MULTIVARIATE DATA ANALYSIS

SEVENTH EDITION

MULTIVARIATE DATA ANALYSIS
A Global Perspective

Joseph F. Hair, Jr.
Kennesaw State University

William C. Black
Louisiana State University

Barry J. Babin
University of Southern Mississippi

Rolph E. Anderson
Drexel University

Upper Saddle River Boston Columbus San Francisco New York
Indianapolis London Toronto Sydney Singapore Tokyo Montreal
Dubai Madrid Hong Kong Mexico City Munich Paris Amsterdam Cape Town

Acquisitions Editor: Julian Partridge
Editorial Director: Sally Yagan
Product Development Manager: Ashley Santora
Editorial Assistant: Elizabeth Walne
Executive Marketing Manager: Patrick Leow
Marketing Manager: Anne Fahlgren
Permissions Project Manager: Charles Morris
Senior Managing Editor: Judy Leale
Production Project Manager: Ana Jankowski
Senior Operations Specialist: Arnold Vila
Cover Designer: Jodi Notowitz
Cover Illustration/Photo: Photo Disk/Photolibrary Group
Composition: Integra Software Services
Full-Service Project Management: Jennifer Welsch/BookMasters, Inc.
Printer/Binder: Hamilton Printing Co./Lehigh-Phoenix Color/Hagerstown
Typeface: 10/12 Times

Credits and acknowledgments borrowed from other sources and reproduced, with permission, in this textbook appear on appropriate page within text.

If you purchased this book within the United States or Canada you should be aware that it has been wrongfully imported without the approval of the Publisher or the Author.

www.pearsonglobaleditions.com

10 9 8 7 6 5 4 3 2 1
ISBN-13: 978-0-13-515309-3
ISBN-10:   0-13-515309-3

To my loving and supportive wife Dale and to my son Joe III and his wife Kerrie

—*Joseph F. Hair, Jr., Kennesaw, GA*

To Deb and Steve for their love and support

—*William C. Black, Baton Rouge, LA*

For Laurie, Amie, and James, and my mother Barbara

—*Barry J. Babin, Hattiesburg, MS*

To Sallie, Rachel, and Stuart for their continuing love and support

—*Rolph E. Anderson, Philadelphia, PA*

BRIEF CONTENTS

CONTENTS

PREFACE

More than 30 years ago when the first edition of *Multivariate Data Analysis* was published, we could not have imagined the applications of multivariate statistics would be as pervasive as they are today. During this time, we have seen phenomenal changes in the environment faced by both academic and applied researchers. First, developing technology has provided desktop analytical capabilities that no one could have anticipated just a few years ago. In a little more than 3 decades, we have gone from punch cards to speech recognition, revolutionizing the way we can interact with and use computers and information. At the same time, we have seen tremendous advances in statistical software, particularly in its ease of use, ranging from completely integrated computer packages such as SPSS and SAS to specialized programs for such techniques as neural networks and conjoint analysis. Today, researchers can find almost any conceivable technique in an accessible, easy-to-use format and often at a reasonable price.

On the statistical front, we have seen widespread application of new techniques, such as structural equation modeling and partial least squares. These advances, however, have been matched by an ever-increasing need for more analytical capability and better metrics. The information explosion has not only challenged our ability to physically handle and analyze the available information, but also required a reassessment of data analysis approaches. Finally, the complexity of the topics being addressed and the increased role of theory and measurement in research design have combined to require more rigorous and sophisticated techniques to perform the necessary confirmatory analyses.

These events have all contributed to the acceptance of the past six editions of this text and the demand for this 7th edition. In approaching this revision, we have tried to embrace both academic and applied researchers with a presentation strongly grounded in statistical techniques, but focusing on design, estimation, and interpretation. We continually strive to reduce our reliance on statistical notation and terminology and instead to identify the fundamental concepts which affect application of these techniques and then express them in simple terms—the result being an applications-oriented introduction to multivariate analysis for the non-statistician. Our commitment remains to provide a firm understanding of the statistical and managerial principles underlying multivariate analysis so as to develop a "comfort zone" not only for the statistical but also the practical issues involved.

NEW FEATURES

First, the authors are continuously working to simplify and streamline coverage of the techniques, and the 7th edition is no exception. This edition is shorter and simpler in its organization, with chapters focusing on a single topic. Moreover, all chapters have been revised to incorporate advances in technology, and several chapters have undergone more extensive change. For example, the initial discussion of topics focuses on a basic understanding of a technique and how to apply it. More advanced issues and concerns are addressed either later in the chapter or in a separate chapter, such as with structural equations modeling. Two chapters, cluster analysis and conjoint, were extensively revised to more effectively demonstrate straightforward approaches to obtain solutions.

Metrics increasingly are relied upon in both scholarly and business applications. This edition updates and expands coverage of important metrics, such as power and effect size. Based on much positive feedback, the "Rules of Thumb" for the application and interpretation of the various techniques have been expanded in this edition, including important issues like sample size. The rules of thumb are highlighted throughout the chapters to facilitate their use. We are confident these guidelines will facilitate your utilization of the techniques.

Another major change is the expansion and reorganization in coverage of structural equations modeling. Chapter 11 provides an overview of structural equation modeling. Chapter 12 then focuses on confirmatory factor analysis, issues in estimating and testing structural models, and advanced topics in both confirmatory factor analysis and structural equations modeling, such as testing higher-order factor models, group models, moderating and mediating variables and PLS. We also worked to eliminate and minimize the use of technical terms and mathematical and statistical notation that often is confusing. These chapters provide a comprehensive overview and explanation of this technique.

Special thanks are due to Pei-ju Lucy Ting and Hsin-Ju Stephanie Tsai, both from University of Manchester, for the revision of the chapter on canonical correlation analysis (Chapter 5). They updated this chapter with an example using the HBAT database, added recently published material, and reorganized it to facilitate understanding.

An important development is the expansion of a Web site (www.mvstats.com) devoted to multivariate analysis, titled "Great Ideas in Teaching Multivariate Statistics." This Web site acts as a resource center for individuals interested in multivariate analysis, providing links to resources for each technique as well as a forum for identifying new topics or statistical methods. In this way, we can provide more timely feedback to researchers other than if they were to wait for a new edition of the book. The Web site also represents a clearinghouse for materials on teaching multivariate statistics, including exercises, datasets, and project ideas.

Each of these changes, and others not mentioned, will assist readers in gaining a more thorough understanding of both the statistical and applied issues underlying these techniques.

ACKNOWLEDGMENTS

We would like to acknowledge the assistance of the following individuals on prior editions of the text: Bruce Alford, Louisiana Tech University; David Andrus, Kansas State University; Jill Attaway, Illinois State University; Jim Boles, Georgia State University; David Booth, Kent State University; Alvin C. Burns, Louisiana State University; Alan J. Bush, University of Memphis; Robert Bush, Louisiana State University at Alexandria; Rabikar Chatterjee, University of Michigan; Kerri Curtis, Golden Gate University; Chaim Ehrman, University of Illinois at Chicago; Joel Evans, Hofstra University; Thomas L. Gillpatrick, Portland State University; Andreas Herrman, University of St. Gallen; Dipak Jain, Northwestern University; Stavros Kalafatis, Kingston University; John Lastovicka, University of Kansas; Margaret Liebman, La Salle University; Arthur Money, Henley Management College; Peter McGoldrick, University of Manchester; Richard Netemeyer, University of Virginia; Ossi Pesamaa, Jonkoping University; Robert Peterson, University of Texas; Torsten Pieper, Kennesaw State University; Scott Roach, Northeast Louisiana University; Phillip Samouel, Kingston University; Marcus Schmidt, Copenhagen Business School; Muzaffar Shaikh, Florida Institute of Technology; Dan Sherrell, University of Memphis; Walter A. Smith, Tulsa University; Goren Svensson, University of Oslo; Ronald D. Taylor, Mississippi State University; Jerry L. Wall, University of Louisiana-Monroe; and Arch Woodside, Boston College. Hans Eibe Sørensen, University of Southern Denmark; Koo Rijkema, Eindhoven University of Technology; James Sallis, Uppsala Universitet; Iain Weir, University of the West of England.

J.F.H.

W.C.B.

B.J.B.

R.E.A.

ABOUT THE AUTHORS

Joseph F. Hair, Jr. Dr. Hair is Professor of Marketing at Kennesaw State University. He previously held the Copeland Endowed Chair of Entrepreneurship and was Director, Entrepreneurship Institute, Ourso College of Business Administration, Louisiana State University. He was a United States Steel Foundation Fellow at the University of Florida, Gainesville, where he earned his Ph.D. in Marketing in 1971. He has authored over 40 books, including Marketing, South-Western Publishing Company, 10th edition 2010; Marketing Essentials, South-Western Publishing Company, 6th edition 2009; MKTG, South-Western Publishing Company, 3rd edition 2009; Essentials of Business Research Methods, Wiley, 2003; Research Methods for Business, Wiley, UK, 2007; Marketing Research, McGraw-Hill/Irwin, 4th edition 2010; Essentials of Marketing Research, McGraw-Hill/Irwin, 2008; and Sales Management: Building Partnerships; Houghton-Mifflin, 2008 He also has published numerous articles in professional journals such as the *Journal of Marketing Research, Journal of Academy of Marketing Science, Journal of Business/Chicago, Journal of Advertising Research, Journal of Business Research, Management Decision, Journal of Marketing Theory and Practice, European Business Review, Journal of Personal Selling and Sales Management, Industrial Marketing Management, Business Horizons, Journal of Retailing, Marketing Education Review, Journal of Marketing Education, Multivariate Behavioral Research*, and others. He is a Distinguished Fellow of the Academy of Marketing Sciences, the Society for Marketing Advances, and Southwestern Marketing Association. He also has served as President of the Academy of Marketing Sciences, the Society for Marketing Advances, the Southern Marketing Association, and other scholarly organizations. He was recognized as the Innovative Marketer of the Year in 2007 by the Marketing Management Association, received the Academy of Marketing Science Outstanding Marketing Teaching Excellence Award in 2004, and the Louisiana State University Entrepreneurship Institute under his leadership was recognized nationally by Entrepreneurship Magazine as one of the top 12 programs in the USA, and also was ranked #3 in the USA by Forbes Magazine/Princeton Review.

William C. Black Dr. Black is the Piccadilly, Inc. Business Administration Business Partnership Professor in the Department of Marketing, E. J. Ourso College of Business at Louisiana State University. He received his M.B.A. in 1976 and Ph.D. in 1980, both from the University of Texas at Austin. He held positions at the University of Arizona from 1980 to 1985, and has been at LSU since 1985. He has also published numerous articles in professional journals such as the *Journal of Marketing, Journal of Marketing Research, Journal of Consumer Research, Journal of Retailing, Growth and Change, Transportation Research, Journal of Real Estate Research, Journal of General Management, Leisure Sciences, Economic Geography,* and others, along with a number of chapters in scholarly books. His teaching interests are in the areas of multivariate statistics and the application of information technology, especially the evolution of marketing principles involved in e-commerce. He is a member of the Editorial Review Board for the *Journal of Business Research.*

Barry J. Babin Dr. Babin is Max P. Watson, Jr. Professor of Business and Chair of Marketing & Quantitative Analysis, Louisiana Tech University. He received his Ph.D. in Business Administration from Louisiana State University in 1991. His research appears in the *Journal of Retailing, Journal of the Academy of Marketing Science, Journal of Business Research, Journal of Marketing, Journal of Consumer Research, Psychological Reports, Psychology and Marketing*, and numerous other professional and trade periodicals. His research focuses on various aspects of retail and service management with an emphasis on the role of value. He has given frequent national and international presentations on the meaning of wine and wine marketing history. Barry has lectured internationally on research related topics in particular on matching theory and analysis and on the use of structural equations modeling. He also lectures on topics related to wine marketing and wine history. He has been recognized for contributions in teaching, service, and research. Included among these awards is his recognition as a Distinguished Fellow of both the Academy of Marketing Science and the Society for Marketing Advances. He is currently Marketing Editor of the *Journal of Business Research* and Immediate Past President of the Academy of Marketing Science.

Rolph E. Anderson Dr. Anderson is the Royal H. Gibson Sr. Professor of Business Administration and former Head of the Department of Marketing at Drexel University. He earned his Ph.D. from the University of Florida, and his M.B.A. and B.A. degrees from Michigan State University. His primary research and publication areas are personal selling and sales management, customer relationship management, and customer loyalty. He is author or co-author of 18 textbooks, including most recently: *Personal Selling: Achieving Customer Satisfaction and Loyalty* (Houghton Mifflin, 2002) and *Professional Sales Management* (Thompson Learning, 3rd ed., 1999). His research has been widely published in the major professional journals in his field, including articles in the *Journal of Marketing Research, Journal of Marketing, Journal of Retailing, Journal of the Academy of Marketing Science, Journal of Experimental Education, Business Horizons, Journal of Global Marketing, Journal of Marketing Education, European Journal of Marketing, Psychology & Marketing, Journal of Business-to-Business Marketing, Marketing Education Review, Industrial Marketing Management, Journal of Business & Industrial Marketing, Journal of Personal Selling & Sales Management,* and numerous others. Dr. Anderson has been selected twice by Drexel's LeBow College of Business students to receive the Faculty Appreciation Award, and serves as a distinguished fellow in the Center for Teaching Excellence. In 1995, he was recipient of the national Excellence in Reviewing Award from the editor of the *Journal of Personal Selling & Sales Management.* In 1998, he received the American Marketing Association Sales Special Interest Group's inaugural Excellence in Sales Scholarship Award. For 2000–2001, he received Drexel University's LeBow College of Business Research Achievement award. Dr. Anderson has served professional organizations as an officer, including: President, Southeast Institute for Decision Sciences (IDS); Board of Directors, American Marketing Association (Philadelphia Chapter); and Secretary and Board of Directors, Academy of Marketing Science. He serves on the editorial boards of five academic journals and on the Faculty Advisory Board of the Fisher Institute for Professional Selling.

CHAPTER 1

Introduction: Methods and Model Building

LEARNING OBJECTIVES

Upon completing this chapter, you should be able to do the following:

- Explain what multivariate analysis is and when its application is appropriate.
- Discuss the nature of measurement scales and their relationship to multivariate techniques.
- Understand the nature of measurement error and its impact on multivariate analysis.
- Determine which multivariate technique is appropriate for a specific research problem.
- Define the specific techniques included in multivariate analysis.
- Discuss the guidelines for application and interpretation of multivariate analyses.
- Understand the six-step approach to multivariate model building.

CHAPTER PREVIEW

Chapter 1 presents a simplified overview of multivariate analysis. It stresses that multivariate analysis methods will increasingly influence not only the analytical aspects of research but also the design and approach to data collection for decision making and problem solving. Although multivariate techniques share many characteristics with their univariate and bivariate counterparts, several key differences arise in the transition to a multivariate analysis. To illustrate this transition, this chapter presents a classification of multivariate techniques. It then provides general guidelines for the application of these techniques as well as a structured approach to the formulation, estimation, and interpretation of multivariate results. The chapter concludes with a discussion of the databases utilized throughout the text to illustrate application of the techniques.

KEY TERMS

Before starting the chapter, review the key terms to develop an understanding of the concepts and terminology used. Throughout the chapter, the key terms appear in boldface. Other points of emphasis in the chapter are italicized. Also, cross-references within the key terms appear in italics.

Alpha (α) See *Type I error*.
Beta (β) See *Type II error*.
Bivariate partial correlation Simple (two-variable) correlation between two sets of residuals (unexplained variances) that remain after the association of other independent variables is removed.

Bootstrapping   An approach to validating a multivariate model by drawing a large number of sub-samples and estimating models for each subsample. Estimates from all the subsamples are then combined, providing not only the "best" estimated coefficients (e.g., means of each estimated coefficient across all the subsample models), but their expected variability and thus their likelihood of differing from zero; that is, are the estimated coefficients statistically different from zero or not? This approach does not rely on statistical assumptions about the population to assess statistical significance, but instead makes its assessment based solely on the sample data.

Composite measure   See *summated scales.*

Dependence technique   Classification of statistical techniques distinguished by having a variable or set of variables identified as the *dependent variable(s)* and the remaining variables as *independent*. The objective is prediction of the dependent variable(s) by the independent variable(s). An example is regression analysis.

Dependent variable   Presumed effect of, or response to, a change in the *independent variable(s).*

Dummy variable   *Nonmetrically* measured variable transformed into a *metric* variable by assigning a 1 or a 0 to a subject, depending on whether it possesses a particular characteristic.

Effect size   Estimate of the degree to which the phenomenon being studied (e.g., correlation or difference in means) exists in the population.

Independent variable   Presumed cause of any change in the *dependent variable.*

Indicator   Single variable used in conjunction with one or more other variables to form a *composite measure.*

Interdependence technique   Classification of statistical techniques in which the variables are not divided into *dependent* and *independent* sets; rather, all variables are analyzed as a single set (e.g., factor analysis).

Measurement error   Inaccuracies of measuring the "true" variable values due to the fallibility of the measurement instrument (i.e., inappropriate response scales), data entry errors, or respondent errors.

Metric data   Also called quantitative data, interval data, or ratio data, these measurements identify or describe subjects (or objects) not only on the possession of an attribute but also by the amount or degree to which the subject may be characterized by the attribute. For example, a person's age and weight are metric data.

Multicollinearity   Extent to which a variable can be explained by the other variables in the analysis. As multicollinearity increases, it complicates the interpretation of the *variate* because it is more difficult to ascertain the effect of any single variable, owing to their interrelationships.

Multivariate analysis   Analysis of multiple variables in a single relationship or set of relationships.

Multivariate measurement   Use of two or more variables as *indicators* of a single *composite measure*. For example, a personality test may provide the answers to a series of individual questions (indicators), which are then combined to form a single score (*summated scale*) representing the personality trait.

Nonmetric data   Also called qualitative data, these are attributes, characteristics, or categorical properties that identify or describe a subject or object. They differ from *metric data* by indicating the presence of an attribute, but not the amount. Examples are occupation (physician, attorney, professor) or buyer status (buyer, nonbuyer). Also called nominal data or ordinal data.

Power   Probability of correctly rejecting the null hypothesis when it is false; that is, correctly finding a hypothesized relationship when it exists. Determined as a function of (1) the statistical significance level set by the researcher for a *Type I error (α)*, (2) the sample size used in the analysis, and (3) the *effect size* being examined.

Practical significance   Means of assessing multivariate analysis results based on their substantive findings rather than their statistical significance. Whereas statistical significance determines whether the result is attributable to chance, practical significance assesses whether the result is useful (i.e., substantial enough to warrant action) in achieving the research objectives.

Reliability   Extent to which a variable or set of variables is consistent in what it is intended to measure. If multiple measurements are taken, the reliable measures will all be consistent in their values. It differs from *validity* in that it relates not to what should be measured, but instead to how it is measured.

Specification error   Omitting a key variable from the analysis, thus affecting the estimated effects of included variables.

Summated scales   Method of combining several variables that measure the same concept into a single variable in an attempt to increase the *reliability* of the measurement through *multivariate measurement*. In most instances, the separate variables are summed and then their total or average score is used in the analysis.

Treatment   Independent variable the researcher manipulates to see the effect (if any) on the dependent variable(s), such as in an experiment (e.g., testing the appeal of color versus black-and-white advertisements).

Type I error   Probability of incorrectly rejecting the null hypothesis—in most cases, it means saying a difference or correlation exists when it actually does not. Also termed *alpha* (α). Typical levels are 5 or 1 percent, termed the .05 or .01 level, respectively.

Type II error   Probability of incorrectly failing to reject the null hypothesis—in simple terms, the chance of not finding a correlation or mean difference when it does exist. Also termed *beta* (β), it is inversely related to *Type I error*. The value of 1 minus the Type II error ($1 - \beta$) is defined as *power*.

Univariate analysis of variance (ANOVA)   Statistical technique used to determine, on the basis of one dependent measure, whether samples are from populations with equal means.

Validity   Extent to which a measure or set of measures correctly represents the concept of study—the degree to which it is free from any systematic or nonrandom error. Validity is concerned with how well the concept is defined by the measure(s), whereas *reliability* relates to the consistency of the measure(s).

Variate   Linear combination of variables formed in the multivariate technique by deriving empirical weights applied to a set of variables specified by the researcher.

WHAT IS MULTIVARIATE ANALYSIS?

Today businesses must be more profitable, react quicker, and offer higher-quality products and services, and do it all with fewer people and at lower cost. An essential requirement in this process is effective knowledge creation and management. There is no lack of information, but there is a dearth of knowledge. As Tom Peters said in his book *Thriving on Chaos*, "We are drowning in information and starved for knowledge" [7].

The information available for decision making exploded in recent years, and will continue to do so in the future, probably even faster. Until recently, much of that information just disappeared. It was either not collected or discarded. Today this information is being collected and stored in data warehouses, and it is available to be "mined" for improved decision making. Some of that information can be analyzed and understood with simple statistics, but much of it requires more complex, multivariate statistical techniques to convert these data into knowledge.

A number of technological advances help us to apply multivariate techniques. Among the most important are the developments in computer hardware and software. The speed of computing equipment has doubled every 18 months while prices have tumbled. User-friendly software packages brought data analysis into the point-and-click era, and we can quickly analyze mountains of complex data with relative ease. Indeed, industry, government, and university-related research centers throughout the world are making widespread use of these techniques.

Throughout the text we use the generic term *researcher* when referring to a data analyst within either the practitioner or academic communities. We feel it inappropriate to make any distinction between these two areas, because research in both relies on theoretical and quantitative bases. Although the research objectives and the emphasis in interpretation may vary, a researcher within either area must address all of the issues, both conceptual and empirical, raised in the discussions of the statistical methods.

MULTIVARIATE ANALYSIS IN STATISTICAL TERMS

Multivariate analysis techniques are popular because they enable organizations to create knowledge and thereby improve their decision making. **Multivariate analysis** refers to all statistical techniques that simultaneously analyze multiple measurements on individuals or objects under investigation. Thus, any simultaneous analysis of more than two variables can be loosely considered multivariate analysis.

Many multivariate techniques are extensions of univariate analysis (analysis of single-variable distributions) and bivariate analysis (cross-classification, correlation, analysis of variance, and simple regression used to analyze two variables). For example, simple regression (with one predictor variable) is extended in the multivariate case to include several predictor variables. Likewise, the single dependent variable found in analysis of variance is extended to include multiple dependent variables in multivariate analysis of variance. Some multivariate techniques (e.g., multiple regression and multivariate analysis of variance) provide a means of performing in a single analysis what once took multiple univariate analyses to accomplish. Other multivariate techniques, however, are uniquely designed to deal with multivariate issues, such as factor analysis, which identifies the structure underlying a set of variables, or discriminant analysis, which differentiates among groups based on a set of variables.

Confusion sometimes arises about what multivariate analysis is because the term is not used consistently in the literature. Some researchers use *multivariate* simply to mean examining relationships between or among more than two variables. Others use the term only for problems in which all the multiple variables are assumed to have a multivariate normal distribution. To be considered truly multivariate, however, all the variables must be random and interrelated in such ways that their different effects cannot meaningfully be interpreted separately. Some authors state that the purpose of multivariate analysis is to measure, explain, and predict the degree of relationship among variates (weighted combinations of variables). Thus, the multivariate character lies in the multiple variates (multiple combinations of variables), and not only in the number of variables or observations. For the purposes of this book, we do not insist on a rigid definition of multivariate analysis. Instead, multivariate analysis will include both multivariable techniques and truly multivariate techniques, because we believe that knowledge of multivariable techniques is an essential first step in understanding multivariate analysis.

SOME BASIC CONCEPTS OF MULTIVARIATE ANALYSIS

Although the roots of multivariate analysis lie in univariate and bivariate statistics, the extension to the multivariate domain introduces additional concepts and issues of particular relevance. These concepts range from the need for a conceptual understanding of the basic building block of multivariate analysis—the variate—to specific issues dealing with the types of measurement scales used and the statistical issues of significance testing and confidence levels. Each concept plays a significant role in the successful application of any multivariate technique.

The Variate

As previously mentioned, the building block of multivariate analysis is the **variate,** a linear combination of variables with empirically determined weights. The variables are specified by the researcher, whereas the weights are determined by the multivariate technique to meet a specific objective. A variate of n weighted variables (X_1 to X_n) can be stated mathematically as:

$$\text{Variate value} = w_1X_1 + w_2X_2 + w_3X_3 + \cdots + w_nX_n$$

where X_n is the observed variable and w_n is the weight determined by the multivariate technique.

The result is a single value representing a combination of the *entire set* of variables that best achieves the objective of the specific multivariate analysis. In multiple regression, the variate is determined in a manner that maximizes the correlation between the multiple independent variables and the single dependent variable. In discriminant analysis, the variate is formed so as to create scores for each observation that maximally differentiates between groups of observations. In factor analysis, variates are formed to best represent the underlying structure or patterns of the variables as represented by their intercorrelations.

In each instance, the variate captures the multivariate character of the analysis. Thus, in our discussion of each technique, the variate is the focal point of the analysis in many respects. We must understand not only its collective impact in meeting the technique's objective but also each separate variable's contribution to the overall variate effect.

Measurement Scales

Data analysis involves the identification and measurement of variation in a set of variables, either among themselves or between a dependent variable and one or more independent variables. The key word here is *measurement* because the researcher cannot identify variation unless it can be measured. Measurement is important in accurately representing the concept of interest and is instrumental in the selection of the appropriate multivariate method of analysis. Data can be classified into one of two categories—nonmetric (qualitative) and metric (quantitative)—based on the type of attributes or characteristics they represent.

The researcher must define the measurement type—nonmetric or metric—for each variable. To the computer, the values are only numbers. As we will see in the following section, defining data as either metric or nonmetric has substantial impact on what the data can represent and how it can be analyzed.

NONMETRIC MEASUREMENT SCALES   **Nonmetric data** describe differences in type or kind by indicating the presence or absence of a characteristic or property. These properties are discrete in that by having a particular feature, all other features are excluded; for example, if a person is male, he cannot be female. An "amount" of gender is not possible, just the state of being male or female. Nonmetric measurements can be made with either a nominal or an ordinal scale.

Nominal Scales. A nominal scale assigns numbers as a way to label or identify subjects or objects. The numbers assigned to the objects have no quantitative meaning beyond indicating the presence or absence of the attribute or characteristic under investigation. Therefore, nominal scales, also known as categorical scales, can only provide the number of occurrences in each class or category of the variable being studied.

For example, in representing gender (male or female) the researcher might assign numbers to each category (e.g., 2 for females and 1 for males). With these values, however, we can only tabulate the number of males and females; it is nonsensical to calculate an average value of gender.

Nominal data only represent categories or classes and do not imply amounts of an attribute or characteristic. Commonly used examples of nominally scaled data include many demographic attributes (e.g., individual's sex, religion, occupation, or political party affiliation), many forms of behavior (e.g., voting behavior or purchase activity), or any other action that is discrete (happens or not).

Ordinal Scales. Ordinal scales are the next "higher" level of measurement precision. In the case of ordinal scales, variables can be ordered or ranked in relation to the amount of the attribute possessed. Every subject or object can be compared with another in terms of a "greater than" or "less than" relationship. The numbers utilized in ordinal scales, however, are really nonquantitative because they indicate only relative positions in an ordered series. Ordinal scales provide no measure of the actual amount or magnitude in absolute terms, only the order of the values. The researcher knows the order, but not the amount of difference between the values.

For example, different levels of an individual consumer's satisfaction with several new products can be illustrated, first using an ordinal scale. The following scale shows a respondent's view of three products.

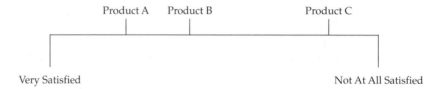

| Product A | Product B | | | Product C |

Very Satisfied   Not At All Satisfied

When we measure this variable with an ordinal scale, we "rank order" the products based on satisfaction level. We want a measure that reflects that the respondent is more satisfied with Product A than Product B and more satisfied with Product B than Product C, based solely on their position on the scale. We could assign "rank order" values (1 = most satisfied, 2 = next most satisfied, etc.) of 1 for Product A (most satisfaction), 2 for Product B, and 3 for Product C.

When viewed as ordinal data, we know that Product A has the most satisfaction, followed by Product B and then Product C. However, we cannot make any statements on the amount of the differences between products (e.g., we cannot answer the question whether the difference between Products A and B is greater than the difference between Products B and C). We have to use an interval scale (see next section) to assess what is the magnitude of differences between products.

In many instances a researcher may find it attractive to use ordinal measures, but the implications for the types of analyses that can be performed are substantial. The analyst cannot perform any arithmetic operations (no sums, averages, multiplication or division, etc.), thus nonmetric data are quite limited in their use in estimating model coefficients. For this reason, many multivariate techniques are devised solely to deal with nonmetric data (e.g., correspondence analysis) or to use nonmetric data as an independent variable (e.g., discriminant analysis with a nonmetric dependent variable or multivariate analysis of variance with nonmetric independent variables). Thus, the analyst must identify all nonmetric data to ensure that they are used appropriately in the multivariate techniques.

METRIC MEASUREMENT SCALES   In contrast to nonmetric data, **metric data** are used when subjects differ in amount or degree on a particular attribute. Metrically measured variables reflect relative quantity or degree and are appropriate for attributes involving amount or magnitude, such as the level of satisfaction or commitment to a job. The two different metric measurement scales are interval and ratio scales.

Interval Scales. Interval scales and ratio scales (both metric) provide the highest level of measurement precision, permitting nearly any mathematical operation to be performed. These two scales have constant units of measurement, so differences between any two adjacent points on any part of the scale are equal.

In the preceding example in measuring satisfaction, metric data could be obtained by measuring the distance from one end of the scale to each product's position. Assume that Product A was 2.5 units from the left end, Product B was 6.0 units, and Product C was 12 units. Using these values as a measure of satisfaction, we could not only make the same statements as we made with the ordinal data (e.g., the rank order of the products), but we could also see that the difference between Products A and B was much smaller (6.0 − 2.5 = 3.5) than was the difference between Products B and C (12.0 − 6.0 = 6.0).

The only real difference between interval and ratio scales is that interval scales use an arbitrary zero point, whereas ratio scales include an absolute zero point. The most familiar interval scales are the Fahrenheit and Celsius temperature scales. Each uses a different arbitrary zero point, and neither indicates a zero amount or lack of temperature, because we can register temperatures below the zero point on each scale. Therefore, it is not possible to say that any value on an interval scale is a multiple of some other point on the scale.

For example, an 80°F day cannot correctly be said to be twice as hot as a 40°F day, because we know that 80°F, on a different scale, such as Celsius, is 26.7°C. Similarly, 40°F on a Celsius scale is 4.4°C. Although 80°F is indeed twice 40°F, one cannot state that the heat of 80°F is twice the heat of 40°F because, using different scales, the heat is not twice as great; that is, $4.4°C \times 2 \neq 26.7°C$.

Ratio Scales. Ratio scales represent the highest form of measurement precision because they possess the advantages of all lower scales plus an absolute zero point. All mathematical operations are permissible with ratio-scale measurements. The bathroom scale or other common weighing machines are examples of these scales, because they have an absolute zero point and can be spoken of in terms of multiples when relating one point on the scale to another; for example, 100 pounds is twice as heavy as 50 pounds.

THE IMPACT OF CHOICE OF MEASUREMENT SCALE   Understanding the different types of measurement scales is important for two reasons:

1. The researcher must identify the measurement scale of each variable used, so that nonmetric data are not incorrectly used as metric data, and vice versa (as in our earlier example of representing gender as 1 for male and 2 for female). If the researcher incorrectly defines this measure as metric, then it may be used inappropriately (e.g., finding the mean value of gender).
2. The measurement scale is also critical in determining which multivariate techniques are the most applicable to the data, with considerations made for both independent and dependent variables. In the discussion of the techniques and their classification in later sections of this chapter, the metric or nonmetric properties of independent and dependent variables are the determining factors in selecting the appropriate technique.

Measurement Error and Multivariate Measurement

The use of multiple variables and the reliance on their combination (the variate) in multivariate techniques also focuses attention on a complementary issue—measurement error. **Measurement error** is the degree to which the observed values are not representative of the "true" values. Measurement error has many sources, ranging from data entry errors to the imprecision of the measurement (e.g., imposing 7-point rating scales for attitude measurement when the researcher knows the respondents can accurately respond only to a 3-point rating) to the inability of respondents to accurately provide information (e.g., responses as to household income may be reasonably accurate but rarely totally precise). *Thus, all variables used in multivariate techniques must be assumed to have some degree of measurement error.* The measurement error adds "noise" to the observed or measured variables. Thus, the observed value obtained represents both the "true" level and the "noise." When used to compute correlations or means, the "true" effect is partially masked by the measurement error, causing the correlations to weaken and the means to be less precise. The specific impact of measurement error and its accommodation in dependence relationships is covered in more detail in Chapter 11.

VALIDITY AND RELIABILITY   The researcher's goal of reducing measurement error can follow several paths. In assessing the degree of measurement error present in any measure, the researcher must address two important characteristics of a measure:

- **Validity** is the degree to which a measure accurately represents what it is supposed to. For example, if we want to measure discretionary income, we should not ask about total household income. Ensuring validity starts with a thorough understanding of what is to be measured and then making the measurement as "correct" and accurate as possible. However, accuracy does not ensure validity. In our income example, the researcher could precisely define total household income, but it would still be "wrong" (i.e., an invalid measure) in measuring discretionary income because the "correct" question was not being asked.

• If validity is assured, the researcher must still consider the reliability of the measurements. **Reliability** is the degree to which the observed variable measures the "true" value and is "error free"; thus, it is the opposite of measurement error. If the same measure is asked repeatedly, for example, more reliable measures will show greater consistency than less reliable measures. The researcher should always assess the variables being used and, if valid alternative measures are available, choose the variable with the higher reliability.

EMPLOYING MULTIVARIATE MEASUREMENT   In addition to reducing measurement error by improving individual variables, the researcher may also choose to develop **multivariate measurements**, also known as **summated scales**, for which several variables are joined in a **composite measure** to represent a concept (e.g., multiple-item personality scales or summed ratings of product satisfaction). The objective is to avoid the use of only a single variable to represent a concept and instead to use several variables as **indicators**, all representing differing facets of the concept to obtain a more well-rounded perspective. The use of multiple indicators enables the researcher to more precisely specify the desired responses. It does not place total reliance on a single response, but instead on the "average" or typical response to a set of related responses.

For example, in measuring satisfaction, one could ask a single question, "How satisfied are you?" and base the analysis on the single response. Or a summated scale could be developed that combined several responses of satisfaction (e.g., finding the average score among three measures—overall satisfaction, the likelihood to recommend, and the probability of purchasing again). The different measures may be in different response formats or in differing areas of interest assumed to comprise overall satisfaction.

The guiding premise is that multiple responses reflect the "true" response more accurately than does a single response. The researcher should assess reliability and incorporate scales into the analysis. For a more detailed introduction to multiple measurement models and scale construction, see further discussion in Chapter 3 (Factor Analysis) and Chapter 11 (SEM: An Introduction) or additional resources [8]. In addition, compilations of scales that can provide the researcher a "ready-to-go" scale with demonstrated reliability have been published in recent years [1, 4].

THE IMPACT OF MEASUREMENT ERROR   The impact of measurement error and poor reliability cannot be directly seen because they are embedded in the observed variables. The researcher must therefore always work to increase reliability and validity, which in turn will result in a more accurate portrayal of the variables of interest. Poor results are not always due to measurement error, but the presence of measurement error is guaranteed to distort the observed relationships and make multivariate techniques less powerful. Reducing measurement error, although it takes effort, time, and additional resources, may improve weak or marginal results and strengthen proven results as well.

STATISTICAL SIGNIFICANCE VERSUS STATISTICAL POWER

All the multivariate techniques, except for cluster analysis and perceptual mapping, are based on the statistical inference of a population's values or relationships among variables from a randomly drawn sample of that population. A census of the entire population makes statistical inference unnecessary, because any difference or relationship, however small, is true and does exist. Researchers very seldom use a census. Therefore, researchers are often interested in drawing inferences from a sample.

Types of Statistical Error and Statistical Power

Interpreting statistical inferences requires the researcher to specify the acceptable levels of statistical error that result from using a sample (known as *sampling error*). The most common approach is to specify the level of **Type I error**, also known as **alpha (α)**. Type I error is the probability of rejecting the null hypothesis when it is actually true—generally referred to as a *false positive*. By specifying an alpha level, the researcher sets the acceptable limits for error and indicates the probability of concluding that significance exists when it really does not.

When specifying the level of Type I error, the researcher also determines an associated error, termed **Type II error, or beta (β).** The Type II error is the probability of not rejecting the null hypothesis when it is actually false. An extension of Type II error is $1 - \beta$, referred to as the **power** of the statistical inference test. Power is the probability of correctly rejecting the null hypothesis when it should be rejected. Thus, power is the probability that statistical significance will be indicated if it is present. The relationship of the different error probabilities in testing for the difference in two means is shown here:

		Reality	
		No Difference	Difference
Statistical Decision	H_0: No Difference	$1 - \alpha$	β **Type II error**
	H_a: Difference	α **Type I error**	$1 - \beta$ **Power**

Although specifying alpha establishes the level of acceptable statistical significance, it is the level of power that dictates the probability of success in finding the differences if they actually exist. Why not set both alpha and beta at acceptable levels? Because the Type I and Type II errors are inversely related. Thus, Type I error becomes more restrictive (moves closer to zero) as the probability of a Type II error increases. That is, reducing Type I errors reduces the power of the statistical test. Thus, researchers must strike a balance between the level of alpha and the resulting power.

Impacts on Statistical Power

But why can't high levels of power always be achieved? Power is not solely a function of alpha. Power is determined by three factors:

1. *Effect size*—The probability of achieving statistical significance is based not only on statistical considerations, but also on the actual size of the effect. Thus, the **effect size** helps researchers determine whether the observed relationship (difference or correlation) is meaningful. For example, the effect size could be a difference in the means between two groups or the correlation between variables. If a weight loss firm claims its program leads to an average weight loss of 25 pounds, the 25 pounds is the effect size. Similarly, if a university claims its MBA graduates get a starting salary that is 50 percent higher than the average, the percent is the effect size attributed to earning the degree. When examining effect sizes, a larger effect is more likely to be found than a smaller effect and is thus more likely to impact the power of the statistical test.

   To assess the power of any statistical test, the researcher must first understand the effect being examined. Effect sizes are defined in standardized terms for ease of comparison. Mean differences are stated in terms of standard deviations, thus an effect size of .5 indicates that the mean difference is one-half of a standard deviation. For correlations, the effect size is based on the actual correlation between the variables.

2. *Alpha* (α)—As alpha becomes more restrictive, power decreases. Therefore, as the researcher reduces the chance of incorrectly saying an effect is significant when it is not, the probability of correctly finding an effect decreases. Conventional guidelines suggest alpha levels of .05 or .01. Researchers should consider the impact of a particular alpha level on the power before selecting the alpha level. The relationship of these two probabilities is illustrated in later discussions.

3. *Sample size*—At any given alpha level, increased sample sizes always produce greater power for the statistical test. As sample sizes increase, researchers must decide if the power is too high. By "too high" we mean that by increasing sample size, smaller and smaller effects (e.g., correlations) will be found to be statistically significant, until at very large sample sizes almost any effect is significant. The researcher must always be aware that sample size can affect the statistical test either by making it insensitive (at small sample sizes) or overly sensitive (at very large sample sizes).

The relationships among alpha, sample size, effect size, and power are complicated, but a number of sources are available for consideration. Cohen [5] examines power for most statistical inference tests and provides guidelines for acceptable levels of power, suggesting that studies be designed to achieve alpha levels of at least .05 with power levels of 80 percent. To achieve such power levels, all three factors—alpha, sample size, and effect size—must be considered simultaneously. These interrelationships can be illustrated by a simple example.

The example involves testing for the difference between the mean scores of two groups. Assume that the effect size is thought to range between small (.2) and moderate (.5). The researcher must now determine the necessary alpha level and sample size of each group. Table 1-1 illustrates the impact of both sample size and alpha level on power. Note that with a moderate effect size power reaches acceptable levels at sample sizes of 100 or more for alpha levels of both .05 and .01. But when the effect size is small, statistical tests have little power, even with more flexible alpha levels or samples sizes of 200 or more. For example, if the effect size is small a sample of 200 with an alpha of .05 still has only a 50 percent chance of significant differences being found. This suggests that if the researcher expects that the effect sizes will be small the study must have much larger sample sizes and/or less restrictive alpha levels (e.g., .10).

TABLE 1-1   **Power Levels for the Comparison of Two Means: Variations by Sample Size, Significance Level, and Effect Size**

Sample Size	alpha (α) = .05 Effect Size (ES)		alpha (α) = .01 Effect Size (ES)	
	Small (.2)	Moderate (.5)	Small (.2)	Moderate (.5)
20	.095	.338	.025	.144
40	.143	.598	.045	.349
60	.192	.775	.067	.549
80	.242	.882	.092	.709
100	.290	.940	.120	.823
150	.411	.990	.201	.959
200	.516	.998	.284	.992

Source: *SOLO Power Analysis*, BMDP Statistical Software, Inc. [2]

Using Power with Multivariate Techniques

Researchers can use power analysis either in the study design or after data is collected. In designing research studies, the sample size and alpha level are selected to achieve the desired power. Power also is examined after analysis is completed to determine the actual power achieved so the results can be interpreted. Are the results due to effect sizes, sample sizes, or significance levels? Each of these factors is assessed to determine their impact on the significance or nonsignificance of the results. Researchers can refer to published studies for specifics on power determination [5] or access Web sites that assist in planning studies to achieve the desired power or calculate the power of actual results [2, 3]. Specific guidelines for multiple regression and multivariate analysis of variance—the most common applications of power analysis—are discussed in more detail in Chapters 4 and 8.

Having addressed the issues in extending multivariate techniques from their univariate and bivariate origins, we present a classification scheme to assist in the selection of the appropriate technique by specifying the research objectives (independence or dependence relationship) and the data type (metric or nonmetric). We then briefly introduce each multivariate method discussed in the text.

A CLASSIFICATION OF MULTIVARIATE TECHNIQUES

To assist you in becoming familiar with the specific multivariate techniques, we present a classification of multivariate methods in Figure 1-1. This classification is based on three judgments the researcher must make about the research objective and nature of the data:

1. Can the variables be divided into independent and dependent classifications based on some theory?
2. If they can, how many variables are treated as dependent in a single analysis?
3. How are the variables, both dependent and independent, measured?

Selection of the appropriate multivariate technique depends on the answers to these three questions.

When considering the application of multivariate statistical techniques, the answer to the first question—Can the data variables be divided into independent and dependent classifications?—indicates whether a dependence or interdependence technique should be utilized. Note that in Figure 1-1, the dependence techniques are on the left side and the interdependence techniques are on the right. A **dependence technique** may be defined as one in which a variable or set of variables is identified as the **dependent variable** to be predicted or explained by other variables

---

RULES OF THUMB 1-1

Statistical Power Analysis

- Researchers should design studies to achieve a power level of .80 at the desired significance level.
- More stringent significance levels (e.g., .01 instead of .05) require larger samples to achieve the desired power level.
- Conversely, power can be increased by choosing a less stringent alpha level (e.g., .10 instead of .05).
- Smaller effect sizes require larger sample sizes to achieve the desired power.
- An increase in power is most likely achieved by increasing the sample size.

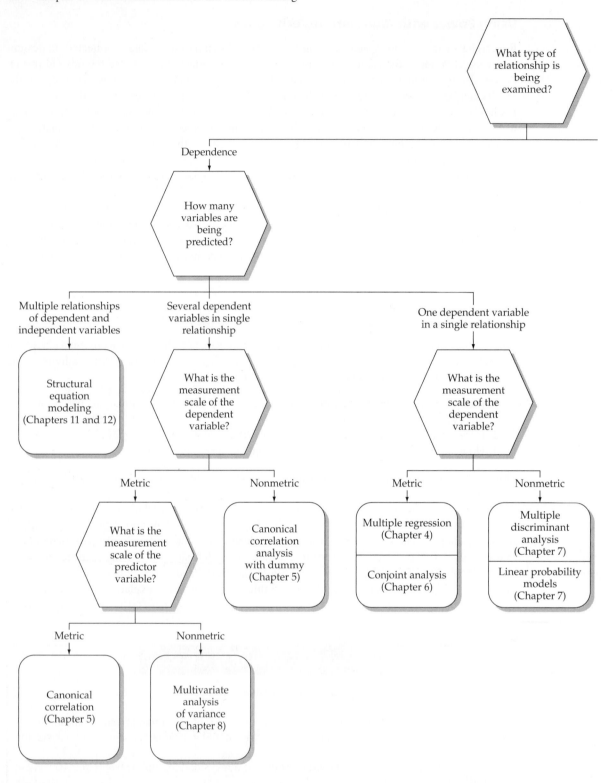

FIGURE 1-1 Selecting a Multivariate Technique

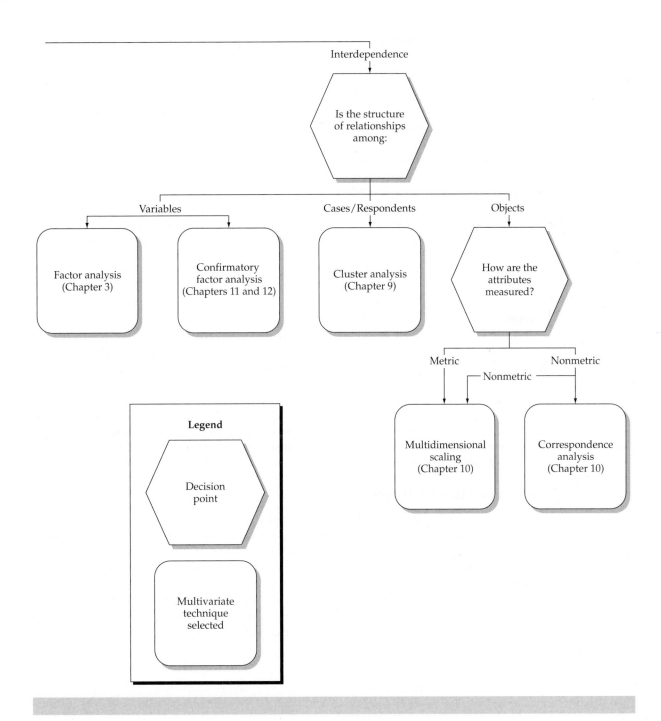

known as **independent variables**. An example of a dependence technique is multiple regression analysis. In contrast, an **interdependence technique** is one in which no single variable or group of variables is defined as being independent or dependent. Rather, the procedure involves the simultaneous analysis of all variables in the set. Factor analysis is an example of an interdependence technique. Let us focus on dependence techniques first and use the classification in Figure 1-1 to select the appropriate multivariate method.

Dependence Techniques

The different dependence techniques can be categorized by two characteristics: (1) the number of dependent variables and (2) the type of measurement scale employed by the variables. First, regarding the number of dependent variables, dependence techniques can be classified as those having a single dependent variable, several dependent variables, or even several dependent/independent relationships. Second, dependence techniques can be further classified as those with either metric (quantitative/numerical) or nonmetric (qualitative/categorical) dependent variables. If the analysis involves a single dependent variable that is metric, the appropriate technique is either multiple regression analysis or conjoint analysis. Conjoint analysis is a special case. It involves a dependence procedure that may treat the dependent variable as either nonmetric or metric, depending on the type of data collected. In contrast, if the single dependent variable is nonmetric (categorical), then the appropriate techniques are multiple discriminant analysis and linear probability models.

When the research problem involves several dependent variables, four other techniques of analysis are appropriate. If the several dependent variables are metric, we must then look to the independent variables. If the independent variables are nonmetric, the technique of multivariate analysis of variance (MANOVA) should be selected. If the independent variables are metric, canonical correlation is appropriate. If the several dependent variables are nonmetric, then they can be transformed through **dummy variable** coding (0–1) and canonical analysis can again be used.[1] Finally, if a set of dependent/independent variable relationships is postulated, then structural equation modeling is appropriate.

A close relationship exists between the various dependence procedures, which can be viewed as a family of techniques. Table 1-2 defines the various multivariate dependence techniques in terms of the nature and number of dependent and independent variables. As we can see, canonical correlation can be considered to be the general model upon which many other multivariate techniques are based, because it places the least restrictions on the type and number of variables in both the dependent and independent variates. As restrictions are placed on the variates, more precise conclusions can be reached based on the specific scale of data measurement employed. Thus, multivariate techniques range from the general method of canonical analysis to the specialized technique of structural equation modeling.

Interdependence Techniques

Interdependence techniques are shown on the right side of Figure 1-1. Readers will recall that with interdependence techniques the variables cannot be classified as either dependent or independent. Instead, all the variables are analyzed simultaneously in an effort to find an underlying structure to the entire set of variables or subjects. If the structure of variables is to be analyzed, then factor analysis or confirmatory factor analysis is the appropriate technique. If cases or respondents are to be grouped to represent structure, then cluster analysis is selected. Finally, if the interest is in the structure of objects, the techniques of perceptual mapping should be applied. As with dependence techniques, the measurement properties of the techniques should be considered. Generally, factor analysis and cluster

---

[1] Dummy variables (see Key Terms) are discussed in greater detail later. Briefly, dummy variable coding is a means of transforming nonmetric data into metric data. It involves the creation of so-called dummy variables, in which 1s and 0s are assigned to subjects, depending on whether they possess a characteristic in question. For example, if a subject is male, assign him a 0, if the subject is female, assign her a 1, or the reverse.

TABLE 1-2   The Relationship Between Multivariate Dependence Methods

Canonical Correlation

$$Y_1 + Y_2 + Y_3 + \cdots + Y_n \quad = \quad X_1 + X_2 + X_3 + \cdots + X_n$$
$$\text{(metric, nonmetric)} \qquad\qquad \text{(metric, nonmetric)}$$

Multivariate Analysis of Variance

$$Y_1 + Y_2 + Y_3 + \cdots + Y_n \quad = \quad X_1 + X_2 + X_3 + \cdots + X_n$$
$$\text{(metric)} \qquad\qquad \text{(nonmetric)}$$

Analysis of Variance

$$Y_1 \quad = \quad X_1 + X_2 + X_3 + \cdots + X_n$$
$$\text{(metric)} \qquad\qquad \text{(nonmetric)}$$

Multiple Discriminant Analysis

$$Y_1 \quad = \quad X_1 + X_2 + X_3 + \cdots + X_n$$
$$\text{(nonmetric)} \qquad\qquad \text{(metric)}$$

Multiple Regression Analysis

$$Y_1 \quad = \quad X_1 + X_2 + X_3 + \cdots + X_n$$
$$\text{(metric)} \qquad\qquad \text{(metric, nonmetric)}$$

Conjoint Analysis

$$Y_1 \quad = \quad X_1 + X_2 + X_3 + \cdots + X_n$$
$$\text{(nonmetric, metric)} \qquad\qquad \text{(nonmetric)}$$

Structural Equation Modeling

$$Y_1 = X_{11} + X_{12} + X_{13} + \cdots + X_{1n}$$
$$Y_2 = X_{21} + X_{22} + X_{23} + \cdots + X_{2n}$$
$$Y_m = X_{m1} + X_{m2} + X_{m3} + \cdots + X_{mn}$$
$$\text{(metric)} \qquad\qquad \text{(metric, nonmetric)}$$

analysis are considered to be metric interdependence techniques. However, nonmetric data may be transformed through dummy variable coding for use with special forms of factor analysis and cluster analysis. Both metric and nonmetric approaches to perceptual mapping have been developed. If the interdependencies of objects measured by nonmetric data are to be analyzed, correspondence analysis is also an appropriate technique.

TYPES OF MULTIVARIATE TECHNIQUES

Multivariate analysis is an ever-expanding set of techniques for data analysis that encompasses a wide range of possible research situations as evidenced by the classification scheme just discussed. The more established as well as emerging techniques include the following:

1. Principal components and common factor analysis
2. Multiple regression and multiple correlation
3. Multiple discriminant analysis and logistic regression
4. Canonical correlation analysis
5. Multivariate analysis of variance and covariance

   6. Conjoint analysis
   7. Cluster analysis
   8. Perceptual mapping, also known as multidimensional scaling
   9. Correspondence analysis
   10. Structural equation modeling and confirmatory factor analysis

Here we introduce each of the multivariate techniques and briefly define the technique and the objective for its application.

Principal Components and Common Factor Analysis

Factor analysis, including both principal component analysis and common factor analysis, is a statistical approach that can be used to analyze interrelationships among a large number of variables and to explain these variables in terms of their common underlying dimensions (factors). The objective is to find a way of condensing the information contained in a number of original variables into a smaller set of variates (factors) with a minimal loss of information. By providing an empirical estimate of the structure of the variables considered, factor analysis becomes an objective basis for creating summated scales.

A researcher can use factor analysis, for example, to better understand the relationships between customers' ratings of a fast-food restaurant. Assume you ask customers to rate the restaurant on the following six variables: food taste, food temperature, freshness, waiting time, cleanliness, and friendliness of employees. The analyst would like to combine these six variables into a smaller number. By analyzing the customer responses, the analyst might find that the variables food taste, temperature, and freshness combine together to form a single factor of food quality, whereas the variables waiting time, cleanliness, and friendliness of employees combine to form another single factor, service quality.

Multiple Regression

Multiple regression is the appropriate method of analysis when the research problem involves a single metric dependent variable presumed to be related to two or more metric independent variables. The objective of multiple regression analysis is to predict the changes in the dependent variable in response to changes in the independent variables. This objective is most often achieved through the statistical rule of least squares.

Whenever the researcher is interested in predicting the amount or size of the dependent variable, multiple regression is useful. For example, monthly expenditures on dining out (dependent variable) might be predicted from information regarding a family's income, its size, and the age of the head of household (independent variables). Similarly, the researcher might attempt to predict a company's sales from information on its expenditures for advertising, the number of salespeople, and the number of stores carrying its products.

Multiple Discriminant Analysis and Logistic Regression

Multiple discriminant analysis (MDA) is the appropriate multivariate technique if the single dependent variable is dichotomous (e.g., male–female) or multichotomous (e.g., high–medium–low) and therefore nonmetric. As with multiple regression, the independent variables are assumed to be metric. Discriminant analysis is applicable in situations in which the total sample can be divided into groups based on a nonmetric dependent variable characterizing several known classes. The primary objectives of multiple discriminant analysis are to understand group differences and to predict the likelihood that an entity (individual or object) will belong to a particular class or group based on several metric independent variables.

Discriminant analysis might be used to distinguish innovators from noninnovators according to their demographic and psychographic profiles. Other applications include distinguishing heavy product users from light users, males from females, national-brand buyers from private-label buyers, and good credit risks from poor credit risks. Even the Internal Revenue Service uses discriminant analysis to compare selected federal tax returns with a composite, hypothetical, normal taxpayer's return (at different income levels) to identify the most promising returns and areas for audit.

Logistic regression models, often referred to as *logit analysis*, are a combination of multiple regression and multiple discriminant analysis. This technique is similar to multiple regression analysis in that one or more independent variables are used to predict a single dependent variable. What distinguishes a logistic regression model from multiple regression is that the dependent variable is nonmetric, as in discriminant analysis. The nonmetric scale of the dependent variable requires differences in the estimation method and assumptions about the type of underlying distribution, yet in most other facets it is quite similar to multiple regression. Thus, once the dependent variable is correctly specified and the appropriate estimation technique is employed, the basic factors considered in multiple regression are used here as well. Logistic regression models are distinguished from discriminant analysis primarily in that they accommodate all types of independent variables (metric and nonmetric) and do not require the assumption of multivariate normality. However, in many instances, particularly with more than two levels of the dependent variable, discriminant analysis is the more appropriate technique.

Assume financial advisors were trying to develop a means of selecting emerging firms for start-up investment. To assist in this task, they reviewed past records and placed firms into one of two classes: successful over a five-year period, and unsuccessful after five years. For each firm, they also had a wealth of financial and managerial data. They could then use a logistic regression model to identify those financial and managerial data that best differentiated between the successful and unsuccessful firms in order to select the best candidates for investment in the future.

Canonical Correlation

Canonical correlation analysis can be viewed as a logical extension of multiple regression analysis. Recall that multiple regression analysis involves a single metric dependent variable and several metric independent variables. With canonical analysis the objective is to correlate simultaneously several metric dependent variables and several metric independent variables. Whereas multiple regression involves a single dependent variable, canonical correlation involves multiple dependent variables. The underlying principle is to develop a linear combination of each set of variables (both independent and dependent) in a manner that maximizes the correlation between the two sets. Stated in a different manner, the procedure involves obtaining a set of weights for the dependent and independent variables that provides the maximum simple correlation between the set of dependent variables and the set of independent variables.

Assume a company conducts a study that collects information on its service quality based on answers to 50 metrically measured questions. The study uses questions from published service quality research and includes benchmarking information on perceptions of the service quality of "world-class companies" as well as the company for which the research is being conducted. Canonical correlation could be used to compare the perceptions of the world-class companies on the 50 questions with the perceptions of the company. The research could then conclude whether the perceptions of the company are correlated with those of world-class companies. The technique would provide information on the overall correlation of perceptions as well as the correlation between each of the 50 questions.

Multivariate Analysis of Variance and Covariance

Multivariate analysis of variance (MANOVA) is a statistical technique that can be used to simultaneously explore the relationship between several categorical independent variables (usually referred to as **treatments**) and two or more metric dependent variables. As such, it represents an extension of **univariate analysis of variance (ANOVA)**. Multivariate analysis of covariance (MANCOVA) can be used in conjunction with MANOVA to remove (after the experiment) the effect of any uncontrolled metric independent variables (known as covariates) on the dependent variables. The procedure is similar to that involved in **bivariate partial correlation**, in which the effect of a third variable is removed from the correlation. MANOVA is useful when the researcher designs an experimental situation (manipulation of several nonmetric treatment variables) to test hypotheses concerning the variance in group responses on two or more metric dependent variables.

Assume a company wants to know if a humorous ad will be more effective with its customers than a nonhumorous ad. It could ask its ad agency to develop two ads—one humorous and one nonhumorous—and then show a group of customers the two ads. After seeing the ads, the customers would be asked to rate the company and its products on several dimensions, such as modern versus traditional or high quality versus low quality. MANOVA would be the technique to use to determine the extent of any statistical differences between the perceptions of customers who saw the humorous ad versus those who saw the nonhumorous one.

Conjoint Analysis

Conjoint analysis is an emerging dependence technique that brings new sophistication to the evaluation of objects, such as new products, services, or ideas. The most direct application is in new product or service development, allowing for the evaluation of complex products while maintaining a realistic decision context for the respondent. The market researcher is able to assess the importance of attributes as well as the levels of each attribute while consumers evaluate only a few product profiles, which are combinations of product levels.

Assume a product concept has three attributes (price, quality, and color), each at three possible levels (e.g., red, yellow, and blue). Instead of having to evaluate all 27 ($3 \times 3 \times 3$) possible combinations, a subset (9 or more) can be evaluated for their attractiveness to consumers, and the researcher knows not only how important each attribute is but also the importance of each level (e.g., the attractiveness of red versus yellow versus blue). Moreover, when the consumer evaluations are completed, the results of conjoint analysis can also be used in product design simulators, which show customer acceptance for any number of product formulations and aid in the design of the optimal product.

Cluster Analysis

Cluster analysis is an analytical technique for developing meaningful subgroups of individuals or objects. Specifically, the objective is to classify a sample of entities (individuals or objects) into a small number of mutually exclusive groups based on the similarities among the entities. In cluster analysis, unlike discriminant analysis, the groups are not predefined. Instead, the technique is used to identify the groups.

Cluster analysis usually involves at least three steps. The first is the measurement of some form of similarity or association among the entities to determine how many groups really exist in the sample. The second step is the actual clustering process, whereby entities are partitioned into groups (clusters). The final step is to profile the persons or variables to determine their composition. Many times this profiling may be accomplished by applying discriminant analysis to the groups identified by the cluster technique.

As an example of cluster analysis, let's assume a restaurant owner wants to know whether customers are patronizing the restaurant for different reasons. Data could be collected on perceptions of pricing, food quality, and so forth. Cluster analysis could be used to determine whether some subgroups (clusters) are highly motivated by low prices versus those who are much less motivated to come to the restaurant based on price considerations.

Perceptual Mapping

In perceptual mapping (also known as *multidimensional scaling*), the objective is to transform consumer judgments of similarity or preference (e.g., preference for stores or brands) into distances represented in multidimensional space. If objects A and B are judged by respondents as being the most similar compared with all other possible pairs of objects, perceptual mapping techniques will position objects A and B in such a way that the distance between them in multidimensional space is smaller than the distance between any other pairs of objects. The resulting perceptual maps show the relative positioning of all objects, but additional analyses are needed to describe or assess which attributes predict the position of each object.

As an example of perceptual mapping, let's assume an owner of a Burger King franchise wants to know whether the strongest competitor is McDonald's or Wendy's. A sample of customers is given a survey and asked to rate the pairs of restaurants from most similar to least similar. The results show that the Burger King is most similar to Wendy's, so the owners know that the strongest competitor is the Wendy's restaurant because it is thought to be the most similar. Follow-up analysis can identify what attributes influence perceptions of similarity or dissimilarity.

Correspondence Analysis

MPI 3

Correspondence analysis is a recently developed interdependence technique that facilitates the perceptual mapping of objects (e.g., products, persons) on a set of nonmetric attributes. Researchers are constantly faced with the need to "quantify the qualitative data" found in nominal variables. Correspondence analysis differs from the interdependence techniques discussed earlier in its ability to accommodate both nonmetric data and nonlinear relationships.

In its most basic form, correspondence analysis employs a contingency table, which is the cross-tabulation of two categorical variables. It then transforms the nonmetric data to a metric level and performs dimensional reduction (similar to factor analysis) and perceptual mapping. Correspondence analysis provides a multivariate representation of interdependence for nonmetric data that is not possible with other methods.

As an example, respondents' brand preferences can be cross-tabulated on demographic variables (e.g., gender, income categories, occupation) by indicating how many people preferring each brand fall into each category of the demographic variables. Through correspondence analysis, the association, or "correspondence," of brands and the distinguishing characteristics of those preferring each brand are then shown in a two- or three-dimensional map of both brands and respondent characteristics. Brands perceived as similar are located close to one another. Likewise, the most distinguishing characteristics of respondents preferring each brand are also determined by the proximity of the demographic variable categories to the brand's position.

Structural Equation Modeling and Confirmatory Factor Analysis

Structural equation modeling (SEM) is a technique that allows separate relationships for each of a set of dependent variables. In its simplest sense, structural equation modeling provides the appropriate and most efficient estimation technique for a series of separate multiple regression equations estimated simultaneously. It is characterized by two basic components: (1) the structural model and (2) the measurement model. The structural model is the *path* model, which relates independent to dependent variables. In such situations, theory, prior experience, or other guidelines enable the researcher to distinguish which independent variables predict each dependent variable. Models discussed previously that accommodate multiple dependent variables—multivariate analysis of variance and canonical correlation—are not applicable in this situation because they allow only a single relationship between dependent and independent variables.

The measurement model enables the researcher to use several variables (indicators) for a single independent or dependent variable. For example, the dependent variable might be a concept represented by a summated scale, such as self-esteem. In a confirmatory factor analysis the researcher can assess the contribution of each scale item as well as incorporate how well the scale measures the concept (reliability). The scales are then integrated into the estimation of the relationships between dependent and independent variables in the structural model. This procedure is similar to performing a factor analysis (discussed in a later section) of the scale items and using the factor scores in the regression.

A study by management consultants identified several factors that affect worker satisfaction: supervisor support, work environment, and job performance. In addition to this relationship, they noted a separate relationship wherein supervisor support and work environment were unique predictors of job performance. Hence, they had two separate, but interrelated relationships. Supervisor support and the work environment not only affected worker satisfaction directly, but had possible indirect effects through the relationship with job performance, which was also a predictor of worker satisfaction. In attempting to assess these relationships, the consultants also developed multi-item scales for each construct (supervisor support, work environment, job performance, and worker satisfaction). SEM provides a means of not only assessing each of the relationships simultaneously rather than in separate analyses, but also incorporating the multi-item scales in the analysis to account for measurement error associated with each of the scales.

GUIDELINES FOR MULTIVARIATE ANALYSES AND INTERPRETATION

As demonstrated throughout this chapter, multivariate analyses' diverse character leads to quite powerful analytical and predictive capabilities. This power becomes especially tempting when the researcher is unsure of the most appropriate analysis design and relies instead on the multivariate technique as a substitute for the necessary conceptual development. And even when applied correctly, the strengths of accommodating multiple variables and relationships create substantial complexity in the results and their interpretation.

Faced with this complexity, we caution the researcher to proceed only when the requisite conceptual foundation to support the selected technique has been developed. We have already discussed several issues particularly applicable to multivariate analyses, and although no single "answer" exists, we find that analysis and interpretation of any multivariate problem can be aided by following a set of general guidelines. By no means an exhaustive list of considerations, these guidelines represent more of a "philosophy of multivariate analysis" that has served us well. The following sections discuss these points in no particular order and with equal emphasis on all.

Establish Practical Significance as Well as Statistical Significance

The strength of multivariate analysis is its seemingly magical ability of sorting through a myriad number of possible alternatives and finding those with statistical significance. However, with this power must come caution. Many researchers become myopic in focusing solely on the achieved significance of the results without understanding their interpretations, good or bad. A researcher must instead look not only at the statistical significance of the results but also at their **practical significance**. Practical significance asks the question, "So what?" For any managerial application, the results must offer a demonstrable effect that justifies action. In academic settings, research is becoming more focused not only on the statistically significant results but also on their substantive and theoretical implications, which are many times drawn from their practical significance.

For example, a regression analysis is undertaken to predict repurchase intentions, measured as the probability between 0 and 100 that the customer will shop again with the firm. The study is conducted and the results come back significant at the .05 significance level. Executives rush to embrace the results and modify firm strategy accordingly. What goes unnoticed, however, is that even though the relationship was significant, the predictive ability was poor—so poor that the estimate of repurchase probability could vary by as much as ±20 percent at the .05 significance level. The "statistically significant" relationship could thus have a range of error of 40 percentage points!

A customer predicted to have a 50 percent chance of return could really have probabilities from 30 percent to 70 percent, representing unacceptable levels upon which to take action. Had researchers and managers probed the practical or managerial significance of the results, they would have concluded that the relationship still needed refinement before it could be relied upon to guide strategy in any substantive sense.

Recognize That Sample Size Affects All Results

The discussion of statistical power demonstrated the substantial impact sample size plays in achieving statistical significance, both in small and large sample sizes. For smaller samples, the sophistication and complexity of the multivariate technique may easily result in either (1) too little statistical power for the test to realistically identify significant results or (2) too easily "overfitting" the data such that the results are artificially good because they fit the sample yet provide no generalizability.

A similar impact also occurs for large sample sizes, which as discussed earlier, can make the statistical tests overly sensitive. Any time sample sizes exceed 400 respondents, the researcher should examine all significant results to ensure they have practical significance due to the increased statistical power from the sample size.

Sample sizes also affect the results when the analyses involve groups of respondents, such as discriminant analysis or MANOVA. Unequal sample sizes among groups influence the results and require additional interpretation or analysis. Thus, a researcher or user of multivariate techniques should always assess the results in light of the sample used in the analysis.

Know Your Data

Multivariate techniques, by their very nature, identify complex relationships that are difficult to represent simply. As a result, the tendency is to accept the results without the typical examination one undertakes in univariate and bivariate analyses (e.g., scatterplots of correlations and boxplots of mean comparisons). Such shortcuts can be a prelude to disaster, however. Multivariate analyses require an even more rigorous examination of the data because the influence of outliers, violations of assumptions, and missing data can be compounded across several variables to create substantial effects.

A wide-ranging set of diagnostic techniques enables discovery of these multivariate relationships in ways quite similar to the univariate and bivariate methods. The multivariate researcher must take the time to utilize these diagnostic measures for a greater understanding of the data and the basic relationships that exist. With this understanding, the researcher grasps not only "the big picture," but also knows where to look for alternative formulations of the original model that can aid in model fit, such as nonlinear and interactive relationships.

Strive for Model Parsimony

Multivariate techniques are designed to accommodate multiple variables in the analysis. This feature, however, should not substitute for conceptual model development before the multivariate techniques are applied. Although it is always more important to avoid omitting a critical predictor variable, termed **specification error**, the researcher must also avoid inserting variables indiscriminately and letting the multivariate technique "sort out" the relevant variables for two fundamental reasons:

1. Irrelevant variables usually increase a technique's ability to fit the sample data, but at the expense of overfitting the sample data and making the results less generalizable to the population. We address this issue in more detail when the concept of degrees of freedom is discussed in Chapter 4.
2. Even though irrelevant variables typically do not bias the estimates of the relevant variables, they can mask the true effects due to an increase in multicollinearity. **Multicollinearity** represents the degree to which any variable's effect can be predicted or accounted for by the other

variables in the analysis. As multicollinearity rises, the ability to define any variable's effect is diminished. The addition of irrelevant or marginally significant variables can only increase the degree of multicollinearity, which makes interpretation of all variables more difficult.

Thus, including variables that are conceptually not relevant can lead to several potentially harmful effects, even if the additional variables do not directly bias the model results.

Look at Your Errors

Even with the statistical prowess of multivariate techniques, rarely do we achieve the best prediction in the first analysis. The researcher is then faced with the question, "Where does one go from here?" The best answer is to look at the errors in prediction, whether they are the residuals from regression analysis, the misclassification of observations in discriminant analysis, or outliers in cluster analysis. In each case, the researcher should use the errors in prediction not as a measure of failure or merely something to eliminate, but as a starting point for diagnosing the validity of the obtained results and an indication of the remaining unexplained relationships.

Validate Your Results

The ability of multivariate analyses to identify complex interrelationships also means that results can be found that are specific only to the sample and not generalizable to the population. The researcher must always ensure there are sufficient observations per estimated parameter to avoid "overfitting" the sample, as discussed earlier. Just as important, however, are the efforts to validate the results by one of several methods:

1. Splitting the sample and using one subsample to estimate the model and the second subsample to estimate the predictive accuracy.
2. Gathering a separate sample to ensure that the results are appropriate for other samples.
3. Employing a **bootstrapping** technique [6], which validates a multivariate model by drawing a large number of subsamples, estimating models for each subsample, and then determining the values for the parameter estimates from the set of models by calculating the mean of each estimated coefficient across all the subsample models. This approach also does not rely on statistical assumptions to assess whether a parameter differs from zero (i.e., Are the estimated coefficients statistically different from zero or not?). Instead it examines the actual values from the repeated samples to make this assessment.

Whenever a multivariate technique is employed, the researcher must strive not only to estimate a significant model but to ensure that it is representative of the population as a whole. Remember, the objective is not to find the best "fit" just to the sample data but instead to develop a model that best describes the population as a whole.

A STRUCTURED APPROACH TO MULTIVARIATE MODEL BUILDING

As we discuss the numerous multivariate techniques available to the researcher and the myriad set of issues involved in their application, it becomes apparent that the successful completion of a multivariate analysis involves more than just the selection of the correct method. Issues ranging from problem definition to a critical diagnosis of the results must be addressed. To aid the researcher or user in applying multivariate methods, a six-step approach to multivariate analysis is presented. The intent is not to provide a rigid set of procedures to follow but, instead, to provide a series of guidelines that emphasize a model-building approach. This model-building approach focuses the analysis on a well-defined research plan, starting with a conceptual model detailing the relationships to be examined. Once defined in conceptual terms, the empirical issues can be addressed, including the selection of the specific multivariate technique and the implementation issues. After obtaining significant results, we focus on their interpretation, with special attention directed toward the variate. Finally, the diagnostic measures ensure that the model is valid not only for the sample data but that it is as generalizable as possible. The following discussion briefly describes each step in this approach.

This six-step model-building process provides a framework for developing, interpreting, and validating any multivariate analysis. Each researcher must develop criteria for "success" or "failure" at each stage, but the discussions of each technique provide guidelines whenever available. Emphasis on a model-building approach here, rather than just the specifics of each technique, should provide a broader base of model development, estimation, and interpretation that will improve the multivariate analyses of practitioner and academician alike.

Stage 1: Define the Research Problem, Objectives, and Multivariate Technique to Be Used

The starting point for any multivariate analysis is to define the research problem and analysis objectives in conceptual terms before specifying any variables or measures. The role of conceptual model development, or theory, cannot be overstated. No matter whether in academic or applied research, the researcher must first view the problem in conceptual terms by defining the concepts and identifying the fundamental relationships to be investigated.

A conceptual model need not be complex and detailed; instead, it can be just a simple representation of the relationships to be studied. If a dependence relationship is proposed as the research objective, the researcher needs to specify the dependent and independent concepts. For an application of an interdependence technique, the dimensions of structure or similarity should be specified. Note that a concept (an idea or topic), rather than a specific variable, is defined in both dependence and interdependence situations. This sequence minimizes the chance that relevant concepts will be omitted in the effort to develop measures and to define the specifics of the research design. Readers interested in conceptual model development should see Chapter 11.

With the objective and conceptual model specified, the researcher has only to choose the appropriate multivariate technique based on the measurement characteristics of the dependent and independent variables. Variables for each concept are specified prior to the study in its design, but may be respecified or even stated in a different form (e.g., transformations or creating dummy variables) after the data have been collected.

Stage 2: Develop the Analysis Plan

With the conceptual model established and the multivariate technique selected, attention turns to the implementation issues. The issues include general considerations such as minimum or desired sample sizes and allowable or required types of variables (metric versus nonmetric) and estimation methods.

Stage 3: Evaluate the Assumptions Underlying the Multivariate Technique

With data collected, the first task is not to estimate the multivariate model but to evaluate its underlying assumptions, both statistical and conceptual, that substantially affect their ability to represent multivariate relationships. For the techniques based on statistical inference, the assumptions of multivariate normality, linearity, independence of the error terms, and equality of variances must all be met. Assessing these assumptions is discussed in more detail in Chapter 2. Each technique also involves a series of conceptual assumptions dealing with such issues as model formulation and the types of relationships represented. Before any model estimation is attempted, the researcher must ensure that both statistical and conceptual assumptions are met.

Stage 4: Estimate the Multivariate Model and Assess Overall Model Fit

With the assumptions satisfied, the analysis proceeds to the actual estimation of the multivariate model and an assessment of overall model fit. In the estimation process, the researcher may choose among options to meet specific characteristics of the data (e.g., use of covariates in MANOVA) or to maximize the fit to the data (e.g., rotation of factors or discriminant functions). After the model is estimated, the overall model fit is evaluated to ascertain whether it achieves acceptable levels on statistical criteria (e.g., level of significance), identifies the proposed relationships, and achieves practical significance. Many times, the model will be respecified in an attempt to achieve better levels of overall fit and/or explanation. In all cases, however, an acceptable model must be obtained before proceeding.

No matter what level of overall model fit is found, the researcher must also determine whether the results are unduly affected by any single or small set of observations that indicate the results may be unstable or not generalizable. Ill-fitting observations may be identified as outliers, influential observations, or other disparate results (e.g., single-member clusters or seriously misclassified cases in discriminant analysis).

Stage 5: Interpret the Variate(s)

With an acceptable level of model fit, interpreting the variate(s) reveals the nature of the multivariate relationship. The interpretation of effects for individual variables is made by examining the estimated coefficients (weights) for each variable in the variate. Moreover, some techniques also estimate multiple variates that represent underlying dimensions of comparison or association. The interpretation may lead to additional respecifications of the variables and/or model formulation, wherein the model is reestimated and then interpreted again. The objective is to identify empirical evidence of multivariate relationships in the sample data that can be generalized to the total population.

Stage 6: Validate the Multivariate Model

Before accepting the results, the researcher must subject them to one final set of diagnostic analyses that assess the degree of generalizability of the results by the available validation methods. The attempts to validate the model are directed toward demonstrating the generalizability of the results to the total population. These diagnostic analyses add little to the interpretation of the results but can be viewed as "insurance" that the results are the most descriptive of the data, yet generalizable to the population.

A Decision Flowchart

For each multivariate technique, the six-step approach to multivariate model building will be portrayed in a decision flowchart partitioned into two sections. The first section (stages 1 through 3) deals with the issues addressed while preparing for actual model estimation (i.e., research objectives, research design considerations, and testing for assumptions). The second section of the decision flowchart (stages 4 through 6) deals with the issues pertaining to model estimation, interpretation, and validation. The decision flowchart provides the researcher with a simplified but systematic method of applying the structural approach to multivariate model building to any application of the multivariate technique.

DATABASES

To explain and illustrate each of the multivariate techniques more fully, we use hypothetical data sets throughout the book. The data sets are for HBAT Industries (HBAT), a manufacturer of paper products. Each data set is assumed to be based on surveys of HBAT customers completed on a secure Web site managed by an established marketing research company. The research company contacts purchasing managers and encourages them to participate. To do so, managers log onto the Web site and complete the survey. The data sets are supplemented by other information compiled and stored in HBAT's data warehouse and accessible through its decision support system.

Primary Database

The primary database, consisting of 100 observations on 18 separate variables, is based on a market segmentation study of HBAT customers. HBAT sells paper products to two market segments: the newsprint industry and the magazine industry. Also, paper products are sold to these market segments either directly to the customer or indirectly through a broker. Two types of information were collected in the surveys. The first type of information was perceptions of HBAT's performance on 13 attributes. These attributes, developed through focus groups, a pretest, and use in previous studies, are considered to be the most influential in the selection of suppliers in the paper industry. Respondents included purchasing managers of firms buying from HBAT, and they rated HBAT on each of the 13 attributes using a 0–10 scale, with 10 being "Excellent" and 0 being "Poor." The second type of information relates to purchase outcomes and business relationships (e.g., satisfaction with HBAT and whether the firm would consider a strategic alliance/ partnership with HBAT). A third type of information is available from HBAT's data warehouse and includes information such as size of customer and length of purchase relationship.

By analyzing the data, HBAT can develop a better understanding of both the characteristics of its customers and the relationships between their perceptions of HBAT, and their actions toward HBAT (e.g., satisfaction and likelihood to recommend). From this understanding of its customers, HBAT will be in a good position to develop its marketing plan for next year. Brief descriptions of the database variables are provided in Table 1-3, in which the variables are classified as either

TABLE 1-3   Description of Database Variables

Variable Description	Variable Type
Data Warehouse Classification Variables	
X_1 Customer Type	Nonmetric
X_2 Industry Type	Nonmetric
X_3 Firm Size	Nonmetric
X_4 Region	Nonmetric
X_5 Distribution System	Nonmetric
Performance Perceptions Variables	
X_6 Product Quality	Metric
X_7 E-Commerce Activities/Web Site	Metric
X_8 Technical Support	Metric
X_9 Complaint Resolution	Metric
X_{10} Advertising	Metric
X_{11} Product Line	Metric
X_{12} Salesforce Image	Metric
X_{13} Competitive Pricing	Metric
X_{14} Warranty and Claims	Metric
X_{15} New Products	Metric
X_{16} Ordering and Billing	Metric
X_{17} Price Flexibility	Metric
X_{18} Delivery Speed	Metric
Outcome/Relationship Measures	
X_{19} Satisfaction	Metric
X_{20} Likelihood of Recommendation	Metric
X_{21} Likelihood of Future Purchase	Metric
X_{22} Current Purchase/Usage Level	Metric
X_{23} Consider Strategic Alliance/Partnership in Future	Nonmetric

independent or dependent, and either metric or nonmetric. Also, a complete listing and electronic copy of the database are available on the Web at www.pearsonglobaleditions.com/hair or www.mvstats.com. A definition of each variable and an explanation of its coding are provided in the following sections.

DATA WAREHOUSE CLASSIFICATION VARIABLES   As respondents were selected for the sample to be used by the marketing research firm, five variables also were extracted from HBAT's data warehouse to reflect the basic firm characteristics and their business relationship with HBAT. The five variables are as follows:

X_1	Customer Type	Length of time a particular customer has been buying from HBAT: 1 = less than 1 year 2 = between 1 and 5 years 3 = longer than 5 years
X_2	Industry Type	Type of industry that purchases HBAT's paper products: 0 = magazine industry 1 = newsprint industry
X_3	Firm Size	Employee size: 0 = small firm, fewer than 500 employees 1 = large firm, 500 or more employees
X_4	Region	Customer location: 0 = USA/North America 1 = outside North America
X_5	Distribution System	How paper products are sold to customers: 0 = sold indirectly through a broker 1 = sold directly

PERCEPTIONS OF HBAT   Each respondent's perceptions of HBAT on a set of business functions were measured on a graphic rating scale, where a 10-centimeter line was drawn between the endpoints, labeled "Poor" and "Excellent," shown here.

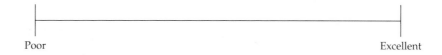

Poor   Excellent

As part of the survey, respondents indicated their perceptions by making a mark anywhere on the line. The location of the mark was electronically observed and the distance from 0 (in centimeters) was recorded in the database for that particular survey. The result was a scale ranging from 0 to 10, rounded to a single decimal place. The 13 HBAT attributes rated by each respondent were as follows:

X_6	Product Quality	Perceived level of quality of HBAT's paper products
X_7	E-Commerce Activities/Web Site	Overall image of HBAT's Web site, especially user-friendliness
X_8	Technical Support	Extent to which technical support is offered to help solve product/service issues
X_9	Complaint Resolution	Extent to which any complaints are resolved in a timely and complete manner

X_{10}	Advertising	Perceptions of HBAT's advertising campaigns in all types of media
X_{11}	Product Line	Depth and breadth of HBAT's product line to meet customer needs
X_{12}	Salesforce Image	Overall image of HBAT's salesforce
X_{13}	Competitive Pricing	Extent to which HBAT offers competitive prices
X_{14}	Warranty and Claims	Extent to which HBAT stands behind its product/service warranties and claims
X_{15}	New Products	Extent to which HBAT develops and sells new products
X_{16}	Ordering and Billing	Perception that ordering and billing is handled efficiently and correctly
X_{17}	Price Flexibility	Perceived willingness of HBAT sales reps to negotiate price on purchases of paper products
X_{18}	Delivery Speed	Amount of time it takes to deliver the paper products once an order has been confirmed

PURCHASE OUTCOMES   Five specific measures were obtained that reflected the outcomes of the respondent's purchase relationships with HBAT. These measures include the following:

X_{19}	Customer Satisfaction	Customer satisfaction with past purchases from HBAT, measured on a 10-point graphic rating scale
X_{20}	Likelihood of Recommending HBAT	Likelihood of recommending HBAT to other firms as a supplier of paper products, measured on a 10-point graphic rating scale
X_{21}	Likelihood of Future Purchases from HBAT	Likelihood of purchasing paper products from HBAT in the future, measured on a 10-point graphic rating scale
X_{22}	Percentage of Purchases from HBAT	Percentage of the responding firm's paper needs purchased from HBAT, measured on a 100-point percentage scale
X_{23}	Perception of Future Relationship with HBAT	Extent to which the customer/respondent perceives his or her firm would engage in strategic alliance/partnership with HBAT:

0 = Would not consider

1 = Yes, would consider strategic alliance or partnership

Other Databases

Five other specialized databases are used in the text. First, Chapter 8 uses an expanded version of the HBAT database containing 200 respondents (HBAT200) that provides sufficient sample sizes for more complex MANOVA analyses. Chapter 2 uses a smaller database (HATMISS) to illustrate the handling of missing data. Chapter 10 on MDS and Correspondence Analysis and the SEM chapters (11 and 12) use different databases that meet the unique data requirements for those techniques. In each instance, the database is described more fully in those chapters. All of the databases used in the text are available at www.pearsonglobaleditions.com/hair or www.mvstats.com.

ORGANIZATION OF THE REMAINING CHAPTERS

The remaining chapters of the text are organized into five sections, each addressing a separate stage in performing a multivariate analysis.

Section I: Understanding and Preparing For Multivariate Analysis

The initial section addresses issues that must be resolved before a multivariate analysis can be performed. It begins with Chapter 2, which covers the topics of accommodating missing data, assurance of meeting the underlying statistical assumptions, and identifying outliers that might disproportionately affect the results. Chapter 3 covers factor analysis, a technique particularly suited to examining the relationships among variables and the opportunities for creating summated scales. These two chapters combine to provide the researcher not only the diagnostic tools necessary for preparing the data for analysis, but also the means for data reduction and scale construction that can be included in other multivariate techniques.

Section II: Analysis Using Dependence Techniques

This section covers six dependence techniques—multiple regression (Chapter 4), canonical correlation (Chapter 5), conjoint analysis (Chapter 6), discriminant analysis and logistic regression (Chapter 7), and multivariate analysis of variance (Chapter 8). Dependence techniques, as noted earlier, enable the researcher to assess the degree of relationship between dependent and independent variables. The dependence techniques vary in the type and character of the relationship as reflected in the measurement properties of the dependent and independent variables. Each technique is examined for its unique perspective on assessing a dependence relationship and its ability to address a particular type of research objective.

Section III: Analysis Using Interdependence Techniques

Two chapters (Chapters 9 and 10) cover the techniques of cluster analysis and perceptual mapping. These techniques present the researcher with tools particularly suited to assessing structure by focusing on the portrayal of the relationships among and between objects, whether they are respondents (cluster analysis) or objects such as firms, products, and so forth (perceptual mapping). It should be noted that one of the primary interdependence techniques, factor analysis and its ability to assess the relationship among variables, is covered in Section I.

Section IV: Structural Equations Modeling

This section introduces the researcher to a widely used advanced multivariate technique, structural equation modeling. Chapter 11 provides an overview of structural equation modeling, focusing on the application of a decision process to SEM analyses. Chapter 12 extends the SEM discussion to the two most widely used applications: confirmatory factor analysis (CFA) and structural modeling.

Summary

Multivariate data analysis is a powerful tool for researchers. Proper application of these techniques reveals relationships that otherwise would not be identified. This chapter introduces you to the major concepts and helps you to do the following:

Explain what multivariate analysis is and when its application is appropriate.   Multivariate analysis techniques are popular because they enable organizations to create knowledge and thereby improve their decision making. Multivariate analysis refers to all statistical techniques that simultaneously analyze multiple measurements on individuals or objects under investigation. Thus, any simultaneous analysis of more than two variables can be considered multivariate analysis.

Some confusion may arise about what multivariate analysis is because the term is not used consistently in the literature. Some researchers use *multivariate* simply to mean examining relationships between or among more than two variables. Others use the term only for problems in which all the multiple variables are assumed to have a multivariate normal distribution. In this book, we do not insist on a rigid definition of multivariate analysis. Instead, multivariate analysis includes both multivariable techniques and truly multivariate techniques, because we believe that knowledge of multivariable techniques is an essential first step in understanding multivariate analysis.

Discuss the nature of measurement scales and their relationship to multivariate techniques.   Data analysis involves the identification and measurement of variation in a set of variables, either among themselves or between a dependent variable and one or more independent variables. The key word here is *measurement* because the researcher cannot identify variation unless it can be measured. Measurement is important in accurately representing the research concepts being studied and is instrumental in the selection of the appropriate multivariate method of analysis. Data can be classified into one of two categories—nonmetric (qualitative) and metric (quantitative)—based on the type of attributes or characteristics they represent. The researcher must define the measurement type for each variable. To the computer, the values are only numbers. Whether data are metric or nonmetric substantially affects what the data can represent, how it can be analyzed, and the appropriate multivariate techniques to use.

Understand the nature of measurement error and its impact on multivariate analysis.   Use of multiple variables and reliance on their combination (the variate) in multivariate methods focuses attention on a complementary issue: measurement error. Measurement error is the degree to which the observed values are not representative of the "true" values. Measurement error has many sources, ranging from data entry errors to the imprecision of the measurement and the inability of respondents to accurately provide information. Thus, all variables used in multivariate techniques must be assumed to have some degree of measurement error. When variables with measurement error are used to compute correlations or means, the "true" effect is partially masked by the measurement error, causing the correlations to weaken and the means to be less precise.

Determine which multivariate technique is appropriate for a specific research problem.   The multivariate techniques can be classified based on three judgments the researcher must make about the research objective and nature of the data: (1) Can the variables be divided into independent and dependent classifications based on some

theory? (2) If they can, how many variables are treated as dependent in a single analysis? and (3) How are the variables, both dependent and independent, measured? Selection of the appropriate multivariate technique depends on the answers to these three questions.

Define the specific techniques included in multivariate analysis.   Multivariate analysis is an ever-expanding set of techniques for data analysis that encompasses a wide range of possible research situations. Among the more established and emerging techniques are principal components and common factor analysis; multiple regression and multiple correlation; multiple discriminant analysis and logistic regression; canonical correlation analysis; multivariate analysis of variance and covariance; conjoint analysis; cluster analysis; perceptual mapping, also known as multidimensional scaling; correspondence analysis; and structural equation modeling (SEM), which includes confirmatory factor analysis.

Discuss the guidelines for application and interpretation of multivariate analyses.   Multivariate analyses have powerful analytical and predictive capabilities. The strengths of accommodating multiple variables and relationships create substantial complexity in the results and their interpretation. Faced with this complexity, the researcher is cautioned to use multivariate methods only when the requisite conceptual foundation to support the selected technique has been developed. The following guidelines represent a "philosophy of multivariate analysis" that should be followed in their application:

1. Establish practical significance as well as statistical significance.
2. Recognize that sample size affects all results.
3. Know your data.
4. Strive for model parsimony.
5. Look at your errors.
6. Validate your results.

Understand the six-step approach to multivariate model building.   The six-step model-building process provides a framework for developing, interpreting, and validating any multivariate analysis.

1. Define the research problem, objectives, and multivariate technique to be used.
2. Develop the analysis plan.
3. Evaluate the assumptions.
4. Estimate the multivariate model and evaluate fit.
5. Interpret the variates.
6. Validate the multivariate model.

This chapter introduced the exciting, challenging topic of multivariate data analysis. The following chapters discuss each of the techniques in sufficient detail to enable the novice researcher to understand what a particular technique can achieve, when and how it should be applied, and how the results of its application are to be interpreted.

Questions

1. In your own words, define *multivariate analysis*.
2. Name the most important factors contributing to the increased application of techniques for multivariate data analysis in the last decade.
3. List and describe the multivariate data analysis techniques described in this chapter. Cite examples for which each technique is appropriate.
4. Explain why and how the various multivariate methods can be viewed as a family of techniques.
5. Why is knowledge of measurement scales important to an understanding of multivariate data analysis?
6. What are the differences between statistical and practical significance? Is one a prerequisite for the other?
7. What are the implications of low statistical power? How can the power be improved if it is deemed too low?
8. Detail the model-building approach to multivariate analysis, focusing on the major issues at each step.

Suggested Readings

A list of suggested readings illustrating issues and applications of multivariate techniques in general is available on the Web at www.pearsonglobaleditions.com/hair or www.mvstats.com.

References

1. Bearden, William O., and Richard G. Netemeyer. 1999. *Handbook of Marketing Scales, Multi-Item Measures for Marketing and Consumer Behavior,* 2nd ed. Thousand Oaks, CA: Sage.
2. BMDP Statistical Software, Inc. 1991. *SOLO Power Analysis.* Los Angeles.
3. Brent, Edward E., Edward J. Mirielli, and Alan Thompson. 1993. *Ex-Sample™: An Expert System to Assist in Determining Sample Size, Version 3.0.* Columbia, MO: Idea Works.
4. Brunner, Gordon C., Karen E. James, and Paul J. Hensel. 2001. *Marketing Scales Handbook,* Vol. 3, *A Compilation of Multi-Item Measures.* Chicago: American Marketing Association.
5. Cohen, J. 1988. *Statistical Power Analysis for the Behavioral Sciences,* 2nd ed. Hillsdale, NJ: Lawrence Erlbaum Publishing.
6. Mooney, Christopher Z., and Robert D. Duval. 1993. *Bootstrapping: A Nonparametric Approach to Statistical Inference.* Thousand Oaks, CA: Sage.
7. Peters, Tom. 1988. *Thriving on Chaos.* New York: Harper and Row.
8. Sullivan, John L., and Stanley Feldman. 1979. *Multiple Indicators: An Introduction.* Thousand Oaks, CA: Sage.

Understanding and Preparing For Multivariate Analysis

OVERVIEW

Section I provides a set of tools and analyses that help to prepare the researcher for the increased complexity of a multivariate analysis. The prudent researcher appreciates the need for a higher level of understanding of the data, both in statistical and conceptual terms. Although the multivariate techniques discussed in this text present the researcher with a powerful set of analytical tools, they also pose the risk of further separating the researcher from a solid understanding of the data and leading to the misplaced notions that the analyses present a "quick and easy" means of identifying relationships. As the researcher relies more heavily on these techniques to find the answer and less on a conceptual basis and understanding of the fundamental properties of the data, the risk increases for serious problems in the misapplication of techniques, violation of statistical properties, or the inappropriate inference and interpretation of the results. These risks can never be totally eliminated, but the tools and analyses discussed in this section will improve the researcher's ability to recognize many of these problems as they occur and to apply the appropriate remedy.

CHAPTERS IN SECTION I

This section begins with Chapter 2, Examining Your Data, which covers the topics of accommodating missing data, meeting the underlying statistical assumptions, and identifying outliers that might disproportionately affect the results. These analyses provide simple empirical assessments that detail the critical statistical properties of the data. Chapter 3, Factor Analysis, presents a discussion of an interdependence technique particularly suited to examining the relationships among variables and the creation of summated scales. The "search for structure" with factor analysis can reveal substantive interrelationships among variables and provide an objective basis for both conceptual model development and improved parsimony among the variables in a multivariate analysis. Thus, the two chapters in this section combine to provide the researcher not only the diagnostic tools necessary for preparing data for analysis, but also the means for data reduction and scale construction that can markedly improve other multivariate techniques.

CHAPTER 2

Cleaning and Transforming Data

LEARNING OBJECTIVES

Upon completing this chapter, you should be able to do the following:

- Select the appropriate graphical method to examine the characteristics of the data or relationships of interest.
- Assess the type and potential impact of missing data.
- Understand the different types of missing data processes.
- Explain the advantages and disadvantages of the approaches available for dealing with missing data.
- Identify univariate, bivariate, and multivariate outliers.
- Test your data for the assumptions underlying most multivariate techniques.
- Determine the best method of data transformation given a specific problem.
- Understand how to incorporate nonmetric variables as metric variables.

CHAPTER PREVIEW

Data examination is a time-consuming, but necessary, initial step in any analysis that researchers often overlook. Here the researcher evaluates the impact of missing data, identifies outliers, and tests for the assumptions underlying most multivariate techniques. The objective of these data examination tasks is as much to reveal what is not apparent as it is to portray the actual data, because the "hidden" effects are easily overlooked. For example, the biases introduced by nonrandom missing data will never be known unless explicitly identified and remedied by the methods discussed in a later section of this chapter. Moreover, unless the researcher reviews the results on a case-by-case basis, the existence of outliers will not be apparent, even if they substantially affect the results. Violations of the statistical assumption may cause biases or nonsignificance in the results that cannot be distinguished from the true results.

Before we discuss a series of empirical tools to aid in data examination, the introductory section of this chapter offers a summary of various graphical techniques available to the researcher as a means of representing data. These techniques provide the researcher with a set of simple yet comprehensive ways to examine both the individual variables and the relationships among them. The graphical techniques are not meant to replace the empirical tools, but rather provide a complementary means of portraying the data and its relationships. As you will see, a histogram can graphically show the shape of a data distribution, just as we can reflect that same distribution with skewness and kurtosis values. The empirical measures quantify the distribution's characteristics, whereas the histogram portrays them in a simple and visual manner. Likewise, other graphical techniques (i.e., scatterplot and boxplot) show relationships between variables represented by the correlation coefficient and means difference test, respectively.

With the graphical techniques addressed, the next task facing the researcher is how to assess and overcome pitfalls resulting from the research design (e.g., questionnaire design) and data collection practices. Specifically, this chapter addresses the following:

- Evaluation of missing data
- Identification of outliers
- Testing of the assumptions underlying most multivariate techniques

Missing data are a nuisance to researchers and primarily result from errors in data collection or data entry or from the omission of answers by respondents. Classifying missing data and the reasons underlying their presence are addressed through a series of steps that not only identify the impacts of the missing data, but that also provide remedies for dealing with it in the analysis. *Outliers,* or extreme responses, may unduly influence the outcome of any multivariate analysis. For this reason, methods to assess their impact are discussed. Finally, the *statistical assumptions* underlying most multivariate analyses are reviewed. Before applying any multivariate technique, the researcher must assess the fit of the sample data with the statistical assumptions underlying that multivariate technique. For example, researchers wishing to apply regression analysis (Chapter 4) would be particularly interested in assessing the assumptions of normality, homoscedasticity, independence of error, and linearity. Each of these issues should be addressed to some extent for each application of a multivariate technique.

In addition, this chapter introduces the researcher to methods of incorporating nonmetric variables in applications that require metric variables through the creation of a special type of metric variable known as *dummy variables.* The applicability of using dummy variables varies with each data analysis project.

KEY TERMS

Before starting the chapter, review the key terms to develop an understanding of the concepts and terminology used. Throughout the chapter the key terms appear in **boldface.** Other points of emphasis in the chapter and key term cross-references are *italicized.*

All-available approach   *Imputation* method for missing data that computes values based on all-available valid observations, also known as the pairwise approach.

Boxplot   Method of representing the distribution of a variable. A box represents the major portion of the distribution, and the extensions—called whiskers—reach to the extreme points of the distribution. This method is useful in making comparisons of one or more variables across groups.

Censored data   Observations that are incomplete in a systematic and known way. One example occurs in the study of causes of death in a sample in which some individuals are still living. Censored data are an example of *ignorable missing data.*

Comparison group   See *reference category.*

Complete case approach   Approach for handling *missing data* that computes values based on data from complete cases, that is, cases with no missing data. Also known as the listwise approach.

Data transformations   A variable may have an undesirable characteristic, such as nonnormality, that detracts from its use in a multivariate technique. A transformation, such as taking the logarithm or square root of the variable, creates a transformed variable that is more suited to portraying the relationship. Transformations may be applied to either the dependent or independent variables, or both. The need and specific type of transformation may be based on theoretical reasons (e.g., transforming a known nonlinear relationship) or empirical reasons (e.g., problems identified through graphical or statistical means).

Dummy variable   Special metric variable used to represent a single category of a nonmetric variable. To account for L levels of a nonmetric variable, $L - 1$ dummy variables are needed. For example, gender is measured as male or female and could be represented by two dummy variables (X_1 and X_2). When the respondent is male, $X_1 = 1$ and $X_2 = 0$. Likewise, when the respondent is female, $X_1 = 0$ and $X_2 = 1$. However, when $X_1 = 1$, we know that X_2 must equal 0. Thus, we need only one variable, either X_1 or X_2, to represent the variable gender. If a nonmetric variable has three levels, only two dummy variables are needed. We always have one dummy variable less than the number of levels for the nonmetric variable. The omitted category is termed the *reference category*.

Effects coding   Method for specifying the *reference category* for a set of *dummy variables* where the reference category receives a value of minus one (-1) across the set of dummy variables. With this type of coding, the dummy variable coefficients represent group deviations from the mean of all groups, which is in contrast to *indicator coding*.

Heteroscedasticity   See *homoscedasticity*.

Histogram   Graphical display of the distribution of a single variable. By forming frequency counts in categories, the shape of the variable's distribution can be shown. Used to make a visual comparison to the *normal distribution*.

Homoscedasticity   When the variance of the error terms (e) appears constant over a range of predictor variables, the data are said to be homoscedastic. The assumption of equal variance of the population error E (where E is estimated from e) is critical to the proper application of many multivariate techniques. When the error terms have increasing or modulating variance, the data are said to be *heteroscedastic*. Analysis of *residuals* best illustrates this point.

Ignorable missing data   *Missing data process* that is explicitly identifiable and/or is under the control of the researcher. Ignorable missing data do not require a remedy because the missing data are explicitly handled in the technique used.

Imputation   Process of estimating the *missing data* of an observation based on valid values of the other variables. The objective is to employ known relationships that can be identified in the valid values of the sample to assist in representing or even estimating the replacements for missing values.

Indicator coding   Method for specifying the *reference category* for a set of *dummy variables* where the reference category receives a value of zero across the set of dummy variables. The dummy variable coefficients represent the category differences from the reference category. Also see *effects coding*.

Kurtosis   Measure of the peakedness or flatness of a distribution when compared with a *normal distribution*. A positive value indicates a relatively peaked distribution, and a negative value indicates a relatively flat distribution.

Linearity   Used to express the concept that the model possesses the properties of additivity and homogeneity. In a simple sense, linear models predict values that fall in a straight line by having a constant unit change (slope) of the dependent variable for a constant unit change of the independent variable. In the population model $Y = b_0 + b_1 X_1 + e$, the effect of a change of 1 in X_1 is to add b_1 (a constant) units to Y.

Missing at random (MAR)   Classification of *missing data* applicable when missing values of Y depend on X, but not on Y. When missing data are MAR, observed data for Y are a truly random sample for the X values in the sample, but not a random sample of all Y values due to missing values of X.

Missing completely at random (MCAR)   Classification of *missing data* applicable when missing values of Y are not dependent on X. When missing data are MCAR, observed values of Y are a truly random sample of all Y values, with no underlying process that lends bias to the observed data.

Missing data   Information not available for a subject (or case) about whom other information is available. Missing data often occur when a respondent fails to answer one or more questions in a survey.

Missing data process   Any systematic event external to the respondent (such as data entry errors or data collection problems) or any action on the part of the respondent (such as refusal to answer a question) that leads to *missing data.*

Multivariate graphical display   Method of presenting a multivariate profile of an observation on three or more variables. The methods include approaches such as glyphs, mathematical transformations, and even iconic representations (e.g., faces).

Normal distribution   Purely theoretical continuous probability distribution in which the horizontal axis represents all possible values of a variable and the vertical axis represents the probability of those values occurring. The scores on the variable are clustered around the mean in a symmetrical, unimodal pattern known as the bell-shaped, or normal, curve.

Normal probability plot   Graphical comparison of the form of the distribution to the *normal distribution.* In the normal probability plot, the normal distribution is represented by a straight line angled at 45 degrees. The actual distribution is plotted against this line so that any differences are shown as deviations from the straight line, making identification of differences quite apparent and interpretable.

Normality   Degree to which the distribution of the sample data corresponds to a *normal distribution.*

Outlier   An observation that is substantially different from the other observations (i.e., has an extreme value) on one or more characteristics (variables). At issue is its representativeness of the population.

Reference category   The category of a nonmetric variable that is omitted when creating *dummy variables* and acts as a reference point in interpreting the dummy variables. In *indicator coding,* the reference category has values of zero (0) for all dummy variables. With *effects coding,* the reference category has values of minus one (−1) for all dummy variables.

Residual   Portion of a dependent variable not explained by a multivariate technique. Associated with dependence methods that attempt to predict the dependent variable, the residual represents the unexplained portion of the dependent variable. Residuals can be used in diagnostic procedures to identify problems in the estimation technique or to identify unspecified relationships.

Robustness   The ability of a statistical technique to perform reasonably well even when the underlying statistical assumptions have been violated in some manner.

Scatterplot   Representation of the relationship between two metric variables portraying the joint values of each observation in a two-dimensional graph.

Skewness   Measure of the symmetry of a distribution; in most instances the comparison is made to a *normal distribution.* A positively skewed distribution has relatively few large values and tails off to the right, and a negatively skewed distribution has relatively few small values and tails off to the left. Skewness values falling outside the range of −1 to +1 indicate a substantially skewed distribution.

Variate   Linear combination of variables formed in the multivariate technique by deriving empirical weights applied to a set of variables specified by the researcher.

INTRODUCTION

The tasks involved in examining your data may seem mundane and inconsequential, but they are an essential part of any multivariate analysis. Multivariate techniques place tremendous analytical power in the researcher's hands. But they also place a greater burden on the researcher to ensure that the statistical and theoretical underpinnings on which they are based also are supported. By examining the data before the application of any multivariate technique, the researcher gains several critical insights into the characteristics of the data:

- First and foremost, the researcher attains a basic understanding of the data and relationships between variables. Multivariate techniques place greater demands on the researcher to understand, interpret, and articulate results based on relationships that are more complex than encountered before. A thorough knowledge of the variable interrelationships can aid immeasurably in the specification and refinement of the multivariate model as well as provide a reasoned perspective for interpretation of the results.

- Second, the researcher ensures that the data underlying the analysis meet all of the requirements for a multivariate analysis. Multivariate techniques demand much more from the data in terms of larger data sets and more complex assumptions than encountered with univariate analyses. Missing data, outliers, and the statistical characteristics of the data are all much more difficult to assess in a multivariate context. Thus, the analytical sophistication needed to ensure that these requirements are met forces the researcher to use a series of data examination techniques that are as complex as the multivariate techniques themselves.

Both novice and experienced researchers may be tempted to skim or even skip this chapter to spend more time in gaining knowledge of a multivariate technique(s). The time, effort, and resources devoted to the data examination process may seem almost wasted because many times no corrective action is warranted. The researcher should instead view these techniques as "investments in multivariate insurance" that ensure the results obtained from the multivariate analysis are truly valid and accurate. Without such an "investment" it is quite easy, for example, for several unidentified outliers to skew the results, for missing data to introduce a bias in the correlations between variables, or for nonnormal variables to invalidate the results. And yet the most troubling aspect of these problems is that they are "hidden," because in most instances the multivariate techniques will go ahead and provide results. Only if the researcher has made the "investment" will the potential for catastrophic problems be recognized and corrected *before* the analyses are performed. These problems can be avoided by following these analyses each and every time a multivariate technique is applied. These efforts will more than pay for themselves in the long run; the occurrence of one serious and possibly fatal problem will make a convert of any researcher. We encourage you to embrace these techniques before problems that arise during analysis force you to do so.

GRAPHICAL EXAMINATION OF THE DATA

As discussed earlier, the use of multivariate techniques places an increased burden on the researcher to understand, evaluate, and interpret complex results. This complexity requires a thorough understanding of the basic characteristics of the underlying data and relationships. When univariate analyses are considered, the level of understanding is fairly simple. As the researcher moves to more complex multivariate analyses, however, the need and level of understanding increase dramatically and require even more powerful empirical diagnostic measures. The researcher can be aided immeasurably in gaining a fuller understanding of what these diagnostic measures mean through the use of graphical techniques, portraying the basic characteristics of individual variables and relationships between variables in a simple "picture." For example, a simple scatterplot represents in a single picture not only the two basic elements of a correlation coefficient, namely the type of relationship (positive or negative) and the strength of the relationship (the dispersion of the cases), but also a simple visual means for assessing linearity that would require a much more detailed analysis if attempted strictly by empirical means. Correspondingly, a boxplot illustrates not only the overall level of differences across groups shown in a *t*-test or analysis of variance, but also the differences between pairs of groups and the existence of outliers that would otherwise take more empirical analysis to detect if the graphical method was not employed. The objective in using graphical techniques is not to replace the empirical measures, but to use them as a complement to provide a visual representation of the basic relationships so that researchers can feel confident in their understanding of these relationships.

The advent and widespread use of statistical programs designed for the personal computer increased access to such methods. Most statistical programs provide comprehensive modules of graphical techniques available for data examination that are augmented with more detailed statistical measures of data description. The following sections detail some of the more widely used techniques for examining the characteristics of the distribution, bivariate relationships, group differences, and even multivariate profiles.

Univariate Profiling: Examining the Shape of the Distribution

The starting point for understanding the nature of any variable is to characterize the shape of its distribution. A number of statistical measures are discussed in a later section on normality, but many times the researcher can gain an adequate perspective of the variable through a **histogram.** A histogram is a graphical representation of a single variable that represents the frequency of occurrences (data values) within data categories. The frequencies are plotted to examine the shape of the distribution of values. If the integer values ranged from 1 to 10, the researcher could construct a histogram by counting the number of responses for each integer value. For continuous variables, categories are formed within which the frequency of data values is tabulated. If examination of the distribution is to assess its normality (see section on testing assumptions for details on this issue), the normal curve can be superimposed on the distribution to assess the correspondence of the actual distribution to the desired (normal) distribution. The histogram can be used to examine any type of metric variable.

For example, the responses for X_6 from the database introduced in Chapter 1 are represented in Figure 2-1. The height of the bars represents the frequencies of data values within each category. The normal curve is also superimposed on the distribution. As will be shown in a later section, empirical measures indicate that the distribution of X_6 deviates significantly from the normal distribution. But how does it differ? The empirical measure that differs most is the kurtosis, representing the peakedness or flatness of the distribution. The values indicate that the distribution is flatter than expected. What does the histogram show? The middle of the distribution falls below the superimposed normal curve, while both tails are higher than expected. Thus, the distribution shows no appreciable skewness to one side or the other, just a shortage of observations in the center of the distribution. This comparison also provides guidance on the type of transformation that would be effective if applied as a remedy for nonnormality. All of this information about the distribution is shown through a single histogram.

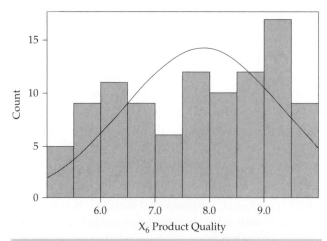

FIGURE 2-1   Graphical Representation of Univariate Distribution

Bivariate Profiling: Examining the Relationship Between Variables

Whereas examining the distribution of a variable is essential, many times the researcher is also interested in examining relationships between two or more variables. The most popular method for examining bivariate relationships is the **scatterplot,** a graph of data points based on two metric variables. One variable defines the horizontal axis and the other variable defines the vertical axis. Variables may be any metric value. The points in the graph represent the corresponding joint values of the variables for any given case. The pattern of points represents the relationship between the variables. A strong organization of points along a straight line characterizes a linear relationship or correlation. A curved set of points may denote a nonlinear relationship, which can be accommodated in many ways (see later discussion on linearity). Or a seemingly random pattern of points may indicate no relationship.

Of the many types of scatterplots, one format particularly suited to multivariate techniques is the scatterplot matrix, in which the scatterplots are represented for all combinations of variables in the lower portion of the matrix. The diagonal contains histograms of the variables. Scatterplot matrices and individual scatterplots are now available in all popular statistical programs. A variant of the scatterplot is discussed in the following section on outlier detection, where an ellipse representing a specified confidence interval for the bivariate normal distribution is superimposed to allow for outlier identification.

Figure 2-2 presents the scatterplots for a set of five variables from the HBAT database (X_6, X_7, X_8, X_{12}, and X_{13}). For example, the highest correlation can be easily identified as between X_7 and X_{12},

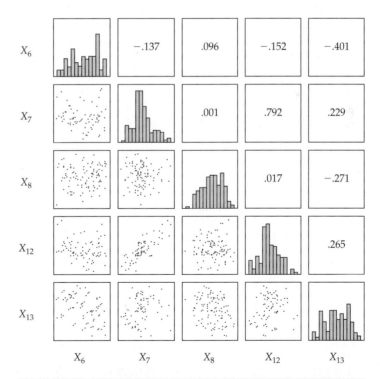

FIGURE 2-2   Bivariate Profiling of Relationships Between Variables: Scatterplot Matrix of Selected Metric Variables (X_6, X_7, X_8, X_{12}, and X_{13})

Note: Values above the diagonal are bivariate correlations, with corresponding scatterplot below the diagonal. Diagonal portrays the distribution of each variable.

as indicated by the observations closely aligned in a well-defined linear pattern. In the opposite extreme, the correlation just above (X_7 versus X_8) shows an almost total lack of relationship as evidenced by the widely dispersed pattern of points and the correlation .001. Finally, an inverse or negative relationship is seen for several combinations, most notably the correlation of X_6 and X_{13} (−.401). Moreover, no combination seems to exhibit a nonlinear relationship that would not be represented in a bivariate correlation.

The scatterplot matrix provides a quick and simple method of not only assessing the strength and magnitude of any bivariate relationship, but also a means of identifying any nonlinear patterns that might be hidden if only the bivariate correlations, which are based on a linear relationship, are examined.

Bivariate Profiling: Examining Group Differences

The researcher also is faced with understanding the extent and character of differences of one or more metric variables across two or more groups formed from the categories of a nonmetric variable. Assessing group differences is done through univariate analyses such as t-tests and analysis of variance and the multivariate techniques of discriminant analysis and multivariate analysis of variance. Another important aspect is to identify outliers (described in more detail in a later section) that may become apparent only when the data values are separated into groups.

The graphical method used for this task is the **boxplot**, a pictorial representation of the data distribution of a metric variable for each group (category) of a nonmetric variable (see example in Figure 2-3). First, the upper and lower quartiles of the data distribution form the upper and lower boundaries of the box, with the box length being the distance between the 25th percentile and the 75th percentile. The box contains the middle 50 percent of the data values and the larger the box, the greater the spread (e.g., standard deviation) of the observations. The median is depicted by a solid line within the box. If the median lies near one end of the box, skewness in the opposite direction is indicated. The lines extending from each box (called *whiskers*) represent the distance to the smallest and the largest observations that are less than one quartile range from the box. Outliers (observations that range between 1.0 and 1.5 quartiles away from the box) and extreme values (observations greater than 1.5 quartiles away from the end of the box) are depicted by symbols outside the whiskers. In using boxplots, the objective is to portray not only the information that is given in the

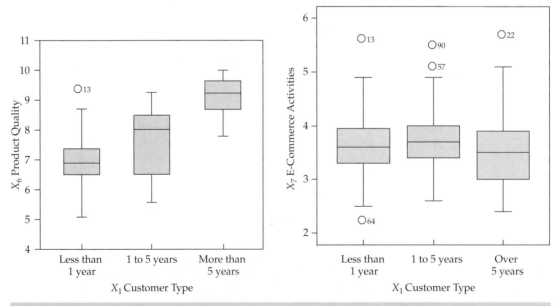

FIGURE 2-3   Bivariate Profiling of Group Differences: Boxplots of X_6 (Product Quality) and X_7 (E-Commerce Activities) with X_1 (Customer Type)

statistical tests (Are the groups different?), but also additional descriptive information that adds to our understanding of the group differences.

Figure 2-3 shows the boxplots for X_6 and X_7 for each of the three groups of X_1 (Customer Type). Before examining the boxplots for each variable, let us first see what the statistical tests tell us about the differences across these groups for each variable. For X_6, a simple analysis of variance test indicates a highly significant statistical difference (F value of 36.6 and a significance level of .000) across the three groups. For X_7, however, the analysis of variance test shows no statistically significant difference (significance level of .419) across the groups of X_1.

Using boxplots, what can we learn about these same group differences? As we view the boxplot of X_6, we do see substantial differences across the groups that confirm the statistical results. We can also see that the primary differences are between groups 1 and 2 versus group 3. Essentially, groups 1 and 2 seem about equal. If we performed more statistical tests looking at each pair of groups separately, the tests would confirm that the only statistically significant differences are group 1 versus 3 and group 2 versus 3. Also, we can see that group 2 has substantially more dispersion (a larger box section in the boxplot), which prevents its difference from group 1. The boxplots thus provide more information about the extent of the group differences of X_6 than just the statistical test.

For X_7, we can see that the three groups are essentially equal, as verified by the nonsignificant statistical test. We can also see a number of outliers in each of the three groups (as indicated by the notations at the upper portion of each plot beyond the whiskers). Although the outliers do not impact the group differences in this case, the researcher is alerted to their presence by the boxplots. The researcher could examine these observations and consider the possible remedies discussed in more detail later in this chapter.

Multivariate Profiles

To this point the graphical methods have been restricted to univariate or bivariate portrayals. In many instances, however, the researcher may desire to compare observations characterized on a multivariate profile, whether it be for descriptive purposes or as a complement to analytical procedures. To address this need, a number of **multivariate graphical displays** center around one of three types of graphs [10]. The first graph type is a direct portrayal of the data values, either by (a) glyphs, or metroglyphs, which are some form of circle with radii that correspond to a data value; or (b) multivariate profiles, which portray a barlike profile for each observation. A second type of multivariate display involves a mathematical transformation of the original data into a mathematical relationship, which can then be portrayed graphically. The most common technique of this type is Andrew's Fourier transformation [1]. The final approach is the use of graphical displays with iconic representativeness, the most popular being a face [5]. The value of this type of display is the inherent processing capacity humans have for their interpretation. As noted by Chernoff [5]:

> I believe that we learn very early to study and react to real faces. Our library of responses to faces exhausts a large part of our dictionary of emotions and ideas. We perceive the faces as a gestalt and our built-in computer is quick to pick out the relevant information and to filter out the noise when looking at a limited number of faces.

Facial representations provide a potent graphical format but also give rise to a number of considerations that affect the assignment of variables to facial features, unintended perceptions, and the quantity of information that can actually be accommodated. Discussion of these issues is beyond the scope of this text, and interested readers are encouraged to review them before attempting to use these methods [24, 25].

The researcher can employ any of these methods when examining multivariate data to provide a format that is many times more insightful than just a review of the actual data values. Moreover, the multivariate methods enable the researcher to use a single graphical portrayal to represent a large number of variables, instead of using a large number of the univariate or bivariate methods to portray the same number of variables.

MISSING DATA

Missing data, where valid values on one or more variables are not available for analysis, are a fact of life in multivariate analysis. In fact, rarely does the researcher avoid some form of missing data problem. The researcher's challenge is to address the issues raised by missing data that affect the generalizability of the results. To do so, the researcher's *primary concern is to identify the patterns and relationships underlying the missing data in order to maintain as close as possible the original distribution of values when any remedy is applied.* The extent of missing data is a secondary issue in most instances, affecting the type of remedy applied. These patterns and relationships are a result of a **missing data process,** which is any systematic event external to the respondent (such as data entry errors or data collection problems) or any action on the part of the respondent (such as refusal to answer) that leads to missing values. The need to focus on the reasons for missing data comes from the fact that the researcher must understand the processes leading to the missing data in order to select the appropriate course of action.

The Impact of Missing Data

The effects of some missing data processes are known and directly accommodated in the research plan as will be discussed later in this section. More often, the missing data processes, particularly those based on actions by the respondent (e.g., nonresponse to a question or set of questions), are rarely known beforehand. To identify any patterns in the missing data that would characterize the missing data process, the researcher asks such questions as (1) Are the missing data scattered randomly throughout the observations or are distinct patterns identifiable? and (2) How prevalent are the missing data? If distinct patterns are found and the extent of missing data is sufficient to warrant action, then it is assumed that some missing data process is in operation.

Why worry about the missing data processes? Can't the analysis be performed with the valid values we do have? Although it might seem prudent to proceed just with the valid values, both substantive and practical considerations necessitate an examination of the missing data processes.

- The *practical impact* of missing data is the reduction of the sample size available for analysis. For example, if remedies for missing data are not applied, any observation with missing data on any of the variables will be excluded from the analysis. In many multivariate analyses, particularly survey research applications, missing data may eliminate so many observations that what was an adequate sample is reduced to an inadequate sample. For example, it has been shown that if 10 percent of the data is randomly missing in a set of five variables, on average almost 60 percent of the cases will have at least one missing value [17]. Thus, when complete data are required, the sample is reduced to 40 percent of the original size. In such situations, the researcher must either gather additional observations or find a remedy for the missing data in the original sample.
- From a *substantive perspective,* any statistical results based on data with a nonrandom missing data process could be biased. This bias occurs when the missing data process "causes" certain data to be missing and these missing data lead to erroneous results. For example, what if we found that individuals who did not provide their household income tended to be almost exclusively those in the higher income brackets? Wouldn't you be suspect of the results

knowing this specific group of people were excluded? The effects of missing data are sometimes termed *hidden* due to the fact that we still get results from the analyses even without the missing data. The researcher could consider these biased results as valid unless the underlying missing data processes are identified and understood.

The concern for missing data processes is similar to the need to understand the causes of nonresponse in the data collection process. Just as we are concerned about who did not respond during data collection and any subsequent biases, we must also be concerned about the nonresponse or missing data among the collected data. The researcher thus needs to not only remedy the missing data if possible, but also understand any underlying missing data processes and their impacts. Yet, too often, researchers either ignore the missing data or invoke a remedy without regard to the effects of the missing data. The next section employs a simple example to illustrate some of these effects and some simple, yet effective, remedies. Then, a four-step process of identifying and remedying missing data processes is presented. Finally, the four-step process is applied to a small data set with missing data.

A Simple Example of a Missing Data Analysis

To illustrate the substantive and practical impacts of missing data, Table 2-1 contains a simple example of missing data among 20 cases. As is typical of many data sets, particularly in survey research, the number of missing data varies widely among both cases and variables.

In this example, we can see that all of the variables (V_1 to V_5) have some missing data, with V_3 missing more than one-half (55%) of all values. Three cases (3, 13, and 15) have more

TABLE 2-1   Hypothetical Example of Missing Data

Case ID	V_1	V_2	V_3	V_4	V_5	Missing Data by Case Number	Percent
1	1.3	9.9	6.7	3.0	2.6	0	0
2	4.1	5.7			2.9	2	40
3		9.9		3.0		3	60
4	.9	8.6		2.1	1.8	1	20
5	.4	8.3		1.2	1.7	1	20
6	1.5	6.7	4.8		2.5	1	20
7	.2	8.8	4.5	3.0	2.4	0	0
8	2.1	8.0	3.0	3.8	1.4	0	0
9	1.8	7.6		3.2	2.5	1	20
10	4.5	8.0		3.3	2.2	1	20
11	2.5	9.2		3.3	3.9	1	20
12	4.5	6.4	5.3	3.0	2.5	0	9
13					2.7	4	80
14	2.8	6.1	6.4		3.8	1	20
15	3.7			3.0		3	60
16	1.6	6.4	5.0		2.1	1	20
17	.5	9.2		3.3	2.8	1	20
18	2.8	5.2	5.0		2.7	1	20
19	2.2	6.7		2.6	2.9	1	20
20	1.8	9.0	5.0	2.2	3.0	0	0
Missing Data by Variable						Total Missing Values	
Number	2	2	11	6	2	Number: 23	
Percent	10	10	55	30	10	Percent: 23	

than 50 percent missing data and only five cases have complete data. Overall, 23 percent of the data values are missing.

From a *practical standpoint,* the missing data in this example can become quite problematic in terms of reducing the sample size. For example, if a multivariate analysis was performed that required complete data on all five variables, the sample would be reduced to only the five cases with no missing data (cases 1, 7, 8, 12, and 20). This sample size is too few for any type of analysis. Among the remedies for missing data that will be discussed in detail in later sections, an obvious option is the elimination of variables and/or cases. In our example, assuming that the conceptual foundations of the research are not altered substantially by the deletion of a variable, eliminating V_3 is one approach to reducing the number of missing data. By just eliminating V_3, seven additional cases, for a total of 12, now have complete information. If the three cases (3, 13, 15) with exceptionally high numbers of missing data are also eliminated, the total number of missing data is now reduced to only five instances, or 7.4 percent of all values.

The *substantive impact,* however, can be seen in these five that are still missing data; all occur in V_4. By comparing the values of V_2 for the remaining five cases with missing data for V_4 (cases 2, 6, 14, 16, and 18) versus those cases having valid V_4 values, a distinct pattern emerges. The five cases with missing values for V_4 have the five lowest values for V_2, indicating that missing data for V_4 are strongly associated with lower scores on V_2. This systematic association between missing and valid data directly affects any analysis in which V_4 and V_2 are both included. For example, the mean score for V_2 will he higher if cases with missing data on V_4 are excluded (mean = 8.4) than if those five cases are included (mean = 7.8). In this instance, the researcher must always scrutinize results including both V_4 and V_2 for the possible impact of this missing data process on the results.

As we have seen in the example, finding a remedy for missing data (e.g., deleting cases or variables) can be a practical solution for missing data. Yet the researcher must guard against applying such remedies without diagnosis of the missing data processes. Avoiding the diagnosis may address the practical problem of sample size, but only cover up the substantive concerns. What is needed is a structured process of first identifying the presence of missing data processes and then applying the appropriate remedies. In the next section we discuss a four-step process to address both the practical and substantive issues arising from missing data.

A Four-Step Process for Identifying Missing Data and Applying Remedies

As seen in the previous discussions, missing data can have significant impacts on any analysis, particularly those of a multivariate nature. Moreover, as the relationships under investigation become more complex, the possibility also increases of not detecting missing data processes and their effects. These factors combine to make it essential that any multivariate analysis begin with an examination of the missing data processes. To this end, a four-step process (see Figure 2-4) is presented, which addresses the types and extent of missing data, identification of missing data processes, and available remedies for accommodating missing data into multivariate analyses.

STEP 1: DETERMINE THE TYPE OF MISSING DATA   The first step in any examination of missing data is to determine the type of missing data involved. Here the researcher is concerned whether the missing data are part of the research design and under the control of the researcher or whether the "causes" and impacts are truly unknown. Let's start with the missing data that are part of the research design and can be handled directly by the researcher.

Ignorable Missing Data. Many times, missing data are expected and part of the research design. In these instances, the missing data are termed **ignorable missing data,** meaning that specific remedies for missing data are not needed because the allowances for missing data are inherent in the technique used [18, 22]. The justification for designating missing data as ignorable is that the missing data process is operating at random (i.e., the observed values are a random sample of the total set

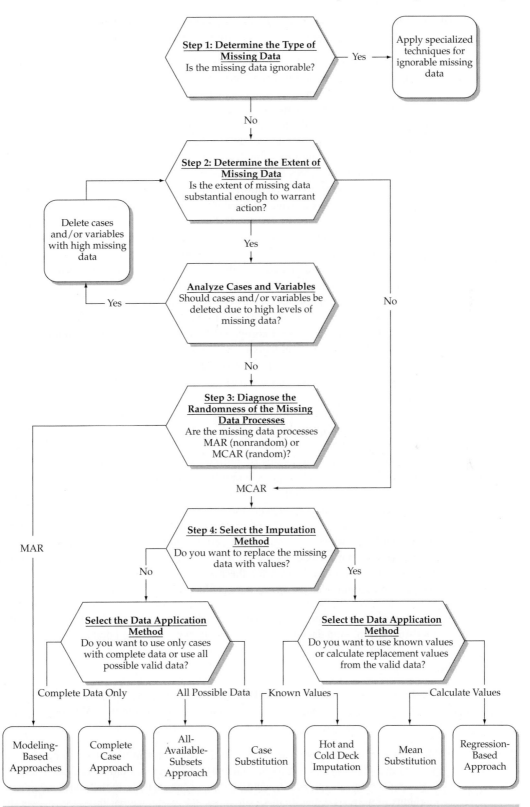

FIGURE 2-4 A Four-Step Process for Identifying Missing Data and Applying Remedies

of values, observed and missing) or explicitly accommodated in the technique used. There are three instances in which a researcher most often encounters ignorable missing data.

- The first example encountered in almost all surveys and most other data sets is the ignorable missing data process resulting from taking a sample of the population rather than gathering data from the entire population. In these instances, the missing data are those observations in a population that are not included when taking a sample. The purpose of multivariate techniques is to generalize from the sample observations to the entire population, which is really an attempt to overcome the missing data of observations not in the sample. The researcher makes these missing data ignorable by using probability sampling to select respondents. Probability sampling enables the researcher to specify that the missing data process leading to the omitted observations is random and that the missing data can be accounted for as sampling error in the statistical procedures. Thus, the missing data of the nonsampled observations are ignorable.

- A second instance of ignorable missing data is due to the specific design of the data collection process. Certain nonprobability sampling plans are designed for specific types of analysis that accommodate the nonrandom nature of the sample. Much more common are missing data due to the design of the data collection instrument, such as through skip patterns where respondents skip sections of questions that are not applicable.

   For example, in examining customer complaint resolution, it might be appropriate to require that individuals make a complaint before asking questions about how complaints are handled. For those respondents not making a complaint, they do not answer the questions on the process and thus create missing data. The researcher is not concerned about these missing data, because they are part of the research design and would be inappropriate to attempt to remedy.

- A third type of ignorable missing data occurs when the data are censored. **Censored data** are observations not complete because of their stage in the missing data process. A typical example is an analysis of the causes of death. Respondents who are still living cannot provide complete information (i.e., cause or time of death) and are thus censored. Another interesting example of censored data is found in the attempt to estimate the heights of the U.S. general population based on the heights of armed services recruits (as cited in [18]). The data are censored because in certain years the armed services had height restrictions that varied in level and enforcement. Thus, the researchers face the task of estimating the heights of the entire population when it is known that certain individuals (i.e., all those below the height restrictions) are not included in the sample. In both instances the researcher's knowledge of the missing data process allows for the use of specialized methods, such as event history analysis, to accommodate censored data [18].

In each instance of an ignorable missing data process, the researcher has an explicit means of accommodating the missing data into the analysis. It should be noted that it is possible to have both ignorable and nonignorable missing data in the same data set when two different missing data processes are in effect.

Missing Data Processes That Are Not Ignorable. Missing data that cannot be classified as ignorable occur for many reasons and in many situations. In general, these missing data fall into two classes based on their source: known versus unknown processes.

- Many missing data processes are *known* to the researcher in that they can be identified due to procedural factors, such as errors in data entry that create invalid codes, disclosure restrictions (e.g., small counts in U.S. Census data), failure to complete the entire questionnaire, or even the morbidity of the respondent. In these situations, the researcher has little control over the missing data processes, but some remedies may be applicable if the missing data are found to be random.

- *Unknown* missing data processes are less easily identified and accommodated. Most often these instances are related directly to the respondent. One example is the refusal to respond to certain questions, which is common in questions of a sensitive nature (e.g., income or controversial issues) or when the respondent has no opinion or insufficient knowledge to answer the question. The researcher should anticipate these problems and attempt to minimize them in the research design and data collection stages of the research. However, they still may occur, and the researcher must now deal with the resulting missing data. But all is not lost. When the missing data occur in a random pattern, remedies may be available to mitigate their effect.

In most instances, the researcher faces a missing data process that cannot be classified as ignorable. Whether the source of this nonignorable missing data process is known or unknown, the researcher must still proceed to the next step of the process and assess the extent and impact of the missing data.

STEP 2: DETERMINE THE EXTENT OF MISSING DATA   Given that some of the missing data are not ignorable, the researcher must next examine the patterns of the missing data and determine the extent of the missing data for individual variables, individual cases, and even overall. The primary issue in this step of the process is to *determine whether the extent or amount of missing data is low enough to not affect the results, even if it operates in a nonrandom manner.* If it is sufficiently low, then any of the approaches for remedying missing data may be applied. If the missing data level is not low enough, then we must first determine the randomness of the missing data process before selecting a remedy (step 3). The unresolved issue at this step is this question: What is low enough? In making the assessment as to the extent of missing data, the researcher may find that the deletion of cases and/or variables will reduce the missing data to levels that are low enough to allow for remedies without concern for creating biases in the results.

Assessing the Extent and Patterns of Missing Data. The most direct means of assessing the extent of missing data is by tabulating (1) the percentage of variables with missing data for each case and (2) the number of cases with missing data for each variable. This simple process identifies not only the extent of missing data, but any exceptionally high levels of missing data that occur for individual cases or observations. The researcher should look for any nonrandom patterns in the data, such as concentration of missing data in a specific set of questions, attrition in not completing the questionnaire, and so on. Finally, the researcher should determine the number of cases with no missing data on any of the variables, which will provide the sample size available for analysis if remedies are not applied.

With this information in hand, the important question is: Is the missing data so high as to warrant additional diagnosis? At issue is the possibility that either ignoring the missing data or using some remedy for substituting values for the missing data can create a bias in the data that will markedly affect the results. Even though most discussions of this issue require researcher judgment, the two guidelines in Rules of Thumb 2-1 apply.

RULES OF THUMB 2-1

How Much Missing Data Is Too Much?

- Missing data under 10 percent for an individual case or observation can generally be ignored, except when the missing data occurs in a specific nonrandom fashion (e.g., concentration in a specific set of questions, attrition at the end of the questionnaire, etc.) [19, 20]
- The number of cases with no missing data must be sufficient for the selected analysis technique if replacement values will not be substituted (imputed) for the missing data

If it is determined that the extent is acceptably low and no specific nonrandom patterns appear, then the researcher can employ any of the imputation techniques (step 4) without biasing the results in any appreciable manner. If the level of missing data is too high, then the researcher must consider specific approaches to diagnosing the randomness of the missing data processes (step 3) before proceeding to apply a remedy.

Deleting Individual Cases and/or Variables.   Before proceeding to the formalized methods of diagnosing randomness in step 3, the researcher should consider the simple remedy of deleting offending case(s) and/or variable(s) with excessive levels of missing data. The researcher may find that the missing data are concentrated in a small subset of cases and/or variables, with their exclusion substantially reducing the extent of the missing data. Moreover, in many cases where a nonrandom pattern of missing data is present, this solution may be the most efficient. Again, no firm guidelines exist on the necessary level for exclusion (other than the general suggestion that the extent should be "large"), but any decision should be based on both empirical and theoretical considerations, as listed in Rules of Thumb 2-2.

Ultimately the researcher must compromise between the gains from deleting variables and/or cases with missing data versus the reduction in sample size and variables to represent the concepts in the study. Obviously, variables or cases with 50 percent or more missing data should be deleted, but as the level of missing data decreases, the researcher must employ more judgment and "trial and error." As we will see when discussing imputation methods, assessing multiple approaches for dealing with missing data is preferable.

STEP 3: DIAGNOSE THE RANDOMNESS OF THE MISSING DATA PROCESSES   Having determined that the extent of missing data is substantial enough to warrant action, the next step is to ascertain the degree of randomness present in the missing data, which then determines the appropriate remedies available. Assume for the purposes of illustration that information on two variables (X and Y) is collected. X has no missing data, but Y has some missing data. A nonrandom missing data process is present between X and Y when significant differences in the values of X occur between cases that have valid data for Y versus those cases with missing data on Y. Any analysis must explicitly accommodate any nonrandom missing data process between X and Y or else bias is introduced into the results.

Levels of Randomness of the Missing Data Process. Of the two levels of randomness when assessing missing data, one requires special methods to accommodate a nonrandom component (Missing At Random, or MAR). A second level (Missing Completely At Random, or MCAR) is sufficiently random to accommodate any type of missing data remedy [18]. Although the titles of

---

RULES OF THUMB 2-2

Deletions Based on Missing Data

- Variables with as little as 15 percent missing data are candidates for deletion [15], but higher levels of missing data (20% to 30%) can often be remedied
- Be sure the overall decrease in missing data is large enough to justify deleting an individual variable or case
- Cases with missing data for dependent variable(s) typically are deleted to avoid any artificial increase in relationships with independent variables
- When deleting a variable, ensure that alternative variables, hopefully highly correlated, are available to represent the intent of the original variable
- Always consider performing the analysis both with and without the deleted cases or variables to identify any marked differences

both levels seem to indicate that they reflect random missing data patterns, only MCAR allows for the use of any remedy desired. The distinction between these two levels is in the generalizability to the population, as described here:

- Missing data are termed **missing at random (MAR)** if the missing values of Y depend on X, but not on Y. In other words, the observed Y values represent a random sample of the actual Y values for each value of X, but the observed data for Y do not necessarily represent a truly random sample of all Y values. Even though the missing data process is random in the sample, its values are not generalizable to the population. Most often, the data are missing randomly within subgroups, but differ in levels between subgroups. The researcher must determine the factors determining the subgroups and the varying levels between groups.

  For example, assume that we know the gender of respondents (the X variable) and are asking about household income (the Y variable). We find that the missing data are random for both males and females but occur at a much higher frequency for males than females. Even though the missing data process is operating in a random manner within the gender variable, any remedy applied to the missing data will still reflect the missing data process because gender affects the ultimate distribution of the household income values.

- A higher level of randomness is termed **missing completely at random (MCAR).** In these instances the observed values of Y are truly a random sample of all Y values, with no underlying process that lends bias to the observed data. In simple terms, the cases with missing data are indistinguishable from cases with complete data.

  From our earlier example, this situation would be shown by the fact that the missing data for household income were randomly missing in equal proportions for both males and females. In this missing data process, any of the remedies can be applied without making allowances for the impact of any other variable or missing data process.

Diagnostic Tests for Levels of Randomness. As previously noted, the researcher must ascertain whether the missing data process occurs in a completely random manner. When the data set is small, the researcher may be able to visually see such patterns or perform a set of simple calculations (such as in our simple example at the beginning of the chapter). However, as sample size and the number of variables increases, so does the need for empirical diagnostic tests. Some statistical programs add techniques specifically designed for missing data analysis (e.g., Missing Value Analysis in SPSS), which generally include one or both diagnostic tests.

- The first diagnostic assesses the missing data process of a single variable Y by forming two groups: observations with missing data for Y and those with valid values of Y. Statistical tests are then performed to determine whether significant differences exist between the two groups on other variables of interest. Significant differences indicate the possibility of a nonrandom missing data process.

  Let us use our earlier example of household income and gender. We would first form two groups of respondents, those with missing data on the household income question and those who answered the question. We would then compare the percentages of gender for each group. If one gender (e.g., males) was found in greater proportion in the missing data group, we would suspect a nonrandom missing data process. If the variable being compared is metric (e.g., an attitude or perception) instead of categorical (gender), then t-tests are performed to determine the statistical significance of the difference in the variable's mean between the two groups. The researcher should examine a number of variables to see whether any consistent pattern emerges. Remember that some differences will occur by chance, but either a large number or a systematic pattern of differences may indicate an underlying nonrandom pattern.

- A second approach is an overall test of randomness that determines whether the missing data can be classified as MCAR. This test analyzes the pattern of missing data on all variables and compares it with the pattern expected for a random missing data process. If no significant differences are found, the missing data can be classified as MCAR. If significant differences are found, however, the researcher must use the approaches described previously to identify the specific missing data processes that are nonrandom.

As a result of these tests, the missing data process is classified as either MAR or MCAR, which then determines the appropriate types of potential remedies. Even though achieving the level of MCAR requires a completely random pattern in the missing data, it is the preferred type because it allows for the widest range of potential remedies.

STEP 4: SELECT THE IMPUTATION METHOD   At this step of the process, the researcher must select the approach used for accommodating missing data in the analysis. This decision is based primarily on whether the missing data are MAR or MCAR, but in either case the researcher has several options for imputation [14, 18, 21, 22]. **Imputation** is the process of estimating the missing value based on valid values of other variables and/or cases in the sample. The objective is to employ known relationships that can be identified in the valid values of the sample to assist in estimating the missing values. However, the researcher should carefully consider the use of imputation in each instance because of its potential impact on the analysis [8]:

> The idea of imputation is both seductive and dangerous. It is seductive because it can lull the user into the pleasurable state of believing that the data are complete after all, and it is dangerous because it lumps together situations where the problem is sufficiently minor that it can be legitimately handled in this way and situations where standard estimators applied to the real and imputed data have substantial biases.

All of the imputation methods discussed in this section are used primarily with metric variables; nonmetric variables are left as missing unless a specific modeling approach is employed. Nonmetric variables are not amenable to imputation because even though estimates of the missing data for metric variables can be made with such values as a mean of all valid values, no comparable measures are available for nonmetric variables. As such, nonmetric variables require an estimate of a specific value rather than an estimate on a continuous scale. It is different to estimate a missing value for a metric variable, such as an attitude or perception—even income—than it is to estimate the respondent's gender when missing.

Imputation of a MAR Missing Data Process. If a nonrandom or MAR missing data process is found, the researcher should apply only one remedy—the specifically designed modeling approach [18]. Application of any other method introduces bias into the results. This set of procedures explicitly incorporates the missing data into the analysis, either through a process specifically designed for missing data estimation or as an integral portion of the standard multivariate analysis. The first approach involves maximum likelihood estimation techniques that attempt to model the processes underlying the missing data and to make the most accurate and reasonable estimates possible [12, 18]. One example is the EM approach [11]. It is an iterative two-stage method (the E and M stages) in which the E stage makes the best possible estimates of the missing data and the M stage then makes estimates of the parameters (means, standard deviations, or correlations) assuming the missing data were replaced. The process continues going through the two stages until the change in the estimated values is negligible and they replace the missing data. This approach has been shown to work quite effectively in instances of nonrandom missing data processes, but has seen limited application due to the need for specialized software. Its inclusion in recent versions of the popular software programs (e.g., the Missing Value Analysis module of SPSS) may increase its use. Comparable procedures employ structural equation modeling (Chapter 11) to estimate the missing data [2, 4, 9], but detailed discussion of these methods is beyond the scope of this chapter.

The second approach involves the inclusion of missing data directly into the analysis, defining observations with missing data as a select subset of the sample. This approach is most applicable for dealing with missing values on the independent variables of a dependent relationship. Its premise has best been characterized by Cohen et al. [6]:

> We thus view missing data as a pragmatic fact that must be investigated, rather than a disaster to be mitigated. Indeed, implicit in this philosophy is the idea that like all other aspects of sample data, missing data are a property of the population to which we seek to generalize.

When the missing values occur on a nonmetric variable, the researcher can easily define those observations as a separate group and then include them in any analysis. When the missing data are present on a metric independent variable in a dependence relationship, the observations are incorporated directly into the analysis while maintaining the relationships among the valid values [6]. This procedure is best illustrated in the context of regression analysis, although it can be used in other dependence relationships as well. The first step is to code all observations having missing data with a dummy variable (where the cases with missing data receive a value of one for the dummy variable and the other cases have a value of zero as discussed in the last section of this chapter). Then, the missing values are imputed by the mean substitution method (see next section for a discussion of this method). Finally, the relationship is estimated by normal means. The dummy variable represents the difference for the dependent variable between those observations with missing data and those observations with valid data. The dummy variable coefficient assesses the statistical significance of this difference. The coefficient of the original variable represents the relationship for all cases with nonmissing data. This method enables the researcher to retain all the observations in the analysis for purposes of maintaining the sample size. It also provides a direct test for the differences between the two groups along with the estimated relationship between the dependent and independent variables.

The primary disadvantage for either of these two approaches is the complexity involved for researchers in implementation or interpretation. Most researchers are unfamiliar with these options, much less even the necessity for diagnosing missing data processes. Yet many of the remedies discussed in the next section for MCAR missing data are directly available from the statistical programs, thus their more widespread application even when inappropriate. Hopefully, the increasing availability of the specialized software needed, as well as the awareness of the implications for nonrandom missing data processes, will enable these more suitable methods to be applied where necessary to accommodate MAR missing data.

Imputation of a MCAR Missing Data Process. If the researcher determines that the missing data process can be classified as MCAR, either of two basic approaches be used: using only valid data or defining replacement values for the missing data. We will first discuss the two methods that use only valid data, and then follow with a discussion of the methods based on using replacement values for the missing data.

Imputation Using Only Valid Data. Some researchers may question whether using only valid data is actually a form of imputation, because no data values are actually replaced. The intent of this approach is to represent the entire sample with those observations or cases with valid data. As seen in the two following approaches, this representation can be done in several ways. The underlying assumption in both is that the missing data are in a random pattern and that the valid data are an adequate representation.

- *Complete Case Approach:* The simplest and most direct approach for dealing with missing data is to include only those observations with complete data, also known as the **complete case approach.** This method, also known as the LISTWISE method in SPSS, is available in all statistical programs and is the default method in many programs. Yet the complete case approach has two distinct disadvantages. First, it is most affected by any nonrandom missing data processes, because the cases with any missing data are deleted from the analysis. Thus, even though only valid observations are used, the results are not generalizable to the population. Second, this approach also results in the greatest reduction in sample size, because missing data on any variable eliminates the entire case. It has been shown that with only 2 percent randomly missing data, more than 18 percent of the cases will have some missing data. Thus, in many situations with even very small amounts of missing data, the resulting sample size is reduced to an inappropriate size when this approach is used. As a result, the complete case approach is best suited for instances in which the extent of missing data is small, the sample is sufficiently large to allow for deletion of the cases with missing data, and the relationships in the data are so strong as to not be affected by any missing data process.

- *Using All-Available Data:* The second imputation method using only valid data also does not actually replace the missing data, but instead imputes the distribution characteristics (e.g., means or standard deviations) or relationships (e.g., correlations) from every valid value. For example, assume that there are three variables of interest (V_1, V_2, and V_3). To estimate the mean of each variable, all of the valid values are used for each respondent. If a respondent is missing data for V_3, the valid values for V_1 and V_2 are still used to calculate the means. Correlations are calculated in the same manner, using all valid pairs of data. Assume that one respondent has valid data for only V_1 and V_2, whereas a second respondent has valid data for V_2 and V_3. When calculating the correlation between V_1 and V_2, the values from the first respondent will be used, but not for correlations of V_1 and V_3 or V_2 and V_3. Likewise, the second respondent will contribute data for calculating the correlation of V_2 and V_3, but not the other correlations.

Known as the **all-available approach,** this method (e.g., the PAIRWISE option in SPSS) is primarily used to estimate correlations and maximize the pairwise information available in the sample. The distinguishing characteristic of this approach is that the characteristic of a variable (e.g., mean, standard deviation) or the correlation for a pair of variables is based on a potentially unique set of observations. It is to be expected that the number of observations used in the calculations will vary for each correlation. The imputation process occurs not by replacing the missing data, but instead by using the obtained correlations on just the cases with valid data as representative for the entire sample.

Even though the all-available method maximizes the data utilized and overcomes the problem of missing data on a single variable eliminating a case from the entire analysis, several problems can arise. First, correlations may be calculated that are "out of range" and inconsistent with the other correlations in the correlation matrix [17]. Any correlation between X and Y is constrained by their correlation to a third variable Z, as shown in the following formula:

$$\text{Range of } r_{XY} = r_{XZ}r_{YZ} \pm \sqrt{(1 - r_{XZ}^2)(1 - r_{YZ}^2)}$$

The correlation between X and Y can range only from -1 to $+1$ if both X and Y have zero correlation with all other variables in the correlation matrix. Yet rarely are the correlations with other variables zero. As the correlations with other variables increase, the range of the correlation between X and Y decreases, which increases the potential for the correlation in a unique set of cases to be inconsistent with correlations derived from other sets of cases. For example, if X and Y have correlations of .6 and .4, respectively, with Z, then the possible range of correlation between X and Y is .24 ± .73, or from $-.49$ to .97. Any value outside this range is mathematically inconsistent, yet may occur if the correlation is obtained with a differing number and set of cases for the two correlations in the all-available approach.

An associated problem is that the eigenvalues in the correlation matrix can become negative, thus altering the variance properties of the correlation matrix. Although the correlation matrix can be adjusted to eliminate this problem, many procedures do not include this adjustment process. In extreme cases, the estimated variance/covariance matrix is not positive definite [17]. Both of these problems must be considered when selecting the all-available approach.

Imputation by Using Replacement Values. The second form of imputation involves replacing missing values with estimated values based on other information available in the sample. The principal advantage is that once the replacement values are substituted, all observations are available for use in the analysis. The available options vary from the direct substitution of values to estimation processes based on relationships among the variables. The following discussion focuses on the four most widely used methods, although many other forms of imputation are available [18, 21, 22]. These methods can be classified as to whether they use a known value as a replacement or calculate a replacement value from other observations.

- *Using Known Replacement Values:* The common characteristic in these methods is to identify a known value, most often from a single observation, that is used to replace the missing data. The observation may be from the sample or even external to the sample. A primary consideration is identifying the appropriate observation through some measure of similarity. The observation with missing data is "matched" to a similar case, which provides the replacement values for the missing data. The trade-off in assessing similarity is between using more variables to get a better "match" versus the complexity in calculating similarity.
  - **Hot or Cold Deck Imputation.** In this approach, the researcher substitutes a value from another source for the missing values. In the "hot deck" method, the value comes from another observation in the sample that is deemed similar. Each observation with missing data is paired with another case that is similar on a variable(s) specified by the researcher. Then, missing data are replaced with valid values from the similar observation. "Cold deck" imputation derives the replacement value from an external source (e.g., prior studies, other samples, etc.). Here the researcher must be sure that the replacement value from an external source is more valid than an internally generated value. Both variants of this method provide the researcher with the option of replacing the missing data with actual values from similar observations that may be deemed more valid than some calculated value from all cases, such as the mean of the sample.
  - **Case Substitution.** In this method, entire observations with missing data are replaced by choosing another nonsampled observation. A common example is to replace a sampled household that cannot be contacted or that has extensive missing data with another household not in the sample, preferably similar to the original observation. This method is most widely used to replace observations with complete missing data, although it can be used to replace observations with lesser amounts of missing data as well. At issue is the ability to obtain these additional observations not included in the original sample.
- *Calculating Replacement Values:* The second basic approach involves calculating a replacement value from a set of observations with valid data in the sample. The assumption is that a value derived from all other observations in the sample is the most representative replacement value. These methods, particularly mean substitution, are more widely used due to their ease in implementation versus the use of known values discussed previously.
  - **Mean Substitution.** One of the most widely used methods, mean substitution replaces the missing values for a variable with the mean value of that variable calculated from all valid responses. The rationale of this approach is that the mean is the best single replacement value. This approach, although it is used extensively, has several disadvantages. First, it understates the variance estimates by using the mean value for all missing data. Second, the actual distribution of values is distorted by substituting the mean for the missing values.

Third, this method depresses the observed correlation because all missing data will have a single constant value. It does have the advantage, however, of being easily implemented and providing all cases with complete information. A variant of this method is group mean substitution, where observations with missing data are grouped on a second variable, and then mean values for each group are substituted for the missing values within the group.

• **Regression Imputation.** In this method, regression analysis (described in Chapter 4) is used to predict the missing values of a variable based on its relationship to other variables in the data set. First, a predictive equation is formed for each variable with missing data and estimated from all cases with valid data. Then, replacement values for each missing value are calculated from that observation's values on the variables in the predictive equation. Thus, the replacement value is derived based on that observation's values on other variables shown to relate to the missing value.

Although it has the appeal of using relationships already existing in the sample as the basis of prediction, this method also has several disadvantages. First, it reinforces the relationships already in the data. As the use of this method increases, the resulting data become more characteristic of the sample and less generalizable. Second, unless stochastic terms are added to the estimated values, the variance of the distribution is understated. Third, this method assumes that the variable with missing data has substantial correlations with the other variables. If these correlations are not sufficient to produce a meaningful estimate, then other methods, such as mean substitution, are preferable. Fourth, the sample must be large enough to allow for a sufficient number of observations to be used in making each prediction. Finally, the regression procedure is not constrained in the estimates it makes. Thus, the predicted values may not fall in the valid ranges for variables (e.g., a value of 11 may be predicted for a 10-point scale) and require some form of additional adjustment.

Even with all of these potential problems, the regression method of imputation holds promise in those instances for which moderate levels of widely scattered missing data are present and for which the relationships between variables are sufficiently established so that the researcher is confident that using this method will not affect the generalizability of the results.

The range of possible imputation methods varies from the conservative (complete data method) to those that attempt to replicate the missing data as much as possible (e.g., regression imputation or model-based methods). What should be recognized is that each method has advantages and disadvantages, such that the researcher must examine each missing data situation and select the most appropriate imputation method. Table 2-2 provides a brief comparison of the imputation method, but a quick review shows that no single method is best in all situations. However, some general suggestions (see Rules of Thumb 2-3) can be made based on the extent of missing data.

Given the many imputation methods available, the researcher should also strongly consider following a multiple imputation strategy, whereby a combination of several methods is used. In this approach, two or more methods of imputation are used to derive a composite estimate—usually the mean of the various estimates—for the missing value. The rationale is that the use of multiple approaches minimizes the specific concerns with any single method and the composite will be the best possible estimate. The choice of this approach is primarily based on the trade-off between the researcher's perception of the potential benefits versus the substantially higher effort required to make and combine the multiple estimates.

An Illustration of Missing Data Diagnosis with the Four-Step Process

To illustrate the four-step process of diagnosing the patterns of missing data and the application of possible remedies, a new data set is introduced (a complete listing of the observations and an electronic copy are available at www.pearsonglobaleditions.com/hair and www.mvstats.com). This data set was collected during the pretest of a questionnaire used to collect the data described in Chapter 1.

TABLE 2-2   Comparison of Imputation Techniques for Missing Data

Imputation Method	*Advantages*	*Disadvantages*	*Best Used When:*
Imputation Using Only Valid Data			
Complete Data	• Simplest to implement • Default for many statistical programs	• Most affected by nonrandom processes • Greatest reduction in sample size • Lowers statistical power	• Large sample size • Strong relationships among variables • Low levels of missing data
All Available Data	• Maximizes use of valid data • Results in largest sample size possible without replacing values	• Varying sample sizes for every imputation • Can generate "out of range" values for correlations and eigenvalues	• Relatively low levels of missing data • Moderate relationships among variables
Imputation Using Known Replacement Values			
Case Substitution	• Provides realistic replacement values (i.e., another actual observation) rather than calculated values	• Must have additional cases not in the original sample • Must define similarity measure to identify replacement case	• Additional cases are available • Able to identify appropriate replacement cases
Hot and Cold Deck Imputation	• Replaces missing data with actual values from the most similar case or best known value	• Must define suitably similar cases or appropriate external values	• Established replacement values are known, or • Missing data process indicates variables upon which to base similarity
Imputation by Calculating Replacement Values			
Mean Substitution	• Easily implemented • Provides all cases with complete information	• Reduces variance of the distribution • Distorts distribution of the data • Depresses observed correlations	• Relatively low levels of missing data • Relatively strong relationships among variables
Regression Imputation	• Employs actual relationships among the variables • Replacement values calculated based on an observation's own values on other variables • Unique set of predictors can be used for each variable with missing data	• Reinforces existing relationships and reduces generalizability • Must have sufficient relationships among variables to generate valid predicted values • Understates variance unless error term added to replacement value • Replacement values may be "out of range"	• Moderate to high levels of missing data • Relationships sufficiently established so as to not impact generalizability • Software availability
Model-Based Methods for MAR Missing Data Processes			
Model-Based Methods	• Accommodates both nonrandom and random missing data processes • Best representation of original distribution of values with least bias	• Complex model specification by researcher • Requires specialized software • Typically not available directly in software programs (except EM method in SPSS)	• Only method that can accommodate nonrandom missing data processes • High levels of missing data require least biased method to ensure generalizability

---

RULES OF THUMB 2-3

Imputation of Missing Data

- **Under 10%** Any of the imputation methods can be applied when missing data are this low, although the complete case method has been shown to be the least preferred

- **10% to 20%** The increased presence of missing data makes the all-available, hot deck case substitution, and regression methods most preferred for MCAR data, whereas model-based methods are necessary with MAR missing data processes

- **Over 20%** If it is deemed necessary to impute missing data when the level is over 20 percent, the preferred methods are:
  - The regression method for MCAR situations
  - Model-based methods when MAR missing data occur

The pretest involved 70 individuals and collected responses on 14 variables (9 metric variables, V_1 to V_9, and 5 nonmetric variables, V_{10} to V_{14}). The variables in this pretest do not coincide directly with those in the HBAT data set, so they will be referred to just by their variable designation (e.g., V_3).

In the course of pretesting, however, missing data occurred. The following sections detail the diagnosis of the missing data through the four-step process. A number of software programs add analyses of missing data, among them BMDP and SPSS. The analyses described in these next sections were performed with the Missing Value Analysis module in SPSS, but all of the analyses can be replicated by data manipulation and conventional analysis. Examples are provided at www.pearsonglobaleditions.com/hair and www.mvstats.com.

STEP 1: DETERMINE THE TYPE OF MISSING DATA   All the missing data in this example are unknown and not ignorable because they are due to nonresponse by the respondent. As such, the researcher is forced to proceed in the examination of the missing data processes.

STEP 2: DETERMINE THE EXTENT OF MISSING DATA   The objective in this step is to determine whether the extent of the missing data is sufficiently high enough to warrant a diagnosis of randomness of the missing data (step 3), or is it at a low enough level to proceed directly to the remedy (step 4). The researcher is interested in the level of missing data on a case and variable basis, plus the overall extent of missing data across all cases.

Table 2-3 contains the descriptive statistics for the observations with valid values, including the percentage of cases with missing data on each variable. Viewing the metric variables (V_1 to V_9), we see that the lowest amount of missing data is six cases for V_6 (9% of the sample), ranging up to 30 percent missing (21 cases) for V_1. This frequency makes V_1 and V_3 possible candidates for deletion in an attempt to reduce the overall amount of missing data. All of the nonmetric variables (V_{10} to V_{14}) have low levels of missing data and are acceptable.

Moreover, the amount of missing data per case is also tabulated. Although 26 cases have no missing data, it is also apparent that 6 cases have 50 percent missing data, making them likely to be deleted because of an excessive number of missing values. Table 2-4 shows the missing data patterns for all the cases with missing data, and these six cases are listed at the bottom of the table. As we view the patterns of missing data, we see that all the missing data for the nonmetric variables occurs in these six cases, such that after their deletion there will be only valid data for these variables.

Even though it is obvious that deleting the six cases will improve the extent of missing data, the researcher must also consider the possibility of deleting a variable(s) if the missing data level is high. The two most likely variables for deletion are V_1 and V_3, with 30 percent and 24 percent missing data, respectively. Table 2-5 on page 59 provides insight into the impact of deleting one or both by

TABLE 2-3   Summary Statistics of Missing Data for Original Sample

Variable	Number of Cases	Mean	Standard Deviation	Missing Data Number	Missing Data Percent
V_1	49	4.0	.93	21	30
V_2	57	1.9	.93	13	19
V_3	53	8.1	1.41	17	24
V_4	63	5.2	1.17	7	10
V_5	61	2.9	.78	9	13
V_6	64	2.6	.72	6	9
V_7	61	6.8	1.68	9	13
V_8	61	46.0	9.36	9	13
V_9	63	4.8	.83	7	10
V_{10}	68	NA	NA	2	3
V_{11}	68	NA	NA	2	3
V_{12}	68	NA	NA	2	3
V_{13}	69	NA	NA	1	1
V_{14}	68	NA	NA	2	3

NA = Not applicable to nonmetric variables

Summary of Cases

Number of Missing Data per Case	Number of Cases	Percent of Sample
0	26	37
1	15	21
2	19	27
3	4	6
7	6	9
Total	70	100%

examining the patterns of missing data and assessing the extent that missing data will be decreased. For example, the first pattern (first row) shows no missing data for the 26 cases. The pattern of the second row shows missing data only on V_3 and indicates that only one case has this pattern. The far right column indicates the number of cases having complete information if this pattern is eliminated (i.e., these variables deleted or replacement values imputed). In the case of this first pattern, we see that the number of cases with complete data would increase by 1, to 27, by deleting V_3 because only 1 case was missing data on only V_3. If we look at the fourth row, we see that 6 cases are missing data on only V_1, so that if we delete V_1 32 cases will have complete data. Finally, row 3 denotes the pattern of missing data on both V_1 and V_3, and if we delete both variables the number of cases with complete data will increase to 37. Thus, deleting just V_3 adds 1 case with complete data, just deleting V_1 increases the total by 6 cases, and deleting both variables increases the cases with complete data by 11, to a total of 37.

For purposes of illustration, we will delete just V_1, leaving V_3 with a fairly high amount of missing data to demonstrate its impact in the imputation process. The result is a sample of 64 cases with now only eight metric variables. Table 2-6 on page 60 contains the summary statistics on this reduced sample. The extent of missing data decreased markedly just by deleting six cases (less than 10% of the sample) and one variable. Now, one-half of the sample has complete data, only two variables have more than 10 percent missing data, and the nonmetric variables now have all complete data. Moreover, the largest number of missing values for any case is two, which indicates that imputation should not affect any case in a substantial manner.

TABLE 2-4   Patterns of Missing Data by Case

Case	# Missing	% Missing	V_1	V_2	V_3	V_4	V_5	V_6	V_7	V_8	V_9	V_{10}	V_{11}	V_{12}	V_{13}	V_{14}
			\multicolumn{14}{c}{*Missing Data Patterns*}												
205	1	7.1			S											
202	2	14.3	S		S											
250	2	14.3	S		S											
255	2	14.3	S		S											
269	2	14.3	S		S											
238	1	7.1	S													
240	1	7.1	S							-						
253	1	7.1	S													
256	1	7.1	S													
259	1	7.1	S													
260	1	7.1	S													
228	2	14.3	S			S										
246	1	7.1				S										
225	2	14.3			S	S										
267	2	14.3			S	S										
222	2	14.3			S		S									
241	2	14.3			S		S									
229	1	7.1					S									
216	2	14.3	S				S									
218	2	14.3	S				S									
232	2	14.3	S	S												
248	2	14.3	S	S												
237	1	7.1		S												
249	1	7.1		S												
220	1	7.1		S												
213	2	14.3		S	S											
257	2	14.3		S	S											
203	2	14.3		S						S						
231	1	7.1								S						
219	2	14.3								S	S					
244	1	7.1								S						
227	2	14.3		S						S						
224	3	21.4	S	S						S						
268	1	7.1										S				
235	2	14.3						S				S				
204	3	21.4	S		S							S				
207	3	21.4	S		S							S				
221	3	21.4	S		S				S							
245	7	50.0	S		S		S		S	S				S		S
233	7	50.0		S	S		S	S			S			S		S
261	7	50.0		S	S			S	S	S	S		S			
210	7	50.0				S	S	S	S	S	S	S				
263	7	50.0		S		S	S	S	S	S		S				
214	7	50.0	S			S		S	S	S			S		S	

Note: Only cases with missing data are shown.

S = missing data.

Having deleted six cases and one variable, the extent of missing data is still high enough to justify moving to step 3 and diagnosing the randomness of the missing data patterns. This analysis will be limited to the metric variables because the nonmetric variables now have no missing data.

TABLE 2-5 Missing Data Patterns

	Missing Data Patterns														
Number of Cases	V_1	V_2	V_3	V_4	V_5	V_6	V_7	V_8	V_9	V_{10}	V_{11}	V_{12}	V_{13}	V_{14}	Number of Complete Cases if Variables Missing in Pattern Are Not Used
26															26
1			X												27
4	X		X												37
6	X														32
1	X			X											34
1				X											27
2			X	X											30
2			X		X										30
1					X										27
2	X				X										35
2	X	X													37
3		X													29
2		X	X												32
1		X					X								31
1							X								27
1							X	X							29
1								X							27
1		X						X							31
1	X	X						X							40
1									X						27
1						X			X						28
2	X		X				X								40
1	X		X				X								39
1	X		X		X			X				X		X	47
1		X	X		X	X			X			X		X	38
1		X	X			X	X	X	X		X				40
1				X	X	X	X	X	X	X					34
1		X		X	X	X	X	X		X					37
1	X			X		X	X	X			X		X		38

Notes: Represents the number of cases with each missing data pattern. For example, reading down the column for the first three values (26, 1, and 4), 26 cases are not missing data on any variable. Then, one case is missing data on V_3. Then, four cases are missing data on two variables (V_1 and V_3).

STEP 3: DIAGNOSING THE RANDOMNESS OF THE MISSING DATA PROCESS   The next step is an empirical examination of the patterns of missing data to determine whether the missing data are distributed randomly across the cases and the variables. Hopefully the missing data will be judged MCAR, thus allowing a wider range of remedies in the imputation process. We will first employ a test of comparison between groups of missing and nonmissing cases and then conduct an overall test for randomness.

The first test for assessing randomness is to compare the observations with and without missing data for each variable on the other variables. For example, the observations with missing data on V_2 are placed in one group and those observations with valid responses for V_2 are placed in another group. Then, these two groups are compared to identify any differences on the remaining metric variables (V_3 through V_9). Once comparisons have been made on all of the variables, new groups are

TABLE 2-6   Summary Statistics for Reduced Sample (Six Cases and V_1 Deleted)

	Number of Cases	Mean	Standard Deviation	Missing Data Number	Missing Data Percent
V_2	54	1.9	.86	10	16
V_3	50	8.1	1.32	14	22
V_4	60	5.1	1.19	4	6
V_5	59	2.8	.75	5	8
V_6	63	2.6	.72	1	2
V_7	60	6.8	1.68	4	6
V_8	60	46.0	9.42	4	6
V_9	60	4.8	.82	4	6
V_{10}	64			0	0
V_{11}	64			0	0
V_{12}	64			0	0
V_{13}	64			0	0
V_{14}	64			0	0

Summary of Cases

Number of Missing Data per Case	Number of Cases	Percent of Sample
0	32	50
1	18	28
2	14	22
Total	64	100

formed based on the missing data for the next variable (V_3) and the comparisons are performed again on the remaining variables. This process continues until each variable (V_2 through V_9; remember V_1 has been excluded) has been examined for any differences. The objective is to identify any systematic missing data process that would be reflected in patterns of significant differences.

Table 2-7 contains the results for this analysis of the 64 remaining observations. The only noticeable pattern of significant t-values occurs for V_2, for which three of the eight comparisons (V_4, V_5, and V_6) found significant differences between the two groups. Moreover, only one other instance (groups formed on V_4 and compared on V_2) showed a significant difference. This analysis indicates that although significant differences can be found due to the missing data on one variable (V_2), the effects are limited to only this variable, making it of marginal concern. If later tests of randomness indicate a nonrandom pattern of missing data, these results would then provide a starting point for possible remedies.

The final test is an overall test of the missing data for being missing completely at random (MCAR). The test makes a comparison of the actual pattern of missing data with what would be expected if the missing data were totally randomly distributed. The MCAR missing data process is indicated by a *nonsignificant* statistical level (e.g., greater than .05), showing that the observed pattern *does not* differ from a random pattern. This test is performed in the Missing Value Analysis module of SPSS as well as several other software packages dealing with missing value analysis.

In this instance, Little's MCAR test has a significance level of .583, indicating a nonsignificant difference between the observed missing data pattern in the reduced sample and a random pattern. This result, coupled with the earlier analysis showing minimal differences in a nonrandom

TABLE 2-7   **Assessing the Randomness of Missing Data Through Group Comparisons of Observations with Missing Versus Valid Data**

Groups Formed by Missing Data on:		V_2	V_3	V_4	V_5	V_6	V_7	V_8	V_9
V_2	t-value	.	.7	−2.2	−4.2	−2.4	−1.2	−1.1	−1.2
	Significance		.528	.044	.001	.034	.260	.318	.233
	Number of cases (valid data)	54	42	50	49	53	51	52	50
	Number of cases (missing data)	0	8	10	10	10	9	8	10
	Mean of cases (valid data)	1.9	8.2	5.0	2.7	2.5	6.7	45.5	4.8
	Mean cases (missing data)	.	7.9	5.9	3.5	3.1	7.4	49.2	5.0
V_3	t-value	1.4	.	1.1	2.0	.2	.0	1.9	.9
	Significance	.180	.	.286	.066	.818	.965	.073	.399
	Number of cases (valid data)	42	50	48	47	49	47	46	48
	Number of cases (missing data)	12	0	12	12	14	13	14	12
	Mean of cases (valid data)	2.0	8.1	5.2	2.9	2.6	6.8	47.0	4.8
	Mean cases (missing data)	1.6	.	4.8	2.4	2.6	6.8	42.5	4.6
V_4	t-value	2.6	−.3	.	.2	1.4	1.5	.2	−2.4
	Significance	.046	.785	.	.888	.249	.197	.830	.064
	Number of cases (valid data)	50	48	60	55	59	56	56	56
	Number of cases (missing data)	4	2	0	4	4	4	4	4
	Mean of cases (valid data)	1.9	8.1	5.1	2.8	2.6	6.8	46.0	4.8
	Mean cases (missing data)	1.3	8.4	.	2.8	2.3	6.2	45.2	5.4
V_5	t-value	−.3	.8	.4	.	−.9	−.4	.5	.6
	Significance	.749	.502	.734	.	.423	.696	.669	.605
	Number of cases (valid data)	49	47	55	59	58	55	55	55
	Number of cases (missing data)	5	3	5	0	5	5	5	5
	Mean of cases (valid data)	1.9	8.2	5.2	2.8	2.6	6.8	46.2	4.8
	Mean cases (missing data)	2.0	7.1	5.0	.	2.9	7.1	43.6	4.6
V_7	t-value	.9	.2	−2.1	.9	−1.5	.	.5	.4
	Significance	.440	.864	.118	.441	.193	.	.658	.704
	Number of cases (valid data)	51	47	56	55	59	60	57	56
	Number of cases (missing data)	3	3	4	4	4	0	3	4
	Mean of cases (valid data)	1.9	8.1	5.1	2.9	2.6	6.8	46.1	4.8
	Mean cases (missing data)	1.5	8.0	6.2	2.5	2.9	.	42.7	4.7
V_8	t-value	−1.4	2.2	−1.1	−.9	−1.8	1.7	.	1.6
	Significance	.384	.101	.326	.401	.149	.128	.	.155
	Number of cases (valid data)	52	46	56	55	59	57	60	56
	Number of cases (missing data)	2	4	4	4	4	3	0	4
	Mean of cases (valid data)	1.9	8.3	5.1	2.8	2.6	6.8	46.0	4.8
	Mean cases (missing data)	3.0	6.6	5.6	3.1	3.0	6.3	.	4.5
V_9	t-value	.8	−2.1	2.5	2.7	1.3	.9	2.4	.
	Significance	.463	.235	.076	.056	.302	.409	.066	.
	Number of cases (valid data)	50	48	56	55	60	56	56	60
	Number of cases (missing data)	4	2	4	4	3	4	4	0
	Mean of cases (valid data)	1.9	8.1	5.2	2.9	2.6	6.8	46.4	4.8
	Mean cases (missing data)	1.6	9.2	3.9	2.1	2.2	6.3	39.5	.

Notes: Each cell contains six values: (1) t-value for the comparison of the means of the column variable across the groups formed between group a (cases with valid data on the row variable) and group b (observations with missing data on the row variable; (2) significance of the t-value for group comparisons; (3) and (4) number of cases for group a (valid data) and group b (missing data); (5) and (6) mean of column variable for group a (valid data on row variable) and group b (missing data on row variable).

Interpretation of the table: The upper right cell indicates that a t-value for the comparison of V_9 between group a (valid data on V_2) and group b (missing data on V_2) is −1.2, which has a significance level of .233. The sample sizes of groups a and b are 50 and 10, respectively. Finally, the mean of group a (valid data on V_2) is 4.8, whereas the mean of group b (missing data on V_2) is 5.0.

pattern, allow for the missing data process to be considered MCAR. As a result, the researcher may employ any of the remedies for missing data, because no potential biases exist in the patterns of missing data.

STEP 4: SELECTING AN IMPUTATION METHOD   As discussed earlier, numerous imputation methods are available for both MAR and MCAR missing data processes. In this instance, the MCAR missing data process allows for use of any imputation methods. The other factor to consider is the extent of missing data. As the missing data level increases, methods such as the complete information method become less desirable due to restrictions on sample size, and the all-available method, regression, and model-based methods become more preferred.

The first option is to use only observations with complete data. The advantage of this approach in maintaining consistency in the correlation matrix is offset in this case, however, by its reduction of the sample to such a small size (32 cases) that it is not useful in further analyses. The next options are to still use only valid data through the all-available method or calculate replacement values through such methods as the mean substitution, the regression-based method, or even a model-building approach (e.g., EM method). Because the missing data are MCAR, all of these methods will be employed and then compared to assess the differences that arise between methods. They could also form the basis for a multiple imputation strategy where all the results are combined into a single overall result.

Table 2-8 details the results of estimating means and standard deviations by four imputation methods (mean substitution, all-available, regression, and EM). In comparing the means, we find a general consistency between the methods, with no noticeable patterns. For the standard deviations, however, we can see the variance reduction associated with the mean substitution method. Across all variables, it consistently provides the smallest standard deviation, attributable to the substitution on the constant value. The other three methods again show a consistency in the results, indicative of the lack of bias in any of the methods since the missing data process was deemed MCAR.

Finally, Table 2-9 contains the correlations obtained from the complete case, all-available, mean substitution, and EM imputation approaches. In most instances the correlations are similar, but several substantial differences arise. First is a consistency between the correlations obtained with the all-available, mean substitution, and EM approaches. Consistent differences occur, however, between these values and the values from the complete case approach. Second, the notable differences are concentrated in the correlations with V_2 and V_3, the two variables with the greatest amount of missing data in the reduced sample (refer back to Table 2-6). These differences may

TABLE 2-8   Comparing the Estimates of the Mean and Standard Deviation Across Four Imputation Methods

	Estimated Means							
Imputation Method	V_2	V_3	V_4	V_5	V_6	V_7	V_8	V_9
Mean Substitution	1.90	8.13	5.15	2.84	2.60	6.79	45.97	4.80
All Available	1.90	8.13	5.15	2.84	2.60	6.79	45.97	4.80
Regression	1.99	8.11	5.14	2.83	2.58	6.84	45.81	4.77
EM	2.00	8.34	5.17	2.88	2.54	6.72	47.72	4.85

	Estimated Standard Deviations							
Imputation Method	V_2	V_3	V_4	V_5	V_6	V_7	V_8	V_9
Mean Substitution	.79	1.16	1.15	.72	.71	1.62	9.12	.79
All Available	.86	1.32	1.19	.75	.72	1.67	9.42	.82
Regression	.87	1.26	1.16	.75	.73	1.67	9.28	.81
EM	.84	1.21	1.11	.69	.72	1.69	9.67	.88

TABLE 2-9   Comparison of Correlations Obtained with the Complete Case (LISTWISE), All-Available (PAIRWISE), Mean Substitution, and EM Methods of Imputation

	V_2	V_3	V_4	V_5	V_6	V_7	V_8	V_9
V_2	1.00							
	1.00							
	1.00							
	1.00							
V_3	−.29	1.00						
	−.36	1.00						
	−.29	1.00						
	−.32	1.00						
V_4	.28	−.07	1.00					
	.30	−.07	1.00					
	.24	−.06	1.00					
	.30	−.09	1.00					
V_5	.29	.25	.26	1.00				
	.44	.05	.43	1.00				
	.38	.04	.42	1.00				
	.48	.07	.41	1.00				
V_6	.35	−.09	.82	.31	1.00			
	.26	−.06	.81	.34	1.00			
	.22	−.03	.77	.32	1.00			
	.30	−.07	.80	.38	1.00			
V_7	.34	−.41	.42	−.03	.54	1.00		
	.35	−.36	.40	.07	.40	1.00		
	.31	−.29	.37	.06	.40	1.00		
	.35	−.30	.40	.03	.41	1.00		
V_8	.01	.72	.20	.71	.26	−.27	1.00	
	.15	.60	.22	.71	.27	−.20	1.00	
	.13	.50	.21	.66	.26	−.20	1.00	
	.17	.54	.20	.68	.27	−.19	1.00	
V_9	−.27	.77	.21	.46	.09	−.43	.71	1.00
	−.18	.70	.38	.53	.23	−.26	.67	1.00
	−.17	.63	.34	.48	.23	−.25	.65	1.00
	−.08	.61	.36	.55	.24	−.24	.67	1.00

Interpretation: The top value is the correlation obtained with the *complete case* method, the second value is obtained with the *all-available* method, the third value is the correlation from the *mean substitution* method, and the fourth correlation results from the *EM* method of imputation.

indicate the impact of a missing data process, even though the overall randomness test showed no significant pattern. Although the researcher has no proof of greater validity for any of the approaches, these results demonstrate the marked differences sometimes obtained between the approaches. Whichever approach is chosen, the researcher should examine the correlations obtained by alternative methods to understand the range of possible values.

The task for the researcher is to coalesce the missing data patterns with the strengths and weaknesses of each approach and then select the most appropriate method. In the instance of differing estimates, the more conservative approach of combining the estimates into a single estimate (the multiple imputation approach) may be the most appropriate choice. Whichever approach is used, the data set with replacement values should be saved for further analysis.

A RECAP OF THE MISSING VALUE ANALYSIS   Evaluation of the issues surrounding missing data in the data set can be summarized in four conclusions:

1. **The missing data process is MCAR.** All of the diagnostic techniques support the conclusion that no systematic missing data process exists, making the missing data MCAR (missing completely at random). Such a finding provides two advantages to the researcher. First, it should not involve any hidden impact on the results that need to be considered when interpreting the results. Second, any of the imputation methods can be applied as remedies for the missing data. Their selection need not be based on their ability to handle nonrandom processes, but instead on the applicability of the process and its impact on the results.

2. **Imputation is the most logical course of action.** Even given the benefit of deleting cases and variables, the researcher is precluded from the simple solution of using the complete case method, because it results in an inadequate sample size. Some form of imputation is therefore needed to maintain an adequate sample size for any multivariate analysis.

3. **Imputed correlations differ across techniques.** When estimating correlations among the variables in the presence of missing data, the researcher can choose from four commonly employed techniques: the complete case method, the all-available information method, the mean substitution method, and the EM method. The researcher is faced in this situation, however, with differences in the results among these methods. The all-available information, mean substitution, and EM approaches lead to generally consistent results. Notable differences, however, are found between these approaches and the complete information approach. Even though the complete information approach would seem the most "safe" and conservative, in this case it is not recommended due to the small sample used (only 26 observations) and its marked differences from the other two methods. The researcher should, if necessary, choose among the other approaches.

4. **Multiple methods for replacing the missing data are available and appropriate.** As already mentioned, mean substitution is one acceptable means of generating replacement values for the missing data. The researcher also has available the regression and EM imputation methods, each of which give reasonably consistent estimates for most variables. The presence of several acceptable methods also enables the researcher to combine the estimates into a single composite, hopefully mitigating any effects strictly due to one of the methods.

In conclusion, the analytical tools and the diagnostic processes presented in the earlier section provide an adequate basis for understanding and accommodating the missing data found in the pretest data. As this example demonstrates, the researcher need not fear that missing data will always preclude a multivariate analysis or always limit the generalizability of the results. Instead, the possibly hidden impact of missing data can be identified and actions taken to minimize the effect of missing data on the analyses performed.

OUTLIERS

Outliers are observations with a *unique combination of characteristics identifiable as distinctly different* from the other observations. What constitutes a unique characteristic? Typically it is judged to be an unusually high or low value on a variable or a unique combination of values across several variables that make the observation stand out from the others. In assessing the impact of outliers, we must consider the practical and substantive considerations:

- From a *practical* standpoint, outliers can have a marked effect on any type of empirical analysis. For example, assume that we sample 20 individuals to determine the average household income. In our sample we gather responses that range between $20,000 and $100,000, so that the average is $45,000. But assume that the 21st person has an income of $1 million. If we include this value in the analysis, the average income increases to more than $90,000. Obviously, the outlier is a valid case, but what is the better estimate of the average household income: $45,000 or $90,000? The researcher must assess whether the outlying value is retained or eliminated due to its undue influence on the results.

- In *substantive* terms, the outlier must be viewed in light of how representative it is of the population. Again, using our example of household income, how representative of the more wealthy segment is the millionaire? If the researcher feels that it is a small, but viable segment in the population, then perhaps the value should be retained. If, however, this millionaire is the only one in the entire population and truly far above everyone else (i.e., unique) and represents an extreme value, then it may be deleted.

Outliers cannot be categorically characterized as either beneficial or problematic, but instead must be viewed within the context of the analysis and should be evaluated by the types of information they may provide. When beneficial, outliers—although different from the majority of the sample—may be indicative of characteristics of the population that would not be discovered in the normal course of analysis. In contrast, problematic outliers are not representative of the population, are counter to the objectives of the analysis, and can seriously distort statistical tests. Owing to the varying impact of outliers, it is imperative that the researcher examine the data for the presence of outliers and ascertain their type of influence. The reader is also referred to the discussions in Chapter 4 that relate to the topic of influential observations. In these discussions, outliers are placed in a framework particularly suited for assessing the influence of individual observations and determining whether this influence is helpful or harmful.

Why do outliers occur? Outliers can be classified into one of four classes based on the source of their uniqueness:

- The first class arises from a *procedural error,* such as a data entry error or a mistake in coding. These outliers should be identified in the data cleaning stage, but if overlooked they should be eliminated or recorded as missing values.
- The second class of outlier is the observation that occurs as the result of an *extraordinary event,* which accounts for the uniqueness of the observation. For example, assume we are tracking average daily rainfall, when we have a hurricane that lasts for several days and records extremely high rainfall levels. These rainfall levels are not comparable to anything else recorded in the normal weather patterns. If included, they will markedly change the pattern of the results. The researcher must decide whether the extraordinary event fits the objectives of the research. If so, the outlier should be retained in the analysis. If not, it should be deleted.
- The third class of outlier comprises *extraordinary observations* for which the researcher has no explanation. In these instances, a unique and markedly different profile emerges. Although these outliers are the most likely to be omitted, they may be retained if the researcher feels they represent a valid element of the population. Perhaps they represent an emerging element, or an untapped element previously not identified. Here the researcher must use judgment in the retention/deletion decision.
- The fourth and final class of outlier contains observations that fall within the ordinary range of values on each of the variables. These observations are not particularly high or low on the variables, but are *unique in their combination* of values across the variables. In these situations, the researcher should retain the observation unless specific evidence is available that discounts the outlier as a valid member of the population.

Detecting and Handling Outliers

The following sections detail the methods used in detecting outliers in univariate, bivariate, and multivariate situations. Once identified, they may be profiled to aid in placing them into one of the four classes just described. Finally, the researcher must decide on the retention or exclusion of each outlier, judging not only from the characteristics of the outlier but also from the objectives of the analysis.

METHODS OF DETECTING OUTLIERS   Outliers can be identified from a univariate, bivariate, or multivariate perspective based on the number of variables (characteristics) considered. The researcher should utilize as many of these perspectives as possible, looking for a consistent pattern across perspectives to identify outliers. The following discussion details the processes involved in each of the three perspectives.

Univariate Detection.   The univariate identification of outliers examines the distribution of observations for each variable in the analysis and selects as outliers those cases falling at the outer ranges (high or low) of the distribution. The primary issue is establishing the threshold for designation of an outlier. The typical approach first converts the data values to standard scores, which have a mean of 0 and a standard deviation of 1. Because the values are expressed in a standardized format, comparisons across variables can be made easily.

In either case, the researcher must recognize that a certain number of observations may occur normally in these outer ranges of the distribution. The researcher should strive to identify only those truly distinctive observations and designate them as outliers.

Bivariate Detection.   In addition to the univariate assessment, pairs of variables can be assessed jointly through a scatterplot. Cases that fall markedly outside the range of the other observations will be seen as isolated points in the scatterplot. To assist in determining the expected range of observations in this two-dimensional portrayal, an ellipse representing a bivariate normal distribution's confidence interval (typically set at the 90% or 95% level) is superimposed over the scatterplot. This ellipse provides a graphical portrayal of the confidence limits and facilitates identification of the outliers. A variant of the scatterplot is termed the influence plot, with each point varying in size in relation to its influence on the relationship.

Each of these methods provides an assessment of the uniqueness of each observation in relationship to the other observation based on a specific pair of variables. A drawback of the bivariate method in general is the potentially large number of scatterplots that arise as the number of variables increases. For three variables, it is only three graphs for all pairwise comparisons. But for five variables, it takes 10 graphs, and for 10 variables it takes 45 scatterplots! As a result, the researcher should limit the general use of bivariate methods to specific relationships between variables, such as the relationship of the dependent versus independent variables in regression. The researcher can then examine the set of scatterplots and identify any general pattern of one or more observations that would result in their designation as outliers.

Multivariate Detection.   Because most multivariate analyses involve more than two variables, the bivariate methods quickly become inadequate for several reasons. First, they require a large number of graphs, as discussed previously, when the number of variables reaches even moderate size. Second, they are limited to two dimensions (variables) at a time. Yet when more than two variables are considered, the researcher needs a means to objectively measure the *multidimensional* position of each observation relative to some common point. This issue is addressed by the Mahalanobis D^2 measure, a multivariate assessment of each observation across a set of variables. This method measures each observation's distance in multidimensional space from the mean center of all observations, providing a single value for each observation no matter how many variables are considered. Higher D^2 values represent observations farther removed from the general distribution of observations in this multidimensional space. This method, however, also has the drawback of only providing an overall assessment, such that it provides no insight as to which particular variables might lead to a high D^2 value.

For interpretation purposes, the Mahalanobis D^2 measure has statistical properties that allow for significance testing. The D^2 measure divided by the number of variables involved (D^2/df) is approximately distributed as a t-value. Given the nature of the statistical tests, it is suggested that conservative levels of significance (e.g., .005 or .001) be used as the threshold value for designation as an outlier. Thus, observations having a D^2/df value exceeding 2.5 in small samples and 3 or 4 in large samples can be designated as possible outliers. Once identified as a potential outlier on the D^2 measure, an observation can be reexamined in terms of the univariate and bivariate methods discussed earlier to more fully understand the nature of its uniqueness.

OUTLIER DESIGNATION   With these univariate, bivariate, and multivariate diagnostic methods, the researcher has a complementary set of perspectives with which to examine observations as to their status as outliers. Each of these methods can provide a unique perspective on the observations and be used in a concerted manner to identify outliers (see Rules of Thumb 2-4).

When observations have been identified by the univariate, bivariate, and multivariate methods as possible outliers, the researcher must then select only observations that demonstrate real uniqueness in comparison with the remainder of the population across as many perspectives as possible. The researcher must refrain from designating too many observations as outliers and not succumb to the temptation of eliminating those cases not consistent with the remaining cases just because they are different.

OUTLIER DESCRIPTION AND PROFILING   Once the potential outliers are identified, the researcher should generate profiles of each outlier observation and identify the variable(s) responsible for its being an outlier. In addition to this visual examination, the researcher can also employ multivariate techniques such as discriminant analysis (Chapter 7) or multiple regression (Chapter 4) to identify the differences between outliers and the other observations. If possible the researcher should assign the outlier to one of the four classes described earlier to assist in the retention or deletion decision to be made next. The researcher should continue this analysis until satisfied with understanding the aspects of the case that distinguish the outlier from the other observations.

RETENTION OR DELETION OF THE OUTLIER   After the outliers are identified, profiled, and categorized, the researcher must decide on the retention or deletion of each one. Many philosophies among researchers offer guidance as to how to deal with outliers. Our belief is that they should be retained unless demonstrable proof indicates that they are truly aberrant and not representative of any observations in the population. If they do portray a representative element or segment of the population, they should be retained to ensure generalizability to the entire population. As outliers are deleted, the researcher runs the risk of improving the multivariate analysis but limiting its generalizability. If outliers are problematic in a particular technique,

---

RULES OF THUMB 2-4

Outlier Detection

- Univariate methods: Examine all metric variables to identify unique or extreme observations
  - For small samples (80 or fewer observations), outliers typically are defined as cases with standard scores of 2.5 or greater
  - For larger sample sizes, increase the threshold value of standard scores up to 4
  - If standard scores are not used, identify cases falling outside the ranges of 2.5 versus 4 standard deviations, depending on the sample size
- Bivariate methods: Focus their use on specific variable relationships, such as the independent versus dependent variables
  - Use scatterplots with confidence intervals at a specified alpha level
- Multivariate methods: Best suited for examining a complete variate, such as the independent variables in regression or the variables in factor analysis
  - Threshold levels for the D^2/df measure should be conservative (.005 or .001), resulting in values of 2.5 (small samples) versus 3 or 4 in larger samples

many times they can be accommodated in the analysis in a manner in which they do not seriously distort the analysis.

An Illustrative Example of Analyzing Outliers

As an example of outlier detection, the observations of the HBAT database introduced in Chapter 1 are examined for outliers. The variables considered in the analysis are the metric variables X_6 through X_{19}, with the context of our examination being a regression analysis, where X_{19} is the dependent variable and X_6 through X_{18} are the independent variables. The outlier analysis will include univariate, bivariate, and multivariate diagnoses. When candidates for outlier designation are found, they are examined, and a decision on retention or deletion is made.

OUTLIER DETECTION The first step is examination of all the variables from a univariate perspective. Bivariate methods will then be employed to examine the relationships between the dependent variable (X_{19}) and each of the independent variables. From each of these scatterplots, observations that fall outside the typical distribution can be identified and their impact on that relationship ascertained. Finally, a multivariate assessment will be made on all of the independent variables collectively. Comparison of observations across the three methods will hopefully provide the basis for the deletion/retention decision.

Univariate Detection. The first step is to examine the observations on each of the variables individually. Table 2-10 contains the observations with standardized variable values exceeding ±2.5 on each of the variables (X_6 to X_{19}). From this univariate perspective, only observations 7, 22, and 90 exceed the threshold on more than a single variable. Moreover, none of these observations had values so extreme as to affect any of the overall measures of the variables, such as the mean or standard deviation. We should note that the dependent variable had one outlying observation (22), which may affect the bivariate scatterplots because the dependent variable appears in each

TABLE 2-10 Univariate, Bivariate, and Multivariate Outlier Detection Results

UNIVARIATE OUTLIERS		BIVARIATE OUTLIERS		MULTIVARIATE OUTLIERS		
Cases with Standardized Values Exceeding ± 2.5		*Cases Outside the 95% Confidence Interval Ellipse*		*Cases with a Value of D^2/df Greater than 2.5 (df = 13)[a]*		
		X_{19} **with:**		**Case**	D^2	D^2/df
X_6	No cases	X_6	44, **90**	98	40.0	3.08
X_7	13, 22, **90**	X_7	13, **22**, 24, 53, **90**	36	36.9	2.84
X_8	8, **7**	X_8	**22**, 87			
X_9	No cases	X_9	**2, 22**, 45, 52			
X_{10}	No cases	X_{10}	**22, 24**, 85			
X_{11}	**7**	X_{11}	**2**, 7, **22**, 45			
X_{12}	**90**	X_{12}	**22**, 44, **90**			
X_{13}	No cases	X_{13}	**22**, 57			
X_{14}	77	X_{14}	**22**, 77, 84			
X_{15}	6, 53	X_{15}	6, **22**, 53			
X_{16}	24	X_{16}	**22, 24**, 48, 62, 92			
X_{17}	No cases	X_{17}	**22**			
X_{18}	**7**, 84	X_{18}	**2**, 7, **22**, 84			
X_{19}	**22**					

[a]*Mahalanobis D^2 value based on the 13 HBAT perceptions (X_6 to X_{18}).*

scatterplot. The three observations will be noted to see whether they appear in the subsequent bivariate and multivariate assessments.

Bivariate Detection. For a bivariate perspective, 13 scatterplots are formed for each of the independent variables (X_6 through X_{18}) with the dependent variable (X_{19}). An ellipse representing the 95% confidence interval of a bivariate normal distribution is then superimposed on the scatterplot. Figure 2-5 contains examples of two such scatterplots involving X_6 and X_7. As we can see in the scatterplot for X_6 with X_{19}, the two outliers fall just outside the ellipse and do not have the most extreme values on either variable. This result is in contrast to the scatterplot of X_7 with X_{19}, where observation 22 is markedly different from the other observations and shows the highest values on both X_7 and X_{19}. The second part of Table 2-10 contains a compilation of the observations falling outside this ellipse for each variable. Because it is a 95% confidence interval, we would expect some observations normally to fall outside the ellipse. Only four observations (2, 22, 24, and 90) fall outside the ellipse more than two times. Observation 22 falls outside in 12 of the 13 scatterplots, mostly because it is an outlier on the dependent variable. Of the remaining three observations, only observation 90 was noted in the univariate detection.

Multivariate Detection. The final diagnostic method is to assess multivariate outliers with the Mahalanobis D^2 measure (see Table 2-10). This analysis evaluates the position of each observation compared with the center of all observations on a set of variables. In this case, all the metric independent

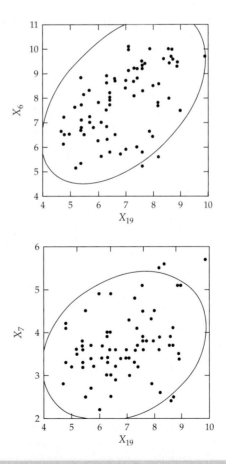

FIGURE 2-5 Selected Scatterplots for Bivariate Detection of Outliers: X_6 (Product Quality) and X_7 (E-Commerce Activities) with X_{19} (Customer Satisfaction)

variables were used. The calculation of the D^2/df value ($df = 13$) allows for identification of outliers through an approximate test of statistical significance. Because the sample has only 100 observations, a threshold value of 2.5 will be used rather than the value of 3.5 or 4.0 used in large samples. With this threshold, two observations (98 and 36) are identified as significantly different. It is interesting that these observations were not seen in earlier univariate and bivariate analyses but appear only in the multivariate tests. This result indicates they are not unique on any single variable but instead are unique in combination.

RETENTION OR DELETION OF THE OUTLIERS   As a result of these diagnostic tests, no observations demonstrate the characteristics of outliers that should be eliminated. Each variable has some observations that are extreme, and they should be considered if that variable is used in an analysis. No observations are extreme on a sufficient number of variables to be considered unrepresentative of the population. In all instances, the observations designated as outliers, even with the multivariate tests, seem similar enough to the remaining observations to be retained in the multivariate analyses. However, the researcher should always examine the results of each specific multivariate technique to identify observations that may become outliers in that particular application. In the case of regression analysis, Chapter 4 will provide additional methods to assess the relative influence of each observation and provide more insight into the possible deletion of an observation as an outlier.

TESTING THE ASSUMPTIONS OF MULTIVARIATE ANALYSIS

The final step in examining the data involves testing for the assumptions underlying the statistical bases for multivariate analysis. The earlier steps of missing data analysis and outlier detection attempted to clean the data to a format most suitable for multivariate analysis. Testing the data for compliance with the statistical assumptions underlying the multivariate techniques now deals with the foundation upon which the techniques make statistical inferences and results. Some techniques are less affected by violating certain assumptions, which is termed **robustness,** but in all cases meeting some of the assumptions will be critical to a successful analysis. Thus, it is necessary to understand the role played by each assumption for every multivariate technique.

The need to test the statistical assumptions is increased in multivariate applications because of two characteristics of multivariate analysis. First, the complexity of the relationships, owing to the typical use of a large number of variables, makes the potential distortions and biases more potent when the assumptions are violated, particularly when the violations compound to become even more detrimental than if considered separately. Second, the complexity of the analyses and results may mask the indicators of assumption violations apparent in the simpler univariate analyses. In almost all instances, the multivariate procedures will estimate the multivariate model and produce results even when the assumptions are severely violated. Thus, the researcher must be aware of any assumption violations and the implications they may have for the estimation process or the interpretation of the results.

Assessing Individual Variables Versus the Variate

Multivariate analysis requires that the assumptions underlying the statistical techniques be tested twice: first for the separate variables, akin to the tests for a univariate analysis, and second for the multivariate model **variate,** which acts collectively for the variables in the analysis and thus must meet the same assumptions as individual variables. This chapter focuses on the examination of individual variables for meeting the assumptions underlying the multivariate procedures. Discussions in each chapter address the methods used to assess the assumptions underlying the variate for each multivariate technique.

Four Important Statistical Assumptions

Multivariate techniques and their univariate counterparts are all based on a fundamental set of assumptions representing the requirements of the underlying statistical theory. Although many assumptions or requirements come into play in one or more of the multivariate techniques we discuss in the text, four of them potentially affect every univariate and multivariate statistical technique.

NORMALITY The most fundamental assumption in multivariate analysis is **normality,** referring to the shape of the data distribution for an individual metric variable and its correspondence to the **normal distribution,** the benchmark for statistical methods. *If the variation from the normal distribution is sufficiently large, all resulting statistical tests are invalid, because normality is required to use the* F *and* t *statistics.* Both the univariate and the multivariate statistical methods discussed in this text are based on the assumption of univariate normality, with the multivariate methods also assuming multivariate normality.

Univariate normality for a single variable is easily tested, and a number of corrective measures are possible, as shown later. In a simple sense, multivariate normality (the combination of two or more variables) means that the individual variables are normal in a univariate sense and that their combinations are also normal. Thus, *if a variable is multivariate normal, it is also univariate normal. However, the reverse is not necessarily true (two or more univariate normal variables are not necessarily multivariate normal).* Thus, a situation in which all variables exhibit univariate normality will help gain, although not guarantee, multivariate normality. Multivariate normality is more difficult to test [13, 23], but specialized tests are available in the techniques most affected by departures from multivariate normality. In most cases assessing and achieving univariate normality for all variables is sufficient, and we will address multivariate normality only when it is especially critical. Even though large sample sizes tend to diminish the detrimental effects of nonnormality, the researcher should always assess the normality for all metric variables included in the analysis.

Assessing the Impact of Violating the Normality Assumption. The severity of nonnormality is based on two dimensions: the shape of the offending distribution and the sample size. As we will see in the following discussion, the researcher must not only judge the extent to which the variable's distribution is nonnormal, but also the sample sizes involved. What might be considered unacceptable at small sample sizes will have a negligible effect at larger sample sizes.

Impacts Due to the Shape of the Distribution. How can we describe the distribution if it differs from the normal distribution? The shape of any distribution can be described by two measures: kurtosis and skewness. **Kurtosis** refers to the "peakedness" or "flatness" of the distribution compared with the normal distribution. Distributions that are taller or more peaked than the normal distribution are termed *leptokurtic,* whereas a distribution that is flatter is termed *platykurtic.* Whereas kurtosis refers to the height of the distribution, **skewness** is used to describe the balance of the distribution; that is, is it unbalanced and shifted to one side (right or left) or is it centered and symmetrical with about the same shape on both sides? If a distribution is unbalanced, it is skewed. A positive skew denotes a distribution shifted to the left, whereas a negative skewness reflects a shift to the right.

Knowing how to describe the distribution is followed by the issue of how to determine the extent or amount to which it differs on these characteristics? Both skewness and kurtosis have empirical measures that are available in all statistical programs. In most programs, the skewness and kurtosis of a normal distribution are given values of zero. Then, values above or below zero denote departures from normality. For example, negative kurtosis values indicate a platykurtic (flatter) distribution, whereas positive values denote a leptokurtic (peaked) distribution. Likewise, positive skewness values indicate the distribution shifted to the left, and the negative values denote a rightward shift. To judge the "Are they large enough to worry about?" question for these values, the following discussion on statistical tests shows how the kurtosis and skewness values can be transformed to reflect the statistical significance of the differences and provide guidelines as to their severity.

Impacts Due to Sample Size. Even though it is important to understand how the distribution departs from normality in terms of shape and whether these values are large enough to warrant attention, the researcher must also consider the affects of sample size. As discussed in Chapter 1, sample size has the effect of increasing statistical power by reducing sampling error. It results in a similar effect here, in that larger sample sizes *reduce* the detrimental effects of nonnormality. In small samples of 50 or fewer observations, and especially if the sample size is less than 30 or so, significant departures from normality can have a substantial impact on the results. For sample sizes of 200 or more, however, these same effects may be negligible. Moreover, when group comparisons are made, such as in ANOVA, the differing sample sizes between groups, if large enough, can even cancel out the detrimental effects. Thus, in most instances, as the sample sizes become large, the researcher can be less concerned about nonnormal variables, except as they might lead to other assumption violations that do have an impact in other ways (e.g., see the following discussion on homoscedasticity).

Graphical Analyses of Normality. The simplest diagnostic test for normality is a visual check of the histogram that compares the observed data values with a distribution approximating the normal distribution (see Figure 2-1). Although appealing because of its simplicity, this method is problematic for smaller samples, where the construction of the histogram (e.g., the number of categories or the width of categories) can distort the visual portrayal to such an extent that the analysis is useless. A more reliable approach is the **normal probability plot,** which compares the cumulative distribution of actual data values with the cumulative distribution of a normal distribution. The normal distribution forms a straight diagonal line, and the plotted data values are compared with the diagonal. If a distribution is normal, the line representing the actual data distribution closely follows the diagonal.

Figure 2-6 shows several departures from normality and their representation in the normal probability in terms of kurtosis and skewness. First, departures from the normal distribution in terms of kurtosis are easily seen in the normal probability plots. When the line falls below the diagonal, the distribution is flatter than expected. When it goes above the diagonal, the distribution is more peaked than the normal curve. For example, in the normal probability plot of a peaked distribution (Figure 2-6d), we see a distinct S-shaped curve. Initially the distribution is flatter, and the plotted line falls below the diagonal. Then the peaked part of the distribution rapidly moves the plotted line above the diagonal, and eventually the line shifts to below the diagonal again as the distribution flattens. A nonpeaked distribution has the opposite pattern (Figure 2-6c). Skewness is also easily seen, most often represented by a simple arc, either above or below the diagonal. A negative skewness (Figure 2-6e) is indicated by an arc below the diagonal, whereas an arc above the diagonal represents a positively skewed distribution (Figure 2-6f). An excellent source for interpreting normal probability plots, showing the various patterns and interpretations, is Daniel and Wood [7]. These specific patterns not only identify nonnormality but also tell us the form of the original distribution and the appropriate remedy to apply.

Statistical Tests of Normality. In addition to examining the normal probability plot, one can also use statistical tests to assess normality. A simple test is a rule of thumb based on the skewness and kurtosis values (available as part of the basic descriptive statistics for a variable computed by all statistical programs). The statistic value (z) for the skewness value is calculated as:

$$z_{skewness} = \frac{skewness}{\sqrt{\dfrac{6}{N}}}$$

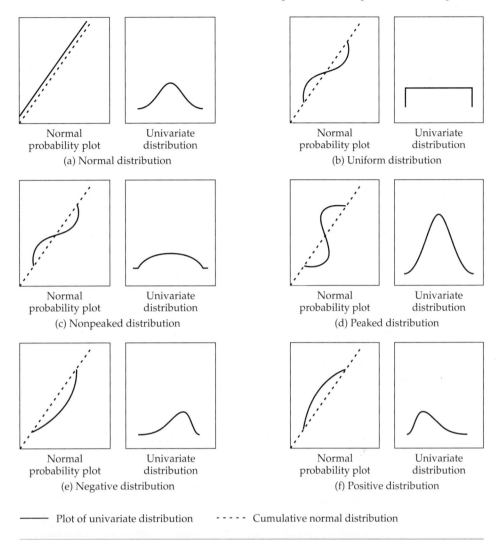

Normal
probability plot

Univariate
distribution

(a) Normal distribution

Normal
probability plot

Univariate
distribution

(b) Uniform distribution

Normal
probability plot

Univariate
distribution

(c) Nonpeaked distribution

Normal
probability plot

Univariate
distribution

(d) Peaked distribution

Normal
probability plot

Univariate
distribution

(e) Negative distribution

Normal
probability plot

Univariate
distribution

(f) Positive distribution

——— Plot of univariate distribution   - - - - - Cumulative normal distribution

FIGURE 2-6   Normal Probability Plots and Corresponding Univariate Distributions

where N is the sample size. A z value can also be calculated for the kurtosis value using the following formula:

$$z_{kurtosis} = \frac{kurtosis}{\sqrt{\frac{24}{N}}}$$

If either calculated z value exceeds the specified critical value, then the distribution is nonnormal in terms of that characteristic. The critical value is from a z distribution, based on the significance level we desire. The most commonly used critical values are ±2.58 (.01 significance level) and ±1.96, which corresponds to a .05 error level. With these simple tests, the researcher can easily assess the degree to which the skewness and peakedness of the distribution vary from the normal distribution.

Specific statistical tests for normality are also available in all the statistical programs. The two most common are the Shapiro-Wilks test and a modification of the Kolmogorov-Smirnov test. Each calculates the level of significance for the differences from a normal distribution. The researcher should always remember that tests of significance are less useful in small samples (fewer than 30) and quite sensitive in large samples (exceeding 1,000 observations). Thus, the researcher should always use both the graphical plots and any statistical tests to assess the actual degree of departure from normality.

Remedies for Nonnormality. A number of data transformations available to accommodate nonnormal distributions are discussed later in the chapter. This chapter confines the discussion to univariate normality tests and transformations. However, when we examine other multivariate methods, such as multivariate regression or multivariate analysis of variance, we discuss tests for multivariate normality as well. Moreover, many times when nonnormality is indicated, it also contributes to other assumption violations; therefore, remedying normality first may assist in meeting other statistical assumptions as well. (For those interested in multivariate normality, see references [13, 16, 25].)

HOMOSCEDASTICITY   The next assumption is related primarily to dependence relationships between variables. **Homoscedasticity** refers to the assumption that dependent variable(s) exhibit equal levels of variance across the range of predictor variable(s). Homoscedasticity is desirable because *the variance of the dependent variable being explained in the dependence relationship should not be concentrated in only a limited range of the independent values.* In most situations, we have many different values of the dependent variable at each value of the independent variable. For this relationship to be fully captured, the dispersion (variance) of the dependent variable values must be relatively equal at each value of the predictor variable. If this dispersion is unequal across values of the independent variable, the relationship is said to be **heteroscedastic.** Although the dependent variables must be metric, this concept of an equal spread of variance across independent variables can be applied when the independent variables are either metric or nonmetric:

- *Metric independent variables.* The concept of homoscedasticity is based on the spread of dependent variable variance across the range of independent variable values, which is encountered in techniques such as multiple regression. The dispersion of values for the dependent variable should be as large for small values of the independent values as it is for moderate and large values. In a scatterplot, it is seen as an elliptical distribution of points
- *Nonmetric independent variables.* In these analyses (e.g., ANOVA and MANOVA) the focus now becomes the equality of the variance (single dependent variable) or the variance/covariance matrices (multiple dependent variables) across the groups formed by the nonmetric independent variables. The equality of variance/covariance matrices is also seen in discriminant analysis, but in this technique the emphasis is on the spread of the independent variables across the groups formed by the nonmetric dependent measure.

In each of these instances, the purpose is the same: to ensure that the variance used in explanation and prediction is distributed across the range of values, thus allowing for a "fair test" of the relationship across all values of the nonmetric variables. The two most common sources of heteroscedasticity are the following:

- *Variable type.* Many types of variables have a natural tendency toward differences in dispersion. For example, as a variable increases in value (e.g., units ranging from near zero to millions) a naturally wider range of answers is possible for the larger values. Also, when percentages are used the natural tendency is for many values to be in the middle range, with few in the lower or higher values.
- *Skewed distribution of one or both variables.* In Figure 2-7a, the scatterplots of data points for two variables (V_1 and V_2) with normal distributions exhibit equal dispersion across all data values (i.e., homoscedasticity). However, in Figure 2-7b we see unequal dispersion (heteroscedasticity) caused by skewness of one of the variables (V_3). For the different values of V_3, there are different patterns of dispersion for V_1.

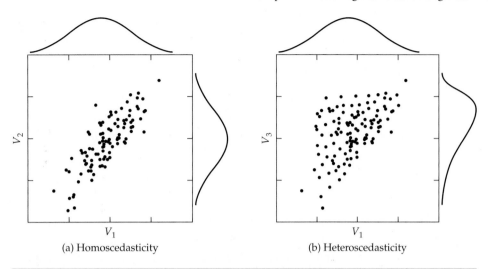

(a) Homoscedasticity   (b) Heteroscedasticity

FIGURE 2-7   Scatterplots of Homoscedastic and Heteroscedastic Relationships

The result of heteroscedasticity is to cause the predictions to be better at some levels of the independent variable than at others. This variability affects the standard errors and makes hypothesis tests either too stringent or too insensitive. The effect of heteroscedasticity is also often related to sample size, especially when examining the variance dispersion across groups. For example, in ANOVA or MANOVA the impact of heteroscedasticity on the statistical test depends on the sample sizes associated with the groups of smaller and larger variances. In multiple regression analysis, similar effects would occur in highly skewed distributions where there were disproportionate numbers of respondents in certain ranges of the independent variable.

Graphical Tests of Equal Variance Dispersion. The test of homoscedasticity for two metric variables is best examined graphically. Departures from an equal dispersion are shown by such shapes as cones (small dispersion at one side of the graph, large dispersion at the opposite side) or diamonds (a large number of points at the center of the distribution). The most common application of graphical tests occurs in multiple regression, based on the dispersion of the dependent variable across the values of either the metric independent variables. We will defer our discussion of graphical methods until we reach Chapter 4, which describes these procedures in much more detail.

Boxplots work well to represent the degree of variation between groups formed by a categorical variable. The length of the box and the whiskers each portray the variation of data within that group. Thus, heteroscedasticity would be portrayed by substantial differences in the length of the boxes and whiskers between groups representing the dispersion of observations in each group.

Statistical Tests for Homoscedasticity. The statistical tests for equal variance dispersion assess the equality of variances within groups formed by nonmetric variables. The most common test, the Levene test, is used to assess whether the variances of a single metric variable are equal across any number of groups. If more than one metric variable is being tested, so that the comparison involves the equality of variance/covariance matrices, the Box's M test is applicable. The Box's M test is available in both multivariate analysis of variance and discriminant analysis and is discussed in more detail in later chapters pertaining to these techniques.

Remedies for Heteroscedasticity. Heteroscedastic variables can be remedied through data transformations similar to those used to achieve normality. As mentioned earlier, many times heteroscedasticity is the result of nonnormality of one of the variables, and correction of the nonnormality also remedies the unequal dispersion of variance. A later section discusses data transformations of the variables to "spread" the variance and make all values have a potentially equal effect in prediction.

LINEARITY　An implicit assumption of all multivariate techniques based on correlational measures of association, including multiple regression, logistic regression, factor analysis, and structural equation modeling, is **linearity.** Because correlations represent only the linear association between variables, nonlinear effects will not be represented in the correlation value. This omission results in an underestimation of the actual strength of the relationship. It is always prudent to examine all relationships to identify any departures from linearity that may affect the correlation.

Identifying Nonlinear Relationships.　The most common way to assess linearity is to examine scatterplots of the variables and to identify any nonlinear patterns in the data. Many scatterplot programs can show the straight line depicting the linear relationship, enabling the researcher to better identify any nonlinear characteristics. An alternative approach is to run a simple regression analysis (the specifics of this technique are covered in Chapter 4) and to examine the **residuals.** The residuals reflect the unexplained portion of the dependent variable; thus, any nonlinear portion of the relationship will show up in the residuals. A third approach is to explicitly model a nonlinear relationship by the testing of alternative model specifications (also know as curve fitting) that reflect the nonlinear elements. A discussion of this approach and residual analysis is found in Chapter 4.

Remedies for Nonlinearity.　If a nonlinear relationship is detected, the most direct approach is to transform one or both variables to achieve linearity. A number of available transformations are discussed later in this chapter. An alternative to data transformation is the creation of new variables to represent the nonlinear portion of the relationship. The process of creating and interpreting these additional variables, which can be used in all linear relationships, is discussed in Chapter 4.

ABSENCE OF CORRELATED ERRORS　Predictions in any of the dependence techniques are not perfect, and we will rarely find a situation in which they are. However, we do attempt to ensure that any prediction errors are uncorrelated with each other. For example, if we found a pattern that suggests every other error is positive while the alternative error terms are negative, we would know that some unexplained systematic relationship exists in the dependent variable. If such a situation exists, we cannot be confident that our prediction errors are independent of the levels at which we are trying to predict. Some other factor is affecting the results, but is not included in the analysis.

Identifying Correlated Errors.　One of the most common violations of the assumption that errors are uncorrelated is due to the data collection process. Similar factors that affect one group may not affect the other. If the groups are analyzed separately, the effects are constant within each group and do not impact the estimation of the relationship. But if the observations from both groups are combined, then the final estimated relationship must be a compromise between the two actual relationships. This combined effect leads to biased results because an unspecified cause is affecting the estimation of the relationship.

Another common source of correlated errors is time series data. As we would expect, the data for any time period is highly related to the data at time periods both before and afterward. Thus, any predictions and any prediction errors will necessarily be correlated. This type of data led to the creation of specialized programs specifically for time series analysis and this pattern of correlated observations.

To identify correlated errors, the researcher must first identify possible causes. Values for a variable should be grouped or ordered on the suspected variable and then examined for any patterns. In our earlier example of grouped data, once the potential cause is identified the researcher could see whether differences did exist between the groups. Finding differences in the prediction errors in the two groups would then be the basis for determining that an unspecified effect was "causing" the correlated errors. For other types of data, such as time series data, we can see any trends or patterns when we order the data (e.g., by time period for time series data). This ordering variable (time in this case), if not included in the analysis in some manner, would cause the errors to be correlated and create substantial bias in the results.

Remedies for Correlated Errors. Correlated errors must be corrected by including the omitted causal factor into the multivariate analysis. In our earlier example, the researcher would add a variable indicating in which class the respondents belonged. The most common remedy is the addition of a variable(s) to the analysis that represents the omitted factor. The key task facing the researcher is not the actual remedy, but rather the identification of the unspecified effect and a means of representing it in the analysis.

OVERVIEW OF TESTING FOR STATISTICAL ASSUMPTIONS The researcher is faced with what may seem to be an impossible task: satisfy all of these statistical assumptions or risk a biased and flawed analysis. We want to note that even though these statistical assumptions are important, the researcher must use judgment in how to interpret the tests for each assumption and when to apply remedies. Even analyses with small sample sizes can withstand small, but significant, departures from normality. What is more important for the researcher is to understand the implications of each assumption with regard to the technique of interest, striking a balance between the need to satisfy the assumptions versus the robustness of the technique and research context. The following guidelines in Rules of Thumb 2-5 attempt to portray the most pragmatic aspects of the assumptions and the reactions that can be taken by researchers.

Data Transformations

Data transformations provide a means of modifying variables for one of two reasons: (1) to correct violations of the statistical assumptions underlying the multivariate techniques or (2) to improve the relationship (correlation) between variables. Data transformations may be based on reasons that are either *theoretical* (transformations whose appropriateness is based on the nature of the data) or *data derived* (where the transformations are suggested strictly by an examination of the data). Yet in either case the researcher must proceed many times by trial and error, monitoring the improvement versus the need for additional transformations.

All the transformations described here are easily carried out by simple commands in the popular statistical packages. We focus on transformations that can be computed in this manner, although more sophisticated and complicated methods of data transformation are available (e.g., see Box and Cox [3]).

RULES OF THUMB 2-5

Testing Statistical Assumptions

- Normality can have serious effects in small samples (fewer than 50 cases), but the impact effectively diminishes when sample sizes reach 200 cases or more
- Most cases of heteroscedasticity are a result of nonnormality in one or more variables; thus, remedying normality may not be needed due to sample size, but may be needed to equalize the variance
- Nonlinear relationships can be well defined, but seriously understated unless the data are transformed to a linear pattern or explicit model components are used to represent the nonlinear portion of the relationship
- Correlated errors arise from a process that must be treated much like missing data; that is, the researcher must first define the causes among variables either internal or external to the dataset; if they are not found and remedied, serious biases can occur in the results, many times unknown to the researcher

TRANSFORMATIONS TO ACHIEVE NORMALITY AND HOMOSCEDASTICITY   Data transformations provide the principal means of correcting nonnormality and heteroscedasticity. In both instances, patterns of the variables suggest specific transformations. For nonnormal distributions, the two most common patterns are flat distributions and skewed distributions. For the flat distribution, the most common transformation is the inverse (e.g., $1/Y$ or $1/X$). Skewed distributions can be transformed by taking the square root, logarithms, squared, or cubed (X^2 or X^3) terms or even the inverse of the variable. Usually negatively skewed distributions are best transformed by employing a squared or cubed transformation, whereas the logarithm or square root typically works best on positive skewness. In many instances, the researcher may apply all of the possible transformations and then select the most appropriate transformed variable.

Heteroscedasticity is an associated problem, and in many instances "curing" this problem will deal with normality problems as well. Heteroscedasticity is also due to the distribution of the variable(s). When examining the scatterplot, the most common pattern is the cone-shaped distribution. If the cone opens to the right, take the inverse; if the cone opens to the left, take the square root. Some transformations can be associated with certain types of data. For example, frequency counts suggest a square root transformation; proportions are best transformed by the arcsin transformation $\left(X_{new} = 2 \arcsin \sqrt{X_{old}}\right)$; and proportional change is best handled by taking the logarithm of the variable. In all instances, once the transformations have been performed, the transformed data should be tested to see whether the desired remedy was achieved.

TRANSFORMATIONS TO ACHIEVE LINEARITY   Numerous procedures are available for achieving linearity between two variables, but most simple nonlinear relationships can be placed in one of four categories (see Figure 2-8). In each quadrant, the potential transformations for both dependent

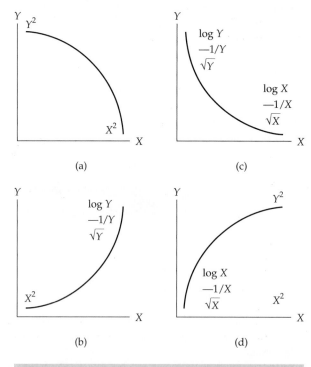

FIGURE 2-8   Selecting Transformations to Achieve Linearity

Source: F. Mosteller and J. W. Tukey, *Data Analysis and Regression.* Reading, MA: Addison-Wesley, 1977.

and independent variables are shown. For example, if the relationship looks like Figure 2-8a, then either variable can be squared to achieve linearity. When multiple transformation possibilities are shown, start with the top method in each quadrant and move downward until linearity is achieved. An alternative approach is to use additional variables, termed *polynomials,* to represent the nonlinear components. This method is discussed in more detail in Chapter 4.

GENERAL GUIDELINES FOR TRANSFORMATIONS   Many possibilities exist for transforming the data to achieve the required statistical assumptions. Apart from the technical issues of the type of transformation, several points to remember when performing data transformations are presented in Rules of Thumb 2-6.

An Illustration of Testing the Assumptions Underlying Multivariate Analysis

To illustrate the techniques involved in testing the data for meeting the assumptions underlying multivariate analysis and to provide a foundation for use of the data in the subsequent chapters, the data set introduced in Chapter 1 will be examined. In the course of this analysis, the assumptions of normality, homoscedasticity, and linearity will be covered. The fourth basic assumption, the absence of correlated errors, can be addressed only in the context of a specific multivariate model; this assumption will be covered in later chapters for each multivariate technique. Emphasis will be placed on examining the metric variables, although the nonmetric variables will be assessed where appropriate.

NORMALITY   The assessment of normality of the metric variables involves both empirical measures of a distribution's shape characteristics (skewness and kurtosis) and the normal probability plots. The empirical measures provide a guide as to the variables with significant deviations from normality, and the normal probability plots provide a visual portrayal of the shape of the distribution. The two portrayals complement each other when selecting the appropriate transformations.

Table 2-11 and Figure 2-9 contain the empirical measures and normal probability plots for the metric variables in our data set. Our first review concerns the empirical measures reflecting the shape of the distribution (skewness and kurtosis) as well as a statistical test for normality

---

RULES OF THUMB 2-6

Transforming Data

- To judge the potential impact of a transformation, calculate the ratio of the variable's mean to its standard deviation:
  - Noticeable effects should occur when the ratio is less than 4
  - When the transformation can be performed on either of two variables, select the variable with the smallest ratio
- Transformations should be applied to the independent variables except in the case of heteroscedasticity
- Heteroscedasticity can be remedied only by the transformation of the dependent variable in a dependence relationship; if a heteroscedastic relationship is also nonlinear, the dependent variable, and perhaps the independent variables, must be transformed
- Transformations may change the interpretation of the variables; for example, transforming variables by taking their logarithm translates the relationship into a measure of proportional change (elasticity); always be sure to explore thoroughly the possible interpretations of the transformed variables
- Use variables in their original (untransformed) format when profiling or interpreting results

TABLE 2-11 Distributional Characteristics, Testing for Normality, and Possible Remedies

| Variable | SHAPE DESCRIPTORS | | | | Tests of Normality | | Description of the Distribution | Applicable Remedies | |
| | Skewness | | Kurtosis | | | | | | |
	Statistic	z value	Statistic	z value	Statistic	Significance		Transformation	Significance After Remedy
Firm Characteristics									
X_6	−.245	−1.01	−1.132	**−2.37**	.109	**.005**	Almost uniform distribution	Squared term	.015
X_7	.660	**2.74**	.735	1.54	.122	**.001**	Peaked with positive skew	Logarithm	.037
X_8	−.203	−.84	−.548	−1.15	.060	.200[a]	Normal distribution		
X_9	−.136	−.56	−.586	−1.23	.051	.200[a]	Normal distribution		
X_{10}	.044	.18	−.888	−1.86	.065	.200[a]	Normal distribution		
X_{11}	−.092	−.38	−.522	−1.09	.060	.200[a]	Normal distribution		
X_{12}	.377	1.56	.410	.86	.111	**.004**	Slight positive skew and peakedness	None	—
X_{13}	−.240	−1.00	−.903	−1.89	.106	**.007**	Peaked	Cubed term	.022
X_{14}	.008	.03	−.445	−.93	.064	.200[a]	Normal distribution		
X_{15}	.299	1.24	.016	.03	.074	.200[a]	Normal distribution		
X_{16}	−.334	−1.39	.244	.51	.129	**.000**	Negative skewness	Squared term	.066
X_{17}	.323	1.34	−.816	−1.71	.101	**.013**	Peaked, positive skewness	Inverse	.187
X_{18}	−.463	−1.92	.218	.46	.084	.082	Normal distribution		
Performance Measures									
X_{19}	.078	.32	−.791	−1.65	.078	.137	Normal distribution		
X_{20}	.044	.18	−.089	−.19	.077	.147	Normal distribution		
X_{21}	−.093	−.39	−.090	−.19	.073	.200[a]	Normal distribution		
X_{22}	−.132	−.55	−.684	−1.43	.075	.180	Normal distribution		

[a]Lower bound of true significance.

Note: The z values are derived by dividing the statistics by the appropriate standard errors of .241 (skewness) and .478 (kurtosis). The equations for calculating the standard errors are given in the text.

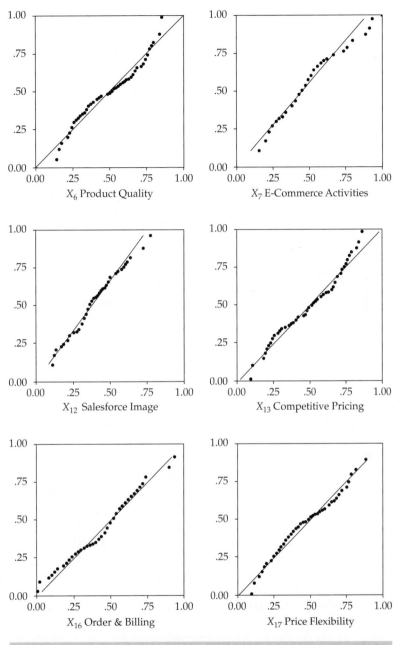

FIGURE 2-9   Normal Probability Plots (NPP) of Nonnormal Metric Variables
$(X_6, X_7, X_{12}, X_{13}, X_{16},$ and $X_{17})$

(the modified Kolmogorov-Smirnov test). Of the 17 metric variables, only 6 ($X_6, X_7, X_{12}, X_{13}, X_{16},$ and X_{17}) show any deviation from normality in the overall normality tests. When viewing the shape characteristics, significant deviations were found for skewness (X_7) and kurtosis (X_6). One should note that only two variables were found with shape characteristics significantly different from the normal curve, while six variables were identified with the overall tests. The overall test provides no insight as to the transformations that might be best, whereas the shape characteristics provide guidelines for possible transformations. The researcher can also use the normal probability plots to identify the shape of the distribution. Figure 2-9 contains the normal probability plots for the six variables found to have the nonnormal distributions. By combining information, from the empirical and graphical methods, the researcher can characterize the nonnormal distribution in anticipation of selecting a transformation (see Table 2-11 for a description of each nonnormal distribution).

Table 2-11 also suggests the appropriate remedy for each of the variables. Two variables (X_6 and X_{16}) were transformed by taking the square root. X_7 was transformed by logarithm, whereas X_{17} was squared and X_{13} was cubed. Only X_{12} could not be transformed to improve on its distributional characteristics. For the other five variables, their tests of normality were now either nonsignificant (X_{16} and X_{17}) or markedly improved to more acceptable levels (X_6, X_7, and X_{13}). Figure 2-10 demonstrates the effect of the transformation on X_{17} in achieving normality. The transformed X_{17} appears markedly more normal in the graphical portrayals, and the statistical descriptors are also improved. The researcher should always examine the transformed variables as rigorously as the original variables in terms of their normality and distribution shape.

In the case of the remaining variable (X_{12}), none of the transformations could improve the normality. This variable will have to be used in its original form. In situations where the normality of the variables is critical, the transformed variables can be used with the assurance that they meet the assumptions of normality. But the departures from normality are not so extreme in any of the original variables that they should never be used in any analysis in their original form. If the technique has a robustness to departures from normality, then the original variables may be preferred for the comparability in the interpretation phase.

HOMOSCEDASTICITY   All statistical packages have tests to assess homoscedasticity on a univariate basis (e.g., the Levene test in SPSS) where the variance of a metric variable is compared across levels of a nonmetric variable. For our purposes, we examine each of the metric variables across the five nonmetric variables in the data set. These analyses are appropriate in preparation for analysis of variance or multivariate analysis of variance, in which the nonmetric variables are the independent variables, or for discriminant analysis, in which the nonmetric variables are the dependent measures.

Table 2-12 contains the results of the Levene test for each of the nonmetric variables. Among the performance factors, only X_4 (Region) has notable problems with heteroscedasticity. For the 13 firm characteristic variables, only X_6 and X_{17} show patterns of heteroscedasticity on more than one of the nonmetric variables. Moreover, in no instance do any of the nonmetric variables have more than two problematic metric variables. The actual implications of these instances of heteroscedasticity must be examined whenever group differences are examined using these nonmetric variables as independent variables and these metric variables as dependent variables. The relative lack of either numerous problems or any consistent patterns across one of the nonmetric variables suggests that heteroscedasticity problems will be minimal. If the assumption violations are found, variable transformations are available to help remedy the variance dispersion.

The ability for transformations to address the problem of heteroscedasticity for X_{17}, if desired, is also shown in Figure 2-10. Before a logarithmic transformation was applied, heteroscedastic conditions were found on three of the five nonmetric variables. The transformation not only corrected the nonnormality problem, but also eliminated the problems with heteroscedasticity. It should be noted, however, that several transformations "fixed" the normality problem, but only the logarithmic transformation also addressed the heteroscedasticity, which demonstrates the relationship between normality and heteroscedasticity and the role of transformations in addressing each issue.

The tests for homoscedasticity of two metric variables, encountered in methods such as multiple regression, are best accomplished through graphical analysis, particularly an analysis of the residuals. The interested reader is referred to Chapter 4 for a complete discussion of residual analysis and the patterns of residuals indicative of heteroscedasticity.

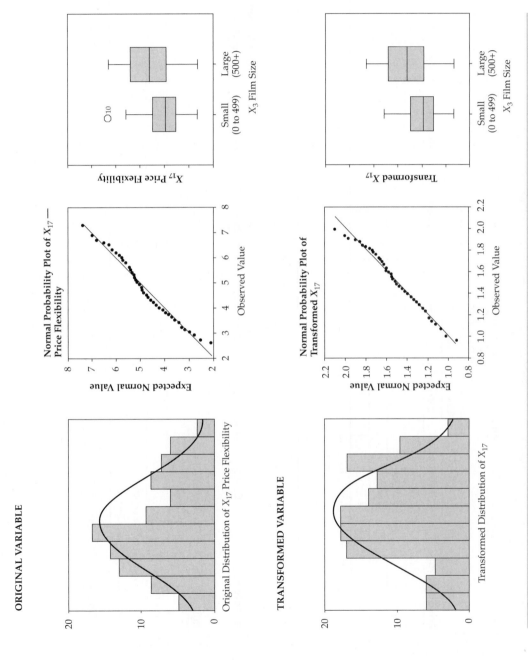

FIGURE 2-10 Transformation of X_{17} (Price Flexibility) to Achieve Normality and Homoscedasticity

Distribution Characteristics Before and After Transformation

Variable Form	SHAPE DESCRIPTORS				Test of Normality	
	Skewness		Kurtosis			
	Statistic	z value[a]	Statistic	z value[a]	Statistic	Significance
Original X_{17}	.323	1.34	−.816	−1.71	.101	.013
Transformed X_{17}[b]	−.121	.50	−.803	−1.68	.080	.117

[a] The z values are derived by dividing the statistics by the appropriate standard errors of .241 (skewness) and .478 (kurtosis). The equations for calculating the standard errors are given in the text.
[b] Logarithmic transformation.

Variable Form	Levene Test Statistic				
	X_1 Customer Type	X_2 Industry Type	X_3 Firm Size	X_4 Region	X_5 Distribution System
Original X_{17}	5.56**	2.84	4.19*	16.21**	.62
Transformed X_{17}	2.76	2.23	1.20	3.11**	.01

* Significant at .05 significance level.
** Significant at .01 significance level.

FIGURE 2-10 Continued

TABLE 2-12 Testing for Homoscedasticity

| | NONMETRIC/CATEGORICAL VARIABLE | | | | | | | | | |
| | X_1 Customer Type | | X_2 Industry Type | | X_3 Firm Size | | X_4 Region | | X_5 Distribution System | |
Metric Variable	Levene Statistic	Sig.	Levene Statistic	Sig.	Levene Statistic	Sig.	Levene Statistic	Sig.	Levene Statistic	Sig.
Firm Characteristics										
X_6	**17.47**	**.00**	.01	.94	.02	.89	**17.86**	**.00**	.48	.49
X_7	.58	.56	.09	.76	.09	.76	.05	.83	2.87	.09
X_8	.37	.69	.48	.49	1.40	.24	.72	.40	.11	.74
X_9	.43	.65	.02	.88	.17	.68	.58	.45	1.20	.28
X_{10}	.74	.48	.00	.99	.74	.39	1.19	.28	.69	.41
X_{11}	.05	.95	.15	.70	.09	.76	3.44	.07	1.72	.19
X_{12}	2.46	.09	.36	.55	.06	.80	1.55	.22	1.55	.22
X_{13}	.84	.43	**4.43**	**.04**	1.71	.19	.24	.63	2.09	.15
X_{14}	2.39	.10	2.53	.11	**4.55**	**.04**	.25	.62	.16	.69
X_{15}	1.13	.33	.47	.49	1.05	.31	.01	.94	.59	.45
X_{16}	1.65	.20	.83	.37	.31	.58	2.49	.12	**4.60**	**.03**
X_{17}	**5.56**	**.01**	2.84	.10	**4.19**	**.04**	**16.21**	**.00**	.62	.43
X_{18}	.87	.43	.30	.59	.18	.67	2.25	.14	**4.27**	**.04**
Performance Measures										
X_{19}	**3.40**	**.04**	.00	.96	.73	.39	**8.57**	**.00**	.18	.67
X_{20}	1.64	.20	.03	.86	.03	.86	**7.07**	**.01**	.46	.50
X_{21}	1.05	.35	.11	.74	.68	.41	**11.54**	**.00**	2.67	.10
X_{22}	.15	.86	.30	.59	.74	.39	.00	.99	1.47	.23

Notes: Values represent the Levene statistic value and the statistical significance in assessing the variance dispersion of each metric variable across the levels of the nonmetric/categorical variables. Values in bold are statistically significant at the .05 level or less.

LINEARITY   The final assumption to be examined is the linearity of the relationships. In the case of individual variables, this linearity relates to the patterns of association between each pair of variables and the ability of the correlation coefficient to adequately represent the relationship. If nonlinear relationships are indicated, then the researcher can either transform one or both of the variables to achieve linearity or create additional variables to represent the nonlinear components. For our purposes, we rely on the visual inspection of the relationships to determine whether non-linear relationships are present. The reader can also refer to Figure 2-2, the scatterplot matrix containing the scatterplot for selected metric variables in the data set. Examination of the scatter-plots does not reveal any apparent nonlinear relationships. Review of the scatterplots not shown in Figure 2-2 also did not reveal any apparent nonlinear patterns. Thus, transformations are not deemed necessary. The assumption of linearity will also be checked for the entire multivariate model, as is done in the examination of residuals in multiple regression.

SUMMARY   The series of graphical and statistical tests directed toward assessing the assumptions underlying the multivariate techniques revealed relatively little in terms of violations of the assumptions. Where violations were indicated, they were relatively minor and should not present any serious problems in the course of the data analysis. The researcher is encouraged always to perform these simple, yet revealing, examinations of the data to ensure that potential problems can be identified and resolved before the analysis begins.

INCORPORATING NONMETRIC DATA WITH DUMMY VARIABLES

A critical factor in choosing and applying the correct multivariate technique is the measurement properties of the dependent and independent variables (see Chapter 1 for a more detailed discussion of selecting multivariate techniques). Some of the techniques, such as discriminant analysis or multivariate analysis of variance, specifically require nonmetric data as dependent or independent variables. In many instances, metric variables must be used as independent variables, such as in regression analysis, discriminant analysis, and canonical correlation. Moreover, the interdependence techniques of factor and cluster analysis generally require metric variables. To this point, all discussions assumed metric measurement for variables. What can we do when the variables are nonmetric, with two or more categories? Are nonmetric variables, such as gender, marital status, or occupation, precluded from use in many multivariate techniques? The answer is no, and we now discuss how to incorporate nonmetric variables into many of these situations that require metric variables.

The researcher has available a method for using dichotomous variables, known as **dummy variables,** which act as replacement variables for the nonmetric variable. *A dummy variable is a dichotomous variable that represents one category of a nonmetric independent variable.* Any non-metric variable with k categories can be represented as $k - 1$ dummy variables. The following example will help clarify this concept.

First, assume we wish to include gender, which has two categories, female and male. We also have measured household income level by three categories (see Table 2-13). To represent the nonmetric variable gender, we would create two new dummy variables (X_1 and X_2), as shown in Table 2-13. X_1 would represent those individuals who are female with a value of 1 and would give all males a value of 0. Likewise, X_2 would represent all males with a value of 1 and give females a value of 0. Both variables (X_1 and X_2) are not necessary, however, because when $X_1 = 0$, gender must be female by definition. Thus, we need include only one of the variables (X_1 or X_2) to test the effect of gender.

Correspondingly, if we had also measured household income with three levels, as shown in Table 2-13, we would first define three dummy variables (X_3, X_4, and X_5). In the case of gender, we would not need the entire set of dummy variables, and instead use $k - 1$ dummy variables, where k

TABLE 2-13   Representing Nonmetric Variables with Dummy Variables

Nonmetric Variable with Two Categories (Gender)		Nonmetric Variable with Three Categories (Household Income Level)	
Gender	**Dummy Variables**	**Household Income Level**	**Dummy Variables**
Female	$X_1 = 1$, else $X_1 = 0$	if $<\$15,000$	$X_3 = 1$, else $X_3 = 0$
Male	$X_2 = 1$, else $X_2 = 0$	if $>\$15,000$ & $\leq\$25,000$	$X_4 = 1$, else $X_4 = 0$
		if $>\$25,000$	$X_5 = 1$, else $X_5 = 0$

is the number of categories. Thus, we would use two of the dummy variables to represent the effects of household income.

In constructing dummy variables, two approaches can be used to represent the categories, and more importantly, the category that is omitted, known as the **reference category** or **comparison group.**

- The first approach is known as **indicator coding,** uses three ways to represent the household income levels with two dummy variables, as shown in Table 2-14. *An important consideration is the reference category or comparison group, the category that received all zeros for the dummy variables.* For example, in regression analysis, the regression coefficients for the dummy variables represent *deviations from the comparison group on the dependent variable.* The deviations represent the differences between the dependent variable mean score for each group of respondents (represented by a separate dummy variable) and the comparison group. This form is most appropriate in a logical comparison group, such as in an experiment. In an experiment with a control group acting as the comparison group, the coefficients are the mean differences on the dependent variable for each treatment group from the control group. Any time dummy variable coding is used, we must be aware of the comparison group and remember the impacts it has in our interpretation of the remaining variables.
- An alternative method of dummy variable coding is termed **effects coding.** It is the same as indicator coding except that the comparison group (the group that got all zeros in indicator coding) is now given the value of –1 instead of 0 for the dummy variables. Now the coefficients represent differences for any group from the mean of all groups rather than from the omitted group. Both forms of dummy variable coding will give the same results; the only differences will be in the interpretation of the dummy variable coefficients.

Dummy variables are used most often in regression and discriminant analysis, where the coefficients have direct interpretation. Their use in other multivariate techniques is more limited, especially for those that rely on correlation patterns, such as factor analysis, because the correlation of a binary variable is not well represented by the traditional Pearson correlation coefficient. However, special considerations can be made in these instances, as discussed in the appropriate chapters.

TABLE 2-14   Alternative Dummy Variable Coding Patterns for a Three-Category Nonmetric Variable

Household Income Level	Pattern 1		Pattern 2		Pattern 3	
	X_1	X_2	X_1	X_2	X_1	X_2
If $<\$15,000$	1	0	1	0	0	0
If $>\$15,000$ & $\leq\$25,000$	0	1	0	0	1	0
If $>\$25,000$	0	0	0	1	0	1

Summary

Researchers should examine and explore the nature of the data and the relationships among variables before the application of any of the multivariate techniques. This chapter helps the researcher to do the following:

Select the appropriate graphical method to examine the characteristics of the data or relationships of interest. Use of multivariate techniques places an increased burden on the researcher to understand, evaluate, and interpret the more complex results. It requires a thorough understanding of the basic characteristics of the underlying data and relationships. The first task in data examination is to determine the character of the data. A simple, yet powerful, approach is through graphical displays, which can portray the univariate, bivariate, and even multivariate qualities of the data in a visual format for ease of display and analysis. The starting point for understanding the nature of a single variable is to characterize the shape of its distribution, which is accomplished with a histogram. The most popular method for examining bivariate relationships is the scatterplot, a graph of data points based on two variables. Researchers also should examine multivariate profiles. Three types of graphs are used. The first graph type is a direct portrayal of the data values, either by glyphs that display data in circles or multivariate profiles that provide a barlike profile for each observation. A second type of multivariate display involves a transformation of the original data into a mathematical relationship, which can then be portrayed graphically. The most common technique of this type is the Fourier transformation. The third graphical approach is iconic representativeness, the most popular being the Chernoff face.

Assess the type and potential impact of missing data. Although some missing data can be ignored, missing data is still one of the most troublesome issues in most research settings. At its best, it is a nuisance that must be remedied to allow for as much of the sample to be analyzed as possible. In more problematic situations, however, it can cause serious biases in the results if not correctly identified and accommodated in the analysis. The four-step process for identifying missing data and applying remedies is as follows:

1. Determine the type of missing data and whether or not it can be ignored.
2. Determine the extent of missing data and decide whether respondents or variables should be deleted.
3. Diagnose the randomness of the missing data.
4. Select the imputation method for estimating missing data.

Understand the different types of missing data processes. A missing data process is the underlying cause for missing data, whether it be something involving the data collection process (poorly worded questions, etc.) or the individual (reluctance or inability to answer, etc.). When missing data are not ignorable, the missing data process can be classified into one of two types. The first is MCAR, which denotes that the effects of the missing data process are randomly distributed in the results and can be remedied without incurring bias. The second is MAR, which denotes that the underlying process results in a bias (e.g., lower response by a certain type of consumer) and any remedy must be sure to not only "fix" the missing data, but not incur bias in the process.

Explain the advantages and disadvantages of the approaches available for dealing with missing data. The remedies for missing data can follow one of two approaches: using only valid data or calculating replacement data for the missing data. Even though using only valid data seems a reasonable approach, the researcher must remember that doing so assures the full effect of any biases due to non-random (MAR) data processes. Therefore, such approaches can be used only when random (MCAR) data processes are present, and then only if the sample is not too depleted for the analysis in question (remember, missing data excludes a case from use in the analysis). The calculation of replacement values attempts to impute a value for each missing value, based on criteria ranging from the sample's overall mean score for that variable to specific characteristics of the case used in a predictive relationship. Again, the researcher must first consider whether the effects are MCAR or MAR, and then select a remedy balancing the specificity of the remedy versus the extent of the missing data and its effect on generalizability.

Identify univariate, bivariate, and multivariate outliers. Outliers are observations with a unique combination of characteristics indicating they are distinctly different from the other observations. These differences can be on a single variable (univariate outlier), a relationship between two variables (bivariate outlier), or across an entire set of variables (multivariate outlier). Although the causes for outliers are varied, the primary issue to be resolved is their representativeness and whether the observation or variable should be deleted or included in the sample to be analyzed.

Test your data for the assumptions underlying most multivariate techniques. Because our analyses involve the use of a sample and not the population, we must be concerned with meeting the assumptions of the statistical inference process that is the foundation for all multivariate statistical techniques. The most important assumptions include normality, homoscedasticity, linearity, and absence of correlated errors. A wide range of tests, from graphical portrayals to empirical measures, is available to determine whether assumptions are met. Researchers are faced with what may seem to be an impossible task: satisfy all of these statistical assumptions or risk a biased and flawed analysis. These statistical assumptions are important, but judgment must be used in how to interpret the tests for each assumption and when to apply remedies. Even analyses with small sample sizes can withstand small, but significant, departures from normality. What is more important for the researcher is to understand the implications of each assumption with regard to the technique of interest, striking a balance between the need to satisfy the assumptions versus the robustness of the technique and research context.

Determine the best method of data transformation given a specific problem. When the statistical assumptions are not met, it is not necessarily a "fatal" problem that prevents further analysis. Instead, the researcher may be able to apply any number of transformations to the data in question that will solve the problem and enable the assumptions to be met. Data transformations provide a means of modifying variables for one of two reasons: (1) to correct violations of the statistical assumptions underlying the multivariate techniques, or (2) to improve the relationship (correlation) between variables. Most of the transformations involve modifying one or more variables (e.g., compute the square root, logarithm, or inverse) and then using the transformed value in the analysis. It should be noted that the underlying data are still intact, just their distributional character is changed so as to meet the necessary statistical assumptions.

Understand how to incorporate nonmetric variables as metric variables. An important consideration in choosing and applying the correct multivariate technique is the measurement properties of the dependent and independent variables. Some of the techniques, such as discriminant analysis or multivariate analysis of variance, specifically require nonmetric data as dependent or independent variables. In many instances, the multivariate methods require that metric variables be used. Yet nonmetric variables are often of considerable interest to the researcher in a particular analysis. A method is available to represent a nonmetric variable with a set of dichotomous variables, known as dummy variables, so that it may be included in many of the analyses requiring only metric variables. A dummy variable is a dichotomous variable that has been converted to a metric distribution and represents one category of a nonmetric independent variable.

Considerable time and effort can be expended in these activities, but the prudent researcher wisely invests the necessary resources to thoroughly examine the data to ensure that the multivariate methods are applied in appropriate situations and to assist in a more thorough and insightful interpretation of the results.

Questions

1. Explain how graphical methods can complement the empirical measures when examining data.
2. List potential underlying causes of outliers. Be sure to include attributions to both the respondent and the researcher.
3. Discuss why outliers might be classified as beneficial and as problematic.
4. Distinguish between data that are missing at random (MAR) and missing completely at random (MCAR). Explain how each type affects the analysis of missing data.
5. Describe the conditions under which a researcher would delete a case with missing data versus the conditions under which a researcher would use an imputation method.
6. Evaluate the following statement: In order to run most multivariate analyses, it is not necessary to meet all the assumptions of normality, linearity, homoscedasticity, and independence.
7. Discuss the following statement: Multivariate analyses can be run on any data set, as long as the sample size is adequate.

Suggested Readings

A list of suggested readings illustrating the issues in examining data in specific applications is available on the Web at www.pearsonglobaleditions.com/hair or www.mvstats.com.

References

1. Anderson, Edgar. 1969. A Semigraphical Method for the Analysis of Complex Problems. *Technometrics* 2 (August): 387–91.
2. Arbuckle, J. 1996. Full Information Estimation in the Presence of Incomplete Data. In *Advanced Structural Equation Modeling: Issues and Techniques,* G. A. Marcoulides and R. E. Schumacher (eds.). Mahwah, NJ: LEA.
3. Box, G. E. P., and D. R. Cox. 1964. An Analysis of Transformations. *Journal of the Royal Statistical Society B* 26: 211–43.
4. Brown, R. L. 1994. Efficacy of the Indirect Approach for Estimating Structural Equation Models with Missing Data: A Comparison of Five Methods. *Structural Equation Modeling* 1: 287–316.
5. Chernoff, Herman. 1978. Graphical Representation as a Discipline. In *Graphical Representation of Multivariate Data,* Peter C. C. Wang (ed.). New York: Academic Press, pp. 1–11.
6. Cohen, Jacob, Stephen G. West, Leona Aiken, and Patricia Cohen. 2002. *Applied Multiple Regression/Correlation Analysis for the Behavioral Sciences,* 3rd ed. Hillsdale, NJ: Lawrence Erlbaum Associates.
7. Daniel, C., and F. S. Wood. 1999. *Fitting Equations to Data,* 2nd ed. New York: Wiley-Interscience.
8. Dempster, A. P., and D. B. Rubin. 1983. Overview. In *Incomplete Data in Sample Surveys: Theory and Annotated Bibliography,* Vol. 2., Madow, Olkin, and Rubin (eds.). New York: Academic Press.
9. Duncan, T. E., R. Omen, and S. C. Duncan. 1994. Modeling Incomplete Data in Exercise Behavior Using Structural Equation Methodology. *Journal of Sport and Exercise Psychology* 16: 187–205.
10. Feinberg, Stephen. 1979. Graphical Methods in Statistics. *American Statistician* 33 (November): 165–78.
11. Graham, J. W., and S. W. Donaldson. 1993. Evaluating Interventions with Differential Attrition: The Importance of Nonresponse Mechanisms and Use of Follow-up Data. *Journal of Applied Psychology* 78: 119–28.
12. Graham, J. W., S. M. Hofer, and D. P. MacKinnon. 1996. Maximizing the Usefulness of Data Obtained with Planned Missing Value Patterns: An Application of Maximum Likelihood Procedures. *Multivariate Behavioral Research* 31(2): 197–218.
13. Gnanedesikan, R. 1977. *Methods for Statistical Analysis of Multivariate Distributions.* New York: Wiley.
14. Heitjan, D. F. 1997. Annotation: What Can Be Done About Missing Data? Approaches to Imputation. *American Journal of Public Health* 87(4): 548–50.
15. Hertel, B. R. 1976. Minimizing Error Variance Introduced by Missing Data Routines in Survey Analysis. *Sociological Methods and Research* 4: 459–74.
16. Johnson, R. A., and D. W. Wichern. 2002. *Applied Multivariate Statistical Analysis.* 5th ed. Upper Saddle River, NJ: Prentice Hall.
17. Kim, J. O., and J. Curry. 1977. The Treatment of Missing Data in Multivariate Analysis. *Sociological Methods and Research* 6: 215–41.
18. Little, Roderick J. A., and Donald B. Rubin. 2002. *Statistical Analysis with Missing Data.* 2nd ed. New York: Wiley.
19. Malhotra, N. K. 1987. Analyzing Marketing Research Data with Incomplete Information on the Dependent Variables. *Journal of Marketing Research* 24: 74–84.
20. Raymonds, M. R., and D. M. Roberts. 1987. A Comparison of Methods for Treating Incomplete Data in Selection Research. *Educational and Psychological Measurement* 47: 13–26.
21. Roth, P. L. 1994. Missing Data: A Conceptual Review for Applied Psychologists. *Personnel Psychology* 47: 537–60.
22. Schafer, J. L. 1997. *Analysis of Incomplete Multivariate Data.* London: Chapman and Hall.
23. Stevens, J. 2001. *Applied Multivariate Statistics for the Social Sciences,* 4th ed. Hillsdale, NJ: Lawrence Erlbaum Publishing.
24. Wang, Peter C. C. (ed.). 1978. *Graphical Representation of Multivariate Data.* New York: Academic Press.
25. Weisberg, S. 1985. *Applied Linear Regression.* New York: Wiley.
26. Wilkinson, L. 1982. An Experimental Evaluation of Multivariate Graphical Point Representations. In *Human Factors in Computer Systems: Proceedings.* New York: ACM Press, pp. 202–9.

CHAPTER 3

Factor Analysis

LEARNING OBJECTIVES

Upon completing this chapter, you should be able to do the following:

▪ Differentiate factor analysis techniques from other multivariate techniques.

▪ Distinguish between exploratory and confirmatory uses of factor analytic techniques.

▪ Understand the seven stages of applying factor analysis.

▪ Distinguish between R and Q factor analysis.

▪ Identify the differences between component analysis and common factor analysis models.

▪ Describe how to determine the number of factors to extract.

▪ Explain the concept of rotation of factors.

▪ Describe how to name a factor.

▪ Explain the additional uses of factor analysis.

▪ State the major limitations of factor analytic techniques.

CHAPTER PREVIEW

Use of the multivariate statistical technique of factor analysis increased during the past decade in all fields of business-related research. As the number of variables to be considered in multivariate techniques increases, so does the need for increased knowledge of the structure and interrelationships of the variables. This chapter describes factor analysis, a technique particularly suitable for analyzing the patterns of complex, multidimensional relationships encountered by researchers. It defines and explains in broad, conceptual terms the fundamental aspects of factor analytic techniques. Factor analysis can be utilized to examine the underlying patterns or relationships for a large number of variables and to determine whether the information can be condensed or summarized in a smaller set of factors or components. To further clarify the methodological concepts, basic guidelines for presenting and interpreting the results of these techniques are also included.

KEY TERMS

Before starting the chapter, review the key terms to develop an understanding of the concepts and terminology used. Throughout the chapter the key terms appear in **boldface.** Other points of emphasis in the chapter and key term cross-references are *italicized.*

Anti-image correlation matrix   Matrix of the partial correlations among variables after factor analysis, representing the degree to which the factors explain each other in the results. The diagonal contains the *measures of sampling adequacy* for each variable, and the off-diagonal values are partial correlations among variables.

Bartlett test of sphericity   Statistical test for the overall significance of all correlations within a correlation matrix.

Cluster analysis   Multivariate technique with the objective of grouping respondents or cases with similar profiles on a defined set of characteristics. Similar to *Q factor analysis.*

Common factor analysis   Factor model in which the factors are based on a reduced correlation matrix. That is, *communalities* are inserted in the diagonal of the *correlation matrix,* and the extracted factors are based only on the *common variance,* with *specific* and *error variance* excluded.

Common variance   Variance shared with other variables in the factor analysis.

Communality   Total amount of variance an original variable shares with all other variables included in the analysis.

Component analysis   Factor model in which the factors are based on the total variance. With component analysis, unities (1s) are used in the diagonal of the *correlation matrix;* this procedure computationally implies that all the variance is common or shared.

Composite measure   See *summated scales.*

Conceptual definition   Specification of the theoretical basis for a concept that is represented by a factor.

Content validity   Assessment of the degree of correspondence between the items selected to constitute a *summated scale* and its *conceptual definition.*

Correlation matrix   Table showing the intercorrelations among all variables.

Cronbach's alpha   Measure of *reliability* that ranges from 0 to 1, with values of .60 to .70 deemed the lower limit of acceptability.

Cross-loading   A variable has two more *factor loadings* exceeding the threshold value deemed necessary for inclusion in the factor interpretation process.

Dummy variable   Binary metric variable used to represent a single category of a nonmetric variable.

Eigenvalue   Column sum of squared loadings for a factor; also referred to as the *latent root.* It represents the amount of variance accounted for by a factor.

EQUIMAX   One of the *orthogonal factor rotation* methods that is a "compromise" between the *VARIMAX* and *QUARTIMAX* approaches, but is not widely used.

Error variance   Variance of a variable due to errors in data collection or measurement.

Face validity   See *content validity.*

Factor   Linear combination (variate) of the original variables. Factors also represent the underlying dimensions (constructs) that summarize or account for the original set of observed variables.

Factor indeterminacy   Characteristic of *common factor analysis* such that several different *factor scores* can be calculated for a respondent, each fitting the estimated factor model. It means the factor scores are not unique for each individual.

Factor loadings   Correlation between the original variables and the factors, and the key to understanding the nature of a particular factor. Squared factor loadings indicate what percentage of the variance in an original variable is explained by a factor.

Factor matrix   Table displaying the *factor loadings* of all variables on each factor.

Factor pattern matrix   One of two factor matrices found in an *oblique rotation* that is most comparable to the *factor matrix* in an *orthogonal rotation.*

Factor rotation   Process of manipulation or adjusting the factor axes to achieve a simpler and pragmatically more meaningful factor solution.

Factor score   Composite measure created for each observation on each factor extracted in the factor analysis. The factor weights are used in conjunction with the original variable values to calculate each observation's score. The factor score then can be used to represent the factor(s) in subsequent analyses. Factor scores are standardized to have a mean of 0 and a standard deviation of 1.

Factor structure matrix   A *factor matrix* found in an *oblique rotation* that represents the simple correlations between variables and factors, incorporating the unique variance and the correlations between factors. Most researchers prefer to use the *factor pattern matrix* when interpreting an oblique solution.

Indicator   Single variable used in conjunction with one or more other variables to form a *composite measure.*

Latent root   See *eigenvalue.*

Measure of sampling adequacy (MSA)   Measure calculated both for the entire correlation matrix and each individual variable evaluating the appropriateness of applying factor analysis. Values above .50 for either the entire matrix or an individual variable indicate appropriateness.

Measurement error   Inaccuracies in measuring the "true" variable values due to the fallibility of the measurement instrument (i.e., inappropriate response scales), data entry errors, or respondent errors.

Multicollinearity   Extent to which a variable can be explained by the other variables in the analysis.

Oblique factor rotation   *Factor rotation* computed so that the extracted factors are correlated. Rather than arbitrarily constraining the factor rotation to an *orthogonal* solution, the oblique rotation identifies the extent to which each of the factors is correlated.

Orthogonal   Mathematical independence (no correlation) of factor axes to each other (i.e., at right angles, or 90 degrees).

Orthogonal factor rotation   *Factor rotation* in which the factors are extracted so that their axes are maintained at 90 degrees. Each factor is independent of, or *orthogonal* to, all other factors. The correlation between the factors is determined to be 0.

Q factor analysis   Forms groups of respondents or cases based on their similarity on a set of characteristics (also see the discussion of *cluster analysis* in Chapter 9).

QUARTIMAX   A type of *orthogonal factor rotation* method focusing on simplifying the columns of a factor matrix. Generally considered less effective than the *VARIMAX* rotation.

R factor analysis   Analyzes relationships among variables to identify groups of variables forming latent dimensions (factors).

Reliability   Extent to which a variable or set of variables is consistent in what it is intended to measure. If multiple measurements are taken, reliable measures will all be consistent in their values. It differs from *validity* in that it does not relate to what should be measured, but instead to how it is measured.

Reverse scoring   Process of reversing the scores of a variable, while retaining the distributional characteristics, to change the relationships (correlations) between two variables. Used in *summated scale* construction to avoid a canceling out between variables with positive and negative *factor loadings* on the same factor.

Specific variance   Variance of each variable unique to that variable and not explained or associated with other variables in the factor analysis.

Summated scales   Method of combining several variables that measure the same concept into a single variable in an attempt to increase the *reliability* of the measurement. In most instances, the separate variables are summed and then their total or average score is used in the analysis.

Surrogate variable   Selection of a single variable with the highest *factor loading* to represent a factor in the data reduction stage instead of using a *summated scale* or *factor score.*

Trace   Represents the total amount of variance on which the factor solution is based. The trace is equal to the number of variables, based on the assumption that the variance in each variable is equal to 1.

Unique variance   See *specific variance.*

Validity   Extent to which a measure or set of measures correctly represents the concept of study—the degree to which it is free from any systematic or nonrandom error. Validity is concerned with how well the concept is defined by the measure(s), whereas *reliability* relates to the consistency of the measure(s).

Variate   Linear combination of variables formed by deriving empirical weights applied to a set of variables specified by the researcher.

VARIMAX   The most popular *orthogonal factor rotation* methods focusing on simplifying the columns in a *factor matrix*. Generally considered superior to other orthogonal factor rotation methods in achieving a simplified factor structure.

WHAT IS FACTOR ANALYSIS?

Factor analysis is an interdependence technique, as defined in Chapter 1, whose *primary purpose is to define the underlying structure among the variables in the analysis.* Obviously, variables play a key role in any multivariate analysis. Whether we are making a sales forecast with regression, predicting success or failure of a new firm with discriminant analysis, or using any of the other multivariate techniques discussed in Chapter 1, we must have a set of variables upon which to form relationships (e.g., What variables best predict sales or success/failure?). As such, variables are the building blocks of relationships.

As we employ multivariate techniques, by their very nature, the number of variables increases. Univariate techniques are limited to a single variable, but multivariate techniques can have tens, hundreds, or even thousands of variables. But how do we describe and represent all of these variables? Certainly, if we have only a few variables, they may all be distinct and different. As we add more and more variables, more and more overlap (i.e., correlation) is likely among the variables. In some instances, such as when we are using multiple measures to overcome measurement error by multivariable measurement (see Chapter 1 for more a more detailed discussion), the researcher even strives for correlation among the variables. As the variables become correlated, the researcher now needs ways in which to manage these variables—grouping highly correlated variables together, labeling or naming the groups, and perhaps even creating a new composite measure that can represent each group of variables.

We introduce factor analysis as our first multivariate technique because it can play a unique role in the application of other multivariate techniques. Broadly speaking, factor analysis provides the tools for analyzing the structure of the interrelationships (correlations) among a large number of variables (e.g., test scores, test items, questionnaire responses) by defining sets of variables that are highly interrelated, known as **factors.** These groups of variables (factors), which are by definition highly intercorrelated, are assumed to represent dimensions within the data. If we are only concerned with reducing the number of variables, then the dimensions can guide in creating new composite measures. However, if we have a conceptual basis for understanding the relationships between variables, then the dimensions may actually have meaning for what they collectively represent. In the latter case, these dimensions may correspond to concepts that cannot be adequately described by a single measure (e.g., store atmosphere is defined by many sensory components that must be measured separately but are all interrelated). We will see that factor analysis presents several ways of representing these groups of variables for use in other multivariate techniques.

We should note at this point that factor analytic techniques can achieve their purposes from either an exploratory or confirmatory perspective. A continuing debate concerns the appropriate role for factor analysis. Many researchers consider it only exploratory, useful in searching for structure among a set of variables or as a data reduction method. In this perspective, factor analytic techniques "take what the data give you" and do not set any a priori constraints on the estimation of components or the number of components to be extracted. For many—if not most—applications, this use of factor analysis is appropriate. However, in other situations, the researcher has preconceived thoughts on the actual structure of the data, based on theoretical support or prior research. For example, the researcher may wish to test hypotheses involving issues such as which variables should be grouped together on a factor or the precise number of factors. In these instances, the researcher requires that factor analysis take a confirmatory approach—that is, assess the degree to which the data meet the expected structure. The methods we discuss in this chapter do not directly provide the necessary structure for formalized hypothesis testing. We explicitly address the confirmatory perspective of factor analysis in Chapter 12. In this chapter, however, we view factor analytic techniques principally from an exploratory or nonconfirmatory viewpoint.

A HYPOTHETICAL EXAMPLE OF FACTOR ANALYSIS

Assume that through qualitative research a retail firm identified 80 different characteristics of retail stores and their service that consumers mentioned as affecting their patronage choice among stores. The retailer wants to understand how consumers make decisions but feels that it cannot evaluate 80 separate characteristics or develop action plans for this many variables, because they are too specific. Instead, it would like to know if consumers think in more general evaluative dimensions rather than in just the specific items. For example, consumers may consider salespersons to be a more general evaluative dimension that is composed of many more specific characteristics, such as knowledge, courtesy, likeability, sensitivity, friendliness, helpfulness, and so on.

To identify these broader dimensions, the retailer could commission a survey asking for consumer evaluations on each of the 80 specific items. Factor analysis would then be used to identify the broader underlying evaluative dimensions. Specific items that correlate highly are assumed to be a member of that broader dimension. These dimensions become composites of specific variables, which in turn allow the dimensions to be interpreted and described. In our example, the factor analysis might identify such dimensions as product assortment, product quality, prices, store personnel, service, and store atmosphere as the broader evaluative dimensions used by the respondents. Each of these dimensions contains specific items that are a facet of the broader evaluative dimension. From these findings, the retailer may then use the dimensions (factors) to define broad areas for planning and action.

An illustrative example of a simple application of factor analysis is shown in Figure 3-1, which represents the correlation matrix for nine store image elements. Included in this set are measures of the product offering, store personnel, price levels, and in-store service and experiences. The question a researcher may wish to address is: Are all of these elements separate in their evaluative properties or do they group into some more general areas of evaluation? For example, do all of the product elements group together? Where does price level fit, or is it separate? How do the in-store features (e.g., store personnel, service, and atmosphere) relate to one another? Visual inspection of the original correlation matrix (Figure 3-1, part 1) does not easily reveal any specific pattern. Among scattered high correlations, variable groupings are not apparent. The application of factor analysis results in the grouping of variables as reflected in part 2 of Figure 3-1. Here some interesting patterns emerge. First, four variables all relating to the in-store experience of shoppers are grouped together. Then, three variables describing the product assortment and availability are grouped together. Finally, product quality and price levels are grouped. Each group represents a set of highly interrelated variables that may reflect a more general evaluative dimension. In this case, we might label the three variable groupings by the labels in-store experience, product offerings, and value.

This simple example of factor analysis demonstrates its basic objective of grouping highly intercorrelated variables into distinct sets (factors). In many situations, these factors can provide a wealth of information about the interrelationships of the variables. In this example, factor analysis identified for store management a smaller set of concepts to consider in any strategic or tactical

PART 1: ORIGINAL CORRELATION MATRIX

	V_1	V_2	V_3	V_4	V_5	V_6	V_7	V_8	V_9
V_1 Price Level	1.000								
V_2 Store Personnel	.427	1.000							
V_3 Return Policy	.302	.771	1.000						
V_4 Product Availability	.470	.497	.427	1.000					
V_5 Product Quality	.765	.406	.307	.427	1.000				
V_6 Assortment Depth	.281	.445	.423	.713	.325	1.000			
V_7 Assortment Width	.345	.490	.471	.719	.378	.724	1.000		
V_8 In-store Service	.242	.719	.733	.428	.240	.311	.435	1.000	
V_9 Store Atmosphere	.372	.737	.774	.479	.326	.429	.466	.710	1.000

PART 2: CORRELATION MATRIX OF VARIABLES AFTER GROUPING ACCORDING TO FACTOR ANALYSIS

	V_3	V_8	V_9	V_2	V_6	V_7	V_4	V_1	V_5
V_3 Return Policy	1.000								
V_8 In-store Service	.773	1.000							
V_9 Store Atmosphere	.771	.710	1.000						
V_2 Store Personnel	.771	.719	.737	1.000					
V_6 Assortment Depth	.423	.311	.429	.445	1.000				
V_7 Assortment Width	.471	.435	.466	.490	.724	1.000			
V_4 Product Availability	.427	.428	.479	.497	.713	.719	1.000		
V_1 Price Level	.302	.242	.372	.427	.281	.354	.470	1.000	
V_5 Product Quality	.307	.240	.326	.406	.325	.378	.427	.765	1.000

FIGURE 3-1   Illustrative Example of the Use of Factor Analysis to Identify Structure within a Group of Variables

Note: Shaded areas represent variables grouped together by factor analysis.

marketing plans, while still providing insight into what constitutes each general area (i.e., the individual variables defining each factor).

FACTOR ANALYSIS DECISION PROCESS

We center the discussion of factor analysis on the six-stage model-building paradigm introduced in Chapter 1. Figure 3-2 shows the first three stages of the structured approach to multivariate model building, and Figure 3-4 details the final three stages, plus an additional stage (stage 7) beyond the estimation, interpretation, and validation of the factor models, which aids in selecting surrogate variables, computing factor scores, or creating summated scales for use in other multivariate techniques. A discussion of each stage follows.

STAGE 1: OBJECTIVES OF FACTOR ANALYSIS

The starting point in factor analysis, as with other statistical techniques, is the research problem. The general purpose of factor analytic techniques is to find a way to condense (summarize) the information contained in a number of original variables into a smaller set of new, composite dimensions or variates (factors) with a minimum loss of information—that is, to search for and define the fundamental constructs or dimensions assumed to underlie the original variables [18, 33]. In meeting its objectives, factor analysis is keyed to four issues: specifying the unit of analysis, achieving data summarization and/or data reduction, variable selection, and using factor analysis results with other multivariate techniques.

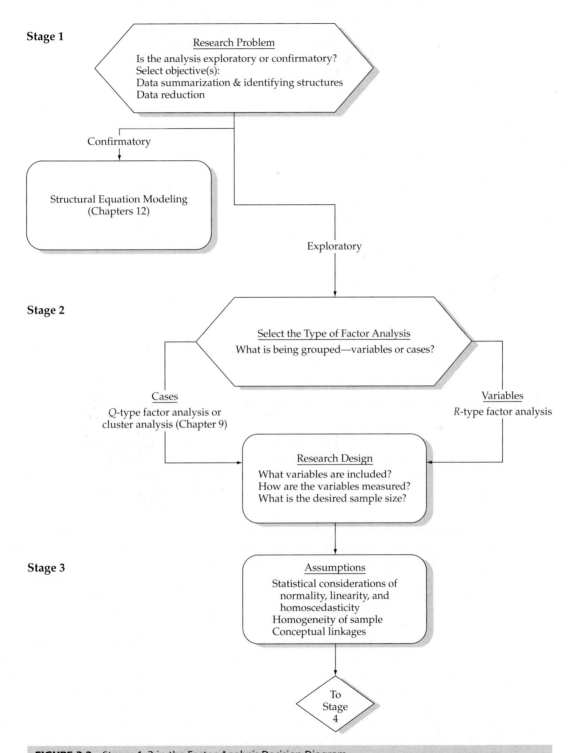

FIGURE 3-2 Stages 1–3 in the Factor Analysis Decision Diagram

Specifying the Unit of Analysis

Up to this time, we defined factor analysis solely in terms of identifying structure among a set of variables. Factor analysis is actually a more general model in that it can identify the structure of relationships among either variables or respondents by examining either the correlations between the variables or the correlations between the respondents.

- If the objective of the research were to summarize the characteristics, factor analysis would be applied to a **correlation matrix** of the variables. This most common type of factor analysis, referred to as *R* **factor analysis,** analyzes a set of variables to identify the dimensions that are latent (not easily observed).
- Factor analysis also may be applied to a correlation matrix of the individual respondents based on their characteristics. Referred to as *Q* **factor analysis,** this method combines or condenses large numbers of people into distinctly different groups within a larger population. The *Q* factor analysis approach is not utilized frequently because of computational difficulties. Instead, most researchers utilize some type of **cluster analysis** (see Chapter 9) to group individual respondents. Also see Stewart [36] for other possible combinations of groups and variable types.

Thus, the researcher must first select the unit of analysis for factor analysis: variables or respondents. Even though we will focus primarily on structuring variables, the option of employing factor analysis among respondents as an alternative to cluster analysis is also available. The implications in terms of identifying similar variables or respondents will be discussed in stage 2 when the correlation matrix is defined.

Achieving Data Summarization Versus Data Reduction

Factor analysis provides the researcher with two distinct, but interrelated, outcomes: data summarization and data reduction. In summarizing the data, factor analysis derives underlying dimensions that, when interpreted and understood, describe the data in a much smaller number of concepts than the original individual variables. Data reduction extends this process by deriving an empirical value (factor score) for each dimension (factor) and then substituting this value for the original values.

DATA SUMMARIZATION   The fundamental concept involved in data summarization is the definition of *structure*. Through structure, the researcher can view the set of variables at various levels of generalization, ranging from the most detailed level (individual variables themselves) to the more generalized level, where individual variables are grouped and then viewed not for what they represent individually, but for what they represent collectively in expressing a concept.

For example, variables at the individual level might be: "I shop for specials," "I usually look for the lowest possible prices," "I shop for bargains," "National brands are worth more than store brands." Collectively, these variables might be used to identify consumers who are "price conscious" or "bargain hunters."

Factor analysis, as an interdependence technique, differs from the dependence techniques discussed in the next section (i.e., multiple regression, discriminant analysis, multivariate analysis of variance, or conjoint analysis) where one or more variables are explicitly considered the criterion or dependent variables and all others are the predictor or independent variables. In factor analysis, all variables are simultaneously considered with no distinction as to dependent or independent variables. Factor analysis still employs the concept of the **variate,** the linear composite of variables, but in factor analysis, the variates (factors) are formed to maximize their explanation of the entire variable set, not to predict a dependent variable(s). The goal of data summarization is achieved by defining a small number of factors that adequately represent the original set of variables.

If we were to draw an analogy to dependence techniques, it would be that each of the observed (original) variables is a dependent variable that is a function of some underlying and latent set of factors (dimensions) that are themselves made up of all other variables. Thus, each variable is predicted by all of the factors, and indirectly by all the other variables. Conversely, one can look at each factor (variate) as a dependent variable that is a function of the entire set of observed variables. Either analogy illustrates the differences in purpose between dependence (prediction) and interdependence (identification of structure) techniques. Structure is defined by the interrelatedness among variables allowing for the specification of a smaller number of dimensions (factors) representing the original set of variables.

DATA REDUCTION   Factor analysis can also be used to achieve data reduction by (1) identifying representative variables from a much larger set of variables for use in subsequent multivariate analyses, or (2) creating an entirely new set of variables, much smaller in number, to partially or completely replace the original set of variables. In both instances, the purpose is to retain the nature and character of the original variables, but reduce their number to simplify the subsequent multivariate analysis. Even though the multivariate techniques were developed to accommodate multiple variables, the researcher is always looking for the most parsimonious set of variables to include in the analysis. As discussed in Chapter 1, both conceptual and empirical issues support the creation of composite measures. Factor analysis provides the empirical basis for assessing the structure of variables and the potential for creating these composite measures or selecting a subset of representative variables for further analysis.

Data summarization makes the identification of the underlying dimensions or factors ends in themselves. Thus, estimates of the factors and the contributions of each variable to the factors (termed *loadings*) are all that is required for the analysis. Data reduction relies on the factor loadings as well, but uses them as the basis for either identifying variables for subsequent analysis with other techniques or making estimates of the factors themselves (factor scores or summated scales), which then replace the original variables in subsequent analyses. The method of calculating and interpreting factor loadings is discussed later.

Variable Selection

Whether factor analysis is used for data reduction and/or summarization, the researcher should always consider the conceptual underpinnings of the variables and use judgment as to the appropriateness of the variables for factor analysis.

- In both uses of factor analysis, the researcher implicitly specifies the potential dimensions that can be identified through the character and nature of the variables submitted to factor analysis. For example, in assessing the dimensions of store image, if no questions on store personnel were included, factor analysis would not be able to identify this dimension.
- The researcher also must remember that factor analysis will always produce factors. Thus, factor analysis is always a potential candidate for the "garbage in, garbage out" phenomenon. If the researcher indiscriminately includes a large number of variables and hopes that factor analysis will "figure it out," then the possibility of poor results is high. The quality and meaning of the derived factors reflect the conceptual underpinnings of the variables included in the analysis.

Obviously, the use of factor analysis as a data summarization technique is based on having a conceptual basis for any variables analyzed. But even if used solely for data reduction, factor analysis is most efficient when conceptually defined dimensions can be represented by the derived factors.

Using Factor Analysis with Other Multivariate Techniques

Factor analysis, by providing insight into the interrelationships among variables and the underlying structure of the data, is an excellent starting point for many other multivariate techniques. From the data summarization perspective, factor analysis provides the researcher with a clear understanding of which variables may act in concert and how many variables may actually be expected to have impact in the analysis.

- Variables determined to be highly correlated and members of the same factor would be expected to have similar profiles of differences across groups in multivariate analysis of variance or in discriminant analysis.
- Highly correlated variables, such as those within a single factor, affect the stepwise procedures of multiple regression and discriminant analysis that sequentially enter variables based on their incremental predictive power over variables already in the model. As one variable from a factor is entered, it becomes less likely that additional variables from that same factor would also be included due to their high correlations with variable(s) already in the model, meaning they have little incremental predictive power. It does not mean that the other variables of the factor are less important or have less impact, but instead their effect is already represented by the included variable from the factor. Thus, knowledge of the structure of the variables by itself would give the researcher a better understanding of the reasoning behind the entry of variables in this technique.

The insight provided by data summarization can be directly incorporated into other multivariate techniques through any of the data reduction techniques. Factor analysis provides the basis for creating a new set of variables that incorporate the character and nature of the original variables in a much smaller number of new variables, whether using representative variables, factor scores, or summated scales. In this manner, problems associated with large numbers of variables or high intercorrelations among variables can be substantially reduced by substitution of the new variables. The researcher can benefit from both the empirical estimation of relationships and the insight into the conceptual foundation and interpretation of the results.

STAGE 2: DESIGNING A FACTOR ANALYSIS

The design of a factor analysis involves three basic decisions: (1) calculation of the input data (a correlation matrix) to meet the specified objectives of grouping variables or respondents; (2) design of the study in terms of number of variables, measurement properties of variables, and the types of allowable variables; and (3) the sample size necessary, both in absolute terms and as a function of the number of variables in the analysis.

Correlations Among Variables or Respondents

The first decision in the design of a factor analysis focuses on calculating the input data for the analysis. We discussed earlier the two forms of factor analysis: R-type versus Q-type factor analysis. Both types of factor analysis utilize a correlation matrix as the basic data input. With R-type factor analysis, the researcher would use a traditional correlation matrix (correlations among variables) as input. But the researcher could also elect to derive the correlation matrix from the correlations between the individual respondents. In this Q-type factor analysis, the results would be a factor matrix that would identify similar individuals.

For example, if the individual respondents are identified by number, the resulting factor pattern might tell us that individuals 1, 5, 6, and 7 are similar. Similarly, respondents 2, 3, 4, and 8 would perhaps load together on another factor, and we would label these individuals as similar.

From the results of a *Q* factor analysis, we could identify groups or clusters of individuals that demonstrate a similar pattern on the variables included in the analysis.

A logical question at this point would be: How does *Q*-type factor analysis differ from cluster analysis, because both approaches compare the pattern of responses across a number of variables and place the respondents in groups? The answer is that *Q*-type factor analysis is based on the inter-correlations between the respondents, whereas cluster analysis forms groupings based on a distance-based similarity measure between the respondents' scores on the variables being analyzed.

To illustrate this difference, consider Figure 3-3, which contains the scores of four respondents over three different variables. A *Q*-type factor analysis of these four respondents would yield two groups with similar covariance structures, consisting of respondents A and C versus B and D. In contrast, the clustering approach would be sensitive to the actual distances among the respondents' scores and would lead to a grouping of the closest pairs. Thus, with a cluster analysis approach, respondents A and B would be placed in one group and C and D in the other group.

If the researcher decides to employ *Q*-type factor analysis, these distinct differences from traditional cluster analysis techniques should be noted. With the availability of other grouping techniques and the widespread use of factor analysis for data reduction and summarization, the remaining discussion in this chapter focuses on *R*-type factor analysis, the grouping of variables rather than respondents.

Variable Selection and Measurement Issues

Two specific questions must be answered at this point: (1) What type of variables can be used in factor analysis? and (2) How many variables should be included? In terms of the types of variables included, the primary requirement is that a correlation value can be calculated among all variables. Metric variables are easily measured by several types of correlations. Nonmetric variables, however, are more problematic because they cannot use the same types of correlation measures used by metric variables. Although some specialized methods calculate correlations among nonmetric variables, the most prudent approach is to avoid nonmetric variables. If a nonmetric variable must be included, one approach is to define **dummy variables** (coded 0–1) to represent categories of nonmetric variables. If all the variables are dummy variables, then specialized forms of factor analysis, such as Boolean factor analysis, are more appropriate [5].

The researcher should also attempt to minimize the number of variables included but still maintain a reasonable number of variables per factor. If a study is being designed to assess a proposed structure, the researcher should be sure to include several variables (five or

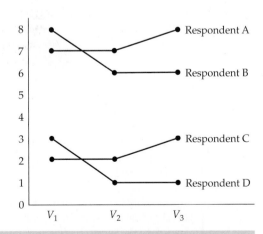

		Variables	
Respondent	V_1	V_2	V_3
A	7	7	8
B	8	6	6
C	2	2	3
D	3	1	1

FIGURE 3-3 Comparisons of Score Profiles for *Q*-type Factor Analysis and Cluster Analysis

more) that may represent each proposed factor. The strength of factor analysis lies in finding patterns among groups of variables, and it is of little use in identifying factors composed of only a single variable. Finally, when designing a study to be factor analyzed, the researcher should, if possible, identify several key variables (sometimes referred to as key indicants or marker variables) that closely reflect the hypothesized underlying factors. This identification will aid in validating the derived factors and assessing whether the results have practical significance.

Sample Size

Regarding the sample size question, the researcher generally would not factor analyze a sample of fewer than 50 observations, and preferably the sample size should be 100 or larger. As a general rule, the minimum is to have at least five times as many observations as the number of variables to be analyzed, and the more acceptable sample size would have a 10:1 ratio. Some researchers even propose a minimum of 20 cases for each variable. One must remember, however, that 30 variables, for example, requires computing 435 correlations in the factor analysis. At a .05 significance level, perhaps even 20 of those correlations would be deemed significant and appear in the factor analysis just by chance. The researcher should always try to obtain the highest cases-per-variable ratio to minimize the chances of overfitting the data (i.e., deriving factors that are sample-specific with little generalizability). In order to do so, the researcher may employ the most parsimonious set of variables, guided by conceptual and practical considerations, and then obtain an adequate sample size for the number of variables examined. When dealing with smaller sample sizes and/or a lower cases-to-variable ratio, the researcher should always interpret any findings cautiously. The issue of sample size will also be addressed in a later section on interpreting factor loadings.

Summary

Issues in the design of factor analysis are of equally critical importance whether an exploratory or confirmatory perspective is taken. In either perspective, the researcher is relying on the technique to provide insights into the structure of the data, but the structure revealed in the analysis is dependent on decisions by the researcher in such areas as variables included in the analysis, sample size, and so on. As such, several key considerations are listed in Rules of Thumb 3-1.

RULES OF THUMB 3-1

Factor Analysis Design

- Factor analysis is performed most often only on metric variables, although specialized methods exist for the use of dummy variables; a small number of "dummy variables" can be included in a set of metric variables that are factor analyzed
- If a study is being designed to reveal factor structure, strive to have at least five variables for each proposed factor
- For sample size:
  - The sample must have more observations than variables
  - The minimum absolute sample size should be 50 observations
  - Strive to maximize the number of observations per variable, with a desired ratio of 5 observations per variable

---

STAGE 3: ASSUMPTIONS IN FACTOR ANALYSIS

The critical assumptions underlying factor analysis are more conceptual than statistical. The researcher is always concerned with meeting the statistical requirement for any multivariate technique, but in factor analysis the overriding concerns center as much on the character and composition of the variables included in the analysis as on their statistical qualities.

Conceptual Issues

The conceptual assumptions underlying factor analysis relate to the set of variables selected and the sample chosen. A basic assumption of factor analysis is that some *underlying structure does exist* in the set of selected variables. The presence of correlated variables and the subsequent definition of factors do not guarantee relevance, even if they meet the statistical requirements. It is the responsibility of the researcher to ensure that the observed patterns are conceptually valid and appropriate to study with factor analysis, because the technique has no means of determining appropriateness other than the correlations among variables. For example, mixing dependent and independent variables in a single factor analysis and then using the derived factors to support dependence relationships is inappropriate.

The researcher must also ensure that the sample is homogeneous with respect to the underlying factor structure. It is inappropriate to apply factor analysis to a sample of males and females for a set of items known to differ because of gender. When the two subsamples (males and females) are combined, the resulting correlations and factor structure will be a poor representation of the unique structure of each group. Thus, whenever differing groups are expected in the sample, separate factor analyses should be performed, and the results should be compared to identify differences not reflected in the results of the combined sample.

Statistical Issues

From a statistical standpoint, departures from normality, homoscedasticity, and linearity apply only to the extent that they diminish the observed correlations. Only normality is necessary if a statistical test is applied to the significance of the factors, but these tests are rarely used. In fact, some degree of **multicollinearity** is desirable, because the objective is to identify interrelated sets of variables.

Assuming the researcher has met the conceptual requirements for the variables included in the analysis, the next step is to ensure that the variables are sufficiently intercorrelated to produce representative factors. As we will see, we can assess this degree of interrelatedness from both overall and individual variable perspectives. The following are several empirical measures to aid in diagnosing the factorability of the correlation matrix.

OVERALL MEASURES OF INTERCORRELATION   In addition to the statistical bases for the correlations of the data matrix, the researcher must also ensure that the data matrix has sufficient correlations to justify the application of factor analysis. If it is found that all of the correlations are low, or that all of the correlations are equal (denoting that no structure exists to group variables), then the researcher should question the application of factor analysis. To this end, several approaches are available:

- If visual inspection reveals no substantial number of correlations greater than .30, then factor analysis is probably inappropriate. The correlations among variables can also be analyzed by computing the partial correlations among variables. A partial correlation is the correlation that is unexplained when the effects of other variables are taken into account. If "true" factors exist in the data, the partial correlation should be small, because the variable can be explained by the variables loading on the factors. If the partial correlations are high, indicating no underlying factors, then factor analysis is inappropriate. The researcher is looking for a pattern of high partial correlations, denoting a variable not correlated with a large number of other variables in the analysis.

The one exception regarding high correlations as indicative of a poor correlation matrix occurs when two variables are highly correlated and have substantially higher loadings than other variables on that factor. Then, their partial correlation may be high because they are not explained to any great extent by the other variables, but do explain each other. This exception is also to be expected when a factor has only two variables loading highly.

A high partial correlation is one with practical and statistical significance, and a rule of thumb would be to consider partial correlations above .7 as high. SPSS and SAS provide the **anti-image correlation matrix**, which is just the negative value of the partial correlation, whereas BMDP directly provides the partial correlations. In each case, larger partial or anti-image correlations are indicative of a data matrix perhaps not suited to factor analysis.

- Another method of determining the appropriateness of factor analysis examines the entire correlation matrix. The **Bartlett test of sphericity**, a statistical test for the presence of correlations among the variables, is one such measure. It provides the statistical significance that the correlation matrix has significant correlations among at least some of the variables. The researcher should note, however, that increasing the sample size causes the Bartlett test to become more sensitive in detecting correlations among the variables.

- A third measure to quantify the degree of intercorrelations among the variables and the appropriateness of factor analysis is the **measure of sampling adequacy (MSA)**. This index ranges from 0 to 1, reaching 1 when each variable is perfectly predicted without error by the other variables. The measure can be interpreted with the following guidelines: .80 or above, meritorious; .70 or above, middling; .60 or above, mediocre; .50 or above, miserable; and below .50, unacceptable [22, 23]. The MSA increases as (1) the sample size increases, (2) the average correlations increase, (3) the number of variables increases, or (4) the number of factors decreases [23]. The researcher should always have an overall MSA value of above .50 before proceeding with the factor analysis. If the MSA value falls below .50, then the variable-specific MSA values (see the following discussion) can identify variables for deletion to achieve an overall value of .50.

VARIABLE-SPECIFIC MEASURES OF INTERCORRELATION   In addition to a visual examination of a variable's correlations with the other variables in the analysis, the MSA guidelines can be extended to individual variables. The researcher should examine the MSA values for each variable and exclude those falling in the unacceptable range. In deleting variables, the researcher should first delete the variable with the lowest MSA and then recalculate the factor analysis. Continue this process of deleting the variable with the lowest MSA value under .50 until all variables have an acceptable MSA value. Once the individual variables achieve an acceptable level, then the overall MSA can be evaluated and a decision made on continuance of the factor analysis.

Summary

Factor analysis, as an interdependence technique, is in many ways more affected by not meeting its underlying conceptual assumptions than by the statistical assumptions. The researcher must be sure to thoroughly understand the implications of not only ensuring that the data meet the statistical requirements for a proper estimation of the factor structure, but that the set of variables has the conceptual foundation to support the results. In doing so, the researcher should consider several key guidelines as listed in Rules of Thumb 3-2.

```
                    RULES OF THUMB 3-2

              Testing Assumptions of Factor Analysis

• A strong conceptual foundation needs to support the assumption that a structure does exist
  before the factor analysis is performed
• A statistically significant Bartlett's test of sphericity (sig. < .05) indicates that sufficient correlations
  exist among the variables to proceed
• Measure of sampling adequacy (MSA) values must exceed .50 for both the overall test and each
  individual variable; variables with values less than .50 should be omitted from the factor analysis
  one at a time, with the smallest one being omitted each time
```

STAGE 4: DERIVING FACTORS AND ASSESSING OVERALL FIT

Once the variables are specified and the correlation matrix is prepared, the researcher is ready to apply factor analysis to identify the underlying structure of relationships (see Figure 3-4). In doing so, decisions must be made concerning (1) the method of extracting the factors (common factor analysis versus components analysis) and (2) the number of factors selected to represent the underlying structure in the data.

Selecting the Factor Extraction Method

The researcher can choose from two similar, yet unique, methods for defining (extracting) the factors to represent the structure of the variables in the analysis. This decision on the method to use must combine the objectives of the factor analysis with knowledge about some basic characteristics of the relationships between variables. Before discussing the two methods available for extracting factors, a brief introduction to partitioning a variable's variance is presented.

PARTITIONING THE VARIANCE OF A VARIABLE   In order to select between the two methods of factor extraction, the researcher must first have some understanding of the variance for a variable and how it is divided or partitioned. First, remember that variance is a value (i.e., the square of the standard deviation) that represents the total amount of dispersion of values for a single variable about its mean. When a variable is correlated with another variable, we many times say it *shares* variance with the other variable, and the amount of sharing between just two variables is simply the squared correlation. For example, if two variables have a correlation of .50, each variable shares 25 percent ($.50^2$) of its variance with the other variable.

In factor analysis, we group variables by their correlations, such that variables in a group (factor) have high correlations with each other. Thus, for the purposes of factor analysis, it is important to understand how much of a variable's variance is shared with other variables in that factor versus what cannot be shared (e.g., unexplained). The total variance of any variable can be divided (partitioned) into three types of variance:

1. **Common variance** is defined as that variance in a variable that is shared with all other variables in the analysis. This variance is accounted for (shared) based on a variable's correlations with all other variables in the analysis. A variable's **communality** is the estimate of its shared, or common, variance among the variables as represented by the derived factors.
2. **Specific variance** (also known as **unique variance**) is that variance associated with only a specific variable. This variance cannot be explained by the correlations to the other variables but is still associated uniquely with a single variable.
3. **Error variance** is also variance that cannot be explained by correlations with other variables, but it is due to unreliability in the data-gathering process, measurement error, or a random component in the measured phenomenon.

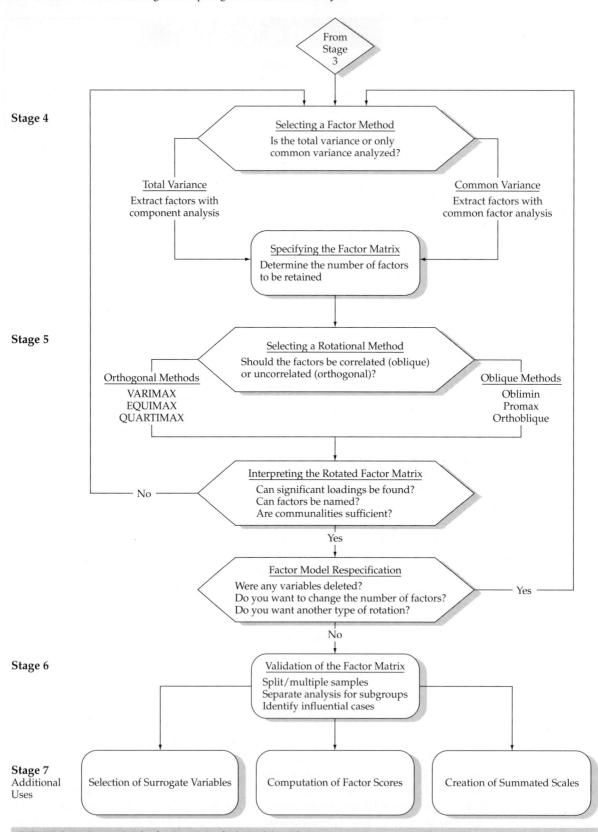

FIGURE 3-4 Stages 4–7 in the Factor Analysis Decision Diagram

Thus, the total variance of any variable is composed of its common, unique, and error variances. As a variable is more highly correlated with one or more variables, the common variance (communality) increases. However, if unreliable measures or other sources of extraneous error variance are introduced, then the amount of possible common variance and the ability to relate the variable to any other variable are reduced.

COMMON FACTOR ANALYSIS VERSUS COMPONENT ANALYSIS With a basic understanding of how variance can be partitioned, the researcher is ready to address the differences between the two methods, known as **common factor analysis** and **component analysis.** The selection of one method over the other is based on two criteria: (1) the objectives of the factor analysis and (2) the amount of prior knowledge about the variance in the variables. Component analysis is used when the objective is to summarize most of the original information (variance) in a minimum number of factors for prediction purposes. In contrast, common factor analysis is used primarily to identify underlying factors or dimensions that reflect what the variables share in common. The most direct comparison between the two methods is by their use of the explained versus unexplained variance:

- Component analysis, also known as principal components analysis, *considers the total variance and derives factors that contain small proportions of unique variance and, in some instances, error variance.* However, the first few factors do not contain enough unique or error variance to distort the overall factor structure. Specifically, with component analysis, unities (values of 1.0) are inserted in the diagonal of the correlation matrix, so that the full variance is brought into the factor matrix. Figure 3-5 portrays the use of the total variance in component analysis and the differences when compared to common factor analysis.
- Common factor analysis, in contrast, considers only the common or shared variance, *assuming that both the unique and error variance are not of interest in defining the structure of the variables.* To employ only common variance in the estimation of the factors, communalities (instead of unities) are inserted in the diagonal. Thus, factors resulting from common factor analysis are based only on the common variance. As shown in Figure 3-5, common factor analysis excludes a portion of the variance included in a component analysis.

How is the researcher to choose between the two methods? First, the common factor and component analysis models are both widely used. As a practical matter, the components model is the typical default method of most statistical programs when performing factor analysis. Beyond the program defaults, distinct instances indicate which of the two methods is most appropriate:

Component factor analysis is most appropriate when:

- *Data reduction is a primary concern,* focusing on the minimum number of factors needed to account for the maximum portion of the *total variance* represented in the original set of variables, and
- Prior knowledge suggests that specific and error variance represent a *relatively small proportion* of the total variance.

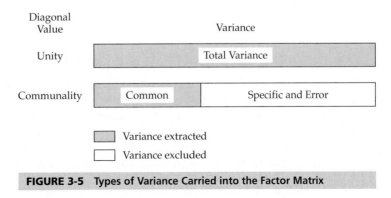

FIGURE 3-5 Types of Variance Carried into the Factor Matrix

Common factor analysis is most appropriate when:

- *The primary objective is to identify the latent dimensions or constructs* represented in the original variables, and
- The researcher has *little knowledge about the amount of specific and error variance* and therefore wishes to eliminate this variance.

Common factor analysis, with its more restrictive assumptions and use of only the latent dimensions (shared variance), is often viewed as more theoretically based. Although theoretically sound, however, common factor analysis has several problems. First, common factor analysis suffers from **factor indeterminacy,** which means that for any individual respondent, several different factor scores can be calculated from a single factor model result [26]. No single unique solution is found, as in component analysis, but in most instances the differences are not substantial. The second issue involves the calculation of the estimated communalities used to represent the shared variance. Sometimes the communalities are not estimable or may be invalid (e.g., values greater than 1 or less than 0), requiring the deletion of the variable from the analysis.

Does the choice of one model or the other really affect the results? The complications of common factor analysis have contributed to the widespread use of component analysis. But the base of proponents for the common factor model is strong. Cliff [13] characterizes the dispute between the two sides as follows:

> Some authorities insist that component analysis is the only suitable approach, and that the common factor methods just superimpose a lot of extraneous mumbo jumbo, dealing with fundamentally unmeasurable things, the common factors. Feelings are, if anything, even stronger on the other side. Militant common-factorists insist that components analysis is at best a common factor analysis with some error added and at worst an unrecognizable hodgepodge of things from which nothing can be determined. Some even insist that the term "factor analysis" must not be used when a components analysis is performed.

Although considerable debate remains over which factor model is the more appropriate [6, 19, 25, 35], empirical research demonstrates similar results in many instances [37]. In most applications, *both component analysis and common factor analysis arrive at essentially identical results if the number of variables exceeds 30 [18] or the communalities exceed .60 for most variables.* If the researcher is concerned with the assumptions of components analysis, then common factor analysis should also be applied to assess its representation of structure.

When a decision has been made on the factor model, the researcher is ready to extract the initial unrotated factors. By examining the unrotated factor matrix, the researcher can explore the potential for data reduction and obtain a preliminary estimate of the number of factors to extract. Final determination of the number of factors must wait, however, until the results are rotated and the factors are interpreted.

Criteria for the Number of Factors to Extract

How do we decide on the number of factors to extract? Both factor analysis methods are interested in the best linear combination of variables—best in the sense that the particular combination of original variables accounts for more of the variance in the data as a whole than any other linear combination of variables. Therefore, the first factor may be viewed as the single best summary of linear relationships exhibited in the data. The second factor is defined as the second-best linear combination of the variables, subject to the constraint that it is orthogonal to the first factor. To be

orthogonal to the first factor, the second factor must be derived from the variance remaining after the first factor has been extracted. Thus, the second factor may be defined as the linear combination of variables that accounts for the most variance that is still unexplained after the effect of the first factor has been removed from the data. The process continues extracting factors accounting for smaller and smaller amounts of variance until all of the variance is explained. For example, the components method actually extracts n factors, where n is the number of variables in the analysis. Thus, if 30 variables are in the analysis, 30 factors are extracted.

So, what is gained by factor analysis? Although our example contains 30 factors, a few of the first factors can represent a substantial portion of the total variance across all the variables. Hopefully, the researcher can retain or use only a small number of the variables and still adequately represent the entire set of variables. Thus, the key question is: *How many factors to extract or retain?*

In deciding when to stop factoring (i.e., how many factors to extract), the researcher must combine a conceptual foundation (How many factors should be in the structure?) with some empirical evidence (How many factors can be reasonably supported?). The researcher generally begins with some predetermined criteria, such as the general number of factors plus some general thresholds of practical relevance (e.g., required percentage of variance explained). These criteria are combined with empirical measures of the factor structure. An exact quantitative basis for deciding the number of factors to extract has not been developed. However, the following stopping criteria for the number of factors to extract are currently being utilized.

LATENT ROOT CRITERION   The most commonly used technique is the latent root criterion. This technique is simple to apply to either components analysis or common factor analysis. The rationale for the latent root criterion is that any individual factor should account for the variance of at least a single variable if it is to be retained for interpretation. With component analysis each variable contributes a value of 1 to the total eigenvalue. Thus, only the factors having **latent roots** or **eigenvalues** greater than 1 are considered significant; all factors with latent roots less than 1 are considered insignificant and are disregarded. Using the eigenvalue for establishing a cutoff is most reliable when the number of variables is between 20 and 50. If the number of variables is less than 20, the tendency is for this method to extract a conservative number of factors (too few); whereas if more than 50 variables are involved, it is not uncommon for too many factors to be extracted.

A PRIORI CRITERION   The a priori criterion is a simple yet reasonable criterion under certain circumstances. When applying it, the researcher already knows how many factors to extract before undertaking the factor analysis. The researcher simply instructs the computer to stop the analysis when the desired number of factors has been extracted. This approach is useful when testing a theory or hypothesis about the number of factors to be extracted. It also can be justified in attempting to replicate another researcher's work and extract the same number of factors that was previously found.

PERCENTAGE OF VARIANCE CRITERION   The percentage of variance criterion is an approach based on achieving a specified cumulative percentage of total variance extracted by successive factors. The purpose is to ensure practical significance for the derived factors by ensuring that they explain at least a specified amount of variance. No absolute threshold has been adopted for all applications. However, in the natural sciences the factoring procedure usually should not be stopped until the extracted factors account for at least 95 percent of the variance or until the last factor accounts for only a small portion (less than 5%). In contrast, in the social sciences, where information is often less precise, it is not uncommon to consider a solution that accounts for 60 percent of the total variance (and in some instances even less) as satisfactory.

A variant of this criterion involves selecting enough factors to achieve a prespecified communality for each of the variables. If theoretical or practical reasons require a certain communality for each variable, then the researcher will include as many factors as necessary to adequately represent each of the original variables. This approach differs from focusing on just the total amount of variance explained, which neglects the degree of explanation for the individual variables.

SCREE TEST CRITERION   Recall that with the component analysis factor model the later factors extracted contain both common and unique variance. Although all factors contain at least some unique variance, the proportion of unique variance is substantially higher in later factors. The scree test is used to identify the optimum number of factors that can be extracted before the amount of unique variance begins to dominate the common variance structure [9].

The scree test is derived by plotting the latent roots against the number of factors in their order of extraction, and the shape of the resulting curve is used to evaluate the cutoff point. Figure 3-6 plots the first 18 factors extracted in a study. Starting with the first factor, the plot slopes steeply downward initially and then slowly becomes an approximately horizontal line. The point at which the curve first begins to straighten out is considered to indicate the maximum number of factors to extract. In the present case, the first 10 factors would qualify. Beyond 10, too large a proportion of unique variance would be included; thus these factors would not be acceptable. Note that in using the latent root criterion only 8 factors would have been considered. In contrast, using the scree test provides us with 2 more factors. As a general rule, the scree test results in at least one and sometimes two or three more factors being considered for inclusion than does the latent root criterion [9].

HETEROGENEITY OF THE RESPONDENTS   Shared variance among variables is the basis for both common and component factor models. An underlying assumption is that shared variance extends across the entire sample. If the sample is heterogeneous with regard to at least one subset of the variables, then the first factors will represent those variables that are more homogeneous across the entire sample. Variables that are better discriminators between the subgroups of the sample

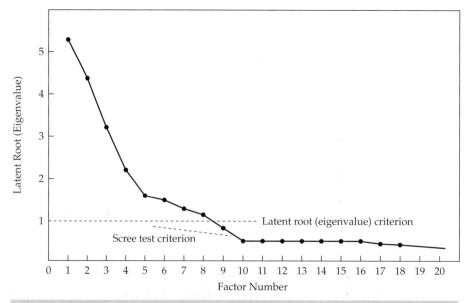

FIGURE 3-6 Eigenvalue Plot for Scree Test Criterion

will load on later factors, many times those not selected by the criteria discussed previously [17]. When the objective is to identify factors that discriminate among the subgroups of a sample, the researcher should extract additional factors beyond those indicated by the methods just discussed and examine the additional factors' ability to discriminate among the groups. If they prove less beneficial in discrimination, the solution can be run again and these later factors eliminated.

SUMMARY OF FACTOR SELECTION CRITERIA   In practice, most researchers seldom use a single criterion in determining how many factors to extract. Instead, they initially use a criterion such as the latent root as a guideline for the first attempt at interpretation. After the factors are interpreted, as discussed in the following sections, the practicality of the factors is assessed. Factors identified by other criteria are also interpreted. Selecting the number of factors is interrelated with an assessment of structure, which is revealed in the interpretation phase. Thus, several factor solutions with differing numbers of factors are examined before the structure is well defined. In making the final decision on the factor solution to represent the structure of the variables, the researcher should remember the considerations listed in Rules of Thumb 3-3.

One word of caution in selecting the final set of factors: Negative consequences arise from selecting either too many or too few factors to represent the data. If too few factors are used, then the correct structure is not revealed, and important dimensions may be omitted. If too many factors are retained, then the interpretation becomes more difficult when the results are rotated (as discussed in the next section). Although the factors are independent, you can just as easily have too many factors as too few. By analogy, choosing the number of factors is something like focusing a microscope. Too high or too low an adjustment will obscure a structure that is obvious when the adjustment is just right. Therefore, by examining a number of different factor structures derived from several trial solutions, the researcher can compare and contrast to arrive at the best representation of the data.

As with other aspects of multivariate models, parsimony is important. The notable exception is when factor analysis is used strictly for data reduction and a set level of variance to be extracted is specified. The researcher should always strive to have the most representative and parsimonious set of factors possible.

RULES OF THUMB 3-3

Choosing Factor Models and Number of Factors

- Although both component and common factor analysis models yield similar results in common research settings (30 or more variables or communalities of .60 for most variables):
  - The component analysis model is most appropriate when data reduction is paramount
  - The common factor model is best in well-specified theoretical applications
- Any decision on the number of factors to be retained should be based on several considerations:
  - Use of several stopping criteria to determine the initial number of factors to retain:
    - Factors with eigenvalues greater than 1.0
    - A predetermined number of factors based on research objectives and/or prior research
    - Enough factors to meet a specified percentage of variance explained, usually 60% or higher
    - Factors shown by the scree test to have substantial amounts of common variance (i.e., factors before inflection point)
    - More factors when heterogeneity is present among sample subgroups
  - Consideration of several alternative solutions (one more and one less factor than the initial solution) to ensure the best structure is identified

STAGE 5: INTERPRETING THE FACTORS

Although no unequivocal processes or guidelines determine the interpretation of factors, the researcher with a strong conceptual foundation for the anticipated structure and its rationale has the greatest chance of success. We cannot state strongly enough the importance of a strong conceptual foundation, whether it comes from prior research, theoretical paradigms, or commonly accepted principles. As we will see, the researcher must repeatedly make subjective judgments in such decisions as to the number of factors, what are sufficient relationships to warrant grouping variables, and how can these groupings be identified. As the experienced researcher can attest, almost anything can be uncovered if one tries long and hard enough (e.g., using different factor models, extracting different numbers of factors, using various forms of rotation). It is therefore left up to the researcher to be the final arbitrator as to the form and appropriateness of a factor solution, and such decisions are best guided by conceptual rather than empirical bases.

To assist in the process of interpreting a factor structure and selecting a final factor solution, three fundamental processes are described. Within each process, several substantive issues (factor rotation, factor-loading significance, and factor interpretation) are encountered. Thus, after each process is briefly described, each of these processes will be discussed in more detail.

The Three Processes of Factor Interpretation

Factor interpretation is circular in nature. The researcher first evaluates the initial results, then makes a number of judgments in viewing and refining these results, with the distinct possibility that the analysis is respecified, requiring a return to the evaluative step. Thus, the researcher should not be surprised to engage in several iterations until a final solution is achieved.

ESTIMATE THE FACTOR MATRIX   First, the initial unrotated **factor matrix** is computed, containing the factor loadings for each variable on each factor. **Factor loadings** are the correlation of each variable and the factor. Loadings indicate the degree of correspondence between the variable and the factor, with higher loadings making the variable representative of the factor. Factor loadings are the means of interpreting the role each variable plays in defining each factor.

FACTOR ROTATION   Unrotated factor solutions achieve the objective of data reduction, but the researcher must ask whether the unrotated factor solution (which fulfills desirable mathematical requirements) will provide information that offers the most adequate interpretation of the variables under examination. In most instances the answer to this question is no, because factor rotation (a more detailed discussion follows in the next section) should simplify the factor structure. Therefore, the researcher next employs a rotational method to achieve simpler and theoretically more meaningful factor solutions. In most cases rotation of the factors improves the interpretation by reducing some of the ambiguities that often accompany initial unrotated factor solutions.

FACTOR INTERPRETATION AND RESPECIFICATION   As a final process, the researcher evaluates the (rotated) factor loadings for each variable in order to determine that variable's role and contribution in determining the factor structure. In the course of this evaluative process, the need may arise to respecify the factor model owing to (1) the deletion of a variable(s) from the analysis, (2) the desire to employ a different rotational method for interpretation, (3) the need to extract a different number of factors, or (4) the desire to change from one extraction method to another. Respecification of a factor model involves returning to the extraction stage (stage 4), extracting factors, and then beginning the process of interpretation once again.

Rotation of Factors

Perhaps the most important tool in interpreting factors is **factor rotation.** The term *rotation* means exactly what it implies. Specifically, the reference axes of the factors are turned about the origin until some other position has been reached. As indicated earlier, unrotated factor solutions extract factors in the order of their variance extracted. The first factor tends to be a general factor with almost every variable loading significantly, and it accounts for the largest amount of variance. The second and subsequent factors are then based on the residual amount of variance. Each accounts for successively smaller portions of variance. The ultimate effect of rotating the factor matrix is to redistribute the variance from earlier factors to later ones to achieve a simpler, theoretically more meaningful factor pattern.

The simplest case of rotation is an **orthogonal factor rotation,** in which the axes are maintained at 90 degrees. It is also possible to rotate the axes and not retain the 90-degree angle between the reference axes. When not constrained to being orthogonal, the rotational procedure is called an **oblique factor rotation.** Orthogonal and oblique factor rotations are demonstrated in Figures 3-7 and 3-8, respectively.

Figure 3-7, in which five variables are depicted in a two-dimensional factor diagram, illustrates factor rotation. The vertical axis represents unrotated factor II, and the horizontal axis represents unrotated factor I. The axes are labeled with 0 at the origin and extend outward to +1.0 or −1.0. The numbers on the axes represent the factor loadings. The five variables are labeled V_1, V_2, V_3, V_4, and V_5. The factor loading for variable 2 (V_2) on unrotated factor II is determined by drawing a dashed line horizontally from the data point to the vertical axis for factor II. Similarly, a vertical line is drawn from variable 2 to the horizontal axis of unrotated factor I to determine the loading of variable 2 on factor I. A similar procedure followed for the remaining variables determines the factor loadings for the unrotated and rotated solutions, as displayed in Table 3-1 for comparison purposes. On the unrotated first factor, all the variables load fairly high. On the unrotated second factor, variables 1 and 2 are very high in the positive direction. Variable 5 is moderately high in the negative direction, and variables 3 and 4 have considerably lower loadings in the negative direction.

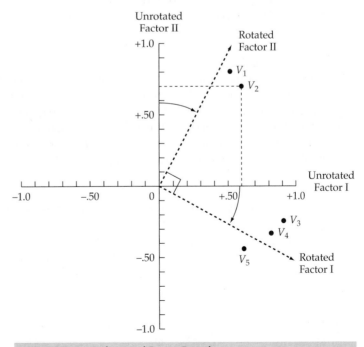

FIGURE 3-7 Orthogonal Factor Rotation

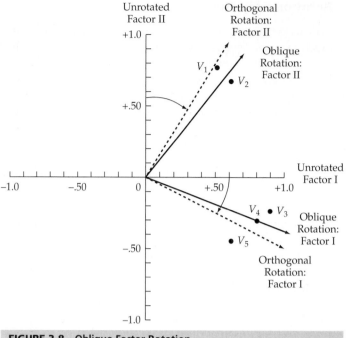

FIGURE 3-8 Oblique Factor Rotation

From visual inspection of Figure 3-7, two clusters of variables are obvious. Variables 1 and 2 go together, as do variables 3, 4, and 5. However, such patterning of variables is not so obvious from the unrotated factor loadings. By rotating the original axes clockwise, as indicated in Figure 3-7, we obtain a completely different factor-loading pattern. Note that in rotating the factors, the axes are maintained at 90 degrees. This procedure signifies that the factors are mathematically independent and that the rotation has been orthogonal. After rotating the factor axes, variables 3, 4, and 5 load high on factor I, and variables 1 and 2 load high on factor II. Thus, the clustering or patterning of these variables into two groups is more obvious after the rotation than before, even though the relative position or configuration of the variables remains unchanged.

The same general principles of orthogonal rotations pertain to oblique rotations. The oblique rotational method is more flexible, however, because the factor axes need not be orthogonal. It is also more realistic because the theoretically important underlying dimensions are not assumed to be uncorrelated with each other. In Figure 3-8 the two rotational methods are compared. Note that the oblique factor rotation represents the clustering of variables more accurately. This accuracy is a result of the fact that each rotated factor axis is now closer to the respective group of variables.

TABLE 3-1 Comparison Between Rotated and Unrotated Factor Loadings

Variables	Unrotated Factor Loadings		Rotated Factor Loadings	
	I	II	I	II
V_1	.50	.80	.03	.94
V_2	.60	.70	.16	.90
V_3	.90	−.25	.95	.24
V_4	.80	−.30	.84	.15
V_5	.60	−.50	.76	−.13

Also, the oblique solution provides information about the extent to which the factors are actually correlated with each other.

Most researchers agree that most unrotated solutions are not sufficient. That is, in most cases rotation will improve the interpretation by reducing some of the ambiguities that often accompany the preliminary analysis. The major option available is to choose an orthogonal or oblique rotation method. The ultimate goal of any rotation is to obtain some theoretically meaningful factors and, if possible, the simplest factor structure. Orthogonal rotational approaches are more widely used because all computer packages with factor analysis contain orthogonal rotation options, whereas the oblique methods are not as widespread. Orthogonal rotations are also utilized more frequently because the analytical procedures for performing oblique rotations are not as well developed and are still subject to some controversy. Several different approaches are available for performing either orthogonal or oblique rotations. However, only a limited number of oblique rotational procedures are available in most statistical packages. Thus, the researcher will probably have to accept the one that is provided.

ORTHOGONAL ROTATION METHODS   In practice, the objective of all methods of rotation is to simplify the rows and columns of the factor matrix to facilitate interpretation. In a factor matrix, columns represent factors, with each row corresponding to a variable's loading across the factors. By simplifying the rows, we mean making as many values in each row as close to zero as possible (i.e., maximizing a variable's loading on a single factor). By simplifying the columns, we mean making as many values in each column as close to zero as possible (i.e., making the number of high loadings as few as possible). Three major orthogonal approaches have been developed:

1. The ultimate goal of a **QUARTIMAX** rotation is to simplify the rows of a factor matrix; that is, QUARTIMAX focuses on rotating the initial factor so that a variable loads high on one factor and as low as possible on all other factors. In these rotations, many variables can load high or near high on the same factor because the technique centers on simplifying the rows. The QUARTIMAX method has not proved especially successful in producing simpler structures. Its difficulty is that it tends to produce a general factor as the first factor on which most, if not all, of the variables have high loadings. Regardless of one's concept of a simpler structure, inevitably it involves dealing with clusters of variables; a method that tends to create a large general factor (i.e., QUARTIMAX) is not in line with the goals of rotation.

2. In contrast to QUARTIMAX, the **VARIMAX** criterion centers on simplifying the columns of the factor matrix. With the VARIMAX rotational approach, the maximum possible simplification is reached if there are only 1s and 0s in a column. That is, the VARIMAX method maximizes the sum of variances of required loadings of the factor matrix. Recall that in QUARTIMAX approaches, many variables can load high or near high on the same factor because the technique centers on simplifying the rows. With the VARIMAX rotational approach, some high loadings (i.e., close to −1 or +1) are likely, as are some loadings near 0 in each column of the matrix. The logic is that interpretation is easiest when the variable-factor correlations are (1) close to either +1 or −1, thus indicating a clear positive or negative association between the variable and the factor; or (2) close to 0, indicating a clear lack of association. This structure is fundamentally simple. Although the QUARTIMAX solution is analytically simpler than the VARIMAX solution, VARIMAX seems to give a clearer separation of the factors. In general, Kaiser's experiment [22, 23] indicates that the factor pattern obtained by VARIMAX rotation tends to be more invariant than that obtained by the QUARTIMAX method when different subsets of variables are analyzed. The VARIMAX method has proved successful as an analytic approach to obtaining an orthogonal rotation of factors.

3. The **EQUIMAX** approach is a compromise between the QUARTIMAX and VARIMAX approaches. Rather than concentrating either on simplification of the rows or on simplification of the columns, it tries to accomplish some of each. EQUIMAX has not gained widespread acceptance and is used infrequently.

OBLIQUE ROTATION METHODS   Oblique rotations are similar to orthogonal rotations, except that oblique rotations allow correlated factors instead of maintaining independence between the rotated factors. Where several choices are available among orthogonal approaches, however, most statistical packages typically provide only limited choices for oblique rotations. For example, SPSS provides OBLIMIN; SAS has PROMAX and ORTHOBLIQUE; and BMDP provides DQUART, DOBLIMIN, and ORTHOBLIQUE. The objectives of simplification are comparable to the orthogonal methods, with the added feature of correlated factors. With the possibility of correlated factors, the factor researcher must take additional care to validate obliquely rotated factors, because they have an additional way (nonorthogonality) of becoming specific to the sample and not generalizable, particularly with small samples or a low cases-to-variable ratio.

SELECTING AMONG ROTATIONAL METHODS   No specific rules have been developed to guide the researcher in selecting a particular orthogonal or oblique rotational technique. In most instances, the researcher simply utilizes the rotational technique provided by the computer program. Most programs have the default rotation of VARIMAX, but all the major rotational methods are widely available. However, no compelling analytical reason suggests favoring one rotational method over another. The choice of an orthogonal or oblique rotation should be made on the basis of the particular needs of a given research problem. To this end, several considerations (in Rules of Thumb 3-4) should guide in selecting the rotational method.

Judging the Significance of Factor Loadings

In interpreting factors, a decision must be made regarding the factor loadings worth consideration and attention. The following discussion details issues regarding practical and statistical significance, as well as the number of variables, that affect the interpretation of factor loadings.

ENSURING PRACTICAL SIGNIFICANCE   The first guideline is not based on any mathematical proposition but relates more to practical significance by making a preliminary examination of the factor matrix in terms of the factor loadings. Because a factor loading is the correlation of the variable and the factor, the squared loading is the amount of the variable's total variance accounted for by the factor. Thus, a .30 loading translates to approximately 10 percent explanation, and a .50 loading denotes that 25 percent of the variance is accounted for by the factor. The loading must exceed .70 for the factor to account for 50 percent of the variance of a variable. Thus, the larger the absolute size of the factor loading, the more important the loading in

---

RULES OF THUMB 3-4

Choosing Factor Rotation Methods

- Orthogonal rotation methods:
  - Are the most widely used rotational methods
  - Are the preferred method when the research goal is data reduction to either a smaller number of variables or a set of uncorrelated measures for subsequent use in other multivariate techniques
- Oblique rotation methods:
  - Are best suited to the goal of obtaining several theoretically meaningful factors or constructs, because, realistically, few constructs in the real world are uncorrelated

interpreting the factor matrix. Using practical significance as the criteria, we can assess the loadings as follows:

- Factor loadings in the range of ±.30 to ±.40 are considered to meet the minimal level for interpretation of structure.
- Loadings ±.50 or greater are considered practically significant.
- Loadings exceeding 1.70 are considered indicative of well-defined structure and are the goal of any factor analysis.

The researcher should realize that extremely high loadings (.80 and above) are not typical and that the practical significance of the loadings is an important criterion. These guidelines are applicable when the sample size is 100 or larger and where the emphasis is on practical, not statistical, significance.

ASSESSING STATISTICAL SIGNIFICANCE   As previously noted, a factor loading represents the correlation between an original variable and its factor. In determining a significance level for the interpretation of loadings, an approach similar to determining the statistical significance of correlation coefficients could be used. However, research [14] has demonstrated that factor loadings have substantially larger standard errors than typical correlations. Thus, factor loadings should be evaluated at considerably stricter levels. The researcher can employ the concept of statistical power discussed in Chapter 1 to specify factor loadings considered significant for differing sample sizes. With the stated objective of obtaining a power level of 80 percent, the use of a .05 significance level, and the proposed inflation of the standard errors of factor loadings, Table 3-2 contains the sample sizes necessary for each factor loading value to be considered significant.

For example, in a sample of 100 respondents, factor loadings of .55 and above are significant. However, in a sample of 50, a factor loading of .75 is required for significance. In comparison with the prior rule of thumb, which denoted all loadings of .30 as having practical significance, this approach would consider loadings of .30 significant only for sample sizes of 350 or greater.

TABLE 3-2 Guidelines for Identifying Significant Factor Loadings Based on Sample Size	
Factor Loading	**Sample Size Needed for Significance**[a]
.30	350
.35	250
.40	200
.45	150
.50	120
.55	100
.60	85
.65	70
.70	60
.75	50

[a] Significance is based on a .05 significance level (α), a power level of 80 percent, and standard errors assumed to be twice those of conventional correlation coefficients.

Source: Computations made with SOLO *Power Analysis,* BMDP Statistical Software, Inc., 1993.

These guidelines are quite conservative when compared with the guidelines of the previous section or even the statistical levels associated with conventional correlation coefficients. Thus, these guidelines should be used as a starting point in factor-loading interpretation, with lower loadings considered significant and added to the interpretation based on other considerations. The next section details the interpretation process and the role that other considerations can play.

ADJUSTMENTS BASED ON THE NUMBER OF VARIABLES   A disadvantage of both of the prior approaches is that the number of variables being analyzed and the specific factor being examined are not considered. It has been shown that as the researcher moves from the first factor to later factors, the acceptable level for a loading to be judged significant should increase. The fact that unique variance and error variance begin to appear in later factors means that some upward adjustment in the level of significance should be included [22]. The number of variables being analyzed is also important in deciding which loadings are significant. As the number of variables being analyzed increases, the acceptable level for considering a loading significantly decreases. Adjustment for the number of variables is increasingly important as one moves from the first factor extracted to later factors.

Rules of Thumb 3-5 summarizes the criteria for the practical or statistical significance of factor loadings.

Interpreting a Factor Matrix

The task of interpreting a factor-loading matrix to identify the structure among the variables can at first seem overwhelming. The researcher must sort through all the factor loadings (remember, each variable has a loading on each factor) to identify those most indicative of the underlying structure. Even a fairly simple analysis of 15 variables on four factors necessitates evaluating and interpreting 60 factor loadings. Using the criteria for interpreting loadings described in the previous section, the researcher finds those distinctive variables for each factor and looks for a correspondence to the conceptual foundation or the managerial expectations for the research to assess practical significance. Thus, interpreting the complex interrelationships represented in a factor matrix requires a combination of applying objective criteria with managerial judgment. By following the five-step procedure outlined next, the process can be simplified considerably. After the process is discussed, a brief example will be used to illustrate the process.

STEP 1: EXAMINE THE FACTOR MATRIX OF LOADINGS   The factor-loading matrix contains the factor loading of each variable on each factor. They may be either rotated or unrotated loadings, but as discussed earlier, rotated loadings are usually used in factor interpretation unless data reduction is the sole objective. Typically, the factors are arranged as columns; thus, each column of numbers represents

---

RULES OF THUMB 3-5

Assessing Factor Loadings

- Although factor loadings of ±.30 to ±.40 are minimally acceptable, values greater than ±.50 are generally considered necessary for practical significance
- To be considered significant:
  - A smaller loading is needed given either a larger sample size or a larger number of variables being analyzed
  - A larger loading is needed given a factor solution with a larger number of factors, especially in evaluating the loadings on later factors
- Statistical tests of significance for factor loadings are generally conservative and should be considered only as starting points needed for including a variable for further consideration

the loadings of a single factor. If an oblique rotation has been used, two matrices of factor loadings are provided. The first is the **factor pattern matrix,** which has loadings that represent the unique contribution of each variable to the factor. The second is the **factor structure matrix,** which has simple correlations between variables and factors, but these loadings contain both the unique variance between variables and factors and the correlation among factors. As the correlation among factors becomes greater, it becomes more difficult to distinguish which variables load uniquely on each factor in the factor structure matrix. Thus, most researchers report the results of the factor pattern matrix.

STEP 2: IDENTIFY THE SIGNIFICANT LOADING(S) FOR EACH VARIABLE   The interpretation should start with the first variable on the first factor and move horizontally from left to right, looking for the highest loading for that variable on any factor. When the highest loading (largest absolute factor loading) is identified, it should be underlined if significant as determined by the criteria discussed earlier. Attention then focuses on the second variable and, again moving from left to right horizontally, looking for the highest loading for that variable on any factor and underlining it. This procedure should continue for each variable until all variables have been reviewed for their highest loading on a factor.

Most factor solutions, however, do not result in a simple structure solution (a single high loading for each variable on only one factor). Thus, the researcher will, after underlining the highest loading for a variable, continue to evaluate the factor matrix by underlining all significant loadings for a variable on all the factors. The process of interpretation would be greatly simplified if each variable had only one significant variable. In practice, however, the researcher may find that one or more variables each has moderate-size loadings on several factors, all of which are significant, and the job of interpreting the factors is much more difficult. When a variable is found to have more than one significant loading, it is termed a **cross-loading.**

The difficulty arises because a variable with several significant loadings (a cross-loading) must be used in labeling all the factors on which it has a significant loading. Yet how can the factors be distinct and potentially represent separate concepts when they "share" variables? Ultimately, the objective is to minimize the number of significant loadings on each row of the factor matrix (i.e., make each variable associate with only one factor). The researcher may find that different rotation methods eliminate any cross-loadings and thus define a simple structure. If a variable persists in having cross-loadings, it becomes a candidate for deletion.

STEP 3: ASSESS THE COMMUNALITIES OF THE VARIABLES   Once all the significant loadings have been identified, the researcher should look for any variables that are not adequately accounted for by the factor solution. One simple approach is to identify any variable(s) lacking at least one significant loading. Another approach is to examine each variable's communality, representing the amount of variance accounted for by the factor solution for each variable. The researcher should view the communalities to assess whether the variables meet acceptable levels of explanation. For example, a researcher may specify that at least one-half of the variance of each variable must be taken into account. Using this guideline, the researcher would identify all variables with communalities less than .50 as not having sufficient explanation.

STEP 4: RESPECIFY THE FACTOR MODEL IF NEEDED   Once all the significant loadings have been identified and the communalities examined, the researcher may find any one of several problems: (a) a variable has no significant loadings; (b) even with a significant loading, a variable's communality is deemed too low; or (c) a variable has a cross-loading. In this situation, the researcher can take any combination of the following remedies, listed from least to most extreme:

- *Ignore those problematic variables* and interpret the solution as is, which is appropriate if the objective is solely data reduction, but the researcher must still note that the variables in question are poorly represented in the factor solution.

- Evaluate each of those variables for *possible deletion,* depending on the variable's overall contribution to the research as well as its communality index. If the variable is of minor importance to the study's objective or has an unacceptable communality value, it may be eliminated and then the factor model respecified by deriving a new factor solution with those variables eliminated.
- Employ an *alternative rotation method,* particularly an oblique method if only orthogonal methods had been used.
- *Decrease/increase the number of factors retained* to see whether a smaller/larger factor structure will represent those problematic variables.
- *Modify the type of factor model used* (component versus common factor) to assess whether varying the type of variance considered affects the factor structure.

No matter which of these options are chosen by the researcher, the ultimate objective should always be to obtain a factor structure with both empirical and conceptual support. As we have seen, many "tricks" can be used to improve upon the structure, but the ultimate responsibility rests with the researcher and the conceptual foundation underlying the analysis.

STEP 5: LABEL THE FACTORS   When an acceptable factor solution has been obtained in which all variables have a significant loading on a factor, the researcher attempts to assign some meaning to the pattern of factor loadings. Variables with higher loadings are considered more important and have greater influence on the name or label selected to represent a factor. Thus, the researcher will examine all the significant variables for a particular factor and, placing greater emphasis on those variables with higher loadings, will attempt to assign a name or label to a factor that accurately reflects the variables loading on that factor. The signs are interpreted just as with any other correlation coefficients. On each factor, like signs mean the variables are positively related, and opposite signs mean the variables are negatively related. In orthogonal solutions the factors are independent of one another. Therefore, the signs for factor loading relate only to the factor on which they appear, not to other factors in the solution.

This label is not derived or assigned by the factor analysis computer program; rather, the label is intuitively developed by the researcher based on its appropriateness for representing the underlying dimensions of a particular factor. This procedure is followed for each extracted factor. The final result will be a name or label that represents each of the derived factors as accurately as possible.

As discussed earlier, the selection of a specific number of factors and the rotation method are interrelated. Several additional trial rotations may be undertaken, and by comparing the factor interpretations for several different trial rotations the researcher can select the number of factors to extract. In short, the ability to assign some meaning to the factors, or to interpret the nature of the variables, becomes an extremely important consideration in determining the number of factors to extract.

AN EXAMPLE OF FACTOR INTERPRETATION   To serve as an illustration of factor interpretation, nine measures were obtained in a pilot test based on a sample of 202 respondents. After estimation of the initial results, further analysis indicated a three-factor solution was appropriate. Thus, the researcher now has the task of interpreting the factor loadings of the nine variables.

Table 3-3 contains a series of factor-loading matrices. The first to be considered is the unrotated factor matrix (part a). We will examine the unrotated and rotated factor-loading matrices through the five-step process described earlier.

Steps 1 and 2: Examine the Factor-Loading Matrix and Identify Significant Loadings. Given the sample size of 202, factor loadings of .40 and higher will be considered significant for interpretative purposes. Using this threshold for the factor loadings, we can see that the unrotated matrix does little to identify any form of simple structure. Five of the nine variables have cross-loadings, and for many of the other variables the significant loadings are fairly low. In this situation, rotation may improve our understanding of the relationship among variables.

As shown in Table 3-3b, the VARIMAX rotation improves the structure considerably in two noticeable ways. First, the loadings are improved for almost every variable, with the loadings more closely aligned to the objective of having a high loading on only a single factor. Second, now only one variable (V_1) has a cross-loading.

Step 3: Assess Communalities.   Only V_3 has a communality that is low (.299). For our purposes V_3 will be retained, but a researcher may consider deletion of such variables in other research contexts. It illustrates the instance in which a variable has a significant loading, but may still be poorly accounted for by the factor solution.

Step 4: Respecify the Factor Model if Needed.   If we set a threshold value of .40 for loading significance and rearrange the variables according to loadings, the pattern shown in Table 3-3c emerges. Variables V_7, V_9, and V_8 all load highly on factor 1; factor 2 is characterized by variables V_5, V_2, and V_3; and factor 3 has two distinctive characteristics (V_4 and V_6). Only V_1 is problematic, with significant loadings on both factors 1 and 3. Given that at least two variables are given on both of these factors, V_1 is deleted from the analysis and the loadings recalculated.

Step 5: Label the Factors.   As shown in Table 3-3d, the factor structure for the remaining eight variables is now very well defined, representing three distinct groups of variables that the researcher may now utilize in further research.

TABLE 3-3   Interpretation of a Hypothetical Factor-Loading Matrix

(a) Unrotated Factor-Loading Matrix

	Factor 1	2	3
V_1	.611	.250	−.204
V_2	.614	−.446	.264
V_3	.295	−.447	.107
V_4	.561	−.176	−.550
V_5	.589	−.467	.314
V_6	.630	−.102	−.285
V_7	.498	.611	.160
V_8	.310	.300	.649
V_9	.492	.597	−.094

(b) VARIMAX Rotated Factor-Loading Matrix

	Factor 1	2	3	Communality
V_1	.462	.099	.505	.477
V_2	.101	.778	.173	.644
V_3	−.134	.517	.114	.299
V_4	−.005	.184	.784	.648
V_5	.087	.801	.119	.664
V_6	.180	.302	.605	.489
V_7	.795	−.032	.120	.647
V_8	.623	.293	−.366	.608
V_9	.694	−.147	.323	.608

(c) Simplified Rotated Factor-Loading Matrix[1]

	Component 1	2	3
V_7	.795		
V_9	.694		
V_8	.623		
V_5		.801	
V_2		.778	
V_3		.517	
V_4			.784
V_6			.605
V_1	.462		.505

[1]Loadings less than .40 are not shown and variables are sorted by highest loading.

(d) Rotated Factor-Loading Matrix with V_1 Deleted[2]

	Factor 1	2	3
V_2	.807		
V_5	.803		
V_3	.524		
V_7		.802	
V_9		.686	
V_8		.655	
V_4			.851
V_6			.717

[2]V_1 deleted from the analysis, loadings less than .40 are not shown, and variables are sorted by highest loading.

RULES OF THUMB 3-6

Interpreting the Factors

- An optimal structure exists when all variables have high loadings only on a single factor
- Variables that cross-load (load highly on two or more factors) are usually deleted unless theoretically justified or the objective is strictly data reduction
- Variables should generally have communalities of greater than .50 to be retained in the analysis
- Respecification of a factor analysis can include such options as the following:
  - Deleting a variable(s)
  - Changing rotation methods
  - Increasing or decreasing the number of factors

As the preceding example shows, the process of factor interpretation involves both objective and subjective judgments. The researcher must consider a wide range of issues, all the time never losing sight of the ultimate goal of defining the best structure of the set of variables. Although many details are involved, some of the general principles are found in Rules of Thumb 3-6.

STAGE 6: VALIDATION OF FACTOR ANALYSIS

The sixth stage involves assessing the degree of generalizability of the results to the population and the potential influence of individual cases or respondents on the overall results. The issue of generalizability is critical for each of the multivariate methods, but it is especially relevant for the interdependence methods because they describe a data structure that should be representative of the population as well. In the validation process, the researcher must address a number of issues in the area of research design and data characteristics as discussed next.

Use of a Confirmatory Perspective

The most direct method of validating the results is to move to a confirmatory perspective and assess the replicability of the results, either with a split sample in the original data set or with a separate sample. The comparison of two or more factor model results has always been problematic. However, several options exist for making an objective comparison. The emergence of confirmatory factor analysis (CFA) through structural equation modeling has provided one option, but it is generally more complicated and requires additional software packages, such as LISREL or EQS [4, 21]. Chapters 11 and 12 discuss confirmatory factor analysis in greater detail. Apart from CFA, several other methods have been proposed, ranging from a simple matching index [10] to programs (FMATCH) designed specifically to assess the correspondence between factor matrices [34]. These methods have had sporadic use, owing in part to (1) their perceived lack of sophistication, and (2) the unavailability of software or analytical programs to automate the comparisons. Thus, when CFA is not appropriate, these methods provide some objective basis for comparison.

Assessing Factor Structure Stability

Another aspect of generalizability is the stability of the factor model results. Factor stability is primarily dependent on the sample size and on the number of cases per variable. The researcher is always encouraged to obtain the largest sample possible and develop parsimonious models to increase the cases-to-variables ratio. If sample size permits, the researcher may wish to randomly split the sample into two subsets and estimate factor models for each subset. Comparison of the two resulting factor matrices will provide an assessment of the robustness of the solution across the sample.

Detecting Influential Observations

In addition to generalizability, another issue of importance to the validation of factor analysis is the detection of influential observations. Discussions in Chapter 2 on the identification of outliers and in Chapter 4 on the influential observations in regression both have applicability in factor analysis. The researcher is encouraged to estimate the model with and without observations identified as outliers to assess their impact on the results. If omission of the outliers is justified, the results should have greater generalizability. Also, as discussed in Chapter 4, several measures of influence that reflect one observation's position relative to all others (e.g., covariance ratio) are applicable to factor analysis as well. Finally, the complexity of methods proposed for identifying influential observations specific to factor analysis [11] limits the application of these methods.

STAGE 7: ADDITIONAL USES OF FACTOR ANALYSIS RESULTS

Depending upon the objectives for applying factor analysis, the researcher may stop with factor interpretation or further engage in one of the methods for data reduction. If the objective is simply to identify logical combinations of variables and better understand the interrelationships among variables, then factor interpretation will suffice. It provides an empirical basis for judging the structure of the variables and the impact of this structure when interpreting the results from other multivariate techniques. If the objective, however, is to identify appropriate variables for subsequent application to other statistical techniques, then some form of data reduction will be employed. The two options include the following:

- *Selecting the variable with the highest factor loading* as a surrogate representative for a particular factor dimension
- *Replacing the original set of variables* with an entirely new, smaller set of variables created either from *summated scales* or *factor scores*

Either option will provide new variables for use, for example, as the independent variables in a regression or discriminant analysis, as dependent variables in multivariate analysis of variance, or even as the clustering variables in cluster analysis. We discuss each of these options for data reduction in the following sections.

Selecting Surrogate Variables for Subsequent Analysis

If the researcher's objective is simply to identify appropriate variables for subsequent application with other statistical techniques, the researcher has the option of examining the factor matrix and selecting the variable with the highest factor loading on each factor to act as a **surrogate variable** that is representative of that factor. This approach is simple and direct only when one variable has a factor loading that is substantially higher than all other factor loadings. In many instances, however, the selection process is more difficult because two or more variables have loadings that are significant and fairly close to each other, yet only one is chosen as representative of a particular dimension. This decision should be based on the researcher's a priori knowledge of theory that may suggest that one variable more than the others would logically be representative of the dimension. Also, the researcher may have knowledge suggesting that a variable with a loading slightly lower is in fact more reliable than the highest-loading variable. In such cases, the researcher may choose the variable that is loading slightly lower as the best variable to represent a particular factor.

The approach of selecting a single surrogate variable as representative of the factor—although simple and maintaining the original variable—has several potential disadvantages.

- It does *not address the issue of measurement error* encountered when using single measures (see the following section for a more detailed discussion).

- It also runs the *risk of potentially misleading results by selecting only a single variable to represent a perhaps more complex result.* For example, assume that variables representing price competitiveness, product quality, and value were all found to load highly on a single factor. The selection of any one of these separate variables would create substantially different interpretations in any subsequent analysis, yet all three may be so closely related as to make any definitive distinction impossible.

In instances where several high loadings complicate the selection of a single variable, the researcher may have no choice but to employ factor analysis as the basis for calculating a summed scale or factor scores instead of the surrogate variable. The objective, just as in the case of selecting a single variable, is to best represent the basic nature of the factor or component.

Creating Summated Scales

Chapter 1 introduced the concept of a **summated scale,** which is formed by combining several individual variables into a single **composite measure.** In simple terms, all of the variables loading highly on a factor are combined, and the total—or more commonly the average score of the variables—is used as a replacement variable. A summated scale provides two specific benefits.

- First, it provides a *means of overcoming to some extent the measurement error* inherent in all measured variables. **Measurement error** is the degree to which the observed values are not representative of the actual values due to any number of reasons, ranging from actual errors (e.g., data entry errors) to the inability of individuals to accurately provide information. The impact of measurement error is to partially mask any relationships (e.g., correlations or comparison of group means) and make the estimation of multivariate models more difficult. The summated scale reduces measurement error by using multiple **indicators** (variables) to reduce the reliance on a single response. By using the average or typical response to a set of related variables, the measurement error that might occur in a single question will be reduced.
- A second benefit of the summated scale is its *ability to represent the multiple aspects of a concept in a single measure.* Many times we employ more variables in our multivariate models in an attempt to represent the many facets of a concept that we know is quite complex. But in doing so, we complicate the interpretation of the results because of the redundancy in the items associated with the concept. Thus, we would like not only to accommodate the richer descriptions of concepts by using multiple variables, but also to maintain parsimony in the number of variables in our multivariate models. The summated scale, when properly constructed, does combine the multiple indicators into a single measure representing what is held in common across the set of measures.

The process of scale construction has theoretical and empirical foundations in a number of disciplines, including psychometric theory, sociology, and marketing. Although a complete treatment of the techniques and issues involved are beyond the scope of this text, a number of excellent sources are available for further reading on this subject [2, 12, 20, 30, 31]. Additionally, a series of compilations of existing scales may be applied in a number of situations [3, 7, 32]. We discuss here, however, four issues basic to the construction of any summated scale: conceptual definition, dimensionality, reliability, and validity.

CONCEPTUAL DEFINITION   The starting point for creating any summated scale is its **conceptual definition.** The conceptual definition specifies the theoretical basis for the summated scale by defining the concept being represented in terms applicable to the research context. In academic research, theoretical definitions are based on prior research that defines the character and nature of a concept. In a managerial setting, specific concepts may be defined that relate to proposed objectives, such as image, value, or satisfaction. In either instance, creating a summated scale is always guided by the conceptual definition specifying the type and character of the items that are candidates for inclusion in the scale.

Content validity is the assessment of the correspondence of the variables to be included in a summated scale and its conceptual definition. This form of validity, also known as **face validity,** subjectively assesses the correspondence between the individual items and the concept through ratings by expert judges, pretests with multiple subpopulations, or other means. The objective is to ensure that the selection of scale items extends past just empirical issues to also include theoretical and practical considerations [12, 31].

DIMENSIONALITY   An underlying assumption and essential requirement for creating a summated scale is that the items are unidimensional, meaning that they are strongly associated with each other and represent a single concept [20, 24]. Factor analysis plays a pivotal role in making an empirical assessment of the dimensionality of a set of items by determining the number of factors and the loadings of each variable on the factor(s). The test of unidimensionality is that each summated scale should consist of items loading highly on a single factor [1, 20, 24, 28]. If a summated scale is proposed to have multiple dimensions, each dimension should be reflected by a separate factor. The researcher can assess unidimensionality with either exploratory factor analysis, as discussed in this chapter, or confirmatory factor analysis, as described in Chapters 11 and 12.

RELIABILITY   **Reliability** is an assessment of the degree of consistency between multiple measurements of a variable. One form of reliability is test–retest, by which consistency is measured between the responses for an individual at two points in time. The objective is to ensure that responses are not too varied across time periods so that a measurement taken at any point in time is reliable. A second and more commonly used measure of reliability is internal consistency, which applies to the consistency among the variables in a summated scale. The rationale for internal consistency is that the individual items or indicators of the scale should all be measuring the same construct and thus be highly intercorrelated [12, 28].

Because no single item is a perfect measure of a concept, we must rely on a series of diagnostic measures to assess internal consistency.

- The first *measures we consider relate to each separate item,* including the item-to-total correlation (the correlation of the item to the summated scale score) and the inter-item correlation (the correlation among items). Rules of thumb suggest that the item-to-total correlations exceed .50 and that the inter item correlations exceed .30 [31].
- The second type of diagnostic measure is the *reliability coefficient,* which assesses the consistency of the entire scale, with **Cronbach's alpha** [15, 28, 29] being the most widely used measure. The generally agreed upon lower limit for Cronbach's alpha is .70 [31, 32], although it may decrease to .60 in exploratory research [31]. One issue in assessing Cronbach's alpha is its positive relationship to the number of items in the scale. Because increasing the number of items, even with the same degree of intercorrelation, will increase the reliability value, researchers must place more stringent requirements for scales with large numbers of items.
- Also available are *reliability measures derived from confirmatory factor analysis.* Included in these measures are the composite reliability and the average variance extracted, both discussed in greater detail in Chapter 12.

Each of the major statistical programs now has reliability assessment modules or programs, such that the researcher is provided with a complete analysis of both item-specific and overall reliability measures. Any summated scale should be analyzed for reliability to ensure its appropriateness before proceeding to an assessment of its validity.

VALIDITY Having ensured that a scale (1) conforms to its conceptual definition, (2) is unidimensional, and (3) meets the necessary levels of reliability, the researcher must make one final assessment: scale validity. **Validity** is the extent to which a scale or set of measures accurately represents the concept of interest. We already described one form of validity—content or face validity—in the discussion of conceptual definitions. Other forms of validity are measured empirically by the correlation between theoretically defined sets of variables. The three most widely accepted forms of validity are convergent, discriminant, and nomological validity [8, 30].

- Convergent validity assesses the *degree to which two measures of the same concept are correlated.* Here the researcher may look for alternative measures of a concept and then correlate them with the summated scale. High correlations here indicate that the scale is measuring its intended concept.
- Discriminant validity is the *degree to which two conceptually similar concepts are distinct.* The empirical test is again the correlation among measures, but this time the summated scale is correlated with a similar, but conceptually distinct, measure. Now the correlation should be low, demonstrating that the summated scale is sufficiently different from the other similar concept.
- Finally, nomological validity refers to the *degree that the summated scale makes accurate predictions of other concepts in a theoretically based model.* The researcher must identify theoretically supported relationships from prior research or accepted principles and then assess whether the scale has corresponding relationships. In summary, convergent validity confirms that the scale is correlated with other known measures of the concept; discriminant validity ensures that the scale is sufficiently different from other similar concepts to be distinct; and nomological validity determines whether the scale demonstrates the relationships shown to exist based on theory or prior research.

A number of differing methods are available for assessing validity, ranging from the multitrait, multimethod (MTMM) matrices to structural equation-based approaches. Although beyond the scope of this text, numerous available sources address both the range of methods available and the issues involved in the specific techniques [8, 21, 30].

CALCULATING SUMMATED SCALES Calculating a summated scale is a straightforward process whereby the items comprising the summated scale (i.e., the items with high loadings from the factor analysis) are summed or averaged. The most common approach is to take the average of the items in the scale, which provides the researcher with complete control over the calculation and facilitates ease of use in subsequent analyses.

Whenever variables have both positive and negative loadings within the same factor, either the variables with the positive or the negative loadings must have their data values reversed. Typically, the variables with the negative loadings are reverse scored so that the correlations, and the loadings, are now all positive within the factor. **Reverse scoring** is the process by which the data values for a variable are reversed so that its correlations with other variables are reversed (i.e., go from negative to positive). For example, on our scale of 0 to 10, we would reverse score a variable by subtracting the original value from 10 (i.e., reverse score = 10 – original value). In this way, original scores of 10 and 0 now have the reversed scores of 0 and 10. All distributional characteristics are retained; only the distribution is reversed.

The purpose of reverse scoring is to prevent a canceling out of variables with positive and negative loadings. Let us use as an example of two variables with a negative correlation.

We are interested in combining V_1 and V_2, with V_1 having a positive loading and V_2 a negative loading. If 10 is the top score on V_1, the top score on V_2 would be 0. Now assume two cases. In case 1,

V_1 has a value of 10 and V_2 has a value of 0 (the best case). In the second case, V_1 has a value of 0 and V_2 has a value of 10 (the worst case). If V_2 is not reverse scored, then the scale score calculated by adding the two variables for both cases 1 and 2 is 10, showing no difference, whereas we know that case 1 is the best and case 2 is the worst. If we reverse score V_2, however, the situation changes. Now case 1 has values of 10 and 10 on V_1 and V_2, respectively, and case 2 has values of 0 and 0. The summated scale scores are now 20 for case 1 and 0 for case 2, which distinguishes them as the best and worst situations.

SUMMARY   Summated scales, one of the recent developments in academic research, experienced increased application in applied and managerial research as well. The ability of the summated scale to portray complex concepts in a single measure while reducing measurement error makes it a valuable addition in any multivariate analysis. Factor analysis provides the researcher with an empirical assessment of the interrelationships among variables, essential in forming the conceptual and empirical foundation of a summated scale through assessment of content validity and scale dimensionality (see Rules of Thumb 3-7).

Computing Factor Scores

The third option for creating a smaller set of variables to replace the original set is the computation of factor scores. **Factor scores** are also composite measures of each factor computed for each subject. Conceptually the factor score represents the degree to which each individual scores high on the group of items with high loadings on a factor. Thus, higher values on the variables with high loadings on a factor will result in a higher factor score. The one key characteristic that differentiates a factor score from a summated scale is that the *factor score is computed based on the factor loadings of all variables on the factor, whereas the summated scale is calculated by combining only selected variables.* Therefore, although the researcher is able to characterize a factor by the variables with the highest loadings, consideration must also be given to the loadings of other variables, albeit lower, and their influence on the factor score.

Most statistical programs can easily compute factor scores for each respondent. By selecting the factor score option, these scores are saved for use in subsequent analyses. The one disadvantage of factor scores is that they are *not easily replicated across studies because they are based on the*

RULES OF THUMB 3-7

Summated Scales

- A summated scale is only as good as the items used to represent the construct; even though it may pass all empirical tests, it is useless without theoretical justification
- Never create a summated scale without first assessing its unidimensionality with exploratory or confirmatory factor analysis
- Once a scale is deemed unidimensional, its reliability score, as measured by Cronbach's alpha:
  - Should exceed a threshold of .70, although a .60 level can be used in exploratory research
  - The threshold should be raised as the number of items increases, especially as the number of items approaches 10 or more
- With reliability established, validity should be assessed in terms of:
  - Convergent validity scale correlates with other like scales
  - Discriminant validity scale is sufficiently different from other related scales
  - Nomological validity scale "predicts" as theoretically suggested

factor matrix, which is derived separately in each study. Replication of the same factor matrix across studies requires substantial computational programming.

Selecting Among the Three Methods

To select among the three data reduction options, the researcher must make a series of decisions, weighing the advantages and disadvantages of each approach with the research objectives. The guidelines in Rules of Thumb 3-8 address the fundamental trade-offs associated with each approach.

The decision rule, therefore, would be as follows:

- If data are used only in the original sample or orthogonality must be maintained, factor scores are suitable.
- If generalizability or transferability is desired, then summated scales or surrogate variables are more appropriate. If the summated scale is a well-constructed, valid, and reliable instrument, then it is probably the best alternative.
- If the summated scale is untested and exploratory, with little or no evidence of reliability or validity, surrogate variables should be considered if additional analysis is not possible to improve the summated scale.

RULES OF THUMB 3-8

Representing Factor Analysis in Other Analyses

- **The Single Surrogate Variable**

  Advantages:
  - Simple to administer and interpret

  Disadvantages:
  - Does not represent all "facets" of a factor
  - Prone to measurement error

- **Factor Scores**

  Advantages
  - Represent all variables loading on the factor
  - Best method for complete data reduction
  - Are by default orthogonal and can avoid complications caused by multicollinearity

  Disadvantages
  - Interpretation more difficult because all variables contribute through loadings
  - Difficult to replicate across studies

- **Summated Scales**

  Advantages
  - Compromise between the surrogate variable and factor score options
  - Reduce measurement error
  - Represent multiple facets of a concept
  - Easily replicated across studies

  Disadvantages
  - Include only the variables that load highly on the factor and excludes those having little or marginal impact
  - Not necessarily orthogonal
  - Require extensive analysis of reliability and validity issues

---

AN ILLUSTRATIVE EXAMPLE

In the preceding sections, the major questions concerning the application of factor analysis were discussed within the model-building framework introduced in Chapter 1. To clarify these topics further, we use an illustrative example of the application of factor analysis based on data from the database presented in Chapter 1. Our discussion of the empirical example also follows the six-stage model-building process. The first three stages, common to either component or common factor analysis, are discussed first. Then, stages 4 through 6 for component analysis will be discussed, along with examples of the additional use of factor results. We conclude with an examination of the differences for common factor analysis in stages 4 and 5.

Stage 1: Objectives of Factor Analysis

Factor analysis can identify the structure of a set of variables as well as provide a process for data reduction. In our example, the perceptions of HBAT on 13 attributes (X_6 to X_{18}) are examined for the following reasons:

- *Understand whether these perceptions can be "grouped."* Even the relatively small number of perceptions examined here presents a complex picture of 78 separate correlations. By grouping the perceptions, HBAT will be able to see the big picture in terms of understanding its customers and what the customers think about HBAT.
- *Reduce the 13 variables to a smaller number.* If the 13 variables can be represented in a smaller number of composite variables, then the other multivariate techniques can be made more parsimonious. Of course, this approach assumes that a certain degree of underlying order exists in the data being analyzed.

Either or both objectives may be found in a research question, making factor analysis applicable to a wide range of research questions. Moreover, as the basis for summated scale development, it has gained even more use in recent years.

Stage 2: Designing a Factor Analysis

Understanding the structure of the perceptions of variables requires R-type factor analysis and a correlation matrix between variables, not respondents. All the variables are metric and constitute a homogeneous set of perceptions appropriate for factor analysis.

The sample size in this example is an 8:1 ratio of observations to variables, which falls within acceptable limits. Also, the sample size of 100 provides an adequate basis for the calculation of the correlations between variables.

Stage 3: Assumptions in Factor Analysis

The underlying statistical assumptions influence factor analysis to the extent that they affect the derived correlations. Departures from normality, homoscedasticity, and linearity can diminish correlations between variables. These assumptions are examined in Chapter 2, and the reader is encouraged to review the findings. The researcher must also assess the factorability of the correlation matrix. The first step is a visual examination of the correlations, identifying those that are statistically significant.

Table 3-4 shows the correlation matrix for the 13 perceptions of HBAT. Inspection of the correlation matrix reveals that 29 of the 78 correlations (37%) are significant at the .01 level, which provides an adequate basis for proceeding to an empirical examination of adequacy for factor analysis on both an overall basis and for each variable. Tabulating the number of significant correlations per variable finds a range from 0 (X_{15}) to 9 (X_{17}). Although no limits are placed on what is too high or low, variables that have no significant correlations may not be part of any factor, and if a variable

TABLE 3–4 Assessing the Appropriateness of Factor Analysis: Correlations, Measures of Sampling Adequacy, and Partial Correlations Among Variables

Correlations Among Variables

	X_6	X_7	X_8	X_9	X_{10}	X_{11}	X_{12}	X_{13}	X_{14}	X_{15}	X_{16}	X_{17}	X_{18}	Correlations Significant at .01 Level
X_6 Product Quality	1.000													3
X_7 E-Commerce	−.137	1.000												3
X_8 Technical Support	.096	.001	1.000											2
X_9 Complaint Resolution	.106	.140	.097	1.000										5
X_{10} Advertising	−.053	**.430**	−.063	.197	1.000									4
X_{11} Product Line	.477	−.053	.193	**.561**	−.012	1.000								7
X_{12} Salesforce Image	−.152	**.792**	.017	**.230**	**.542**	−.061	1.000							6
X_{13} Competitive Pricing	**−.401**	.229	**−.271**	−.128	.134	**−.495**	**.265**	1.000						6
X_{14} Warranty & Claims	.088	.052	**.797**	.140	.011	**.273**	.107	**−.245**	1.000					3
X_{15} Packaging	.027	−.027	−.074	.059	.084	.046	.032	.023	.035	1.000				0
X_{16} Order & Billing	.104	.156	.080	**.757**	.184	**.424**	.195	−.115	.197	.069	1.000			4
X_{17} Price Flexibility	**−.493**	**.271**	−.186	**.395**	**.334**	**−.378**	**.352**	**.471**	−.170	.094	**.407**	1.000		9
X_{18} Delivery Speed	.028	.192	.025	**.865**	**.276**	**.602**	**.272**	−.073	.109	.106	**.751**	**.497**	1.000	6

Note: Bolded values indicate correlations significant at the .01 significance level.
Overall Measure of Sampling Adequacy: .609
Bartlett Test of Sphericity: 948.9
Significance: .000

Measures of Sampling Adequacy and Partial Correlations

	X_6	X_7	X_8	X_9	X_{10}	X_{11}	X_{12}	X_{13}	X_{14}	X_{15}	X_{16}	X_{17}	X_{18}
X_6 Product Quality	.873												
X_7 E-Commerce	.038	.620											
X_8 Technical Support	-.049	-.060	.527										
X_9 Complaint Resolution	-.082	.117	-.150	.890									
X_{10} Advertising	-.122	.002	.049	.092	.807								
X_{11} Product Line	-.023	-.157	.067	-.152	-.101	.448							
X_{12} Salesforce Image	-.006	-.729	.077	-.154	-.333	.273	.586						
X_{13} Competitive Pricing	.054	-.018	.125	.049	.090	-.088	-.138	.879					
X_{14} Warranty & Claims	.124	.091	-.792	.123	-.020	-.103	-.172	-.019	.529				
X_{15} Packaging	-.076	.091	.143	.061	-.026	-.118	-.054	.015	-.138	.314			
X_{16} Order & Billing	-.189	-.105	.160	-.312	.044	.044	.100	.106	-.250	.031	.859		
X_{17} Price Flexibility	.135	-.134	.031	-.143	-.151	.953	.241	-.212	-.029	-.137	-.037	.442	
X_{18} Delivery Speed	.013	.136	-.028	-.081	.064	-.941	-.254	.126	.070	.090	-.109	-.922	.533

Note: Measures of sampling adequacy (MSA) are on the diagonal, partial correlations in the off-diagonal.

has a large number of correlations, it may be part of several factors. We can note these patterns and see how they are reflected as the analysis proceeds.

The researcher can assess the overall significance of the correlation matrix with the Bartlett test and the factorability of the overall set of variables and individual variables using the measure of sampling adequacy (MSA). Because factor analysis will always derive factors, the objective is to ensure a base level of statistical correlation within the set of variables, such that the resulting factor structure has some objective basis.

In this example, the Bartlett's test finds that the correlations, when taken collectively, are significant at the .0001 level (see Table 3-4). This test only indicates the presence of nonzero correlations, not the pattern of these correlations. The measure of sampling adequacy (MSA) looks not only at the correlations, but also at patterns between variables. In this situation the overall MSA value falls in the acceptable range (above .50) with a value of .609. Examination of the values for each variable, however, identifies three variables (X_{11}, X_{15}, and X_{17}) that have MSA values under .50. Because X_{15} has the lowest MSA value, it will be omitted in the attempt to obtain a set of variables that can exceed the minimum acceptable MSA levels. Recalculating the MSA values finds that X_{17} still has an individual MSA value below .50, so it is also deleted from the analysis. We should note at this point that X_{15} and X_{17} were the two variables with the lowest and highest number of significant correlations, respectively.

Table 3-5 contains the correlation matrix for the revised set of variables (X_{15} and X_{17} deleted) along with the measures of sampling adequacy and the Bartlett test value. In the reduced correlation matrix, 20 of the 55 correlations are statistically significant. As with the full set of variables, the Bartlett test shows that nonzero correlations exist at the significance level of .0001. The reduced set of variables collectively meets the necessary threshold of sampling adequacy with an MSA value of .653. Each of the variables also exceeds the threshold value, indicating that the reduced set of variables meets the fundamental requirements for factor analysis. Finally, examining the partial correlations shows only five with values greater than .50 (X_6–X_{11}, X_7–X_{12}, X_8–X_{14}, X_9–X_{18}, and X_{11}–X_{18}), which is another indicator of the strength of the interrelationships among the variables in the reduced set. It is of note that both X_{11} and X_{18} are involved in two of the high partial correlations. Collectively, these measures all indicate that the reduced set of variables is appropriate for factor analysis, and the analysis can proceed to the next stages.

Component Factor Analysis: Stages 4 Through 7

As noted earlier, factor analysis procedures are based on the initial computation of a complete table of intercorrelations among the variables (correlation matrix). The correlation matrix is then transformed through estimation of a factor model to obtain a factor matrix containing factor loadings for each variable on each derived factor. The loadings of each variable on the factors are then interpreted to identify the underlying structure of the variables, in this case perceptions of HBAT. These steps of factor analysis, contained in stages 4 through 7, are examined first for component analysis. Then, a common factor analysis is performed and comparisons made between the two factor models.

STAGE 4: DERIVING FACTORS AND ASSESSING OVERALL FIT   Given that the components method of extraction will be used first, the next decision is to select the number of components to be retained for further analysis. As discussed earlier, the researcher should employ a number of different criteria in determining the number of factors to be retained for interpretation, ranging from the more subjective (e.g., selecting a number of factors a priori or specifying the percentage of variance extracted) to the more objective (latent root criterion or scree test) criteria.

Table 3-6 contains the information regarding the 11 possible factors and their relative explanatory power as expressed by their eigenvalues. In addition to assessing the importance of each component, we can also use the eigenvalues to assist in selecting the number of factors.

TABLE 3-5 Assessing the Appropriateness of Factor Analysis for the Revised Set of Variables (X_{15} and X_{17} Deleted): Correlations, Measures of Sampling Adequacy, and Partial Correlations Among Variables

Correlations Among Variables

	X_6	X_7	X_8	X_9	X_{10}	X_{11}	X_{12}	X_{13}	X_{14}	X_{16}	X_{18}	Correlations Significant at .01 Level
X_6 Product Quality	1.000	-.137	.096	.106	-.053	**.477**	-.152	**-.401**	.088	.104	.028	2
X_7 E-Commerce		1.000	.001	.140	**.430**	-.053	**.792**	.229	.052	.156	.192	2
X_8 Technical Support			1.000	.097	-.063	.193	.017	**-.271**	**.797**	.080	.025	2
X_9 Complaint Resolution				1.000	.197	**.561**	.230	-.128	.140	**.757**	**.865**	4
X_{10} Advertising					1.000	-.012	**.542**	.134	.011	.184	**.276**	3
X_{11} Product Line						1.000	-.061	**-.495**	**.273**	**.424**	**.602**	6
X_{12} Salesforce Image							1.000	**.265**	.107	.195	**.272**	5
X_{13} Competitive Pricing								1.000	**-.245**	-.115	-.073	5
X_{14} Warranty & Claims									1.000	.197	.109	3
X_{16} Order & Billing										1.000	**.751**	3
X_{18} Delivery Speed											1.000	5

Note: Bolded values indicate correlations significant at the .01 significance level.
Overall Measure of Sampling Adequacy: .653
Bartlett's Test of Sphericity: 619.3
Significance: .000

Measures of Sampling Adequacy and Partial Correlations

	X_6	X_7	X_8	X_9	X_{10}	X_{11}	X_{12}	X_{13}	X_{14}	X_{16}	X_{18}
X_6 Product Quality	.509										
X_7 E-Commerce	.061	.626									
X_8 Technical Support	-.045	-.068	.519								
X_9 Complaint Resolution	-.062	.097	-.156	.787							
X_{10} Advertising	-.107	-.015	.062	.074	.779						
X_{11} Product Line	-.503	-.101	.117	-.054	.143	.622					
X_{12} Salesforce Image	-.042	-.725	.076	-.124	-.311	.148	.622				
X_{13} Competitive Pricing	.085	-.047	.139	.020	.060	.386	-.092	.753			
X_{14} Warranty & Claims	.122	.100	-.787	.127	-.032	-.246	-.175	-.028	.511		
X_{16} Order & Billing	-.184	-.113	.160	-.322	.040	.261	.113	.101	-.250	.760	
X_{18} Delivery Speed	.355	.040	.017	-.555	-.202	-.529	-.087	-.184	.100	-.369	.666

Note: Measures of sampling adequacy (MSA) are on the diagonal, partial correlations in the off-diagonal.

TABLE 3-6   Results for the Extraction of Component Factors

	Eigenvalues		
Component	Total	% of Variance	Cumulative %
1	3.43	31.2	31.2
2	2.55	23.2	54.3
3	1.69	15.4	69.7
4	1.09	9.9	79.6
5	.61	5.5	85.1
6	.55	5.0	90.2
7	.40	3.7	93.8
8	.25	2.2	96.0
9	.20	1.9	97.9
10	.13	1.2	99.1
11	.10	.9	100.0

The researcher is not bound by preconceptions as to the number of factors that should be retained, but practical reasons of desiring multiple measures per factor dictate that between three and five factors would be best given the 11 variables to be analyzed. If we apply the latent root criterion of retaining factors with eigenvalues greater than 1.0, four factors will be retained. The scree test (Figure 3-9), however, indicates that five factors may be appropriate when considering the changes in eigenvalues (i.e., identifying the "elbow" in the eigenvalues). In viewing the eigenvalue for the fifth factor, its low value (.61) relative to the latent root criterion value of 1.0 precluded its inclusion. If the eigenvalue had been quite close to 1, then it might be considered for inclusion as well. The four factors retained represent 79.6 percent of the variance of the 11 variables, deemed sufficient in terms of total variance explained. Combining all these criteria together leads to the conclusion to retain four factors for further analysis. More important, these results illustrate the need for multiple decision criteria in deciding the number of components to be retained.

STAGE 5: INTERPRETING THE FACTORS   With four factors to be analyzed, the researcher now turns to interpreting the factors. Once the factor matrix of loadings has been calculated, the interpretation

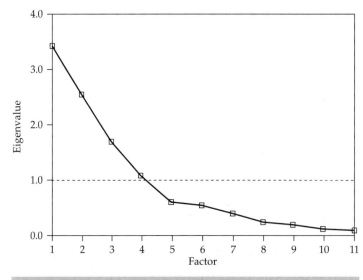

FIGURE 3-9   Scree Test for Component Analysis

process proceeds by examining the unrotated and then rotated factor matrices for significant factor loadings and adequate communalities. If deficiencies are found, respecification of the factors is considered. Once the factors are finalized, they can be described based on the significant factor loadings characterizing each factor.

Step 1: Examine the Factor Matrix of Loadings for the Unrotated Factor Matrix. Factor loadings, in either the unrotated or rotated factor matrices, represent the degree of association (correlation) of each variable with each factor. The loadings take on a key role in interpretation of the factors, particularly if they are used in ways that require characterization as to the substantive meaning of the factors (e.g., as predictor variables in a dependence relationship). The objective of factor analysis in these instances is to maximize the association of each variable with a single factor, many times through rotation of the factor matrix. The researcher must make a judgment as to the adequacy of the solution in this stage and its representation of the structure of the variables and ability to meet the goals of the research. We will first examine the unrotated factor solution and determine whether the use of the rotated solution is necessary.

Table 3-7 presents the unrotated component analysis factor matrix. To begin the analysis, let us explain the numbers included in the table. Five columns of numbers are shown. The first four are the results for the four factors that are extracted (i.e., factor loadings of each variable on each of the factors). The fifth column provides summary statistics detailing how well each variable is explained by the four components, which are discussed in the next section. The first row of numbers at the bottom of each column is the column sum of squared factor loadings (*eigenvalues*) and indicates the relative importance of each factor in accounting for the variance associated with the set of variables. Note that the sums of squares for the four factors are 3.427, 2.551, 1.691, and 1.087, respectively. As expected, the factor solution extracts the factors in the order of their importance, with factor 1 accounting for the most variance, factor 2 slightly less, and so on through all 11 factors. At the far right-hand side of the row is the number 8.756, which represents the total of the four eigenvalues (3.427 + 2.551 + 1.691 + 1.087). The total eigenvalues represents the total amount of variance extracted by the factor solution.

TABLE 3-7   Unrotated Component Analysis Factor Matrix

Variables	Factor				Communality
	1	2	3	4	
X_6 Product Quality	.248	−.501	−.081	.670	.768
X_7 E-Commerce	.307	.713	.306	.284	.777
X_8 Technical Support	.292	−.369	.794	−.202	.893
X_9 Complaint Resolution	.871	.031	−.274	−.215	.881
X_{10} Advertising	.340	.581	.115	.331	.576
X_{11} Product Line	.716	−.455	−.151	.212	.787
X_{12} Salesforce Image	.377	.752	.314	.232	.859
X_{13} Competitive Pricing	−.281	.660	−.069	−.348	.641
X_{14} Warranty & Claims	.394	−.306	.778	−.193	.892
X_{16} Order & Billing	.809	.042	−.220	−.247	.766
X_{18} Delivery Speed	.876	.117	−.302	−.206	.914
					Total
Sum of Squares (eigenvalue)	3.427	2.551	1.691	1.087	8.756
Percentage of trace[a]	31.15	23.19	15.37	9.88	79.59

[a]*Trace = 11.0 (sum of eigenvalues)*

The total amount of variance explained by either a single factor or the overall factor solution can be compared to the total variation in the set of variables as represented by the **trace** of the factor matrix. The trace is the total variance to be explained and is equal to the sum of the eigenvalues of the variable set. In components analysis, the trace is equal to the number of variables because each variable has a possible eigenvalue of 1.0. By adding the percentages of trace for each of the factors (or dividing the total eigenvalues of the factors by the trace), we obtain the total percentage of trace extracted for the factor solution. This total is used as an index to determine how well a particular factor solution accounts for what all the variables together represent. If the variables are all very different from one another, this index will be low. If the variables fall into one or more highly redundant or related groups, and if the extracted factors account for all the groups, the index will approach 100 percent.

The percentages of trace explained by each of the four factors (31.15%, 23.19%, 15.37%, and 9.88%, respectively) are shown as the last row of values of Table 3-7. The percentage of trace is obtained by dividing each factor's sum of squares (eigenvalues) by the trace for the set of variables being analyzed. For example, dividing the sum of squares of 3.427 for factor 1 by the trace of 11.0 results in the percentage of trace of 31.154 percent for factor 1. The index for the overall solution shows that 79.59 percent of the total variance $(8.756 \div 411.0)$ is represented by the information contained in the factor matrix of the four-factor solution. Therefore, the index for this solution is high, and the variables are in fact highly related to one another.

Step 2: Identify the Significant Loadings in the Unrotated Factor Matrix. Having defined the various elements of the unrotated factor matrix, let us examine the factor-loading patterns. As discussed earlier, the factor loadings allow for the description of each factor and the structure in the set of variables.

As anticipated, the first factor accounts for the largest amount of variance in Table 3-7. The second factor is somewhat of a general factor, with half of the variables having a high loading (high loading is defined as greater than .40). The third factor has two high loadings, whereas the fourth factor only has one high loading. Based on this factor-loading pattern with a relatively large number of high loadings on factor 2 and only one high loading on factor 4, interpretation would be difficult and theoretically less meaningful. Therefore, the researcher should proceed to rotate the factor matrix to redistribute the variance from the earlier factors to the later factors. Rotation should result in a simpler and theoretically more meaningful factor pattern. However, before proceeding with the rotation process, we must examine the communalities to see whether any variables have communalities so low that they should be eliminated.

Step 3: Assess the Communalities of the Variables in the Unrotated Factor Matrix. The row sum of squared factor loadings, known as *communalities*, show the amount of variance in a variable that is accounted for by the two factors taken together. The size of the communality is a useful index for assessing how much variance in a particular variable is accounted for by the factor solution. Higher communality values indicate that a large amount of the variance in a variable has been extracted by the factor solution. Small communalities show that a substantial portion of the variable's variance is not accounted for by the factors. Although no statistical guidelines indicate exactly what is "large" or "small," practical considerations dictate a lower level of .50 for communalities in this analysis.

The communalities in Table 3-7 are shown at the far right side of table. For instance, the communality figure of .576 for variable X_{10} indicates that it has less in common with the other variables included in the analysis than does variable X_8, which has a communality of .893. Both variables, however, still share more than one-half of their variance with the four factors. All of the communalities are sufficiently high to proceed with the rotation of the factor matrix.

Applying an Orthogonal (VARIMAX) Rotation. Given that the unrotated factor matrix did not have a completely clean set of factor loadings (i.e., had substantial cross-loadings or did not maximize the loadings of each variable on one factor), a rotation technique can be applied to hopefully improve the interpretation. In this case, the VARIMAX rotation is used and its impact on the overall factor solution and the factor loadings are described next.

The VARIMAX-rotated component analysis factor matrix is shown in Table 3-8. Note that the total amount of variance extracted is the same in the rotated solution as it was in the unrotated one, 79.6 percent. Also, the communalities for each variable do not change when a rotation technique is applied. Still, two differences do emerge. First, the variance is redistributed so that the factor-loading pattern and the percentage of variance for each of the factors are slightly different. Specifically, in the VARIMAX-rotated factor solution, the first factor accounts for 26.3 percent of the variance, compared to 31.2 percent in the unrotated solution. Likewise, the other factors also change, the

TABLE 3-8   VARIMAX-Rotated Component Analysis Factor Matrices: Full and Reduced Sets of Variables

| | VARIMAX-ROTATED LOADINGS[a] Factor | | | | |
Full Set of Variables	1	2	3	4	Communality
X_{18} Delivery Speed	.938				.914
X_9 Complaint Resolution	.926				.881
X_{16} Order & Billing	.864				.766
X_{12} Salesforce Image		.900			.859
X_7 E-Commerce		.871			.777
X_{10} Advertising		.742			.576
X_8 Technical Support			.939		.893
X_{14} Warranty & Claims			.931		.892
X_6 Product Quality				.876	.768
X_{13} Competitive Pricing				−.723	.641
X_{11} Product Line	.591			.642	.787
					Total
Sum of Squares (eigenvalue)	2.893	2.234	1.855	1.774	8.756
Percentage of trace	26.30	20.31	16.87	16.12	79.59

[a]*Factor loadings less than .40 have not been printed and variables have been sorted by loadings on each factor.*

| | VARIMAX-ROTATED LOADINGS[a] Factor | | | | |
Reduced Set of Variables (X_{11} deleted)	1	2	3	4	Communality
X_9 Complaint Resolution	.933				.890
X_{18} Delivery Speed	.931				.894
X_{16} Order & Billing	.886				.806
X_{12} Salesforce Image		.898			.860
X_7 E-Commerce		.868			.780
X_{10} Advertising		.743			.585
X_8 Technical Support			.940		.894
X_{14} Warranty & Claims			.933		.891
X_6 Product Quality				.892	.798
X_{13} Competitive Pricing				−.730	.661
					Total
Sum of Squares (eigenvalue)	2.589	2.216	1.846	1.406	8.057
Percentage of trace	25.89	22.16	18.46	14.06	80.57

[a]*Factor loadings less than .40 have not been printed and variables have been sorted by loadings on each factor.*

largest change being the fourth factor, increasing from 9.9 percent in the unrotated solution to 16.1 percent in the rotated solution. Thus, the explanatory power shifted slightly to a more even distribution because of the rotation. Second, the interpretation of the factor matrix is simplified. As will be discussed in the next section, the factor loadings for each variable are maximized for each variable on one factor, except in any instances of cross-loadings.

Steps 2 and 3: Assess the Significant Factor Loading(s) and Communalities of the Rotated Factor Matrix. With the rotation complete, the researcher now examines the *rotated factor matrix* for the patterns of significant factor loadings hoping to find a simplified structure. If any problems remain (i.e., nonsignificant loadings for one or more variables, cross-loadings, or unacceptable communalities), the researcher must consider respecification of the factor analysis through the set of options discussed earlier.

In the rotated factor solution (Table 3-8) each of the variables has a significant loading (defined as a loading above .40) on only one factor, except for X_{11}, which cross-loads on two factors (factors 1 and 4). Moreover, all of the loadings are above .70, meaning that more than one-half of the variance is accounted for by the loading on a single factor. With all of the communalities of sufficient size to warrant inclusion, the only remaining decision is to determine the action to be taken for X_{11}.

Step 4: Respecify the Factor Model if Needed. Even though the rotated factor matrix improved upon the simplicity of the factor loadings, the cross-loading of X_{11} on factors 1 and 4 requires action. The possible actions include ignoring the cross-loading, deleting X_{11} to eliminate the cross-loading, using another rotation technique, or decreasing the number of factors. The following discussion addresses these options and the course of action chosen.

Examining the correlation matrix in Table 3-5 shows that X_{11} has high correlations with X_6 (part of factor 4), X_9 (part of factor 1), and X_{12} (part of factor 2). Thus, it is not surprising that it may have several high loadings. With the loadings of .642 (factor 4) and .591 (factor 1) almost identical, the cross-loading is so substantial as to not be ignorable. As for employing another rotation technique, additional analysis showed that the other orthogonal methods (QUARTIMAX and EQUIMAX) still had this fundamental problem. Also, the number of factors should not be decreased due to the relatively large explained variance (16.1%) for the fourth factor.

Thus, the course of action taken is to delete X_{11} from the analysis, leaving 10 variables in the analysis. The rotated factor matrix and other information for the reduced set of 10 variables are also shown in Table 3-8. As we see, the factor loadings for the 10 variables remain almost identical, exhibiting both the same pattern and almost the same values for the loadings. The amount of explained variance increases slightly to 80.6 percent. With the simplified pattern of loadings (all at significant levels), all communalities above 50 percent (and most much higher), and the overall level of explained variance high enough, the 10-variable/four-factor solution is accepted, with the final step being to describe the factors.

Step 5: Naming the Factors. When a satisfactory factor solution has been derived, the researcher next attempts to assign some meaning to the factors. The process involves substantive interpretation of the pattern of factor loadings for the variables, including their signs, in an effort to name each of the factors. Before interpretation, a minimum acceptable level of significance for factor loadings must be selected. Then, all significant factor loadings typically are used in the interpretation process. Variables with higher loadings influence to a greater extent the name or label selected to represent a factor.

Let us look at the results in Table 3-8 to illustrate this procedure. The factor solution was derived from component analysis with a VARIMAX rotation of 10 perceptions of HBAT. The cutoff point for interpretation purposes in this example is all loadings ±.40 or above (see Table 3-2). The cutoff point was set somewhat low to illustrate the factor interpretation process with as many significant loadings as possible. In this example, however, all the loadings are substantially above or below this threshold, making interpretation quite straightforward.

Substantive interpretation is based on the significant loadings. In Table 3-8, loadings below .40 have not been printed and the variables are sorted by their loadings on each factor. A marked pattern of variables with high loadings for each factor is evident. Factors 1 and 2 have three variables with significant loadings and factors 3 and 4 have two. Each factor can be named based on the variables with significant loadings:

1. *Factor 1 Postsale Customer Service:* X_9, complaint resolution; X_{18}, delivery speed; and X_{16}, order and billing
2. *Factor 2 Marketing:* X_{12}, salesforce image; X_7, e-commerce presence; and X_{10}, advertising
3. *Factor 3 Technical Support:* X_8, technical support; and X_{14}, warranty and claims
4. *Factor 4 Product Value:* X_6, product quality; and X_{13}, competitive pricing

One particular issue should be noted: In factor 4, competitive pricing (X_{13}) and product quality (X_6) have opposite signs. It means that product quality and competitive pricing vary together, but move in directions opposite to each other. Perceptions are more positive whether product quality increases or price decreases. This fundamental trade-off leads to naming the factor product value. When variables have differing signs, the researcher needs to be careful in understanding the relationships between variables before naming the factors and must also make special actions if calculating summated scales (see earlier discussion on reverse scoring).

Three variables (X_{11}, X_{15}, and X_{17}) were not included in the final factor analysis. When the factor-loading interpretations are presented, it must be noted that these variables were not included. If the results are used in other multivariate analyses, these three could be included as separate variables, although they would not be assured to be orthogonal to the factor scores.

The process of naming factors is based primarily on the subjective opinion of the researcher. Different researchers in many instances will no doubt assign different names to the same results because of differences in their backgrounds and training. For this reason, the process of labeling factors is subject to considerable criticism. If a logical name can be assigned that represents the underlying nature of the factors, it usually facilitates the presentation and understanding of the factor solution and therefore is a justifiable procedure.

Applying an Oblique Rotation. The VARIMAX rotation is orthogonal, meaning that the factors remain uncorrelated throughout the rotation process. In many situations, however, the factors need not be uncorrelated and may even be conceptually linked, which requires correlation between the factors. The researcher should always consider applying a nonorthogonal rotation method and assess its comparability to the orthogonal results.

In our example, it is quite reasonable to expect that perceptual dimensions would be correlated; thus the application of the nonorthogonal oblique rotation is justified. Table 3-9 contains the pattern and structure matrices with the factor loadings for each variable on each factor. As discussed earlier, the pattern matrix is typically used for interpretation purposes, especially if the factors have a substantial correlation between them. In this case, the highest correlation between the factors is only 2.241 (factors 1 and 2), so that the pattern and structure matrices have quite comparable loadings. By examining the variables loading highly on each factor, we note that the interpretation is exactly the same as found with the VARIMAX rotation. The only difference is that all three loadings on factor 2 are negative, so that if the variables are reverse coded the correlations between factors will reverse signs as well.

STAGE 6: VALIDATION OF FACTOR ANALYSIS Validation of any factor analysis result is essential, particularly when attempting to define underlying structure among the variables. Optimally, we would always follow our use of factor analysis with some form of confirmatory factor analysis, such as structural equation modeling (see Chapter 12), but this type of follow-up is often not feasible. We must look to other means, such as split sample analysis or application to entirely new samples.

TABLE 3-9   Oblique Rotation of Components Analysis Factor Matrix

PATTERN MATRIX

		OBLIQUE ROTATED LOADINGS[a] Factor				
		1	**2**	**3**	**4**	*Communality[b]*
X_9	Complaint Resolution	.943				.890
X_{18}	Delivery Speed	.942				.894
X_{16}	Order & Billing	.895				.806
X_{12}	Salesforce Image		−.897			.860
X_7	E-Commerce		−.880			.780
X_{10}	Advertising		−.756			.585
X_8	Technical Support			.946		.894
X_{14}	Warranty & Claims			.936		.891
X_6	Product Quality				.921	.798
X_{13}	Competitive Pricing				−.702	.661

STRUCTURE MATRIX

		OBLIQUE ROTATED LOADINGS[a] Factor			
		1	**2**	**3**	**4**
X_9	Complaint Resolution	.943			
X_{18}	Delivery Speed	.942			
X_{16}	Order & Billing	.897			
X_{12}	Salesforce Image		−.919		
X_7	E-Commerce		−.878		
X_{10}	Advertising		−.750		
X_8	Technical Support			.944	
X_{14}	Warranty & Claims			.940	
X_6	Product Quality				.884
X_{13}	Competitive Pricing				−.773

FACTOR CORRELATION MATRIX

Factor	**1**	**2**	**3**	**4**
1	1.000			
2	−.241	1.000		
3	.118	.021	1.000	
4	.121	.190	.165	1.000

[a]*Factor loadings less than .40 have not been printed and variables have been sorted by loadings on each factor.*
[b]*Communality values are not equal to the sum of the squared loadings due to the correlation of the factors.*

In this example, we split the sample into two equal samples of 50 respondents and reestimate the factor models to test for comparability. Table 3-10 contains the VARIMAX rotations for the two-factor models, along with the communalities. As can be seen, the two VARIMAX rotations are quite comparable in terms of both loadings and communalities for all six perceptions. The only notable occurrence is the presence of a slight cross-loading for X_{13} in subsample 1, although the fairly large difference in loadings (.445 versus 2.709) makes assignment of X_{13} only to factor 4 appropriate.

TABLE 3-10 Validation of Component Factor Analysis by Split-Sample Estimation with VARIMAX Rotation

VARIMAX-ROTATED LOADINGS

	Factor				
Split-Sample 1	1	2	3	4	Communality
X_9 Complaint Resolution	.924				.901
X_{18} Delivery Speed	.907				.878
X_{16} Order & Billing	.901				.841
X_{12} Salesforce Image		.885			.834
X_7 E-Commerce		.834			.733
X_{10} Advertising		.812			.668
X_8 Technical Support			.927		.871
X_{14} Warranty & Claims			.876		.851
X_6 Product Quality				.884	.813
X_{13} Competitive Pricing		.445		−.709	.709

VARIMAX-ROTATED LOADINGS

	Factor				
Split-Sample 2	1	2	3	4	Communality
X_9 Complaint Resolution	.943				.918
X_{18} Delivery Speed	.935				.884
X_{16} Order & Billing	.876				.807
X_{12} Salesforce Image		.902			.886
X_7 E-Commerce		.890			.841
X_{10} Advertising		.711			.584
X_8 Technical Support			.958		.932
X_{14} Warranty & Claims			.951		.916
X_6 Product Quality				.889	.804
X_{13} Competitive Pricing				−.720	.699

With these results we can be reasonably assured that the results are stable within our sample. If possible, we would always like to perform additional work by gathering additional respondents and ensuring that the results generalize across the population or generating new subsamples for analysis and assessment of comparability.

STAGE 7: ADDITIONAL USES OF THE FACTOR ANALYSIS RESULTS   The researcher has the option of using factor analysis not only as a data summarization tool, as seen in the prior discussion, but also as a data-reduction tool. In this context, factor analysis would assist in reducing the number of variables, either through selection of a set of surrogate variables, one per factor, or by creating new composite variables for each factor. The following sections detail the issues in data reduction for this example.

Selecting Surrogate Variables for Subsequent Analysis. Let us first clarify the procedure for selecting surrogate variables. In selecting a single variable to represent an entire factor, it is preferable to use an orthogonal rotation so as to ensure that, to the extent possible, the selected variables be uncorrelated with each other. Thus, on this analysis the orthogonal solution (Table 3-8) will be used instead of the oblique results.

Assuming we want to select only a single variable for further use, attention is on the magnitude of the factor loadings (Table 3-8), irrespective of the sign (positive or negative). Focusing on the factor loadings for factors 1 and 3, we see that the first and second highest loadings are essentially identical (.933 for X_9 and .931 for X_{18} on factor 1, .940 for X_8 and .933 for X_{14} on factor 3). If we have no a priori evidence to suggest that the reliability or validity for one of the variables is better than for the other, and if none would be theoretically more meaningful for the factor interpretation, we would select the variable with the highest loading (X_9 and X_8 for factors 1 and 3, respectively). However, the researcher must be cautious to not let these single measures provide the sole interpretation for the factor, because each factor is a much more complex dimension than could be represented in any single variable. The difference between the first and second highest loadings for factors 2 and 4 are much greater, making selection of variables X_{12} (factor 2) and X_6 (factor 4) easier and more direct. For all four factors, however, no single variable represents the component best; thus factor scores or a summated scale would be more appropriate if possible.

Creating Summated Scales. A summated scale is a composite value for a set of variables calculated by such simple procedures as taking the average of the variables in the scale. It is much like the variates in other multivariate techniques, except that the weights for each variable are assumed to be equal in the averaging procedure. In this way, each respondent would have four new variables (summated scales for factors 1, 2, 3, and 4) that could be substituted for the original 13 variables in other multivariate techniques. Factor analysis assists in the construction of the summated scale by identifying the dimensionality of the variables (defining the factors), which then form the basis for the composite values if they meet certain conceptual and empirical criteria. After the actual construction of the summated scales, which includes reverse scoring of opposite-signed variables (see earlier discussion), the scales should also be evaluated for reliability and validity if possible.

In this example, the four-factor solution suggests that four summated scales should be constructed. The four factors, discussed earlier, correspond to dimensions that can be named and related to concepts with adequate content validity. The dimensionality of each scale is supported by the clean interpretation of each factor, with high factor loadings of each variable on only one factor. The reliability of the summated scales is best measured by Cronbach's alpha, which in this case is .90 for scale 1, .78 for scale 2, .80 for scale 3, and .57 for scale 4. Only scale 4, representing the Product Value factor, has a reliability below the recommended level of .70. A primary reason for the low reliability value is that the scale only has two variables. Future research should strive to find additional items that measure this concept. It will be retained for further use with the caveat of a somewhat lower reliability and the need for future development of additional measures to represent this concept. Also remember that because X_{13} has a negative relationship (loading) it should be reverse-scored before creating the summated scale.

Although no direct test is available to assess the validity of the summated scales in the HBAT database, one approach is to compare the summated scales with the surrogate variables to see whether consistent patterns emerge. Table 3-11 illustrates the use of summated scales as replacements for the original variables by comparing the differences in the surrogate variables across the two regions (USA/North America versus outside North America) of X_4 to those differences of the corresponding summated scales and factor scores.

When viewing the two groups of and factor scores X_4, we can see that the pattern of differences is consistent. X_{12} and X_6 (the surrogate variables for factors 2 and 4) and scales 2 and 4 (the summated scales for factors 2 and 4) all have significant differences between the two regions, whereas the measures for the first and third factors (X_9 and X_8, scales 1 and 3, and factor scores 1 and 3) all show no difference. The summated scales and the surrogate variables all show the same patterns of differences between the two regions, as seen for the factor scores, demonstrating a level of convergent validity between these three measures.

USE OF FACTOR SCORES   Instead of calculating summated scales, we could calculate factor scores for each of the four factors in our component analysis. The factor scores differ from the summated scales in that factor scores are based directly on the factor loadings, meaning that every

TABLE 3-11 Evaluating the Replacement of the Original Variables by Factor Scores or Summated Scales

	MEAN DIFFERENCE BETWEEN GROUPS OF RESPONDENTS BASED ON X_4 – REGION			
Statistical Test	*Mean Scores*		*t-test*	
Measure	**Group 1: USA/North America**	**Group 2: Outside North America**	**t value**	**Significance**
Representative Variables from Each Factor				
X_9 – Complaint Resolution	5.456	5.433	.095	.925
X_{12} – Salesforce Image	4.587	5.466	−4.341	.000
X_8 – Technical Support	5.697	5.152	1.755	.082
X_6 – Product Quality	8.705	7.238	5.951	.000
Factor Scores				
Factor 1 – Customer Service	−.031	.019	−.248	.805
Factor 2 – Marketing	−.308	.197	−2.528	.013
Factor 3 – Technical Support	.154	−.098	1.234	.220
Factor 4 – Product Value	.741	−.474	7.343	.000
Summated Scales				
Scale 1 – Customer Service	4.520	4.545	−.140	.889
Scale 2 – Marketing	3.945	4.475	−3.293	.001
Scale 3 – Technical Support	5.946	5.549	1.747	.084
Scale 4 – Product Value	6.391	4.796	8.134	.000

	Correlations Between Summated Scales			
	Scale 1	**Scale 2**	**Scale 3**	**Scale 4**
Scale 1	1.000			
Scale 2	.260**	1.000		
Scale 3	.113	.010	1.000	
Scale 4	.126	−.225*	.228*	1.000

	Correlations Between Factor Scores and Summated Scales			
	Factor 1	**Factor 2**	**Factor 3**	**Factor 4**
Scale 1	.987**	.127	.057	.060
Scale 2	.147	.976**	.008	−.093
Scale 3	.049	.003	.984**	.096
Scale 4	.082	−.150	.148	.964**

* Significant at .05 level.
** Significant at .01 level.

variable contributes to the factor score based on the size of its loading (rather than calculating the summated scale score as the mean of selected variables with high loadings).

The first test of comparability of factor scores is similar to that performed with summated scales in assessing the pattern of differences found on X_4 for the surrogate variables and now the factor scores. Just as seen with the summated scales, the patterns of differences were identical, with differences on factor scores 2 and 4 corresponding to the differences in the surrogate variables for factors 2 and 4, with no differences for the others.

The consistency between factor scores and summated scales is also seen in the correlations in Table 3-11. We know that the factor scores, since rotated with a VARIMAX technique, are orthogonal (uncorrelated). But how closely do the scales correspond to the factor scores? The second portion of Table 3-11 shows the correlations between the summated scales and factor scores. The first portion of the table shows that the scales are relatively uncorrelated among themselves (the highest correlation is .260), which matches fairly closely to an orthogonal solution. This pattern also closely matches the oblique solution shown in Table 3-9 (note that the second factor in the oblique solution had all negative loadings, thus the difference between positive and negative correlations among the factors). Finally, the second correlation matrix shows a high degree of similarity between the factor scores and the summated scores, with correlations ranging from .964 to .987. These results further support the use of the summated scales as valid substitutes for factor scores if desired.

SELECTING BETWEEN DATA REDUCTION METHODS If the original variables are to be replaced by surrogate variables, factor scores, or summated scales, a decision must be made on which to use. This decision is based on the need for simplicity (which favors surrogate variables) versus replication in other studies (which favors use of summated scales) versus the desire for orthogonality of the measures (which favors factor scores). Although it may be tempting to employ surrogate variables, the preference among researchers today is the use of summated scales or, to a lesser degree, factor scores. From an empirical perspective, the two composite measures are essentially identical. The correlations in Table 3-11 demonstrate the high correspondence of factor scores to summated scales and the low correlations among summated scales, approximating the orthogonality of the factor scores. The final decision, however, rests with the researcher and the need for orthogonality versus replicability in selecting factor scores versus summated scales.

Common Factor Analysis: Stages 4 and 5

Common factor analysis is the second major factor analytic model that we discuss. The primary distinction between component analysis and common factor analysis is that the latter considers only the common variance associated with a set of variables. This aim is accomplished by factoring a "reduced" correlation matrix with estimated initial communalities in the diagonal instead of unities. The differences between component analysis and common factor analysis occur only at the factor estimation and interpretation stages (stages 4 and 5). Once the communalities are substituted on the diagonal, the common factor model extracts factors in a manner similar to component analysis. The researcher uses the same criteria for factor selection and interpretation. To illustrate the differences that can occur between common factor and component analysis, the following sections detail the extraction and interpretation of a common factor analysis of the 13 HBAT perceptions used in the component analysis.

STAGE 4: DERIVING FACTORS AND ASSESSING OVERALL FIT The reduced correlation matrix with communalities on the diagonal was used in the common factor analysis. Recalling the procedures employed in the component analysis, the original 13 variables were reduced to 11 due to low MSA values for X_{15} and X_{17}.

The first step is to determine the number of factors to retain for examination and possible rotation. Table 3-12 shows the extraction statistics. If we were to employ the latent root criterion with a cutoff value of 1.0 for the eigenvalue, four factors would be retained. However, the scree analysis indicates that five factors be retained (see Figure 3-10). In combining these two criteria, we will retain four factors for further analysis because of the low eigenvalue for the fifth factor and to maintain comparability with the component analysis. Note that this same set of circumstances was encountered in the component analysis. As with the component analysis examined earlier, the researcher should employ a combination of criteria in determining the number of factors to retain and may even wish to examine the three-factor solution as an alternative.

TABLE 3-12 Results for the Extraction of Common Factors: Extraction Method—Principal Axis Factoring

Factor	Initial Eigenvalues			Extraction Sums of Squared Loadings		
	Total	% of Variance	Cumulative %	Total	% of Variance	Cumulative %
1	3.427	31.154	31.154	3.215	29.231	29.231
2	2.551	23.190	54.344	2.225	20.227	49.458
3	1.691	15.373	69.717	1.499	13.630	63.088
4	1.087	9.878	79.595	.678	6.167	69.255
5	.609	5.540	85.135			
6	.552	5.017	90.152			
7	.402	3.650	93.802			
8	.247	2.245	96.047			
9	.204	1.850	97.898			
10	.133	1.208	99.105			
11	.098	.895	100.000			

Because the final common factor model sometimes differs from the initial extraction estimates (e.g., see discussion of Table 3-12 that follows), the researcher should be sure to evaluate the extraction statistics for the final common factor model. Remember that in common factor analysis, only the "common" or shared variance is used. Thus, the trace (sum of all eigenvalues) and the eigenvalues for all factors will be lower when only the common variance is considered. As such, a researcher may wish to be more liberal in making judgments on such issues as variance extracted or the latent root criterion threshold. If the researcher was dissatisfied with the total variance explained, for example, the remedies discussed earlier are still available (such as extracting one or more additional factors to increase explained variance). Also, communalities should also be examined to ensure an adequate level is maintained after extraction.

As also shown in Table 3-12, the eigenvalues for extracted factors can be restated in terms of the common factor extraction process. As shown in Table 3-12, the values for the extracted factors still support four factors because the percentage of total variance explained is still 70 percent.

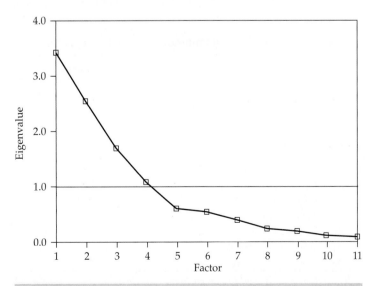

FIGURE 3-10 Scree Test for Common Factor Analysis

The only substantive difference is for the eigenvalue of factor 4, which falls below the 1.0 threshold. It is retained for this analysis, however, because the scree test still supports the four factors and to maintain comparability with the component analysis.

The unrotated factor matrix (Table 3-13) shows that the communalities of each variable are comparable to those found in component analysis. Because several variables fell below a communality of .50, a five-factor model could be made in an attempt to increase the communalities, as well as the overall variance explained. For our purposes here, however, we interpret the four-factor solution.

STAGE 5: INTERPRETING THE FACTORS With the factors extracted and the number of factors finalized, the proceeds to interpretation of the factors.

By examining the unrotated loadings (see Table 3-13), we note the need for a factor matrix rotation just as we found in the component analysis. Factor loadings were generally not as high as desired, and two variables (X_6 and X_{11}) exhibited cross-loadings. Turning then to the VARIMAX-rotated common factor analysis factor matrix (Table 3-14), the information provided is the same as in the component analysis solution (e.g., sums of squares, percentages of variance, communalities, total sums of squares, and total variances extracted).

Comparison of the information provided in the rotated common factor analysis factor matrix and the rotated component analysis factor matrix shows remarkable similarity. X_{11} has substantial cross-loadings on both factors 1 and 4 in both analyses (Tables 3-8 and 3-14). When X_{11} is deleted from the analysis, the four-factor solution is almost identical to the component analysis. The primary differences between the component analysis and common factor analysis are the generally lower loadings in the common factor analysis, owing primarily to the lower communalities of the variables used in common factor analysis. Even with these slight differences in the patterns of loadings, though, the basic interpretations are identical between the component analysis and the common factor analysis.

A Managerial Overview of the Results

Both the component and common factor analyses provide the researcher with several key insights into the structure of the variables and options for data reduction. First, concerning the structure of the variables, clearly four separate and distinct dimensions of evaluation are used by the HBAT

TABLE 3-13 Unrotated Common Factor-Loadings Matrix

		Factor[a]				Communality
		1	**2**	**3**	**4**	
X_{18}	Delivery Speed	.895				.942
X_9	Complaint Resolution	.862				.843
X_{16}	Order & Billing	.747				.622
X_{11}	Product Line	.689	−.454			.800
X_{12}	Salesforce Image		.805			.990
X_7	E-Commerce Presence		.657			.632
X_{13}	Competitive Pricing		.553			.443
X_{10}	Advertising		.457			.313
X_8	Technical Support			.739		.796
X_{14}	Warranty & Claims			.735		.812
X_6	Product Quality		−.408		.463	.424

[a]*Factor loadings less than .40 have not been printed and variables have been sorted by loadings on each factor.*

**TABLE 3-14   VARIMAX-Rotated Common Factor Matrix: Full and Reduced
Sets of Variables**

		Factor[a]				Communality
Full Set of 11 Variables		**1**	**2**	**3**	**4**	**Communality**
X_{18}	Delivery Speed	.949				.942
X_9	Complaint Resolution	.897				.843
X_{16}	Order & Billing	.768				.622
X_{12}	Salesforce Image		.977			.990
X_7	E-Commerce		.784			.632
X_{10}	Advertising		.529			.313
X_{14}	Warranty & Claims			.884		.812
X_8	Technical Support			.884		.796
X_{11}	Product Line	.525			.712	.800
X_6	Product Quality				.647	.424
X_{13}	Competitive Pricing				−.590	.443
						Total
Sum of Squared Loadings (eigenvalue)		2.635	1.971	1.641	1.371	7.618
Percentage of Trace		23.95	17.92	14.92	12.47	69.25

		Factor[a]				Communality
Reduced Set of 10 Variables		**1**	**2**	**3**	**4**	**Communality**
X_{18}	Delivery Speed	.925				.885
X_9	Complaint Resolution	.913				.860
X_{16}	Order & Billing	.793				.660
X_{12}	Salesforce Image		.979			.993
X_7	E-Commerce		.782			.631
X_{10}	Advertising		.531			.316
X_8	Technical Support			.905		.830
X_{14}	Warranty & Claims			.870		.778
X_6	Product Quality				.788	.627
X_{13}	Competitive Pricing				−.480	.353
						Total
Sum of Squared Loadings (eigenvalue)		2.392	1.970	1.650	.919	6.932
Percentage of Trace		23.92	19.70	16.50	9.19	69.32

[a]*Factor loadings less than .40 have not been printed and variables have been sorted by loadings on each factor.*

customers. These dimensions encompass a wide range of elements in the customer experience, from the tangible product attributes (Product Value) to the relationship with the firm (Customer Service and Technical Support) to even the outreach efforts (Marketing) by HBAT. Business planners within HBAT can now discuss plans revolving around these four areas instead of having to deal with all of the separate variables.

Factor analysis also provides the basis for data reduction through either summated scales or factor scores. The researcher now has a method for combining the variables within each factor into a single score that can replace the original set of variables with four new composite variables. When looking for differences, such as between regions, these new composite variables can be used so that only differences for composite scores, rather than the individual variables, are analyzed.

Summary

The multivariate statistical technique of factor analysis has been presented in broad conceptual terms. Basic guidelines for interpreting the results were included to clarify further the methodological concepts. An example of the application of factor analysis was presented based on the HBAT database. This chapter helps you to do the following:

Differentiate factor analysis techniques from other multivariate techniques. Exploratory factor analysis (EFA) can be a highly useful and powerful multivariate statistical technique for effectively extracting information from large bodies of interrelated data. When variables are correlated, the researcher needs ways to manage these variables: grouping highly correlated variables together, labeling or naming the groups, and perhaps even creating a new composite measure that can represent each group of variables. The primary purpose of exploratory factor analysis is to define the underlying structure among the variables in the analysis. As an interdependence technique, factor analysis attempts to identify groupings among variables (or cases) based on relationships represented in a correlation matrix. It is a powerful tool to better understand the structure of the data, and also can be used to simplify analyses of a large set of variables by replacing them with composite variables. When it works well, it points to interesting relationships that might not have been obvious from examination of the raw data alone, or even the correlation matrix.

Distinguish between exploratory and confirmatory uses of factor analysis techniques. Factor analysis as discussed in this chapter is primarily an exploratory technique because the researcher has little control over the specification of the structure (e.g., number of factors, loadings of each variable, etc.). Although the methods discussed in this chapter provide insights into the data, any attempt at confirmation will most likely require the use of specific methods discussed in the chapters on structural equation modeling.

Understand the seven stages of applying factor analysis. The seven stages of applying factor analysis include the following:

1. Clarifying the objectives of factor analysis
2. Designing a factor analysis, including selection of variables and sample size
3. Assumptions of factor analysis

4. Deriving factors and assessing overall fit, including which factor model to use and the number of factors
5. Rotating and interpreting the factors
6. Validation of factor analysis solutions
7. Additional uses of factor analysis results, such as selecting surrogate variables, creating summated scales, or computing factor scores

Distinguish between R and Q factor analysis. The principal use of factor analysis is to develop a structure among variables, referred to as R factor analysis. Factor analysis also can be used to group cases and is then referred to as Q factor analysis. Q factor analysis is similar to cluster analysis. The primary difference is that Q factor analysis uses correlation as the measure of similarity whereas cluster analysis is based on a distance measure.

Identify the differences between component analysis and common factor analysis models. Three types of variance are considered when applying factor analysis: common variance, unique variance, and error variance. When you add the three types of variance together, you get total variance. Each of the two methods of developing a factor solution uses different types of variance. Component analysis, also known as *principal components analysis*, considers the total variance and derives factors that contain small proportions of unique variance and, in some instances, error variance. Component analysis is preferred when data reduction is a primary goal. Common factor analysis is based only on common (shared) variance and assumes that both the unique and error variance are not of interest in defining the structure of the variables. It is more useful in identifying latent constructs and when the researcher has little knowledge about the unique and error variance. The two methods achieve essentially the same results in many research situations.

Describe how to determine the number of factors to extract. A critical decision in factor analysis is the number of factors to retain for interpretation and further use. In deciding when to stop factoring (i.e., how many factors to extract), the researcher must combine a conceptual foundation (How many factors should be in the structure?) with some empirical evidence (How many factors can be reasonably supported?). The researcher generally begins with some predetermined criteria, such as the general number of factors, plus some general thresholds of practical relevance (e.g., required percentage of variance explained). These criteria are combined with empirical measures of the factor structure. An exact quantitative basis for deciding the number of factors to extract has not been developed. Stopping criteria for the number of factors to extract include latent root or eigenvalue, a priori, percentage of variance, and scree test. These empirical criteria must be balanced against any theoretical bases for establishing the number of factors.

Explain the concept of rotation of factors. Perhaps the most important tool in interpreting factors is factor rotation. The term *rotation* means the reference axes of the factors are turned about the origin until some other position has been reached. Two types of rotation are orthogonal and oblique. Unrotated factor solutions extract factors in the order of their importance, with the first factor being a general factor with almost every variable loading significantly and accounting for the largest amount of variance. The second and subsequent factors are based on the residual amount of variance, with each accounting for successively smaller portions of variance. The ultimate effect of rotating the factor matrix is to redistribute the variance from earlier factors to later ones to achieve a simpler, theoretically more meaningful factor pattern. Factor rotation assists in the interpretation of the factors by simplifying the structure through maximizing the significant loadings of a variable on a single factor. In this manner, the variables most useful in defining the character of each factor can be easily identified.

Describe how to name a factor. Factors represent a composite of many variables. When an acceptable factor solution has been obtained in which all variables have a significant loading on a factor, the researcher attempts to assign some meaning to the pattern of factor loadings. Variables with higher loadings are considered more important and have greater influence on the name or label selected to represent a factor. The significant variables for a particular factor are examined and, placing greater emphasis on those variables with higher loadings, a name or label is assigned to a factor that accurately reflects the variables loading on that factor. The researcher identifies the variables with the greatest contribution to a factor and assigns a "name" to represent the factor's conceptual meaning.

Explain the additional uses of factor analysis. Depending on the objectives for applying factor analysis, the researcher may stop with factor interpretation or further engage in one of the methods for data reduction. If the objective is simply to identify logical combinations of variables and better understand the interrelationships among variables, then factor interpretation will suffice. If the objective, however, is to identify appropriate variables for subsequent application to other statistical techniques, then some form of data reduction will be employed. One of the data reduction options of factor analysis is to select a single (surrogate) variable with the highest factor loading. In doing so, the researcher identifies a single variable as the best representative for all variables in the factor. A second option for data reduction is to calculate a summated scale, where variables with the highest factor loadings are summated. A single summated score represents the factor, but only selected variables contribute to the composite score. A third option for data reduction is to calculate factor scores for each factor, where each variable contributes to the score based on its factor loading. This single measure is a composite variable that reflects the relative contributions of all variables to the factor. If the summated scale is valid and reliable, it is probably the best of these three data reduction alternatives.

State the major limitations of factor analytic techniques. Three of the most frequently cited limitations are as follows:

1. Because many techniques for performing exploratory factor analyses are available, controversy exists over which technique is the best.
2. The subjective aspects of factor analysis (i.e., deciding how many factors to extract, which technique should be used to rotate the factor axes, which factor loadings are significant) are all subject to many differences in opinion.
3. The problem of reliability is real.

Like any other statistical procedure, a factor analysis starts with a set of imperfect data. When the data vary because of changes in the sample, the data-gathering process, or the numerous kinds of measurement errors, the results of the analysis also may change. The results of any single analysis are therefore less than perfectly dependable.

The potential applications of exploratory factor analysis to problem solving and decision making in business research are numerous. Factor analysis is a much more complex and involved subject than might be indicated here. This problem is especially critical because the results of a single-factor analytic solution frequently look plausible. It is important to emphasize that plausibility is no guarantee of validity or stability.

Questions

1. What are the differences between the objectives of data summarization and data reduction?
2. How can factor analysis help the researcher improve the results of other multivariate techniques?
3. What guidelines can you use to determine the number of factors to extract? Explain each briefly.
4. How do you use the factor-loading matrix to interpret the meaning of factors?
5. How and when should you use factor scores in conjunction with other multivariate statistical techniques?
6. What are the differences between factor scores and summated scales? When is each most appropriate?
7. What is the difference between Q-type factor analysis and cluster analysis?
8. When would the researcher use an oblique rotation instead of an orthogonal rotation? What are the basic differences between them?

Suggested Readings

A list of suggested readings illustrating issues and applications of factor analysis is available on the Web at www.pearsonglobaleditions.com/hair or www.mvstats.com.

References

1. Anderson, J. C., D. W. Gerbing, and J. E. Hunter. 1987. On the Assessment of Unidimensional Measurement: Internal and External Consistency and Overall Consistency Criteria. *Journal of Marketing Research* 24 (November): 432–37.
2. American Psychological Association. 1985. *Standards for Educational and Psychological Tests.* Washington, DC: APA.
3. Bearden, W. O., and R. G. Netemeyer. 1999. *Handbook of Marketing Scales: Multi-Item Measures for Marketing and Consumer Behavior,* 2nd ed. Newbury Park, CA: Sage.
4. Bentler, Peter M. 1995. *EQS Structural Equations Program Manual.* Los Angeles: BMDP Statistical Software.
5. BMDP Statistical Software, Inc. 1992. *BMDP Statistical Software Manual, Release 7, vols. 1 and 2.* Los Angeles: BMDP Statistical Software.
6. Borgatta, E. F., K. Kercher, and D. E. Stull. 1986. A Cautionary Note on the Use of Principal Components Analysis. *Sociological Methods and Research* 15: 160–68.
7. Bruner, G. C., Karen E. James, and P. J. Hensel. 2001. *Marketing Scales Handbook,* Vol. 3, *A Compilation of Multi-Item Measures.* Chicago: American Marketing Association.
8. Campbell, D. T., and D. W. Fiske. 1959. Convergent and Discriminant Validity by the Multitrait-Multimethod Matrix. *Psychological Bulletin* 56 (March): 81–105.
9. Cattell, R. B. 1966. The Scree Test for the Number of Factors. *Multivariate Behavioral Research* 1 (April): 245–76.
10. Cattell, R. B., K. R. Balcar, J. L. Horn, and J. R. Nesselroade. 1969. Factor Matching Procedures: An Improvement of the s Index; with Tables. *Educational and Psychological Measurement* 29: 781–92.
11. Chatterjee, S., L. Jamieson, and F. Wiseman. 1991. Identifying Most Influential Observations in Factor Analysis. *Marketing Science* 10 (Spring): 145–60.
12. Churchill, G. A. 1979. A Paradigm for Developing Better Measures of Marketing Constructs. *Journal of Marketing Research* 16 (February): 64–73.
13. Cliff, N. 1987. *Analyzing Multivariate Data.* San Diego: Harcourt Brace Jovanovich.
14. Cliff, N., and C. D. Hamburger. 1967. The Study of Sampling Errors in Factor Analysis by Means of Artificial Experiments. *Psychological Bulletin* 68: 430–45.
15. Cronbach, L. J. 1951. Coefficient Alpha and the Internal Structure of Tests. *Psychometrika* 31: 93–96.
16. Dillon, W. R., and M. Goldstein. 1984. *Multivariate Analysis: Methods and Applications.* New York: Wiley.
17. Dillon, W. R., N. Mulani, and D. G. Frederick. 1989. On the Use of Component Scores in the Presence of Group Structure. *Journal of Consumer Research* 16: 106–12.
18. Gorsuch, R. L. 1983. *Factor Analysis.* Hillsdale, NJ: Lawrence Erlbaum Associates.
19. Gorsuch, R. L. 1990. Common Factor Analysis Versus Component Analysis: Some Well and Little Known Facts. *Multivariate Behavioral Research* 25: 33–39.
20. Hattie, J. 1985. Methodology Review: Assessing Unidimensionality of Tests and Items. *Applied Psychological Measurement* 9: 139–64.
21. Jöreskog, K. G., and D. Sörbom. 1993. *LISREL 8: Structural Equation Modeling with the SIMPLIS Command Language.* Mooresville, IN: Scientific Software International.
22. Kaiser, H. F. 1970. A Second-Generation Little Jiffy. *Psychometrika* 35: 401–15.

23. Kaiser, H. F. 1974. Little Jiffy, Mark IV. *Educational and Psychology Measurement* 34: 111–17.

24. McDonald, R. P. 1981. The Dimensionality of Tests and Items. *British Journal of Mathematical and Social Psychology* 34: 100–117.

25. Mulaik, S. A. 1990. Blurring the Distinction Between Component Analysis and Common Factor Analysis. *Multivariate Behavioral Research* 25: 53–59.

26. Mulaik, S. A., and R. P. McDonald. 1978. The Effect of Additional Variables on Factor Indeterminacy in Models with a Single Common Factor. *Psychometrika* 43: 177–92.

27. Nunnally, J. L. 1978. *Psychometric Theory,* 2nd ed. New York: McGraw-Hill.

28. Nunnally, J. L. 1979. *Psychometric Theory.* New York: McGraw-Hill.

29. Peter, J. P. 1979. Reliability: A Review of Psychometric Basics and Recent Marketing Practices. *Journal of Marketing Research* 16 (February): 6–17.

30. Peter, J. P. 1981. Construct Validity: A Review of Basic Issues and Marketing Practices. *Journal of Marketing Research* 18 (May): 133–45.

31. Robinson, J. P., P. R. Shaver, and L. S. Wrightsman. 1991. Criteria for Scale Selection and Evaluation. In *Measures of Personality and Social Psychological Attitudes,* J. P. Robinson, P. R. Shaver, and L. S. Wrightsman (eds.). San Diego, CA: Academic Press.

32. Robinson, J. P., P. R. Shaver, and L. S. Wrightman. 1991. *Measures of Personality and Social Psychological Attitudes.* San Diego: Academic Press.

33. Rummel, R. J. 1970. *Applied Factor Analysis.* Evanston, IL: Northwestern University Press.

34. Smith, Scott M. 1989. *PC-MDS: A Multidimensional Statistics Package.* Provo, UT: Brigham Young University Press.

35. Snook, S. C., and R. L. Gorsuch. 1989. Principal Component Analysis Versus Common Factor Analysis: A Monte Carlo Study. *Psychological Bulletin* 106: 148–54.

36. Stewart, D. W. 1981. The Application and Misapplication of Factor Analysis in Marketing Research. *Journal of Marketing Research* 18 (February): 51–62.

37. Velicer, W. F., and D. N. Jackson. 1990. Component Analysis Versus Common Factor Analysis: Some Issues in Selecting an Appropriate Procedure. *Multivariate Behavioral Research* 25: 1–28.

Analysis Using Dependence Techniques

OVERVIEW

Whereas Section I focused on the preparation of data for multivariate analysis, Section II deals with what many would term the essence of multivariate analysis: dependence techniques. As noted in Chapter 1, dependence techniques are based on the use of a set of independent variables to predict and explain one or more dependent variables. The researcher, whether faced with dependent variables of a metric or nonmetric nature, has a variety of dependence methods available to assist in the process of relating independent variables to dependent variables. Given the multivariate nature of these methods, all of the dependence techniques accommodate multiple independent variables while also allowing multiple dependent variables in certain situations. Thus, the researcher has a set of techniques that should allow for the analysis of almost any type of research question involving a dependence relationship. They also provide the opportunity not only for increased prediction capability, but also for enhanced explanation of the dependent variable's relationship to the independent variables. Explanation becomes increasingly important as the research questions begin to address issues concerning how the relationship between independent and dependent variables operates.

CHAPTERS IN SECTION II

Section II covers five dependence techniques: multiple regression, discriminant analysis, logistic regression, multivariate analysis of variance, and conjoint analysis, in Chapters 4 through 8, respectively. Dependence techniques, as noted earlier, enable the researcher to assess the degree of relationship between the dependent and independent variables. Dependence techniques vary in the type and character of the relationship, as reflected in the measurement properties of the dependent and independent variables discussed in Chapter 1. For example, multiple regression and discriminant analysis accommodate multiple metric independent variables but vary in the type of dependent variable (regression analysis—single metric and discriminant analysis—single nonmetric). Chapter 4, Multiple Regression Analysis, focuses on what is perhaps the most fundamental of all multivariate techniques and a building block for our discussion of the other dependence methods. Whether assessing the conformity to underlying statistical assumptions, measuring predictive accuracy, or interpreting the variate of independent variables, the issues discussed in Chapter 4 will be seen as crucial in many of the other techniques as well. Chapter 5, Canonical Correlation, addresses the most generalized form of multivariate analysis, which accommodates multiple dependent and independent variables. In situations in which variates exist for both the dependent and independent variables, canonical correlation provides a flexible method for both prediction and explanation. Chapter 6, Conjoint Analysis, presents us with a technique unlike any of the other multivariate methods in that the researcher determines the values of the independent nonmetric variables in a quasi-experimental fashion. Once designed, the respondent provides information regarding only the dependent variable. Although it places more responsibility on the researcher, conjoint analysis provides a powerful tool for understanding complex decision processes. Chapter 7, Multiple Discriminant Analysis and Logistic Regression, investigates a unique form of dependence relationship—a dependent variable that is not metric but rather is nonmetric. In this situation, the researcher is attempting to classify observations into groups. This classification into groups can be accomplished through either discriminant analysis or logistic regression, a variant of regression designed to specifically deal with nonmetric dependent variables. In Chapter 8, Multivariate Analysis of Variance, the discussion differs in several ways from the prior techniques; it is suited to the analysis of multiple metric dependent variables and nonmetric independent variables. Although this technique is a direct extension of simple analysis of variance, the multiple metric dependent variables make both prediction and explanation more difficult.

This section provides the researcher with exposure to the wide range of dependence techniques available, each suited to a specific task and relationship. When you complete this section, the issues regarding selecting from these methods should be apparent, and you should feel comfortable in selecting from these techniques and analyzing their results.

CHAPTER 4

Simple and Multiple Regression

LEARNING OBJECTIVES

Upon completing this chapter, you should be able to do the following:

- Determine when regression analysis is the appropriate statistical tool in analyzing a problem.
- Understand how regression helps us make predictions using the least squares concept.
- Use dummy variables with an understanding of their interpretation.
- Be aware of the assumptions underlying regression analysis and how to assess them.
- Select an estimation technique and explain the difference between stepwise and simultaneous regression.
- Interpret the results of regression.
- Apply the diagnostic procedures necessary to assess influential observations.

CHAPTER PREVIEW

This chapter describes multiple regression analysis as it is used to solve important research problems, particularly in business. Regression analysis is by far the most widely used and versatile dependence technique, applicable in every facet of business decision making. Its uses range from the most general problems to the most specific, in each instance relating a factor (or factors) to a specific outcome. For example, regression analysis is the foundation for business forecasting models, ranging from the econometric models that predict the national economy based on certain inputs (income levels, business investment, etc.) to models of a firm's performance in a market if a specific marketing strategy is followed. Regression models are also used to study how consumers make decisions or form impressions and attitudes. Other applications include evaluating the determinants of effectiveness for a program (e.g., what factors aid in maintaining quality) and determining the feasibility of a new product or the expected return for a new stock issue. Even though these examples illustrate only a small subset of all applications, they demonstrate that regression analysis is a powerful analytical tool designed to explore all types of dependence relationships.

Multiple regression analysis is a general statistical technique used to analyze the relationship between a single dependent variable and several independent variables. As noted in Chapter 1, its basic formulation is

$$Y_1 = X_1 + X_2 + \cdots + X_n$$
$$\text{(metric)} \qquad \text{(metric)}$$

This chapter presents guidelines for judging the appropriateness of multiple regression for various types of problems. Suggestions are provided for interpreting the results of its application from a managerial as well as a statistical viewpoint. Possible transformations of the data to remedy violations of various model assumptions are examined, along with a series of diagnostic procedures that identify observations with particular influence on the results. Readers who are already knowledgeable about multiple regression procedures can skim the early portions of the chapter, but for those who are less familiar with the subject, this chapter provides a valuable background for the study of multivariate data analysis.

KEY TERMS

Before beginning this chapter, review the key terms to develop an understanding of the concepts and terminology used. Throughout the chapter the key terms appear in **boldface.** Other points of emphasis in the chapter and key term cross-references are italicized.

Adjusted coefficient of determination (adjusted R^2)   Modified measure of the *coefficient of determination* that takes into account the number of independent variables included in the regression equation and the sample size. Although the addition of independent variables will always cause the coefficient of determination to rise, the adjusted coefficient of determination may fall if the added independent variables have little explanatory power or if the *degrees of freedom* become too small. This statistic is quite useful for comparison between equations with different numbers of independent variables, differing sample sizes, or both.

All-possible-subsets regression   Method of selecting the variables for inclusion in the regression model that considers all possible combinations of the independent variables. For example, if the researcher specifies four potential independent variables, this technique would estimate all possible regression models with one, two, three, and four variables. The technique would then identify the model(s) with the best predictive accuracy.

Backward elimination   Method of selecting variables for inclusion in the regression model that starts by including all independent variables in the model and then eliminating those variables not making a significant contribution to prediction.

Beta coefficient   Standardized regression coefficient (see *standardization*) that allows for a direct comparison between coefficients as to their relative explanatory power of the dependent variable. Whereas *regression coefficients* are expressed in terms of the units of the associated variable, thereby making comparisons inappropriate, beta coefficients use standardized data and can be directly compared.

Coefficient of determination (R^2)   Measure of the proportion of the variance of the dependent variable about its mean that is explained by the independent, or predictor, variables. The coefficient can vary between 0 and 1. If the regression model is properly applied and estimated, the researcher can assume that the higher the value of R^2, the greater the explanatory power of the regression equation, and therefore the better the prediction of the dependent variable.

Collinearity   Expression of the relationship between two (collinearity) or more (multicollinearity) independent variables. Two independent variables are said to exhibit complete collinearity if their correlation coefficient is 1, and complete lack of collinearity if their correlation coefficient is 0. *Multicollinearity* occurs when any single independent variable is highly correlated with a set of other independent variables. An extreme case of collinearity/multicollinearity is *singularity,* in which an independent variable is perfectly predicted (i.e., correlation of 1.0) by another independent variable (or more than one).

Correlation coefficient (r)   Coefficient that indicates the strength of the association between any two metric variables. The sign ($+$ or $-$) indicates the direction of the relationship. The value can range from $+1$ to -1, with $+1$ indicating a perfect positive relationship, 0 indicating no relationship, and −1 indicating a perfect negative or reverse relationship (as one variable grows larger, the other variable grows smaller).

Criterion variable (Y)   See *dependent variable.*

Degrees of freedom (df)   Value calculated from the total number of observations minus the number of estimated *parameters.* These parameter estimates are restrictions on the data because, once made, they define the population from which the data are assumed to have been drawn. For example, in estimating a regression model with a single independent variable, we estimate two parameters, the *intercept* (b_0) and a *regression coefficient* for the independent variable (b_1). In estimating the random error, defined as the sum of the *prediction errors* (actual minus predicted dependent values) for all cases, we would find ($n − 2$) degrees of freedom. Degrees of freedom provide a measure of how restricted the data are to reach a certain level of prediction. If the number of degrees of freedom is small, the resulting prediction may be less generalizable because all but a few observations were incorporated in the prediction. Conversely, a large degrees-of-freedom value indicates the prediction is fairly robust with regard to being representative of the overall sample of respondents.

Dependent variable (Y)   Variable being predicted or explained by the set of independent variables.

Dummy variable   Independent variable used to account for the effect that different levels of a nonmetric variable have in predicting the dependent variable. To account for L levels of a non-metric independent variable, $L − 1$ dummy variables are needed. For example, gender is measured as male or female and could be represented by two dummy variables, X_1 and X_2. When the respondent is male, $X_1 = 1$ and $X_2 = 0$. Likewise, when the respondent is female, $X_1 = 0$ and $X_2 = 1$. However, when $X_1 = 1$, we know that X_2 must equal 0. Thus, we need only one variable, either X_1 or X_2, to represent gender. We need not include both variables because one is perfectly predicted by the other (a *singularity*) and the *regression coefficients* cannot be estimated. If a variable has three levels, only two dummy variables are needed. Thus, the number of dummy variables is one less than the number of levels of the nonmetric variable. The two most common methods of determining the values of the dummy values are *indicator coding* and *effects coding.*

Effects coding   Method for specifying the *reference category* for a set of *dummy variables* in which the reference category receives a value of −1 across the set of dummy variables. In our example of dummy variable coding for gender, we coded the dummy variable as either 1 or 0. But with effects coding, the value of −1 is used instead of 0. With this type of coding, the coefficients for the dummy variables become group deviations on the dependent variable from the mean of the dependent variable across all groups. Effects coding contrasts with *indicator coding,* in which the reference category is given the value of zero across all dummy variables and the coefficients represent group deviations on the dependent variable from the reference group.

Forward addition   Method of selecting variables for inclusion in the regression model by starting with no variables in the model and then adding one variable at a time based on its contribution to prediction.

Heteroscedasticity   See *homoscedasticity.*

Homoscedasticity   Description of data for which the variance of the error terms (e) appears constant over the range of values of an independent variable. The assumption of equal variance of the population error ε (where ε is estimated from the sample value e) is critical to the proper application of linear regression. When the error terms have increasing or modulating variance, the data are said to be *heteroscedastic.* The discussion of *residuals* in this chapter further illustrates this point.

Independent variable   Variable(s) selected as predictors and potential explanatory variables of the dependent variable.

Indicator coding   Method for specifying the *reference category* for a set of *dummy variables* where the reference category receives a value of 0 across the set of dummy variables. The *regression coefficients* represent the group differences in the dependent variable from the reference category. Indicator coding differs from *effects coding,* in which the reference category is given the value of −1 across all dummy variables and the regression coefficients represent group deviations on the dependent variable from the overall mean of the dependent variable.

Influential observation   An observation that has a disproportionate influence on one or more aspects of the regression estimates. This influence may be based on extreme values of the independent or dependent variables, or both. Influential observations can either be "good," by reinforcing the pattern of the remaining data, or "bad," when a single or small set of cases unduly affects the regression estimates. It is not necessary for the observation to be an *outlier,* although many times outliers can be classified as influential observations as well.

Intercept (b_0)   Value on the Y axis (dependent variable axis) where the line defined by the regression equation $Y = b_0 + b_1X_1$ crosses the axis. It is described by the constant term b_0 in the regression equation. In addition to its role in prediction, the intercept may have a managerial interpretation. If the complete absence of the independent variable has meaning, then the intercept represents that amount. For example, when estimating sales from past advertising expenditures, the intercept represents the level of sales expected if advertising is eliminated. But in many instances the constant has only predictive value because in no situation are all independent variables absent. An example is predicting product preference based on consumer attitudes. All individuals have some level of attitude, so the intercept has no managerial use, but it still aids in prediction.

Least squares   Estimation procedure used in simple and multiple regression whereby the *regression coefficients* are estimated so as to minimize the total sum of the squared *residuals.*

Leverage points   Type of *influential observation* defined by one aspect of influence termed *leverage.* These observations are substantially different on one or more independent variables, so that they affect the estimation of one or more *regression coefficients.*

Linearity   Term used to express the concept that the model possesses the properties of additivity and homogeneity. In a simple sense, linear models predict values that fall in a straight line by having a constant unit change (slope) of the dependent variable for a constant unit change of the *independent variable.* In the population model $Y = b_0 + b_1X_1 + \varepsilon$, the effect of changing X_1 by a value of 1.0 is to add b_1 (a constant) units of Y.

Measurement error   Degree to which the data values do not truly measure the characteristic being represented by the variable. For example, when asking about total family income, many sources of measurement error (e.g., reluctance to answer full amount, error in estimating total income) make the data values imprecise.

Moderator effect   Effect in which a third independent variable (the moderator variable) causes the relationship between a dependent/independent variable pair to change, depending on the value of the moderator variable. It is also known as an interactive effect and is similar to the interaction effect seen in analysis of variance methods.

Multicollinearity   See *collinearity.*

Multiple regression   Regression model with two or more independent variables.

Normal probability plot   Graphical comparison of the shape of the sample distribution to the normal distribution. In the graph, the normal distribution is represented by a straight line angled at 45 degrees. The actual distribution is plotted against this line, so any differences are shown as deviations from the straight line, making identification of differences quite simple.

Null plot   Plot of *residuals* versus the predicted values that exhibits a random pattern. A null plot is indicative of no identifiable violations of the assumptions underlying regression analysis.

Outlier   In strict terms, an observation that has a substantial difference between the actual value for the dependent variable and the predicted value. Cases that are substantially different with regard to either the dependent or independent variables are often termed *outliers* as well. In all instances, the objective is to identify observations that are inappropriate representations of the population from which the sample is drawn, so that they may be discounted or even eliminated from the analysis as unrepresentative.

Parameter   Quantity (measure) characteristic of the population. For example, μ and σ^2 are the symbols used for the population parameters mean (μ) and variance (σ^2). They are typically estimated from sample data in which the arithmetic average of the sample is used as a measure of the population average and the variance of the sample is used to estimate the variance of the population.

Part correlation   Value that measures the strength of the relationship between a dependent and a single independent variable when the predictive effects of the other independent variables in the regression model are removed. The objective is to portray the unique predictive effect due to a single independent variable among a set of independent variables. Differs from the *partial correlation coefficient,* which is concerned with incremental predictive effect.

Partial correlation coefficient   Value that measures the strength of the relationship between the criterion or dependent variable and a single independent variable when the effects of the other independent variables in the model are held constant. For example, rY, X_2, X_1 measures the variation in Y associated with X_2 when the effect of X_1 on both X_2 and Y is held constant. This value is used in sequential variable selection methods of regression model estimation (e.g., *stepwise, forward addition,* or *backward elimination*) to identify the independent variable with the greatest incremental predictive power beyond the independent variables already in the regression model.

Partial F (or t) values   The partial F-test is simply a statistical test for the additional contribution to prediction accuracy of a variable above that of the variables already in the equation. When a variable (X_a) is added to a regression equation after other variables are already in the equation, its contribution may be small even though it has a high correlation with the dependent variable. The reason is that X_a is highly correlated with the variables already in the equation. The partial F value is calculated for all variables by simply pretending that each, in turn, is the last to enter the equation. It gives the additional contribution of each variable above all others in the equation. A low or insignificant partial F value for a variable not in the equation indicates its low or insignificant contribution to the model as already specified. A t value may be calculated instead of F values in all instances, with the t value being approximately the square root of the F value.

Partial regression plot   Graphical representation of the relationship between the dependent variable and a single independent variable. The scatterplot of points depicts the partial correlation between the two variables, with the effects of other independent variables held constant (see *partial correlation coefficient*). This portrayal is particularly helpful in assessing the form of the relationship (linear versus nonlinear) and the identification of *influential observations.*

Polynomial   Transformation of an independent variable to represent a curvilinear relationship with the dependent variable. By including a squared term (X^2), a single inflection point is estimated. A cubic term estimates a second inflection point. Additional terms of a higher *power* can also be estimated.

Power   Probability that a significant relationship will be found if it actually exists. Complements the more widely used *significance level alpha* (α).

Prediction error   Difference between the actual and predicted values of the dependent variable for each observation in the sample (see *residual*).

Predictor variable (X_n)   See *independent variable.*

PRESS statistic   Validation measure obtained by eliminating each observation one at a time and predicting this dependent value with the regression model estimated from the remaining observations.

Reference category   The omitted level of a nonmetric variable when a *dummy variable* is formed from the nonmetric variable.

Regression coefficient (b_n)   Numerical value of the parameter estimate directly associated with an independent variable; for example, in the model $Y = b_0 + b_1X_1$ the value b_1 is the regression coefficient for the variable X_1. The regression coefficient represents the amount of change in the dependent variable for a one-unit change in the independent variable. In the multiple predictor model (e.g., $Y = b_0 + b_1X_1 + b_2X_2$), the regression coefficients are partial coefficients because each takes into account not only the relationships between Y and X_1 and between Y and X_2, but also between X_1 and X_2. The coefficient is not limited in range, because it is based on both the degree of association and the scale units of the independent variable. For instance, two variables with the same association to Y would have different coefficients if one independent variable was measured on a 7-point scale and another was based on a 100-point scale.

Regression variate   Linear combination of weighted independent variables used collectively to predict the dependent variable.

Residual (*e* or ε)   Error in predicting our sample data. Seldom will our predictions be perfect. We assume that random error will occur, but we assume that this error is an estimate of the true random error in the population (ε), not just the error in prediction for our sample (*e*). We assume that the error in the population we are estimating is distributed with a mean of 0 and a constant (*homoscedastic*) variance.

Sampling error   The expected variation in any estimated parameter (*intercept* or *regression coefficient*) that is due to the use of a sample rather than the population. Sampling error is reduced as the sample size is increased and is used to statistically test whether the estimated parameter differs from zero.

Significance level (alpha)   Commonly referred to as the level of statistical significance, the significance level represents the probability the researcher is willing to accept that the estimated coefficient is classified as different from zero when it actually is not. This is also known as Type I error. The most widely used level of significance is .05, although researchers use levels ranging from .01 (more demanding) to .10 (less conservative and easier to find significance).

Simple regression   Regression model with a single independent variable, also known as bivariate regression.

Singularity   The extreme case of *collinearity* or *multicollinearity* in which an independent variable is perfectly predicted (a correlation of ±1.0) by one or more independent variables. Regression models cannot be estimated when a singularity exists. The researcher must omit one or more of the independent variables involved to remove the singularity.

Specification error   Error in predicting the dependent variable caused by excluding one or more relevant independent variables. This omission can bias the estimated coefficients of the included variables as well as decrease the overall predictive power of the regression model.

Standard error   Expected distribution of an estimated regression coefficient. The standard error is similar to the standard deviation of any set of data values, but instead denotes the expected range of the coefficient across multiple samples of the data. It is useful in statistical tests of significance that test to see whether the coefficient is significantly different from zero (i.e., whether the expected range of the coefficient contains the value of zero at a given level of confidence). The *t* value of a *regression coefficient* is the coefficient divided by its standard error.

Standard error of the estimate (SE$_E$)   Measure of the variation in the predicted values that can be used to develop confidence intervals around any predicted value. It is similar to the standard deviation of a variable around its mean, but instead is the expected distribution of predicted values that would occur if multiple samples of the data were taken.

Standardization   Process whereby the original variable is transformed into a new variable with a mean of 0 and a standard deviation of 1. The typical procedure is to first subtract the variable mean from each observation's value and then divide by the standard deviation. When all the variables in a *regression variate* are standardized, the b_0 term (the *intercept*) assumes a value of 0 and the *regression coefficients* are known as *beta coefficients,* which enable the researcher to compare directly the relative effect of each independent variable on the dependent variable.

Statistical relationship   Relationship based on the correlation of one or more independent variables with the dependent variable. Measures of association, typically correlations, represent the degree of relationship because there is more than one value of the dependent variable for each value of the independent variable.

Stepwise estimation   Method of selecting variables for inclusion in the regression model that starts by selecting the best predictor of the dependent variable. Additional independent variables are selected in terms of the incremental explanatory power they can add to the regression model. Independent variables are added as long as their partial *correlation coefficients* are statistically significant. Independent variables may also be dropped if their predictive power drops to a non-significant level when another independent variable is added to the model.

Studentized residual   The most commonly used form of standardized *residual*. It differs from other methods in how it calculates the standard deviation used in *standardization*. To minimize the effect of any observation on the standardization process, the standard deviation of the residual for observation i is computed from regression estimates omitting the ith observation in the calculation of the regression estimates.

Sum of squared errors (SS_E)   Sum of the squared prediction errors (*residuals*) across all observations. It is used to denote the variance in the dependent variable not yet accounted for by the regression model. If no independent variables are used for prediction, it becomes the squared errors using the mean as the predicted value and thus equals the *total sum of squares.*

Sum of squares regression (SS_R)   Sum of the squared differences between the mean and predicted values of the dependent variable for all observations. It represents the amount of improvement in explanation of the dependent variable attributable to the independent variable(s).

Suppression effect   The instance in which the expected relationships between independent and dependent variables are hidden or suppressed when viewed in a bivariate relationship. When additional independent variables are entered, the *multicollinearity* removes "unwanted" shared variance and reveals the "true" relationship.

Tolerance   Commonly used measure of *collinearity* and *multicollinearity*. The tolerance of variable i (TOL_i) is $1 - R^{2*}i$, where $R^{2*}i$ is the coefficient of determination for the prediction of variable i by the other independent variables in the *regression variate*. As the tolerance value grows smaller, the variable is more highly predicted by the other independent variables (collinearity).

Total sum of squares (SS_T)   Total amount of variation that exists to be explained by the independent variables. This baseline value is calculated by summing the squared differences between the mean and actual values for the dependent variable across all observations.

Transformation   A variable may have an undesirable characteristic, such as nonnormality, that detracts from the ability of the *correlation coefficient* to represent the relationship between it and another variable. A transformation, such as taking the logarithm or square root of the variable, creates a new variable and eliminates the undesirable characteristic, allowing for a better measure of the relationship. Transformations may be applied to either the dependent or independent variables, or both. The need and specific type of transformation may be based on theoretical reasons (such as transforming a known nonlinear relationship) or empirical reasons (identified through graphical or statistical means).

Variance inflation factor (VIF)   Indicator of the effect that the other independent variables have on the standard error of a *regression coefficient*. The variance inflation factor is directly related to the *tolerance* value ($VIF_i = 1/TOL_i$). Large VIF values also indicate a high degree of *collinearity* or *multicollinearity* among the independent variables.

WHAT IS MULTIPLE REGRESSION ANALYSIS?

Multiple regression analysis is a statistical technique that can be used to analyze the relationship between a single **dependent (criterion) variable** and several **independent (predictor) variables.** The objective of multiple regression analysis is to use the independent variables whose values are known to predict the single dependent value selected by the researcher. Each independent variable is weighted by the regression analysis procedure to ensure maximal prediction from the set of independent variables. The weights denote the relative contribution of the independent variables to the overall prediction and facilitate interpretation as to the influence of each variable in making the prediction, although correlation among the independent variables complicates the interpretative process. The set of weighted independent variables forms the **regression variate,** a linear combination of the independent variables that best predicts the dependent variable (Chapter 1 contains a more detailed explanation of the variate). The regression variate, also referred to as the *regression equation* or *regression model*, is the most widely known example of a variate among the multivariate techniques.

As noted in Chapter 1, multiple regression analysis is a dependence technique. Thus, to use it you must be able to divide the variables into dependent and independent variables. Regression analysis is also a statistical tool that should be used only when both the dependent and independent variables are metric. However, under certain circumstances it is possible to include nonmetric data either as independent variables (by transforming either ordinal or nominal data with dummy variable coding) or the dependent variable (by the use of a binary measure in the specialized technique of logistic regression, see Chapter 7). In summary, to apply multiple regression analysis: (1) the data must be metric or appropriately transformed, and (2) before deriving the regression equation, the researcher must decide which variable is to be dependent and which remaining variables will be independent.

AN EXAMPLE OF SIMPLE AND MULTIPLE REGRESSION

The objective of regression analysis is to predict a single dependent variable from the knowledge of one or more independent variables. When the problem involves a single independent variable, the statistical technique is called **simple regression.** When the problem involves two or more independent variables, it is termed **multiple regression.** The following example will demonstrate the application of both simple and multiple regression. Readers interested in a more detailed discussion of the underlying foundations and basic elements of regression analysis are referred to the Basic Stats appendix on the text's Web sites (www.pearsonglobaleditions.com/hair or www.mvstats.com.).

To illustrate the basic principles involved, results from a small study of eight families regarding their credit card usage are provided. The purpose of the study was to determine which factors affected the number of credit cards used. Three potential factors were identified (family size, family income, and number of automobiles owned), and data were collected from each of the eight families (see Table 4-1). In the terminology of regression analysis, the dependent variable (Y) is the number of credit cards used and the three independent variables (V_1, V_2, and V_3) are family size, family income, and number of automobiles owned, respectively.

Prediction Using a Single Independent Variable: Simple Regression

As researchers, a starting point in any regression analysis is identifying the single independent variable that achieves the best prediction of the dependent measure. Based on the concept of minimizing the sum

TABLE 4-1   Credit Card Usage Survey Results

Family ID	Number of Credit Cards Used (Y)	Family Size (V_1)	Family Income ($000) ($V_2$)	Number of Automobiles Owned (V_3)
1	4	2	14	1
2	6	2	16	2
3	6	4	14	2
4	7	4	17	1
5	8	5	18	3
6	7	5	21	2
7	8	6	17	1
8	10	6	25	2

of squared errors of prediction (see the Basic Stats appendix on the text's Web sites [www.pearson globaleditions.com/hair or www.mvstats.com.] for more detail), we can select the "best" independent variable based on the **correlation coefficient,** because the higher the correlation coefficient, the stronger the relationship and the greater the predictive accuracy. In the regression equation, we represent the **intercept** as b_0. The amount of change in the dependent variable due to the independent variable is represented by the term b_1, also known as a **regression coefficient.** Using a mathematical procedure known as **least squares** [8, 11, 15], we can estimate the values of b_0 and b_1 such that the **sum of squared errors** (**SS**$_E$) of prediction is minimized. The **prediction error,** the difference between the actual and predicted values of the dependent variable, is termed the **residual** (*e* or ε).

Table 4-2 contains a correlation matrix depicting the association between the dependent (Y) variable and independent (V_1, V_2, or V_3) variables that can be used in selecting the best independent variable. Looking down the first column, we can see that V_1, family size, has the highest correlation with the dependent variable and is thus the best candidate for our first simple regression. The correlation matrix also contains the correlations among the independent variables, which we will see is important in multiple regression (two or more independent variables).

We can now estimate our first simple regression model for the sample of eight families and see how well the description fits our data. The regression model can be stated as follows:

$$\text{Predicted number of} = \text{Intercept} + \text{Change in number of credit} \times \text{Family size}$$
credit cards used   cards used associated with
                                     a unit change in family size

or

$$\hat{Y} = b_0 + b_1 V_1$$

For this example, the appropriate values are a constant (b_0) of 2.87 and a regression coefficient (b_1) of .97 for family size.

INTERPRETING THE SIMPLE REGRESSION MODEL With the intercept and regression coefficient estimated by the least squares procedure, attention now turns to interpretation of these two values:

- *Regression coefficient.* The estimated change in the dependent variable for a unit change of the independent variable. If the regression coefficient is found to be statistically significant (i.e., the coefficient is significantly different from zero), the value of the regression coefficient indicates the extent to which the independent variable is associated with the dependent variable.
- *Intercept.* Interpretation of the intercept is somewhat different. The intercept has explanatory value only within the range of values for the independent variable(s). Moreover, its interpretation is based on the characteristics of the independent variable:
  - In simple terms, the intercept has interpretive value only if zero is a conceptually valid value for the independent variable (i.e., the independent variable can have a value of zero

TABLE 4-2 **Correlation Matrix for the Credit Card Usage Study**

Variable	Y	V_1	V_2	V_3
Y Number of Credit Cards Used	1.000			
V_1 Family Size	.866	1.000		
V_2 Family Income	.829	.673	1.000	
V_3 Number of Automobiles	.342	.192	.301	1.000

and still maintain its practical relevance). For example, assume that the independent variable is advertising dollars. If it is realistic that, in some situations, no advertising is done, then the intercept will represent the value of the dependent variable when advertising is zero.

- If the independent value represents a measure that never can have a true value of zero (e.g., attitudes or perceptions), the intercept aids in improving the prediction process, but has no explanatory value.

For some special situations where the specific relationship is known to pass through the origin, the intercept term may be suppressed (called *regression through the origin*). In these cases, the interpretation of the residuals and the regression coefficients changes slightly.

Our regression model predicting credit card holdings indicates that for each additional family member, the credit card holdings are higher on average by .97. The constant 2.87 can be interpreted only within the range of values for the independent variable. In this case, a family size of zero is not possible, so the intercept alone does not have practical meaning. However, this impossibility does not invalidate its use, because it aids in the prediction of credit card usage for each possible family size (in our example from 1 to 5). The simple regression equation and the resulting predictions and residuals for each of the eight families are shown in Table 4-3.

Because we used only a sample of observations for estimating a regression equation, we can expect that the regression coefficients will vary if we select another sample of observations and estimate another regression equation. We do not want to take repeated samples, so we need an empirical test to see whether the regression coefficient we estimated has any real value (i.e., is it different from zero?) or could we possibly expect it to equal zero in another sample. To address this issue, regression analysis allows for the statistical testing of the intercept and regression coefficient(s) to determine whether they are significantly different from zero (i.e., they do have an impact that we can expect with a specified probability to be different from zero across any number of samples of observations). Later in the chapter, we will discuss in more detail the concept of significance testing for specific coefficients.

ASSESSING PREDICTION ACCURACY   The most commonly used measure of predictive accuracy for the regression model is the **coefficient of determination (R^2).** Calculated as the squared correlation between the actual and predicted values of the dependent variable, it represents the combined effects of the entire variate (one or more independent variables plus the intercept) in predicting the

TABLE 4-3   Simple Regression Results Using Family Size as the Independent Variable

Regression Variate: $Y = b_0 + b_1V_1$
Prediction Equation: $Y = 2.87 + .97V_1$

Family ID	Number of Credit Cards Used	Family Size (V_1)	Simple Regression Prediction	Prediction Error	Prediction Error Squared
1	4	2	4.81	−.81	.66
2	6	2	4.81	1.19	1.42
3	6	4	6.75	−.75	.56
4	7	4	6.75	.25	.06
5	8	5	7.72	.28	.08
6	7	5	7.72	−.72	.52
7	8	6	8.69	−.69	.48
8	10	6	8.69	1.31	1.72
Total					5.50

dependent variable. It ranges from 1.0 (perfect prediction) to 0.0 (no prediction). Because it is the squared correlation of actual and predicted values, it also represents the amount of variance in the dependent variable explained by the independent variable(s).

Another measure of predictive accuracy is the expected variation in the predicted values, termed the **standard error of the estimate (SE_E).** Defined simply as the standard deviation of the predicted values, it allows the researcher to understand the confidence interval that can be expected for any prediction from the regression model. Obviously smaller confidence intervals denote greater predictive accuracy.

The interested reader is referred again to the Basic Stats appendix on the text's Web sites (www.pearsonglobaleditions.com/hair or www.mvstats.com.) where all of these basic concepts are described in more detail and calculations are provided using the credit card example.

In our example, the simple regression model has a total prediction error of 5.5 (see Table 4-3), meaning that it accounted for 16.5 (22.0 − 5.5 = 16.5) of the total prediction error of 22.0. Because the coefficient of determination is the amount of variation accounted for by the regression model, the simple regression model with one independent variable has a R^2 of 75 percent (16.5 / 22.0 = .75).

We can also calculate the standard error of the estimate (SE_E) as .957, giving a 95% confidence interval of 2.34, which was obtained by multiplying the SE_E by the t-value, in this case 2.477 (see the Basic Stats appendix on the text's Web sites [www.pearsonglobaleditions.com/hair or www.mvstats.com.] for a description of these calculations).

Both of these measures of predictive accuracy can now be used to assess not only this simple regression model, but also the improvement made when more independent variables are added in a multiple regression model.

Prediction Using Several Independent Variables: Multiple Regression

We previously demonstrated how simple regression can help improve our prediction of a dependent variable (e.g., by using data on family size, we predicted the number of credit cards a family would use much more accurately than we could by simply using the arithmetic average). This result raises the question of whether we could improve our prediction even further by using additional independent variables (e.g., other data obtained from the families). Would our prediction be improved if we used not only data on family size, but data on another variable, perhaps family income or number of automobiles owned by the family?

THE IMPACT OF MULTICOLLINEARITY   The ability of an additional independent variable to improve the prediction of the dependent variable is related not only to its correlation to the dependent variable, but also to the correlation(s) of the additional independent variable to the independent variable(s) already in the regression equation. **Collinearity** is the association, measured as the correlation, between two independent variables. **Multicollinearity** refers to the correlation among three or more independent variables (evidenced when one is regressed against the others). Although a precise distinction separates these two concepts in statistical terms, it is rather common practice to use the terms interchangeably.

As might be expected, correlation among the independent variables can have a marked impact on the regression model:

- *The impact of multicollinearity is to reduce any single independent variable's predictive power by the extent to which it is associated with the other independent variables.* As collinearity increases, the unique variance explained by each independent variable decreases and the shared prediction percentage rises. Because this shared prediction can count only once, the overall prediction increases much more slowly as independent variables with high multicollinearity are added.

> • *To maximize the prediction from a given number of independent variables, the researcher should look for independent variables that have low multicollinearity with the other independent variables but also have high correlations with the dependent variable.*

We revisit the issues of collinearity and multicollinearity in later sections to discuss their implications for the selection of independent variables and the interpretation of the regression variate.

THE MULTIPLE REGRESSION EQUATION   As noted earlier, multiple regression is the use of two or more independent variables in the prediction of the dependent variable. *The task for the researcher is to expand upon the simple regression model by adding independent variable(s) that have the greatest additional predictive power.* Even though we can determine any independent variable's association with the dependent variable through the correlation coefficient, the extent of the incremental predictive power for any additional variable is many times as much determined by its multicollinearity with other variables already in the regression equation. We can look to our credit card example to demonstrate these concepts.

To improve further our prediction of credit card holdings, let us use additional data obtained from our eight families. The second independent variable to include in the regression model is family income (V_2), which has the next highest correlation with the dependent variable. Although V_2 does have a fair degree of correlation with V_1 already in the equation, it is still the next best variable to enter because V_3 has a much lower correlation with the dependent variable. We simply expand our simple regression model to include two independent variables as follows:

$$\text{Predicted number of credit cards used} = b_0 + b_1 V_1 + b_2 V_2 + e$$

where

b_0 = constant number of credit cards independent of family size and income
b_1 = change in credit card usage associated with unit change in family size
b_2 = change in credit card usage associated with unit change in family income
V_1 = family size
V_2 = family income
e = prediction error (residual)

The multiple regression model with two independent variables, when estimated with the least squares procedure, provides a constant of .482 with regression coefficients of .63 and .216 for V_1 and V_2, respectively. We can again find our residuals by predicting Y and subtracting the prediction from the actual value. We then square the resulting prediction error, as in Table 4-4. The sum of squared errors for the multiple regression model with family size and family income is 3.04. This result can be compared to the simple regression model value of 5.50 (Table 4-3), which uses only family size for prediction.

When family income is added to the regression analysis, R^2 also increases to .86.

$$R^2_{(\text{family size + family income})} = \frac{22.0 - 3.04}{22.0} = \frac{18.96}{22.0} = .86$$

The inclusion of family income in the regression analysis increases the prediction by 11 percent (.86 − .75), all due to the unique incremental predictive power of family income.

ADDING A THIRD INDEPENDENT VARIABLE   We have seen an increase in prediction accuracy gained in moving from the simple to multiple regression equation, but we must also note that at

TABLE 4-4   Multiple Regression Results Using Family Size and Family Income as Independent Variables

Regression Variate: $Y = b_0 + b_1V_1 + b_2V_2$
Prediction Equation: $Y = .482 + .63V_1 + .216V_2$

Family ID	Number of Credit Cards Used	Family Size (V_1)	Family Income (V_2)	Multiple Regression Prediction	Prediction Error	Prediction Error Squared
1	4	2	14	4.76	−.76	.58
2	6	2	16	5.20	.80	.64
3	6	4	14	6.03	−.03	.00
4	7	4	17	6.68	.32	.10
5	8	5	18	7.53	.47	.22
6	7	5	21	8.18	−1.18	1.39
7	8	6	17	7.95	.05	.00
8	10	6	25	9.67	.33	.11
Total						3.04

some point the addition of independent variables will become less advantageous and even in some instances counterproductive. The addition of more independent variables is based on trade-offs between increased predictive power versus overly complex and even potentially misleading regression models.

The survey of credit card usage provides one more possible addition to the multiple regression equation, the number of automobiles owned (V_3). If we now specify the regression equation to include all three independent variables, we can see some improvement in the regression equation, but not nearly of the magnitude seen earlier. The R^2 value increases to .87, only a .01 increase over the previous multiple regression model. Moreover, as we discuss in a later section, the regression coefficient for V_3 is not statistically significant. Therefore, in this instance, the researcher is best served by employing the multiple regression model with two independent variables (family size and income) and not employing the third independent variable (number of automobiles owned) in making predictions.

Summary

Regression analysis is a simple and straightforward dependence technique that can provide both prediction and explanation to the researcher. The prior example illustrated the basic concepts and procedures underlying regression analysis in an attempt to develop an understanding of the rationale and issues of this procedure in its most basic form. The following sections discuss these issues in much more detail and provide a decision process for applying regression analysis to any appropriate research problem.

A DECISION PROCESS FOR MULTIPLE REGRESSION ANALYSIS

In the previous sections we discussed examples of simple regression and multiple regression. In those discussions, many factors affected our ability to find the best regression model. To this point, however, we examined these issues only in simple terms, with little regard to how they combine in an overall approach to multiple regression analysis. In the following sections, the six-stage model-building process introduced in Chapter 1 will be used as a framework for discussing the factors that affect the creation, estimation, interpretation, and validation of a regression analysis. The process begins with specifying the objectives of the regression analysis, including the selection of the dependent and independent variables. The researcher then proceeds to design the regression analysis, considering such factors as sample size and the need for variable transformations. With the regression model formulated, the assumptions underlying regression analysis are first tested for the individual variables. If all assumptions are met, then the model is estimated. Once results are obtained, diagnostic analyses are performed to ensure that the overall model meets the regression assumptions and that no observations have undue influence on the results. The next stage is the interpretation of the regression variate; it examines the role played by each independent variable in the prediction of the dependent measure. Finally, the results are validated to ensure generalizability to the population. Figures 4-1 and 4-6 represent stages 1–3 and 4–6, respectively, in providing a graphical representation of the model-building process for multiple regression, and the following sections discuss each step in detail.

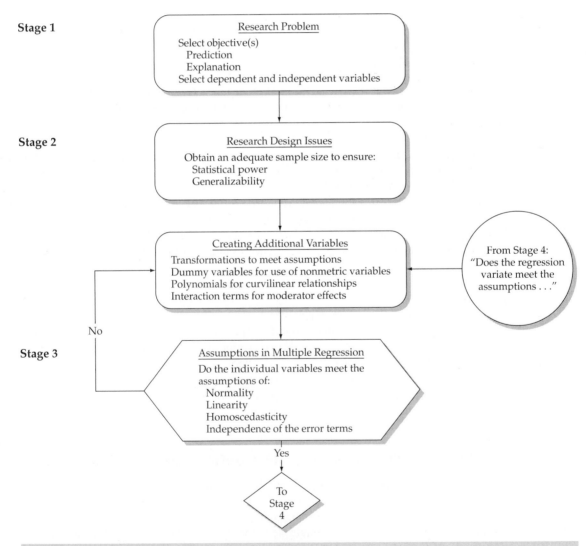

FIGURE 4-1   Stages 1–3 in the Multiple Regression Decision Diagram

STAGE 1: OBJECTIVES OF MULTIPLE REGRESSION

Multiple regression analysis, a form of general linear modeling, is a multivariate statistical technique used to examine the relationship between a single dependent variable and a set of independent variables. The necessary starting point in multiple regression, as with all multivariate statistical techniques, is the research problem. The flexibility and adaptability of multiple regression allow for its use with almost any dependence relationship. In selecting suitable applications of multiple regression, the researcher must consider three primary issues:

1. The appropriateness of the research problem
2. Specification of a statistical relationship
3. Selection of the dependent and independent variables

Research Problems Appropriate for Multiple Regression

Multiple regression is by far the most widely used multivariate technique of those examined in this text. With its broad applicability, multiple regression has been used for many purposes. The ever-widening applications of multiple regression fall into two broad classes of research problems: prediction and explanation. Prediction involves the extent to which the regression variate (one or more independent variables) can predict the dependent variable. Explanation examines the regression coefficients (their magnitude, sign, and statistical significance) for each independent variable and attempts to develop a substantive or theoretical reason for the effects of the independent variables. These research problems are not mutually exclusive, and an application of multiple regression analysis can address either or both types of research problem.

PREDICTION WITH MULTIPLE REGRESSION   One fundamental purpose of multiple regression is to predict the dependent variable with a set of independent variables. In doing so, multiple regression fulfills one of two objectives:

- *The first objective is to maximize the overall predictive power of the independent variables as represented in the variate.* As shown in our earlier example of predicting credit card usage, the variate is formed by estimating regression coefficients for each independent variable so as to be the optimal predictor of the dependent measure. Predictive accuracy is always crucial to ensuring the validity of the set of independent variables. Measures of predictive accuracy are developed and statistical tests are used to assess the significance of the predictive power. In all instances, whether or not the researcher intends to interpret the coefficients of the variate, the regression analysis must achieve acceptable levels of predictive accuracy to justify its application. The researcher must ensure that both statistical and practical significance are considered (see the discussion of stage 4).

    In certain applications focused solely on prediction, the researcher is primarily interested in achieving maximum prediction, and interpreting the regression coefficients is relatively unimportant. Instead, the researcher employs the many options in both the form and the specification of the independent variables that may modify the variate to increase its predictive power, often maximizing prediction at the expense of interpretation. One specific example is a variant of regression, time series analysis, in which the sole purpose is prediction and the interpretation of results is useful only as a means of increasing predictive accuracy.
- Multiple regression can also achieve a *second objective of comparing two or more sets of independent variables to ascertain the predictive power of each variate.* Illustrative of a confirmatory approach to modeling, this use of multiple regression is concerned with the comparison of results across two or more alternative or competing models. The primary

focus of this type of analysis is the relative predictive power among models, although in any situation the prediction of the selected model must demonstrate both statistical and practical significance.

EXPLANATION WITH MULTIPLE REGRESSION   Multiple regression also provides a means of objectively assessing the degree and character of the relationship between dependent and independent variables by forming the variate of independent variables and then examining the magnitude, sign, and statistical significance of the regression coefficient for each independent variable. In this manner, the independent variables, in addition to their collective prediction of the dependent variable, may also be considered for their individual contribution to the variate and its predictions. Interpretation of the variate may rely on any of three perspectives: the importance of the independent variables, the types of relationships found, or the interrelationships among the independent variables.

- *The most direct interpretation of the regression variate is a determination of the relative importance of each independent variable in the prediction of the dependent measure.* In all applications, the selection of independent variables should be based on their theoretical relationships to the dependent variable. Regression analysis then provides a means of objectively assessing the magnitude and direction (positive or negative) of each independent variable's relationship. The character of multiple regression that differentiates it from its univariate counterparts is the simultaneous assessment of relationships between each independent variable and the dependent measure. In making this simultaneous assessment, the relative importance of each independent variable is determined.
- *In addition to assessing the importance of each variable, multiple regression also affords the researcher a means of assessing the nature of the relationships between the independent variables and the dependent variable.* The assumed relationship is a linear association based on the correlations among the independent variables and the dependent measure. Transformations or additional variables are available to assess whether other types of relationships exist, particularly curvilinear relationships. This flexibility ensures that the researcher may examine the true nature of the relationship beyond the assumed linear relationship.
- *Finally, multiple regression provides insight into the relationships among independent variables in their prediction of the dependent measure.* These interrelationships are important for two reasons. First, correlation among the independent variables may make some variables redundant in the predictive effort. As such, they are not needed to produce the optimal prediction given the other independent variable(s) in the regression equation. In such instances, the independent variable will have a strong individual relationship with the dependent variable (substantial bivariate correlations with the dependent variable), but this relationship is markedly diminished in a multivariate context (the partial correlation with the dependent variable is low when considered with other variables in the regression equation). What is the "correct" interpretation in this situation? Should the researcher focus on the strong bivariate correlation to assess importance, or should the diminished relationship in the multivariate context form the basis for assessing the variable's relationship with the dependent variable?

    Here the researcher must rely on the theoretical bases for the regression analysis to assess the "true" relationship for the independent variable. *In such situations, the researcher must guard against determining the importance of independent variables based solely on the derived variate, because relationships among the independent variables may mask or confound relationships that are not needed for predictive purposes but represent substantive findings nonetheless.* The interrelationships among variables can extend not only to their predictive power but also to interrelationships among their estimated effects, which is best seen when the effect of one independent variable is contingent on another independent variable.

Multiple regression provides diagnostic analyses that can determine whether such effects exist based on empirical or theoretical rationale. Indications of a high degree of inter-relationships (multicollinearity) among the independent variables may suggest the use of summated scales, as discussed in Chapter 3.

Specifying a Statistical Relationship

Multiple regression is appropriate when the researcher is interested in a statistical, not a functional, relationship. For example, let us examine the following relationship:

$$\text{Total cost} = \text{Variable cost} + \text{Fixed cost}$$

If the variable cost is $2 per unit, the fixed cost is $500, and we produce 100 units, we assume that the total cost will be exactly $700 and that any deviation from $700 is caused by our inability to measure cost because the relationship between costs is fixed. It is called a *functional relationship* because we expect no error in our prediction. As such, we always know the impact of each variable in calculating the outcome measure.

But in our earlier example dealing with sample data representing human behavior, we assumed that our description of credit card usage was only approximate and not a perfect prediction. It was defined as a **statistical relationship** because some random component is always present in the relationship being examined. A statistical relationship is characterized by two elements:

1. When multiple observations are collected, more than one value of the dependent value will usually be observed for any value of an independent variable.
2. Based on the use of a random sample, the error in predicting the dependent variable is also assumed to be random, and for a given independent variable we can only hope to estimate the average value of the dependent variable associated with it.

For example, in our simple regression example, we found two families with two members, two with four members, and so on, who had different numbers of credit cards. The two families with four members held an average of 6.5 credit cards, and our prediction was 6.75. Our prediction is not as accurate as we would like, but it is better than just using the average of 7 credit cards. The error is assumed to be the result of random behavior among credit card holders.

In summary, a functional relationship calculates an exact value, whereas a statistical relationship estimates an average value. Both of these relationships are displayed in Figure 4-2. Throughout this book, we are concerned with statistical relationships. Our ability to employ just a sample of observations and then use the estimation methods of the multivariate techniques and assess the significance of the independent variables is based on statistical theory. In doing so, we must be sure to meet the statistical assumptions underlying each multivariate technique, because they are critical to our ability to make unbiased predictions of the dependent variable and valid interpretations of the independent variables.

Selection of Dependent and Independent Variables

The ultimate success of any multivariate technique, including multiple regression, starts with the selection of the variables to be used in the analysis. Because multiple regression is a dependence technique, the researcher must specify which variable is the dependent variable and which variables are the independent variables. Although many times the options seem apparent, the researcher should always consider three issues that can affect any decision: strong theory, measurement error, and specification error.

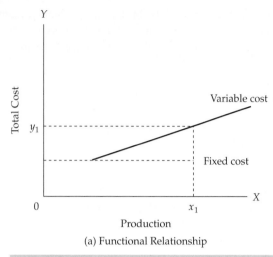

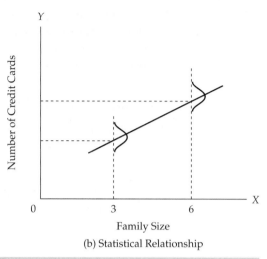

FIGURE 4-2 **Comparison of Functional and Statistical Relationships**

STRONG THEORY The selection of both types of variables should be based principally on conceptual or theoretical grounds, even when the objective is solely for prediction. Chapters 1 and 12 discuss the role of theory in multivariate analysis, and those issues strongly apply to multiple regression. The researcher should always perform the fundamental decisions of variable selection, even though many options and program features are available to assist in model estimation. If the researcher does not exert judgment during variable selection, but instead (1) selects variables indiscriminately or (2) allows for the selection of an independent variable to be based solely on empirical bases, several of the basic tenets of model development will be violated.

MEASUREMENT ERROR The selection of a dependent variable is many times dictated by the research problem. In all instances, the researcher must be aware of the **measurement error,** especially in the dependent variable. Measurement error refers to the degree to which the variable is an accurate and consistent measure of the concept being studied. If the variable used as the dependent measure has substantial measurement error, then even the best independent variables may be unable to achieve acceptable levels of predictive accuracy. Although measurement error can come from several sources (see Chapter 1 for a more detailed discussion), multiple regression has no direct means of correcting for known levels of measurement error for the dependent or independent variables.

Measurement error that is problematic may be addressed through either of two approaches:

- *Summated scales,* employ multiple variables to reduce the reliance on any single variable as the sole representative of a concept.
- *Structural equation modeling* directly accommodates measurement error in estimating the effects of the independent variables in any specified dependence relationship.

Summated scales can be directly incorporated into multiple regression by replacing either dependent or independent variables with the summated scale values, whereas structural equation modeling requires the use of an entirely different technique generally regarded as much more difficult to implement. Thus, summated scales are recommended as the first choice as a remedy for measurement error where possible.

SPECIFICATION ERROR Perhaps the most problematic issue in independent variable selection is **specification error,** which concerns the inclusion of irrelevant variables or the omission of relevant

> ### RULES OF THUMB 4-1
>
> **Meeting Multiple Regression Objectives**
>
> - Only structural equation modeling (SEM) can directly accommodate measurement error, but using summated scales can mitigate it when using multiple regression
> - When in doubt, include potentially irrelevant variables (they can only confuse interpretation) rather than possibly omitting a relevant variable (which can bias all regression estimates)

variables from the set of independent variables. Both types of specification error can have substantial impacts on any regression analysis, although in quite different ways:

- Although the *inclusion of irrelevant variables* does not bias the results for the other independent variables, it does impact the regression variate. First, it reduces model parsimony, which may be critical in the interpretation of the results. Second, the additional variables may mask or replace the effects of more useful variables, especially if some sequential form of model estimation is used (see the discussion of stage 4 for more detail). Finally, the additional variables may make the testing of statistical significance of the independent variables less precise and reduce the statistical and practical significance of the analysis.
- Given the problems associated with adding irrelevant variables, should the researcher be concerned with *excluding relevant variables?* The answer is definitely yes, because the exclusion of relevant variables can seriously bias the results and negatively affect any interpretation of them. In the simplest case, the omitted variables are uncorrelated with the included variables, and the only effect is to reduce the overall predictive accuracy of the analysis. When correlation exists between the included and omitted variables, the effects of the included variables become biased to the extent that they are correlated with the omitted variables. The greater the correlation, the greater the bias. The estimated effects for the included variables now represent not only their actual effects but also the effects that the included variables share with the omitted variables. These effects can lead to serious problems in model interpretation and the assessment of statistical and managerial significance.

The researcher must be careful in the selection of the variables to avoid both types of specification error. Perhaps most troublesome is the omission of relevant variables, because the variables' effect cannot be assessed without their inclusion (see Rules of Thumb 4-1). Their potential influence on any results heightens the need for theoretical and practical support for all variables included or excluded in a multiple regression analysis.

STAGE 2: RESEARCH DESIGN OF A MULTIPLE REGRESSION ANALYSIS

Adaptability and flexibility are two principal reasons for multiple regression's widespread usage across a wide variety of applications. As you will see in the following sections, multiple regression can represent a wide range of dependence relationships. In doing so, the researcher incorporates three features:

1. *Sample size.* Multiple regression maintains the necessary levels of statistical power and practical/statistical significance across a broad range of sample sizes.
2. *Unique elements of the dependence relationship.* Even though independent variables are assumed to be metric and have a linear relationship with the dependent variable, both assumptions can be relaxed by creating additional variables to represent these special aspects of the relationship.

3. *Nature of the independent variables.* Multiple regression accommodates metric independent variables that are assumed to be fixed in nature as well as those with a random component.

Each of these features plays a key role in the application of multiple regression to many types of research questions while maintaining the necessary levels of statistical and practical significance.

Sample Size

The sample size used in multiple regression is perhaps the single most influential element under the control of the researcher in designing the analysis. The effects of sample size are seen most directly in the statistical power of the significance testing and the generalizability of the result. Both issues are addressed in the following sections.

STATISTICAL POWER AND SAMPLE SIZE   The size of the sample has a direct impact on the appropriateness and the statistical power of multiple regression. Small samples, usually characterized as having fewer than 30 observations, are appropriate for analysis only by simple regression with a single independent variable. Even in these situations, only strong relationships can be detected with any degree of certainty. Likewise, large samples of 1,000 observations or more make the statistical significance tests overly sensitive, often indicating that almost any relationship is statistically significant. With such large samples the researcher must ensure that the criterion of practical significance is met along with statistical significance.

Power Levels in Various Regression Models.   In multiple regression **power** refers to the probability of detecting as statistically significant a specific level of R^2 or a regression coefficient at a specified significance level for a specific sample size (see Chapter 1 for a more detailed discussion). Sample size plays a role in not only assessing the power of a current analysis, but also in anticipating the statistical power of a proposed analysis.

Table 4-5 illustrates the *interplay among the sample size, the significance level (α) chosen, and the number of independent variables* in detecting a significant R^2. The table values are the minimum R^2 that the specified sample size will detect as statistically significant at the specified alpha (α) level with a power (probability) of .80.

For example, if the researcher employs five independent variables, specifies a .05 significance level, and is satisfied with detecting the R^2 80 percent of the time it occurs (corresponding to a power of .80), a sample of 50 respondents will detect R^2 values of 23 percent and greater. If the sample is increased to 100 respondents, then R^2 values of 12 percent and above will be detected.

TABLE 4-5 **Minimum R^2 That Can Be Found Statistically Significant with a Power of .80 for Varying Numbers of Independent Variables and Sample Sizes**

Sample Size	Significance Level (α) = .01 No. of Independent Variables				Significance Level (α) = .05 No. of Independent Variables			
	2	5	10	20	2	5	10	20
20	45	56	71	NA	39	48	64	NA
50	23	29	36	49	19	23	29	42
100	13	16	20	26	10	12	15	21
250	5	7	8	11	4	5	6	8
500	3	3	4	6	3	4	5	9
1,000	1	2	2	3	1	1	2	2

Note: Values represent percentage of variance explained.
NA = not applicable.

But if 50 respondents are all that are available, and the researcher wants a .01 significance level, the analysis will detect R^2 values only in excess of 29 percent.

Sample Size Requirements for Desired Power. The researcher can also consider the *role of sample size in significance testing before collecting data.* If weaker relationships are expected, the researcher can make informed judgments as to the necessary sample size to reasonably detect the relationships, if they exist.

For example, Table 4-5 demonstrates that sample sizes of 100 will detect fairly small R^2 values (10% to 15%) with up to 10 independent variables and a significance level of .05. However, if the sample size falls to 50 observations in these situations, the minimum R^2 that can be detected doubles.

The researcher can also determine the sample size needed to detect effects for individual independent variables given the expected effect size (correlation), the α level, and the power desired. The possible computations are too numerous for presentation in this discussion, and the interested reader is referred to texts dealing with power analyses [5] or to a computer program to calculate sample size or power for a given situation [3].

Summary. The researcher must always be aware of the anticipated power of any proposed multiple regression analysis. It is critical to understand the elements of the research design, particularly sample size, that can be changed to meet the requirements for an acceptable analysis [9].

GENERALIZABILITY AND SAMPLE SIZE In addition to its role in determining statistical power, sample size also affects the generalizability of the results by the ratio of observations to independent variables. A general rule is that the ratio should never fall below 5:1, meaning that five observations are made for each independent variable in the variate. Although the minimum ratio is 5:1, the desired level is between 15 to 20 observations for each independent variable. When this level is reached, the results should be generalizable if the sample is representative. However, if a stepwise procedure is employed, the recommended level increases to 50:1 because this technique selects only the strongest relationships within the data set and suffers from a greater tendency to become sample-specific [16]. In cases for which the available sample does not meet these criteria, the researcher should be certain to validate the generalizability of the results.

Defining Degrees of Freedom. As this ratio falls below 5:1, the researcher encounters the risk of overfitting the variate to the sample, making the results too specific to the sample and thus lacking generalizability. In understanding the concept of overfitting, we need to address the statistical concept of **degrees of freedom.** In any statistical estimation procedure, the researcher is making estimates of parameters from the sample data. In the case of regression, the parameters are the regression coefficients for each independent variable and the constant term. As described earlier, the regression coefficients are the weights used in calculating the regression variate and indicate each independent variable's contribution to the predicted value. What, then, is the relationship between the number of observations and variables? Let us look at a simple view of estimating parameters for some insight into this problem.

Each observation represents a separate and independent unit of information (i.e., one set of values for each independent variable). In a simplistic view, the researcher could dedicate a single variable to perfectly predicting only one observation, a second variable to another observation, and so forth. If the sample is relatively small, then predictive accuracy could be quite high, and many of the observations would be perfectly predicted. As a matter of fact, if the number of estimated **parameters** (regression coefficients and the constant) equals the sample size, perfect prediction will occur even if all the variable values are random numbers. This scenario would be totally unacceptable and considered extreme overfitting because the estimated parameters have no generalizability, but relate only to the sample data. *Moreover, anytime a variable is added to the regression equation, the* R^2 *value will increase.*

Degrees of Freedom as a Measure of Generalizability. What happens to generalizability as the sample size increases? We can perfectly predict one observation with a single variable, but what about all the other observations? Thus, the researcher is searching for the best regression model, one with the highest predictive accuracy for the largest (most generalizable) sample. The degree of generalizability is represented by the degrees of freedom, calculated as:

$$\text{Degrees of freedom } (df) = \text{Sample size} - \text{Number of estimated parameters}$$

or

$$\text{Degrees of freedom } (df) = N - (\text{Number of independent variables} + 1)$$

The larger the degrees of freedom, the more generalizable are the results. Degrees of freedom increase for a given sample by reducing the number of independent variables. Thus, the objective is to achieve the highest predictive accuracy with the most degrees of freedom. In our preceding example, where the number of estimated parameters equals the sample size, we have perfect prediction, but *zero degrees of freedom!* The researcher must reduce the number of independent variables (or increase the sample size), lowering the predictive accuracy but also increasing the degrees of freedom. No specific guidelines determine how large the degrees of freedom are, just that they are indicative of the generalizability of the results and give an idea of the overfitting for any regression model as shown in Rules of Thumb 4-2.

Creating Additional Variables

The basic relationship represented in multiple regression is the *linear* association between metric dependent and independent variables based on the product-moment correlation. One problem often faced by researchers is the desire to incorporate nonmetric data, such as gender or occupation, into a regression equation. Yet, as we already discussed, regression is limited to metric data. Moreover, regression's inability to directly model nonlinear relationships may constrain the researcher when faced with situations in which a nonlinear relationship (e.g., U-shaped) is suggested by theory or detected when examining the data.

USING VARIABLE TRANSFORMATIONS   In these situations, new variables must be created by **transformations,** because multiple regression is totally reliant on creating new variables in the model to incorporate nonmetric variables or represent any effects other than linear relationships. We also will encounter the use of transformations discussed in Chapter 2 as a means of remedying violations of some statistical assumptions, but our purpose here is to provide the researcher with a means of modifying either the dependent or independent variables for one of two reasons:

1. Improve or modify the relationship between independent and dependent variables
2. Enable the use of nonmetric variables in the regression variate

---

RULES OF THUMB 4-2

Sample Size Considerations

- Simple regression can be effective with a sample size of 20, but maintaining power at .80 in multiple regression requires a minimum sample of 50 and preferably 100 observations for most research situations
- The minimum ratio of observations to variables is 5:1, but the preferred ratio is 15:1 or 20:1, which should increase when stepwise estimation is used
- Maximizing the degrees of freedom improves generalizability and addresses both model parsimony and sample size concerns

Data transformations may be based on reasons that are either *theoretical* (transformations whose appropriateness is based on the nature of the data) or *data derived* (transformations that are suggested strictly by an examination of the data). In either case the researcher must proceed many times by trial and error, constantly assessing the improvement versus the need for additional transformations. We explore these issues with discussions of data transformations that enable the regression analysis to best represent the actual data and a discussion of the creation of variables to supplement the original variables.

All the transformations we describe are easily carried out by simple commands in all the popular statistical packages. Although we focus on transformations that can be computed in this manner, other more sophisticated and complicated methods of data transformation are available (e.g., see Box and Cox [4]).

INCORPORATING NONMETRIC DATA WITH DUMMY VARIABLES   One common situation faced by researchers is the desire to utilize nonmetric independent variables. Yet, to this point, all our illustrations assumed metric measurement for both independent and dependent variables. When the dependent variable is measured as a dichotomous (0, 1) variable, either discriminant analysis or a specialized form of regression (logistic regression), discussed in later chapters, is appropriate. What can we do when the independent variables are nonmetric and have two or more categories? Chapter 2 introduced the concept of dichotomous variables, known as **dummy variables,** which can act as replacement independent variables. Each dummy variable represents one category of a nonmetric independent variable, and any nonmetric variable with k categories can be represented as $k - 1$ dummy variables.

Indicator Coding: The Most Common Format.   Of the two forms of dummy variable coding, the most common is **indicator coding** in which each category of the nonmetric variable is represented by either 1 or 0. *The regression coefficients for the dummy variables represent differences on the dependent variable for each group of respondents from the **reference category*** (i.e., the omitted group that received all zeros). *These group differences can be assessed directly, because the coefficients are in the same units as the dependent variable.*

This form of dummy-variable coding can be depicted as differing intercepts for the various groups, with the reference category represented in the constant term of the regression model (see Figure 4-3). In this example, a three-category nonmetric variable is represented by two dummy variables (D_1 and D_2) representing groups 1 and 2, with group 3 the reference category. The regression coefficients are 2.0 for D_1 and -3.0 for D_2. These coefficients translate into three parallel lines. The reference group (in this case group 3) is defined by the regression equation with both dummy variables equaling zero. Group 1's line is two units above the line for the reference group. Group 2's line is three units below the line for reference group 3. The parallel lines indicate that dummy variables do not change the nature of the relationship, but only provide for differing intercepts among the groups.

This form of coding is most appropriate when a logical reference group is present, such as in an experiment. Any time dummy variable coding is used, we must be aware of the comparison group and remember that the coefficients represent the differences in group means from this group.

Effects Coding.   An alternative method of dummy-variable coding is termed **effects coding.** It is the same as indicator coding except that the comparison or omitted group (the group that got all zeros) is now given the value of -1 instead of 0 for the dummy variables. *Now the coefficients represent differences for any group from the mean of all groups rather than from the omitted group.* Both forms of dummy-variable coding will give the same predictive results, coefficient of determination, and regression coefficients for the continuous variables. The only differences will be in the interpretation of the dummy-variable coefficients.

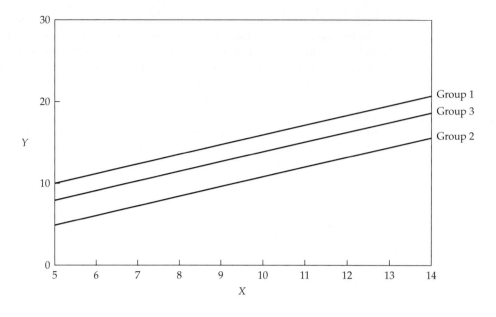

FIGURE 4-3   Incorporating Nonmetric Variables Through Dummy Variables

REPRESENTING CURVILINEAR EFFECTS WITH POLYNOMIALS   Several types of data transformations are appropriate for linearizing a curvilinear relationship. Direct approaches, discussed in Chapter 2, involve modifying the values through some arithmetic transformation (e.g., taking the square root or logarithm of the variable). However, such transformations are subject to the following limitations:

- They are applicable only in a simple curvilinear relationship (a relationship with only one turning or inflection point).
- They do not provide any statistical means for assessing whether the curvilinear or linear model is more appropriate.
- They accommodate only univariate relationships and not the interaction between variables when more than one independent variable is involved.

We now discuss a means of creating new variables to explicitly model the curvilinear components of the relationship and address each of the limitations inherent in data transformations.

Specifying a Curvilinear Effect. Power transformations of an independent variable that add a nonlinear component for each additional power of the independent variable are known as **polynomials.** The power of 1 (X^1) represents the linear component and is the form discussed so far in this chapter. The power of 2, the variable squared (X^2), represents the quadratic component. In graphical terms, X^2 represents the first inflection point of a curvilinear relationship. A cubic component, represented by the variable cubed (X^3), adds a second inflection point. With these variables, and even higher powers, we can accommodate more complex relationships than are possible with only transformations. For example, in a simple regression model, a curvilinear model with one turning point can be modeled with the equation

$$Y = b_0 + b_1X_1 + b_2X_1^2$$

where

$$b_0 = \text{intercept}$$
$$b_1X_1 = \text{linear effect of } X_1$$
$$b_2X_1^2 = \text{curvilinear effect of } X_1$$

Although any number of nonlinear components may be added, the cubic term is usually the highest power used. Multivariate polynomials are created when the regression equation contains two or more independent variables. We follow the same procedure for creating the polynomial terms as before, but must also create an additional term, the interaction term (X_1X_2), which is needed for each variable combination to represent fully the multivariate effects. In graphical terms, a two-variable multivariate polynomial is portrayed by a surface with one peak or valley. For higher-order polynomials, the best form of interpretation is obtained by plotting the surface from the predicted values.

Interpreting a Curvilinear Effect. As each new variable is entered into the regression equation, we can also perform a direct statistical test of the nonlinear components, which we cannot do with data transformations. However, multicollinearity can create problems in assessing the statistical significance of the individual coefficients to the extent that the researcher should assess incremental effects as a measure of any polynomial terms in a three-step process:

1. Estimate the original regression equation.
2. Estimate the curvilinear relationship (original equation plus polynomial term).
3. Assess the change in R^2. If it is statistically significant, then a significant curvilinear effect is present. The focus is on the incremental effect, not the significance of individual variables.

Three (two nonlinear and one linear) relationships are shown in Figure 4-4. For interpretation purposes, the positive quadratic term indicates a ∪-shaped curve, whereas a negative coefficient indicates a ∩-shaped curve. The use of a cubic term can represent such forms as the **S**-shaped or growth curve quite easily, but it is generally best to plot the values to interpret the actual shape.

How many terms should be added? Common practice is to start with the linear component and then sequentially add higher-order polynomials until nonsignificance is achieved. The use of polynomials, however, also has potential problems. First, each additional term requires a degree of freedom, which may be particularly restrictive with small sample sizes. This limitation does not occur with data transformations. Also, multicollinearity is introduced by the additional terms and makes statistical significance testing of the polynomial terms inappropriate. Instead, the researcher must compare the R^2 values from the equation model with linear terms to the R^2 for the equation with the polynomial terms. Testing for the statistical significance of the incremental R^2 is the appropriate manner of assessing the impact of the polynomials.

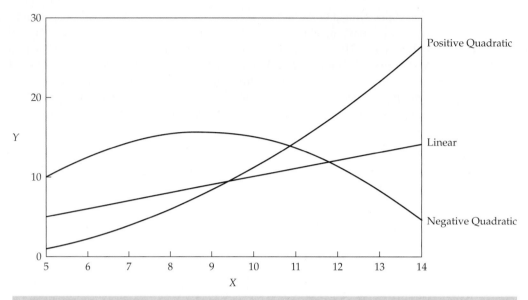

FIGURE 4-4 **Representing Nonlinear Relationships with Polynomials**

REPRESENTING INTERACTION OR MODERATOR EFFECTS The nonlinear relationships just discussed require the creation of an additional variable (e.g., the squared term) to represent the changing slope of the relationship over the range of the independent variable. This variable focuses on the relationship between a single independent variable and the dependent variable. But what if an independent–dependent variable relationship is affected by another independent variable? This situation is termed a **moderator effect,** which occurs when the moderator variable, a second independent variable, changes the *form* of the relationship between another independent variable and the dependent variable. It is also known as an *interaction effect* and is similar to the interaction term found in analysis of variance and multivariate analysis of variance (see Chapter 8 for more detail on interaction terms).

Examples of Moderator Effects. The most common moderator effect employed in multiple regression is the *quasi* or *bilinear moderator,* in which the slope of the relationship of one independent variable (X_1) changes across values of the moderator variable (X_2) [7, 14].

In our earlier example of credit card usage, assume that family income (X_2) was found to be a positive moderator of the relationship between family size (X_1) and credit card usage (Y). It would mean that the expected change in credit card usage based on family size (b_1, the regression coefficient for X_1) might be lower for families with low incomes and higher for families with higher incomes. Without the moderator effect, we assumed that family size had a constant effect on the number of credit cards used, but the interaction term tells us that this relationship changes, depending on family income level. Note that it does not necessarily mean the effects of family size or family income by themselves are unimportant, but instead that the interaction term complements their explanation of credit card usage.

Adding the Moderator Effect. The moderator effect is represented in multiple regression by a term quite similar to the polynomials described earlier to represent nonlinear effects. The moderator term is a compound variable formed by multiplying X_1 by the moderator X_2, which is entered into the regression equation. In fact, the nonlinear term can be viewed as a form of interaction, where the independent variable "moderates" itself, thus the squared term (X_iX_i). The moderated relationship is represented as

$$Y = b_0 + b_1X_1 + b_2X_2 + b_3X_1X_2$$

where

$$b_0 = \text{intercept}$$
$$b_1X_1 = \text{linear effect of } X_1$$
$$b_2X_2 = \text{linear effect of } X_2$$
$$b_3X_1X_2 = \text{moderator effect of } X_2 \text{ on } X_1$$

Because of the multicollinearity among the old and new variables, an approach similar to testing for the significance of polynomial (nonlinear) effects is employed. To determine whether the moderator effect is significant, the researcher follows a three-step process:

1. Estimate the original (unmoderated) equation.
2. Estimate the moderated relationship (original equation plus moderator variable).
3. Assess the change in R^2: If it is statistically significant, then a significant moderator effect is present. Only the incremental effect is assessed, not the significance of individual variables.

Interpreting Moderator Effects. The interpretation of the regression coefficients changes slightly in moderated relationships. *The b_3 coefficient, the moderator effect, indicates the unit change in the effect of X_1 as X_2 changes. The b_1 and b_2 coefficients now represent the effects of X_1 and X_2, respectively, when the other independent variable is zero.* In the unmoderated relationship, the b_1 coefficient represents the effect of X_1 across all levels of X_2, and similarly for b_2. Thus, in unmoderated regression, the regression coefficients b_1 and b_2 are averaged across levels of the other independent variables, whereas in a moderated relationship they are separate from the other independent variables. To determine the total effect of an independent variable, the separate and moderated effects must be combined. The overall effect of X_1 for any value of X_2 can be found by substituting the X_2 value into the following equation:

$$b_{\text{total}} = b_1 + b_3X_2$$

For example, assume a moderated regression resulted in the following coefficients: $b_1 = 2.0$ and $b_3 = .5$. If the value of X_2 ranges from 1 to 7, the researcher can calculate the total effect of X_1 at any value of X_2. When X_2 equals 3, the total effect of X_1 is 3.5 [2.0 + .5(3)]. When X_2 increases to 7, the total effect of X_1 is now 5.5 [2.0 + .5(7)].

We can see the moderator effect at work, making the relationship of X_1 and the dependent variable change, given the level of X_2. Excellent discussions of moderated relationships in multiple regression are available in a number of sources [5, 7, 14].

SUMMARY The creation of new variables provides the researcher with great flexibility in representing a wide range of relationships within regression models (see Rules of Thumb 4-3). Yet too often the desire for better model fit leads to the inclusion of these special relationships without theoretical support. In those instances the researcher is running a much greater risk of finding results with little or no generalizability. Instead, in using these additional variables, the researcher must be guided by theory that is supported by empirical analysis. In this manner, both practical and statistical significance can hopefully be met.

STAGE 3: ASSUMPTIONS IN MULTIPLE REGRESSION ANALYSIS

We have shown how improvements in prediction of the dependent variable are possible by adding independent variables and even transforming them to represent aspects of the relationship that are not linear. To do so, however, we must make several assumptions about the relationships between the dependent and independent variables that affect the statistical procedure (least squares) used for

---

RULES OF THUMB 4-3

Variable Transformations

- Nonmetric variables can only be included in a regression analysis by creating dummy variables
- Dummy variables can only be interpreted in relation to their reference category
- Adding an additional polynomial term represents another inflection point in the curvilinear relationship
- Quadratic and cubic polynomials are generally sufficient to represent most curvilinear relationships
- Assessing the significance of a polynomial or interaction term is accomplished by evaluating incremental R^2, not the significance of individual coefficients, due to high multicollinearity

---

multiple regression. In the following sections we discuss testing for the assumptions and corrective actions to take if violations occur.

The basic issue is whether, in the course of calculating the regression coefficients and predicting the dependent variable, the assumptions of regression analysis are met. Are the errors in prediction a result of an actual absence of a relationship among the variables, or are they caused by some characteristics of the data not accommodated by the regression model? The assumptions to be examined are in four areas:

1. Linearity of the phenomenon measured
2. Constant variance of the error terms
3. Independence of the error terms
4. Normality of the error term distribution

Assessing Individual Variables Versus the Variate

Before addressing the individual assumptions, we must first understand that the assumptions underlying multiple regression analysis apply both to the individual variables (dependent and independent) and to the relationship as a whole. Chapter 2 examined the available methods for assessing the assumptions for individual variables. In multiple regression, once the variate is derived, it acts collectively in predicting the dependent variable, *which necessitates assessing the assumptions not only for individual variables but also for the variate itself.* This section focuses on examining the variate and its relationship with the dependent variable for meeting the assumptions of multiple regression. These analyses actually must be performed *after* the regression model has been estimated in stage 4. Thus, testing for assumptions must occur not only in the initial phases of the regression, but also after the model has been estimated.

A common question is posed by many researchers: Why examine the individual variables when we can just examine the variate and avoid the time and effort expended in assessing the individual variables? The answer rests in the insight gained in examining the individual variables in two areas:

- Have assumption violations for individual variable(s) caused their relationships to be misrepresented?
- What are the sources and remedies of any assumptions violations for the variate?

Only with a thorough examination of the individual variables will the researcher be able to address these two key questions. If only the variate is assessed, then the researcher not only has little insight into how to correct any problems, but perhaps more importantly does not know what opportunities were missed for better representations of the individual variables and, ultimately, the variate.

Methods of Diagnosis

The principal measure of prediction error for the variate is the *residual*—the difference between the observed and predicted values for the dependent variable. When examining residuals, some form of standardization is recommended to make the residuals directly comparable. (In their original form, larger predicted values naturally have larger residuals.) The most widely used is the **studentized residual,** whose values correspond to *t* values. This correspondence makes it quite easy to assess the statistical significance of particularly large residuals.

Plotting the residuals versus the independent or predicted variables is a basic method of identifying assumption violations for the overall relationship. Use of the residual plots, however, depends on several key considerations:

- The most common residual plot involves the residuals (r_i) versus the predicted dependent values (Y_i). For a simple regression model, the residuals may be plotted against either the dependent or independent variables, because they are directly related. In multiple regression, however, only the predicted dependent values represent the total effect of the regression variate. Thus, unless the residual analysis intends to concentrate on only a single variable, the predicted dependent variables are used.
- Violations of each assumption can be identified by specific patterns of the residuals. Figure 4-5 contains a number of residual plots that address the basic assumptions discussed in the following sections. One plot of special interest is the **null plot** (Figure 4-5a), the plot of residuals when all assumptions are met. The null plot shows the residuals falling randomly, with relatively equal dispersion about zero and no strong tendency to be either greater or less than zero. Likewise, no pattern is found for large versus small values of the independent variable. The remaining residual plots will be used to illustrate methods of examining for violations of the assumptions underlying regression analysis. In the following sections, we examine a series of statistical tests that can complement the visual examination of the residual plots.

Linearity of the Phenomenon

The **linearity** of the relationship between dependent and independent variables represents the degree to which the change in the dependent variable is associated with the independent variable. The regression coefficient is constant across the range of values for the independent variable. The concept of correlation is based on a linear relationship, thus making it a critical issue in regression analysis. Linearity of any bivariate relationship is easily examined through residual plots. Figure 4-5b shows a typical pattern of residuals indicating the existence of a nonlinear relationship not represented in the current model. Any consistent curvilinear pattern in the residuals indicates that corrective action will increase both the predictive accuracy of the model and the validity of the estimated coefficients. Corrective action can take one of three forms:

- Transforming the data values (e.g., logarithm, square root, etc.) of one or more independent variables to achieve linearity is discussed in Chapter 2 [10]
- Directly including the nonlinear relationships in the regression model, such as through the creation of polynomial terms as discussed in stage 2
- Using specialized methods such as nonlinear regression specifically designed to accommodate the curvilinear effects of independent variables or more complex nonlinear relationships

IDENTIFYING THE INDEPENDENT VARIABLES FOR ACTION   How do we determine which independent variable(s) to select for corrective action? In multiple regression with more than one independent variable, an examination of the residuals shows only the combined effects of all independent variables, but we cannot examine any independent variable separately in a residual plot. To do so, we use what are called **partial regression plots,** which show the relationship of a single

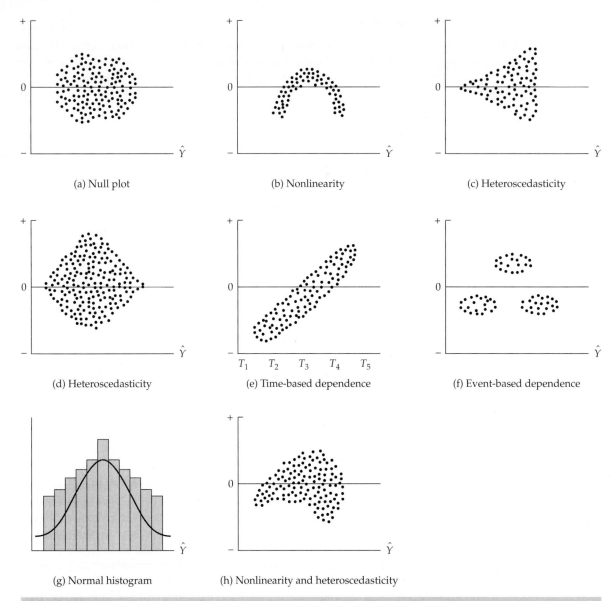

FIGURE 4-5   **Graphical Analysis of Residuals**

independent variable to the dependent variable, controlling for the effects of all other independent variables. As such, the partial regression plot portrays the unique relationship between dependent and independent variables. These plots differ from the residual plots just discussed because the line running through the center of the points, which was horizontal in the earlier plots (refer to Figure 4-5), will now slope up or down depending on whether the regression coefficient for that independent variable is positive or negative.

Examining the observations around this line is done exactly as before, but now the curvilinear pattern indicates a nonlinear relationship between a specific independent variable and the dependent variable. This method is more useful when several independent variables are involved, because we can tell which specific variables violate the assumption of linearity and apply the needed remedies only to them. Also, the identification of outliers or influential observations is facilitated on the basis of one independent variable at a time.

Constant Variance of the Error Term

The presence of unequal variances (**heteroscedasticity**) is one of the most common assumption violations. Diagnosis is made with residual plots or simple statistical tests. Plotting the residuals (studentized) against the predicted dependent values and comparing them to the null plot (see Figure 4-5a) shows a consistent pattern if the variance is not constant. Perhaps the most common pattern is triangle-shaped in either direction (Figure 4-5c). A diamond-shaped pattern (Figure 4-5d) can be expected in the case of percentages where more variation is expected in the midrange than at the tails. Many times, a number of violations occur simultaneously, such as in nonlinearity and heteroscedasticity (Figure 4-5h). Remedies for one of the violations often correct problems in other areas as well.

Each statistical computer program has statistical tests for heteroscedasticity. For example, SPSS provides the Levene test for homogeneity of variance, which measures the equality of variances for a single pair of variables. Its use is particularly recommended because it is less affected by departures from normality, another frequently occurring problem in regression.

If heteroscedasticity is present, two remedies are available. If the violation can be attributed to a single independent variable through the analysis of residual plots discussed earlier, then the procedure of weighted least squares can be employed. More direct and easier, however, are a number of variance-stabilizing transformations discussed in Chapter 2 that allow the transformed variables to exhibit **homoscedasticity** (equality of variance) and be used directly in our regression model.

Independence of the Error Terms

We assume in regression that each predicted value is independent, which means that the predicted value is not related to any other prediction; that is, they are not sequenced by any variable. We can best identify such an occurrence by plotting the residuals against any possible sequencing variable. If the residuals are independent, the pattern should appear random and similar to the null plot of residuals. Violations will be identified by a consistent pattern in the residuals. Figure 4-5e displays a residual plot that exhibits an association between the residuals and time, a common sequencing variable. Another frequent pattern is shown in Figure 4-5f. This pattern occurs when basic model conditions change but are not included in the model. For example, swimsuit sales are measured monthly for 12 months, with two winter seasons versus a single summer season, yet no seasonal indicator is estimated. The residual pattern will show negative residuals for the winter months versus positive residuals for the summer months. Data transformations, such as first differences in a time series model, inclusion of indicator variables, or specially formulated regression models, can address this violation if it occurs.

Normality of the Error Term Distribution

Perhaps the most frequently encountered assumption violation is nonnormality of the independent or dependent variables or both [13]. The simplest diagnostic for the set of independent variables in the equation is a histogram of residuals, with a visual check for a distribution approximating the normal distribution (see Figure 4-5g). Although attractive because of its simplicity, this method is particularly difficult in smaller samples, where the distribution is ill-formed. A better method is the use of **normal probability plots.** They differ from residual plots in that the standardized residuals are compared with the normal distribution. The normal distribution makes a straight diagonal line, and the plotted residuals are compared with the diagonal. If a distribution is normal, the residual line closely follows the diagonal. The same procedure can compare the dependent or independent variables separately to the normal distribution [6]. Chapter 2 provides a more detailed discussion of the interpretation of normal probability plots.

RULES OF THUMB 4-4

Assessing Statistical Assumptions

- Testing assumptions must be done not only for each dependent and independent variable, but for the variate as well
- Graphical analyses (i.e., partial regression plots, residual plots, and normal probability plots) are the most widely used methods of assessing assumptions for the variate
- Remedies for problems found in the variate must be accomplished by modifying one or more independent variables as described in Chapter 2

Summary

Analysis of residuals, whether with the residual plots or statistical tests, provides a simple yet powerful set of analytical tools for examining the appropriateness of our regression model. Too often, however, these analyses are not made, and the violations of assumptions are left intact. Thus, users of the results are unaware of the potential inaccuracies that may be present, which range from inappropriate tests of the significance of coefficients (either showing significance when it is not present, or vice versa) to the biased and inaccurate predictions of the dependent variable. We strongly recommend that these methods be applied for each set of data and regression model (see Rules of Thumb 4-4). Application of the remedies, particularly transformations of the data, will increase confidence in the interpretations and predictions from multiple regression.

STAGE 4: ESTIMATING THE REGRESSION MODEL AND ASSESSING OVERALL MODEL FIT

Having specified the objectives of the regression analysis, selected the independent and dependent variables, addressed the issues of research design, and assessed the variables for meeting the assumptions of regression, the researcher is now ready to estimate the regression model and assess the overall predictive accuracy of the independent variables (see Figure 4-6). In this stage, the researcher must accomplish three basic tasks:

1. Select a method for specifying the regression model to be estimated.
2. Assess the statistical significance of the overall model in predicting the dependent variable.
3. Determine whether any of the observations exert an undue influence on the results.

Selecting an Estimation Technique

In most instances of multiple regression, the researcher has a number of possible independent variables from which to choose for inclusion in the regression equation. Sometimes the set of independent variables is exactly specified and the regression model is essentially used in a confirmatory approach. In other instances, the researcher may use the estimation technique to pick and choose among the set of independent variables with either sequential search methods or combinatorial processes. Each is designed to assist the researcher in finding the best regression model through different approaches. All three approaches to specifying the regression model are discussed next.

CONFIRMATORY SPECIFICATION The simplest, yet perhaps most demanding, approach for specifying the regression model is to employ a confirmatory perspective wherein the researcher specifies the exact set of independent variables to be included. As compared with the sequential or

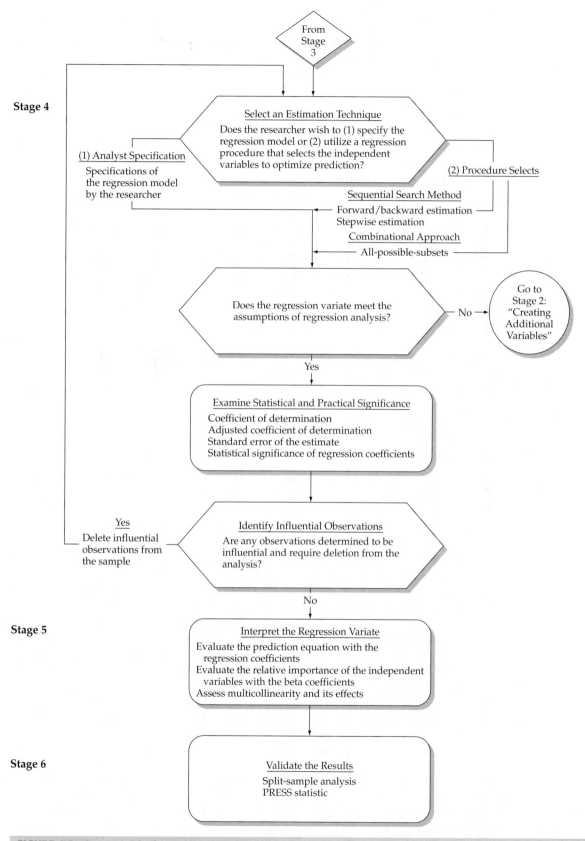

FIGURE 4-6 Stages 4–6 in the Multiple Regression Decision Diagram

combinatorial approaches to be discussed next, the researcher has total control over the variable selection. Although the confirmatory specification is simple in concept, the researcher is completely responsible for the trade-offs between more independent variables and greater predictive accuracy versus model parsimony and concise explanation. Particularly problematic are specification errors of either omission or inclusion. Guidelines for model development are discussed in Chapters 1 and 10. The researcher must avoid being guided by empirical information and instead rely heavily on theoretical justification for a truly confirmatory approach.

SEQUENTIAL SEARCH METHODS   In a marked contrast to the prior approach, sequential search methods have in common the general approach of estimating the regression equation by considering a set of variables defined by the researcher, and then selectively adding or deleting among these variables until some overall criterion measure is achieved. This approach provides an objective method for selecting variables that maximizes the prediction while employing the smallest number of variables. Two types of sequential search approaches are (1) stepwise estimation and (2) forward addition and backward elimination. In each approach, variables are individually assessed for their contribution to prediction of the dependent variable and added to or deleted from the regression model based on their relative contribution. The stepwise procedure is discussed and then contrasted with the forward addition and backward elimination procedures.

Stepwise Estimation.   Perhaps the most popular sequential approach to variable selection is **stepwise estimation.** This approach enables the researcher to examine the contribution of each independent variable to the regression model. Each variable is considered for inclusion prior to developing the equation. The independent variable with the greatest contribution is added first. Independent variables are then selected for inclusion based on their incremental contribution over the variable(s) already in the equation. The stepwise procedure is illustrated in Figure 4-7. The specific issues at each stage are as follows:

1. Start with the simple regression model by selecting the one independent variable that is the most highly correlated with the dependent variable. The equation would be $Y = b_0 + b_1X_1$.
2. Examine the **partial correlation coefficients** to find an additional independent variable that explains the *largest statistically significant* portion of the unexplained (error) variance remaining from the first regression equation.
3. Recompute the regression equation using the two independent variables, and examine the **partial F value** for the original variable in the model to see whether it still makes a significant contribution, given the presence of the new independent variable. If it does not, eliminate the variable. This ability to eliminate variables already in the model distinguishes the stepwise model from the forward addition/backward elimination models. If the original variable still makes a significant contribution, the equation would be $Y = b_0 + b_1X_1 + b_2X_2$.
4. Continue this procedure by examining all independent variables not in the model to determine whether one would make a *statistically significant addition to the current equation* and thus should be included in a revised equation. If a new independent variable is included, examine all independent variables previously in the model to judge whether they should be kept.
5. Continue adding independent variables until none of the remaining candidates for inclusion would contribute a statistically significant improvement in predictive accuracy. This point occurs when all of the remaining partial regression coefficients are nonsignificant.

A potential bias in the stepwise procedure results from considering only one variable for selection at a time. Suppose variables X_3 and X_4 together would explain a significant portion of the variance (each given the presence of the other), but neither is significant by itself. In this situation, neither would be considered for the final model. Also, as will be discussed later, multicollinearity among the independent variables can substantially affect all sequential estimation methods.

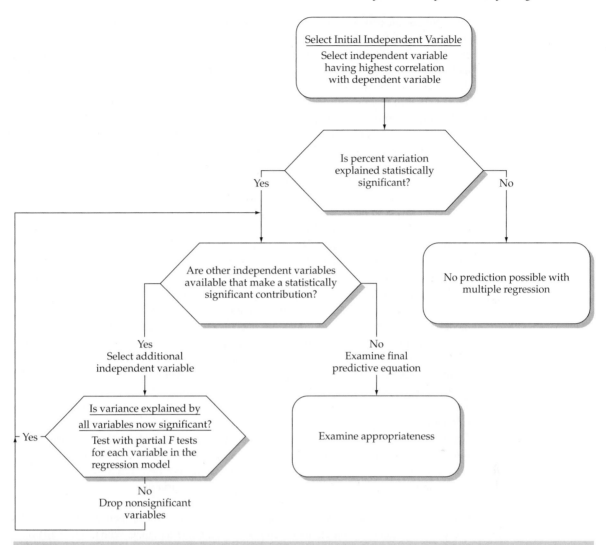

FIGURE 4-7 Flowchart of the Stepwise Estimation Method

Forward Addition and Backward Elimination. The procedures of **forward addition** and **backward elimination** are largely trial-and-error processes for finding the best regression estimates. The forward addition model is similar to the stepwise procedure in that it builds the regression equation starting with a single independent variable, whereas the backward elimination procedure starts with a regression equation including all the independent variables and then deletes independent variables that do not contribute significantly. *The primary distinction of the stepwise approach from the forward addition and backward elimination procedures is its ability to add or delete variables at each stage. Once a variable is added or deleted in the forward addition or backward elimination schemes, the action cannot be reversed at a later stage.* Thus, the ability of the stepwise method to add and delete makes it the preferred method among most researchers.

Caveats to Sequential Search Methods. To many researchers, the sequential search methods seem the perfect solution to the dilemma faced in the confirmatory approach by achieving the maximum predictive power with only those variables that contribute in a statistically significant amount. Yet in the selection of variables for inclusion in the regression variate, three critical caveats markedly affect the resulting regression equation.

1. The multicollinearity among independent variables has substantial impact on the final model specification. Let us examine the situation with two independent variables that have almost equal correlations with the dependent variable that are also highly correlated. The criterion for inclusion or deletion in these approaches is to maximize the incremental predictive power of the additional variable. *If one of these variables enters the regression model, it is highly unlikely that the other variable will also enter because these variables are highly correlated and separately show little unique variance (see the later discussion on multicollinearity).* For this reason, the researcher must assess the effects of multicollinearity in model interpretation by not only examining the final regression equation, but also examining the direct correlations of all potential independent variables. This knowledge will help the researcher to avoid concluding that the independent variables that do not enter the model are inconsequential when in fact they may be highly related to the dependent variable, but also correlated with variables already in the model. Although the sequential search approaches will maximize the predictive ability of the regression model, the researcher must be careful in using these methods in establishing the impact of independent variables without considering multicollinearity among independent variables.

2. All sequential search methods create a loss of control on the part of the researcher. Even though the researcher does specify the variables to be considered for the regression variate, it is the estimation technique, interpreting the empirical data, that specifies the final regression model. In many situations, complications such as multicollinearity can result in a final regression model that achieves the highest levels of predictive accuracy, but has little managerial relevance in terms of variables included and so on. Yet in such instances, what recourse does the researcher have? The ability to specify the final regression model has been relinquished by the researcher. Use of these estimation techniques must consider trade-offs between the advantages found in these methods versus the lack of control in establishing the final regression model.

3. The third caveat pertains primarily to the stepwise procedure. In this approach, multiple significance tests are performed in the model estimation process. To ensure that the overall error rate across all significance tests is reasonable, the researcher should employ more conservative thresholds (e.g., .01) in adding or deleting variables.

The sequential estimation methods have become widely used because of their efficiency in selecting that subset of independent variables that maximizes the predictive accuracy. With this benefit comes the potential for misleading results in explanation where only one of a set of highly correlated variables is entered into the equation and a loss of control in model specification. These potential issues do not suggest that sequential search methods should be avoided, just that the researcher must realize the issues (pro and con) involved in their use.

COMBINATORIAL APPROACH   The third basic type of estimation technique is the combinatorial approach, primarily a generalized search process across all possible combinations of independent variables. The best-known procedure is **all-possible-subsets regression,** which is exactly as the name suggests. All possible combinations of the independent variables are examined, and the best-fitting set of variables is identified. For example, a model with 10 independent variables has 1,024 possible regressions (1 equation with only the constant, 10 equations with a single independent variable, 45 equations with all combinations of two variables, and so on). With computerized estimation procedures, this process can be managed today for even rather large problems, identifying the best overall regression equation for any number of measures of predictive fit.

Usage of this approach has decreased due to criticisms of its (1) atheoretical nature and (2) lack of consideration of such factors as multicollinearity, the identification of outliers and influentials, and the interpretability of the results. When these issues are considered, the "best" equation may involve serious problems that affect its appropriateness, and another model may ultimately be selected. This approach can, however, provide insight into the number of regression models that are roughly equivalent in predictive power, yet possess quite different combinations of independent variables.

REVIEW OF THE MODEL SELECTION APPROACHES   Whether a confirmatory sequential search, or combinatorial method is chosen, the most important criterion is the researcher's substantive knowledge of the research context and any theoretical foundation that allows for an objective and informed perspective as to the variables to be included as well as the expected signs and magnitude of their coefficients (see Rules of Thumb 4-5). Without this knowledge, the regression results can have high predictive accuracy but little managerial or theoretical relevance. Each estimation method has advantages and disadvantages, such that no single method is always preferred over the other approaches. As such, the researcher should never totally rely on any one of these approaches without understanding how the implications of the estimation method relate to the researcher's objectives for prediction and explanation and the theoretical foundation for the research. Many times the use of two or more methods in combination may provide a more balanced perspective for the researcher versus using only a single method and trying to address all of the issues affecting the results.

Testing the Regression Variate for Meeting the Regression Assumptions

With the independent variables selected and the regression coefficients estimated, the researcher must now assess the estimated model for meeting the assumptions underlying multiple regression. As discussed in stage 3, the individual variables must meet the assumptions of linearity, constant variance, independence, and normality. In addition to the individual variables, the regression variate must also meet these assumptions. The diagnostic tests discussed in stage 3 can be applied to assessing the collective effect of the variate through examination of the residuals. If substantial violations are found, the researcher must take corrective actions on one or more of the independent variables and then reestimate the regression model.

RULES OF THUMB 4-5

Estimation Techniques

- No matter which estimation technique is chosen, theory must be a guiding factor in evaluating the final regression model because:
  - Confirmatory specification, the only method to allow direct testing of a prespecified model, is also the most complex from the perspectives of specification error, model parsimony, and achieving maximum predictive accuracy
  - Sequential search (e.g., stepwise), although maximizing predictive accuracy, represents a completely "automated" approach to model estimation, leaving the researcher almost no control over the final model specification
  - Combinatorial estimation, while considering all possible models, still removes control from the researcher in terms of final model specification even though the researcher can view the set of roughly equivalent models in terms of predictive accuracy

- No single method is best, and the prudent strategy is to employ a combination of approaches to capitalize on the strengths of each to reflect the theoretical basis of the research question

Examining the Statistical Significance of Our Model

If we were to take repeated random samples of respondents and estimate a regression equation for each sample, we would not expect to get exactly the same values for the regression coefficients each time. Nor would we expect the same overall level of model fit. Instead, some amount of random variation due to sampling error should cause differences among many samples. From a researcher's perspective, we take only one sample and base our predictive model on it. With only this one sample, we need to test the hypothesis that our regression model can represent the population rather than just our one sample. These statistical tests take two basic forms: a test of the variation explained (coefficient of determination) and a test of each regression coefficient.

SIGNIFICANCE OF THE OVERALL MODEL: TESTING THE COEFFICIENT OF DETERMINATION   To test the hypothesis that the amount of variation explained by the regression model is more than the baseline prediction (i.e., that R^2 is significantly greater than zero), the F ratio is calculated as:

$$F \text{ ratio} = \frac{\dfrac{SS_{regression}}{df_{regression}}}{\dfrac{SS_{residual}}{df_{residual}}}$$

where

$$df_{regression} = \text{Number of estimated coefficients (including intercept)} - 1$$
$$df_{residual} = \text{Sample size} - \text{Number of estimated coefficients (including intercept)}$$

Three important features of this ratio should be noted:

1. Dividing each sum of squares by its appropriate degrees of freedom (df) results in an estimate of the variance. The top portion of the F ratio is the variance explained by the regression model, while the bottom portion is the unexplained variance.
2. Intuitively, if the ratio of the explained variance to the unexplained is high, the regression variate must be of significant value in explaining the dependent variable. Using the F distribution, we can make a statistical test to determine whether the ratio is different from zero (i.e., statistically significant). In those instances in which it is statistically significant, the researcher can feel confident that the regression model is not specific to just this sample, but would be expected to be significant in multiple samples from this population.
3. Although larger R^2 values result in higher F values, the researcher must base any assessment of practical significance separate from statistical significance. Because statistical significance is really an assessment of the impact of sampling error, the researcher must be cautious of always assuming that statistically significant results are also practically significant. This caution is particularly relevant in the case of large samples where even small R^2 values (e.g., 5% or 10%) can be statistically significant, but such levels of explanation would not be acceptable for further action on a practical basis.

In our example of credit card usage, the F ratio for the simple regression model is $(16.5 \div 1)/(5.50 \div 6) = 18.0$. The tabled F statistic of 1 with 6 degrees of freedom at a significance level of .05 yields 5.99. Because the F ratio is greater than the table value, we reject the hypothesis that the reduction in error we obtained by using family size to predict credit card usage was a chance occurrence. This outcome means that, considering the sample used for estimation, we can explain 18 times more variation than when using the average, which is not likely to happen by chance (less than 5% of the time). Likewise, the F ratio for the multiple regression model with two independent variables is $(18.96 \div 2)/(3.04 \div 5) = 15.59$. The multiple regression model is also statistically significant, indicating that the additional independent variable was substantial in adding to the regression model's predictive ability.

ADJUSTING THE COEFFICIENT OF DETERMINATION   As was discussed earlier in defining degrees of freedom, the addition of a variable will always increase the R^2 value. This increase then creates concern with generalizability because R^2 will increase even if nonsignificant predictor variables are added. The impact is most noticeable when the sample size is close in size to the number of predictor variables (termed *overfitting*—when the degrees of freedom is small). With this impact minimized when the sample size greatly exceeds the number of independent variables, a number of guidelines have been proposed as discussed earlier (e.g., 10 to 15 observations per independent variable to a minimum of 5 observations per independent variable). Yet what is needed is a more objective measure relating the level of overfitting to the R^2 achieved by the model.

This measure involves an adjustment based on the number of independent variables relative to the sample size. In this way, adding nonsignificant variables just to increase the R^2 can be discounted in a systematic manner. As part of all regression programs, an **adjusted coefficient of determination (adjusted R^2)** is given along with the coefficient of determination. Interpreted the same as the unadjusted coefficient of determination, the adjusted R^2 decreases as we have fewer observations per independent variable. The adjusted R^2 value is particularly useful in comparing across regression equations involving different numbers of independent variables or different sample sizes because it makes allowances for the degrees of freedom for each model.

In our example of credit card usage, R^2 for the simple regression model is .751, and the adjusted R^2 is .709. As we add the second independent variable, R^2 increases to .861, but the adjusted R^2 increases to only .806. When we add the third variable, R^2 increases to only .872, and the adjusted R^2 decreases to .776. Thus, while we see the R^2 always increased when adding variables, the decrease in the adjusted R^2 when adding the third variable indicates an overfitting of the data. When we discuss assessing the statistical significance of regression coefficients in the next section, we will see that the third variable was not statistically significant. The adjusted R^2 not only reflects overfitting, but also the addition of variables that do not contribute significantly to predictive accuracy.

SIGNIFICANCE TESTS OF REGRESSION COEFFICIENTS   Statistical significance testing for the estimated coefficients in regression analysis is appropriate and necessary when the analysis is based on a sample of the population rather than a census. When using a sample, the researcher is not just interested in the estimated regression coefficients for that sample, but is also interested in how the coefficients are expected to vary across repeated samples. The interested reader can find more detailed discussion of the calculations underlying significance tests for regression coefficients in the Basic Stats appendix on the text's Web sites (www.pearsonglobaleditions.com/hair or www.mvstats.com.).

Establishing a Confidence Interval. Significance testing of regression coefficients is a statistically based probability estimate of whether the estimated coefficients across a large number of samples of a certain size will indeed be different from zero. To make this judgment, a confidence interval must be established around the estimated coefficient. If the confidence interval does not include the value of zero, then it can be said that the coefficient's difference from zero is statistically significant. To make this judgment, the researcher relies on three concepts:

- Establishing the **significance level (alpha)** denotes the chance the researcher is willing to take of being wrong about whether the estimated coefficient is different from zero. A value typically used is .05. As the researcher desires a smaller chance of being wrong and sets the significance level smaller (e.g., .01 or .001), the statistical test becomes more demanding. Increasing the significance level to a higher value (e.g., .10) allows for a larger chance of being wrong, but also makes it easier to conclude that the coefficient is different from zero.
- **Sampling error** is the cause for variation in the estimated regression coefficients for each sample drawn from a population. For small sample sizes, the sampling error is larger

and the estimated coefficients will most likely vary widely from sample to sample. As the size of the sample increases, the samples become more representative of the population (i.e., sampling error decreases), and the variation in the estimated coefficients for these large samples become smaller. This relationship holds true until the analysis is estimated using the population. Then the need for significance testing is eliminated because the sample is equal to, and thus perfectly representative of, the population (i.e., no sampling error).

- The **standard error** is the expected variation of the estimated coefficients (both the constant and the regression coefficients) due to sampling error. The standard error acts like the standard deviation of a variable by representing the expected dispersion of the *coefficients* estimated from repeated samples of this size.

With the significance level selected and the standard error calculated, we can establish a confidence interval for a regression coefficient based on the standard error just as we can for a mean based on the standard deviation. For example, setting the significance level at .05 would result in a confidence interval of $\pm 1.96 \times$ standard error, denoting the outer limits containing 95 percent of the coefficients estimated from repeated samples. With the confidence interval in hand, the researcher now must ask three questions about the statistical significance of any regression coefficient:

1. *Was statistical significance established?* The researcher sets the significance level from which the confidence interval is derived (e.g., a significance level of 5% for a large sample establishes the confidence interval at $\pm 1.96 \times$ standard error). A coefficient is deemed statistically significant if the confidence interval does not include zero.

2. *How does the sample size come into play?* If the sample size is small, sampling error may cause the standard error to be so large that the confidence interval includes zero. However, if the sample size is larger, the test has greater precision because the variation in the coefficients becomes less (i.e., the standard error is smaller). Larger samples do not guarantee that the coefficients will not equal zero, but instead make the test more precise.

3. *Does it provide practical significance in addition to statistical significance?* As we saw in assessing the R^2 value for statistical significance, just because a coefficient is statistically significant does not guarantee that it also is practically significant. Be sure to evaluate the sign and size of any significant coefficient to ensure that it meets the research needs of the analysis.

Significance testing of regression coefficients provides the researcher with an empirical assessment of their "true" impact. Although it is not a test of validity, it does determine whether the impacts represented by the coefficients are generalizable to other samples from this population.

Identifying Influential Observations

Up to now, we focused on identifying general patterns within the entire set of observations. Here we shift our attention to individual observations, with the objective of finding the observations that

- lie outside the general patterns of the data set or
- strongly influence the regression results.

These observations are not necessarily "bad" in the sense that they must be deleted. In many instances they represent the distinctive elements of the data set. However, we must first identify them and assess their impact before proceeding. This section introduces the concept of influential observations and their potential impact on the regression results. A more detailed discussion of the procedures for identifying influential observations is available on the Web at www.pearsonglobaleditions.com/hair or www.mvstats.com.

TYPES OF INFLUENTIAL OBSERVATIONS   **Influential observations,** in the broadest sense, include all observations that have a disproportionate effect on the regression results. The three basic types are based upon the nature of their impact on the regression results:

- **Outliers** are observations that have large residual values and can be identified only with respect to a specific regression model. Outliers were traditionally the only form of influential observation considered in regression models, and specialized regression methods (e.g., robust regression) were even developed to deal specifically with outliers' impact on the regression results [1, 12]. Chapter 2 provides additional procedures for identifying outliers.
- **Leverage points** are observations that are distinct from the remaining observations based on their independent variable values. Their impact is particularly noticeable in the estimated coefficients for one or more independent variables.
- Influential observations are the broadest category, including all observations that have a disproportionate effect on the regression results. Influential observations potentially include outliers and leverage points but may include other observations as well. Also, not all outliers and leverage points are necessarily influential observations.

IDENTIFYING INFLUENTIAL OBSERVATIONS   Influential observations many times are difficult to identify through the traditional analysis of residuals when looking for outliers. Their patterns of residuals would go undetected because the residual for the influential points (the perpendicular distance from the point of the estimated regression line) would not be so large as to be classified as an outlier. Thus, focusing only on large residuals would generally ignore these influential observations.

Figure 4-8 illustrates several general forms of influential observations and their corresponding pattern of residuals:

- *Reinforcing:* In Figure 4-8a, the influential point is a "good" one, reinforcing the general pattern of the data and lowering the standard error of the prediction and coefficients. It is a leverage point but has a small or zero residual value, because it is predicted well by the regression model.
- *Conflicting:* Influential points can have an effect that is *contrary* to the general pattern of the remaining data but still have small residuals (see Figures 4-8b and 4-8c). In Figure 4-8b, two influential observations almost totally account for the observed relationship, because without them no real pattern emerges from the other data points. They also would not be identified if only large residuals were considered, because their residual value would be small. In Figure 4-8c, an even more profound effect is seen in which the influential observations counteract the general pattern of all the remaining data. In this case, the real data would have larger residuals than the bad influential points.
- Multiple influential points may also work toward the same result. In Figure 4-8e, two influential points have the same relative position, making detection somewhat harder. In Figure 4-8f, influentials have quite different positions but a similar effect on the results.
- *Shifting:* Influential observations may affect all of the results in a similar manner. One example is shown in Figure 4-8d, where the slope remains constant but the intercept is shifted. Thus, the relationship among all observations remains unchanged except for the shift in the regression model. Moreover, even though all residuals would be affected, little in the way of distinguishing features among them would assist in diagnosis.

These examples illustrate that we must develop a bigger toolkit of methods for identifying these influential cases. Procedures for identifying all types of influential observations are becoming quite widespread, yet are still less well known and utilized infrequently in regression analysis. All computer programs provide an analysis of residuals from which those with large values (particularly standardized residuals greater than 2.0) can be easily identified. Moreover, most computer programs now provide at least some of the diagnostic measures for identifying leverage points and other influential observations.

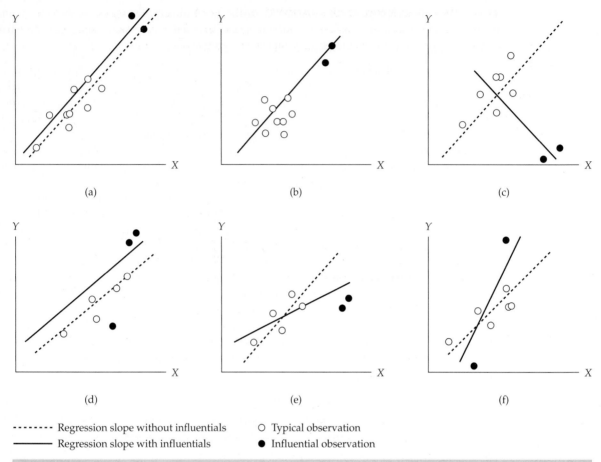

Legend:
- - - - - - Regression slope without influentials   ○ Typical observation
———— Regression slope with influentials   ● Influential observation

FIGURE 4-8   Patterns of Influential Observations

Source: Adapted from Belsley et al. and Mason and Perreault [2, 9].

REMEDIES FOR INFLUENTIALS   The need for additional study of leverage points and influentials is highlighted when we see the substantial extent to which the generalizability of the results and the substantive conclusions (the importance of variables, level of fit, and so forth) can be changed by only a small number of observations. Whether good (accentuating the results) or bad (substantially changing the results), these observations must be identified to assess their impact. Influentials, out-liers, and leverage points are based on one of four conditions, each of which has a specific course of corrective action:

1. *An error in observations or data entry:* Remedy by correcting the data or deleting the case.
2. *A valid but exceptional observation that is explainable by an extraordinary situation:* Remedy by deletion of the case unless variables reflecting the extraordinary situation are included in the regression equation.
3. *An exceptional observation with no likely explanation:* Presents a special problem because it lacks reasons for deleting the case, but its inclusion cannot be justified either, suggesting analyses with and without the observations to make a complete assessment.
4. *An ordinary observation in its individual characteristics but exceptional in its combination of characteristics:* Indicates modifications to the conceptual basis of the regression model and should be retained.

RULES OF THUMB 4-6

Statistical Significance and Influential Observations

- Always ensure practical significance when using large sample sizes, because the model results and regression coefficients could be deemed irrelevant even when statistically significant due just to the statistical power arising from large sample sizes
- Use the adjusted R^2 as your measure of overall model predictive accuracy
- Statistical significance is required for a relationship to have validity, but statistical significance without theoretical support does not support validity
- Although outliers may be easily identifiable, the other forms of influential observations requiring more specialized diagnostic methods can be equal to or have even more impact on the results

In all situations, the researcher is encouraged to delete truly exceptional observations but still guard against deleting observations that, although different, are representative of the population. Remember that the objective is to ensure the most representative model for the sample data so that it will best reflect the population from which it was drawn. This practice extends beyond achieving the highest predictive fit, because some outliers may be valid cases that the model should attempt to predict, even if poorly. The researcher should also be aware of instances in which the results would be changed substantially by deleting just a single observation or a small number of observations.

STAGE 5: INTERPRETING THE REGRESSION VARIATE

The researcher's next task is to interpret the regression variate by evaluating the estimated regression coefficients for their explanation of the dependent variable. The researcher must evaluate not only the regression model that was estimated but also the potential independent variables that were omitted if a sequential search or combinatorial approach was employed. In those approaches, multicollinearity may substantially affect the variables ultimately included in the regression variate. Thus, in addition to assessing the estimated coefficients, the researcher must also evaluate the potential impact of omitted variables to ensure that the managerial significance is evaluated along with statistical significance.

Using the Regression Coefficients

The estimated regression coefficients, termed the b coefficients, represent both the type of relationship (positive or negative) and the strength of the relationship between independent and dependent variables in the regression variate. The sign of the coefficient denotes whether the relationship is positive or negative, and the value of the coefficient indicates the change in the dependent value each time the independent variable changes by one unit.

For example, in the simple regression model of credit card usage with family size as the only independent variable, the coefficient for family size was .971. This coefficient denotes a positive relationship that shows as a family adds a member, credit card usage is expected to increase almost by one (.971) credit card. Moreover, if the family size decreases by one member, credit card usage would also decrease by almost one credit card (−.971).

The regression coefficients play two key functions in meeting the objectives of prediction and explanation for any regression analysis.

PREDICTION Prediction is an integral element in regression analysis, both in the estimation process as well as in forecasting situations. As described in the first section of the chapter, regression involves the use of a variate (the regression model) to estimate a single value for the dependent variable. This process is used not only to calculate the predicted values in the estimation procedure, but also with additional samples used for validation or forecasting purposes.

Estimation. First, in the ordinary least squares (OLS) estimation procedure used to derive the regression variate, a prediction of the dependent variable is made for each observation in the data set. The estimation procedure sets the weights of the regression variate to minimize the residuals (e.g., minimizing the differences between predicted and actual values of the dependent variable). No matter how many independent variables are included in the regression model, a single predicted value is calculated. As such, the predicted value represents the total of all effects of the regression model and allows the residuals, as discussed earlier, to be used extensively as a diagnostic measure for the overall regression model.

Forecasting. Although prediction is an integral element in the estimation procedure, the real benefits of prediction come in forecasting applications. A regression model is used in these instances for prediction with a set of observations not used in estimation. For example, assume that a sales manager developed a forecasting equation to forecast monthly sales of a product line. After validating the model, the sales manager inserts the upcoming month's expected values for the independent variables and calculates an expected sales value.

A simple example of a forecasting application can be shown using the credit card example. Assume that we are using the following regression equation that was developed to estimate the number of credit cards (Y) held by a family:

$$Y = .286 + .635V_1 + .200V_2 + .272V_3$$

Now, suppose that we have a family with the following characteristics: Family size (V_1) of 2 persons, family income (V_2) of 22 ($22,000), and number of autos (V_3) being 3. What would be the expected number of credit cards for this family?

We would substitute the values for V_1, V_2, and V_3 into the regression equation and calculate the predicted value:

$$Y = .286 + .635(2) + .200(22) + .272(3)$$
$$= .286 + 1.270 + 4.40 + .819$$
$$= 6.775$$

Our regression equation would predict that this family would have 6.775 credit cards.

EXPLANATION Many times the researcher is interested in more than just prediction. It is important for a regression model to have accurate predictions to support its validity, but many research questions are more focused on assessing the nature and impact of each independent variable in making the prediction of the dependent variable. In the multiple regression example discussed earlier, an appropriate question is to ask which variable—family size or family income—has the larger effect in predicting the number of credit cards used by a family. Independent variables with larger regression coefficients, all other things equal, would make a greater contribution to the predicted value. Insights into the relationship between independent and dependent variables are gained by examining the relative contributions of each independent variable. In our simple example, a marketer looking to sell additional credit cards and looking for families with higher numbers of cards would know whether to seek out families based on family size or family income.

Interpretation with Regression Coefficients. Thus, for explanatory purposes, the regression coefficients become indicators of the relative impact and importance of the independent variables in their relationship with the dependent variable. Unfortunately, in many instances the regression coefficients do not give us this information directly, the key issue being "all other things equal." As we will see, the scale of the independent variables also comes into play. To illustrate, we use a simple example.

Suppose we want to predict the amount a married couple spends at restaurants during a month. After gathering a number of variables, it was found that two variables, the husband's and wife's annual incomes, were the best predictors. The following regression equation was calculated using a least squares procedure:

$$Y = 30 + 4INC_1 + .004INC_2$$

where

$$INC_1 = \text{Husband's annual income (in \$1,000s)}$$
$$INC_2 = \text{Wife's annual income (in dollars)}$$

If we just knew that INC_1 and INC_2 were annual incomes of the two spouses, then we would probably conclude that the income of the husband was much more important (actually 1,000 times more) than that of the wife. On closer examination, however, we can see that the two incomes are actually equal in importance, the difference being in the way each was measured. The husband's income is in thousands of dollars, such that a \$40,000 income is used in the equation as 40, whereas a wife's \$40,000 income is entered as 40,000. If we predict the restaurant dollars due just to the wife's income, it would be \$160 ($40,000 \times .004$), which would be exactly the same for a husband's income of \$40,000 (40×4). Thus, each spouse's income is equally important, but this interpretation would probably not occur through just an examination of the regression coefficients.

In order to use the regression coefficients for explanatory purposes, we must first ensure that all of the independent variables are on comparable scales. Yet even then, differences in variability from variable to variable can affect the size of the regression coefficient. What is needed is a way to make all independent variables comparable in both scale and variability. We can achieve both these objectives and resolve this problem in explanation by using a modified regression coefficient called the beta coefficient.

Standardizing the Regression Coefficients: Beta Coefficients. The variation in response scale and variability across variables makes direct interpretation problematic. Yet, what if each of our independent variables had been standardized before we estimated the regression equation? **Standardization** converts variables to a common scale and variability, the most common being a mean of zero (0.0) and standard deviation of one (1.0). In this way, we make sure that all variables are comparable. If we still want the original regression coefficients for predictive purposes, is our only recourse to standardize all the variables and then perform a second regression analysis?

Luckily, multiple regression gives us not only the regression coefficients, but also coefficients resulting from the analysis of standardized data termed **beta (β) coefficients.** Their advantage is that they eliminate the problem of dealing with different units of measurement (as illustrated previously) and thus reflect the relative impact on the dependent variable of a change in one standard deviation in either variable. Now that we have a common unit of measurement, we can determine which variable has the most impact. We will return to our credit card example to see the differences between the regression (b) and beta (β) coefficients.

In the credit card example, the regression (b) and beta (β) coefficients for the regression equation with three independent variables (V_1, V_2, and V_3) are shown here:

Variable	Coefficients	
	Regression (*b*)	Beta (β)
V_1 Family Size	.635	.566
V_2 Family Income	.200	.416
V_3 Number of Autos	.272	.108

Interpretation using the regression versus the beta coefficients yields substantially different results. The regression coefficients indicate that V_1 is markedly more important than either V_2 or V_3, which are roughly comparable. The beta coefficients tell a different story. V_1 is still most important, but V_2 is now almost as important, while V_3 now is marginally important at best. These simple results portray the inaccuracies in interpretation that may occur when regression coefficients are used with variables of differing scale and variability.

Although the beta coefficients represent an objective measure of importance that can be directly compared, two cautions must be observed in their use:

- First, they should be used as a *guide to the relative importance of individual independent variables only when collinearity is minimal.* As we will see in the next section, collinearity can distort the contributions of any independent variable even if beta coefficients are used.
- Second, the beta values can be *interpreted only in the context of the other variables in the equation.* For example, a beta value for family size reflects its importance only in relation to family income, not in any absolute sense. If another independent variable were added to the equation, the beta coefficient for family size would probably change, because some relationship between family size and the new independent variable is likely.

In summary, beta coefficients should be used only as a guide to the relative importance of the independent variables included in the equation and only for those variables with minimal multicollinearity.

Assessing Multicollinearity

A key issue in interpreting the regression variate is the correlation among the independent variables. This problem is one of data, not of model specification. The ideal situation for a researcher would be to have a number of independent variables highly correlated with the dependent variable, but with little correlation among themselves. If you refer back to Chapter 3 and our discussion of factor analysis, the use of factor scores that are orthogonal (uncorrelated) was suggested to achieve such a situation.

Yet in most situations, particularly situations involving consumer response data, some degree of multicollinearity is unavoidable. On some other occasions, such as using dummy variables to represent nonmetric variables or polynomial terms for nonlinear effects, the researcher is creating situations of high multicollinearity. The researcher's task includes the following:

- Assess the degree of multicollinearity.
- Determine its impact on the results.
- Apply the necessary remedies if needed.

In the following sections we discuss in detail some useful diagnostic procedures, the effects of multicollinearity, and then possible remedies.

IDENTIFYING MULTICOLLINEARITY   The simplest and most obvious means of identifying collinearity is an examination of the correlation matrix for the independent variables. The presence of high correlations (generally .90 and higher) is the first indication of substantial collinearity. Lack of any high correlation values, however, does not ensure a lack of collinearity. Collinearity may be due to the combined effect of two or more other independent variables (termed *multicollinearity*).

To assess multicollinearity, we need a measure expressing the degree to which each independent variable is explained by the set of other independent variables. *In simple terms, each independent variable becomes a dependent variable and is regressed against the remaining independent variables.* The two most common measures for assessing both pairwise and multiple variable collinearity are tolerance and its inverse, the variance inflation factor.

Tolerance. A direct measure of multicollinearity is **tolerance,** which is defined as the amount of variability of the selected independent variable *not explained by the other independent variables.* Thus, for any regression model with two or more independent variables the tolerance can be simply defined in two steps:

1. Take each independent variable, one at a time, and calculate R^{2*}—the amount of that independent variable that is explained by all of the other independent variables in the regression model. In this process, the selected independent variable is made a dependent variable predicted by all the other remaining independent variables.
2. Tolerance is then calculated as $1 - R^{2*}$. For example, if the other independent variables explain 25 percent of independent variable X_1 ($R^{2*} = .25$), then the tolerance value of X_1 is .75 ($1.0 - .25 = .75$).

The tolerance value should be high, which means a small degree of multicollinearity (i.e., the other independent variables do not collectively have any substantial amount of shared variance). Determining the appropriate levels of tolerance will be addressed in a following section.

Variance Inflation Factor. A second measure of multicollinearity is the **variance inflation factor (VIF),** which is calculated simply as the inverse of the tolerance value. In the preceding example with a tolerance of .75, the VIF would be 1.33 ($1.0 \div .75 = 1.33$). Thus, instances of higher degrees of multicollinearity are reflected in lower tolerance values and higher VIF values. The VIF gets its name from the fact that the square root of the VIF ($\sqrt{VIF}$) is the degree to which the standard error has been increased due to multicollinearity. Let us examine a couple of examples to illustrate the interrelationship of tolerance, VIF, and the impact on the standard error.

For example, if the VIF equals 1.0 (meaning that tolerance equals 1.0 and thus no multicollinearity), then the $\sqrt{VIF} = 1.0$ and the standard error is unaffected. However, assume that the tolerance is .25 (meaning that there is fairly high multicollinearity, because 75% of the variable's variance is explained by other independent variables). In this case the VIF is 4.0 ($1.0 \div .25 = 4$) and the standard error has been doubled ($\sqrt{4} = 2$) due to multicollinearity.

The VIF translates the tolerance value, which directly expresses the degree of multicollinearity, into an impact on the estimation process. As the standard error is increased, it makes the confidence intervals around the estimated coefficients larger, thus making it harder to demonstrate that the coefficient is significantly different from zero.

THE EFFECTS OF MULTICOLLINEARITY The effects of multicollinearity can be categorized in terms of estimation or explanation. In either instance, however, the underlying reason is the same: Multicollinearity creates "shared" variance between variables, thus decreasing the ability to predict the dependent measure as well as ascertain the relative roles of each independent variable. Figure 4-9 portrays the proportions of shared and unique variance for two independent variables in varying instances of collinearity. If the collinearity of these variables is zero, then the individual variables predict 36 and 25 percent of the variance in the dependent variable, for an overall prediction (R^2) of 61 percent. As multicollinearity increases, the total variance explained decreases *(estimation).* Moreover, the amount of unique variance for the independent variables is reduced to levels that make estimation of their individual effects quite problematic *(explanation).* The following sections address these impacts in more detail.

Impacts on Estimation. Multicollinearity can have substantive effects not only on the predictive ability of regression model (as just described), but also on the estimation of the regression coefficients and their statistical significance tests.

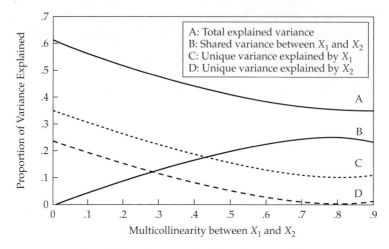

Correlation between dependent and independent variables:
X_1 and dependent (.60), X_2 and dependent (.50)

FIGURE 4-9   **Proportions of Unique and Shared Variance by Levels of Multicollinearity**

1. First, the extreme case of multicollinearity in which two or more variables are perfectly correlated, termed **singularity,** prevents the estimation of any coefficients. Although singularities may occur naturally among the independent variables, many times they are a result of researcher error. A common mistake is to include all of the dummy variables used to represent a nonmetric variable, rather than omitting one as the reference category. Also, such actions as including a summated scale along with the individual variables that created it will result in singularities. For whatever reason, however, the singularity must be removed before the estimation of coefficients can proceed.
2. As multicollinearity increases, the ability to demonstrate that the estimated regression coefficients are significantly different from zero can become markedly impacted due to increases in the standard error as shown in the VIF value. This issue becomes especially problematic at smaller sample sizes, where the standard errors are generally larger due to sampling error.
3. Apart from affecting the statistical tests of the coefficients or the overall model, high degrees of multicollinearity can also result in regression coefficients being incorrectly estimated and even having the wrong signs. Two examples illustrate this point.

Our first example (see Table 4-6) illustrates the situation of reversing signs due to high negative correlation between two variables. In Example A it is clear in examining the correlation matrix and the simple regressions that the relationship between Y and V_1 is positive, whereas the relationship between Y and V_2 is negative. The multiple regression equation, however, does not maintain the relationships from the simple regressions. It would appear to the casual observer examining only the multiple regression coefficients that both relationships (Y and V_1, Y and V_2) are negative, when we know that such is not the case for Y and V_1. The sign of V_1's regression coefficient is wrong in an intuitive sense, but the strong negative correlation between V_1 and V_2 results in the reversal of signs for V_1. Even though these effects on the estimation procedure occur primarily at relatively high levels of multicollinearity (above .80), the possibility of counterintuitive and misleading results necessitates a careful scrutiny of each regression variate for possible multicollinearity.

TABLE 4-6   Regression Estimates with Multicollinear Data

	EXAMPLE A				EXAMPLE B		
	Data				*Data*		
ID	Y	V_1	V_2	ID	Y	Z_1	Z_2
1	5	6	13	1	3.7	3.2	2.9
2	3	8	13	2	3.7	3.3	4.2
3	9	8	11	3	4.2	3.7	4.9
4	9	10	11	4	4.3	3.3	5.1
5	13	10	9	5	5.1	4.1	5.5
6	11	12	9	6	5.2	3.8	6.0
7	17	12	7	7	5.2	2.8	4.9
8	15	14	7	8	5.6	2.6	4.3
				9	5.6	3.6	5.4
				10	6.0	4.1	5.5

	Correlation Matrix				*Correlation Matrix*		
	Y	V_1	V_2		Y	Z_1	Z_2
Y	1.0			Y	1.0		
V_1	.823	1.0		Z_1	.293	1.0	
V_2	−.977	−.913	1.0	Z_2	.631	.642	1.0

Regression Estimates	*Regression Estimates*
Simple Regression (V_1):	Simple Regression (Z_1):
$Y = -4.75 + 1.5V_1$	$Y = 2.996 + .525Z_1$
Simple Regression (V_2):	Simple Regression (Z_2):
$Y = 29.75 - 1.95V_2$	$Y = 1.999 + .587Z_2$
Multiple Regression (V_1, V_2):	Multiple Regression (Z_1, Z_2):
$Y = 44.75 - .75V_1 - 2.7V_2$	$Y = 2.659 - .343Z_1 + .702Z_2$

A similar situation can be seen in Example B of Table 4-6. Here, both Z_1 and Z_2 are positively correlated with the dependent measure (.293 and .631, respectively), but have a higher intercorrelation (.642). In this regression model, even though both bivariate correlations of the independent variables are positive with the dependent variable and the two independent variables are positively intercorrelated, when the regression equation is estimated the coefficient for Z_1 becomes negative (−.343) and the other coefficient is positive (.702). This typifies the case of high multicollinearity reversing the signs of the weaker independent variables (i.e., lower correlations with the dependent variable).

In some instances, this reversal of signs is expected and desirable. Termed a **suppression effect,** it denotes instances when the "true" relationship between the dependent and independent variable(s) has been hidden in the bivariate correlations (e.g., the expected relationships are nonsignificant or even reversed in sign). By adding additional independent variables and inducing multicollinearity, some unwanted shared variance is accounted for and the remaining unique variance allows for the estimated coefficients to be in the expected direction. More detailed descriptions of all of the potential instances of suppression effects are shown in [5].

However, in other instances, the theoretically supported relationships are reversed because of multicollinearity, leaving the researcher to explain why the estimated coefficients are revered from the expected sign. In these instances, the researcher may need to revert to using the bivariate correlations to describe the relationship rather than the estimated coefficients that are impacted by multicollinearity.

The reversal of signs may be encountered in all of the estimation procedures, but is seen more often in confirmatory estimation processes where a set of variables is entered into the regression model and the likelihood of weaker variables being affected by multicollinearity is increased.

Impacts on Explanation. The effects on explanation primarily concern the ability of the regression procedure and the researcher to represent and understand the effects of each independent variable in the regression variate. As multicollinearity occurs (even at the relatively low levels of .30 or so) the process for identifying the unique effects of independent variables becomes increasingly difficult. Remember that the regression coefficients represent the amount of unique variance explained by each independent variable. As multicollinearity results in larger portions of shared variance and lower levels of unique variance, the effects of the individual independent variables become less distinguishable. It is even possible to find those situations in which multicollinearity is so high that none of the independent regression coefficients are statistically significant, yet the overall regression model has a significant level of predictive accuracy.

HOW MUCH MULTICOLLINEARITY IS TOO MUCH?   Because the tolerance value is the amount of a variable unexplained by the other independent variables, small tolerance values (and thus large VIF values because VIF = 1 ÷ tolerance) denote high collinearity. A common cutoff threshold is a tolerance value of .10, which corresponds to a VIF value of 10. However, particularly when samples sizes are smaller, the researcher may wish to be more restrictive due to the increases in the standard errors due to multicollinearity. With a VIF threshold of <u>10</u>, this tolerance would correspond to standard errors being "inflated" more than three times ($\sqrt{10} = 3.16$) what they would be with no multicollinearity.

Each researcher must determine the degree of collinearity that is acceptable, because most defaults or recommended thresholds still allow for substantial collinearity. Some suggested guidelines follow:

- When assessing bivariate correlations, two issues should be considered. First, correlations of even .70 (which represents "shared" variance of 50%) can impact both the explanation and estimation of the regression results. Moreover, even lower correlations can have an impact if the correlation between the two independent variables is greater than either independent variable's correlation with the dependent measure (e.g., the situation in our earlier example of the reversal of signs).
- The suggested cutoff for the tolerance value is. 10 (or a corresponding VIF of 10.0), which corresponds to a multiple correlation of .95 with the other independent variables. When values at this level are encountered, multicollinearity problems are almost certain. However, problems are likely at much lower levels as well. For example, a VIF of 5.3 corresponds to a multiple correlation of .9 between one independent variable and all other independent variables. Even a VIP of 3.0 represents a multiple correlation of .82, which would be considered high if between dependent and independent variables.

Therefore, the researcher should always assess the degree and impact of multicollinearity even when the diagnostic measures are substantially below the suggested cutoff (e.g., VIF values of 3 to 5).

We strongly suggest that the researcher always specify the allowable tolerance values in regression programs, because the default values for excluding collinear variables allow for an extremely high degree of collinearity. For example, the default tolerance value in SPSS for excluding a variable is .0001, which means that until more than 99.99 percent of variance is predicted by the other independent variables, the variable could be included in the regression equation. Estimates of the actual effects of high collinearity on the estimated coefficients are possible but beyond the scope of this text (see Neter et al. [11]).

Even with diagnoses using VIF or tolerance values, we still do not necessarily know which variables are correlated. A procedure development by Belsley et al. [2] allows for the correlated variables to be identified, even if we have correlation among several variables. It provides the researcher greater diagnostic power in assessing the extent and impact of multicollinearity and is described in the supplement to this chapter available on the Web at www.pearsonglobaleditions.com/hair or www.mvstats.com.

REMEDIES FOR MULTICOLLINEARITY   The remedies for multicollinearity range from modification of the regression variate to the use of specialized estimation procedures. Once the degree of collinearity has been determined, the researcher has a number of options:

1. Omit one or more highly correlated independent variables and identify other independent variables to help the prediction. The researcher should be careful when following this option, however, to avoid creating specification error when deleting one or more independent variables.
2. Use the model with the highly correlated independent variables for prediction only (i.e., make no attempt to interpret the regression coefficients), while acknowledging the lowered level of overall predictive ability.
3. Use the simple correlations between each independent variable and the dependent variable to understand the independent–dependent variable relationship.
4. Use a more sophisticated method of analysis such as Bayesian regression (or a special case— ridge regression) or regression on principal components to obtain a model that more clearly reflects the simple effects of the independent variables. These procedures are discussed in more detail in several texts [2, 11].

Each of these options requires that the researcher make a judgment on the variables included in the regression variate, which should always be guided by the theoretical background of the study.

RULES OF THUMB 4-7

Interpreting the Regression Variate

- Interpret the impact of each independent variable relative to the other variables in the model, because model respecification can have a profound effect on the remaining variables:
  - Use beta weights when comparing relative importance among independent variables
  - Regression coefficients describe changes in the dependent variable, but can be difficult in comparing across independent variables if the response formats vary

- Multicollinearity may be considered "good" when it reveals a suppressor effect, but generally it is viewed as harmful because increases in multicollinearity:
  - Reduce the overall R^2 that can be achieved
  - Confound estimation of the regression coefficients
  - Negatively affect the statistical significance tests of coefficients

- Generally accepted levels of multicollinearity (tolerance values up to .10, corresponding to a VIF of 10) almost always indicate problems with multicollinearity, but these problems may also be seen at much lower levels of collinearity and multicollinearity:
  - Bivariate correlations of .70 or higher may result in problems, and even lower correlations may be problematic if they are higher than the correlations between the independent and dependent variables
  - Values much lower than the suggested thresholds (VIF values of even 3 to 5) may result in interpretation or estimation problems, particularly when the relationships with the dependent measure are weaker

STAGE 6: VALIDATION OF THE RESULTS

After identifying the best regression model, the final step is to ensure that it represents the general population (generalizability) and is appropriate for the situations in which it will be used (transferability). The best guideline is the extent to which the regression model matches an existing theoretical model or set of previously validated results on the same topic. In many instances, however, prior results or theory are not available. Thus, we also discuss empirical approaches to model validation.

Additional or Split Samples

The most appropriate empirical validation approach is to test the regression model on a new sample drawn from the general population. A new sample will ensure representativeness and can be used in several ways. First, the original model can predict values in the new sample and predictive fit can be calculated. Second, a separate model can be estimated with the new sample and then compared with the original equation on characteristics such as the significant variables included; sign, size, and relative importance of variables; and predictive accuracy. In both instances, the researcher determines the validity of the original model by comparing it to regression models estimated with the new sample.

Many times the ability to collect new data is limited or precluded by such factors as cost, time pressures, or availability of respondents. Then, the researcher may divide the sample into two parts: an estimation subsample for creating the regression model and the holdout or validation subsample used to test the equation. Many procedures, both random and systematic, are available for splitting the data, each drawing two independent samples from the single data set. All the popular statistical packages include specific options to allow for estimation and validation on separate subsamples. Chapter 5 provides a discussion of the use of estimation and validation subsamples in discriminant analysis.

Whether a new sample is drawn or not, it is likely that differences will occur between the original model and other validation efforts. The researcher's role now shifts to being a mediator among the varying results, looking for the best model across all samples. The need for continued validation efforts and model refinements reminds us that no regression model, unless estimated from the entire population, is the final and absolute model.

Calculating the PRESS Statistic

An alternative approach to obtaining additional samples for validation purposes is to employ the original sample in a specialized manner by calculating the **PRESS statistic,** a measure similar to R^2 used to assess the predictive accuracy of the estimated regression model. It differs from the prior approaches in that not one, but $n - 1$ regression models are estimated. The procedure omits one observation in the estimation of the regression model and then predicts the omitted observation with the estimated model. Thus, the observation cannot affect the coefficients of the model used to calculate its predicted value. The procedure is applied again, omitting another observation, estimating a new model, and making the prediction. The residuals for the observations can then be summed to provide an overall measure of predictive fit.

Comparing Regression Models

When comparing regression models, the most common standard used is overall predictive fit. R^2 provides us with this information, but it has one drawback: As more variables are added, R^2 will always increase. Thus, by including all independent variables, we will never find another model with a higher R^2, but we may find that a smaller number of independent variables result in an almost identical value. Therefore, to compare between models with different numbers of independent variables, we use the adjusted R^2. The adjusted R^2 is also useful in comparing models between different data sets, because it will compensate for the different sample sizes.

Forecasting with the Model

Forecasts can always be made by applying the estimated model to a new set of independent variable values and calculating the dependent variable values. However, in doing so, we must consider several factors that can have a serious impact on the quality of the new predictions:

1. When applying the model to a new sample, we must remember that the predictions now have not only the sampling variations from the original sample, but also those of the newly drawn sample. Thus, we should always calculate the confidence intervals of our predictions in addition to the point estimate to see the expected range of dependent variable values.
2. We must make sure that the conditions and relationships measured at the time the original sample was taken have not changed materially. For instance, in our credit card example, if most companies started charging higher fees for their cards, actual credit card holdings might change substantially, yet this information would not be included in the model.
3. Finally, we must not use the model to estimate beyond the range of independent variables found in the sample. For instance, in our credit card example, if the largest family had 6 members, it might be unwise to predict credit card holdings for families with 10 members. One cannot assume that the relationships are the same for values of the independent variables substantially greater or less than those in the original estimation sample.

ILLUSTRATION OF A REGRESSION ANALYSIS

The issues concerning the application and interpretation of regression analysis have been discussed in the preceding sections by following the six-stage model-building framework introduced in Chapter 1 and discussed earlier in this chapter. To provide an illustration of the important questions at each stage, an illustrative example is presented here detailing the application of multiple regression to a research problem specified by HBAT. Chapter 1 introduced a research setting in which HBAT had obtained a number of measures in a survey of customers. To demonstrate the use of multiple regression, we show the procedures used by researchers to attempt to predict customer satisfaction of the individuals in the sample with a set of 13 independent variables.

Stage 1: Objectives of Multiple Regression

HBAT management has long been interested in more accurately predicting the satisfaction level of its customers. If successful, it would provide a better foundation for their marketing efforts. To this end, researchers at HBAT proposed that multiple regression analysis should be attempted to predict the customer satisfaction based on their perceptions of HBAT's performance. In addition to finding a way to accurately predict satisfaction, the researchers also were interested in identifying the factors that lead to increased satisfaction for use in differentiated marketing campaigns.

To apply the regression procedure, researchers selected customer satisfaction (X_{19}) as the dependent variable (Y) to be predicted by independent variables representing perceptions of HBAT's performance. The following 13 variables were included as independent variables:

X_6	Product Quality	X_{13}	Competitive Pricing
X_7	E-Commerce	X_{14}	Warranty & Claims
X_8	Technical Support	X_{15}	New Products
X_9	Complaint Resolution	X_{16}	Ordering & Billing
X_{10}	Advertising	X_{17}	Price Flexibility
X_{11}	Product Line	X_{18}	Delivery Speed
X_{12}	Salesforce Image		

The relationship among the 13 independent variables and customer satisfaction was assumed to be statistical, not functional, because it involved perceptions of performance and may include levels of measurement error.

Stage 2: Research Design of a Multiple Regression Analysis

The HBAT survey obtained 100 respondents from their customer base. All 100 respondents provided complete responses, resulting in 100 observations available for analysis. The first question to be answered concerning sample size is the level of relationship (R^2) that can be detected reliably with the proposed regression analysis.

Table 4-5 indicates that the sample of 100, with 13 potential independent variables, is able to detect relationships with R^2 values of approximately 23 percent at a power of .80 with the significance level set at .01. If the significance level is relaxed to .05, then the analysis will identify relationships explaining about 18 percent of the variance. The sample of 100 observations also meets the guideline for the minimum ratio of observations to independent variables (5:1) with an actual ratio of 7:1 (100 observations with 13 variables).

The proposed regression analysis was deemed sufficient to identify not only statistically significant relationships but also relationships that had managerial significance. Although HBAT researchers can be reasonably assured that they are not in danger of overfitting the sample, they should still validate the results if at all possible to ensure the generalizability of the findings to the entire customer base, particularly when using a stepwise estimation technique.

Stage 3: Assumptions in Multiple Regression Analysis

Meeting the assumptions of regression analysis is essential to ensure that the results obtained are truly representative of the sample and that we obtain the best results possible. Any serious violations of the assumptions must be detected and corrected if at all possible. Analysis to ensure the research is meeting the basic assumptions of regression analysis involves two steps: (1) testing the individual dependent and independent variables and (2) testing the overall relationship after model estimation. This section addresses the assessment of individual variables. The overall relationship will be examined after the model is estimated.

The three assumptions to be addressed for the individual variables are linearity, constant variance (homoscedasticity), and normality. For purposes of the regression analysis, we summarize the results found in Chapter 2 detailing the examination of the dependent and independent variables.

First, scatterplots of the individual variables did not indicate any nonlinear relationships between the dependent variable and the independent variables. Tests for heteroscedasticity found that only two variables (X_6 and X_{17}) had minimal violations of this assumption, with no corrective action needed. Finally, in the tests of normality, six variables (X_6, X_7, X_{12}, X_{13}, X_{16}, and X_{17}) were found to violate the statistical tests. For all but one variable (X_{12}), transformations were sufficient remedies. In the HBAT example, we will first provide the results using the original variables and then compare these findings with the results obtained using the transformed variables.

Although regression analysis has been shown to be quite robust even when the normality assumption is violated, researchers should estimate the regression analysis with both the original and transformed variables to assess the consequences of nonnormality of the independent variables on the interpretation of the results. To this end, the original variables are used first and later results for the transformed variables are shown for comparison.

Stage 4: Estimating the Regression Model and Assessing Overall Model Fit

With the regression analysis specified in terms of dependent and independent variables, the sample deemed adequate for the objectives of the study, and the assumptions assessed for the individual variables, the model-building process now proceeds to estimation of the regression model and assessing the overall model fit. For purposes of illustration, the stepwise procedure is employed to select variables for inclusion in the regression variate. After the regression model has been estimated, the variate will be assessed for meeting the assumptions of regression analysis. Finally, the observations will be examined to determine whether any observation should be deemed influential. Each of these issues is discussed in the following sections.

STEPWISE ESTIMATION: SELECTING THE FIRST VARIABLE The stepwise estimation procedure maximizes the incremental explained variance at each step of model building. In the first step, the highest bivariate correlation (also the highest partial correlation, because no other variables are in the equation) will be selected. The process for the HBAT example follows.

Table 4-7 displays all the correlations among the 13 independent variables and their correlations with the dependent variable (X_{19}, Customer Satisfaction). Examination of the correlation matrix (looking down the first column) reveals that complaint resolution (X_9) has the highest bivariate correlation with the dependent variable (.603). The first step is to build a regression equation using just this single independent variable. The results of this first step appear as shown in Table 4-8.

Overall Model Fit. From Table 4-8 the researcher can address issues concerning both overall model fit as well as the stepwise estimation of the regression model.

- *Multiple R* Multiple R is the correlation coefficient (at this step) for the simple regression of X_9 and the dependent variable. It has no plus or minus sign because in multiple regression the signs of the individual variables may vary, so this coefficient reflects only the degree of association. In the first step of the stepwise estimation, the Multiple R is the same as the bivariate correlation (.603) because the equation contains only one variable.
- *R Square* R square (R^2) is the correlation coefficient squared ($.603^2 = .364$), also referred to as the *coefficient of determination*. This value indicates the percentage of total variation of Y (X_{19}, Customer Satisfaction) explained by the regression model consisting of X_9.
- *Standard Error of the Estimate* The standard error of the estimate is another measure of the accuracy of our predictions. It is the square root of the sum of the squared errors divided by the degrees of freedom, also represented by the square root of the $MS_{residual}$ ($\sqrt{89.45 \div 98} = .955$). It represents an estimate of the standard deviation of the actual dependent values around the regression line; that is, it is a measure of variation around the regression line. The standard error of the estimate also can be viewed as the standard deviation of the prediction errors; thus it becomes a measure to assess the absolute size of the prediction error. It is used also in estimating the size of the confidence interval for the predictions. See Neter et al. [11] for details regarding this procedure.
- *ANOVA and F Ratio* The ANOVA analysis provides the statistical test for the overall model fit in terms of the F ratio. The total sum of squares ($51.178 + 89.450 = 140.628$) is the squared error that would occur if we used only the mean of Y to predict the dependent variable. Using the values of X_9 reduces this error by 36.4 percent ($51.178 \div 140.628$). This reduction is deemed statistically significant with an F ratio of 56.070 and a significance level of .000.

Variables in the Equation (Step 1). In step 1, a single independent variable (X_9) is used to calculate the regression equation for predicting the dependent variable. For each variable in the equation, several measures need to be defined: the regression coefficient, the standard error of the coefficient, the *t* value of variables in the equation, and the collinearity diagnostics (tolerance and VIF).

TABLE 4-7 Correlation Matrix: HBAT Data

	X_{19}	X_6	X_7	X_8	X_9	X_{10}	X_{11}	X_{12}	X_{13}	X_{14}	X_{15}	X_{16}	X_{17}	X_{18}
Dependent Variable														
X_{19} Customer Satisfaction														
Independent Variables														
X_6 Product Quality	**.486**	1.000												
X_7 E-Commerce	**.283**	−.137	1.000											
X_8 Technical Support	.113	.096	.001	1.000										
X_9 Complaint Resolution	**.603**	.106	.140	.097	1.000									
X_{10} Advertising	**.305**	−.053	**.430**	−.063	**.197**	1.000								
X_{11} Product Line	**.551**	**.477**	−.053	**.193**	**.561**	−.012	1.000							
X_{12} Salesforce Image	**.500**	−.152	**.792**	.017	**.230**	**.542**	−.061	1.000						
X_{13} Competitive Pricing	**−.208**	**−.401**	**.229**	**−.271**	−.128	.134	**−.495**	**.265**	1.000					
X_{14} Warranty & Claims	**.178**	.088	.052	**.797**	.140	.011	**.273**	.107	**−.245**	1.000				
X_{15} New Products	.071	.027	−.027	−.074	.059	.084	.046	.032	.023	.035	1.000			
X_{16} Order & Billing	**.522**	.104	.156	.080	**.757**	**.184**	**.424**	**.195**	−.115	**.197**	.069	1.000		
X_{17} Price Flexibility	.056	**−.493**	**.271**	**−.186**	**.395**	**.334**	**−.378**	**.352**	**.471**	**−.170**	.094	**.407**	1.000	
X_{18} Delivery Speed	**.577**	.028	**.192**	.025	**.865**	**.276**	**.602**	**.272**	−.073	.109	.106	**.751**	**.497**	1.000

Note: Items in bold are significant at .05 level.

TABLE 4-8   Example Output: Step 1 of HBAT Multiple Regression Example

Step 1–Variable Entered: X_9 Complaint Resolution

Multiple R	.603
Coefficient of Determination (R^2)	.364
Adjusted R^2	.357
Standard error of the estimate	.955

Analysis of Variance

	Sum of Squares	df	Mean Square	F	Sig.
Regression	51.178	1	51.178	56.070	.000
Residual	89.450	98	.913		
Total	140.628	99			

Variables Entered into the Regression Model

	Regression Coefficients			Statistical Significance		Correlations			Collinearity Statistics	
Variables Entered	**B**	**Std. Error**	**Beta**	**t**	**Sig.**	**Zero-order**	**Partial**	**Part**	**Tolerance**	**VIF**
(Constant)	3.680	.443		8.310	.000					
X_9 Complaint Resolution	.595	.079	.603	7.488	.000	.603	.603	.603	1.000	1.000

Variables Not Entered into the Regression Model

		Statistical Significance		Partial Correlation	Collinearity Statistics	
	Beta In	**t**	**Sig.**		**Tolerance**	**VIF**
X_6 Product Quality	.427	6.193	.000	.532	.989	1.011
X_7 E-Commerce	.202	2.553	.012	.251	.980	1.020
X_8 Technical Support	.055	.675	.501	.068	.991	1.009
X_{10} Advertising	.193	2.410	.018	.238	.961	1.040
X_{11} Product Line	.309	3.338	.001	.321	.685	1.460
X_{12} Salesforce Image	.382	5.185	.000	.466	.947	1.056
X_{13} Competitive Pricing	−.133	−1.655	.101	−.166	.984	1.017
X_{14} Warranty & Claims	.095	1.166	.246	.118	.980	1.020
X_{15} New Products	.035	.434	.665	.044	.996	1.004
X_{16} Order & Billing	.153	1.241	.218	.125	.427	2.341
X_{17} Price Flexibility	−.216	−2.526	.013	−.248	.844	1.184
X_{18} Delivery Speed	.219	1.371	.173	.138	.252	3.974

- *Regression Coefficients (b and Beta)* The regression coefficient (b) and the standardized coefficient (β) reflect the change in the dependent measure for each unit change in the independent variable. Comparison between regression coefficients allows for a relative assessment of each variable's importance in the regression model.

  The value .595 is the regression coefficient (b_9) for the independent variable (X_9). The predicted value for each observation is the intercept (3.680) plus the regression coefficient (.595) times its value of the independent variable ($Y = 3.680 + .595X_9$). The standardized regression coefficient, or beta value, of .603 is the value calculated from standardized data. With only one independent variable, the squared beta coefficient equals the coefficient of determination. The beta value enables you to compare the effect of X_9 on Y to the effect of other independent variables on Y at each stage, because this value reduces the regression coefficient to a comparable unit, the number of standard deviations. (Note that at this time we have no other variables available for comparison.)

- **Standard Error of the Coefficient** The standard error of the regression coefficient is an estimate of how much the regression coefficient will vary between samples of the same size taken from the same population. In a simple sense, it is the standard deviation of the estimates of b_9 across multiple samples. If one were to take multiple samples of the same sample size from the same population and use them to calculate the regression equation, the standard error is an estimate of how much the regression coefficient would vary from sample to sample. A smaller standard error implies more reliable prediction and therefore smaller confidence intervals.

  The standard error of b_9 is .079, denoting that the 95% confidence interval for b_9 would be $.595 \pm (1.96 \times .079)$, or ranging from a low of .44 to a high of .75. The value of b_9 divided by the standard error $(.595 \div .079 = 7.488)$ is the calculated t value for a t-test of the hypothesis $b_9 = 0$ (see following discussion).

- **t Value of Variables in the Equation** The t value of variables in the equation, as just calculated, measures the significance of the partial correlation of the variable reflected in the regression coefficient. As such, it indicates whether the researcher can confidently say, with a stated level of error, that the coefficient is not equal to zero. F values may be given at this stage rather than t values. They are directly comparable because the t value is approximately the square root of the F value.

  The t value is also particularly useful in the stepwise procedure in helping to determine whether any variable should be dropped from the equation once another independent variable has been added. The calculated level of significance is compared to the threshold level set by the researcher for dropping the variable. In our example, we set a .10 level for dropping variables from the equation. The critical value for a significance level of .10 with 98 degrees of freedom is 1.658. As more variables are added to the regression equation, each variable is checked to see whether it still falls within this threshold. If it falls outside the threshold (significance greater than .10), it is eliminated from the regression equation, and the model is estimated again.

  In our example, the t value (as derived by dividing the regression coefficient by the standard error) is 7.488, which is statistically significant at the .000 level. It gives the researcher a high level of assurance that the coefficient is not equal to zero and can be assessed as a predictor of customer satisfaction.

- **Correlations** Three different correlations are given as an aid in evaluating the estimation process. The zero-order correlation is the simple bivariate correlation between the independent and dependent variable. The partial correlation denotes the incremental predictive effect, controlling for other variables in the regression model on both dependent and independent variables. This measure is used for judging which variable is next added in sequential search methods. Finally, the **part correlation** denotes the unique effect attributable to each independent variable.

  For the first step in a stepwise solution, all three correlations are identical (.603) because no other variables are in the equation. As variables are added, these values will differ, each reflecting their perspective on each independent variable's contribution to the regression model.

- **Collinearity Statistics** Both collinearity measures (tolerance and VIF) are given to provide a perspective on the impact of collinearity on the independent variables in the regression equation. Remember that the tolerance value is the amount of an independent variable's predictive capability that is not predicted by the other independent variables in the equation. Thus, it represents the unique variance remaining for each variable. The VIF is the inverse of the tolerance value.

  In the case of a single variable in the regression model, the tolerance is 1.00, indicating that it is totally unaffected by other independent variables (as it should be since it is the only variable in the model). Also, the VIF is 1.00. Both values indicate a complete lack of multicollinearity.

Variables Not in the Equation. With X_9 included in the regression equation, 12 other potential independent variables remain for inclusion to improve the prediction of the dependent variable. For each of these variables, four types of measures are available to assess their potential contribution to the regression model: partial correlations, collinearity measures, standardized coefficients (Beta), and t values.

- ***Partial Correlation and Collinearity Measures*** The partial correlation is a measure of the variation in Y that can be accounted for by each of these additional variables, controlling for the variables already in the equation (only X_9 in step 1). As such, the sequential search estimation methods use this value to denote the next candidate for inclusion. If the variable with the highest partial correlation exceeds the threshold of statistical significance required for inclusion, it will be added to the regression model at the next step.

    The partial correlation represents the correlation of each variable not in the model with the unexplained portion of the dependent variable. As such, the contribution of the partial correlation (the squared partial correlation) is that percentage of the unexplained variance that is explained by the addition of this independent variable. Assume that the variable(s) in the regression model already account for 60 percent of the dependent measure ($R^2 = .60$ with unexplained variance $= .40$). If a partial correlation has a value of .5, then the *additional* explained variance it accounts for is the square of the partial correlation times the amount of unexplained variance. In this simple example, that is $.5^2 \times .40$, or 10 percent. By adding this variable, we would expect the R^2 value to increase by 10 percent (from .60 to .70).

    For our example, the values of partial correlations range from a high of .532 to a low of .044. X_6, with the highest value of .532, should be the variable next entered if this partial correlation is found to be statistically significant (see later section). It is interesting to note, however, that X_6 had only the sixth highest bivariate correlation with X_{19}. Why was it the second variable to enter the stepwise equation, ahead of the variables with higher correlations? The variables with the second, third, and fourth highest correlations with X_{19} were X_{18} (.577), X_{11} (.551), and X_{16} (.522). Both X_{18} and X_{16} had high correlations with X_9, reflected in their rather low tolerance values of .252 and .427, respectively. It should be noted that this fairly high level of multicollinearity is not unexpected, because these three variables (X_9, X_{16}, and X_{18}) constituted the first factor derived in Chapter 3. X_{11}, even though it did not join this factor, is highly correlated with X_9 (.561) to the extent that the tolerance is only .685. Finally, X_{12}, the fifth highest bivariate correlation with X_{19}, only has a correlation with X_9 of .230, but it was just enough to make the partial correlation slightly lower than that of X_6. The correlation of X_9 and X_6 of only .106 resulted in a tolerance of .989 and transformed the bivariate correlation of .486 into a partial correlation of .532, which was highest among all the remaining 12 variables.

    If X_6 is added, then the R^2 value should increase by the partial correlation squared times the amount of unexplained variance (change in $R^2 = .532^2 \times .636 = .180$). Because 36.4 percent was already explained by X_9, X_6 can explain only 18.0 percent of the remaining variance. A Venn diagram illustrates this concept.

    The shaded area of X_6 as a proportion of the shaded area of Y represents the partial correlation of X_6 with Y given X_9. The shaded area, as a proportion of Y, denotes the incremental variance explained by X_6 given that X_9 is already in the equation.

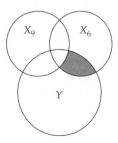

    Note that the total explained variance (all of the overlap areas for Y) is not equal to just the areas associated with the partial correlations of X_6 and X_9. Some explanation is due to the shared effects of both X_6 and X_9. The shared effects are denoted by the middle section where these two variables overlap among themselves and with Y. The calculation of the unique variance associated with adding X_6 can also be determined through the part correlation.

- *Standardized Coefficients* For each variable not in the equation, the standardized coefficient (Beta) that it would have if entered into the equation is calculated. In this manner, the researcher can assess the relative magnitude of this variable if added to those variables already in the equation. Moreover, it allows for an assessment of practical significance in terms of relative predictive power of the added variable.

   In Table 4-8, we see that X_6, the variable with the highest partial correlation, also has the highest Beta coefficient if entered. Even though the magnitude of .427 is substantial, it can also be compared with the beta for the variable now in the model (X_9 with a beta of .603), indicating that X_6 will make a substantive contribution to the explanation of the regression model, as well as to its predictive capability.

- *t Values of Variables Not in the Equation* The t value measures the significance of the partial correlations for variables not in the equation. They are calculated as a ratio of the additional sum of squares explained by including a particular variable and the sum of squares left after adding that same variable. If this t value does not exceed a specified significance level (e.g., .05), the variable will not be allowed to enter the equation. The tabled t value for a significance level of .05 with 97 degrees of freedom is 1.98.

   Looking at the column of t values in Table 4-8, we note that six variables (X_6, X_7, X_{10}, X_{11}, X_{12}, and X_{17}) exceed this value and are candidates for inclusion. Although all are significant, the variable added will be that variable with the highest partial correlation. We should note that establishing the threshold of statistical significance before a variable is added precludes adding variables with no significance even though they increase the overall R^2.

Looking Ahead. With the first step of the stepwise procedure completed, the final task is to evaluate the variables not in the equation and determine whether another variable meets the criteria and can be added to the regression model. As noted earlier, the partial correlation must be great enough to be statistically significant at the specified level (generally .05). If two or more variables meet this criterion, then the variable with the highest partial correlation is selected.

As described earlier, X_6 (Product Quality) has the highest partial correlation at this stage, even though four other variables had higher bivariate correlations with the dependent variable. In each instance, multicollinearity with X_9, entered in the first step, caused the partial correlations to decrease below that of X_6.

We know that a significant portion of the variance in the dependent variable is explained by X_9, but the stepwise procedure indicates that if we add X_6 with the highest partial correlation coefficient with the dependent variable and a t value is significant at the .05 level, we will make a significant increase in the predictive power of the overall regression model. Thus, we can now look at the new model using both X_9 and X_6.

STEPWISE ESTIMATION: ADDING A SECOND VARIABLE (X_6)   The next step in a stepwise estimation is to check and delete any of the variables in the equation that now fall below the significance threshold, and once done, add the variable with the highest statistically significant partial correlation. The following section details the newly formed regression model and the issues regarding its overall model fit, the estimated coefficients, the impact of multicollinearity, and identification of a variable to add in the next step.

Overall Model Fit. As described in the prior section, X_6 was the next variable to be added to the regression model in the stepwise procedure. The multiple R and R^2 values have both increased with the addition of X_6 (see Table 4-9). The R^2 increased by 18.0 percent, the amount we derived in examining the partial correlation coefficient from X_6 of .532 by multiplying the 63.6 percent of variation that was not explained after step 1 by the partial correlation squared ($63.6 \times .532^2 = 18.0$). Then, of the 63.3 percent unexplained with X_9, $(.532)^2$ of this variance was explained by adding X_6, yielding a total variance explained (R^2) of .544. The adjusted R^2 also increased to .535 and the standard error of the estimate decreased from .955 to .813. Both of these measures also demonstrate the improvement in the overall model fit.

Estimated Coefficients. The regression coefficient for X_6 is .364 and the beta weight is .427. Although not as large as the beta for X_9 (.558), X_6 still has a substantial impact in the overall regression

TABLE 4-9   Example Output: Step 2 of HBAT Multiple Regression Example

Step 2 – Variable Entered: X_6 Product Quality

Multiple R	.738
Coefficient of Determination (R^2)	.544
Adjusted R^2	.535
Standard error of the estimate	.813

Analysis of Variance

	Sum of Squares	df	Mean Square	F	Sig.
Regression	76.527	2	38.263	57.902	.000
Residual	64.101	97	.661		
Total	140.628	99			

Variables Entered into the Regression Model

Variables Entered	Regression Coefficients			Statistical Significance		Correlations			Collinearity Statistics	
	B	Std. Error	Beta	t	Sig.	Zero-order	Partial	Part	Tolerance	VIF
(Constant)	1.077	.564		1.909	.059					
X_9 Complaint Resolution	.550	.068	.558	8.092	.000	.603	.635	.555	.989	1.011
X_6 Product Quality	.364	.059	.427	6.193	.000	.486	.532	.425	.989	1.011

Variables Not Entered into the Regression Model

			Statistical Significance			Collinearity Statistics	
		Beta In	t	Sig.	Partial Correlation	Tolerance	VIF
X_7	E-Commerce	.275	4.256	.000	.398	.957	1.045
X_8	Technical Support	.018	.261	.794	.027	.983	1.017
X_{10}	Advertising	.228	3.423	.001	.330	.956	1.046
X_{11}	Product Line	.066	.683	.496	.070	.508	1.967
X_{12}	Salesforce Image	.477	8.992	.000	.676	.916	1.092
X_{13}	Competitive Pricing	.041	.549	.584	.056	.832	1.202
X_{14}	Warranty & Claims	.063	.908	.366	.092	.975	1.026
X_{15}	New Products	.026	.382	.703	.039	.996	1.004
X_{16}	Order & Billing	.129	1.231	.221	.125	.427	2.344
X_{17}	Price Flexibility	.084	.909	.366	.092	.555	1.803
X_{18}	Delivery Speed	.334	2.487	.015	.246	.247	4.041

model. The coefficient is statistically significant and multicollinearity is minimal with X_9 (as described in the earlier section). Thus, tolerance is quite acceptable with a value of .989 indicating that only 1.1 percent of either variable is explained by the other.

Impact of Multicollinearity. The lack of multicollinearity results in little change for either the value of b_9 (.550) or the beta of X_9 (.558) in step 1. It further indicates that variables X_9 and X_6 are relatively independent (the simple correlation between the two variables is .106). If the effect of X_6 on Y were totally independent of the effect of X_9, the b_9 coefficient would not change at all. The t values indicate that both X_9 and X_6 are statistically significant predictors of Y. The t value for X_9 is now 8.092, whereas it was 7.488 in step 1. The t value for X_6 relates to the contribution of this variable given that X_5 is already in the equation. Note that the t value for X_6 (6.193) is the same value shown for X_6 in step 1 under the heading "Variables Not Entered into the Regression Model" (see Table 4-8).

Identifying Variables to Add. Because X_9 and X_6 both make significant contributions, neither will be dropped in the stepwise estimation procedure. We can now ask "Are other predictors available?" To address this question, we can look in Table 4-9 under the section "Variables Not Entered into the Regression Model."

Looking at the partial correlations for the variables not in the equation in Table 4-9, we see that X_{12} has the highest partial correlation (.676), which is also statistically significant at the .000 level. This variable would explain 45.7 percent of the heretofore unexplained variance ($.676^2 = .457$), or 20.9 percent of the total variance ($.676^2 \times .456$). This substantial contribution actually slightly surpasses the incremental contribution of X_6, the second variable entered in the stepwise procedure.

STEPWISE ESTIMATION: A THIRD VARIABLE (X_{12}) IS ADDED   The next step in a stepwise estimation follows the same pattern of (1) first checking and deleting any variables in the equation falling below the significance threshold and then (2) adding the variable with the highest statistically significant partial correlation. The following section details the newly formed regression model and the issues regarding its overall model fit, the estimated coefficients, the impact of multicollinearity, and identification of a variable to add in the next step.

Overall Model Fit. Entering X_{12} into the regression equation gives the results shown in Table 4-10. As predicted, the value of R^2 increases by 20.9 percent ($.753 - .544 = .209$). Moreover, the adjusted R^2 increases to .745 and the standard error of the estimate decreases to .602. Again, as was the case with X_6 in the previous step, the new variable entered (X_{12}) makes substantial contribution to overall model fit.

Estimated Coefficients. The addition of X_{12} brought a third statistically significant predictor of customer satisfaction into the equation. The regression weight of .530 is complemented by a beta weight of .477, second highest among the three variables in the model (behind the .512 of X_6).

Impact of Multicollinearity. It is noteworthy that even with the third variable in the regression equation, multicollinearity is held to a minimum. The lowest tolerance value is for X_{12} (.916), indicating that only 8.4 percent of variance of X_{12} is accounted for by the other two variables. This pattern of variables entering the stepwise procedure should be expected, however, when viewed in light of the factor analysis done in Chapter 2. From those results, we see that the three variables now in the equation (X_9, X_6, and X_{12}) were each members of different factors in that analysis. Because variables within the same factor exhibit a high degree of multicollinearity, it would be expected that when one variable from a factor enters a regression equation, the odds of another variable from that same factor entering the equation are rather low (and if it does, the impact of both variables will be reduced due to multicollinearity).

TABLE 4-10   Example Output: Step 3 of HBAT Multiple Regression Example

Step 3 – Variable Entered: X_{12} Salesforce Image

Multiple R	.868
Coefficient of Determination (R^2)	.753
Adjusted R^2	.745
Standard error of the estimate	.602

Analysis of Variance

	Sum of Squares	df	Mean Square	F	Sig.
Regression	105.833	3	35.278	97.333	.000
Residual	34.794	96	.362		
Total	140.628	99			

Variables Entered into the Regression Model

	Regression Coefficients			Statistical Significance		Correlations			Collinearity Statistics	
Variables Entered	B	Std. Error	Beta	t	Sig.	Zero-order	Partial	Part	Tolerance	VIF
(Constant)	−1.569	.511		−3.069	.003					
X_9 Complaint Resolution	.433	.052	.439	8.329	.000	.603	.648	.423	.927	1.079
X_6 Product Quality	.437	.044	.512	9.861	.000	.486	.709	.501	.956	1.046
X_{12} Salesforce Image	.530	.059	.477	8.992	.000	.500	.676	.457	.916	1.092

Variables Not Entered into the Regression Model

		Statistical Significance			Collinearity Statistics	
	Beta In	t	Sig.	Partial Correlation	Tolerance	VIF
X_7 E-Commerce	−.232	−2.890	.005	−.284	.372	2.692
X_8 Technical Support	.013	.259	.796	.027	.983	1.017
X_{10} Advertising	−.019	−.307	.760	−.031	.700	1.428
X_{11} Product Line	.180	2.559	.012	.254	.494	2.026
X_{13} Competitive Pricing	−.094	−1.643	.104	−.166	.776	1.288
X_{14} Warranty & Claims	.020	.387	.700	.040	.966	1.035
X_{15} New Products	.016	.312	.755	.032	.996	1.004
X_{16} Order & Billing	.101	1.297	.198	.132	.426	2.348
X_{17} Price Flexibility	−.063	−.892	.374	−.091	.525	1.906
X_{18} Delivery Speed	.219	2.172	.032	.217	.243	4.110

Looking Ahead. At this stage in the analysis, only three variables (X_7, X_{11}, and X_{18}) have the statistically significant partial correlations necessary for inclusion in the regression equation. What happened to the other variables' predictive power? By reviewing the bivariate correlations of each variable with X_{19} in Table 4-7, we can see that of the 13 original independent variables, three variables had nonsignificant bivariate correlations with the dependent variable (X_8, X_{15}, and X_{17}). Thus X_{10}, X_{13}, X_{14}, and X_{16} all have significant bivariate correlations, yet their partial correlations are now nonsignificant. For X_{16}, the high bivariate correlation of .522 was reduced markedly by high multicollinearity (tolerance value of .426, denotes that less than half of original predictive power remaining). For the other three variables, X_{10}, X_{13}, and X_{14}, their lower bivariate correlations (.305, −.208, and .178) have been reduced by multicollinearity just enough to be nonsignificant.

At this stage, we will skip to the final regression model and detail the entry of the final two variables (X_7 and X_{11}) in a single stage for purposes of conciseness.

STEPWISE ESTIMATION: FOURTH AND FIFTH VARIABLES (X_7 AND X_{11}) ARE ADDED The final regression model (Table 4-11) is the result of two more variables (X_7 and X_{11}) being added. For purposes of conciseness, we omitted the details involved in adding X_7 and will focus on the final regression model with both variables included.

Overall Model Fit. The final regression model with five independent variables (X_9, X_6, X_{12}, X_7, and X_{11}) explains almost 80 percent of the variance of customer satisfaction (X_{19}). The adjusted

TABLE 4-11 **Example Output: Step 5 of HBAT Multiple Regression Example**

Step 5 – Variable Entered: X_{11} Product Line

Multiple R	.889
Coefficient of Determination (R^2)	.791
Adjusted R^2	.780
Standard error of the estimate	.559

Analysis of Variance

	Sum of Squares	df	Mean Square	F	Sig.
Regression	111.205	5	22.241	71.058	.000
Residual	29.422	94	.313		
Total	140.628	99			

Variables Entered into the Regression Model

Variables Entered	Regression Coefficients B	Std. Error	Beta	Statistical Significance t	Sig.	Correlations Zero-order	Partial	Part	Collinearity Statistics Tolerance	VIF
(Constant)	−1.151	.500		−2.303	.023					
X_9 Complaint Resolution	.319	.061	.323	5.256	.000	.603	.477	.248	.588	1.701
X_6 Product Quality	.369	.047	.432	7.820	.000	.486	.628	.369	.728	1.373
X_{12} Salesforce Image	.775	.089	.697	8.711	.000	.500	.668	.411	.347	2.880
X_7 E-Commerce	−.417	.132	−.245	−3.162	.002	.283	−.310	−.149	.370	2.701
X_{11} Product Line	.174	.061	.192	2.860	.005	.551	.283	.135	.492	2.033

Variables Not Entered into the Regression Model

	Beta In	Statistical Significance t	Sig.	Partial Correlation	Collinearity Statistics Tolerance	VIF
X_8 Technical Support	−.009	−.187	.852	−.019	.961	1.041
X_{10} Advertising	−.009	−.162	.872	−.017	.698	1.432
X_{13} Competitive Pricing	−.040	−.685	.495	−.071	.667	1.498
X_{14} Warranty & Claims	−.023	−.462	.645	−.048	.901	1.110
X_{15} New Products	.002	.050	.960	.005	.989	1.012
X_{16} Order & Billing	.124	1.727	.088	.176	.423	2.366
X_{17} Price Flexibility	.129	1.429	.156	.147	.272	3.674
X_{18} Delivery Speed	.138	1.299	.197	.133	.197	5.075

R^2 of .780 indicates no overfitting of the model and that the results should be generalizable from the perspective of the ratio of observations to variables in the equation (20:1 for the final model). Also, the standard error of the estimate has been reduced to .559, which means that at the 95% confidence level ($\pm 1.96 \times$ standard error of the estimate), the margin of error for any predicted value of X_{19} can be calculated at ± 1.1.

Estimated Coefficients. The five regression coefficients, plus the constant, are all significant at the .05 level, and all except the constant are significant at the .01 level. The next section (stage 5) provides a more detailed discussion of the regression coefficients and beta coefficients as they relate to interpreting the variate.

Impact of Multicollinearity. The impact of multicollinearity, even among just these five variables, is substantial. Of the five variables in the equation, three of them (X_{12}, X_7, and X_{11}) have tolerance values less than .50 indicating that over one-half of their variance is accounted for by the other variables in the equation. Moreover, these variables were the last three to enter in the step-wise process.

If we examine the zero-order (bivariate) and partial correlations, we can see more directly the effects of multicollinearity. For example, X_{11} has the third highest bivariate correlation (.551) among all 13 variables, yet multicollinearity (tolerance of .492) reduces it to a partial correlation of only .135, making it a marginal contributor to the regression equation. In contrast, X_{12} has a bivariate correlation (.500) that even with high multicollinearity (tolerance of .347) still has a partial correlation of .411. Thus, multicollinearity will always affect a variable's contribution to the regression model, but must be examined to assess the actual degree of impact.

If we take a broader perspective, the variables entering the regression equation correspond almost exactly to the factors derived in Chapter 3. X_9 and X_6 are each members of separate factors, with multicollinearity reducing the partial correlations of other members of these factors to a non-significant level. X_{12} and X_7 are both members of a third factor, but multicollinearity caused a change in the sign of the estimated coefficient for X_7 (see a more detailed discussion in stage 5). Finally, X_{11} did not load on any of the factors, but was a marginal contributor in the regression model.

The impact of multicollinearity as reflected in the factor structure becomes more apparent in using a stepwise estimation procedure and will be discussed in more detail in stage 5. Even apart from issues in explanation, however, multicollinearity can have a substantial impact on the overall predictive ability of any set of independent variables.

Looking Ahead. As noted earlier, the regression model at this stage consists of the five independent variables with the addition of X_{11}. Examining the partial correlations of variables not in the model at this stage (see Table 4-11), we see that none of the remaining variables have a significant partial correlation at the .05 level needed for entry. Moreover, all of the variables in the model remain statistically significant, avoiding the need to remove a variable in the stepwise process. Thus, no more variables are considered for entry or exit and the model is finalized.

AN OVERVIEW OF THE STEPWISE PROCESS The stepwise estimation procedure is designed to develop a regression model with the fewest number of statistically significant independent variables and maximum predictive accuracy. In doing so, however, the regression model can be markedly affected by issues such as multicollinearity. Moreover, the researcher relinquishes control over the formation of the regression model and runs a higher risk of decreased generalizability. The following section provides an overview of the estimation of the stepwise regression model discussed earlier from the perspective of overall model fit. Issues relating to interpretation of the variate, other estimation procedures, and alternative model specifications will be addressed in subsequent sections.

Table 4-12 provides a step-by-step summary detailing the measures of overall fit for the regression model used by HBAT in predicting customer satisfaction. Each of the first three variables added to the equation made substantial contributions to the overall model fit, with substantive increases in the R^2 and adjusted R^2 while also decreasing the standard error of the estimate. With only the first three variables, 75 percent of the variation in customer satisfaction is explained with a confidence interval of ±1.2. Two additional variables are added to arrive at the final model, but these variables, although statistically significant, make much smaller contributions. The R^2 increases by 3 percent and the confidence interval decreases to ±1.1, an improvement of .1. The relative impacts of each variable will be discussed in stage 5, but the stepwise procedure highlights the importance of the first three variables in assessing overall model fit.

In evaluating the estimated equation, we considered statistical significance. We must also address two other basic issues: (1) meeting the assumptions underlying regression and (2) identifying the influential data points. We consider each of these issues in the following sections.

EVALUATING THE VARIATE FOR THE ASSUMPTIONS OF REGRESSION ANALYSIS   To this point, we examined the individual variables for meeting the assumptions required for regression analysis. However, we must also evaluate the variate for meeting these assumptions as well. The assumptions to examine are linearity, homoscedasticity, independence of the residuals, and normality. The principal measure used in evaluating the regression variate is the residual—the difference between the actual dependent variable value and its predicted value. For comparison, we use the studentized residuals, a form of standardized residuals (see Key Terms).

The most basic type of residual plot is shown in Figure 4-10, the studentized residuals versus the predicted values. As we can see, the residuals fall within a generally random pattern, similar to the null plot in Figure 4-5a. However, we must make specific tests for each assumption to check for violations.

Linearity.   The first assumption, linearity, will be assessed through an analysis of residuals (testing the overall variate) and partial regression plots (for each independent variable in the analysis).

Figure 4-10 does not exhibit any nonlinear pattern to the residuals, thus ensuring that the overall equation is linear. We must also be certain, when using more than one independent variable, that each independent variable's relationship is also linear to ensure its best representation in the equation. To do

TABLE 4-12   **Model Summary of Stepwise Multiple Regression Model**

Model Summary

| | Overall Model Fit | | | | R^2 Change Statistics | | | | |
Step	R	R^2	Adjusted R^2	Std. Error of the Estimate	R^2 Change	F Value of R^2 Change	df1	df2	Significance of R^2 Change
1	603	.364	.357	.955	.364	56.070	1	98	.000
2	.738	.544	.535	.813	.180	38.359	1	97	.000
3	868	.753	.745	.602	.208	80.858	1	96	.000
4	.879	.773	.763	.580	.020	8.351	1	95	.005
5	.889	.791	.780	.559	.018	8.182	1	94	.005

Step 1: X_9 Complaint Resolution
Step 2: X_9 Complaint Resolution, X_6 Product Quality
Step 3: X_9 Complaint Resolution, X_6 Product Quality, X_{12} Salesforce Image
Step 4: X_9 Complaint Resolution, X_6 Product Quality, X_{12} Salesforce Image, X_7 E-Commerce Activities
Step 5: X_9 Complaint Resolution, X_6 Product Quality, X_{12} Salesforce Image, X_7 E-Commerce Activities, X_{11} Product Line

Note: Constant (intercept term) included in all regression models.

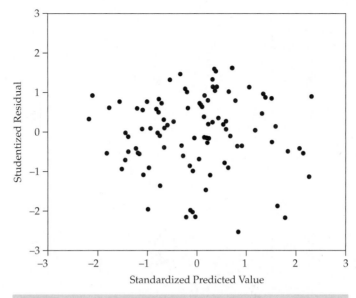

FIGURE 4-10   **Analysis of Standardized Residuals**

so, we use the partial regression plot for each independent variable in the equation. In Figure 4-11 we see that the relationships for X_6, X_9, and X_{12} are reasonably well defined; that is, they have strong and significant effects in the regression equation. Variables X_7 and X_{11} are less well defined, both in slope and scatter of the points, thus explaining their lesser effect in the equation (evidenced by the smaller coefficient, beta value, and significance level). For all five variables, no nonlinear pattern is shown, thus meeting the assumption of linearity for each independent variable.

Homoscedasticity. The next assumption deals with the constancy of the residuals across values of the independent variables. Our analysis is again through examination of the residuals (Figure 4-10), which shows no pattern of increasing or decreasing residuals. This finding indicates homoscedasticity in the multivariate (the set of independent variables) case.

Independence of the Residuals. The third assumption deals with the effect of carryover from one observation to another, thus making the residual not independent. When carryover is found in such instances as time series data, the researcher must identify the potential sequencing variables (such as time in a time series problem) and plot the residuals by this variable. For example, assume that the identification number represents the order in which we collect our responses. We could plot the residuals and see whether a pattern emerges.

In our example, several variables, including the identification number and each independent variable, were tried and no consistent pattern was found. We must use the residuals in this analysis, not the original dependent variable values, because the focus is on the prediction errors, not the relationship captured in the regression equation.

Normality. The final assumption we will check is normality of the error term of the variate with a visual examination of the normal probability plots of the residuals.

As shown in Figure 4-12, the values fall along the diagonal with no substantial or systematic departures; thus, the residuals are considered to represent a normal distribution. The regression variate is found to meet the assumption of normality.

Applying Remedies for Assumption Violations. After testing for violations of the four basic assumptions of multivariate regression for both individual variables and the regression variate, the researcher should assess the impact of any remedies on the results.

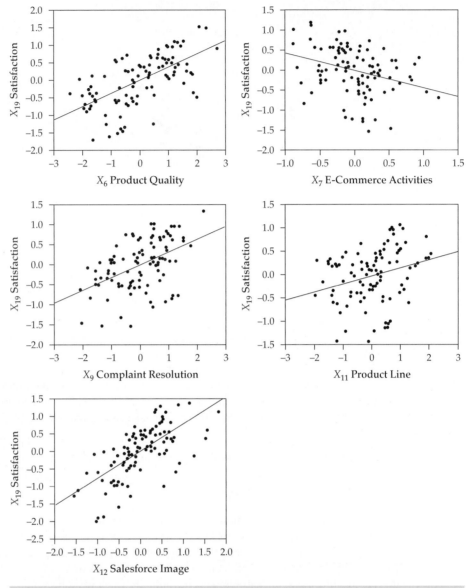

FIGURE 4-11   **Standardized Partial Regression Plots**

In the examination of individual variables in Chapter 2, the only remedies needed were the transformations of X_6, X_7, X_{12}, X_{13}, X_{16}, and X_{17}. A set of differing transformations were used, including the squared term (X_6 and X_{16}), logarithm (X_7), cubed term (X_{13}), and inverse (X_{16}). Only in the case of X_{12} did the transformation not achieve normality. If we substitute these variables for their original values and reestimate the regression equation with a stepwise procedure, we achieve almost identical results. The same variables enter the equation with no substantive differences in either the estimated coefficients or overall model fit as assessed with R^2 and standard error of the estimate. The independent variables not in the equation still show nonsignificant levels for entry— even those that were transformed. Thus, in this case, the remedies for violating the assumptions improved the prediction slightly but did not alter the substantive findings.

IDENTIFYING OUTLIERS AS INFLUENTIAL OBSERVATIONS   For our final analysis, we attempt to identify any observations that are influential (having a disproportionate impact on the regression

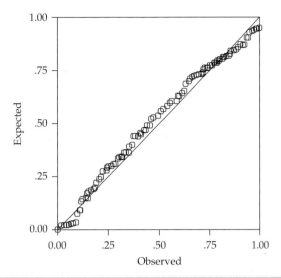

FIGURE 4-12 **Normal Probability Plot: Standardized Residuals**

results) and determine whether they should be excluded from the analysis. Although more detailed procedures are available for identifying outliers as influential observations, we address the use of residuals in identifying outliers in the following section.

The most basic diagnostic tool involves the residuals and identification of any outliers—that is, observations not predicted well by the regression equation that have large residuals. Figure 4-13 shows the studentized residuals for each observation. Because the values correspond to t values, upper and lower limits can be set once the desired confidence interval has been established. Perhaps the most widely used level is the 95% confidence interval ($\alpha = .05$). The corresponding t value is 1.96, thus identifying statistically significant residuals as those with residuals greater than this value (1.96). Seven observations can be seen in Figure 4-13 (2, 10, 20, 45, 52, 80, and 99) to have significant residuals and thus be classified as outliers. Outliers are important because they are observations not represented by the regression equation for one or more reasons, any one of which may be an influential effect on the equation that requires a remedy.

Examination of the residuals also can be done through the partial regression plots (see Figure 4-11). These plots help to identify influential observations for each independent–dependent variable relationship. Consistently across each graph in Figure 4-11, the points at the lower portion are those observations identified as having high negative residuals (observations 2, 10, 20, 45, 52, 80, and 99 in Figure 4-13). These points are not well represented by the relationship and thus affect the partial correlation as well.

More detailed analyses to ascertain whether any of the observations can be classified as influential observations, as well as what may be the possible remedies, are discussed in the supplement to this chapter available on the Web at www.pearsonglobaleditions.com/hair or www.mvstats.com.

Stage 5: Interpreting the Regression Variate

With the model estimation completed, the regression variate specified, and the diagnostic tests that confirm the appropriateness of the results administered, we can now examine our predictive equation based on five independent variables (X_6, X_7, X_9, X_{11}, and X_{12}).

INTERPRETATION OF THE REGRESSION COEFFICIENTS   The first task is to evaluate the regression coefficients for the estimated signs, focusing on those of unexpected direction.

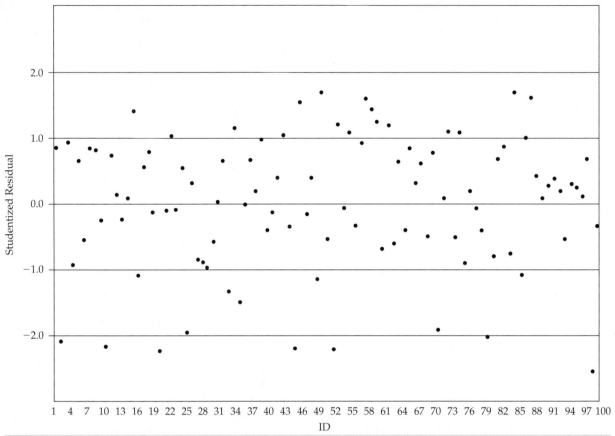

FIGURE 4-13 Plot of Studentized Residuals

The section of Table 4-11 headed "Variables Entered into the Regression Equation" yields the prediction equation from the column labeled "Regression Coefficient: B." From this column, we read the constant term (−1.151) and the coefficients (.319, .369, .775, −.417, and .174) for X_9, X_6, X_{12}, X_7, and X_{11}, respectively. The predictive equation would be written

$$Y = -1.151 + .319X_9 + .369X_6 + .775X_{12} + (-.417)X_7 + .174X_{11}$$

Note: The coefficient of X_7 is included in parentheses to avoid confusion due to the negative value of the coefficient.

With this equation, the expected customer satisfaction level for any customer could be calculated if that customer's evaluations of HBAT are known. For illustration, let us assume that a customer rated HBAT with a value of 6.0 for each of these five measures. The predicted customer satisfaction level for that customer would be

$$\text{Predicted Customer} = -1.151 + .319 \times 6 + .369 \times 6 + .775 \times 6 + (-.417) \times 6 \\ + .174 \times 6$$
$$\text{Satisfaction} = -1.151 + 1.914 + 2.214 + 4.650 - 2.502 + 1.044$$
$$= 6.169$$

We first start with an interpretation of the constant. It is statistically significant (significance = .023), thus making a substantive contribution to the prediction. However, because in our situation it is highly unlikely that any respondent would have zero ratings on all the HBAT perceptions, the constant merely plays a part in the prediction process and provides no insight for interpretation.

In viewing the regression coefficients, the sign is an indication of the relationship (positive or negative) between the independent and dependent variables. All of the variables except one have positive coefficients. Of particular note is the reversed sign for X_7 (E-Commerce), suggesting that an increase in perceptions on this variable has a negative impact on predicted customer satisfaction. All the other variables have positive coefficients, meaning that more positive perceptions of HBAT (higher values) increase customer satisfaction.

Does X_7, then, somehow operate differently from the other variables? In this instance, the bivariate correlation between X_7 and customer satisfaction is positive, indicating that when considered separately, X_7 has a positive relationship with customer satisfaction, just as the other variables. We will discuss in the following section the impact of multicollinearity on the reversal of signs for estimated coefficients.

ASSESSING VARIABLE IMPORTANCE   In addition to providing a basis for predicting customer satisfaction, the regression coefficients also provide a means of assessing the relative importance of the individual variables in the overall prediction of customer satisfaction. When all the variables are expressed in a standardized scale, then the regression coefficients represent relative importance. However, in other instances the beta weight is the preferred measure of relative importance.

In this situation, all the variables are expressed on the same scale, but we will use the beta coefficients for comparison between independent variables. In Table 4-11, the beta coefficients are listed in the column headed "Regression Coefficients: Beta." The researcher can make direct comparisons among the variables to determine their relative importance in the regression variate. For our example, X_{12} (Salesforce Image) was the most important, followed by X_6 (Product Quality), X_9 (Complaint Resolution), X_7 (E-Commerce), and finally X_{11} (Product Line). With a steady decline in size of the beta coefficients across the variables, it is difficult to categorize variables as high, low, or otherwise. However, viewing the relative magnitudes does indicate that, for example, X_{12} (Salesforce Image) shows a more marked effect (three times as much) than X_{11} (Product Line). Thus, to the extent that salesforce image can be increased uniquely from other perceptions, it represents the most direct way, *ceterus paribus,* of increasing customer satisfaction.

MEASURING THE DEGREE AND IMPACT OF MULTICOLLINEARITY   In any interpretation of the regression variate, the researcher must be aware of the impact of multicollinearity. As discussed earlier, highly collinear variables can distort the results substantially or make them quite unstable and thus not generalizable. Two measures are available for testing the impact of collinearity: (1) calculating the tolerance and VIF values and (2) using the condition indices and decomposing the regression coefficient variance (see the supplement to this chapter available on the Web at www.pearsonglobaleditions.com/hair or www.mvstats.com for more details on this process). The tolerance value is 1 minus the proportion of the variable's variance explained by the other independent variables. Thus, a high tolerance value indicates little collinearity, and tolerance values approaching zero indicate that the variable is almost totally accounted for by the other variables (high multicollinearity). The variance inflation factor is the reciprocal of the tolerance value; thus we look for small VIF values as indicative of low correlation among variables.

Diagnosing Multicollinearity.   In our example, tolerance values for the variables in the equation range from .728 (X_6) to .347 (X_{12}), indicating a wide range of multicollinearity effects (see Table 4-11). Likewise, the VIF values range from 1.373 to 2.701. Even though none of these values indicate levels of multicollinearity that should seriously distort the regression variate, we must be careful even with these levels to understand their effects, especially on the stepwise estimation process. The following section will detail some of these effects on both the estimation and interpretation process.

A second approach to identifying multicollinearity and its effects is through the decomposition of the coefficient variance. Researchers are encouraged to explore this technique and the additional insights it offers into the interpretation of the regression equation. Details of this method are discussed in the supplement to this chapter available on the Web at www.pearsonglobaleditions.com/hair or www.mvstats.com.

Impacts Due to Multicollinearity. Although multicollinearity is not so high that the researcher must take corrective action before valid results can be obtained, multicollinearity still has impact on the estimation process, particularly on the composition of the variate and the estimated regression coefficients.

If you examine the bivariate correlations, after X_9 (the first variable added to the regression variate in the stepwise process), the second-highest correlation with the dependent variable is X_{18} (Delivery Speed), followed by X_{11} (Product Line), and X_{16} (Order & Billing). Yet due to collinearity with X_9, the second variable entered was X_6, only the sixth highest bivariate correlation with X_{19}.

The impacts of multicollinearity are seen repeatedly throughout the estimation process, such that the final set of five variables entered into the regression model (X_6, X_7, X_9, X_{11}, and X_{12}) represent the first, sixth, fifth, eighth, and third highest correlations with the dependent variable, respectively. Variables with the second highest correlation (X_{18} at .577) and fourth highest correlation (X_{16} at .522) are never entered into the regression model. Does their exclusion mean they are unimportant? Are they lacking in impact? If a researcher went only by the estimated regression model, multicollinearity would cause serious problems in interpretation. What happened is that X_{16} and X_{18} are highly correlated with X_9, to such an extent that they have little unique explanatory power apart from that shared with X_9. Yet by themselves, or if X_9 was not allowed to enter the model, they would be important predictors of customer satisfaction. The extent of multicollinearity among these three variables is evidenced in Chapter 3, where these three variables were found to all form one of the four factors arising from the HBAT perceptions.

In addition to affecting the composition of the variate, multicollinearity has a distinct impact on the signs of the estimated coefficients. In this situation, it relates to the collinearity between X_{12} (Salesforce Image) and X_7 (E-Commerce). As noted in our earlier discussion about multicollinearity, one possible effect is the reversal of sign for an estimated regression coefficient from the expected direction represented in the bivariate correlation. Here, the high positive correlation between X_{12} and X_7 (correlation = .792) causes the sign for the regression coefficient for X_7 to change from positive (in the bivariate correlation) to a negative sign. If the researcher did not investigate the extent of multicollinearity and its impact, the inappropriate conclusion might be drawn that increases in E-Commerce activities decrease customer satisfaction.

Thus, the researcher must understand the basic relationships supported by the conceptual theory underlying the original model specification and make interpretation based on this theory, not just on the estimated variate.

The researcher must never allow an estimation procedure to dictate the interpretation of the results, but instead must understand the issues of interpretation accompanying each estimation procedure. For example, if all 13 independent variables are entered into the regression variate, the researcher must still contend with the effects of collinearity on the interpretation of the coefficients, but in a different manner than if stepwise were used.

Stage 6: Validating the Results

The final task facing the researcher involves the validation process of the regression model. The primary concern of this process is to ensure that the results are generalizable to the population and not specific to the sample used in estimation. The most direct approach to validation is to obtain another sample from the population and assess the correspondence of the results from the two samples. In the absence of an additional sample, the researcher can assess the validity of the results in several approaches, including an assessment of the adjusted R^2 or estimating the regression model on two or more subsamples of the data.

Examining the adjusted R^2 value reveals little loss in predictive power when compared to the R^2 value (.780 versus .791, see Table 4-11), which indicates a lack of overfitting that would be shown by a more marked difference between the two values. Moreover, with five variables in the model, it maintains an adequate ratio of observations to variables in the variate.

A second approach is to divide the sample into two subsamples, estimate the regression model for each subsample, and compare the results. Table 4-13 contains the stepwise models estimated for two subsamples of 50 observations each. Comparison of the overall model fit demonstrates a high level of similarity of the results in terms of R^2, adjusted R^2, and the standard error of the estimate. Yet in comparing the individual coefficients, some differences do appear. In sample 1, X_9 did not enter in the stepwise results as it did in sample 2 and the overall sample. Instead, X_{16}, highly collinear with X_9, entered. Moreover, X_{12} had a markedly greater beta weight in sample 1 than found in the overall results. In the second sample, four of the variables entered as with the overall results, but X_{11}, the least forceful variable in the overall results, did not enter the model. The omission of X_{11} in one of the subsamples confirms that it was a marginal predictor, as indicated by the low beta and t values in the overall model.

Evaluating Alternative Regression Models

The stepwise regression model examined in the previous discussion provided a solid evaluation of the research problem as formulated. However, a researcher is always well served in evaluating

TABLE 4-13   Split-Sample Validation of Stepwise Estimation

Overall Model Fit

	Sample 1	Sample 2
Multiple R	.910	.888
Coefficient of Determination (R^2)	.828	.788
Adjusted R^2	.808	.769
Standard error of the estimate	.564	.529

Analysis of Variance

	Sample 1					Sample 2				
	Sum of Squares	df	Mean Square	F	Sig.	Sum of Squares	df	Mean Square	F	Sig.
Regression	67.211	5	13.442	2.223	000	46.782	4	11.695	41.747	.000
Residual	14.008	44	.318			12.607	45	.280		
Total	81.219	49				59.389	49			

Variables Entered into the Stepwise Regression Model

	SAMPLE 1					SAMPLE 2				
	Regression Coefficients			Statistical Significance		Regression Coefficients			Statistical Significance	
Variables Entered	B	Std. Error	Beta	t	Sig.	B	Std. Error	Beta	t	Sig.
(Constant)	−1.413	.736		−1.920	.061	−.689	.686		−1.005	.320
X_{12} Salesforce Image	1.069	.151	.916	7.084	.000	.594	.105	.568	5.679	.000
X_6 Product Quality	.343	.066	.381	5.232	.000	.447	.062	.518	7.170	.000
X_7 E-Commerce	−.728	.218	−.416	−3.336	.002	−.349	.165	−.212	−2.115	.040
X_{11} Product Line	.295	.078	.306	3.780	.000					
X_{16} Order & Billing	.285	.115	.194	2.473	.017					
X_9 Complaint Resolution						.421	.070	.445	5.996	.000

alternative regression models in the search of additional explanatory power and confirmation of earlier results. In this section, we examine two additional regression models: one model including all 13 independent variables in a confirmatory approach and a second model adding a nonmetric variable (X_3, Firm Size) through the use of a dummy variable.

CONFIRMATORY REGRESSION MODEL   A primary alternative to the stepwise regression estimation method is the confirmatory approach, whereby the researcher specifies the independent variable to be included in the regression equation. In this manner, the researcher retains complete control over the regression variate in terms of both prediction and explanation. This approach is especially appropriate in situations of replication of prior research efforts or for validation purposes.

In this situation, the confirmatory perspective involves the inclusion of all 13 perceptual measures as independent variables. These same variables are considered in the stepwise estimation process, but in this instance all are directly entered into the regression equation at one time. Here the researcher can judge the potential impacts of multicollinearity on the selection of independent variables and the effect on overall model fit from including all seven variables.

The primary comparison between the stepwise and confirmatory procedures involves examination of the overall model fit of each procedure as well as the interpretations drawn from each set of results.

Impact on Overall Model Fit. The results in Table 4-14 are similar to the final results achieved through stepwise estimation (see Table 4-11), with two notable exceptions:

1. Even though more independent variables are included, the overall model fit decreases. Whereas the coefficient of determination increases (.889 to .897) because of the additional independent variables, the adjusted R^2 decreases slightly (.780 to .774), which indicates the inclusion of several independent variables that were nonsignificant in the regression equation. Although they contribute to the overall R^2 value, they detract from the adjusted R^2. This change illustrates the role of the adjusted R^2 in comparing regression variates with differing numbers of independent variables.
2. Another indication of the overall poorer fit of the confirmatory model is the increase in the standard error of the estimate (SEE) from .559 to .566, which illustrates that overall R^2 should not be the sole criterion for predictive accuracy because it can be influenced by many factors, one being the number of independent variables.

Impact on Variate Interpretation. The other difference is in the regression variate, where multicollinearity affects the number and strength of the significant variables.

1. First, only three variables (X_6, X_7, and X_{12}) are statistically significant, whereas the stepwise model contains two more variables (X_9 and X_{11}). In the stepwise model, X_{11} was the least significant variable, with a significance level of .005. When the confirmatory approach is used, the multicollinearity with other variables (as indicated by its tolerance value of .026) renders it nonsignificant. The same happens to X_9, which was the first variable entered in the stepwise solution, but it now has a nonsignificant coefficient in the confirmatory model. Again, multicollinearity had a sizeable impact, reflected in its tolerance value of .207.
2. The impact of multicollinearity on other variables not in the stepwise model is also substantial. In the confirmatory approach, three variables (X_{11}, X_{17}, and X_{18}) have tolerance values under .05 (with corresponding VIF values of 33.3, 37.9, and 44.0), meaning that 95 percent or more of their variance is accounted for by the other HBAT perceptions. In such situations, it is practically impossible for these variables to be significant predictors. Six other variables have tolerance values under .50, indicating that the regression model variables account for more than half of the variance in these variables.

Thus, whereas multicollinearity provided for the creation of four well-developed factors in Chapter 3, here the inclusion of all variables creates issues in estimation and interpretation.

TABLE 4-14   Multiple Regression Results Using a Confirmatory Estimation Approach with All 13 Independent Variables

Confirmatory Specification with 13 Variables

Multiple R	.897
Coefficient of Determination (R^2)	.804
Adjusted R^2	.774
Standard error of the estimate	.566

Analysis of Variance

	Sum of Squares	df	Mean Square	F	Sig.
Regression	113.044	13	8.696	27.111	.000
Residual	27.584	86	.321		
Total	140.628	99			

Variables Entered into the Regression Model

		Regression Coefficients			Statistical Significance		Correlations			Collinearity Statistics	
Variables Entered		B	Std. Error	Beta	t	Sig.	Zero-order	Partial	Part	Tolerance	VIF
(Constant)		−1.336	1.120		−1.192	.236					
X_6	Product Quality	.377	.053	.442	7.161	.000	.486	.611	.342	.598	1.672
X_7	E-Commerce	−.456	.137	−.268	−3.341	.001	.283	−.339	−.160	.354	2.823
X_8	Technical Support	.035	.065	.045	.542	.589	.113	.058	.026	.328	3.047
X_9	Complaint Resolution	.154	.104	.156	1.489	.140	.603	.159	.071	.207	4.838
X_{10}	Advertising	−.034	.063	−.033	−.548	.585	.305	−.059	−.026	.646	1.547
X_{11}	Product Line	.362	.267	.400	1.359	.178	.551	.145	.065	.026	37.978
X_{12}	Salesforce Image	.827	.101	.744	8.155	.000	.500	.660	.389	.274	3.654
X_{13}	Competitive Pricing	−.047	.048	−.062	−.985	.328	−.208	−.106	−.047	.584	1.712
X_{14}	Warranty & Claims	−.107	.126	−.074	−.852	.397	.178	−.092	−.041	.306	3.268
X_{15}	New Products	−.003	.040	−.004	−.074	.941	.071	−.008	−.004	.930	1.075
X_{16}	Order & Billing	.143	.105	.111	1.369	.175	.522	.146	.065	.344	2.909
X_{17}	Price Flexibility	.238	.272	.241	.873	.385	.056	.094	.042	.030	33.332
X_{18}	Delivery Speed	−.249	.514	−.154	−.485	.629	.577	−.052	−.023	.023	44.004

The confirmatory approach provides the researcher with control over the regression variate, but at the possible cost of a regression equation with poorer prediction and explanation if the researcher does not closely examine the results. The confirmatory and sequential approaches both have strengths and weaknesses that should be considered in their use, but the prudent researcher will employ both approaches in order to address the strengths of each.

INCLUDING A NONMETRIC INDEPENDENT VARIABLE   The prior discussion focused on the confirmatory estimation method as an alternative for possibly increasing prediction and explanation, but the researcher also should consider the possible improvement from the addition of nonmetric independent variables. As discussed in an earlier section and in Chapter 2, nonmetric variables cannot be directly included in the regression equation, but must instead be represented by a series of newly created variables, or dummy variables, that represent the separate categories of the nonmetric variable.

In this example, the variable of firm size (X_3), which has the two categories (large and small firms), will be included in the stepwise estimation process. The variable is already coded in the appropriate form, with large firms (500 or more employees) coded as 1 and small firms as 0.

The variable can be directly included in the regression equation to represent the difference in customer satisfaction between large and small firms, given the other variables in the regression equation. Specifically, because large firms have the value 1, small firms act as the reference category.

The regression coefficient is interpreted as the value for large firms compared to small firms. A positive coefficient indicates that large firms have higher customer satisfaction than small firms, whereas a negative coefficient indicates that small firms have higher customer satisfaction. The amount of the coefficient represents the difference in customer satisfaction between the means of the two groups, controlling for all other variables in the model.

Table 4-15 contains the results of the addition of X_3 in a stepwise model, where it was added with the five variables that formed the stepwise model earlier in this section (see Table 4-11). Examination of the overall fit statistics indicates minimal improvement, with all of the measures (R^2, adjusted R^2, and SEE) increasing over the original stepwise model (see Table 4-11).

When we examine the regression coefficients, note that the coefficient for X_3 is .271 and is significant at the .03 level. The positive value of the coefficient indicates that large firms, given their characteristics on the other five independent variables in the equation, still have a customer satisfaction level that is about a quarter point higher (.271) on the 10-point customer satisfaction question. The use of X_3 increased the prediction only slightly. From an explanatory perspective, though, we now know that large firms enjoy higher customer satisfaction.

This example illustrates the manner in which the researcher can add nonmetric variables to the metric variables in the regression variate and improve both prediction and explanation.

TABLE 4-15 **Multiple Regression Results Adding X_3 (Firm Size) as an Independent Variable by Using a Dummy Variable**

Stepwise Regression with Transformed Variables

Multiple R	.895
Coefficient of Determination (R^2)	.801
Adjusted R^2	.788
Standard error of the estimate	.548

Analysis of Variance

	Sum of Squares	df	Mean Square	F	Sig.
Regression	112.669	6	18.778	62.464	.000
Residual	27.958	93	.301		
Total	140.628	99			

Variables Entered into the Regression Model

	Regression Coefficients			Statistical Significance		Correlations			Collinearity Statistics	
Variables Entered	B	Std. Error	Beta	t	Sig.	Zero-order	Partial	Part	Tolerance	VIF
(Constant)	−1.250	.492		−2.542	.013					
X_9 Complaint Resolution	.300	.060	.304	4.994	.000	.603	.460	.231	.576	1.736
X_6 Product Quality	.365	.046	.427	7.881	.000	.486	.633	.364	.727	1.375
X_{12} Salesforce Image	.701	.093	.631	7.507	.000	.500	.614	.347	.303	3.304
X_7 E-Commerce	−.333	.135	−.196	−2.473	.015	.283	−.248	−.114	.341	2.935
X_{11} Product Line	.203	.061	.224	3.323	.001	.551	.326	.154	.469	2.130
X_3 Firm Size	.271	.123	.114	2.207	.030	.229	.223	.102	.798	1.253

A Managerial Overview of the Results

The regression results, including the complementary evaluation of the confirmatory model and the addition of the nonmetric variable, all assist in addressing the basic research question: What affects customer satisfaction? In formulating a response, the researcher must consider two aspects: prediction and explanation.

In terms of prediction, the regression models all achieve high levels of predictive accuracy. The amount of variance explained equals about 80 percent and the expected error rate for any prediction at the 95% confidence level is about 1.1 points. In this type of research setting, these levels, augmented by the results supporting model validity, provide the highest levels of assurance as to the quality and accuracy of the regression models as the basis for developing business strategies.

In terms of explanation, all of the estimated models arrived at essentially the same results: three strong influences (X_{12}, Salesforce Image; X_6, Product Quality; and X_9, Complaint Resolution). Increases in any of these variables result in increases in customer satisfaction. For example, an increase of 1 point in the customer's perception of Salesforce Image (X_{12}) will result in an average increase of at least seven-tenths (.701) of a point on the 10-point customer satisfaction scale. Similar results are seen for the other variables. Moreover, at least one firm characteristic, firm size, demonstrated a significant effect on customer satisfaction. Larger firms have levels of satisfaction about a quarter of a point (.271) higher than the smaller firms. These results provide management with a framework for developing strategies for improving customer satisfaction. Actions directed toward increasing the perceptions of HBAT can be justified in light of the corresponding increases in customer satisfaction.

The impact of two other variables (X_7, E-Commerce; X_{11}, Product Line) on customer satisfaction is less certain. Even though these two variables were included in the stepwise solution, their combined explained variance was only .038 out of an overall model R^2 of .791. Both variables were not significant in the confirmatory model. Moreover, X_7 had the reversed sign in the stepwise model, which, although due to multicollinearity, still represents a result contrary to managerial strategy development. As a result, the researcher should consider reducing the influence allotted to these variables and even possibly omit them from consideration as influences on customer satisfaction.

In developing any conclusions or business plans from these results, the researcher should also note that the three major influences (X_{12}, X_6, and X_9) are primary components of the perceptual dimensions identified through factor analysis in Chapter 3. These dimensions, which represent broad measures of customers' perceptions of HBAT, should thus also be considered in any conclusions. To state that only these three specific variables are influences on customer satisfaction would be a serious misstatement of the more complex patterns of collinearity among variables. Thus, these variables are better viewed as representatives of the perceptual dimensions, with the other variables in each dimension also considered in any conclusions drawn from these results.

Management now has an objective analysis that confirms not only the specific influences of key variables, but also the perceptual dimensions that must be considered in any form of business planning regarding strategies aimed at affecting customer satisfaction.

Summary

This chapter provided an overview of the fundamental concepts underlying multiple regression analysis. Multiple regression analysis can describe the relationships among two or more intervally scaled variables and is much more powerful than simple regression with a single independent variable. This chapter helps you to do the following:

Determine when regression analysis is the appropriate statistical tool in analyzing a problem. Multiple regression analysis can be used to analyze the relationship between a single dependent (criterion) variable and several independent (predictor) variables. The objective of multiple regression analysis is to use the several independent variables whose values are known to predict the single dependent value. Multiple regression is a dependence technique. To use it you must be able to divide the variables into dependent and independent variables, and both the dependent and independent variables are metric. Under certain circumstances, it is possible to include nonmetric data either as independent variables (by transforming either ordinal or nominal data with dummy variable coding) or the dependent variable (by the use of a binary measure in the specialized technique of logistic regression). Thus, to apply multiple regression analysis: (1) the data must be metric or appropriately transformed, and (2) before deriving the regression equation, the researcher must decide which variable is to be dependent and which remaining variables will be independent.

Understand how regression helps us make predictions using the least squares concept.   The objective of regression analysis is to predict a single dependent variable from the knowledge of one or more independent variables. Before estimating the regression equation, we must calculate the baseline against which we will compare the predictive ability of our regression models. The baseline should represent our best prediction without the use of any independent variables. In regression, the baseline predictor is the simple mean of the dependent variable. Because the mean will not perfectly predict each value of the dependent variable, we must have a way to assess predictive accuracy that can be used with both the baseline prediction and the regression models we create. The customary way to assess the accuracy of any prediction is to examine the errors in predicting the dependent variable. Although we might expect to obtain a useful measure of prediction accuracy by simply adding the errors, this approach is not possible because the errors from using the mean value always sum to zero. To overcome this problem, we square each error and add the results together. This total, referred to as the sum of squared errors (SS_E), provides a measure of prediction accuracy that will vary according to the amount of prediction errors. The objective is to obtain the smallest possible sum of squared errors as our measure of prediction accuracy. Hence, the concept of least squares enables us to achieve the highest accuracy possible.

Use dummy variables with an understanding of their interpretation.   A common situation faced by researchers is the desire to utilize nonmetric independent variables. Many multivariate techniques assume metric measurement for both independent and dependent variables. When the dependent variable is measured as a dichotomous (0, 1) variable, either discriminant analysis or a specialized form of regression (logistic regression), is appropriate. When the independent variables are nonmetric and have two or more categories, we can create dummy variables that act as replacement independent variables. Each dummy variable represents one category of a nonmetric independent variable, and any nonmetric variable with k categories can be represented as $k - 1$ dummy variables. Thus, nonmetric variables can be converted to a metric format for use in most multivariate techniques.

Be aware of the assumptions underlying regression analysis and how to assess them.   Improvements in predicting the dependent variable are possible by adding independent variables and even transforming them to represent nonlinear relationships. To do so, we must make several assumptions about the relationships between the dependent and independent variables that affect the statistical procedure (least squares) used for multiple regression. The basic issue is to know whether in the course of calculating the regression coefficients and predicting the dependent variable the assumptions of regression analysis have been met. We must know whether the errors in prediction are the result of the absence of a relationship among the variables or caused by some characteristics of the data not accommodated by the regression model. The assumptions to be examined include linearity of the phenomenon measured, constant variance of the error terms, independence of the error terms, and normality of the error term distribution. The assumptions underlying multiple regression analysis apply both to the individual variables (dependent and independent) and to the relationship as a whole. Once the variate has been derived, it acts collectively in predicting the dependent variable, which necessitates assessing the assumptions not only for individual variables, but also for the variate. The principal measure of prediction error for the variate is the residual—the difference between the observed and predicted values for the dependent variable. Plotting the residuals versus the independent or predicted variables is a basic method of identifying assumption violations for the overall relationship.

Select an estimation technique and explain the difference between stepwise and simultaneous regression.   In multiple regression, a researcher may choose from a number of possible independent variables for inclusion in the regression equation. Sometimes the set of independent variables are exactly specified and the regression model is essentially used in a confirmatory approach. This approach, referred to as *simultaneous regression*, includes all the variables at the same time. In other instances, the researcher may use the estimation technique to "pick and choose" among the set of independent variables with either sequential search methods or combinatorial processes.

The most popular sequential search method is stepwise estimation, which enables the researcher to examine the contribution of each independent variable to the regression model. The combinatorial approach is a generalized search process across all possible combinations of independent variables. The best-known procedure is all-possible-subsets regression, which is exactly as the name suggests. All possible combinations of the independent variables are examined, and the best-fitting set of variables is identified. Each estimation technique is designed to assist the researcher in finding the best regression model using different approaches.

Interpret the results of regression.   The regression variate must be interpreted by evaluating the estimated regression coefficients for their explanation of the dependent variable. The researcher must evaluate not only the regression model that was estimated, but also the potential independent variables that were omitted if a sequential search or combinatorial approach was employed. In those approaches, multicollinearity may substantially affect the variables ultimately included in the regression variate. Thus, in addition to assessing the estimated coefficients, the researcher must also evaluate the potential impact of omitted variables to ensure that the managerial significance is evaluated along with statistical significance. The estimated regression coefficients, or beta coefficients, represent both the type of relationship (positive or negative) and the strength of the relationship between independent and dependent variables in the regression variate. The sign of the coefficient denotes whether the relationship is positive or negative, whereas the value of the coefficient indicates the change in the dependent value each time the independent variable changes by one unit. Prediction is an integral element in regression analysis, both in the estimation process as well as in forecasting situations. Regression involves the use of a variate to estimate a single value for the dependent variable. This process is used not only to calculate the predicted values in the estimation procedure, but also with additional samples for validation or forecasting purposes. The researcher often is interested not only in prediction, but also explanation. Independent variables with larger regression coefficients make a greater contribution to the predicted value. Insight into the relationship between independent and dependent variables is gained by examining the relative contributions of each independent variable. Thus, for explanatory purposes, the regression coefficients become indicators of the relative impact and importance of the independent variables in their relationship with the dependent variable.

Apply the diagnostic procedures necessary to assess influential observations.   Influential observations include all observations that have a disproportionate effect on the regression results. The three basic types of influentials are as follows:

1. *Outliers.* Observations that have large residual values and can be identified only with respect to a specific regression model.
2. *Leverage points.* Observations that are distinct from the remaining observations based on their independent variable values.
3. *Influential observations.* All observations that have a disproportionate effect on the regression results.

Influentials, outliers, and leverage points are based on one of four conditions:

1. *An error in observations or data entry:* Remedy by correcting the data or deleting the case.
2. *A valid but exceptional observation that is explainable by an extraordinary situation:* Remedy by deletion of the case unless variables reflecting the extraordinary situation are included in the regression equation.
3. *An exceptional observation with no likely explanation:* Presents a special problem because the researcher has no reason for deleting the case, but its inclusion cannot be justified either, suggesting analyses with and without the observations to make a complete assessment.
4. *An ordinary observation in its individual characteristics but exceptional in its combination of characteristics:* Indicates modifications to the conceptual basis of the regression model and should be retained.

The researcher should delete truly exceptional observations but avoid deleting observations that, although different, are representative of the population.

This chapter provides a fundamental presentation of how regression works and what it can achieve. Familiarity with the concepts presented will provide a foundation for regression analyses the researcher might undertake and help you to better understand the more complex and detailed technical presentations in other textbooks on this topic.

Questions

1. How would you explain the relative importance of the independent variables used in a regression equation?
2. Why is it important to examine the assumption of linearity when using regression?
3. How can nonlinearity be corrected or accounted for in the regression equation?
4. Could you find a regression equation that would be acceptable as statistically significant and yet offer no acceptable interpretational value to management?
5. What is the difference in interpretation between regression coefficients associated with interval-scale independent variables and dummy-coded (0, 1) independent variables?
6. What are the differences between interactive and correlated independent variables? Do any of these differences affect your interpretation of the regression equation?
7. Are influential cases always omitted? Give examples of occasions when they should or should not be omitted.

Suggested Readings

A list of suggested readings illustrating issues and applications of multivariate techniques in general is available on the Web at www.pearsonglobaleditions.com/hair or www.mvstats.com.

References

1. Barnett, V., and T. Lewis. 1994. *Outliers in Statistical Data,* 3rd ed. New York: Wiley.
2. Belsley, D. A., E. Kuh, and E. Welsch. 1980. *Regression Diagnostics: Identifying Influential Data and Sources of Collinearity.* New York: Wiley.
3. BMDP Statistical Software, Inc. 1991. *SOLO Power Analysis.* Los Angeles: BMDP.
4. Box, G. E. P., and D. R. Cox. 1964. An Analysis of Transformations. *Journal of the Royal Statistical Society B* 26: 211–43.
5. Cohen, J., Stephen G. West, Leona Aiken, and P. Cohen. 2002. *Applied Multiple Regression/ Correlation Analysis for the Behavioral Sciences,* 3rd ed. Hillsdale, NJ: Lawrence Erlbaum Associates.
6. Daniel, C., and F. S. Wood. 1999. *Fitting Equations to Data,* 2nd ed. New York: Wiley-Interscience.
7. Jaccard, J., R. Turrisi, and C. K. Wan. 2003. *Interaction Effects in Multiple Regression.* 2nd ed. Beverly Hills, CA: Sage Publications.
8. Johnson, R. A., and D. W. Wichern. 2002. *Applied Multivariate Statistical Analysis.* 5th ed. Upper Saddle River, NJ: Prentice Hall.
9. Mason, C. H., and W. D. Perreault, Jr. 1991. Collinearity, Power, and Interpretation of Multiple Regression Analysis. *Journal of Marketing Research* 28 (August): 268–80.
10. Mosteller, F., and J. W. Tukey. 1977. *Data Analysis and Regression.* Reading, MA: Addison-Wesley.
11. Neter, J., M. H. Kutner, W. Wassermann, and C. J. Nachtsheim. 1996. *Applied Linear Regression Models.* 3rd ed. Homewood, IL: Irwin.
12. Rousseeuw, P. J., and A. M. Leroy. 2003. *Robust Regression and Outlier Detection.* New York: Wiley.
13. Seber, G. A. F. 2004. *Multivariate Observations.* New York: Wiley.
14. Sharma, S., R. M. Durand, and O. Gur-Arie. 1981. Identification and Analysis of Moderator Variables. *Journal of Marketing Research* 18 (August): 291–300.
15. Weisberg, S. 1985. *Applied Linear Regression.* 2nd ed. New York: Wiley.
16. Wilkinson, L. 1975. Tests of Significance in Stepwise Regression. *Psychological Bulletin* 86: 168–74.

CHAPTER 5

Canonical Correlation

LEARNING OBJECTIVES

Upon completing this chapter, you should be able to do the following:

- Describe canonical correlation analysis and understand its purpose.
- State the similarities and differences between multiple regression, discriminant analysis, factor analysis, and canonical correlation.
- Summarize the conditions that must be met for application of canonical correlation analysis.
- Define and compare canonical root measures and the redundancy index.
- Compare the advantages and disadvantages of the three methods for interpreting the nature of canonical functions.

CHAPTER PREVIEW

In the previous chapter, we introduced multiple regression, a technique that predicts a single metric dependent variable from a linear function of a set of several independent variables. For some research problems, however, interest may not center on a single dependent variable. Instead, the researcher may be interested in relationships between sets of multiple dependent and multiple independent variables. Canonical correlation analysis is the answer for this kind of research problem. It is a method that enables the assessment of the relationship between two sets of multiple variables.

Application of canonical correlation analysis has increased as the software has become more widely available. Even though it is less popular than many other methods, examples of canonical correlation analysis are found across various disciplines and in many business studies. For example, canonical analysis was used to examine the relationships between product innovation strategies and market orientation [12] and between adoption of outsourcing services and the characteristics of environments in which the services operate [3]. Canonical correlation analysis will continue to be a useful technique for situations involving multiple independent and multiple dependent variables.

In applying canonical analysis, it is helpful to think of one set of variables as independent and the other set as dependent. Use of the terms *independent variables* and *dependent variables,* however, does not imply that they share a causal relationship. Instead, it simply refers to how the two sets of multiple variables correlate. As discussed in Chapter 1, canonical correlation analysis is considered a general model on which many other multivariate techniques are based because it can use both metric and nonmetric data for either the dependent or independent variables. The general form of canonical analysis can be expressed as:

$$Y_1 + Y_2 + Y_3 + \cdots + Y_n = X_1 + X_2 + X_3 + \cdots + X_n$$
$$(\text{metric, nonmetric}) \qquad\qquad (\text{metric, nonmetric})$$

This chapter introduces the researcher to the multivariate statistical technique of canonical correlation analysis. We first describe the nature of canonical correlation analysis and then summarize a six-step procedure and guidelines for judging the appropriateness of the method. We then illustrate the application and interpretation of canonical correlation analysis with an example from the HBAT database. The chapter also summarizes the potential advantages and limitations of the technique.

KEY TERMS

Before starting the chapter, review the key terms to develop an understanding of the concepts and terminology used. Throughout the chapter the key terms appear in **boldface.** Other points of emphasis in the chapter are italicized and cross-references within the Key Terms appear in *italics.*

Canonical correlation coefficient   Measures the strength of the overall relationship between the two linear composites (*canonical variates*), one variate for the independent variables and one for the dependent variables. In effect, it represents the bivariate correlation between the two *canonical variates* in a *canonical function.*

Canonical cross-loadings   Correlation of each observed independent or dependent variable with the opposite canonical variate. For example, the independent variables are correlated with the dependent canonical variate. They can be interpreted like *canonical loadings,* but with the opposite canonical variate.

Canonical function   Relationship (correlational) between two linear composites (*canonical variates*). Each canonical function has two *canonical variates,* one for the set of dependent variables and one for the set of independent variables. The strength of the relationship is given by the *canonical correlation coefficient.*

Canonical loading   Simple linear correlation between the independent variables and their respective *canonical variates.* These can be interpreted like factor loadings; they are also known as *canonical structure correlations.* Each independent variable has different canonical loadings for each *canonical function.*

Canonical roots   Squared *canonical correlation coefficients,* which provide an estimate of the amount of shared variance between the respective *canonical variates* of dependent and independent variables. Also known as *eigenvalues.*

Canonical structure correlations   See *canonical loading.*

Canonical variates   Linear combinations that represent the optimally weighted sum of two or more variables and are formed for both the dependent and independent variables in each *canonical function.* Also referred to as *linear composites,* linear compounds, and linear combinations.

Eigenvalues   See *canonical roots.*

Linear composites   See *canonical variates.*

Orthogonal   Mathematical constraint specifying that the canonical functions are statistically independent of each other. This is similar to orthogonal factor analysis.

Redundancy index   Amount of variance in a canonical variate (dependent or independent) explained by the other canonical variate in the canonical function. It can be computed for both the dependent and the independent *canonical variates* in each canonical function. For example, a redundancy index of the dependent variate represents the amount of variance in the dependent variables explained by the independent canonical variate.

WHAT IS CANONICAL CORRELATION?

Canonical correlation analysis is a multivariate statistical model that facilitates the study of linear interrelationships between two sets of variables. One set of variables is referred to as *independent variables* and the others are considered *dependent variables* [8, 9]; a **canonical variate** is formed for each set. It may be helpful to think of a canonical variate as being like the variate (i.e., **linear composite**) formed from the set of independent variables in a multiple regression analysis. But in canonical correlation there is also a variate formed from several dependent variables whereas multiple regression can accommodate only one dependent variable. Canonical correlation analysis develops a **canonical function** that maximizes the **canonical correlation coefficient** between the two canonical variates. The canonical correlation coefficient measures the strength of the relationship between the two canonical variates. Each canonical variate is interpreted with **canonical loadings,** the correlation of the individual variables and their respective variates. Canonical loadings are similar to the factor loadings of each variable and the factors that were described in factor analysis (see Chapter 3). So in a sense this is analogous to estimating a separate factor for each set of variables in order to maximize the correlation between factors. A canonical function between the sets of independent and dependent variables and their variates is shown in Figure 5-1.

Another unique characteristic of canonical correlation is that it develops multiple canonical functions. Each canonical function is independent (orthogonal) from the other canonical functions so that they represent different relationships found among the sets of dependent and independent variables. The canonical loadings of the individual variables differ in each canonical function and represent that variable's contribution to the specific relationship being depicted. Extending the simple example above, each canonical function consists of a different pair of variates (one for the independent variables and the other for the dependent variables), each function representing a different relationship between the sets of variables. The researcher retains and interprets all of the statistically significant canonical functions. Here we see the development of different canonical functions as somewhat analogous to the discriminant functions in discriminant analysis in which each represents a different dimension of discrimination in the dependent variable (see Chapter 7).

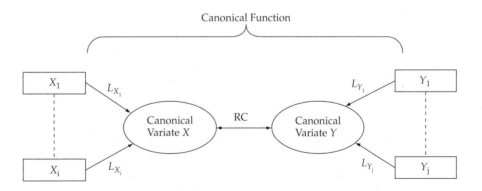

$X_i = i^{th}$ measured variable on canonical variate X
$Y_j = j^{th}$ measured variable on canonical variate Y
$L_{X_i} =$ Loading of ith X measured variable on canonical variate X
$L_{Y_j} =$ Loading of jth Y measured variable on canonical variate Y
RC = Canonical correlation coefficient for the pair of canonical variates in the canonical function

FIGURE 5-1   Relationship of Variables and Canonical Loadings with the Canonical Variates in the Canonical Function

Canonical correlation analysis has several advantages for researchers. First, canonical correlation analysis limits the probability of committing Type I errors. The risk of a Type I error is related to the likelihood of finding a statistically significant result when it does not exist. Increased risk of Type I error results from when the same variables in a data set are used for too many statistical tests. If a researcher wants to see if four X variables can predict six Y variables through multiple regression, then a series of six regression equations are required (i.e., one for each dependent variable). But using separate statistical significance tests for each equation substantially increases the risk of Type I error. Canonical correlation can assess these relationships between the two set of variables (independent and dependent) in a single relationship rather than using separate relationships for each dependent variable. Second, canonical correlation analysis may better reflect the reality of research studies. The complexity of research studies involving human and/or organizational behavior may suggest multiple variables that represent a concept and thus create problems when the variables are examined separately. In the example above, canonical correlation would represent a relationship between the sets of variables rather than individual variables. Moreover, it can identify two or more unique relationships, if they exist. Thus canonical correlation analysis is both technically able to analyze the data involving multiple sets of variables and is theoretically consistent with that purpose [16].

HYPOTHETICAL EXAMPLE OF CANONICAL CORRELATION

To further explain the nature of canonical correlation, let us consider an extension of the regression example used in Chapter 4. Recall the small survey that used family size and income as predictors of the number of credit cards a family would hold. That problem involved examining the relationship between two independent variables and a single dependent variable.

Developing a Variate of Dependent Variables

Suppose the researcher is now interested in the broader concept of credit card usage, especially how consumer characteristics and credit history can predict credit card usage. To measure credit card usage, not only is the number of credit cards held by the family taken into consideration, but also the family's average monthly dollar spending on all credit cards and the outstanding balances of all credit cards. These three measures are believed to give a much better perspective on a family's credit card usage. Readers interested in other approaches to using multiple indicators to represent a concept are referred to Chapter 3 ("Factor Analysis") and Chapter 11 ("Structural Equation Modeling"). The problem now involves predicting three dependent measures simultaneously (number of credit cards, average monthly dollar spending, and outstanding balance). Multiple regression is capable of handling only a single dependent variable. Discriminant analysis is also unsuitable because its dependent variable must be nonmetric. Canonical correlation is therefore the only technique available for examining relationships with multiple dependent variables.

The problem of predicting credit card usage is illustrated in Figure 5-2. The three dependent variables used to measure credit card usage—number of credit cards held by the family, average monthly dollar expenditures on all credit cards, and outstanding balance—are shown on the right. The three independent variables selected to predict credit usage—family size, family income, and credit history—are shown on the left. By using canonical correlation analysis, the researcher creates a composite measure of credit usage that consists of three dependent variables, rather than having to compute a separate regression equation for each of the dependent variables.

Estimating the First Canonical Function

Estimation of the first canonical function (see top portion of Figure 5-3) provides the researcher with two items of particular interest. The first is the two canonical variates representing the optimal linear combinations of the dependent or independent variables and the canonical correlation coefficient

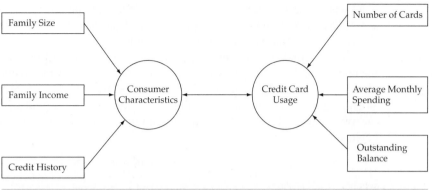

FIGURE 5-2   **Canonical Relationship Between Consumer Characteristics
and Their Credit Card Usage**

(Rc) representing the relationship between them. More about how to interpret the variate with the canonical loadings will be discussed in later sections, but for now we can use high loadings as an indication of the most descriptive variables for a variate. The first function is between the variate of independent variables characterized by family size and family income (e.g., perhaps described as

First Canonical Function

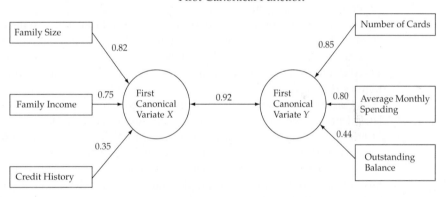

Second Canonical Function

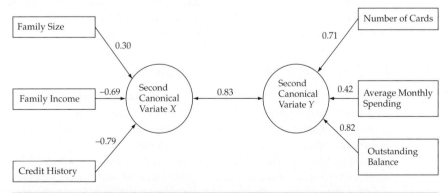

FIGURE 5-3   **Canonical Loadings and Correlations for the Two Canonical Functions
of the Hypothetical Example**

household characteristics) and the variate of dependent variables characterized by the number of cards and average monthly dollar spending. The dependent variate could perhaps be described as the tendency for convenience when using credit cards.

The second item of interest is the canonical correlation coefficient, which measures the correlation between the two variates. In the case of the first function, the canonical correlation coefficient is .92, indicating a fairly high degree of association between the two variates. When squared, the canonical correlation represents the amount of shared variance between the two canonical variates. Squared canonical correlations are called **canonical roots** or **eigenvalues**:

$$Canonical\ Root_n = Rc_n^2$$

where

$$Rc_n = \text{canonical correlation for the } n^{\text{th}} \text{ canonical function}$$

In our credit card example, the canonical correlation coefficient of the first canonical function is 0.92, which corresponds to a canonical root of 0.85 (i.e., $.92^2 = .85$).

The researcher might also be interested in knowing how much variance in the independent variate is explained by the dependent variate, and vice versa. This differs from the canonical root because each variate accounts for only a portion of the variance of its variables. Thus, although two variates might have a squared correlation of .80, it does not mean that each variable is explained in this amount. Instead, a variable is explained only to the degree that it relates to the variate (i.e., its canonical loading). So to determine the "explained" variance in the actual variable, we must take into account not only the canonical root, but also the canonical loadings of the variable. The Stewart-Love **redundancy index** [14] was developed to represent the amount of variance in one set of variables that can be explained by the variables in the other set. This index serves as a measure of accounted-for or explained variance, similar to the R^2 calculation used in multiple regression.

Canonical correlation differs from multiple regression in that it does not deal with a single dependent variable but instead with a composite of the dependent variables, and this composite represents only a portion of each dependent variable's total variance. For this reason, we cannot assume that 100 percent of the variance in the dependent variable set is available to be explained by the independent variable set. The set of independent variables can be expected to account only for the shared variance in the dependent canonical variate. As we will discuss later the amount explained is not symmetrical between variates and one variate may explain more of the other variate than vice versa.

Although we will demonstrate how to calculate the redundancy index in a later section, the results for our example show that the first canonical variate of the independent variables relating to consumer characteristics explains 44 percent of the variance in the variate for the dependent variables. Likewise, the first variate for the dependent variables (Credit Card Usage) explains 38 percent of the consumer characteristics variate (independent variables).

Estimating a Second Canonical Function

In this example, canonical correlation analysis estimated a second statistically significant canonical function (lower portion of Figure 5-3). This second canonical function represents a second unique and independent relationship between the dependent and independent variables. Note that the same variables (independent and dependent) are included in each variate, but the loadings differ from the first canonical function, thus giving each variate a different "meaning" based on the variables with the highest canonical loadings. The second set of variates was characterized on the variate of independent variables by a high canonical loading on credit history, whereas the variate of dependent variables is characterized by a high canonical loading on outstanding balance. This result might be interpreted as usage of credit cards for cash advances. When combined with the relationship for the first variate, this suggests that consumers' credit card usage might be explained by their need for both convenience and cash.

The canonical correlation coefficient of the second canonical correlation coefficient is 0.83, which corresponds to a canonical root is 0.69. Calculation of the redundancy index shows that the variate for the consumer characteristics (independent variables) explains 31 percent of the credit card usage variate. Conversely, the second credit card usage variate explains 28 percent of the consumer characteristics variate. Because the two functions are independent, the explained variance for the two functions can be added. This means that in total consumer characteristics explain over 70 percent of the variance in the dependent variables (credit card usage), whereas the credit card usage variables explain a total of 66 percent of the variance in consumer characteristics.

RELATIONSHIPS OF CANONICAL CORRELATION ANALYSIS TO OTHER MULTIVARIATE TECHNIQUES

Canonical correlation analysis is the most generalized member of the family of multivariate statistical techniques. It is directly related to several dependence methods. Chapter 4 discussed multiple regression analysis, which can predict the value of a single metric dependent variable from a linear function of a set of independent variables. Similar to regression, the goal of canonical correlation is to quantify the strength of the relationship, in this case between the two sets of variables (independent and dependent). Whereas multiple regression predicts a single dependent variable from a set of multiple independent variables, canonical correlation simultaneously predicts multiple dependent variables from multiple independent variables. Canonical correlation analysis also resembles discriminant analysis in its ability to determine independent dimensions (similar to discriminant functions) for each variable set. Discriminant analysis, which will be introduced in Chapter 7, is a method of estimating the relationship between a single nonmetric dependent variable and a set of metric independent variables. In sum, canonical correlation analysis is more general than multiple regression and discriminant analysis because it can handle multiple dependent variables that can be metric or nonmetric.

Canonical correlation analysis is also closely related to principal components analysis, which is included in factor analysis [15]. Chapter 3 discussed factor analysis, the primary purpose of which is to define the underlying structure among the variables in the analysis. Canonical correlation analysis corresponds to principal components analysis and factor analysis in the creation of the optimum structure or dimensionality of each variable set that maximizes the relationship between independent and dependent variable sets. Whereas principal components analysis and factor analysis attempt to explain the linear relationship among a set of observed variables and an unknown number of factors/variates, canonical correlation analysis focuses more on the linear relationship between two variates. As such, it is similar in purpose to PLS, a variant of structural equation modeling, which will be discussed in Chapter 12.

Canonical correlation places the fewest restrictions on the types of data on which it operates and can be used for both metric and nonmetric data. Because the other techniques impose more rigid restrictions, it is generally believed that the information obtained from them is more robust statistically and may be presented in a more interpretable manner. But in situations with multiple dependent and independent variables, canonical correlation is the most appropriate and powerful multivariate technique. It has gained acceptance in many fields and represents a useful tool for multivariate analysis, particularly with the expanding interest in considering relationships between multiple dependent variables.

Our discussion of canonical correlation analysis is organized around the model-building process described in Chapter 1. Figure 5-4 (stages 1–3) and Figure 5-5 (stages 4–6) depict the stages for canonical correlation analysis, which include (1) specifying the objectives of canonical correlation, (2) developing the analysis plan, (3) assessing the assumptions underlying canonical correlation, (4) estimating the canonical model and assessing overall model fit, (5) interpreting the canonical variates, and (6) validating the model.

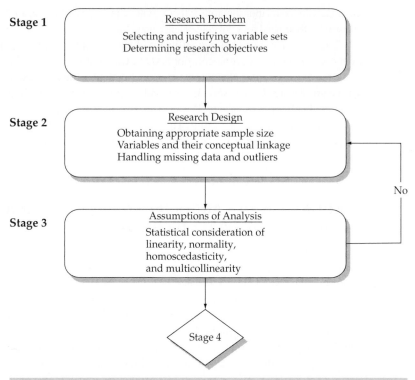

FIGURE 5-4   Stages 1–3 in the Canonical Correlation Analysis Decision Diagram

STAGE 1: OBJECTIVES OF CANONICAL CORRELATION ANALYSIS

As noted earlier, canonical correlation analysis examines the relationships between two sets of variables. Although it places the least number of restrictions on the types of data that can be used, this can also increase the difficulty in interpreting the solutions. Therefore, it is especially important to have clearly defined objectives before using the method. In order to reassure that canonical correlation analysis is appropriately applied, the researcher needs to consider two issues: (1) selection of variable sets and (2) evaluation of research objectives.

Selection of Variable Sets

The appropriate data for canonical correlation analysis are two sets of variables. Each set needs to be given theoretical meaning for why the variables are being treated together, at least to the extent that one set could be defined as the independent variables and the other as the dependent variables.

Canonical correlation analysis solutions are sensitive to the changes of variables from one set to another. The results of canonical correlation analysis are derived both from correlations among variables in each set and correlations between canonical variates. Therefore changing the variables in one variate can noticeably change the composition of the other canonical variate.

Evaluating Research Objectives

Once the theoretical justification has been made for the combination of variables, canonical correlation can address a wide range of objectives. The objectives are usually exploratory and may include any or all of the following:

- Determining whether two sets of variables (measurements made on the same objects) are independent of one another or, conversely, determining the magnitude of the relationships that may exist between the two sets.

- Deriving a set of weights for each set of dependent and independent variables so that the linear combinations of each set are maximally correlated. Additional linear functions that maximize the remaining correlation are independent of the preceding set(s) of linear combinations.
- Explaining the nature of whatever relationships exist between the sets of dependent and independent variables, generally by measuring the relative contribution of each variable to the canonical functions (relationships) that are extracted.

Canonical correlation coefficients can be used to address the first research objective and canonical factor loadings are appropriate for the second objective. The redundancy index and proportion of variance explained is used for the third objective. These indices, especially the redundancy index and proportion of variance, will be discussed more in stage 4.

The inherent flexibility of canonical correlation in terms of the number and types of variables handled, both dependent and independent, makes it a logical alternative to consider for many of the more complex problems addressed with multivariate techniques.

STAGE 2: DESIGNING A CANONICAL CORRELATION ANALYSIS

As the most general form of multivariate analysis, canonical correlation analysis shares basic implementation issues common to all multivariate techniques. Some principal features that are discussed in other chapters (particularly multiple regression, discriminant analysis, and factor analysis) are also relevant to canonical correlation analysis. These include (1) appropriate sample size, (2) variables and their conceptual linkage, and (3) absence of missing data and outliers.

Sample Size

Issues related the impact of sample size (both small and large) and the necessity for a sufficient number of observations per variable are frequently encountered with canonical correlation. Researchers are tempted to include many variables in both the independent and dependent variable set, not realizing the implications for sample size. Sample sizes that are very small will not represent the correlations well, thus obscuring meaningful relationships. Very large samples have a tendency to result in statistical significance in all instances, even where practical significance is not indicated. The appropriate sample size is related to the reliability of the variables (see Chapter 3 for a more detailed discussion). Different disciplines have different expectations regarding reliability but for social science and business researchers reliability is generally expected to be .7 or higher, and they are encouraged to maintain at least 10 observations per independent variable to avoid "overfitting" the data. For exploratory studies, however, these requirements may be relaxed somewhat.

Variables and Their Conceptual Linkage

Canonical correlation analysis is the most liberal form of multivariate analysis in that both metric and nonmetric variables can be included in the latent variables. The classification of variates as dependent or independent is of little importance for the statistical estimation of the canonical functions, however, because the method calculates weights for both variates to maximize the correlation and places no particular emphasis on either variate. Yet because the technique produces variates to maximize the correlation between them, a variable in either set relates to all other variables in both sets. The result is that the addition or deletion of a single variable may affect the entire solution, particularly the other variate. The composition of each variate, either independent or dependent, thus becomes critical. Researchers must have conceptually linked the sets of variables before applying canonical correlation analysis. This makes the specification of dependent versus independent variates essential to establishing a strong conceptual foundation for the variables.

Missing Data and Outliers

Canonical correlation analysis is sensitive to changes in the data set. Therefore, different procedures for handling missing data can create substantial changes in canonical solutions [11]. Similarly, outliers can also substantially impact canonical analysis results. Missing data can be replaced by estimating values or removing cases with missing data. Outliers can be detected by univariate, bivariate, and multivariate diagnostic methods. Consult Chapter 2 for more details on these issues.

STAGE 3: ASSUMPTIONS IN CANONICAL CORRELATION

The generality of canonical correlation analysis also extends to its underlying statistical assumptions. Even though some assumptions are not strictly required, interpretability of canonical solutions is enhanced if they are. Readers unfamiliar with these statistical assumptions, the tests for their diagnosis, or the alternative remedies when the assumptions are not met should refer to Chapter 2. The following section discusses some essential assumptions and their impact on canonical correlation analysis.

Linearity

Linearity is important to canonical correlation analysis and it affects two aspects of canonical correlation results. First, the canonical correlation coefficient between the pair of variates is based on a linear relationship. If the variates relate in a nonlinear manner, the relationship will not be captured by the canonical correlation coefficient. Second, the canonical correlation analysis maximizes the linear correlation between the variates. Thus, although canonical correlation analysis is the most generalized multivariate method, it is still constrained to identifying linear relationships. If the relationship is nonlinear, then one or both variates should be transformed, if possible.

If the relationship is nonlinear, procedures are available to assess nonlinear canonical correlation (e.g., OVERALS in SPSS). When using this technique, the variables can be numerical, ordinal, and nominal, and there can be more than two sets of variables. Nonlinear canonical correlation analysis is beyond the scope of this book. Readers interested in this topic can acquire more information from van der Burg et al. (1994).

Normality

Canonical correlation analysis can accommodate any metric variable without the strict assumption of normality. However, normality is desirable because it allows for the highest correlation among the variables. Indeed, canonical correlation analysis can accommodate nonnormal variables if the distributional form (e.g., highly skewed) does not decrease the correlation with other variables. This allows for transformed nonmetric data (in the form of dummy variables) to be used as well. However, multivariate normality is required for the statistical inference test of the significance of each canonical function. Because tests for multivariate normality are not readily available, the prevailing guideline is to ensure that each variable has univariate normality. Thus, although normality is not strictly required, it is highly recommended that all variables be evaluated for normality and transformed if possible.

Homoscedasticity and Multicollinearity

Canonical correlation analysis best portrays the relationships when they are homoscedastic. Homoscedasticity is important because the opposite, heteroscedasticity, decreases the correlation between variables. Similarly, multicollinearity should be dealt with as well. Multicollinearity occurs when two or more variables are highly correlated. Multicollinearity among either variable set will confound the ability of the technique to isolate the impact of any single variable, making interpretation less reliable.

RULES OF THUMB 5-1

Canonical Correlation Analysis Design

- A strong conceptual foundation is needed to specify the sets of independent and dependent variables
- Variables can be either metric or nonmetric
- The acceptable sample size is at least 10 cases per measured variable, except for exploratory research
- As with all other multivariate methods, the essential assumptions including linearity, normality, homoscedasticity, and multicollinearity should be met or remedied

STAGE 4: DERIVING THE CANONICAL FUNCTIONS AND ASSESSING OVERALL FIT

Once the selection of variables has been justified and the data have been examined, the next step of canonical correlation analysis is to derive one or more canonical functions (Figure 5-5). Each function consists of a pair of variates—one representing the independent variables and the other representing the dependent variables. The maximum number of canonical variates (functions) that can be extracted from the sets of variables equals the number of variables in the smallest variable

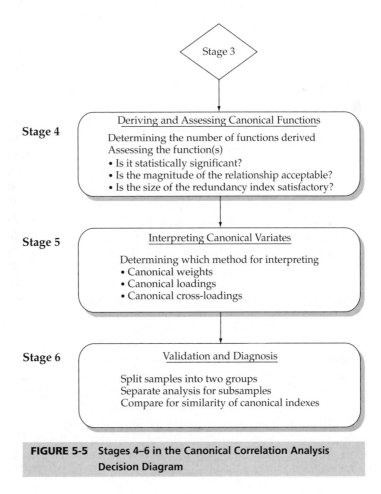

FIGURE 5-5   **Stages 4–6 in the Canonical Correlation Analysis Decision Diagram**

set, either independent or dependent. For example, when the research problem involves five independent variables and three dependent variables, the maximum number of canonical functions that can be extracted is three.

Deriving Canonical Functions

The derivation of successive canonical variates is similar to the procedure used with unrotated factor analysis (see Chapter 3). The first canonical function that is extracted accounts for the maximum amount of variance in the set of variables. The second function is then computed so that it accounts for as much as possible of the variance not accounted for by the first function, and so forth, until all functions have been extracted. Therefore, successive functions are derived from residual or leftover variance from earlier functions. Canonical correlation analysis follows a procedure similar to factor analysis but focuses on accounting for the maximum amount of the relationship between the two sets of variables, rather than within a single set (e.g., a single factor). The result is that the first pair of canonical variates is derived so as to have the highest intercorrelation possible between the two variates. The second pair of canonical variates is then derived so that it exhibits the maximum relationship between the two sets of variables (variates) not accounted for by the first pair of variates. In short, successive canonical functions estimate pairs of canonical variates based on residual variance from the previous canonical functions, and their respective canonical correlations (which reflect the interrelationships between the variates) become smaller as each additional function is extracted. That is, the first pair of canonical variates exhibits the highest intercorrelation, the next pair the second-highest correlation, and so forth.

Because canonical correlation analysis is closely linked with principal components analysis, rotation of canonical variates can be considered as an aid to increase the interpretability of canonical results. Rotation does not change the sums of the squared canonical correlation coefficients but it will lead to a simpler structure. As noted, successive pairs of canonical variates are based on residual variance. Therefore, each pair of variates is **orthogonal** and independent of all other variates derived from the same set of data. The selection of rotation methods is therefore limited to those that are orthogonal and Varimax rotation is the most obvious choice given the close relationship between canonical correlation analysis and principal components analysis [15]. Rotation is only possible when there are at least two canonical functions. It is available in some computer programs, including SPSS and Statit. However, some researchers do not recommend rotation for canonical correlation analysis for two reasons [13]. First, rotation can reduce the optimality of the canonical correlations when each pair of canonical variates is derived to maximize their correlation. Second, rotation introduces correlations among succeeding canonical variates. Therefore, even though rotation may increase the interpretability of the canonical results, this gain may be offset by the increased complexity due to interrelationships among the pairs of canonical variates. Researchers therefore need to be careful when using rotation for canonical correlation analysis.

Which Canonical Functions Should Be Interpreted?

As with other statistical techniques, the most common practice is to analyze functions whose canonical correlation coefficients are statistically significant beyond some level, typically .05 or less. Canonical functions deemed nonsignificant are not interpreted. Interpretation of the canonical variates in a significant function is based on the premise that variables in each set that contribute heavily to shared variances for these functions are considered to be related to each other.

The authors believe that the use of a single criterion, such as the level of significance, is too superficial. Instead, we recommend that three criteria be used in conjunction with one another to decide which canonical functions should be interpreted. The three criteria are (1) level of statistical significance of the function, (2) magnitude of the canonical correlation, and (3) redundancy measure for the percentage of variance accounted for from the two data sets.

LEVEL OF SIGNIFICANCE   The level of significance of a canonical correlation generally considered to be the minimum acceptable for interpretation is the .05 level, which (along with the .01 level) has become the generally accepted level for considering a correlation coefficient statistically significant. This consensus has developed largely because of the availability of tables for these levels. These levels are not necessarily required in all situations, however, and researchers from various disciplines frequently must rely on results based on lower levels of significance. The most widely used test, and the one normally provided by computer packages, is the F-statistic, based on Rao's approximation [5].

In addition to separate tests of each canonical function, a multivariate test of all canonical roots can also be used to evaluate the significance of canonical roots. Many of the measures for assessing the significance of discriminant functions, including Wilks' lambda, Hotelling's trace, Pillai's trace, and Roy's gcr, are also provided. See Chapter 8 for a discussion of these measures.

MAGNITUDE OF THE CANONICAL RELATIONSHIPS   The practical significance of the canonical functions, represented by the size of the canonical correlations, also should be considered when deciding which functions to interpret. No generally accepted guidelines have been established regarding suitable sizes for canonical correlations. Rather, the decision is usually based on the contribution of the findings to better understanding of the research problem being studied. It seems logical that the guidelines suggested for significant factor loadings (see Chapter 3) might be useful with canonical correlations, particularly when one considers that canonical correlations refer to the variance explained in the canonical variates not the original variables.

REDUNDANCY MEASURE OF SHARED VARIANCE   Recall that squared canonical correlations (canonical roots) provide an estimate of the shared variance between the canonical variates. Although this is a simple and appealing measure of the shared variance, it may lead to some misinterpretation because the squared canonical correlations represent the variance shared by the linear composites of the sets of dependent and independent variables, not the variance extracted from the sets of variables [1]. Thus, a relatively strong canonical correlation may be obtained between two linear composites (canonical variates), even though these linear composites may not extract significant portions of variance from their respective sets of variables.

Canonical correlations may be obtained that are considerably larger than previously reported bivariate and multiple correlation coefficients. Thus, there may be a temptation to assume that canonical analysis has uncovered substantial relationships of conceptual and practical significance. Before such conclusions are warranted, however, further analysis involving measures other than canonical correlations must be undertaken to determine the amount of the dependent variable variance accounted for or shared with the independent variables [10].

To overcome the inherent bias and uncertainty in using canonical roots (squared canonical correlations) as a measure of shared variance, a redundancy index has been proposed [14]. It is the equivalent of computing the squared multiple correlation coefficient between the total independent variable set and each variable in the dependent variable set, and then averaging these squared coefficients to arrive at an average R^2. This index provides a summary measure of the ability of a set of independent variables (taken as a set) to explain variation in the dependent variables (taken one at a time). As such, the redundancy measure is analogous to multiple regression's R^2 statistic, and its value as an index is similar.

Calculation of the redundancy index is a three-step process. The first step involves calculating the amount of shared variance from the set of dependent variables included in the dependent canonical variate. The second step involves calculating the amount of variance in the dependent canonical variate that can be explained by the independent canonical variate. The final step is to calculate the redundancy index, which is determined by multiplying these two components.

Step 1: The Amount of Shared Variance.   To calculate the amount of shared variance in a variable set included in its canonical variate, let's first consider how the regression R^2 statistic is calculated. R^2 is the square of the correlation coefficient R, which represents the correlation between the dependent variable and the predicted value. In the canonical case, we are concerned with correlation between the canonical variate and each of its variables. This information can be obtained from the canonical loadings (L_{Xi} or L_{Yi}), which represent the correlation between each variable and its own canonical variate. By squaring each of the canonical loadings (L_i^2), one obtains a measure of the amount of variation in each of the variables explained by the canonical variate. To calculate the amount of shared variance explained by the canonical variate, an average of the squared loadings is used:

$$SV_{xn} = \frac{L_{xn1}^2 + \cdots + L_{xni}^2}{i} \quad \text{and} \quad SV_{yn} = \frac{L_{yn1}^2 + \cdots + L_{ynj}^2}{j}$$

where

SV_{xn} = shared variance of the nth variate of independent (X) variables

SV_{yn} = shared variance of the nth variate of dependent (Y) variables

L_{xni}^2 = squared loading of the ith independent variable in the nth variate of X variables

L_{ynj}^2 = squared loading of the jth dependent variable in the nth variate of Y variables

Thus, the amount of shared variance explained by a set of variables by a canonical variate of the set is the sum of the squared loadings divided by the number of variables in the set.

From our earlier example of credit card usage, the amount of shared variance for the first and the second canonical variates of independent variables can be calculated as:

$$SV_{x1} = \frac{(0.82)^2 + (0.75)^2 + (0.35)^2}{3} \cong 0.45$$

$$SV_{x2} = \frac{(0.30)^2 + (-0.69)^2 + (-0.79)^2}{3} \cong 0.40$$

Likewise, the amount of shared variance for the first and the second canonical variates of dependent variables is:

$$SV_{y1} = \frac{(0.85)^2 + (0.80)^2 + (0.44)^2}{3} \cong 0.52$$

$$SV_{y2} = \frac{(0.71)^2 + (0.42)^2 + (0.82)^2}{3} \cong 0.45$$

The first canonical variate explains 45 percent of the variance in consumer characteristics and the second canonical variate explains 40 percent. Together, the two canonical variates explain nearly 100 percent of the variance in consumer characteristics. For the credit card usage variables, the first canonical variate explains 52 percent of the variance and the second canonical variate explains 45 percent. When summing the two variates, almost 100 percent of variance in credit card usage is explained by the two canonical variates.

Step 2: The Amount of Explained Variance. The second step in calculating redundancy involves an estimate of the account of shared variance between the dependent and independent variates; namely, the canonical root. This is the squared correlation between the independent canonical variate and the dependent canonical variate. Recall that in our example the first pair of canonical variates had a canonical root of 85 percent and the second canonical root was 69 percent.

Step 3: The Redundancy Index. The redundancy index of a variate is then derived by multiplying the two components (shared variance of the variate multiplied by the squared canonical correlation) to find the amount of shared variance explained by the opposite variate. To have a high redundancy index, one must have a high canonical correlation and a high degree of shared variance explained by its own variate. A high canonical correlation alone does not ensure a valuable canonical function. Redundancy indices are calculated for both the dependent and the independent variates, although in most instances the researcher is concerned only with the variance extracted from the dependent variable set, which provides a much more realistic measure of the predictive ability of canonical relationships. The researcher should note that although the canonical correlation is the same for both variates in the canonical function, the redundancy index will most likely vary between the two variates, because each will have a differing amount of shared variance:

$$RI = SV \times Rc^2$$

The redundancy index of a canonical variate is the percentage of variance explained by its own set of variables multiplied by the squared canonical correlation for the pair of variates. Thus, in our example, the redundancy indices for the independent variables explaining the dependent variables are:

$$RI_{x1:\ y} = SV_{y1} \times Rc_1^2 = 0.52 \times 0.85 \cong 0.44$$
$$RI_{x2:\ y} = SV_{y2} \times Rc_2^2 = 0.45 \times 0.69 \cong 0.31$$

The amounts of variance of the independent variables, explained by the dependent variables for the two functions, are:

$$RI_{y1:\ x} = SV_{x1} \times Rc_1^2 = 0.45 \times 0.85 \cong 0.38$$
$$RI_{y2:\ x} = SV_{x2} \times Rc_2^2 = 0.40 \times 0.69 \cong 0.28$$

That is, the first canonical variate for Consumer Characteristics explains 44 percent of the variance in Credit Card Usage and the second canonical variate explains 31 percent. Together they explain over 70 percent of the variance in the dependent variables. The first and the second canonical variates for Credit Card Usage explain 38 percent and 28 percent of the variance in Consumer Characteristics, respectively. Together they explain 66 percent of the variance in Consumer Characteristics.

What is the minimum acceptable redundancy index needed to justify the interpretation of canonical functions? Just as with canonical correlations, no generally accepted guidelines have been established. The researcher must judge each canonical function in light of its theoretical and practical significance to the research problem being investigated to determine whether the redundancy index is sufficient to justify interpretation. A test for the significance of the redundancy index has been developed [2], although it has not been widely utilized.

RULES OF THUMB 5-2

Choosing and Assessing Canonical Functions

- More than one canonical function can be derived, depending on the number of independent or dependent variables included
- The first function has the highest intercorrelation between the variates, the second function has the second-highest, and so forth
- Orthogonal rotation methods (e.g., Varimax) can increase the interpretability of canonical results, but it can also distort the fundamental logic of the method
- The canonical functions can be interpreted based on:
  - Their level of statistical significance
  - The size of the canonical correlation
  - The magnitude of the redundancy index
- The indices used varies based on the aims of the research problem and the guidelines within each discipline
- Canonical correlations can be judged by the same guidelines as factor loadings
- Theoretical and practical significance are more important than statistical significance in determining the minimum acceptable redundancy index

STAGE 5: INTERPRETING THE CANONICAL VARIATE

If the canonical relationship is statistically significant and the magnitudes of the canonical root and the redundancy index are acceptable, the researcher still needs to make substantive interpretations of the results. Making these interpretations involves examining the canonical functions to determine the relative importance of each of the original variables in the canonical relationships. Three methods have been proposed: (1) canonical weights (standardized coefficients), (2) canonical loadings (structure correlations), and (3) canonical cross-loadings.

Canonical Weights

The traditional approach to interpreting canonical functions involves examining the sign and the magnitude of the canonical weight assigned to each variable in its canonical variate. Variables with relatively larger weights contribute more to the variates, and vice versa. Similarly, variables whose weights have opposite signs exhibit an inverse relationship with each other, and variables with weights of the same sign exhibit a direct relationship. However, interpreting the relative importance or contribution of a variable by its canonical weight is subject to the same criticisms associated with the interpretation of beta weights in regression techniques. For example, a small weight may mean either that its corresponding variable is irrelevant in determining a relationship or that it has been partialed out of the relationship because of a high degree of multicollinearity. Another problem with the use of canonical weights is that these weights are subject to considerable instability (variability) from one sample to another. This instability occurs because the computational procedure for canonical analysis yields weights that maximize the canonical correlations for a particular sample of observed dependent and independent variable sets [10]. These problems suggest considerable caution in using canonical weights to interpret the results of a canonical analysis.

Canonical Loadings

Canonical loadings have been increasingly used as a basis for interpretation because of the deficiencies inherent in canonical weights. Canonical loadings, also called **canonical structure correlations,** measure the simple linear correlation between an original variable in the dependent or independent set and the set's canonical variate. The canonical loading reflects the variance that the observed variable shares with the canonical variate and can be interpreted like a factor loading in assessing the relative contribution of each variable to each canonical function. The approach considers each independent canonical function separately and computes the within-set variable-to-variate correlation. The larger the coefficient, the more important it is in deriving the canonical variate. Also, the criteria for determining the significance of canonical structure correlations are the same as with factor loadings (see Chapter 3).

Canonical loadings, like weights, may be subject to considerable variability from one sample to another. This variability suggests that loadings, and hence the relationships ascribed to them, may be sample-specific, resulting from chance or extraneous factors [8]. Also, when using only canonical loadings for interpretation the risks increase that the application is closer to a univariate setting than to a multivariate one [13]. Although canonical loadings are considered relatively more valid than weights as a means of interpreting the nature of canonical relationships, the researcher must be cautious when using loadings for interpreting canonical relationships, particularly with regard to the external validity of the findings.

Canonical Cross-Loadings

The computation of **canonical cross-loadings** has been suggested as an alternative to canonical loadings [7]. This procedure involves correlating each of the variables directly with the other canonical variate, and vice versa. For example, each dependent variable would be correlated separately with the variate for the independent variables. Recall that conventional loadings correlate the variables with their respective variates after the two canonical variates (dependent and independent) are maximally correlated with each other. This may also seem similar to multiple regression, but it differs in that each independent variable, for example, is correlated with the dependent variate instead of a single dependent variable. Thus cross-loadings provide a more direct measure of the dependent–independent variable relationships by eliminating an intermediate step involved in conventional loadings. Some canonical analyses do not compute correlations between the variables and the variates. In such cases, the canonical weights are considered comparable but not equivalent for purposes of our discussion. The cross-loadings can be expressed as:

$$\lambda_{xni:\ y} = Rc_n \times L_{xni}$$
$$\lambda_{ynj:\ x} = Rc_n \times L_{ynj}$$

where

$\lambda_{xni:\ y}$ = cross-loading of the ith independent variable in the nth variate X to the nth variate Y

$\lambda_{ynj:\ x}$ = cross-loading of the jth dependent variable in the nth variate Y to the nth variate X

Rc_n = canonical correlation coefficient for the ith canonical function

L_{xni} = loading of the ith independent variable in the nth variate X

L_{ynj} = loading of the jth dependent variable in the nth variate Y

Thus, each canonical cross-loading is equal to the product of canonical correlation coefficient for the nth canonical function and the canonical loading of the corresponding variable.

Which Interpretation Approach to Use

Several different methods for interpreting the nature of canonical relationships have been discussed. The question remains, however: Which method should the researcher use? Most often the researcher is constrained to the method(s) available in the statistical package being used. Canonical weights perform well in the multivariate setting because they adjust when the combination of the variable sets change [13]. However, they are valid only when collinearity is minimal. The canonical-loadings approach is somewhat more representative than the use of weights, as was seen with factor analysis and discriminant analysis. Cross-loadings and loadings are based on a linear relationship and thus have corresponding results. But cross-loadings facilitate the transformation of a canonical model to a single latent construct which resembles Structural Equation Modeling [4]. Therefore, the cross-loadings approach is preferred and whenever possible the loadings approach is recommended as the best alternative to the canonical cross-loading method. Cross-loadings are provided by many computer programs, but if the cross-loadings are not available, the researcher is forced either to compute the cross-loadings by hand or to rely on the interpretation of canonical loadings.

STAGE 6: VALIDATION AND DIAGNOSIS

As with other multivariate techniques, canonical correlation analysis should be subjected to validation methods to ensure that the results are specific not only to the sample data but can be generalized to the population. The most direct procedure is to create two subsamples of the data (if sample size allows) and perform the analysis on each subsample separately. Then the results can be compared for similarity of canonical functions, variate loadings, and the like. If marked differences are found, the researcher should consider additional investigation to ensure that the final results are representative of the population values, not solely those of a single sample.

Another approach is to assess the sensitivity of the results to the removal of a dependent and/or independent variable. Because the canonical correlation procedure maximizes the correlation and does not optimize the interpretability, the canonical weights and loadings may vary

---

RULES OF THUMB 5-3

Interpreting and Validating Canonical Variates

- Canonical cross-loadings are preferred over canonical weights and canonical loadings when interpreting the results.
- If possible, the data should be split randomly into two subgroups, running canonical correlation analysis separately and comparing the results, in order to ensure the validity of the canonical solution.

---

substantially if one variable is removed from either variate. To ensure the stability of the canonical weights and loadings, the researcher should estimate multiple canonical correlations, each time removing a different independent or dependent variable.

Although few diagnostic procedures have been developed specifically for canonical correlation analysis, the researcher should view the results within the limitations of the technique. Among the limitations that can have the greatest impact on the results and their interpretation are the following:

1. The canonical correlation reflects the variance shared by the linear composites of the sets of variables, not the variance extracted from the variables.
2. Canonical weights derived in computing canonical functions are subject to a great deal of instability.
3. Canonical weights are derived to maximize the correlation between linear composites, not the variance extracted.
4. The interpretation of the canonical variates may be difficult, because they are calculated to maximize the relationship, and aids for interpretation, such as rotation of variates in factor analysis, are limited.
5. It is difficult to identify meaningful relationships between the subsets of independent and dependent variables because precise statistics have not yet been developed to interpret canonical analysis, and we must rely on inadequate measures such as loadings or cross-loadings [10].

These limitations are not meant to discourage the use of canonical correlation. Rather, they are pointed out to enhance the effectiveness of canonical correlation as a research tool.

AN ILLUSTRATIVE EXAMPLE

To illustrate the application of canonical correlation, we use variables drawn from the HBAT database introduced in Chapter 1. Recall that the data consisted of a series of measures obtained on samples of HBAT customers. For this example, we chose the sample involving 200 customers (HBAT_200). The variables included ratings of HBAT on five measures of basic firm characteristics and respondents' business relationship with HBAT (X_1 to X_5), 13 indicators reflecting perceptions of HBAT (X_6 to X_{18}), and five indicators of purchase outcome (X_{19} to X_{23}). SPSS and Statit are used in the design and analysis of this example. Comparable results are obtained with other statistical programs available for commercial and academic use. The syntax and dataset of canonical correlation analysis is available on the text's Web sites at www.pearsonglobaleditions.com/hair or www.mvstats.com.

As with previous chapters, discussion of this application of canonical correlation analysis follows the six-stage process described earlier in the chapter. At each stage, the results illustrating the decisions in that stage are examined.

STAGE 1: OBJECTIVES OF CANONICAL CORRELATION ANALYSIS

In demonstrating the application of canonical correlation, 13 indicators of perceptions and 2 indicators of customers' likelihood to do business with HBAT are used as input data. Both dependent and independent variables were assessed for meeting the basic assumptions underlying multivariate analyses. Among them, X_{14} and X_{18} are highly correlated with other variables. They are thus removed from the analysis due to their violation of multicollinearity. The HBAT ratings (X_6 to X_{13} and X_{15} to X_{17}) are designated as the set of independent variables. The set of dependent variables is defined as two variables—the likelihood of recommending HBAT and future purchases from HBAT (X_{20} and X_{21}). The statistical problem involves identifying any relationships between the variates formed for customer perceptions of HBAT and the likelihood of doing business with HBAT in the future. This canonical model is illustrated in Figure 5-6.

Stages 2 and 3: Designing a Canonical Correlation Analysis and Testing the Assumptions

The designation of the variables includes two metric-dependent and 11 metric-independent variables. The 11 variables resulted in an 18-to-1 ratio of observations to variables, exceeding the guideline of 10 observations per independent variable. The HBAT data set with 200 observations was used in this example. The sample size of 200 should not affect the estimates of sampling error markedly and thus should not impact the statistical significance of the results.

Stage 4: Deriving the Canonical Functions and Assessing Overall Fit

The canonical correlation analysis was restricted to deriving two canonical functions because the dependent variable set contained only two indicators. To determine the number of canonical functions to include in the interpretation stage, the analysis focused on the level of statistical significance, the practical significance of the canonical correlation, and the redundancy indices for each variate.

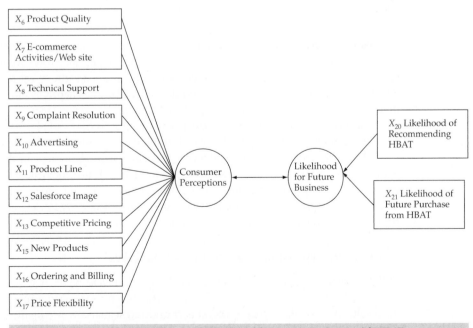

FIGURE 5-6   Canonical Correlation of Likelihood for Future Business with HBAT and Customer's Perception of HBAT

TABLE 5-1 Canonical Correlation Analysis Relating Perceptions of HBAT to Likelihood to Do Business with HBAT

Measures of Overall Model Fit for Canonical Correlation Analysis

Canonical Function	Canonical Correlation	Canonical R^2	F Statistics	Probability
1	.765	.585	9.956	.000
2	.203	.041	.807	.622

Multivariate Tests of Significance

Statistic	Value	Approximate F Statistic	Probability
Wilks' lambda	.398	9.956	.000
Pillai's trace	.626	7.793	.000
Hotelling's trace	1.454	12.291	.000
Roy's gcr	.585		

STATISTICAL AND PRACTICAL SIGNIFICANCE The first statistical significance test is for the canonical correlations of each of the two canonical functions. In this example, only the first canonical correlation is statistically significant (see Table 5-1). In addition to tests of each canonical function separately, multivariate tests of both functions were performed simultaneously. The test statistics employed included Wilks' lambda, Pillai's criterion, Hotelling's trace, and Roy's gcr. Table 5-1 also details the multivariate test statistics, which all indicate that the first canonical function, taken collectively, is statistically significant at the .01 level; the second canonical function fails to achieve significance level at the .05 level.

In addition to statistical significance, the canonical correlations show that only the first function is of sufficient size to be deemed practically significant. Because rotation is available only when there are at least two functions, rotation is not performed in this example.

REDUNDANCY ANALYSIS A redundancy index is calculated for the independent and dependent variates of the first function in Table 5-2. As can be seen, the redundancy index for the dependent variate is substantial (.484). The independent variate, however, has a markedly lower redundancy index (.119), although in this case, because there is a clear delineation between dependent and independent variables, this lower value is not unexpected or problematic. The low redundancy of the independent variate results from the relatively low shared variance in the independent variate (.204), not the canonical R^2. From the redundancy analysis and the statistical significance tests, the first function should be accepted.

The interested researcher should review Chapter 3, with attention to the discussion of scale development. Canonical correlation is in some ways a form of scale development, because the dependent and independent variates represent dimensions of the variable sets, similar to the scales developed with factor analysis. The primary difference is that these dimensions are developed to maximize the relationship between them, whereas factor analysis maximizes the explanation (shared variance) of the variable set.

Stage 5: Interpreting the Canonical Variates

With the canonical relationship deemed statistically significant and the magnitude of the canonical root and the redundancy index acceptable, the researcher proceeds to making substantive interpretations of the results. Because the second function is considered statistically nonsignificant, it is excluded from further analysis and the interpretation phase. These interpretations involve examining the canonical function to determine the relative importance of each of the original variables in deriving the canonical relationships. The three methods for interpretation are (1) canonical weights (standardized coefficients), (2) canonical loadings (structure correlations), and (3) canonical cross-loadings.

TABLE 5-2 Calculation of the Redundancy Indices for the First Canonical Function

Variate/Variables	Canonical Loading	Canonical Loading Squared	Average Loading Squared	Canonical R^2	Redundancy Index
Dependent Variables					
X_{20} Likelihood of Recommending HBAT	.938	.880			
X_{21} Likelihood of Future Purchases from HBAT	.881	.776			
Dependent Variate		1.656	$\cong$.827	.585	.484
Independent Variables					
X_6 Product Quality	.622	.387			
X_7 E-commerce	.393	.154			
X_8 Technical Support	.328	.108			
X_9 Complaint Resolution	.605	.366			
X_{10} Advertising	.336	.113			
X_{11} Product Line	.661	.437			
X_{12} Salesforce Image	.524	.275			
X_{13} Competitive Pricing	−.291	.085			
X_{15} New Products	.150	.023			
X_{16} Ordering and Billing	.540	.292			
X_{17} Pricing Flexibility	.036	.001			
Independent Variate		2.239	$\cong$.204	.585	$\cong$.119

CANONICAL WEIGHTS Table 5-4 contains the standardized canonical weights for each canonical variate of both the dependent and independent variables. As discussed earlier, the magnitude of the weights represents their relative contribution to the variate. The four variables with the highest canonical weights on the independent variate are X_{12} (Salesforce Image), X_6 (Product Quality), X_{17} (Pricing Flexibility), and X_{11} (Product Line). The dependent variable order on the variate is X_{20} (Likelihood of Recommending), then X_{21} (Likelihood of Future Purchase). Recall that the weights obtained are typically unstable unless the collinearity among variables is minimal. In this example, X_7 with X_{12} and X_9 with X_{16} do share moderately high correlation (.788 and .741, respectively). Thus, the interpretation based on canonical weights is likely to be biased and therefore is not recommended in this example.

CANONICAL LOADINGS Table 5-4 contains the canonical loadings for the dependent and independent variates for the first canonical function. The objective of maximizing the variates for the correlation between them results in variates "optimized" not for interpretation, but rather for prediction. This makes identification of relationships more difficult. In the first dependent variate, both variables have loadings exceeding .80, resulting in the high shared variance (.824). This indicates a high degree of intercorrelation among the two variables and suggests that both or either measures are representative of customer intentions to do business with HBAT.

The first independent variate has a quite different pattern, with loadings ranging from .036 to .661, with one independent variable (X_{13}) even having a negative loading, although it is rather small and not of substantive interest. The four variables with the highest loadings on the independent variate are X_{11} (Product Line), X_6 (Product Quality), X_9 (Complaint Resolution), and X_{16} (Ordering and Billing). This variate partly corresponds to the dimensions extracted in factor analysis (see Chapter 3), but it would not be expected to fully match because the variates in canonical correlation are extracted only to maximize predictive objectives. As such, it should correspond more closely to the results from other dependence techniques.

TABLE 5-3 Redundancy Analysis of Dependent and Independent Variates for Both Canonical Functions

Standardized Variance of the Dependent Variables Explained by

Canonical Function	Their Own Canonical Variate (Shared Variance)		Canonical R^2	The Opposite Canonical Variate (Redundancy)	
	Percentage	Cumulative Percentage		Percentage	Cumulative Percentage
1	.827	.827	.585	.484	.484

Standardized Variance of the Independent Variables Explained by

Canonical Function	Their Own Canonical Variate (Shared Variance)		Canonical R^2	The Opposite Canonical Variate (Redundancy)	
	Percentage	Cumulative Percentage		Percentage	Cumulative Percentage
1	.204	.204	.585	.119	.119

There is a close correspondence to the multiple regression results summarized in Chapter 4. Two of these variables (X_6 and X_{11}) were included in the stepwise regression analysis in which X_{21} (one of the two variables in the dependent variate) was the dependent variable. Thus, the first canonical function closely corresponds to the multiple regression results, with the independent variate representing the set of variables best predicting the two dependent measures. The researcher should also perform a sensitivity analysis of the independent variate to see whether the loadings change when an independent variable is deleted (see stage 6).

CANONICAL CROSS-LOADINGS Table 5-4 also includes the cross-loadings for the two canonical functions. In studying the first canonical function, we see that both dependent variables (X_{20} and X_{21}) exhibit high correlations with the independent canonical variate $-$ .717 and .674, respectively. This

TABLE 5-4 Canonical Weights, Loadings, and Cross-Loadings for the Canonical Function

	Canonical Weights	Canonical Loadings	Canonical Cross-Loadings
Independent Variables			
X_6 Product Quality	.631	.622	.476
X_7 E-commerce	−.139	.393	.301
X_8 Technical Support	.161	.328	.251
X_9 Complaint Resolution	.026	.605	.463
X_{10} Advertising	−.120	.336	.257
X_{11} Product Line	.416	.661	.506
X_{12} Salesforce Image	.657	.524	.401
X_{13} Competitive Pricing	−.087	−.291	−.222
X_{15} New Products	−.013	.150	.114
X_{16} Ordering and Billing	−.043	.540	.413
X_{17} Pricing Flexibility	.421	.036	.028
Dependent Variables			
X_{20} Likelihood of Recommending HBAT	.632	.938	.717
X_{21} Likelihood of Future Purchases from HBAT	.462	.881	.674

reflects the high shared variance between these two variables. By squaring these terms, we find the percentage of the variance for each of the variables explained by function 1. The results show that 51.4 percent of the variance in X_{20} and 45.4 percent of the variance in X_{21} is explained by function 1. Looking at the independent variables' cross-loadings, we see that variables X_{11} and X_6 have the highest correlations with the dependent canonical variate − .506 and .476, respectively. From this information, approximately 25.6 percent of the variance in X_{11} and 22.7 percent of variance in X_6 are explained by the dependent variate (the 25.6% is obtained by squaring the correlation coefficient, .506).

The final issue of interpretation is examining the signs of the cross-loadings. All independent variables except X_{13} (Competitive Pricing) have a positive, direct relationship. The four highest cross-loadings of the independent variate correspond to the variables with the highest canonical loadings as well. Thus all the relationships are positive except for one inverse relationship (X_{13}).

Stage 6: Validation and Diagnosis

The last stage should involve a validation of the canonical correlation analyses through one of several procedures. Among the available approaches would be (1) splitting the sample into estimation and validation samples or (2) sensitivity analysis of the independent variables set. Table 5-5 contains the results of a sensitivity analysis in which the canonical loadings are examined for stability when

TABLE 5-5   **Sensitivity Analysis of the Canonical Correlation Results to Removal of an Independent Variable**

	Complete Variate	Results after Deletion of		
		X_6	X_7	X_8
Canonical correlation (*R*)	.765	.668	.762	.756
Canonical root (R^2)	.585	.447	.581	.571
Independent Variate				
Canonical loadings				
X_6 Product Quality	.622	omitted	.624	.630
X_7 E-commerce	.393	.452	omitted	.398
X_8 Technical Support	.328	.376	.329	omitted
X_9 Complaint Resolution	.605	.692	.607	.613
X_{10} Advertising	.336	.384	.337	.340
X_{11} Product Line	.661	.756	.663	.670
X_{12} Salesforce Image	.524	.601	.526	.530
X_{13} Competitive Pricing	−.291	−.331	−.291	−.295
X_{15} New Products	.150	.173	.151	.151
X_{16} Ordering and Billing	.540	.621	.543	.546
X_{17} Pricing Flexibility	.036	.043	.037	.036
Shared variance	.204	.243	.210	.219
Redundancy	.120	.109	.122	.125
Dependent Variate				
Canonical loadings				
X_{20} Likelihood of Recommending HBAT	.938	.944	.941	.935
X_{21} Likelihood of Future Purchases from HBAT	.881	.872	.876	.884
Shared variance	.827	.825	.826	.828
Redundancy	.484	.369	.480	.473

individual independent variables are deleted from the analysis. As seen, the canonical loadings in our example are remarkably stable and consistent in each of the three cases where an independent variable (X_6, X_7, or X_8) is deleted. The overall canonical correlations also remain stable. But the researcher examining the canonical weights (not presented in the table) would find widely varying results, depending on which variable was deleted. This reinforces the procedure of using the canonical loadings and cross-loadings for interpretation purposes.

A Managerial Overview of the Results

The canonical correlation analysis addresses two primary objectives: (1) identification of dimensions among the dependent and independent variable sets and (2) maximization of the relationship between the dimensions. From a managerial perspective, this provides the researcher with insights into the structure of the different variable sets as they relate to a dependence relationship. First, the results indicate that only a single relationship exists, supported by the lack of statistical significance and low practical significance of the second canonical function. In examining this relationship, we first see that the two dependent variables are quite closely related and create a well-defined dimension for representing the outcomes of HBAT's efforts. Second, this outcome dimension is fairly well predicted by the set of independent variables when acting as a set. The redundancy value of .484 would be an acceptable R^2 for a comparable multiple regression. When interpreting the independent variate, we see that three variables, X_{11} (Product Line), X_6 (Product Quality), and X_9 (Complaint Resolution) provide the substantive contributions and thus are the key predictors of the outcome dimension. These should be the focal points in the development of any strategy directed toward impacting the outcomes of HBAT's future actions.

Summary

The concept of canonical correlation analysis and a decision-making procedure for this technique have been introduced in this chapter. Basic guidelines for assessing the results are also included and an example of the application of canonical correlation analysis with the HBAT data set is presented to further clarify the methodological concepts.

Describe canonical correlation analysis and understand its purpose.   Canonical correlation analysis is a useful and powerful technique for exploring the relationships among multiple dependent and independent variables. The technique is primarily descriptive, although it can be used for predictive purposes. Results obtained from a canonical analysis should suggest answers to questions concerning the number of ways in which the two sets of multiple variables are related, the strengths of the relationships, and the nature of the relationships defined. Canonical analysis enables the researcher to combine into a composite measure what otherwise might be an unmanageably large number of bivariate correlations between sets of variables. It is useful for identifying overall relationships between multiple independent and dependent variables, particularly when the researcher has little a priori knowledge about relationships among the sets of variables. Essentially, the researcher can apply canonical correlation analysis to a set of variables, select those variables (both independent and dependent) that appear to be significantly related, and run subsequent canonical correlations with the more significant variables remaining or perform individual regressions with these variables.

State the similarities and differences between multiple regression, discriminant analysis, factor analysis, and canonical correlation.   Canonical correlation analysis is the most general form of the multivariate statistical techniques. It can have multiple dependent variables that can be metric or nonmetric. It is similar to multiple regression in that the goal of canonical correlation analysis is to quantify the strength of the relationships between independent variables and dependent variable(s). It also resembles discriminant analysis in its ability to determine independent dimensions. Like factor analysis, canonical correlation analysis can create an optimized structure for a set of variables. However, canonical correlation analysis is unique in that it can integrate more than one dependent variable, which is not possible with multiple regression and discriminant analysis. Factor analysis gives details about the linear relationship between the measured variables and latent variables, but canonical correlation focuses more on the linear relationship between a pair of canonical variates.

Summarize the conditions that must be met for application of canonical correlation analysis. Strong theoretical support is needed for selecting and grouping variables, as well as determining the research objectives. However, it is of little importance to strictly define the latent variables into independent and dependent canonical variates, because the technique does not differentiate between the two sets when weighting both variates to maximize the correlation. Even though canonical correlation analysis is relatively less demanding in meeting the underlying statistical assumptions, interpretability is improved if the assumptions are satisfied. These include linearly, normality, homoscedasticity, and multicollinearity. Missing data and outliers should also be avoided.

Define and compare canonical root measures and the redundancy index. A canonical root is the squared canonical correlation coefficient. It indicates the amount of shared variance between the two respective optimally weighted canonical variates. It tells the researcher the proportion of variance explained in the canonical variates but it does not differentiate how much variance is explained in each of the two sets of variables themselves. The canonical root is the same for both variates in the canonical function.

The redundancy index is the amount of variance in a canonical variate explained by the opposite canonical variate in the canonical function. It reflects how well the independent canonical variate predicts values of the dependent variables. The redundancy index is similar to a regression R^2—high redundancy means high ability to predict. It suggests the ability of a set of independent variables to explain variation in the dependent variables, or vice versa. Thus, the redundancy indices are different for dependent and independent variates.

Compare the advantages and disadvantages of the three methods for interpreting the nature of canonical functions. The canonical function can be interpreted by the sign and the magnitude of the canonical weights assigned to each variable in its respective canonical variate. Variables with larger weights contribute more to the variates, and vice versa. Because canonical weights are derived to maximize the canonical correlations, they are subject to considerable instability from one sample to another. Also, weights may be distorted due to multicollinearity. Therefore, considerable caution is necessary if interpretation is based on canonical weights.

Canonical loadings measure the correlation between the original observed variables and its canonical variate. They can be interpreted like factor loadings. Variables with larger loadings are more important in deriving the canonical variate. Whereas weights are more suitable for prediction, loadings are better at explaining underlying constructs. Canonical loadings are considered more valid and stable than weights.

Canonical cross-loadings measure the correlation between the original observed variables and their opposite variate (i.e., independent variables correlate to the dependent variate, dependent variables correlate to the independent variate). They offer more direct interpretations by eliminating an intermediate step of conventional loadings. However, computation of cross-loadings is not as widely available as canonical loadings and weights in statistical programs.

Questions

1. Under what circumstances would you select canonical correlation analysis over multiple regression as the appropriate statistical technique?
2. What three criteria should you use in deciding which canonical functions should be interpreted? Explain the role of each.
3. How would you interpret a canonical correlation analysis?
4. What is the relationship among the canonical root, the redundancy index, and multiple regression's R^2?
5. What are the limitations associated with canonical correlation analysis?
6. Why has canonical correlation analysis been used much less frequently than other multivariate techniques?

Suggested Readings

A list of suggested readings illustrating issues and applications of canonical correlation in general is available on the Web at www.pearsonglobaleditions.com/hair or www.mvstats.com.

References

1. Alpert, Mark I., and Robert A. Peterson. 1972. On the Interpretation of Canonical Analysis. *Journal of Marketing Research* 9 (May): 187.

2. Alpert, Mark I., Robert A. Peterson, and Warren S. Martin. 1975. Testing the Significance of Canonical Correlations. *Proceedings,* American Marketing Association 37: 117–19.

3. Aksin, Zeynep O., and Andrea Masini. 2008. Effective Strategies for Internal Outsourcing and Offshoring of Business Services: An Empirical Investigation. *Journal of Operations Management* 26: 239–56.

4. Bagozzi, R. P., C. Fornell, and D. F. Larcker. 1981. Canonical Correlation Analysis as a Special Case of a Structural Relations Model. *Multivariate Behavioral Research* 16(4): 437–54.

5. Bartlett, M. S. 1941. The Statistical Significance of Canonical Correlations. *Biometrika* 32: 29.

6. Van der Burg, Eeke, Jan de Leeuw, and Garmt Dijksterhuis. 1994. OVERALS: Nonlinear Canonical Correlation with *K* Sets of Variables. *Computational Statistics & Data Analysis* 18: 141–63.

7. Dillon, W. R., and M. Goldstein. 1984. *Multivariate Analysis: Methods and Applications.* New York: Wiley.

8. Green, P. E. 1978. *Analyzing Multivariate Data.* Hinsdale, IL: Holt, Rinehart, & Winston.

9. Green, P. E., and J. Douglas Carroll. 1978. *Mathematical Tools for Applied Multivariate Analysis.* New York: Academic Press.

10. Lambert, Z., and R. Durand. 1975. Some Precautions in Using Canonical Analysis. *Journal of Marketing Research* 12 (November): 468–75.

11. Levine, Mark S. 1977. *Canonical Analysis and Factor Comparison.* Newbury Park, CA: Sage.

12. Lukas, Bryan A., and O. C. Ferrell. 2000. The Effect of Market Orientation on Product Innovation. *Journal of Academy of Marketing Science* 28: 239–47.

13. Rencher, Alvin C. 2002. *Methods of Multivariate Analysis,* 2d ed. New York: John Wiley and Sons.

14. Stewart, Douglas, and William Love. 1968. A General Canonical Correlation Index. *Psychological Bulletin* 70: 160–63.

15. Thompson, Bruce. 1984. *Canonical Correlation Analysis: Uses and Interpretation.* Newbury Park, CA: Sage.

16. Thompson, Bruce. 1991. A Primer on the Logic and Use of Canonical Correlation Analysis. *Measurement and Evaluation in Counseling and Development* 24: 80–95.

CHAPTER 6

Conjoint Analysis

LEARNING OBJECTIVES

Upon completing this chapter, you should be able to do the following:

- Explain the managerial uses of conjoint analysis.
- Know the guidelines for selecting the variables to be examined by conjoint analysis.
- Formulate the experimental plan for a conjoint analysis.
- Understand how to create factorial designs.
- Explain the impact of choosing rank choice versus ratings as the measure of preference.
- Assess the relative importance of the predictor variables and each of their levels in affecting consumer judgments.
- Apply a choice simulator to conjoint results for the prediction of consumer judgments of new attribute combinations.
- Compare a main effects model and a model with interaction terms and show how to evaluate the validity of one model versus the other.
- Recognize the limitations of traditional conjoint analysis and select the appropriate alternative methodology (e.g., choice-based or adaptive conjoint) when necessary.

CHAPTER PREVIEW

Since the mid-1970s, conjoint analysis has attracted considerable attention as a method that realistically portrays consumers' decisions as trade-offs among multi-attribute products or services [35]. Conjoint analysis gained widespread acceptance and use in many industries, with usage rates increasing up to tenfold in the 1980s [114]. During the 1990s, the application of conjoint analysis increased even further, spreading to almost every field of study. Marketing's widespread utilization of conjoint analysis in new product development for consumers led to its adoption in many other areas, such as segmentation, industrial marketing, pricing, and advertising [31, 61]. This rise in usage in the United States has been similar in other parts of the world as well, particularly in Europe [119].

Coincident with this continued growth was the development of alternative methods of constructing the choice tasks for consumers and estimating the conjoint models. Most of the multivariate techniques we discuss in this text are established in the statistical field. Conjoint analysis, however, will continue to develop in terms of its design, estimation, and applications within many areas of research [14].

The use of conjoint analysis accelerated with the widespread introduction of computer programs that integrate the entire process, from generating the combinations of independent variable values to be evaluated to creating choice simulators for predicting consumer choices across a wide number of alternative product and service formulations. Today, several widely employed packages can be accessed by any researcher with a personal computer [9, 10, 11, 41, 86, 87, 88, 92, 96, 97]. Moreover, the conversion of even the most advanced research developments into the PC-based programs is continuing [14], and interest in these software programs is increasing [13, 69, 70].

In terms of the basic dependence model discussed in Chapter 1, conjoint analysis can be expressed as

$$Y_1 = X_1 + X_2 + X_3 + \cdots + X_N$$

$$\text{(nonmetric or metric)} \qquad \text{(nonmetric)}$$

With the use of nonmetric independent variables, conjoint analysis closely resembles analysis of variance (ANOVA), which has a foundation in the analysis of experiments. As such, conjoint analysis is closely related to traditional experimentation. Let us compare a traditional experiment with a conjoint analysis.

The use of experiments in studying individuals typically involves designing a series of stimuli and then asking respondents to evaluate a single stimulus (or sometimes multiple stimuli in a repeated measures design). The results are then analyzed with ANOVA (analysis of variance) procedures, such as those discussed in Chapter 8. Conjoint analysis follows this same approach through the design of stimuli (known as *profiles*). It differs in that respondents are always shown multiple profiles (most often 15 or more profiles) to allow for model estimates to be made for each respondent because each respondent provides multiple observations by evaluating multiple profiles.

In both situations, the researcher has a limited number of attributes that can be systematically varied in amount or character. Although we might try to utilize the traditional experimental format to understand consumers' preferences, it requires large numbers of respondents and only makes comparisons between groups (refer to Chapter 8 for design considerations). As an option, conjoint analysis affords the researcher a technique that can be applied to a single individual or group of individuals and provides insights into not only the preferences for each attribute (e.g., fragrance), but also the amount of the attribute (slightly or highly) [28, 30].

Conjoint analysis is actually a family of techniques and methods specifically developed to understand individual preferences that share a theoretical foundation based on the models of information integration and functional measurement [58]. It is best suited for understanding consumers' reactions to and evaluations of predetermined attribute combinations that represent potential products or services. The flexibility and uniqueness of conjoint analysis arise primarily from the following:

- An ability to accommodate either a metric or a nonmetric dependent variable
- The use of only categorical predictor variables
- Quite general assumptions about the relationships of independent variables with the dependent variable

As we will see in the following sections, conjoint analysis provides the researcher with substantial insight into the composition of consumer preferences while maintaining a high degree of realism.

KEY TERMS

Before starting the chapter, review the key terms to develop an understanding of the concepts and terminology to be used. Throughout the chapter the key terms appear in **boldface**. Other points of emphasis in the chapter and key term cross-references are *italicized*.

Adaptive conjoint method   Methodology for conducting a conjoint analysis that relies on respondents providing additional information not in the actual *conjoint task* (e.g., importance of attributes). This information is then used to adapt and simplify the *conjoint task*. Examples are the *self-explicated* and *adaptive,* or *hybrid,* models.

Adaptive model   Technique for simplifying conjoint analysis by combining the *self-explicated model* and *traditional conjoint analysis.* The most widely known example is Adaptive Conjoint Analysis (ACA) from Sawtooth Software.

Additive model   Model based on the additive *composition rule,* which assumes that individuals just "add up" the *part-worths* to calculate an overall or total worth score indicating *utility* or preference. It is also known as a *main effects* model and is the simplest conjoint model in terms of the number of evaluations and the estimation procedure required.

Balanced design   Profile *design* in which each *level* within a *factor* appears an equal number of times across the profiles in the *conjoint task.*

Bayesian analysis   Alternative estimation procedure relying on probability estimates derived from both the individual cases as well as the sample population that are combined to estimate the conjoint model.

Bridging design   Profile *design* for a large number of *factors* (attributes) in which the attributes are broken into a number of smaller groups. Each attribute group has some attributes contained in other groups, enabling the results from each group to be combined, or bridged.

Choice-based conjoint approach   Alternative form of *conjoint task* for collecting responses and estimating the conjoint model. The primary difference is that respondents select a single *full profile* from a set of profiles (known as a *choice set*) instead of rating or ranking each profile separately.

Choice set   Set of profiles constructed through experimental design principles and used in the *choice-based conjoint approach.*

Choice simulator   Procedure that enables the researcher to assess many "what-if" scenarios. Once the conjoint *part-worths* have been estimated for each respondent, the choice simulator analyzes a set of *profiles* and predicts both individual and aggregate choices for each profile in the set. Multiple sets of profiles can be analyzed to represent any scenario (e.g., preferences for hypothetical product or service configurations or the competitive interactions among profiles assumed to constitute a market).

Composition rule   Rule used to represent how respondents combine attributes to produce a judgment of relative value, or *utility,* for a product or service. For illustration, let us suppose a person is asked to evaluate four objects. The person is assumed to evaluate the attributes of the four objects and to create some overall relative value for each. The rule may be as simple as creating a mental weight for each perceived attribute and adding the weights for an overall score (*additive model*), or it may be a more complex procedure involving *interaction effects.*

Compositional model   Class of multivariate models that estimates the dependence relationship based on respondent observations regarding both the dependent and the independent variables. Such models calculate or "compose" the dependent variable from the respondent-supplied values for all of the independent variables. Principal among such methods are regression analysis and discriminant analysis. These models are in direct contrast to *decompositional models.*

Conjoint task   The procedure for gathering judgments on each profile in the conjoint *design* using one of the three types of presentation method (i.e., *full-profile, pairwise comparison,* or *trade-off*).

Conjoint variate   Combination of independent variables (known as *factors*) specified by the researcher that constitute the total worth or *utility* of the profile.

Decompositional model   Class of multivariate models that decompose the individual's responses to estimate the dependence relationship. This class of models presents the respondent with a predefined set of objects (e.g., a hypothetical or actual product or service) and then asks for an overall evaluation or preference of the object. Once given, the evaluation/preference is decomposed by relating the known attributes of the object (which become the independent variables) to the evaluation (dependent variable). Principal among such models is conjoint analysis and some forms of multidimensional scaling (see Chapter 10).

Design   Specific set of conjoint *profiles* created to exhibit the statistical properties of *orthogonality* and *balance.*

Design efficiency   Degree to which a *design* matches an *orthogonal* design. This measure is primarily used to evaluate and compare *nearly orthogonal* designs. Design efficiency values range from 0 to 100, which denotes an *optimal design.*

Environmental correlation   See *interattribute correlation.*

Factor   Independent variable the researcher manipulates that represents a specific attribute. In conjoint analysis, the factors are nonmetric. Factors must be represented by two or more values (known as *levels*), which are also specified by the researcher.

Factorial design   Method of designing *profiles* by generating all possible combinations of *levels.* For example, a three-factor conjoint analysis with three levels per factor ($3 \times 3 \times 3$) would result in 27 combinations that would act as profiles in the *conjoint task.*

Fractional factorial design   Method of designing profiles (i.e., an alternative to *factorial design*) that uses only a subset of the possible profiles needed to estimate the results based on the assumed composition rule. Its primary objective is to reduce the number of evaluations collected while still maintaining *orthogonality* among the levels and subsequent *part-worth* estimates. It achieves this objective by designing profiles that can estimate only a subset of the total possible effects. The simplest design is an *additive model,* in which only *main effects* are estimated. If selected *interaction terms* are included, then additional profiles are created. The design can be created either by referring to published sources or by using computer programs that accompany most conjoint analysis packages.

Full-profile method   Method of gathering respondent evaluations by presenting *profiles* that are described in terms of all *factors.* For example, let us assume that a candy was described by three factors with two levels each: price (15 cents or 25 cents), flavor (citrus or butterscotch), and color (white or red). A full profile would be defined by one level of each factor. One such profile would be a red butterscotch candy costing 15 cents.

Holdout profiles   See *validation profiles.*

Hybrid model   See *adaptive model.*

Interaction effects   Effects of a combination of related features (independent variables), also known as *interaction terms.* In assessing value, a person may assign a unique value to specific combinations of features that runs counter to the additive *composition rule.* For example, let us assume a person is evaluating mouthwash products described by the two factors (attributes) of color and brand. Let us further assume that this person has an average preference for the attributes red and brand X when considered separately. Thus, when this specific combination of levels (red and brand X) is evaluated with the additive composition rule, the red brand X product would have an expected overall preference rating somewhere in the middle of all possible profiles. If, however, the person actually prefers the red brand X mouthwash more than any other profiles, even above other combinations of attributes (color and brand) that had higher evaluations of the individual features, then an interaction is found to exist. This unique evaluation of a combination that is greater (or could be less) than expected based on the separate judgments indicates a two-way interaction. Higher-order (three-way or more) interactions can occur among more combinations of levels.

Interattribute correlation   Also known as *environmental correlation,* it is the correlation among attributes that makes combinations of attributes unbelievable or redundant. A negative correlation depicts the situation in which two attributes are naturally assumed to operate in different directions, such as horsepower and gas mileage. As one increases, the other is naturally assumed to decrease. Thus, because of this correlation, all combinations of these two attributes (e.g., high gas mileage and high horsepower) are not believable. The same effects can be seen for positive correlations, where perhaps price and quality are assumed to be positively correlated. It may not be believable to find a high-price, low-quality product in such a situation. The presence of strong interattribute correlations requires that the researcher closely examine the profiles presented to respondents and avoid unbelievable combinations that are not useful in estimating the *part-worths*.

Level   Specific nonmetric value describing a *factor*. Each factor must be represented by two or more levels, but the number of levels typically never exceeds four or five. If the factor is originally metric, it must be reduced to a small number of nonmetric levels. For example, the many possible values of size and price may be represented by a small number of levels: size (10, 12, or 16 ounces) or price ($1.19, $1.39, or $1.99). If the factor is nonmetric, the original values can be used as in these examples: color (red or blue), brand (X, Y, or Z), or fabric softener additive (present or absent).

Main effects   Direct effect of each *factor* (independent variable) on the dependent variable. May be complemented by *interaction effects* in specific situations.

Monotonic relationship   The assumption by the researcher that a preference order among *levels* should apply to the *part-worth* estimates. Examples may include objective factors (closer distance preferred over farther distance traveled) or more subjective factors (more quality preferred over lower quality). The implication is that the estimated part-worths should have some ordering in the values, and violations (known as *reversals*) should be addressed.

Nearly orthogonal   Characteristic of a profiles design that is not *orthogonal,* but the deviations from orthogonality are slight and carefully controlled in the generation of the profiles. This type of design can be compared with other profiles designs with measures of *design efficiency.*

Optimal design   Profiles design that is *orthogonal* and *balanced.*

Orthogonality   Mathematical constraint requiring that the *part-worth* estimates be independent of each other. In conjoint analysis, *orthogonality* refers to the ability to measure the effect of changing each attribute level and to separate it from the effects of changing other attribute levels and from experimental error.

Pairwise comparison method   Method of presenting a pair of *profiles* to a respondent for evaluation, with the respondent selecting one profile as preferred.

Part-worth   Estimate from conjoint analysis of the overall preference or *utility* associated with each *level* of each *factor* used to define the product or service.

Preference structure   Representation of both the relative importance or worth of each *factor* and the impact of individual *levels* in affecting *utility.*

Profile   By taking one *level* from each *factor,* the researcher creates a specific "object" (also known as a *treatment*) that can be evaluated by respondents. For example, if a soft drink was being defined by three factors, each with two levels (diet versus regular, cola versus non-cola, and caffeine-free or not), then a profile would be one of the combinations with levels from each factor. Some of the possible profiles would be a caffeine-free diet cola, a regular caffeine-free cola, or a diet caffeine-free non-cola. There can be as many profiles as there are unique combinations of levels. One method of defining profiles is the *factorial design,* which creates separate profiles for each combination of all levels. For example, three factors with two levels each would create eight ($2 \times 2 \times 2$) profiles. However, in many conjoint analyses, the total number of combinations is too large for a respondent to evaluate them all. In these instances, some subsets of profiles are created according to a systematic plan, most often a *fractional factorial design.*

Prohibited pair   A specific combination of *levels* from two *factors* that is prohibited from occurring in the creation of profiles. The most common cause is *interattribute correlation* among the factors.

Respondent heterogeneity   The variation in *part-worths* across unique individuals found in disaggregate models. When aggregate models are estimated, modifications in the estimation process can approximate this expected variation in *part-worths*.

Reversal   A violation of a *monotonic relationship,* where the estimated *part-worth* for a level is greater/lower than it should be in relation to another level. For example, in distance traveled to a store, closer stores would always be expected to have more utility than those farther away. A reversal would be when a farther distance has a larger part-worth than a closer distance.

Self-explicated model   *Compositional model* for performing conjoint analysis in which the respondent provides the *part-worth* estimates directly without making choices.

Stimulus   See *profile.*

Trade-off method   Method of presenting profiles to respondents in which *factors* (attributes) are depicted two at a time and respondents rank all combinations of the *levels* in terms of preference.

Traditional conjoint analysis   Methodology that employs the classic principles of conjoint analysis in the *conjoint task,* using an *additive model* of consumer preference and *pairwise comparison* or *full-profile methods* of presentation.

Utility   An individual's subjective preference judgment representing the holistic value or worth of a specific object. In conjoint analysis, utility is assumed to be formed by the combination of *part-worth* estimates for any specified set of levels with the use of an *additive model,* perhaps in conjunction with *interaction effects.*

Validation profiles   Set of *profiles* that are not used in the estimation of *part-worths.* Estimated *part-worths* are then used to predict preference for the validation profiles to assess validity and reliability of the original estimates. Similar in concept to the validation sample of respondents in discriminant analysis.

WHAT IS CONJOINT ANALYSIS?

Conjoint analysis is a multivariate technique developed specifically to understand how respondents develop preferences for any type of object (products, services, or ideas). It is based on the simple premise that consumers evaluate the value of an object (real or hypothetical) by combining the separate amounts of value provided by each attribute. Moreover, consumers can best provide their estimates of preference by judging objects formed by combinations of attributes.

Utility, a subjective judgment of preference unique to each individual, is the most fundamental concept in conjoint analysis and the conceptual basis for measuring value. The researcher using conjoint analysis to study what things determine utility should consider several key issues:

- Utility encompasses all features of the object, both tangible and intangible, and as such is a measure of an individual's overall preference.
- Utility is assumed to be based on the value placed on each of the levels of the attributes. In doing so, respondents react to varying combinations of attribute levels (e.g., different prices, features, or brands) with varying levels of preference.
- Utility is expressed by a relationship reflecting the manner in which the utility is formulated for any combination of attributes. For example, we might sum the utility values associated with each feature of a product or service to arrive at an overall utility. Then we would assume that products or services with higher utility values are more preferred and have a better chance of choice.

To be successful in defining *utility,* the researcher must be able to describe the object in terms of both its attributes and all relevant values for each attribute. To do so, the researcher develops a **conjoint task,** which not only identifies the relevant attributes, but also defines those attributes so that hypothetical choice situations can be constructed. In doing so, the researcher faces four specific questions:

1. *What are the important attributes that could affect preference?* In order to accurately measure preference, the researcher must be able to identify all of the attributes, known as **factors,** that provide utility and form the basis for preference and choice. Factors represent the specific attributes or other characteristics of the product or service.
2. *How will respondents know the meaning of each factor?* In addition to specifying the factors, the researcher must also define each factor in terms of **levels,** which are the possible values for that factor. These values enable the researcher to then describe an object in terms of its levels on the set of factors characterizing it. For example, brand name and price might be two factors in a conjoint analysis. Brand name might have two levels (brand X and brand Y), whereas price might have four levels (39 cents, 49 cents, 59 cents, and 69 cents).

3. *What do the respondents actually evaluate?* After the researcher selects the factors and the levels to describe an object, they are combined (one level from each factor) into a **profile,** which is similar to a **stimulus** in a traditional experiment. Therefore, a profile for our simple example might be brand X at 49 cents.

4. *How many profiles are evaluated?* Conjoint analysis is unique among the multivariate methods, as will be discussed later, in that respondents provide multiple evaluations. In terms of the conjoint task, a respondent will evaluate a number of profiles in order to provide a basis for understanding their preferences. The process of deciding on the actual number of profiles and their composition is contained in the **design.**

These four questions are focused on ensuring that the respondent is able to perform a realistic task—choosing among a set of objects (profiles). Respondents need not tell the researcher anything else, such as how important an individual attribute is to them or how well the object performs on any specific attribute. Because the researcher constructed the hypothetical objects in a specific manner, the influence of each attribute and each value of each attribute on the utility judgment of a respondent can be determined from the respondents' overall ratings.

HYPOTHETICAL EXAMPLE OF CONJOINT ANALYSIS

As an illustration, we examine a simple conjoint analysis for a hypothetical product with three attributes. We first describe the process of defining utility in terms of attributes (factors) and the possible values of each attribute (levels). With the factors specified, the process of collecting preference data through evaluations of profiles is discussed, followed by an overview of the process of estimating the utility associated with each factor and level.

Specifying Utility, Factors, Levels, and Profiles

The first task is to define the attributes that constitute utility for the product being studied. A key issue involves defining the attributes that truly affect preferences and then establishing the most appropriate values for the levels.

Assume that HBAT is trying to develop a new industrial cleanser. After discussions with sales representatives and focus groups, management decides that three attributes are important: cleaning ingredients, form, and brand name. To operationalize these attributes, the researchers create three factors with two levels each:

	Levels	
Factor	*1*	*2*
1. Ingredients	Phosphate-Free	Phosphate-Based
2. Form	Liquid	Powder
3. Brand Name	HBAT	Generic Brand

A profile of a hypothetical cleaning product can be constructed by selecting one level of each attribute. For the three attributes (factors) with two values (levels), eight ($2 \times 2 \times 2$) combinations can be formed. Three examples of the eight possible combinations (profiles) are:

- Profile 1: HBAT phosphate-free powder
- Profile 2: Generic phosphate-based liquid
- Profile 3: Generic phosphate-free liquid

By constructing specific combinations (profiles), the researcher attempts to understand a respondent's **preference structure.** The preference structure depicts not only how important each factor is in the overall decision, but also how the differing levels within a factor influence the formation of an overall preference (utility).

Gathering Preferences from Respondents

With the profiles defined in terms of the attributes giving rise to utility, the next step is to gather preference evaluations from respondents. This process shows why conjoint analysis is also called *trade-off analysis,* because in making a judgment on a hypothetical product respondents must consider both the "good" and "bad" characteristics of the product in forming a preference. Thus, respondents must weigh all attributes simultaneously in making their judgments. Respondents can either rank-order the profiles in terms of preference or rate each combination on a preference scale (perhaps a 1–10 scale).

In our example, conjoint analysis assesses the relative impact of each brand name (HBAT versus generic), each form (powder versus liquid), and the different cleaning ingredients (phosphate-free versus phosphate-based) in determining a person's utility by evaluating the eight profiles. Each respondent was presented with eight descriptions of cleanser products (profiles) and asked to rank them in order of preference for purchase (1 = most preferred, 8 = least preferred). The eight profiles are described in Table 6-1, along with the rank orders given by two respondents.

This utility, which represents the total worth or overall preference of an object, can be thought of as the sum of what the product parts are worth, or **part-worths.** The general form of a conjoint model can be shown as

$$(\text{Total worth for product})_{ij} \ldots n_{ij} = \text{Part worth of level } i \text{ for factor 1}$$
$$+ \text{Part worth of level } j \text{ for factor 2} + \cdots$$
$$+ \text{Part worth of level } n \text{ for factor } m$$

where the product or service has m attributes, each having n levels. The product consists of level i of factor 2, level j of factor 2, and so forth, up to level n for factor m.

In our example, the simplest model would represent the preference structure for the industrial cleanser as determined by adding the three factors (utility = brand effect + ingredient effect + form

TABLE 6-1 **Profile Descriptions and Respondent Rankings for Conjoint Analysis of Industrial Cleanser Example**

| | PROFILE DESCRIPTIONS | | | Respondent Rankings | |
| | *Levels of:* | | | | |
Profile #	Form	Ingredients	Brand	Respondent 1	Respondent 2
1	Liquid	Phosphate-free	HBAT	1	1
2	Liquid	Phosphate-free	Generic	2	2
3	Liquid	Phosphate-based	HBAT	5	3
4	Liquid	Phosphate-based	Generic	6	4
5	Powder	Phosphate-free	HBAT	3	7
6	Powder	Phosphate-free	Generic	4	5
7	Powder	Phosphate-based	HBAT	7	8
8	Powder	Phosphate-based	Generic	8	6

Note: The eight profiles represent all combinations of the three attributes, each with two levels $(2 \times 2 \times 2)$.

effect). This format is known as an *additive model* and will be discussed in more detail in a later section. The preference for a specific cleanser product can be calculated from the part-worth values. For example, the preference for profile 1 previous described above (HBAT phosphate-free powder) is defined as

$$
\begin{aligned}
\text{Utility} = {} & \text{Part-worth of HBAT brand} \\
& + \text{Part-worth of phosphate-free cleaning ingredient} \\
& + \text{Part-worth of powder}
\end{aligned}
$$

With the part-worth estimates, the preference of an individual can be estimated for any combination of factors. Moreover, the preference structure would reveal the factor(s) most important in determining overall utility and product choice. The choices of multiple respondents could also be combined to represent the real-world competitive environment.

Estimating Part-Worths

How do we estimate the part-worths for each level when we have only rankings or ratings of the profiles? The procedure is analogous to multiple regression with dummy variables or ANOVA, although other estimation techniques are also used, such as multinomial logit models. We should note that these calculations are done for each respondent separately. This approach differs markedly from other techniques where we deal with relationships across all respondents or group differences. More detail on the actual estimation process is provided for interested readers in the Basic Stats appendix on the text's Web sites (www.pearsonglobaleditions.com/hair or www.mvstats.com).

Table 6-2 provides the estimated part-worths for two respondents in our example. As we can see, each level has a unique part-worth estimate that reflects that level's contribution to utility when contained in a profile. In viewing part-worths for respondent 1, we can see that *Ingredients* seems to be most important because they have the largest impact on utility (part-worths). This differs from respondent 2, where the largest estimated part-worths relate to *Form.*

TABLE 6-2   Estimated Part-Worths and Factor Importance for Respondents 1 and 2

| Factor Level | Respondent 1 | | | Respondent 2 | | |
| | Estimated Part-Worths | Calculating Factor Importance | | Estimated Part-Worths | Calculating Factor Importance | |
	Estimated Part-Worth	Range of Part-Worths	Factor Importance[a]	Estimated Part-Worth	Range of Part-Worths	Factor Importance[a]
Form						
Liquid	+.756	1.512	28.6%	+1.612	3.224	66.7%
Powder	−.756			−1.612		
Ingredients						
Phosphate-free	+1.511	3.022	57.1%	+.604	1.208	25.0%
Phosphate-based	−1.511			−.604		
Brand						
HBAT	+.378	.756	14.3%	−.20	.400	8.3%
Generic	−.378			+.20		
Sum of Part-Worth Ranges		5.290			4.832	

[a]Factor importance is equal to the range of a factor divided by the sum of ranges across all factors, multiplied by 100 to get a percentage.

Determining Attribute Importance

Because the part-worth estimates are on a common scale, we can compute the relative importance of each factor directly. The importance of a factor is represented by the range of its levels (i.e., the difference between the highest and lowest values) divided by the sum of the ranges across all factors. This calculation provides a relative impact or importance of each attribute based on the size of the range of its part-worth estimates. Factors with a larger range for their part-worths have a greater impact on the calculated utility values and thus are deemed of greater importance. The relative importance scores across all attributes will total 100 percent.

For example, for respondent 1, the ranges of the three attributes are 1.512 [.756 − (−.756)], 3.022 [1.511 − (−1.511)], and .756 [.378 − (−.378)]. The sum total of ranges is 5.290. From these, the relative importance for the three factors (form, ingredients, and brand) is calculated as 1.512/5.290, 3.022/5.290, and .756/5.290, or 28.6 percent, 57.1 percent, and 14.3 percent, respectively. We can follow the same procedure for the second respondent and calculate the importance of each factor, with the results of form (66.7%), ingredients (25%), and brand (8.3%). These calculations for respondents 1 and 2 are also shown in Table 6-2.

Assessing Predictive Accuracy

To examine the ability of this model to predict the actual choices of the respondents, we predict preference order by summing the part-worths for the profiles and then rank-ordering the resulting scores. Comparing the predicted preference order to the respondent's actual preference order assesses predictive accuracy. Note that the total part-worth values have no real meaning except as a means of developing the preference order and, as such, are not compared across respondents.

The calculations for both respondents for all eight profiles are shown in Table 6-3, along with the predicted and actual preference orders. Let us examine the results for these respondents to understand how well their preferences were represented by the part-worth estimates:

- *Respondent 1.* The estimated part-worths predict the preference order perfectly for this respondent. This result indicates that the preference structure was successfully represented in the part-worth estimates and that the respondent made choices consistent with the preference structure.
- *Respondent 2.* The inconsistency in rankings for respondent 2 prohibits a full representation of the preference structure. For example, the average rank for profiles with the generic brand is lower than those profiles with the HBAT brand (refer to Table 6-3). This result indicates that, all things being equal, the profiles with the generic brand will be more preferred. Yet, examining the actual rank orders, this response is not always seen. Profiles 1 and 2 are equal except for brand name, yet HBAT is more preferred. The same thing occurs for profiles 3 and 4. However, the correct ordering (generic preferred over HBAT) is seen for the profile pairs of 5–6 and 7–8. Thus, the preference structure of the part-worths will have a difficult time predicting this choice pattern. When we compare the actual and predicted rank orders (see Table 6-3), we see that respondent 2's choices are often incorrectly predicted, but most often miss by one position due to what is termed an *interaction effect* (discussed in a later section).

As you can see, the results of a conjoint analysis provide a complete understanding of the respondent's preference structure. Estimates are made not only of the utility of each level (e.g., Brand X versus Brand Y) but of the relative importance of factors as well (e.g., Ingredients versus Brand). This provides a unique insight into the choice process and the role of important factors.

TABLE 6-3 Predicted Part-Worth Totals for Each Profile and a Comparison of Actual and Estimated Preference Rankings

Profile	Profile Description			Part-Worth Estimates				Preference Rankings	
Profile	Form	Ingredients	Brand	Form	Ingredients	Brand	Total	Estimated	Actual
Respondent 1									
1	Liquid	Phosphate-free	HBAT	.756	1.511	.378	2.645	1	1
2	Liquid	Phosphate-free	Generic	.756	1.511	−.378	1.889	2	2
3	Liquid	Phosphate-based	HBAT	.756	−1.511	.378	−.377	5	5
4	Liquid	Phosphate-based	Generic	.756	−1.511	−.378	−1.133	6	6
5	Powder	Phosphate-free	HBAT	−.756	1.511	.378	1.133	3	3
6	Powder	Phosphate-free	Generic	−.756	1.511	−.378	.377	4	4
7	Powder	Phosphate-based	HBAT	−.756	−1.511	.378	−1.889	7	7
8	Powder	Phosphate-based	Generic	−.756	−1.511	−.378	−2.645	8	8
Respondent 2									
1	Liquid	Phosphate-free	HBAT	1.612	.604	−.200	2.016	2	1
2	Liquid	Phosphate-free	Generic	1.612	.604	.200	2.416	1	2
3	Liquid	Phosphate-based	HBAT	1.612	−.604	−.200	.808	4	3
4	Liquid	Phosphate-based	Generic	1.612	−.604	.200	1.208	3	4
5	Powder	Phosphate-free	HBAT	−1.612	.604	−.200	−1.208	6	7
6	Powder	Phosphate-free	Generic	−1.612	.604	.200	−.808	5	5
7	Powder	Phosphate-based	HBAT	−1.612	−.604	−.200	−2.416	8	8
8	Powder	Phosphate-based	Generic	−1.612	−.604	.200	−2.016	7	6

THE MANAGERIAL USES OF CONJOINT ANALYSIS

Before discussing the statistical basis of conjoint analysis, we should understand the technique in terms of its role in understanding consumer decision making and providing a basis for strategy development [98]. The simple example we just discussed presents some of the basic benefits of conjoint analysis. The flexibility of conjoint analysis gives rise to its application in almost any area in which decisions are studied. Conjoint analysis assumes that any set of objects (e.g., brands, companies) or concepts (e.g., positioning, benefits, images) is evaluated as a bundle of attributes. Having determined the contribution of each factor to the consumer's overall evaluation, the researcher could then proceed with the following:

1. Define the object or concept with the optimum combination of features.
2. Show the relative contributions of each attribute and each level to the overall evaluation of the object.
3. Use estimates of purchaser or customer judgments to predict preferences among objects with differing sets of features (other things held constant).
4. Isolate groups of potential customers who place differing importance on the features to define high and low potential segments.
5. Identify marketing opportunities by exploring the market potential for feature combinations not currently available.

The knowledge of the preference structure for each individual allows the researcher almost unlimited flexibility in examining both individual and aggregate reactions to a wide range of product- or service-related issues. We examine some of the most popular applications later in this chapter.

COMPARING CONJOINT ANALYSIS WITH OTHER MULTIVARIATE METHODS

Conjoint analysis represents a hybrid type of multivariate technique for estimating dependence relationships. In one sense it combines traditional methods (i.e., regression and ANOVA), yet it is unique in that it is decompositional in nature and the results can be estimated for each respondent separately. It offers the researcher an analysis tool developed specifically to understand consumer decisions and their preference structures. Conjoint analysis differs from other multivariate techniques in four distinct areas: (1) its decompositional nature, (2) specification of the variate, (3) the fact that estimates can be made at the individual level, and (4) its flexibility in terms of relationships between dependent and independent variables.

Compositional Versus Decompositional Techniques

Many of the dependence multivariate techniques we examined in previous chapters are termed **compositional models** (e.g., discriminant analysis and many regression applications). With these techniques, the researcher collects ratings from the respondent on many product characteristics (e.g., favorability toward color, style, specific features) and then relates these ratings to some overall preference rating to develop a predictive model. The researcher does not know beforehand the ratings on the product characteristics, but collects them from the respondent. With regression and discriminant analysis, the respondent's ratings and overall preferences are analyzed to "compose" the overall preference from the respondent's evaluations of the product on each attribute.

Conjoint analysis, which is a type of **decompositional model,** differs in that the researcher needs to know only a respondent's overall preference for a profile. The values of each attribute (levels act as the values of the independent variables) were already specified by the researcher when the profile was created. In this way conjoint analysis can determine (decompose) the value of each attribute using only the overall preference measure. It should be noted that conjoint analysis does share one characteristic with compositional models in that the researcher defines the set of attributes to be included in the analysis. Thus, it differs in this regard with other decompositional models such as MDS (see Chapter 10), which do not require specification of the attributes.

Specifying the Conjoint Variate

Conjoint analysis employs a variate quite similar in form to what is used in other multivariate techniques. The **conjoint variate** is a linear combination of effects of the independent variables (levels of each factor) on a dependent variable. The important difference is that in the conjoint variate the researcher specifies both the independent variables (factors) *and* their values (levels). The only information provided by the respondent is the dependent measure. The levels specified by the researcher are then used by conjoint analysis to decompose the respondent's response into effects for each level, much as is done in regression analysis for each independent variable.

This feature illustrates the common characteristics shared by conjoint analysis and experimentation, whereby designing the project is a critical step to success. For example, if a variable or effect was not anticipated in the research design, then it will not be available for analysis. For this reason, a researcher may be tempted to include a number of variables that might be relevant. However, conjoint analysis is limited in the number of variables it can include, so the researcher cannot just include additional questions to compensate for a lack of clear conceptualization of the problem.

Separate Models for Each Individual

Conjoint analysis differs from almost all other multivariate methods in that it can be carried out at the individual level, meaning that the researcher generates a separate model for predicting the preference structure of each respondent. Most other multivariate methods use each respondent's measures as a single observation and then perform the analysis using all respondents simultaneously. In fact, many methods require that a respondent provide only a single observation (the assumption of independence) and then develop a common model for all respondents, fitting each respondent with varying degrees of accuracy (represented by the errors of prediction for each observation, such as residuals in regression).

The ability to estimate models for each individual comes with the requirement, however, that consumers provide multiple evaluations of differing profiles. And as the number of factors and levels increase, the required number of profiles increases as well (see later section for more detailed discussion). But even the simplest situation, such as the cleanser example earlier, requires a substantial number of responses that quickly increase the difficulty of the conjoint task.

Although we have focused on estimates for the individual (disaggregate), estimates can also be made for groups of individuals representing a market segment or the entire market (aggregate). Each approach has distinct benefits. At the disaggregate level, each respondent must rate enough profiles for the analysis to be performed separately for each person. Predictive accuracy is calculated for each person, rather than only for the total sample. The individual results can then be aggregated to portray an overall (aggregate) model as well.

Many times, however, the researcher selects an aggregate analysis method that performs the estimation of part-worths for the group of respondents as a whole. Aggregate analysis can provide several advantages. First, it is a means for reducing the data collection task so that the number of evaluations per person is reduced (discussed in later sections). Second, methods for estimating interactions between factors (e.g., choice-based conjoint) are easily estimated with aggregate models. Third, greater statistical efficiency is gained by using more observations in the estimation process.

In selecting between aggregate and disaggregate conjoint analyses, the researcher must balance the benefits gained by aggregate methods versus the insights provided by the separate models for each respondent obtained by disaggregate methods.

Flexibility in Types of Relationships

Conjoint analysis is not limited in the types of relationships required between the dependent and independent variables. As discussed in earlier chapters, most dependence methods assume that a linear relationship exists when the dependent variable increases (or decreases) in equal amounts for each unit change in the independent variable. If any type of nonlinear relationship is to be represented, either the model form must be modified or specialized variables must be created (e.g., polynomials).

Conjoint analysis, however, can make separate predictions for the effects of each level of the independent variable and does not assume that the levels are related at all. Conjoint analysis can easily handle nonlinear relationships—even the complex curvilinear relationship, in which one value is positive, the next negative, and the third positive again. Moreover, the types of relationships can vary between attributes. As we discuss later, however, the simplicity and flexibility of conjoint analysis compared with the other multivariate methods are based on a number of assumptions made by the researcher.

DESIGNING A CONJOINT ANALYSIS EXPERIMENT

The researcher applying conjoint analysis must make a number of key decisions in designing the experiment and analyzing its results. Figure 6-1 (stages 1–3) on pages 274–275 and Figure 6-4 (stages 4–7) on page 295 show the general steps followed in the design and execution of a conjoint analysis experiment. The discussion follows the model-building paradigm introduced in Chapter 1.

The decision process is initiated with a specification of the objectives of conjoint analysis. Because conjoint analysis is similar to an experiment, the conceptualization of the research is critical to its success. After the defining the objectives, addressing the issues related to the actual research design, and evaluating the assumptions, the discussion looks at how the decision process then considers the actual estimation of the conjoint results, the interpretation of the results, and the methods used to validate the results. The discussion ends with an examination of the use of conjoint analysis results in further analyses, such as market segmentation and choice simulators.

Each of these decisions stems from the research question and the use of conjoint analysis as a tool in understanding the respondent's preferences and judgment process. We follow this discussion

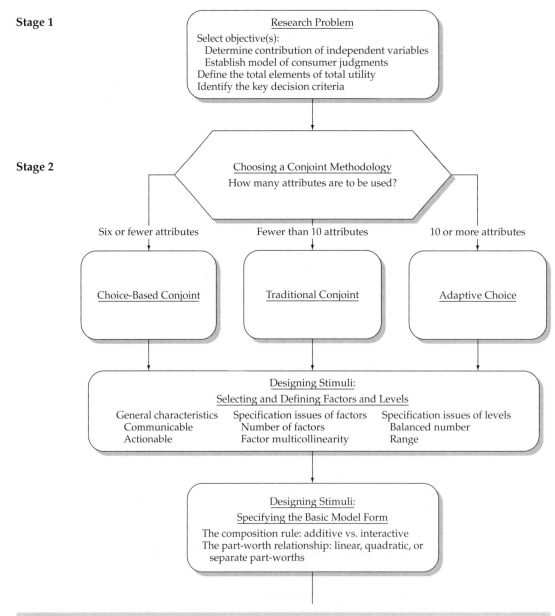

FIGURE 6-1   **Stages 1–3 of the Conjoint Analysis Decision Diagram**

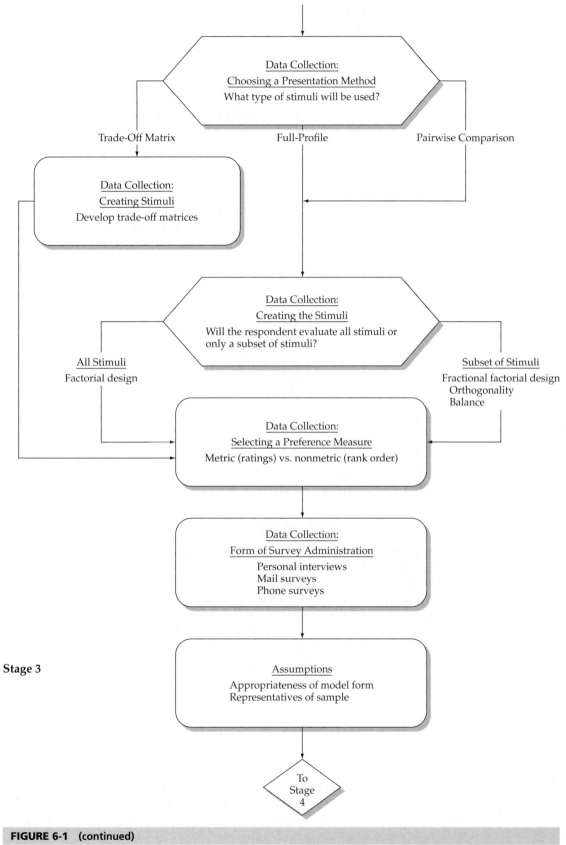

Stage 3

FIGURE 6-1 (continued)

275

of the model-building approach by examining two alternative conjoint methodologies (choice-based and adaptive conjoint), which are then compared to the issues addressed here for traditional conjoint analysis.

STAGE 1: THE OBJECTIVES OF CONJOINT ANALYSIS

As with any statistical analysis, the starting point is the research question. In understanding consumer decisions, conjoint analysis meets two basic objectives:

1. *To determine the contributions of predictor variables and their levels in the determination of consumer preferences.* For example, how much does price contribute to the willingness to buy a product? Which price level is the best? How much change in the willingness to buy soap can be accounted for by differences between the levels of price?
2. *To establish a valid model of consumer judgments.* Valid models enable us to predict the consumer acceptance of any combination of attributes, even those not originally evaluated by consumers. In doing so, the issues addressed include the following: Do the respondents' choices indicate a simple linear relationship between the predictor variables and choices? Is a simple model of adding up the value of each attribute sufficient, or do we need to include more complex evaluations of preference to mirror the judgment process adequately?

The respondent reacts only to what the researcher provides in terms of profiles (attribute combinations). Are these the actual attributes used in making a decision? Are other attributes, particularly attributes of a more qualitative nature such as emotional reactions, important as well? These and other considerations require the research question to be framed around two major issues:

- Is it possible to describe all the attributes that give utility or value to the product or service being studied?
- What are the key attributes involved in the choice process for this type of product or service?

These questions need to be resolved before proceeding into the design phase of a conjoint analysis because they provide critical guidance for key decisions in each stage.

Defining the Total Utility of the Object

The researcher must first be sure to define the total utility of the object. To represent the respondent's judgment process accurately, all attributes that potentially *create* or *detract* from the overall utility of the product or service should be included. It is essential that both positive and negative factors be considered. First, by focusing on only positive factors the research may seriously distort the respondents' judgments. A realistic choice task requires that the "good and bad" attributes be considered. Second, even if the researcher omits the negative factors respondents can subconsciously employ them, either explicitly or through association with attributes that are included. In either instance, the researcher has an incomplete view of the factors that influenced the choice process.

Fortunately, the omission of a single factor typically can have only a minimal impact on the estimates for other factors when using an additive model [84], but the omission of a key attribute may still seriously distort the representation of the preference structure and diminish predictive accuracy.

Specifying the Determinant Factors

In addition, the researcher must be sure to include all determinant factors (drawn from the concept of determinant attributes [5]). The goal is to include the factors that best *differentiate* between the objects. Many attributes may be considered important but may not differentiate in making choices because they do not vary substantially between objects.

RULES OF THUMB 6-1

Objectives of Conjoint Analysis

- Conjoint analysis is unique from other multivariate techniques in that:
  - It is a form of decompositional model that has many elements of an experiment
  - Consumers only provide overall preference rating for objects (stimuli) created by the researcher
  - Stimuli are created by combining one level (value) from each factor (attribute)
  - Each respondent evaluates enough stimuli so that conjoint results are estimated for each individual
- A "successful" conjoint analysis requires that the researcher:
  - Accurately define all of the attributes (factors) that have a positive and negative impact on preference
  - Apply the appropriate model of how consumers combine the values of individual attributes into overall evaluations of an object
- Conjoint analysis results can be used to:
  - Provide estimates of the "utility" of each level within each attribute
  - Define the total utility of any stimuli so that it can be compared to other stimuli to predict consumer choices (e.g., market share)

For example, safety in automobiles is an important attribute, but it may not be determinant in most cases because all cars meet strict government standards and thus are considered safe, at least at an acceptable level. However, other features, such as gas mileage, performance, or price, are important and much more likely to be used to decide among different car models. The researcher should always strive to identify the key *determinant* variables because they are pivotal in the actual judgment decision.

STAGE 2: THE DESIGN OF A CONJOINT ANALYSIS

Having resolved the issues stemming from the research objectives, the researcher shifts attention to the particular issues involved in designing and executing the conjoint analysis experiment. As described in the introductory section, four issues face a researcher in terms of research design:

1. First, which of several alternative conjoint methods should be chosen? Conjoint analysis has three differing approaches to collecting and analyzing data, each with specific benefits and limitations.
2. With the conjoint method selected, the next issue centers on the composition and design of the profiles. What are the factors and levels to be used in defining utility? How are they to be combined into profiles?
3. A key benefit of conjoint analysis is its ability to represent many types of relationships in the conjoint variate. A crucial consideration is the type of effects that are to be included because they require modifications in the research design. **Main effects,** representing the direct impact of each attribute, can be augmented by **interaction effects,** which represent the unique impact of various combinations of attributes.

The last issue relates to data collection, specifically the type of preference measure to be used and the actual conjoint task faced by the respondent.

Note that the design issues are perhaps the most important phase in conjoint analysis. A poorly designed study cannot be "fixed" after administration if design flaws are discovered. Thus, the researcher must pay particular attention to the issues surrounding construction and administration of the conjoint experiment.

Selecting a Conjoint Analysis Methodology

After the researcher determines the basic attributes that constitute the utility of the product or service (object), a fundamental question must be resolved: Which of the three basic conjoint methodologies (traditional conjoint, adaptive conjoint, or choice-based conjoint) should be used [74]?

The choice of conjoint methodologies revolves around the basic characteristics of the proposed research: number of attributes handled, level of analysis, choice task, and the permitted model form. Table 6-4 compares the three methodologies on these considerations. As portrayed in the earlier example, **traditional conjoint analysis** is characterized by a simple additive model generally containing up to nine factors estimated for each individual. A respondent evaluates profiles constructed with selected levels from each attribute. Although this format has been the mainstay of conjoint studies for many years, two additional methodologies have been developed in an attempt to deal with certain design issues. The **adaptive conjoint method** was developed specifically to accommodate a large number of factors (many times up to 30), which would not be feasible in traditional conjoint analysis. It employs a computerized process that adapts the profiles shown to a respondent as the choice task proceeds. Moreover, the profiles can be composed of subsets of attributes, thus allowing for many more attributes. Finally, the **choice-based conjoint approach** employs a unique form of presenting profiles in sets (choose one profile from a set of profiles) rather than one by one. Due to the more complicated task, the number of factors included is more limited, but the approach does allow for inclusion of interactions and can be estimated at the aggregate or individual level.

Many times the research objectives create situations not handled well by traditional conjoint analysis, thus the use of these alternative methodologies. The issues of establishing the number of attributes and selecting the model form are discussed in greater detail in the following section, focusing on traditional conjoint analysis. Then, the unique characteristics of the two other methodologies are addressed in subsequent sections. The researcher should note that the basic issues discussed in this section apply to the two other methodologies as well.

Designing Profiles: Selecting and Defining Factors and Levels

The experimental foundations of conjoint analysis place great importance on the design of the profiles evaluated by respondents. The design involves specifying the conjoint variate by selecting the factors and levels to be included in constructing the profiles. Other issues relate to the general character of both factors, and levels as well as considerations are specific to each. These design

TABLE 6-4   A Comparison of Alternative Conjoint Methodologies

| | Conjoint Methodology | | |
	Traditional Conjoint	Adaptive/Hybrid Conjoint	Choice-Based Conjoint
Characteristic			
Upper Limit on Number of Attributes	9	30	6
Level of Analysis	Individual	Individual	Aggregate or Individual
Model Form	Additive	Additive	Additive + Interaction
Choice Task	Evaluating Full-Profiles One at a Time	Rating Profile Containing Subsets of Attributes	Choice Between Sets of Profiles
Data Collection Format	Any Format	Generally Computer-Based	Any Format

issues are important because they affect the effectiveness of the profiles in the task, the accuracy of the results, and ultimately their managerial relevance.

GENERAL CHARACTERISTICS OF FACTORS AND LEVELS   Before discussing the specific issues relating to factors or levels, characteristics applicable to the specification of factors and levels should be addressed. When operationalizing factors or levels, the researcher should ensure that the measures are both communicable and actionable.

Communicable Measures. First, the factors and levels must be easily communicated for a realistic evaluation. Traditional methods of administration (pencil and paper or computer) limit the types of factors that can be included. For example, it is difficult to describe the actual fragrance of a perfume or the "feel" of a hand lotion. Written descriptions do not capture sensory effects well unless the respondent sees the product firsthand, smells the fragrance, or uses the lotion. If respondents are unsure as to the nature of the attributes being used, then the results are not a true reflection of their preference structure.

One attempt to bring a more realistic portrayal of sensory characteristics that may have been excluded in the past involves specific forms of conjoint developed to employ virtual reality [83] or to engage the entire range of sensory and multimedia effects in describing the product or service [43, 57, 94]. Regardless of whether these approaches are used, the researcher must always be concerned about the communicability of the attributes and levels used.

Actionable Measures. The factors and levels also must be capable of being put into practice, meaning the attributes must be distinct and represent a concept that can be precisely implemented. Researchers should avoid using attributes that are hard to specify or quantify, such as overall quality or convenience. A fundamental aspect of conjoint analysis is that respondents trade off between attributes in evaluating a profile. If they are uncertain as to how one attribute compares to another attribute (e.g., one more precisely defined than the other), then the task cannot reflect the actual preference structure. Likewise, levels should not be specified in imprecise terms such as low, moderate, or high. These specifications are imprecise because of the perceptual differences among individuals as to what they actually mean (as compared with actual differences as to how they feel about them).

If factors cannot be defined more precisely, the researcher may use a two-stage process. A preliminary conjoint study defines profiles in terms of more global or imprecise factors (quality or convenience). Then the factors identified as important in the preliminary study are included in the larger study in more precise terms.

SPECIFICATION ISSUES REGARDING FACTORS   Having selected the attributes to be included as factors and ensured that the measures will be communicable and actionable, the researcher still must address three issues specific to defining factors: the number of factors to be included, multicollinearity among the factors, and the unique role of price as a factor. Specification of factors is a critical phase of research design because once a factor is included in a conjoint analysis choice task, it cannot be removed from the analysis. Respondents always evaluate sets of attributes collectively. Removal of an attribute in the estimation of the part-worths will invalidate the conjoint analysis.

Number of Factors. The number of factors included in the analysis directly affects the statistical efficiency and reliability of the results. Two limits come into play when considering the number of factors to be included in the study.

First, adding factors to a conjoint study always increases the minimum number of profiles in the conjoint design. This requirement is similar to those encountered in regression where the number of observations must exceed the number of estimated coefficients. A conjoint design with only a couple of factors is fairly simple, but the addition of factors can quickly make it a quite complex and arduous task for the respondent. The minimum number of profiles that must be evaluated by each respondent is:

$$\text{Minimum number of profiles} = \text{Total number of levels across all factors} \\ - \text{Number of factors} + 1$$

For example, a conjoint analysis with five factors with three levels each (a total of 15 levels) would need a minimum of 11 $(15 - 5 + 1)$ profiles.

Even though it might seem that increasing the number of factors would reduce the number of profiles required (i.e., the number of factors is subtracted in the preceding equation), remember that each factor must have at least two levels (and many times more), such that an additional factor will always increase the number of profiles. Thus, in the previous example adding one additional factor with three levels would necessitate at least two additional profiles. Some evidence indicates that traditional conjoint analysis techniques can employ a larger number of attributes (20 or so) than originally thought [82]. As we will discuss later, some techniques have been developed to specifically handle large numbers of attributes with specialized designs. Even in these situations, however, the researcher is cautioned to ensure that no matter how many attributes are included, it does not present too complex a task for the respondent.

Second, the number of profiles also must increase when modeling a more complex relationship, such as the case of adding interaction terms. Some reductions in profiles are possible by specialized conjoint designs, but the increased number of parameters to be estimated requires either a larger number of profiles or a reduction in the reliability of parameters.

It is especially important to note that conjoint analysis differs from other multivariate analyses in that the need for more profiles described above cannot be fixed by adding more respondents. In conjoint analysis each respondent generates the required number of observations, and therefore the required number of stimuli is constant no matter how many respondents are analyzed. Specialized forms of estimation estimate aggregate models across individuals, thus requiring fewer stimuli per respondent, but in these cases the fundamental concept of obtaining conjoint estimates for each respondent is eliminated. We will discuss these options in greater detail in a later section.

Interattribute Correlation. A critical issue that many times goes undetected unless the researcher carefully examines all of the profiles in the conjoint design is the correlation among factors (known as **interattribute** or **environmental correlation**). In practical terms, the presence of correlated factors denotes a lack of conceptual independence among the factors. We first examine the effects of interattribute correlation on the conjoint design and then discuss several remedies.

When two or more factors are correlated, two direct outcomes occur. First, as in many other multivariate techniques, particularly multiple regression, the parameter estimates are affected (Chapter 4 contains a discussion of multicollinearity and its impact). Among the more problematic effects is the inability to obtain reliable estimates due to the lack of uniqueness for each level.

Perhaps the more important effect is the creation of unbelievable combinations of two or more factors that can distort the conjoint design. This issue typically occurs in two situations. The first is when two attributes are negatively correlated, such that consumers expect that high levels of one factor should be matched with low levels of another factor. Yet when levels from each are combined in the conjoint task, the profiles are not realistic. The problem lies not in the levels themselves but in the fact that they cannot realistically be paired in all combinations, which is required for parameter estimation. A simple example of unbelievable combinations is for horsepower and gas mileage. Although both attributes are quite valid when considered separately, many combinations of their levels are not believable. What is the realism of an automobile with the highest levels of both horsepower and gas mileage? Moreover, why would anyone consider an automobile with the lowest levels of horsepower and gas mileage?

The second situation where unbelievable combinations are formed occurs when one factor indicates presence/absence of a feature and another attribute indicates amount. In this situation the conjoint task includes profiles denoting that a feature is available/unavailable, with a second factor indicating the amount. Again, each factor is plausible when considered separately, yet when combined create profiles that are not possible and cannot be used in the analysis. An example of the problems caused by a presence/absence factor is when one factor indicates the presence/absence of a price discount and the second factor indicates the amount of the discount. The problem comes whenever the profiles are constructed that indicate the absence of a price discount, yet the second factor specifies an amount. Including a level with the amount of zero only increases the problem, because now included profiles may indicate a price discount with the amount of zero. The result in each situation is an implausible profile.

Even though a researcher would always like to avoid an environmental correlation among factors, in some cases the attributes are essential to the conjoint analysis and must be included. When the correlated factors are retained, the researcher has three basic remedies to overcome the unrealistic profiles included in the conjoint design.

The most direct remedy is to create *superattributes* that combine the aspects of correlated attributes. Here the researcher takes the two factors and creates new levels that represent realistic amounts of both attributes. It is important to note that when these superattributes are added they should be made as actionable and specific as possible. If it is not possible to define the broader factor with the necessary level of specificity, then the researchers may be forced to eliminate one of the original factors from the design.

In our example of horsepower and gas mileage, perhaps a factor of "performance" could be substituted. In this instance levels of performance can be defined in terms of horsepower and gas mileage, but as realistic combinations in a single factor. As an example of positively correlated attributes, factors of store layout, lighting, and decor may be better addressed by a single concept, such as "store atmosphere." This factor designation avoids the unrealistic profiles of high levels of layout and lighting, but low levels of décor (along with other equally unbelievable combinations). When a presence/absence factor is utilized with another factor indicating amount, the most direct approach is to combine them into a single factor, with the levels including zero to indicate the absence of the attribute.

A second approach involves refined experimental designs and estimation techniques that create nearly orthogonal profiles, which can be used to eliminate any unbelievable profiles resulting from interattribute correlation [102]. Here the researcher can specify which combinations of levels (known as **prohibited pairs**) or even profiles of the orthogonal design are to be eliminated from the conjoint design, thus presenting respondents only with believable profiles. However, the danger in this approach is that poorly designed profiles will result in so large a number of unacceptable profiles that one or more of the correlated factors are effectively eliminated from the study, which then affects the part-worth estimates for that and all other factors.

The third remedy is to constrain the estimation of part-worths to conform to a prespecified relationship. These constraints can be between factors as well as pertaining to the levels within any single factor [100, 106]. Again, however, the researcher is placing restrictions on the estimation process, which may produce poor estimates of the preference structures.

Of the three remedies discussed, the creation of superattributes is the conceptually superior approach because it preserves the basic structure of the conjoint analysis. The other two remedies, which add significant complexity to the design and estimation of the conjoint analysis, should be considered only after the more direct remedy has been attempted.

The Unique Role of Price as a Factor. Price is a factor included in many conjoint studies because it represents a distinct component of value for many products or services being studied. Price, however, is not like other factors in its relationship to other factors [50]. We will first discuss the unique features of price and then address the approaches for the inclusion of price into a conjoint analysis.

Price is a principal element in any assessment of value and thus an attribute ideally suited to the trade-off nature of conjoint analysis. However, it is this basic nature of being an inherent trade-off that creates several issues with its inclusion. The most basic issue is that in many, if not most instances, price has a high degree of interattribute correlation with other factors. For many attributes, an increase in the amount of the attribute is associated with an increase in price, and a decreasing price level may be unrealistic (e.g., the price–quality relationship). The result is one or more profiles that are inappropriate for inclusion in the conjoint design (see earlier discussion of interattribute correlation for possible remedies).

Second, many times price is included in the attempt to represent value—the trade-off between the utility you get (the positive factors of quality, reliability, etc.) versus what you must give up; that is, price. Most times utility is defined by many factors whereas price is defined by only one factor. As a result, just due to the disparate number of factors there may be a decrease in the importance of price [77].

Finally, price may interact (i.e., have different effects for differing levels) when combined with other factors, particularly more intangible factors such as brand name. An example is that a certain price level has different meanings for different brands [50, 77], one that applies to a premium brand and another for a discount brand. We discuss the concept of interactions later in this chapter.

All of these unique features of price as a factor should not cause a researcher to avoid the use of price, but instead to anticipate the impacts and adjust the design and interpretation as required. First, explicit forms of conjoint analysis, such as conjoint value analysis (CVA), have been developed for occasions in which the focus is on price [92]. Moreover, if interactions of price and other factors are considered to be important, methods such as choice-based conjoint or multistage analyses [77, 81, 112] provide quantitative estimates of these relationships. Even if no specific adjustment is made, the researcher should consider these issues in the definition of the price levels and in the interpretation of the results.

SPECIFICATION ISSUES REGARDING LEVELS The specification of levels is a critical aspect of conjoint analysis because the levels are the actual measures used to form the profiles. Thus, in addition to being actionable and communicable, research has shown that the number of levels, the balance in number of levels between factors, and the range of the levels within a factor all have distinct effects on the evaluations.

Number and Balance of Levels. The estimated relative importance of a variable tends to increase as the number of levels increases, even if the end points stay the same [52, 71, 110, 117, 118]. Known as the "number of levels effect," the refined categorization calls attention to the attribute and causes consumers to focus on that factor more than on others. Thus, researchers should attempt as best as possible to balance or equalize the number of levels across factors so as to not bias the conjoint task in favor of factors with more levels. If the relative importance of factors is known a priori, then the researcher may wish to expand the levels of the more important factors to avoid a dilution of importance and to capture additional information on the more important factors [116].

Range of Levels. The range (low to high) of the levels should be set somewhat outside existing values but not at an unbelievable level. This range should include all levels of interest because the results should never be extrapolated beyond the levels defined for an attribute [77]. Although this practice helps to reduce interattribute correlation, it also can reduce believability if the levels are set too extreme. Levels that are impractical, unbelievable, or that would never be used in realistic situations can artificially affect the results and should be eliminated.

Before excluding a level, however, the researcher must ensure that it is truly unacceptable, because many times people select products or services that have what they term *unacceptable levels*. If an unacceptable level is found after the experiment has been administered, the recommended solutions are either to eliminate all profiles that have unacceptable levels or to reduce part-worth estimates of the offending level to such a low level that any objects containing that level will not be chosen.

For example, assume that in the normal course of business activity, the range of prices varies about 10 percent around the average market price. If a price level 50 percent lower was included, but would not realistically be offered, its inclusion would markedly distort the results. Respondents would logically be most favorable to such a price level. When the part-worth estimates are made and the importance of price is calculated, price will artificially appear more important than it would actually be in day-to-day decisions.

Specifying the Basic Model Form

For conjoint analysis to explain a respondent's preference structure based only on overall evaluations of a set of profiles, the researcher must make two key decisions regarding the underlying conjoint model: specifying the composition rule to be used and selecting the type of relationships between part-worth estimates. These decisions affect both the design of the profiles and the analysis of respondent evaluations.

THE COMPOSITION RULE: SELECTING AN ADDITIVE VERSUS AN INTERACTIVE MODEL   The most wide-ranging decision by the researcher involves the specification of the respondent's **composition rule.** The composition rule describes how the researcher postulates that the respondent combines the part-worths of the factors to obtain overall worth or utility. It is a critical decision because it defines the basic nature of the preference structure that will be estimated. In the following

---

RULES OF THUMB 6-2

Designing the Conjoint Task

- Researchers must select one of three methodologies based on number of attributes, choice task requirements, and assumed consumer model of choice:
  - Traditional methods are best suited when the number of attributes is fewer than 10, results are desired for each individual, and the simplest model of consumer choice is applicable
  - Adaptive methods are best suited when larger numbers of attributes are involved (up to 30), but require computer-based interviews
  - Choice-based methods are considered the most realistic, can model more complex models of consumer choice, and have become the most popular, but are generally limited to six or fewer attributes
- The researcher faces a fundamental trade-off in the number of factors included:
  - Increase them to better reflect the "utility" of the object
  - Minimize them to reduce the complexity of the respondent's conjoint task and allow use of any of the three methods
- Specifying factors (attributes) and levels (values) of each factor must ensure that:
  - Factors and levels are distinct influences on preference defined in objective terms with minimal ambiguity, thereby generally eliminating emotional or aesthetic elements
  - Factors generally have the same number of levels
  - Interattribute correlations (e.g., acceleration and gas mileage) may be present at minimal levels (.20 or less) for realism, but higher levels must be accommodated by:
    - Creating a "superattribute" (e.g., performance)
    - Specifying prohibited pairs in the analysis to eliminate unrealistic stimuli (e.g., fast acceleration and outstanding gas mileage)
    - Constraining the model estimation to conform to prespecified relationships
  - Price requires special attention because:
    - It generally has interattribute correlations with most other attributes (e.g., price–quality relationship)
    - It uniquely represents in many situations what is traded off in cost for the object
    - Substantial interactions with other variables may require choice-based conjoint or multistage conjoint methods

section we discuss the basic elements of the most common composition rule—the additive model—and then address the issues involved in the addition of other forms of part-worth relationships known as *interaction terms*.

The Additive Model. The most common and basic composition rule is an **additive model.** It assumes the respondent simply adds the values for each attribute (i.e., the part-worths of the levels) to get the total value for a profile. Thus, the total utility of any defined profile is simply the sum of the part-worths.

For example, let us assume that a product has two factors (1 and 2), each with two levels (A, B and C, D). The part-worths of factor 1 have been estimated at 2 and 4 (levels A and B), and factor 2 has part-worth values of 3 and 5 (levels C and D). We can then calculate the total utility of the four possible profiles as follows:

Profile	Levels Defining Profile	Additive Model Part-Worths	Total Utility
1	A and C	2 + 3	5
2	A and D	2 + 5	7
3	B and C	4 + 3	7
4	B and D	4 + 5	9

The additive model typically accounts for the majority (up to 80% or 90%) of the variation in preference in almost all cases, and it suffices for most applications. It is also the basic model underlying both traditional and adaptive conjoint analysis (see Table 6-4).

Adding Interaction Effects. The composition rule using interaction effects is similar to the additive form in that it assumes the consumer sums the part-worths to get an overall total across the set of attributes. It differs in that it allows for certain combinations of levels to be more or less than just their sum. The interactive composition rule corresponds to the statement, "The whole is greater (or less) than the sum of its parts." Let us revisit one of our earlier examples to see how interaction effects impact utility scores.

In our industrial cleanser example, let us examine the results for respondent 2 (refer back to Table 6-3). In the estimated part-worths, the generic brand was preferred over HBAT, phosphate-free over phosphate-based ingredients, and liquid over powder form. But the respondent's results are not always this consistent. As discussed earlier, for profiles 5 through 8 the respondent always preferred profiles with the generic brand over profiles with the HBAT brand, all other things held constant. But the reverse is true with profiles 1 through 4. What differs between these two sets of profiles? Looking at Table 6-3 we see that profiles 1–4 contain the liquid form, whereas profiles 5–8 contain the powder form. Thus, it looks like respondent 2's preferences for brand differ depending on whether the profile contains a liquid or powder form. In this case, we say that the factors of Brand and Form interact, such that one or more combinations of these factors result in much higher or lower ratings than expected. Without including this interaction effect the estimated and actual preference rankings will not match.

With the ability of interaction terms to add generalizability to the composition rule, why not use the interactive model in all cases? The addition of interaction terms does have some drawbacks that must be considered. First, each interaction term requires an additional part-worth estimate with at least one additional profile for each respondent to evaluate. Unless the researcher knows exactly which interaction terms to estimate, the number of profiles increases dramatically. Moreover, if respondents do not utilize an interactive model, estimating the additional interaction terms in the conjoint variate reduces the statistical efficiency (more part-worth estimates) of the estimation process and makes the conjoint task more arduous. Second, from a practical perspective, interactions (when present) predict substantially less variance than the additive effects, most often not exceeding a 5 to 10 percent increase in explained variance. Interaction terms are most likely to be substantial in cases where the attributes are less tangible, particularly when aesthetic or emotional reactions play a large role. Thus, in many instances, the increased predictive power will be minimal.

The researcher must balance the potential for increased explanation from interaction terms with the negative consequences from adding interaction terms. The interaction term is most effective when the researcher can hypothesize that "unexplained" portions of utility are associated with only certain levels of an attribute. Interested readers are referred to the text's Web sites (www.pearsonglobaleditions.com/hair or www.mvstats.com) for a more detailed examination of how to identify interactions terms and their impact on part-worth estimates and predictive accuracy.

Selecting the Model Type. The choice of a composition rule (additive only or with interaction effects) determines the types and number of treatments or profiles that the respondent must evaluate, along with the form of estimation method used. As discussed earlier, trade-offs between the two approaches need to be considered. An additive form requires fewer evaluations from the respondent, and it is easier to obtain estimates for the part-worths. However, the interactive form is a more accurate representation when respondents use more complex decision rules in evaluating a product or service. This choice must be made before the data are collected in order to design the set of profiles correctly.

SELECTING THE PART-WORTH RELATIONSHIP: LINEAR, QUADRATIC, OR SEPARATE PART-WORTHS   The flexibility of conjoint analysis in handling different types of variables comes from the assumptions the researcher makes regarding the relationships of the part-worths within a factor. In making decisions about the composition rule, the researcher decides how factors relate to one another in the respondent's preference structure. In defining the type of part-worth relationship, the researcher focuses on how the levels of a factor are related.

Types of Part-Worth Relationships. Conjoint analysis gives the researcher three alternatives, ranging from the most restrictive (a linear relationship) to the least restrictive (separate part-worths), with the ideal point, or quadratic model, falling in between. Figure 6-2 illustrates the differences among the three types of relationships.

The *linear model* is the simplest yet most restricted form, because we estimate only a single part-worth (similar to a regression coefficient), which is multiplied by the level's value to arrive at a part-worth value for each level. In the *quadratic form,* also known as the *ideal model,* the assumption of strict linearity is relaxed, so that we have a simple curvilinear relationship. The curve can turn either upward or downward. Finally, the *separate part-worth form* (often referred to simply as the *part-worth form*) is the most general, allowing for separate estimates for each level. When using separate part-worths, the number of estimated values is the highest because a separate parameter is estimated for each level.

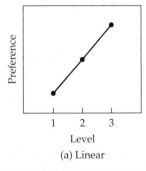
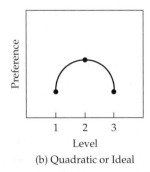
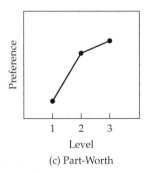

FIGURE 6-2   **Three Basic Types of Relationships Between Factor Levels in Conjoint Analysis**

The form of part-worth relationship can be specified for each factor separately such that each factor takes on a different part-worth relationship. This choice does not affect how the profiles are created and part-worth values are still calculated for each level. It does, however, affect how and what types of part-worths are estimated by conjoint analysis. If we can reduce the number of parameters estimated for any given set of profiles by using a more restricted part-worth relationship (e.g., linear or quadratic form), the calculations will be more efficient and reliable from a statistical estimation perspective.

Selecting a Part-Worth Relationship. The researcher must consider the trade-off between the gains in statistical efficiency by using the linear or quadratic forms versus the potentially more accurate representation of how the consumer actually forms overall preference if we employ less restrictive part-worth relationships. The researcher has several approaches to deciding on the type of relationship for each factor.

The primary support for any part-worth relationship should be from prior research or conceptual models. In this way a researcher may be able to specify a linear or quadratic relationship to achieve not only statistical efficiency, but also consistency with the research question. If adequate conceptual support is not available the researcher may follow a more empirical approach. Here the conjoint model is estimated first as a part-worth model. Then the different part-worth estimates are examined visually to detect whether a linear or a quadratic form is appropriate. In many instances, the general form is apparent, and the model can be reestimated with relationships specified for each variable as justified. When different relationships seem reasonable and with support, then the relationship type maximizing predictive ability would be chosen. An empirical approach is not recommended without at least some theoretical or empirical evidence for the possible type of relationship considered. Without such support, the results may have high predictive ability but offer little use in decision making.

Analyzing and Interpreting the Separate Part-Worth Relationship. The separate part-worth relationship may seem like the logical option in all instances, but the researcher must realize that this flexibility in the form of the relationship may also create difficulties in estimation or interpretation. These problems occur whenever the researcher expects some form of **monotonic relationship** to exist among the levels (i.e., some form of ordered preference is present among the levels) without specifying the actual form of this relationship (e.g., linear or quadratic). Let us look at an example to see where these problems might occur.

Assume that we have a simple conjoint analysis addressing store patronage with two factors (store type and travel distance to store). We can estimate both sets of part-worths with the separate part-worth relationship. For the store type factor, the part-worth estimates represent the relative utility of each type of store with no predefined ordering of which store must be preferred over another. With distance, the most likely assumption is that closer distance would be preferred over a farther distance. At the very least, farther distances should not be more preferred than short distances. Yet when we employ a separate part-worth relationship, the part-worths method lacks the predefined pattern of the linear or quadratic relationship. We may find that the estimated part-worths do not follow the prescribed pattern for one or more levels, due most likely to inconsistencies in the responses. Three miles away, for example, may have a higher part-worth utility than 1 mile away, which seems illogical.

The researcher must always be aware of the possibility of these violations of the monotonic relationship (known as **reversals**) and examine the results to ascertain the severity of any occurrences. We discuss this issue in more detail when discussing estimation (where remedies are possible) and in the interpretation of the part-worths themselves.

Data Collection

Having specified the factors and levels, plus the basic model form and the relationships among part-worth estimates, the researcher must next make three decisions involving data collection: type of presentation method for the factors (trade-off, full-profile, or pairwise comparison), type of

---

RULES OF THUMB 6-3

Specifying Model Form and Part-Worth Relationships

- Researchers can choose between two basic model forms based on the assumed composition rule for individuals:
  - Additive model: Assumes the simplest type of composition rule (utility for each attribute is simply added up to get overall utility) and requires the simplest choice task and estimation procedures
  - Interactive model: Adds interaction terms between attributes to more realistically portray the composition rule, but requires a more complex choice task for the respondent and estimation procedure
  - Additive models generally suffice for most situations and are the most widely used
- Estimating the utility of each level (known as part-worths) can follow one of three relationships:
  - Linear: Requires that the part-worths be linearly related, but may be unrealistic for specific types of attributes
  - Quadratic: Most appropriate when an "ideal point" in the attribute levels is expected
  - Separate: Makes each part-worth estimate independently of other levels, but is most likely to encounter reversals (violations of the hypothesized relationship)

---

response variable, and the method of data collection. The overriding objective is to present the attribute combinations to the respondent in the most realistic and efficient manner possible. Most often they are presented in written descriptions, although physical or pictorial models can be quite useful for aesthetic or sensory attributes.

CHOOSING A PRESENTATION METHOD   Three methods of profile presentation are most widely associated with conjoint analysis. Although they differ markedly in the form and amount of information presented to the respondent (see Figure 6-3), all three are acceptable within the traditional conjoint model. The choice between presentation methods focuses on the assumptions as to the extent of consumer processing being performed during the conjoint task and the type of estimation process being employed.

Full-Profile Method. The most popular presentation method is the **full-profile method** principally because of its perceived realism as well as its ability to reduce the number of comparisons through the use of fractional factorial designs. In this approach, each profile is described separately, most often using a profile card (see Figure 6-3b). This approach elicits fewer judgments, but each is more complex, and the judgments can be either ranked or rated. Its advantages include a more realistic description achieved by defining a profile in terms of a level for each factor and a more explicit portrayal of the trade-offs among all factors and the existing environmental correlations among the attributes. It is also possible to use more types of preference judgments, such as intentions to buy, likelihood of trial, and chances of switching, which are difficult to answer with the other methods.

The full-profile method is not flawless and faces two major limitations based on the respondents' ability and capacity to make reasoned decisions. First, as the number of factors increases, so does the possibility of information overload. The respondent is tempted to simplify the process by focusing on only a few factors, when in an actual situation all factors would be considered. Second, the order in which factors are listed on the profile card may have an impact on the evaluation. Thus, the researcher needs to rotate the factors across respondents when possible to minimize order effects.

The full-profile method is recommended when the number of factors is 6 or fewer. When the number of factors ranges from 7 to 10, the trade-off approach becomes a possible option to the full-profile method. If the number of factors exceeds 10, then alternative methods (adaptive conjoint) are suggested [29].

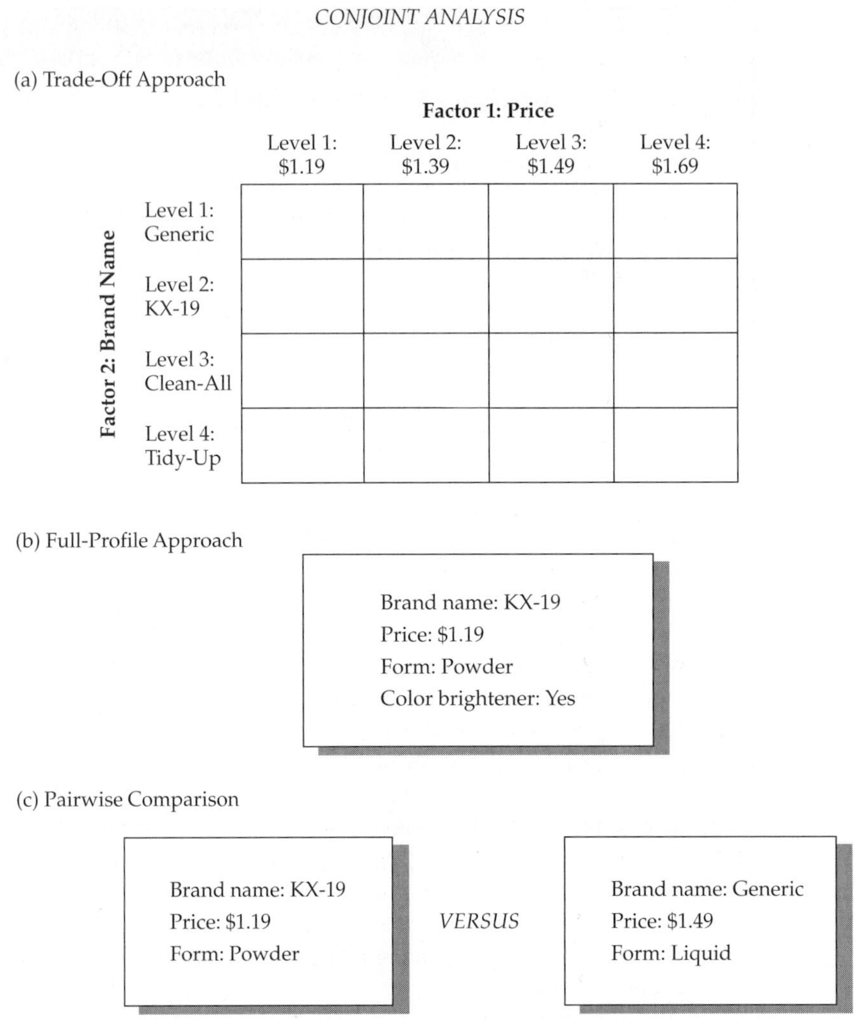

CONJOINT ANALYSIS

(a) Trade-Off Approach

Factor 1: Price

	Level 1: $1.19	Level 2: $1.39	Level 3: $1.49	Level 4: $1.69
Level 1: Generic				
Level 2: KX-19				
Level 3: Clean-All				
Level 4: Tidy-Up				

Factor 2: Brand Name

(b) Full-Profile Approach

> Brand name: KX-19
> Price: $1.19
> Form: Powder
> Color brightener: Yes

(c) Pairwise Comparison

Brand name: KX-19		Brand name: Generic
Price: $1.19	*VERSUS*	Price: $1.49
Form: Powder		Form: Liquid

FIGURE 6-3 **Examples of the Trade-Off and Full-Profile Methods of Presenting Stimuli**

The Pairwise Combination Presentation. The second presentation method, the **pairwise comparison method,** involves a comparison of two profiles (see Figure 6-3c), with the respondent most often using a rating scale to indicate strength of preference for one profile over the other [46]. The distinguishing characteristic of the pairwise comparison is that the profile typically does not contain all the attributes, as does the full-profile method. Instead only a few attributes at a time are selected in constructing profiles in order to simplify the task if the number of attributes is large. The researcher must be careful to not take this characteristic to the extreme and portray profiles with too few attributes to realistically portray the objects.

The pairwise comparison method is also instrumental in many specialized conjoint designs, such as adaptive conjoint analysis (ACA) [87], which is used in conjunction with a large number of attributes (a more detailed discussion of dealing with a large number of attributes appears later in this chapter).

Trade-Off Presentation. The final method is the **trade-off approach,** which compares attributes two at a time by ranking all combinations of levels (see Figure 6-3a). It has the advantages of being simple for the respondent and easy to administer, and it avoids information overload by presenting only two attributes at a time. It was the most widely used form of presentation in the early years of conjoint analysis. Use of this method has decreased dramatically in recent years, however, owing to several limitations. Most limiting is the sacrifice in realism by using only two factors at a time, which also makes a large number of judgments necessary for even a small number of levels. Respondents have a tendency to get confused or follow a routinized response pattern because of fatigue. It is also limited to only nonmetric responses and cannot use fractional factorial designs to reduce the number of necessary comparisons. It is rarely used today in conjoint studies except in specialized designs [118].

CREATING THE PROFILES   Once the factors and levels have been selected and the presentation method chosen, the researcher turns to the task of creating the profiles for evaluation by the respondents. For any presentation method, the researcher always faces increasing the burden on the respondent as the number of profiles increases to handle more factors or levels. The researcher must weigh the benefits of increased task effort versus the additional information gained. The following discussion focuses on creating profiles for the full-profile or pairwise comparison approaches. The trade-off approach is not addressed due to its limited use.

These two methods involve the evaluation of one profile at a time (full-profile) or pairs of profiles (pairwise comparison). In a simple conjoint analysis with a small number of factors and levels (such as those discussed earlier for which three factors with two levels each resulted in eight combinations), the respondent evaluates all possible profiles. This format is known as a **factorial design.**

As the number of factors and levels increases, this design can quickly become impractical. For example if the conjoint task involves four variables with four levels for each variable, 256 profiles (4 levels $\times$ 4 levels $\times$ 4 levels $\times$ 4 levels) would be created in a full factorial design. Even if the number of levels decreases, a moderate number of factors can create a difficult task. For a situation with six factors and two levels each, 64 profiles would be needed. If the number of levels increased just to three for the six factors, then the number of profiles would increase to 729. These situations obviously include too many profiles for one respondent to evaluate and still give consistent, meaningful answers. An even greater number of pairs of profiles would be created for the pairwise combinations of profiles with differing numbers of attributes.

In addition to the limitations of the respondent, the number of profiles must also be large enough to derive stable part-worth estimates. The minimum number of profiles equals the number of parameters to be estimated, calculated as:

$$\text{Number of estimated parameters} = \text{Total number of levels} - \text{Number of attributes} + 1$$

It is suggested that the respondent evaluate a set of profiles equal to a multiple of (two or three times) the number of parameters. Yet as the number of levels and attributes increases, the respondent burden increases quickly. Research has shown that respondents can complete up to 30 choice tasks, but after that point the quality of the data may come into question [92]. The researcher then faces a dilemma: Increasing the complexity of the choice tasks by adding more levels and/or factors increases the number of estimated parameters and the recommended number of choice tasks. Against this the researcher must consider the limit on the number of choice tasks that can be completed by a respondent, which vary by type of presentation method and complexity of the profiles.

DEFINING SUBSETS OF PROFILES   Many times, as discussed previously, the number of profiles in the full factorial design becomes too large and must be reduced. The process of selecting a subset of all possible profiles must be done in a manner to preserve the **orthogonality** (no correlation among levels of an attribute) and **balanced design** aspect (each level in a factor appears the same number of times). There are two approaches for selecting the subset of profiles that still meet these criteria.

Fractional Factorial. A **fractional factorial design** is the most common method for defining a subset of profiles for evaluation. The process designs a sample of possible profiles, with the number of profiles depending on the type of composition rule assumed to be used by respondents. Using the additive model, which assumes only main effects with no interactions, the full-profile method with four factors at four levels requires only 16 profiles to estimate the main effects. Table 6-5 shows two possible sets of 16 profiles. The sets of profiles must be carefully constructed to ensure the correct estimation of the main effects. The two designs in Table 6-5 are **optimal designs,** meaning they are orthogonal and balanced.

The remaining 240 possible profiles in our example that are not in the selected fractional factorial design are used to estimate interaction terms if desired. If the researcher decides that selected interactions are important and should be included in the model estimation, the fractional factorial design must include additional profiles to accommodate the interactions. Published guides for fractional factorial designs or conjoint program components will design the subsets of profiles to maintain orthogonality, making the generation of full-profile profiles quite easy [1, 17, 33, 65].

Bridging Design. If the number of factors becomes too large and the adaptive conjoint methodology is not acceptable, a **bridging design** can be employed [8]. In this design, the factors are divided in subsets of appropriate size, with some attributes overlapping between the sets so that each set has a factor(s) in common with other sets of factors. The profiles are then constructed for each subset so that the respondents never see the original number of factors in a single profile. When the part-worths are estimated, the separate sets of profiles are combined, and a single set of estimates is provided. Computer programs handle the division of the attributes, creation of profiles, and their recombination for estimation [12]. When using pairwise comparisons, the number may be quite large and complex, so that most often interactive computer programs are used that select the optimal sets of pairs as the questioning proceeds.

TABLE 6-5 **Two Alternative Fractional Factorial Designs for an Additive Model (Main Effects Only) with Four Factors at Four Levels Each**

	Design 1: Levels for . . . [a]				Design 2: Levels for . . . [a]			
Profile	Factor 1	Factor 2	Factor 3	Factor 4	Factor 1	Factor 2	Factor 3	Factor 4
1	3	2	3	1	2	3	1	4
2	3	1	2	4	4	1	2	4
3	2	2	1	2	3	3	2	1
4	4	2	2	3	2	2	4	1
5	1	1	1	1	1	1	1	1
6	4	3	4	1	1	4	4	4
7	1	3	2	2	4	2	1	3
8	2	1	4	3	2	4	2	3
9	2	4	2	1	3	2	3	4
10	3	3	1	3	3	4	1	2
11	1	4	3	3	4	3	4	2
12	3	4	4	2	1	3	3	3
13	1	2	4	4	2	1	3	2
14	2	3	3	4	3	1	4	3
15	4	4	1	4	1	2	2	2
16	4	1	3	2	4	4	3	1

[a]The numbers in the columns under factor 1 though factor 4 are the levels of each factor. For example, the first profile in design 1 consists of level 3 for factor 1, level 2 for factor 2, level 3 for factor 3, and level 1 for factor 4.

UNACCEPTABLE PROFILES   The creation of any design, even those with orthogonality and balance, does not mean, however, that all of the profiles in that design will be acceptable for evaluation. We will first discuss the most common reasons for the occurrence of unacceptable profiles and then address the potential remedies.

The most common reason for unacceptable profiles is the creation of "obvious" profiles—profiles whose evaluation is obvious because of their combination of levels. Typical examples of unacceptable profiles are those with all levels at either the highest or lowest values. These profiles really provide little information about choice and can create a perception of unbelievability on the part of the respondent. The second reason is interattribute correlation, which can create profiles with combinations of levels (high gas mileage, high acceleration) that are not realistic. Finally, external constraints may be placed on the combinations of attributes. The research setting may preclude certain combinations as unacceptable (i.e., certain attributes cannot be combined) or inappropriate (e.g., certain levels cannot be combined). In either instance, the attributes and levels are important to the research question, but certain combinations must be excluded.

In any of these instances, the unacceptable profiles present unrealistic choices to the respondent and should be eliminated to ensure a valid estimation process as well as a perception of credibility of the choice task among the respondents. Several courses of action help eliminate unacceptable profiles. First, the researcher can generate another fractional factorial design and assess the acceptability of its profiles. Because many fractional factorial designs are possible from any larger set of profiles, it may be possible to identify one that does not contain any unacceptable profiles. If all designs contain unacceptable profiles and a better alternative design cannot be found, then the unacceptable profile can be deleted. Although the design will not be totally orthogonal (i.e., it will be somewhat correlated and is termed to be **nearly orthogonal**), it will not violate any assumptions of conjoint analysis. Many conjoint programs also have an option to exclude certain combinations of levels (known as *prohibited pairs*). In these instances the program attempts to create a set of profiles that is as close as possible to optimal, but it should be noted that this option cannot overcome design flaws in the specification of factors or levels. In instances in which a systemic problem exists, the researcher should not be comforted by a program that can generate a set of profiles, because the resulting fractional factorial design may still have serious biases (low orthogonality or balance) that can impact the part-worth estimation.

All nearly orthogonal designs should be assessed for **design efficiency,** which is a measure of the correspondence of the design in terms of orthogonality and balance to an optimal design [55]. Typically measured on a 100-point scale (optimal design = 100), alternative nonorthogonal designs can be assessed, and the most efficient design with all acceptable profiles selected. Most conjoint programs for developing nearly orthogonal designs assess the efficiency of the designs [54].

Unacceptable profiles due to interattribute correlations are a unique case and must be accommodated within the development of designs on a conceptual basis. In practical terms, interattribute correlations should be minimized but they do not need to be zero if small correlations (.20 or less) will add to realism. Most problems are found in the case of negative correlations, as between gas mileage and horsepower. Adding uncorrelated factors can reduce the average interattribute correlation, so that with a realistic number of factors (e.g., 6 factors), the average correlation would be close to .20, which has relatively inconsequential effects. The researcher should always assess the believability of the profiles as a measure of practical relevance.

SELECTING A MEASURE OF CONSUMER PREFERENCE   The researcher must also select the measure of preference: rank-ordering versus rating (e.g., a 1–10 scale). Both the pairwise comparison and full-profile methods can evaluate preferences, either by obtaining a rating of preference of one profile over the other or just a binary measure of which is preferred.

Using a Rank-Order Preference Measure.   Each preference measure has certain advantages and limitations. Obtaining a rank-order preference measure (i.e., rank-ordering the profiles from most to least preferred) has two major advantages: (1) it is likely to be more reliable because ranking is easier than rating with a reasonably small number (20 or fewer) of profiles and (2) it provides more flexibility in estimating different types of composition rules.

It has, however, one major drawback: It is difficult to administer, because the ranking process is most commonly performed by sorting profile cards into the preference order, and this sorting can be done only in a personal interview setting.

Measuring Preference by Ratings. The alternative is to obtain a rating of preference on a metric scale. Metric measures are easily analyzed and administered, even by mail, and enable conjoint estimation to be performed by multivariate regression. However, respondents can be less discriminating in their judgments than when they are rank-ordering. Also, given the large number of profiles evaluated, it is useful to expand the number of response categories over that found in most consumer surveys. A rule of thumb is to have 11 categories (i.e., rating from 0 to 10 or 0 to 100 in increments of 10) for 16 or fewer profiles and expand to 21 categories for more than 16 profiles [58].

Choosing the Preference Measure. The decision on the type of preference measure to be used must be based on practical as well as conceptual issues. Many researchers favor the rank-order measure because it depicts the underlying choice process inherent in conjoint analysis: choosing among objects. From a practical perspective, however, the effort of ranking large numbers of profiles becomes overwhelming, particularly when the data collection is done in a setting other than personal interview.

The ratings measure has the inherent advantage of being easy to administer in any type of data collection context, yet it too has drawbacks. If the respondents are not engaged and involved in the choice task, a ratings measure may provide little differentiation among profiles (e.g., all profiles rated about the same). Moreover, as the choice task becomes more involved with additional profiles, the researcher must be concerned with not only task fatigue, but also with the reliability of the ratings across the profiles.

SAMPLE SIZE   Conjoint analysis represents a somewhat unique situation with regard to determining the sample size requirements. First, as mentioned before, improving the accuracy of the part-worth estimates for an individual relates to the number of choice tasks (i.e., profiles rated) performed by each respondent. So theoretically a conjoint analysis can be estimated with one respondent if that respondent provided enough choice tasks (see earlier discussion on number of choice tasks required).

So is sample size irrelevant for conjoint analysis? The answer depends on the research objective being addressed. Although each respondent is estimated separately in a disaggregate approach, the research still must consider the degree to which the respondents are representative of the population of interest. The required sample size needed relates to what the choices reflect (e.g., purchase/no purchase or market share) and how accurate you want that prediction to be. Here the conventional procedures for estimating confidence intervals based on sample size now come into play. If a specific confidence interval is desired (i.e., ± error rate), then estimating the standard error of the estimate provides the necessary sample size. Given the typical applications of conjoint analysis, sample sizes of 200 have been found to provide an acceptable margin of error. We should note that this relates to each group in the population, so if you are expecting to segment the population you should try and have sample sizes of 200 for each group. But small scale studies of as small as 50 respondents can provide a simple glimpse into the preferences of respondents and how they might vary in basic ways.

SURVEY ADMINISTRATION   In the past, the complexity of the conjoint analysis task led most often to the use of personal interviews to obtain the conjoint responses. Personal interviews enable the interviewer to explain the sometimes more difficult tasks associated with conjoint analysis.

RULES OF THUMB 6-4

Data Collection

- Data collection by traditional methods of conjoint analysis:
  - Generally is accomplished with some form of full-profile approach using a stimulus defined on all attributes
  - Increasing the number of factors and/or levels above the simplest task (two or three factors with only two or three levels each) requires some form of fractional factorial design that specifies a statistically valid set of stimuli
- Alternative methodologies (adaptive or choice-based methods) discussed in later sections provide options in terms of the complexity and realism of the choice task that can be accommodated
- Respondents should be limited to evaluating no more than 30 stimuli, regardless of the methodology used
- The estimation of an individual's part-worths is related to the number of choice tasks a respondent completes, not the sample size of respondents
- Sample size impacts the ability of the respondents to represent the population. Fifty respondents is suggested as the minimum sample size, and the recommended sample size is at least 200 per group
- If multiple groups are going to be formed from the respondents (e.g., with cluster analysis to identify segments) then the sample size considerations apply to each group

---

Recent developments in interviewing methods, however, make conducting conjoint analyses feasible both through the mail (with pencil-and-paper questionnaires or computer-based surveys) and by telephone. If the survey is designed to ensure that the respondent can assimilate and process the profiles properly, then all of the interviewing methods produce relatively equal predictive accuracy [2]. The use of computerized interviewing has greatly simplified the conjoint task demands on the respondent and made the administration of full-profile designs feasible [79, 113] while also accommodating even adaptive conjoint analysis [87]. Recent research has even demonstrated the reliability and validity of full-profile conjoint when administered over the Internet [80].

One concern in any conjoint study is the burden placed on the respondent due to the number of conjoint profiles evaluated. Obviously, the respondent could not evaluate all 256 profiles in our earlier factorial design, but what is the appropriate number of tasks in a conjoint analysis? A recent review of commercial conjoint studies found that respondents can easily complete up to 20 or even 30 conjoint evaluations [51, 92]. After that many evaluations, the responses start to become less reliable and less representative of the underlying reference structure. The researcher should always strive to use the fewest possible conjoint evaluations while maintaining efficiency in the estimation process. Yet, in trying to reduce the effort involved in the choice task, the researcher should not make it too simplistic or unrealistic. Also, nothing substitutes for pretesting a conjoint study to assess the respondent burden, the method of administration, and the acceptability of the profiles.

STAGE 3: ASSUMPTIONS OF CONJOINT ANALYSIS

Conjoint analysis has the least restrictive set of assumptions associated with model estimation. The structured experimental design and the generalized nature of the model make most of the tests performed in other dependence methods unnecessary. Therefore, the statistical tests for normality, homoscedasticity, and independence that were performed with other dependence techniques are not necessary for conjoint analysis. The use of statistically based profiles designs also ensures that the estimation is not confounded and that the results are interpretable under the assumed composition rule.

Yet even with fewer statistical assumptions, the conceptual assumptions are probably greater than with any other multivariate technique. As mentioned earlier, the researcher must specify the general form of the model (main effects versus interactive model) before the research is designed. The development of the actual conjoint task builds upon this decision and makes it impossible to test alternative models once the research is designed and the data are collected. Conjoint analysis is not like regression, for example, where additional effects (interaction or non-linear terms) can be easily evaluated after the data are collected. In conjoint analysis the researcher must make a decision regarding model form and then design the research accordingly. Thus, conjoint analysis, although having few statistical assumptions, is theory-driven in its design, estimation, and interpretation.

STAGE 4: ESTIMATING THE CONJOINT MODEL AND ASSESSING OVERALL FIT

The options available to the researcher in terms of estimation techniques have increased dramatically in recent years. Moreover, the development of techniques in conjunction with specialized methods of profile presentation (e.g., the adaptive or choice-based conjoint) is just one improvement of this type. The researcher, in obtaining the results of a conjoint analysis study, has numerous options available when selecting the estimation method and evaluating the results.

Selecting an Estimation Technique

For many years, the type of estimation process was dictated by the choice of preference measure. Recent research, however, focused on developing an alternative estimation approach appropriate for all types of preference measures while also providing a more robust estimation methodology and improvement in both aggregate and disaggregate results.

TRADITIONAL ESTIMATION APPROACHES   Rank-order preference measures were typically estimated using a modified form of analysis of variance specifically designed for ordinal data. Among the most popular and best-known computer programs are MONANOVA (Monotonic Analysis of Variance) [46, 53] and LINMAP [95]. These programs give estimates of attribute part-worths, so that the rank order of their sum (total worth) for each treatment is correlated as closely as possible to the observed rank order.

When a metric measure of preference is used (e.g., ratings rather than rankings), then many methods, even multiple regression, can estimate the part-worths for each level. Most computer programs available today can accommodate either type of evaluation (ratings or rankings), as well as estimate any of the three types of relationships (linear, ideal point, and part-worth). Estimation through the multinomial logit model and its extensions allow for more complicated consumer preference models including interaction terms and cross-attribute effects and the specific model forms discussed below.

EXTENSIONS TO THE BASIC ESTIMATION PROCESS   Up to this point, we discussed only estimation of the basic conjoint model with main and perhaps interaction effects. Although this model formulation is the foundation of all conjoint analysis, extensions of this approach may be warranted in some instances. The following sections discuss extensions applicable to disaggregate and aggregate methods.

One of the primary criticisms of aggregate model estimations is the lack of separate part-worth estimates for each individual versus the single aggregate solution. Yet the researcher is not always able to utilize a disaggregate approach due to any number of design considerations (e.g., type of choice format, number of variables, or sample size). One approach for accounting for

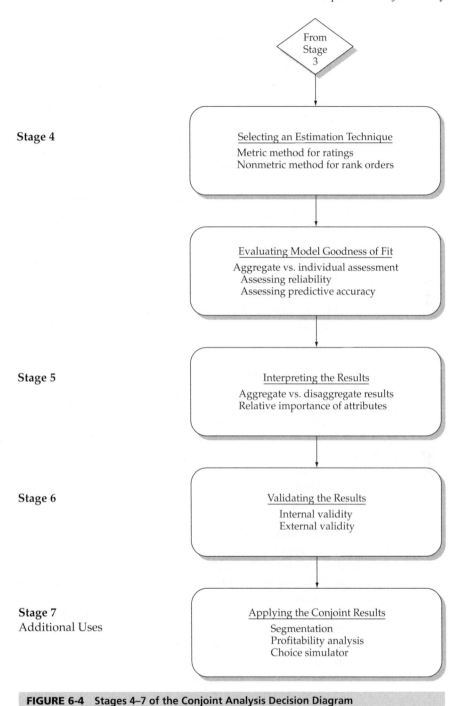

FIGURE 6-4   Stages 4–7 of the Conjoint Analysis Decision Diagram

heterogeneity is the Bayesian estimation approach discussed in the following section [4, 55]. A second is modification of the traditional estimation to introduce a form of **respondent heterogeneity,** which represents the variation expected across individuals if the model was estimated at the disaggregate level [111]. In both approaches, improvements in predictive accuracy have been achieved at levels comparable to those found in disaggregate models [76].

Another extension in the basic conjoint model is the incorporation of additional variables in the estimation process, particularly variables reflecting characteristics of the individual or choice context. Up to this point, we assumed that preferences for the profiles are completely expressed in the levels of the various attributes. But what about other less quantifiable measures, such as attitudes, or even socioeconomic characteristics? Even though we may assume that these individual differences will be reflected in the disaggregated part-worth estimates, in some instances it is beneficial to establish the relationship with these types of variables. Recent research has explored techniques for including socioeconomic and choice context variables as well as attitudinal variables and even latent constructs [142]. These techniques are not widely available yet, but they represent potentially useful approaches for quantifying the impacts of variables outside those used in constructing the profiles.

BAYESIAN ESTIMATION: A RADICALLY NEW APPROACH The estimation procedures just described rely on classical statistical theory, which that is the foundation for all of the multivariate methods discussed in this text. These approaches, however, are being supplanted by a new approach, ***Bayesian analysis*** [22], which is quite different in its basic method in the estimation of the conjoint model. The application of Bayesian analysis is occurring not only in conjoint analysis [3, 56, 62], but in the more traditional methods such as regression analysis as well [4, 93]. Bayesian estimation represents potentially significant improvements over existing methods in terms of both predictive and explanatory ability. Researchers are encouraged to examine Bayesian estimation options in conjoint analysis where available and continue to follow its progress as the issues of implementation discussed below are addressed in the following sections.

The Basics of Bayesian Analysis. The underlying premise of Bayesian analysis is Bayes' theorem, which is based on defining two probability values: the prior probability and the likelihood probability. In a general sense, the likelihood probability is the probability we derive from the actual data observations. The prior probability is an estimate of how likely this particular set of observations (and the associated prior probability) is to occur in the population. By combining these two probabilities we make some estimate of the actual probability of an event (known as the *joint probability*). Interested readers are referred to the Basic Stats appendix on the text's Web sites at www.pearsonglobaleditions.com/hair or www.mvstats.com for a more detailed look at the fundamental principles underlying Bayesian estimation.

Advantages and Disadvantages of Bayesian Estimation. In using Bayesian analysis for estimation of a conjoint model, the researcher does not need to do anything different; these probability values are estimated by the program from the set of observations. The question to be asked, however, is: What are the advantages and disadvantages of employing this technique? Let us examine them in more detail.

Numerous studies have examined Bayesian estimation versus the more traditional methods, and in all instances those studies found Bayesian estimation to be comparable, or even superior, for both part-worth estimation and predictive capability [6]. The advantages go beyond just estimation precision, however. Given the nature of the required probability estimates, Bayesian estimation allows for conjoint models to be estimated at the individual level where previously only aggregate models were possible (i.e., choice-based conjoint and more complex models with interaction terms). To this end, it has been incorporated into all of the basic conjoint models [89, 91].

Bayesian estimation does have some drawbacks. First, it requires a large sample (typically 200 or more respondents) because it is dependent on the sample for the estimates of prior probabilities. This requirement differs from traditional conjoint models that could be estimated solely for one individual. Second, it requires substantially more computing resources because it takes an iterative approach in estimation. Analyses that could be estimated in seconds using traditional means now take several hours [103]. Even though the rapid increases in desktop computing power somewhat mitigated this issue, the researcher must still be aware of the additional resources needed.

Estimated Part-Worths

Once an estimation method is chosen, the responses for each profile are analyzed to estimate the part-worths for each level. The most common method is some form of linear model, depending on whether the dependent measure is metric or nonmetric. As such, the estimated part-worths are essentially regression coefficients estimated with dummy variables, and one level for each attribute is eliminated to avoid singularity (see Chapter 4 for a more detailed discussion of using dummy variables in regression). Thus, the resulting part-worth estimates must be interpreted in a relative sense.

Here is an example of estimated part-worths using ACA [87] for a simple three-attribute design with five and four levels.

Attribute 1		*Attribute 2*		*Attribute 3*	
Level	**Part-Worth**	**Level**	**Part-Worth**	**Level**	**Part-Worth**
1	−.657	1	−.751	1	−.779
2	−.0257	2	−.756	2	−.826
3	−.378	3	.241	3	−.027
4	.098	4	.302	4	.667
5	−.0111				

As we can see, the part-worths must be judged relative to one another, because they have both negative and positive values. For example, for the second attribute, the second level is actually the least desired (most negative at −.756) by a small amount with the fourth level having the highest utility (.302). The levels can also be compared across attributes, but care must be taken to first assess the levels within each attribute to establish their relative position.

To assist in interpretation, many programs convert the part-worth estimates to some common scale (e.g., minimum of 0 to a maximum of 100 points) to allow for comparison both across attributes for an individual and across individuals. Shown next are the scaled part-worths for the example just discussed. As we can see, they are much easier to interpret, both within attributes as well as across attributes. The relative ordering in the original utility values is preserved, but now the lowest level in each attribute is set to zero and all other levels valued relative to that level.

Attribute 1		*Attribute 2*		*Attribute 3*	
Level	**Part-Worth**	**Level**	**Part-Worth**	**Level**	**Part-Worth**
1	0.00	1	.23	1	2.15
2	18.29	2	0.00	2	0.00
3	12.76	3	45.59	3	36.59
4	34.53	4	48.38	4	68.28
5	29.54				

Because the part-worth estimates are always interpreted in a relative perspective (one part-worth versus another) rather than an absolute amount (the actual amount of change in the dependent measure), the researcher should focus on a method of portraying the results that best facilitates both application and interpretation. Scaling the part-worth estimates provides a simple yet effective way of presenting the relative positioning of each level. This format is also conducive to graphical portrayal and also provides a means of more easily using the part-worths in other multivariate techniques such as cluster analysis.

Evaluating Model Goodness-of-Fit

Conjoint analysis results are assessed for accuracy at both the individual and aggregate levels. The objective in both situations is to ascertain how consistently the model predicts the set of preference evaluations. Because any measure of goodness-of-fit may be overfitted when evaluating a single respondent, the researcher must take care to complement any empirical process with additional evaluation through examination of the estimated preference structure as discussed in the next section.

ASSESSING INDIVIDUAL-LEVEL CONJOINT MODELS The role of any goodness-of-fit measure is to assess the quality of the estimated model by comparing actual values of the dependent variable(s) with values predicted by the estimated model. For example, in multiple regression we correlate the actual and predicted values of the dependent variable for the coefficient of determination (R^2) across all respondents. In discriminant analysis we compare the actual and predicted group memberships for each member of the sample in the classification matrix. What distinguishes conjoint analysis from the other multivariate techniques is that separate conjoint models are estimated for each individual, requiring that the goodness-of-fit measure provide information on the estimated part-worths for each respondent. As we will see in the following discussions, this process requires special care in the type of goodness-of-fit measure used and how it is interpreted.

Types of Goodness-of-Fit Measures. For an individual-level model, the goodness-of-fit measure is calculated for each individual. As such, it is based on the number and type of choice tasks performed by each respondent. When the choice tasks involve nonmetric rank-order data, correlations based on the actual and the predicted ranks (e.g., Spearman's rho or Kendall's tau) are used. When the choice tasks involve a rating (e.g., preference on a 0–100 scale), then a simple Pearson correlation, just like that used in regression, is appropriate. In both cases, the estimated part-worths are used to generate predicted preference values (ranks or metric ratings) for each profile. The actual and predicted preferences are then correlated for each person and tested for statistical significance. Individuals who have poor predictive fit should be candidates for deletion from the analysis.

Evaluating the Strength of the Goodness-of-Fit Measure. How high should the goodness-of-fit values be? As with most fit measures, higher values indicate a better fit. In most conjoint experiments, however, the number of profiles does not substantially exceed the number of parameters, and the potential for overfitting the data, and thus overestimating the goodness-of-fit, is always present. Goodness-of-fit measures are not corrected for the degrees of freedom in the estimation model.

Thus, as the number of profiles approaches the number of estimated parameters, the researcher must apply a higher threshold for acceptable goodness-of-fit values. For example, multiple regression is many times used in the metric estimation process. In assessing goodness-of-fit with the coefficient of determination (R^2), the researcher should always calculate the adjusted R^2 value, which compensates for lower degrees of freedom. Thus, in many instances what seem to be acceptable goodness-of-fit values in conjoint analyses may actually reflect markedly lower adjusted values because the number of profiles evaluated is not substantially greater than the number of part-worths (see Chapter 4 for more detailed discussion of the adjustment process). Moreover, values that are exceedingly high (very close to 100%) may not reflect exceedingly good fit, but rather indicate respondents who may not be following the choice tasks correctly and thus are also candidates for deletion.

Using a Validation Sample. Researchers are also strongly encouraged to measure model accuracy not only on the original profiles but also with a set of **validation** or **holdout profiles.** In a procedure similar to a holdout sample in discriminant analysis, the researcher prepares more profile cards than needed for estimation of the part-worths, and the respondent rates all of them at the same time. Parameters from the estimated conjoint model are then used to predict preference for the new set of profiles, which is compared with actual responses to assess model reliability [48].

```
┌─────────────────────────────────────────────────────────────┐
│                    ██ RULES OF THUMB 6-5 ██                   │
│                                                               │
│                   Estimating a Conjoint Model                 │
│                                                               │
│ • The selection of an estimation method is straightforward:   │
│   • The most common method is a regression-based approach,    │
│     applicable with all metric preference measures            │
│   • Using rank-order preference data requires more            │
│     specialized estimation (e.g., MONANOVA)                   │
│   • Bayesian methods are emerging that allow for individual   │
│     level models to be estimated where not previously         │
│     possible, but they require larger samples, are more       │
│     computationally intensive, and are not as widely          │
│     available                                                 │
│ • Goodness-of-fit should be assessed with:                    │
│   • Coefficient of determination ($R^2$) between actual and   │
│     predicted preferences                                     │
│   • Measures based on the rank orders of the predicted and    │
│     actual preferences                                        │
│   • Measures for both the estimation sample and a holdout,    │
│     or validation, sample of additional stimuli not used in   │
│     the estimation process                                    │
└─────────────────────────────────────────────────────────────┘
```

The holdout sample also gives the researcher an opportunity for a direct evaluation of profiles of interest to the research study.

In measuring the goodness-of-fit of a holdout sample, however, the researcher must use extreme caution in evaluating the actual values of the goodness-of-fit measure. In most instances the holdout sample may contain a small number of additional profiles (four to six), thus the values are calculated on a small number of observations. Extremely high values may be suspect in that they do not reflect good fit, but rather fundamental problems in the estimated preference structure of the choice process itself.

AGGREGATE-LEVEL ASSESSMENT If an aggregate estimation technique is used, then the same basic procedures apply, only now aggregated across respondents. Researchers also have the option of selecting a holdout sample of respondents in each group to assess predictive accuracy. In these instances, the aggregate model is applied to individuals and then evaluated in terms of predictive accuracy of their choices. This method is not feasible for disaggregate results because no generalized model is available to apply to the holdout sample, and each respondent in the estimation sample has individualized part-worth estimates.

STAGE 5: INTERPRETING THE RESULTS

The customary approach to interpreting conjoint analysis is at the disaggregate level. That is, each respondent is modeled separately, and the results of the model (part-worth estimates and assessments of attribute importance) are examined for each respondent. Interpretation, however, can also take place with aggregate results. Whether the model estimation is made at the individual level and then aggregated or aggregate estimates are made for a set of respondents, the analysis fits one model to the aggregate of the responses. As one might expect, this process generally yields poor results when predicting what any single respondent would do or when interpreting the part-worths for any single respondent. Unless the researcher is dealing with a population definitely exhibiting homogeneous behavior with respect to the attributes, aggregate analysis should not be used as the only method of analysis. However, many times aggregate analysis more accurately predicts aggregate behavior, such as market share. Thus, the researcher must identify the primary purpose of the study and employ the appropriate level of analysis or a combination of the levels of analysis.

Examining the Estimated Part-Worths

One of the unique elements of conjoint analysis is the ability to represent the preference structure of individuals through part-worths, yet many researchers neglect to validate these preference structures. Much insight can be gained from such examination, plus the potential for improving the overall results by correcting for invalid patterns among the part-worths. The most common method of interpretation is an examination of the part-worth estimates for each factor, assessing their magnitude and pattern. Part-worth estimates are typically scaled so that the higher the part-worth (either positive or negative), the more impact it has on overall utility. As noted earlier, many programs will rescale the part-worths to common scale, such as 0–100 points, so as to allow easier comparison across and within respondents.

ENSURING PRACTICAL RELEVANCE   In evaluating any set of part-worth estimates, the researcher must consider both practical relevance as well as correspondence to any theory-based relationships among levels. In terms of practical relevance, the primary consideration is the degree of differentiation among part-worths within each attribute. For example, part-worth values can be plotted graphically to identify patterns. Relatively flat patterns would indicate a degree of indifference among the levels in that the respondent did not see much difference between the levels as affecting choice. As such, whether it be graphical or empirical comparisons among the levels, it is imperative that the researcher evaluate each set of part-worths to ensure that they are an appropriate representation of the preference structure.

REVERSALS: A CASE OF THEORETICAL INCONSISTENCY   Many times an attribute has a theoretically based structure for the relationships between levels. The most common is a monotonic relationship, such that the part-worths of level C should be greater than those of level B, which should in turn be greater than the part-worths of level A. Common situations in which such a relationship is hypothesized include such attributes as price (lower prices always valued more highly), quality (higher quality always better), or convenience (closer stores preferred over more distant stores). With these and many other attributes, the researcher has a theoretically based relationship to which the part-worth values should correspond.

What happens when the part-worths do not follow the theorized pattern? We introduced the concept of a *reversal* in our earlier discussion of model forms as being when the part-worth values violate the assumed monotonic relationship. In a simple sense, we are referring to the seemingly nonsensical situations in which respondents value paying higher prices, having lower quality, or driving further distances. A reversal represents potentially serious distortions in the representation of a preference structure. Not only does it affect the relationship between adjacent levels, but it may affect the part-worths for the entire attribute.

When reversals create a preference structure that cannot be supported theoretically, the researcher should then consider deletion of the respondent. At issue is the size and frequency of reversals, because they represent illogical or inconsistent patterns in the overall preference structure as measured by the part-worths.

Factors Contributing to Reversals. Given the potentially serious consequences from a reversal, a researcher must be cognizant of factors in the research setting that create the possibility of reversals. These factors should be considered when judging the extent of reversals and making a conclusion as to the validity/invalidity of a preference structure:

- *Respondent effort.* A critical factor in the success of any conjoint analysis is sustained interest in the conjoint tasks in order to accurately assess preference structure. Many factors, however, can diminish this effort, such as respondent fatigue with the conjoint tasks or other parts of the survey or disinterest in the research task. A simple measure of respondent interest is the time spent the conjoint tasks. The researcher should always pretest the conjoint tasks and develop a minimum time period considered necessary to reliably complete the task. Then, for individuals falling under this time threshold, special consideration should be given in examining their part-worths for reversals.

- *Data collection method.* The preferred method of administration is through a personal interview because of the possible complexity of the choice tasks, but recent advances make alternative means of data collection (e.g., Web-, mail-, or phone-based methods) feasible. Although studies support the predictive validity of these alternative methods the researcher must consider that such situations may exhibit a higher level of reversals due to such factors as increased respondent effort required, loss of respondent interest, or even the inability to resolve questions or confusion with the research task.

  The researcher should always include some form of debriefing with the respondent, either through a series of questions administered after the conjoint task or through a series of probes by the survey administrator in a personal interview. The purpose should be to assess the respondent's level of understanding of the factors and levels involved as well as the realism of the choice task.

- *Research context.* A final issue that contributes to the potential level of reversals is the object/concept being studied. Low-involvement products or situations (e.g., commodities, lower-risk ideas or concepts) always run the risk of inconsistencies in the actual choices and resulting part-worths. The researcher must always consider the ability of any set of choice tasks to maintain enough consumer involvement with a decision process when in actuality the consumer may not give the decision the level of thought modeled by the conjoint tasks. Too many times researchers identify too many attributes for consideration, overcomplicating a simple process from the respondent's perspective. When this situation happens, the respondent may view the choice tasks as too complex or unrealistic and provide inconsistent or illogical results.

Identifying Reversals. With the potential influences of the research setting considered, the researcher is still faced with the critical question: What actually is a reversal? Technically, any time a part-worth is hypothesized to be higher than an adjacent level but isn't, it violates the monotonic relationship and could be considered a reversal. Yet what amount of increase is needed to avoid being considered a reversal? What if the two adjacent levels are equal? What if the decline is miniscule?

The first step is to identify possible reversals. A simple yet effective method is to graphically portray the part-worth patterns for each attribute. Illogical patterns can quickly be identified within each attribute. However, as the number of attributes and respondents increases, the need for some empirical measure becomes apparent. It is a simple process to calculate the differences between adjacent levels, which can then be examined for illogical patterns. A miniscule decline might not constitute a reversal, so how large does the difference have to be? As a practical matter, however, some range of differences, even when contrary to the expected pattern, would probably be considered acceptable. In order to establish this range of acceptability, several options exist:

- One approach is to examine the differences and see where some natural break occurs, denoting those truly different values. Again, the researcher is looking for those truly distinctive values that indicate preferences contrary to the hypothesized relationship.
- A second approach would be to try and establish some estimate of a confidence interval that takes into account the established distributional characteristics of the differences. One possibility is to determine the standard error of the differences and then use that to construct a confidence interval around zero to denote truly distinctive differences. We should note that technically the confidence interval should be constructed "within subject" but too few observations are provided on any factor to do so. Thus, use of the standard error calculated across subjects is necessary.

Ultimately, to answer this question the researcher is encouraged to examine the distribution of differences and then identify those deemed outside a reasonable range. The extent of this range should be based not only on the actual differences, but also on the factors discussed earlier (respondent effort, data collection method, and research context), which impact the possibility of reversals.

The objective of any analysis of reversals is to identify consistent patterns of reversals that indicate an invalid representation a preference structure. Although a researcher would hope for no reversals, reversals can occur occasionally and still provide a valid preference structure. It is the job of the researcher to consider all of the factors discussed, along with the extent of the reversals for each respondent, and identify those respondents with an inappropriate number of reversals.

Remedies for Reversals. Even though the presence of reversals does not necessarily invalidate a set of part-worth estimates, the research must strongly consider a series of remedies to ensure both the appropriateness of the results as well as maximize the predictive ability of the part-worths. When faced with a substantial number of reversals, the researcher has several options:

- *Do nothing.* Many times a small number of reversals can be ignored, particularly if the focus is on aggregate results. Many researchers suggest leaving these reversals in as a measure of real-world inconsistency. The reasoning is that the reversals will be compensated for during aggregation.
- *Apply constraints.* Constraints can be applied in the estimation process such that reversals are prohibited [3, 109]. The specificity of these constraints ranges from simple approaches of creating a "tie" for the levels involved (i.e., give them the same part-worth estimate) to monotonicity constraints both within and across attributes [107]. One can also view the linear or ideal point models of part-worths discussed earlier as a type of constraint.

  Even though studies show that the predictive accuracy can be improved with these constraints, the researcher also must assess the degree to which these constraints potentially distort the preferences into predefined relationships. Thus, whereas constraints may be utilized to correct the occasional reversal, it would be inappropriate to utilize constraints to correct for incorrectly specified levels or attributes even if predictive accuracy was improved.
- *Delete respondents.* A final remedy involves the deletion of respondents with substantial numbers of reversals from the analysis. At issue here is the trade-off between reducing representativeness and diversity of the sample through deletion versus the inclusion of invalid preference structures. Again, the researcher must weigh the costs versus benefits in making such a decision.

Care should always be taken any time the researcher directly affects the estimated part-worths. Although the absence of reversals achieves a sense of validity by corresponding to the hypothesized relationships, the researcher must be sure to not impose restrictions that might obscure valid but counterintuitive results. With any remedy for reversals, the researcher also must be cognizant of the implications not only on the individual part-worth estimates, but also on the overall depictions of preference seen in aggregate results or other applications (e.g., segmentation, choice simulators).

Assessing the Relative Importance of Attributes

In addition to portraying the impact of each level with the part-worth estimates, conjoint analysis can assess the relative importance of each factor. Because part-worth estimates are typically converted to a common scale, the greatest contribution to overall utility—and hence the most important factor—is the factor with the greatest range (low to high) of part-worths. The importance values of each factor can be converted to percentages summing to 100 percent by dividing each factor's range by the sum of all range values.

Using our earlier example of estimated part-worths with three attributes, the calculation of importance would be as follows. First, find the range (maximum value minus minimum value) for the attribute. Then divide each range value by the total for the importance value.

Attribute	Minimum	Maximum	Range	Importance
1	−.657	.098	.755	22.8%
2	−.756	.302	1.058	32.0%
3	−.826	.667	1.493	45.2%
Total			3.306	100.0%

In this case, the third attribute accounted for almost one-half of the variation (1.493 ÷ 3.306 = .452) in the utility scores, even though the other two attributes were lower (32.0% and 22.8%). We could then state that for this respondent, attribute 3 was twice as important as attribute 1 in deriving utility scores and preferences.

This approach allows for comparison across respondents using a common scale, as well as giving meaning to the magnitude of the importance score. The researcher must always consider the impact on the importance values of an extreme or practically infeasible level. If such a level is found, it should be deleted from the analysis or the importance values should be reduced to reflect only the range of feasible levels.

STAGE 6: VALIDATION OF THE CONJOINT RESULTS

Conjoint results can be validated both internally and externally. Internal validation involves confirmation that the selected composition rule (i.e., additive versus interactive) is appropriate [19]. The researcher is typically limited to empirically assessing the validity of only the selected model form in a full study, owing to the high demands of collecting data to test both models. This validation process is most efficiently accomplished by comparing alternative models (additive versus interactive) in a pretest study to confirm which model is appropriate. We already discussed the use of holdout profiles to assess the predictive accuracy for each individual or a holdout sample of respondents if the analysis is performed at the aggregate level.

In general, external validation involves in general the ability of conjoint analysis to predict actual choices, and in specific terms the issue of sample representativeness. Although conjoint analysis has been employed in numerous studies over the past 20 years, relatively few studies have focused on its external validity. One study confirmed that conjoint analysis closely corresponded to the results from traditional concept testing, an accepted methodology for predicting customer preference [105], whereas other studies demonstrated the predictive accuracy for purchases of consumer electronics and groceries [37, 76]. Although no evaluation is made of sampling error in the individual-level models, the researcher must always ensure that the sample is representative of the population of study [72]. This representativeness becomes especially important when the conjoint results are used for segmentation or choice-simulation purposes (see the next section for a more detailed discussion of these uses of conjoint results).

MANAGERIAL APPLICATIONS OF CONJOINT ANALYSIS

Typically, conjoint models estimated at the individual level (separate model per individual) are used in one or more of the following areas: segmentation, profitability analysis, and conjoint simulators. In addition to the individual-level results, aggregate conjoint results can represent groups of individuals and also provide a means of predicting their decisions for any number of situations. The unique advantage of conjoint analysis is its ability to represent the preferences for each individual in an objective manner (e.g., part-worth utilities). In the most fundamental sense, conjoint analysis can help identify customers' needs, prioritize those needs, and then translate those needs into actual strategies [67, 90, 98]. The most common managerial and academic applications of conjoint analysis in conjunction with its portrayal of the consumer's preference structure include segmentation, profitability analysis, and conjoint simulators.

RULES OF THUMB 6-6

Interpreting and Validating Conjoint Results

- Results should be estimated for each individual unless:
  - The conjoint model requires aggregate-level estimates (i.e., some forms of choice-based conjoint)
  - The population is known to be homogeneous with no variation between individual preference structures
- Part-worth estimates are generally scaled to a common basis to allow for comparison across respondents
- Theoretically inconsistent patterns of part-worths, known as reversals, can give rise to deletion of a respondent unless:
  - Their occurrence is minimal
  - Constraints are applied to prohibit reversals
- Attribute importance must be derived based on the relative ranges of part-worths for each attribute
- Validation must occur at two levels:
  - Internal validation: Testing whether the appropriate composition rule has been selected (i.e., additive or interactive) and is done in a study pretest
  - External validation: Assessing the predictive validity of the results in an actual setting in which the researcher must always ensure the sample is representative of the population of study

---

Segmentation

One of the most common uses of individual-level conjoint analysis results is to group respondents with similar part-worths or importance values to identify segments. The estimated conjoint part-worth utilities can be used solely or in combination with other variables (e.g., demographics) to derive respondent groupings that are most similar in their preferences [20, 26]. In the industrial cleanser example, we might first group individuals based on their attribute importance scores, finding one group for which brand is the most important feature, whereas another group might value price more highly. Another approach would be to examine the part-worth scores directly, again identifying individuals with similar patterns of scores across each of the levels within one or more attributes.

For the researcher interested in knowing the presence of such groups and their relative magnitude, a number of different approaches to segmentation are available, all with differing strengths and weaknesses [66, 109]. One logical approach would be to apply cluster analysis (see Chapter 9) to the part-worth estimates or the importance scores for each attribute to identify homogeneous subgroups of respondents. Conjoint analysis has also been proposed as a means of validating segmentation analyses formed with other clustering variables, whereby differences in conjoint preference structures are used to demonstrate distinctiveness between the segments [18].

Profitability Analysis

A complement to the product-design decision is a marginal profitability analysis of the proposed product design. If the cost of each feature is known, the cost of each product can be combined with the expected market share and sales volume to predict its viability. This process could identify combinations of attributes that would be profitable even with smaller market shares because of the low cost of particular components. An adjunct to profitability analysis is assessing price sensitivity [45], which can be addressed through either specific research designs [81] or specialized programs [92]. Both individual and aggregate results can be used in this analysis.

Conjoint Simulators

At this point, the researcher still understands only the relative importance of the attributes and the impact of specific levels. So how does conjoint analysis achieve its other primary objective of using what-if analyses to predict the share of preferences that a profile (real or hypothetical) is likely to capture in various competitive scenarios of interest to management? This role is played by **choice simulators,** which enable the researcher to simulate any number of competitive scenarios and then estimate how the respondents would react to each scenario.

The researcher is cautioned in any application of the conjoint simulator in assuming that the share of preference in a conjoint simulation directly translates to market share [15]. The conjoint simulation represents only the product and perhaps price aspects of marketing management, omitting all of the other marketing factors (e.g., advertising and promotion, distribution, competitive responses) that ultimately affect market share. The conjoint simulation does, however, present a view of the product market and the dynamics of preferences that may be seen in the sample under study.

CONDUCTING A SIMULATION   A conjoint simulation is an attempt to understand how the set of respondents would choose among a specified set of profiles. This process provides the researcher with the ability to use the estimated part-worths in evaluating any number of scenarios consisting of differing combinations of profiles. For any given scenario the researcher follows a three-step process.

Step 1: Specify the Scenario(s). After the conjoint model is estimated, the researcher can specify any number of sets of profiles for simulation of consumer choices. Among the possible scenarios that can be assessed are the following:

- Impacts of adding a product to an existing market
- Increased potential from a multiproduct or multibrand strategy, including estimates of cannibalism
- Impacts of deleting a product or brand from the market
- Optimal product design(s) for a specific market setting

In each case, the researcher provides the set of profiles representing the objects (products, services, etc.) available in the market scenario being examined, and the choices of respondents are then simulated. The unique value of using conjoint analysis in the simulation is that multiple scenarios can be evaluated and the results compiled for each respondent through their preference structure of part-worths.

Step 2: Simulate Choices. Once the scenarios are complete, the part-worths for each individual are used to predict the choices across the profiles in each scenario. Choice simulators afford the researcher the ability to evaluate any number of scenarios, but their real benefit involves the ability of the researcher to specify conditions or relationships among the profiles to represent market conditions more realistically. For example, will all objects compete equally with all others? Does similarity among the objects create differing patterns of preference? Can the unmeasured characteristics of the market be included in the simulation? These questions are just a few of the many that can be addressed through a choice simulator in portraying a realistic market within which respondents make choices [37].

The ability of choice simulators to represent these relationships enables researchers to more realistically portray the forces acting among the set of objects being considered in the scenario. Moreover, predictive accuracy is markedly improved along with a better understanding of the underlying market behavior of the respondents [37, 78].

Step 3: Calculate Share of Preference. The final step in conjoint simulation is to predict preferences for each individual and then calculate the share of preferences for each profile by aggregating the individual choices. Choice simulators can use a wide range of choice rules [25] in predicting the choice for any individual:

- *Maximum utility (first choice) model.* This model assumes the respondent chooses the profile with the highest predicted utility score. Share of preference is determined by calculating the number of the individuals preferring each profile. This approach is best suited for situations with individuals of widely different preferences and in situations involving sporadic, nonroutine purchases.
- *Preference probability model.* In this model, predictions of choice probability sum to 100 percent over the set of profiles tested, with each person having some probability of purchasing each profile. Overall share of preference is measured by summing the preference probabilities across all respondents. This approach, which can approximate some elements of product similarity, is best suited to repetitive purchasing situations, for which purchases may be more tied to usage situations over time. The two most common methods of making these predictions are the BTL (Bradford-Terry-Luce) and logit models, which make quite similar predictions in almost all situations [36].
- *Randomized first choice.* Developed by Sawtooth Software [73, 78], this method attempts to combine the best of the two prior approaches. It samples each respondent multiple times, each time adding random variation to the utility estimates for each profile. For each iteration, it applies the first-choice rule and then totals the outcomes for each individual to get a share of preference per respondent. It corrects for product similarity and can be fine-tuned by specifying the amount and type of random variation that best approximates known preference shares [37, 75].

The share of preference, determined by any of the three methods described, provides insight into many factors underlying the actual choices of respondents. Multiple product scenarios can be evaluated, giving rise to not only a perspective of any single scenario, but of the dynamics in share of preference as the profiles change.

ALTERNATIVE CONJOINT METHODOLOGIES

Up to this point we have dealt with conjoint analysis applications involving the traditional conjoint methodology. However, real-world applications many times involve 20 to 30 attributes or require a more realistic choice task than those used in our earlier discussions. Recent research directed toward overcoming these problems encountered in many conjoint studies is leading to the development of two new conjoint methodologies: (1) an adaptive/self-explicated conjoint for dealing with a large number of attributes and (2) a choice-based conjoint for providing more realistic choice tasks. These areas represent the primary focus of current research in conjoint analysis [14, 29, 63].

Adaptive/Self-Explicated Conjoint: Conjoint with a Large Number of Factors

The full-profile method starts to become unmanageable with more than 10 attributes, yet many conjoint studies need to incorporate 20, 30, or even more attributes. In these cases, some adapted or reduced form of conjoint analysis is used to simplify the data collection effort and still represent a realistic choice decision. The two options are the self-explicated models and adaptive or hybrid models.

SELF-EXPLICATED CONJOINT MODELS   In the **self-explicated model,** the respondent provides a rating of the desirability of each level of an attribute and then rates the relative importance of the attribute overall. Part-worths are then calculated by a combination of the two values [99]. In this compositional approach, ratings are made on the components of utility rather than just overall preference. As a major variant of conjoint analysis and closer to traditional multi-attribute models, this model from raises several concerns.

First, can respondents assess the relative importance of attributes accurately? A common problem with self-ratings is the potential for importance to be underestimated in multi-attribute models because respondents want to give socially desirable answers. In such situations, the resulting conjoint model is also biased. Second, interattribute correlations may play a greater role and cause substantial biases in the results due to double-counting of correlated factors. Traditional conjoint models suffer from this problem as well, but the self-explicated approach is particularly affected because respondents must never explicitly consider these attributes in relation to other attributes. Finally, respondents never perform a choice task (rating the set of hypothetical combinations of attributes), and this lack of realism is a critical limitation, particularly in new-product applications.

Recent research demonstrates that this method may offer suitable predictive ability when compared to traditional conjoint methods [27]. This approach is best utilized when aggregate models are preferred because individual idiosyncrasies can be compensated for in the aggregate results. Thus, if the number of factors cannot be reduced to a manageable level acceptable for any of the other conjoint methods, then a self-explicated model may be a viable alternative method.

ADAPTIVE OR HYBRID CONJOINT MODELS   A second approach is the **adaptive** or **hybrid model,** so termed because it combines the self-explicated and part-worth conjoint models [23, 24]. This approach utilizes the self-explicated values to create a small subset of profiles selected from a fractional factorial design. The profiles are then evaluated in a manner similar to traditional conjoint analysis. The sets of profiles differ among respondents based on their self-explicated responses, and although each respondent evaluates only a small number, collectively all profiles are evaluated by a portion of the respondents. The approach of integrating information from the respondent to simplify or augment the choice tasks led to a number of recent research efforts aimed at differing aspects of the research design [3, 44, 101, 106].

One of the most popular variants of this approach is ACA, a computer-administered conjoint program developed by Sawtooth Software [87]. ACA employs self-explicated ratings to reduce the factorial design size and make the process more manageable. It is particularly useful when the study includes a large number of attributes that are not appropriate for the other approaches. The program first collects self-explicated ratings of each factor. These ratings are used in generating the profiles such that the less important factors are quickly eliminated. Moreover, each profile contains just a small number of factors (three to six) to keep the choice task more manageable. This adaptive process can only be accomplished through the associated software, making this approach inappropriate for any type of noninteractive setting (e.g., written survey). Yet its flexibility in accommodating large numbers of attributes with simple choice tasks has made it one of the most widely used approaches. Moreover, its relative predictive ability has been shown to be comparable to traditional conjoint analysis, thus making it a suitable alternative when the number of attributes is large [27, 47, 105, 115, 119].

CHOOSING BETWEEN SELF-EXPLICATED AND ADAPTIVE/HYBRID MODELS   When faced with a number of factors that cannot be accommodated in the conjoint methods discussed to this point, the self-explicated and adaptive/hybrid models preserve at least a portion of the underlying principles of conjoint analysis. In comparing these two extensions, the self-explicated methods have a slightly lower reliability, although recent developments may provide improvement. When the adaptive/hybrid models and self-explicated methods are compared with full-profile methods, the results are mixed, with slightly better performance by the hybrid/adaptive method, particularly ACA [38]. Although more research is needed to confirm the comparisons across methods, the empirical studies indicate that the adaptive/hybrid methods and the newer forms of self-explicated models both offer viable alternatives to traditional conjoint analysis when dealing with a large number of factors.

Choice-Based Conjoint: Adding Another Touch of Realism

In recent years, many researchers in the area of conjoint analysis have directed their efforts toward a new conjoint methodology that provides increased realism in the choice task. With the overriding objective of understanding the respondent's decision-making process and predicting behavior in the marketplace, traditional conjoint analysis assumes that the judgment task, based on ranking or rating, captures the choices of the respondent. Yet researchers argue that this approach is not the most realistic way of depicting a respondent's actual decision process, and others have pointed to the lack of formal theory linking these measured judgments to choice [59].

What emerged is an alternative conjoint methodology, known as **choice-based conjoint (CBC),** with the inherent face validity of asking the respondent to choose a full profile from a set of alternative profiles known as a **choice set.** This method is much more representative of the actual process of selecting a product from a set of competing products. Moreover, choice-based conjoint provides an option of not choosing any of the presented profiles by including a no-choice option in the choice set. Whereas traditional conjoint analysis assumes respondents' preferences will always be allocated among the set of profiles, the choice-based approach allows for market contraction if all the alternatives in a choice set are unattractive.

The advantages of the choice-based approach are the additional realism and the ability to estimate interaction terms. After each respondent has chosen a profile for each choice set, the data can be analyzed either at the disaggregate level (individual respondents) or aggregated across respondents (segments or some other homogeneous groupings of respondents) to estimate the conjoint part-worths for each level and the interaction terms. From these results, we can assess the contributions of each factor and factor-level interaction and estimate the likely market shares of competing profiles.

A SIMPLE ILLUSTRATION OF FULL-PROFILE VERSUS CHOICE-BASED CONJOINT Before discussing some of the more technical details of choice-based conjoint and how it differs from the other conjoint methodologies we will first examine the differences in creating profiles and then review the actual data collection process.

Creating Profiles. The first difference between full-profile and choice-based conjoint is the type of profiles. Both methods use a form of full-profile profiles, but the choice task is quite different. Let's examine a simple example to illustrate the differences.

A wireless phone company wishes to estimate the market potential for three service options that can be added to the base service fee of $14.95 per month and $0.50 per minute of calling time:

ICA	Itemized call accounting with a $2.75-per-month charge
CW	Call waiting with a $3.50-per-month service charge
TWC	Three-way calling with a $3.50-per-month service charge

Traditional conjoint analysis is performed with full-profile profiles representing the various combinations of service, ranging from just the base service to the base service and all three options. The complete set of profiles (factorial design) is shown in Table 6-6. Profile 1 represents the base service with no options, profile 2 is the base service plus itemized call accounting, and so forth, up to profile 8 being the base service plus all three options (itemized call accounting, call waiting, and three-way calling).

TABLE 6-6 **A Comparison of Profile Designs Used in Traditional and Choice-Based Conjoint Analysis**

TRADITIONAL CONJOINT ANALYSIS

	Levels of Factors[a]				Choice-Based Conjoint	
Profile	ICA	CW	TWC	Choice Set	Profiles in Choice Set[b]
1	0	0	0	1	1, 2, 4, 5, 6, and No Choice
2	1	0	0	2	2, 3, 5, 6, 7, and No Choice
3	0	1	0	3	1, 3, 4, 6, 7, 8, and No Choice
4	0	0	1	4	2, 4, 5, 7, 8, and No Choice
5	1	1	0	5	3, 5, 6, 8, and No Choice
6	1	0	1	6	4, 6, 7, and No Choice
7	0	1	1	7	1, 5, 7, 8, and No Choice
8	1	1	1	8	1, 2, 6, 8, and No Choice
				9	1, 2, 3, 7, and No Choice
				10	2, 3, 4, 8, and No Choice
				11	1, 3, 4, 5, and No Choice

[a]Levels: 1 = service option included; 0 = service option not included.
[b]Profiles used in choice sets are those defined in the design for the traditional conjoint analysis.

In a choice-based approach, the respondent is shown a series of choice sets. Each choice set has several full-profile profiles. A choice-based design is also shown in Table 6-6. The first choice set consists of five of the full-profile profiles (profiles 1, 2, 4, 5, and 6) and a "No Choice" option. The respondent then *chooses only one of the profiles in the choice set* ("most preferred" or "most liked") *or "none of these."* An example choice set task for choice set 6 is shown in Table 6-7. The preparation of profiles and choice sets is based on experimental design principles [44, 59] and is the subject of considerable research effort to refine and improve on the choice task [3, 14, 40, 44, 81].

Data Collection. Given the differing ways in which profiles are formed, the choice tasks facing the respondent are quite different. As we will see, the researcher must select between a simpler choice task in the full-profile method versus the choice-based task that is more realistic.

For the full-profile approach, the respondent is asked to rate or rank each of the eight profiles. The respondent evaluates each profile separately and provides a preference rating. The task is relatively simple and can be performed quite quickly after a few warm-up tasks. As discussed earlier, as the number of attributes and levels increases (remember our earlier example of four factors with

TABLE 6-7 **Example of a Choice Set in Choice-Based Conjoint**

	Which Calling System Would You Choose?		
1	**2**	**3**	**4**
Base system at $14.95/month and $.05/minute plus: • TWC: Three-way calling for $3.50/month	Base system at $14.95/month and $.05/minute plus: • ICA: Itemized call accounting for $2.75/month	Base system at $14.95/month and $.05/minute plus: • CW: Call waiting for $3.50/month and • TWC: Three-way calling for $3.50/month	None of these

four levels each generating 256 profiles), the task can become very large and require some form of subset of profiles that still may be fairly substantial.

For the choice-based approach, the number of profiles may or may not vary across choice sets [59]. Also, the number of choices made (one choice for each of 11 choice sets) is actually more in this case than required in this example. As the number of factors and levels increases, however, the choice-based design requires considerably fewer evaluations. But in all situations, the respondent sees multiple full profiles and selects one profile from the choice set.

UNIQUE CHARACTERISTICS OF CHOICE-BASED CONJOINT   The basic nature of choice-based conjoint and its background in the theoretical field of information integration [58] have led to a somewhat more technical perspective than found in the other conjoint methodologies. Even though the other methodologies are based on sound experimental and statistical principles, the additional complexity in both profile designs and estimation has prompted a great deal of developmental efforts in these areas. From these efforts, researchers now have a clearer understanding of the issues involved at each stage. The following sections detail some of the areas and issues in which choice-based conjoint is unique among the conjoint methodologies.

Type of Decision-Making Process Portrayed.   Traditional conjoint has always been associated with an information-intensive approach to decision making because it involves examining the profiles composed of levels from each attribute. But in choice-based conjoint analysis, researchers are coming to the conclusion that the choice task may invoke a different type of decision-making process. In making choices among profiles, consumers seem to choose among a smaller subset of factors upon which comparisons, and ultimately choice, are made [39]. This parallels the types of decisions associated with time-constrained or simplifying strategies, each characterized by a lower depth of processing. Thus, each conjoint methodology provides different insights into the decision-making process. Because researchers may not be willing to select only one methodology, an emerging strategy is to employ both methodologies and draw unique perspectives from each [39, 80].

Choice Set Design.   Perhaps the greatest advantage of choice-based conjoint is the realistic choice process portrayed by the choice set. Recent developments have further enhanced the choice task, allowing for additional relationships within the choice model to be analyzed while increasing the effectiveness of the choice set design.

A recent effort showed how the choice set can be created to ensure balance not just among factor levels, but also among the utilities of the profiles [40]. The most realistic and informative choice is among closely comparable alternatives, rather than the situation in which one or more profiles are markedly inferior or superior. However, the profile-design process is typically focused on achieving orthogonality and balance among the attributes. This approach provides a more realistic task by creating profiles with more comparable utility levels, increasing consumer involvement and providing better results.

Choice-based conjoint also provides the options to include the "No Choice" alternative, in which the respondent has the choice of choosing none of the specified options [32]. This option provides the respondent with an additional level of realism while also providing the researcher with a means of establishing absolute as well as relative effects. Finally, CBC readily accommodates model modifications such as prohibited pairs, level-specific effects, or cross-effects between levels (e.g., brands) that require specially designed choice tasks best accomplished through choice-based conjoint [16, 85]. Moreover, in a method involving additional information from the respondents, choice sets are created that fit the unique preferences of each individual and achieve better predictive accuracy in market-based situations [12].

Estimation Technique.   The conceptual foundation of choice-based conjoint is psychology [60, 104], but it was the development of the multinomial logit estimation technique [64] that provided an operational method for estimating these types of choice models. Although considerable efforts have refined and made the technique widely available, it still represents a more complex methodology than those associated with the other conjoint methodologies.

The choice-based approach was originally estimated only at the aggregate level, but developments have allowed for the formation of segment-level models (known as *latent class models*) and even individual models through Bayesian estimation [6, 56, 91, 103]. This development fostered even more widespread adoption of choice-based methods by making disaggregate models more conducive to use in choice simulators and other applications.

One particular aspect that remains problematic in aggregate models or in the use of choice simulators is the property of IIA (independence of irrelevant alternatives), an assumption that makes the prediction of similar alternatives problematic. Although exploring all of the issues underlying IIA is beyond the scope of this discussion, the researcher is cautioned when using aggregate-level models estimated by choice-based conjoint to understand the ramifications of this assumption.

SOME ADVANTAGES AND LIMITATIONS OF CHOICE-BASED CONJOINT   The growing popularity of choice-based conjoint analysis among marketing research practitioners is primarily due to the belief that obtaining preferences by having respondents choose a single preferred profile from among a set of profiles is more realistic—and thus a better method—for approximating actual decision processes. Yet the added realism of the choice task is accompanied with a number of trade-offs the researcher must consider before selecting choice-based conjoint.

The Choice Task.   Each choice set contains several profiles and each profile contains all of the factors, similar to the full-profile profiles. Therefore, the respondent must process a considerably greater amount of information than the other conjoint methodologies in making a choice in each choice set. Sawtooth Software, developer of a choice-based conjoint (CBC) system, believes that choices involving more than six attributes are likely to confuse and overwhelm the respondent [88]. Although the choice-based method does mimic actual decisions more closely, the inclusion of too many attributes creates a formidable task that ends up with less information than would have been gained through the rating of each profile individually.

Predictive Accuracy.   In practice, all three conjoint methodologies allow for similar types of analyses, simulations, and reporting, even though the estimation processes are different. Choice-based models still have to be subjected to more thorough empirical tests, yet some researchers believe they gain an advantage in predicting choice behavior, particularly when segment-level or aggregate models are desired [108]. However, empirical tests indicate little difference between individual-level ratings-based models adjusted to take the "No Choice" option into account and the generalized multinomial logit choice-based models [68].

In comparing the two approaches (ratings-based or choice-based) in terms of the ability to predict shares in a holdout sample at the individual level [21], both approaches predict holdout sample choices well, with neither approach dominant and the results mixed in different situations. Ultimately, the decision to use one method over the other is dictated by the objectives and scope of the study, the researcher's familiarity with each method, and the available software to properly analyze the data.

Managerial Applications.   Choice-based models estimated at the aggregate level provide the values and statistical significance of all estimates, easily produce realistic market-share predictions for new profiles [44, 108], and offer the added assurances that "choices" among profiles were used to calibrate the model. However, aggregate choice-based conjoint models hinder segmentation of the market segmentation. The development of segment-based or even individual-level models was the response to this need [56, 103, 111]. Their ability to represent interaction terms and complex interattribute relationships does provide greater insight into both the actual choice process as well as the expected aggregate relationships seen through choice simulators. Yet, for most basic choice

> ### RULES OF THUMB 6-7
>
> ### Alternative Conjoint Models
>
> - When 10 or more attributes are included in the conjoint variate, two alternative models are available:
>   - Adaptive models can easily accommodate up to 30 attributes, but require a computer-based interview
>   - Self-explicated models can be done through any form of data collection, but represent a distinct departure from traditional conjoint methods
> - Choice-based conjoint models have become the most popular format of all, even though they generally accommodate no more than six attributes, with popularity based on:
>   - Use of a realistic choice task of selecting most preferred stimulus from a choice set of stimuli, including a "No Choice" option
>   - Ability to more easily estimate interaction effects
>   - Increased availability of software, particularly with Bayesian estimation options

situations, the ratings-based models described earlier are well suited to segmentation studies and the simulation of choice shares. Again, the researcher must decide on the level of realism versus complexity desired in any application of conjoint analysis.

Availability of Computer Programs. The good news is that several choice-based programs are now available for researchers that assist in all phases of research design, model estimation, and interpretation [42, 88]. Moreover, recent research by academicians and applied researchers is being integrated into these commercially available programs. These improvements and enhanced capabilities, after rigorous validation by the research community, should become a standard part of all choice-based programs.

Overview of the Three Conjoint Methodologies

Conjoint analysis evolved past its origins of what we now know as traditional conjoint analysis to develop two additional methodologies, each of which addresses two substantive issues: dealing with large numbers of attributes and making the choice task more realistic [74]. Each methodology provides distinctive features that help define those situations in which it is most applicable (see our earlier discussion in stage 2). Yet, in many situations two or more methodologies are feasible and the researcher has the option of selecting one methodology or, more increasingly, combining the methodologies. Only by being knowledgeable about the strengths and weaknesses of each methodology can the researcher make the more appropriate choice. The advantages of the choice-based approach are making it the most widely used. The adaptive approach also has considerable use given its ability to accommodate large numbers of attributes and levels. But whatever approach is used they all rely on the basic principles of conjoint design. Researchers interested in conjoint analysis are encouraged to continue to monitor the developments of this widely employed multivariate technique.

AN ILLUSTRATION OF CONJOINT ANALYSIS

In this section we examine the steps in an application of conjoint analysis to a product design problem. The discussion follows the model-building process introduced in Chapter 1 and focuses on (1) design of the profiles, (2) estimation and interpretation of the conjoint part-worths, and (3) application of a conjoint simulator to predict market shares for a new product formulation. The CONJOINT module of SPSS is used in the design, analysis, and choice simulator phases of this example [97]. Comparable results are obtained with other conjoint analysis programs available for commercial and academic use. The dataset of conjoint responses is available on the text's Web sites (www.pearsonglobaleditions.com/hair or www.mvstats.com).

Stage 1: Objectives of the Conjoint Analysis

Conjoint analysis, as discussed earlier, has been quite effectively applied to product development situations requiring (1) an understanding of consumer preferences for attributes as well as (2) a method for simulating consumer response to various product designs. Through the application of conjoint analysis, researchers can develop either aggregate (e.g., segment-level) estimates of consumer preferences or estimate disaggregate models (i.e., individual-level) from which segments can be derived.

HBAT was seriously considering designing a new industrial cleanser for use in not only in its own industry, but in many manufacturing facilities. In developing the product concept, HBAT wanted a more thorough understanding of the needs and preferences of its industrial customers. Thus, in an adjunct study to the one described in Chapter 1, HBAT commissioned a conjoint analysis experiment among 86 industrial customers.

Before the actual conjoint study was performed, internal marketing research teams, in consultation with the product development group, identified five factors as the determinant attributes in the targeted segment of the industrial cleanser market. The five attributes are shown in Table 6-8. Focus group research confirmed that these five attributes represented the primary determinants of value in an industrial cleanser for this segment, thus enabling the design phase to proceed with further specification of the attributes and their levels.

Stage 2: Design of the Conjoint Analysis

The decisions at this phase are (1) selecting the conjoint methodology to be used, (2) designing the profiles to be evaluated, (3) specifying the basic model form, and (4) selecting the method of data collection.

SELECTING A CONJOINT METHODOLOGY   The first issue to be resolved is the selection of the conjoint methodology from among the three options: traditional conjoint, adaptive/hybrid conjoint, or choice-based conjoint. The choice of method should be based not only on design considerations (e.g., number of attributes, type of survey administration, etc.), but also on the appropriateness of the choice task to the product decision being studied.

Given the small number of factors (five), all three methodologies would be appropriate. Because the emphasis was on a thorough understanding of the preference structure and the decision was expected to be one of fairly high consumer involvement, the traditional conjoint methodology was chosen as suitable in terms of response burden on the respondent and the depth of information portrayed. Choice-based conjoint was also strongly considered, but the absence of proposed interactions and the desire for reducing the task complexity led to the selection of the traditional conjoint method.

TABLE 6-8   **Attributes and Levels for the HBAT Conjoint Analysis Experiment Involving Product Design of an Industrial Cleanser**

Attribute Description	Levels		
Form of the Product	Premixed liquid	Concentrated liquid	Powder
Number of Applications per Container	50	100	200
Addition of Disinfectant to Cleanser	Yes	No	
Biodegradable Formulation	No	Yes	
Price per Typical Application	35 cents	49 cents	79 cents

The adaptive approach was not strongly considered given the small number of attributes and the desire to use traditional survey-based approaches such as written surveys.

DESIGNING PROFILES   With the traditional full-profile method selected, the next step involves the design of the profiles. Although the attributes have already been selected, the researcher must take great care during this stage in specifying the attribute levels to operationalize the attributes for use in design profiles. Among the considerations to be addressed are the nature of the levels (ensuring they are actionable and communicable), the magnitude and range of the levels for each attribute, and the potential for interattribute correlation.

The first consideration was to ensure that each level was actionable and communicable. Focus group research established specific levels for each attribute (see Table 6-8). The levels were each designed to (1) employ terminology used in the industry and (2) represent aspects of the product routinely specified in buying decisions.

Although the three attributes of *Product Form, Disinfectant,* and *Biodegradability* only portrayed specific characteristics; two attributes needed further examination for appropriateness of the ranges of levels. First, *Number of Applications* ranged from 50 to 200. Given the product form selected, these levels were chosen to result in the typical types of product packaging found in industrial settings, ranging from small containers for individuals to larger containers normally associated with centralized maintenance operations. Next, the three levels of *Price per Application* were determined from examining existing products. As such they were deemed to be realistic and to represent the most common price points in the current market. It should be noted that the price levels are considered monotonic (i.e., have a rank ordering), but not linear, because the intervals (differences between levels) are not consistent.

The product type did not suggest intangible factors that would contribute to interattribute correlation, and the attributes were specifically defined to minimize interattribute correlation. All of the possible combinations of levels were examined to identify any inappropriate combinations and none were found. A small-scale pretest and evaluation study was conducted to ensure that the measures were understood and represented reasonable alternatives when formed into profiles. The results indicated no problems with the levels, thus allowing the process to continue.

SPECIFYING THE BASIC MODEL FORM   With the levels specified, the researcher must next specify the type of model form to be used. In doing so, two critical issues must be addressed: (1) whether interactions are to be represented among the attributes and (2) the type of relationship among the levels (part-worth, linear, or quadratic) for each attribute.

After careful consideration, HBAT researchers felt confident in assuming that an additive composition rule was appropriate. Although research showed that price often has interactions with other factors, it was assumed that all of the other factors were reasonably orthogonal and that interaction terms were not needed. This assumption allowed for the use of aggregate or disaggregate models as needed.

Three of the attributes (*Product Form, Applications per Container,* and *Price per Application*) have more than two levels, thus requiring a decision on the type of part-worth relationship to be used. The *Product Form* attribute represented distinct product types, so separate part-worth estimates are appropriate. The *Application per Container* attribute also had three levels, yet they did not have equal intervals. Thus, separate part-worth estimates were used here as well. Finally, price also was specified with separate part-worth estimates because the intervals were not consistent among levels.

Of these three factors, only *Price per Application* was specified as monotonic, because of the implied relationship for price. *Product Form* represented separate levels with no preconceived order. The factor *Applications per Container* was not considered monotonic, even though the levels are defined in numeric terms (e.g., 50 applications per container). In this situation, no prior knowledge led researchers to propose that the part-worths should either increase or decrease consistently across these levels.

SELECTING THE METHOD OF DATA COLLECTION   The final step in designing the conjoint analysis revolves around the actual collection of preferences from respondents. In doing so, several issues must be addressed, including selection of the presentation method, actual creation of the profiles and identification of any unacceptable profiles, selecting a preference measure, and finalizing the survey administration procedure.

Selection of Presentation Method. To ensure realism and allow for the use of ratings rather than rankings, HBAT decided to use the full-profile method of obtaining respondent evaluations. A hybrid/adaptive method was not needed due to the relatively small number of factors. A choice-based method would have been equally appropriate given the smaller number of attributes and the realism of the choice task, but the full-profile approach was ultimately selected due to the need for disaggregate additive results with the simplest method of estimation.

Profile Subsets. In choosing the additive rule, researchers were also able to use a fractional factorial design to avoid the evaluation of all 108 possible combinations ($3 \times 3 \times 2 \times 2 \times 3$). The profile design component of the computer program generated a set of 18 full-profile descriptions (see Table 6-9), allowing for the estimation of the orthogonal main effects for each factor. Four additional profiles were generated to serve as the validation profiles. None of the profiles were deemed unacceptable after being reviewed for realism and appropriateness to the research question.

TABLE 6-9   Set of 18 Full-Profiles Used in the HBAT Conjoint Analysis Experiment for Designing an Industrial Cleanser

Profile #	Product Form	Number of Applications	Disinfectant Quality	Biodegradable Form	Price per Application
Profiles Used in Estimation of Part-Worths					
1	Concentrate	200	Yes	No	35 cents
2	Powder	200	Yes	No	35 cents
3	Premixed	100	Yes	Yes	49 cents
4	Powder	200	Yes	Yes	49 cents
5	Powder	50	Yes	No	79 cents
6	Concentrate	200	No	Yes	79 cents
7	Premixed	100	Yes	No	79 cents
8	Premixed	200	Yes	No	49 cents
9	Powder	100	No	No	49 cents
10	Concentrate	50	Yes	No	49 cents
11	Powder	100	No	No	35 cents
12	Concentrate	100	Yes	No	79 cents
13	Premixed	200	No	No	79 cents
14	Premixed	50	Yes	No	35 cents
15	Concentrate	100	Yes	Yes	35 cents
16	Premixed	50	No	Yes	35 cents
17	Concentrate	50	No	No	49 cents
18	Powder	50	Yes	Yes	79 cents
Holdout Validation Profiles					
19	Concentrate	100	Yes	No	49 cents
20	Powder	100	No	Yes	35 cents
21	Powder	200	Yes	Yes	79 cents
22	Concentrate	50	No	Yes	35 cents

Sample Size. HBAT researchers considered samples sizes ranging from 50 to 200. Obviously, larger samples would provide a more accurate representation of the population of interest, but practical considerations (relatively small population of customers and fairly high cost of personal interviews) called for smaller sample sizes. Given the relative homogeneity of the respondents it was determined that a sample of approximately 100 would be adequate. After the data collection was completed, a total of 86 respondents completed the entire survey. This was deemed adequate to represent the buyer group in question.

Collecting Respondent Preferences. The conjoint analysis experiment was administered during a personal interview. After collecting some preliminary data, the respondents were handed a set of 22 cards, each containing one profile description. A ratings measure of preference was gathered by presenting each respondent with a foldout form that had seven response categories, ranging from "not at all likely to buy" to "certain to buy." Respondents were instructed to place each card in the response category best describing their purchase intentions. After initially placing the cards, they were asked to review their placements and rearrange any cards, if necessary. The validation profiles were rated at the same time as the other profiles but withheld from the analysis at the estimation stage. Upon completion, the interviewer recorded the category for each card and proceeded with the interview. A total of 86 respondents successfully completed the entire conjoint task.

Stage 3: Assumptions in Conjoint Analysis

The relevant assumption in conjoint analysis is the specification of the composition rule and thus the model form used to estimate the conjoint results. This assessment must be based on conceptual terms as well as practical issues.

In this situation, the nature of the product, the tangibility of the attributes, and the lack of intangible or emotional appeals justifies the use of an additive model. HBAT felt confident in using an additive model for the industrial cleanser decision-making situation. Moreover, it simplified the design of the profiles and facilitated the data collection efforts.

Stage 4: Estimating the Conjoint Model and Assessing Overall Model Fit

With the conjoint tasks specified and responses collected, the next step is to utilize the appropriate estimation approach for deriving the part-worth estimates and then assess overall goodness-of-fit. In doing so, the researcher must consider not only the responses used in estimation, but also those collected for validation purposes.

MODEL ESTIMATION   Given that the preference measure used was a metric rating, either the traditional regression-based approach or the newer Bayesian methodology could be employed. Because the fractional factorial design provided enough profiles for estimation of disaggregate models, the traditional approach was used. It should be noted, however, that Bayesian estimation would have been just as appropriate, particularly because additional interaction effects were desired.

The estimation of part-worths of each attribute was first performed for each respondent separately, and the results were then aggregated to obtain an overall result. Separate part-worth estimates were made for all levels initially, with examination of the individual estimates undertaken to determine the possibility of placing constraints on a factor's relationship form (i.e., employ a linear or quadratic relationship form). Table 6-10 shows the results for the overall sample, as well as for the first five respondents in the data set. Examination of the overall results suggests that perhaps a linear relationship could be estimated for the price variable (i.e., the part-worth values decrease from 1.13 to .08 to −1.21 as the price per application increases from 35 cents to 49 cents to 79 cents). However, review of the individual results shows that only three of the five respondents (107, 123, and 135) had part-worth estimates for the price factors that were of a generally linear pattern.

TABLE 6-10 Conjoint Part-Worth Estimates for the Overall Sample and Five Selected Respondents

					PART-WORTH ESTIMATES							
	Product Form			Number of Applications		Disinfectant		Biodegradable		Price per Application		
Premixed	Concentrate	Powder	50	100	200	Yes	No	No	Yes	$.35	$.49	$.79
Overall Sample												
−.2171	.1667	.0504	−.3450	.0233	.3217	.5102	−.5102	−.1541	.1541	1.1318	.0814	−1.2132
Selected Respondents (107, 110, 123, 129, and 135, respectively)												
−.0556	.6111	−.5556	.4444	.6111	−1.0556	−.2083	.2083	.5417	−.5417	1.4444	.9444	−2.3889
.4444	−.5556	.1111	−.0556	−.3889	.4444	.1667	−.1667	−.5833	.5833	.6111	−.8889	.2778
−.6111	.3889	.2222	−.4444	.2222	.2222	−.4167	.4167	−.5417	.5417	2.5556	.0556	−2.6111
−.0556	.1111	−.0556	−.0556	−.0556	.1111	.4167	−.4167	−.0833	.0833	−.0556	−.0556	.1111
−.2222	−.3889	.6111	−.2222	−.3889	.6111	.1667	−.1667	.1667	−.1667	2.944	−.7222	−2.2222

For respondent 129 the pattern was essentially flat, and respondent 110 had a somewhat illogical pattern with the part-worths actually increasing when going from 49 cents to 79 cents. Thus, application of a linear form for the price factor would severely distort the relationship among levels, and the estimation of separate part-worth values for the *Price per Application* attribute was retained.

ASSESSING GOODNESS-OF-FIT   For both disaggregate and aggregate results, three goodness-of-fit measures were provided. Preference was measured using ratings (metric data); therefore, Pearson correlations were calculated for the estimation sample. The ratings values also were converted to rank orders and a Kendall's tau measure calculated. The holdout sample had only four profiles, so goodness-of-fit, for validation purposes, used only the rank-order measure of Kendall's tau.

Unlike many other multivariate techniques, when evaluating disaggregate results no direct statistical significance test evaluates the goodness-of-fit measures just described. However, we can use generally accepted levels of correlation to assess goodness-of-fit for both the estimation and validation phases. In establishing any threshold for evaluating the goodness-of-fit measures, the researcher must look at both the very low and very high values, because each may indicate respondents for whom the choice task was not applicable.

Assessing Low Goodness-of-Fit Values. In evaluating the lower values, the obvious threshold is some minimum value of correlation between the actual preference scores and the predicted utility values. One way to set a minimum value is to examine the distribution of values for the goodness-of-fit measures. Outlying values may indicate respondents for whom the choice task was not applicable when compared to the other respondents. A second approach is to establish a minimum correlation value based on the small number of profiles for each respondent, similar to the adjusted R^2 measure in multivariate regression (see Chapter 4 for more details).

In this example, the estimation process used 18 profiles and five attributes as independent variables. In a regression model of 18 observations and five independent variables, an adjusted R^2 of zero is found when the R^2 is approximately .300. This establishes a minimum correlation of .55 (the square root of .300) so that the adjusted R^2 would always be above zero. The researcher may also wish to set some minimum threshold that corresponds to a level of fit. For example, if the researcher wanted the estimation process to explain at least 50 percent of the variation, then a correlation of .707 is required.

Using a minimum goodness-of-fit level of .55 and a desired level of .707 for the Pearson correlation (metric-based), only three respondents had values less than .707 and all of them were above the lower threshold of .55 (see Table 6-11). The Kendall's tau values, although generally

TABLE 6-11   Goodness-of-Fit Measures for Conjoint Analysis Results

Respondent	Estimation Sample		Validation Sample	Respondent	Estimation Sample		Validation Sample
	Pearson	Kendall's Tau	Kendall's Tau		Pearson	Kendall's Tau	Kendall's Tau
107	.929	.784	.707	363	.947	.819	.548
110	.756	.636	.408	364	.863	.760	.707
123	.851	.753	.707	366	.828	.751	.548
129	.945	.718	.816	368	.928	.783	.775
135	.957	.876	.816	370	.783	.690	.913
155	.946	.736	.707	372	.950	.813	.183
161	.947	.841	.913	382	.705	.463	.548
162	.880	.828	.667	396	1.000	1.000	1.000
168	.990	.848	.913	399	.948	.766	.913
170	.808	.635	.667	401	.985	.869	.913
171	.792	.648	.548	416	.947	.762	.816
173	.920	.783	.548	421	.887	.732	.548
174	.967	.785	.913	422	.897	.832	1.000
181	.890	.771	.913	425	.945	.743	.707
187	.963	.858	.913	428	.967	.834	.913
193	.946	.820	.816	433	.864	.754	.548
194	.634	.470	.913	440	.903	.778	.816
197	.869	.731	.548	441	.835	.666	.548
211	.960	.839	.707	453	.926	.815	.913
222	.907	.761	.707	454	.894	.661	.816
225	.990	.931	1.000	467	.878	.798	.913
229	.737	.582	.236	471	.955	.840	.707
235	.771	.639	.775	472	.899	.748	.707
236	.927	.843	.707	475	.960	.875	.667
240	.955	.735	.816	476	.722	.538	.775
260	.939	.738	.775	492	.944	.791	.816
261	.965	.847	.707	502	.946	.832	.707
266	.570	.287	.236	507	.857	.746	.548
271	.811	.654	.707	514	.924	.795	.707
277	.843	.718	.707	516	.936	.850	.548
287	.892	.744	.913	518	.902	.803	1.000
300	.961	.885	.707	520	.888	.812	.913
302	.962	.871	.816	522	.957	.903	.548
303	.898	.821	1.000	528	.917	.797	.816
309	.876	.821	.800	535	.883	.748	.816
318	.896	.713	.816	538	.827	.665	1.000
323	.874	.762	.816	557	.948	.854	.913
336	.878	.780	.667	559	.900	.767	.913
348	.949	.747	.816	578	.905	.726	.707
350	.970	.861	.816	580	.714	.614	.913
354	.795	.516	.707	586	.974	.862	1.000
356	.893	.780	.913	589	.934	.679	.913
357	.915	.730	.913	592	.931	.832	.913
Aggregate	.957	.876	.816				

lower in value given their use of rank order rather than ratings, demonstrated the same general pattern. For the validation profiles, four respondents (110, 229, 266, and 372) had particularly low Kendall's tau values (.40 or lower). Although one of these respondents (266) also had low estimation values, the other three had low values only on the validation process.

Assessing Very High Goodness-of-Fit Values. Extremely high goodness-of-fit measures should also be examined; they may indicate that the choice tasks did not capture the decision process, similar to extremely low values. For example, values of 1.0 indicate that the estimated part-worths perfectly captured the choice process, which may occur when the respondent uses only a single or small number of attributes. But it may also indicate a respondent who did not follow the spirit of the task and thus provides unrepresentative results. Although assessing these values requires a degree of researcher judgment, it is important to evaluate the results for every value to ensure that they are truly representative of the choice process.

Three respondents (225, 396, and 586) were identified based on their extremely high goodness-of-fit values for the estimation sample. The goodness-of-fit values for the estimation sample are .990, 1.000, and .974, respectively, and all three have goodness-of-fit values of 1.000 for the validation sample. Thus, all three should be examined to see whether the part-worth estimates represent reasonable preference structures.

When looking at the individual part-worth estimates, quite different preference structures emerge (see Table 6-12). For respondent 225, all of the attributes are valued to some degree, with *Price per Application* and *Disinfectant* being the most important. Yet when we examine respondent 396, we see a totally different pattern. Only *Price per Application* has estimated part-worths, indicating that the decision was made solely on this attribute. Respondent 586 placed some importance on *Product Form* and *Number of Applications,* but *Price per Application* still played a dominant role.

As a result, the researcher must determine whether these respondents are retained based on the appropriateness of their preference structures. In this situation, all three respondents will be retained. For respondent 225, the preference structure seems quite reasonable. For the other two respondents, even though their preference structure is highly concentrated in the *Price per Application* attribute, it still represents a reasonable pattern that would reflect the preferences of specific consumers.

Assessing Validation Sample Goodness-of-Fit Levels. In addition, the researcher must also examine the goodness-of-fit levels for the validation sample. Here the focus is on low values of fit, because the relatively few number of profiles makes higher values quite possible along with the reasonable expectation that the estimated model would perfectly fit the validation profiles.

For the validation profiles four respondents (110, 229, 266, and 372) had very low goodness-of-fit values. Thus, in order to maintain the most appropriate characterization of the preference structures of the sample, these four respondents will be candidates for elimination. The final decision will be made after the part-worths are examined for theoretically consistent patterns.

TABLE 6-12 Examining Part-Worth Estimates for Respondents with Extremely High Goodness-of-Fit Values

PART-WORTH ESTIMATES FOR RESPONDENTS 225, 396, AND 586, RESPECTIVELY

Product Form			Number of Applications			Disinfectant		Biodegradable		Price per Application		
Premixed	Concentrate	Powder	50	100	200	Yes	No	No	Yes	$.35	$.49	$.79
−.4444	.2222	.2222	−.7778	−.4444	1.2222	1.2083	−1.2083	−.0417	.0417	1.0556	.3889	−1.4444
.0000	.0000	.0000	.0000	.0000	.0000	.0000	.0000	.0000	.0000	2.6667	.6667	−3.3333
−.1667	.0000	.1667	.1667	.0000	−.1667	.0000	.0000	.0000	.0000	2.1667	.667	−2.8333

Note: The goodness-of-fit values for the estimation sample are .990, 1.000, and .974, respectively. All three respondents have goodness-of-fit values of 1.000 for the validation sample.

Stage 5: Interpreting the Results

The first task is to examine the part-worths and assess whether reversals (violation of monotonic relationships) exist that would cause deletion of any respondents. To assist in this task, the part-worths will be rescaled to provide a measure of comparison. With any reversals identified, the focus shifts to interpreting the part-worth estimates and examining each respondent's importance score for the attributes.

RESCALING   Comparing part-worth estimates both across attributes and between respondents can sometimes be difficult given the nature of the estimated coefficients. They are centered on zero, making a direct comparison difficult without any obvious reference point. One approach to simplifying the interpretation process is rescaling the part-worths to a common standard, which typically involves a two-step process. First, within each attribute, the minimum part-worth is set to zero and the other part-worth(s) are expressed as values above zero (easily done by adding the minimum part-worth to all levels within each attribute). Then, the part-worths are totaled and rescaled proportionately to equal 100 times the number of attributes. This type of rescaling does not affect the relative magnitude of any part-worth, but provides a common scale across all part-worth values for comparison across attributes and respondents.

Table 6-13 presents the rescaling process and results for respondent 107 in the HBAT study. The process described is used with rescaling such that the sum of the part-worths across the five attributes equals to 500. As shown in the table, step 1 restates each part-worth within each attribute as the difference from the lowest level in the attribute. Then the part-worths are totaled and rescaled to equal 500 (100 × 5). When rescaled, the lowest part-worth on each attribute has a value of zero. Other part-worths can now be compared either within or between respondents knowing that they are all on the same scale.

EXAMINING PART-WORTH ESTIMATES   Now that the part-worths are rescaled, the researcher may examine the part-worth estimates for each respondent to understand not only the differences between levels within a factor or across factors, but also between respondents. The profiles created for each respondent based on the part-worths enable the researcher to quickly categorize the preference

TABLE 6-13   Rescaling Part-Worth Estimates for Respondent 107

	Product Form			Number of Applications			Disinfectant		Biodegradable		Price per Application		
Premixed	Concentrate	Powder	50	100	200	Yes	No	No	Yes	$.35	$.49	$.79
Original Part-Worth Estimates													
−.0556	.6111	−.5556	.4444	.6111	−1.0556	−.2083	.2083	.5417	−.5417	1.4444	.9444	−2.3889
Step 1. Restating Part-Worths in Relationship to Minimum Levels Within Each Attribute:[a]													
.5000	1.1667	0.00	1.500	1.6667	0.00	0.00	.4166	1.0834	0.00	3.8333	3.3333	0.00
Step 2. Rescaling the Part-Worth Estimates:[b]													
18.52	43.21	.00	55.56	61.73	.00	.00	15.43	40.13	.00	141.96	123.46	.00

[a]Minimum part-worth on each attribute added to other part-worths of that attribute [e.g., minimum part-worth of product form is .5556, which when added to premixed value (.5556) equals .5000].

[b]Total of restated part-worths is proportionally rescaled to total 500 [e.g., total of restated part-worths is 13.50; thus, premixed part-worth rescaled to 18.52 (500 ÷ 13.50 × 500)].

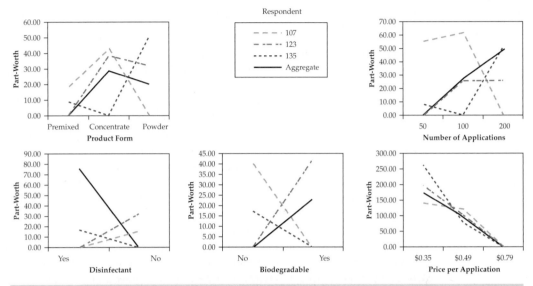

FIGURE 6-5   Part-Worth Estimates for Aggregate Results and Selected Respondents

structure of a respondent, or even sets of respondents. Although more sophisticated techniques could be used, such as cluster analysis (see Chapter 9 for a more detailed discussion), even a visual inspection will identify patterns. If a monotonic relationship is assumed between the levels of an attribute, then the researcher must also identify any reversals (i.e., theoretically inconsistent part-worth patterns) as discussed in the next section.

Figure 6-5 shows the diversity of part-worth estimates across the five attributes for three selected respondents (107, 123, and 135) as well as the aggregate results compiled for all respondents. The aggregate results might be thought of as the average respondent, against which the researcher can view the preference structures of each respondent separately as portrayed by the part-worths to gain unique insights into each individual.

For example, for the attribute *Product Form* the aggregate results indicate that *Concentrate* (part-worth of 28.8) is the most preferred form, followed closely by *Powder* (20.1) and then *Premixed* (0.0).When viewing the three respondents, we can see that respondent 123 has an almost identical pattern, although with slightly higher part-worths for *Concentrate* and *Powder.* For respondent 107, *Concentrate* (43.2) is also the most preferred, but then *Premixed* (18.5) is second most preferred followed by *Powder* (0.0). Respondent 135 has an almost reversed pattern from the aggregate results, with *Powder* (51.7) valued most highly across the entire set of part-worths shown here and *Premixed* (8.6) and *Concentrate* (0.0) valued quite low.

In retrospect, we can see how the aggregate results portray the group overall, but we must also be aware of the differences between respondents. For just these three respondents, we see that two prefer the *Concentrate* over all other forms, yet it is also the lowest valued form for another respondent who values *Powder* most. We can also say that *Premixed* is generally valued low, although it is not the lowest valued level for all respondents as might be surmised if only the aggregate results are viewed.

REVERSALS   A specific form of examining part-worths involves the search for reversals—those patterns of part-worths that are theoretically inconsistent. As noted earlier, some attributes may have implied patterns among the part-worths, typically monotonic relationships that define at least the rank ordering of the levels in terms of preference. For example, in a retail context travel distance should be monotonic, such that stores farther away are preferred less than closer stores.

These relationships are defined by the researcher and then should be reflected in the estimated part-worths.

Identification. The first task is to review all the part-worth patterns and identify any that may reflect reversals. The most direct approach is to examine the differences between adjacent levels that should be monotonically related. For example, if level A is hypothesized to be more preferred than level B, then the difference between the part-worths of level A and level B (i.e., part-worth of level A minus the part-worth of level B) should be positive.

In our example, *Price per Application* was deemed to be monotonic, such that increasing the price per application should decrease preference (and thus estimated part-worths). If we view Figure 6-5 again, we can see that the patterns of part-worths for the aggregate and individual respondents all follow the expected pattern. Although some variability is found at each level, we see that the monotonic pattern (35 cents preferred over 49 cents with 79 cents preferred least) is maintained.

When we scan across the entire set of respondents, however, we do find patterns that seem to indicate a reversal of the monotonic relationship. Figure 6-6 illustrates such patterns as well as an example of the part-worth pattern that follows the monotonic relationship. First, respondent 229 has the expected pattern, with 39 cents the most preferred, then 49 cents, and finally 79 cents. Respondent 382 shows an unexpected pattern between the first two levels (39 cents and 49 cents) where the part-worth actually increases for 49 cents when compared to 35 cents. A second example is the reversal between the levels of 49 cents and 79 cents for respondent 110. Here we find a decrease between 35 cents and 49 cents, but then an increase between 49 cents and 79 cents.

As we look across the entire sample, a number of possible reversals can be identified. Table 6-14 contains all of the part-worth pairs that exhibit part-worth patterns contrary to the monotonic relationship (i.e., the part-worth difference is positive rather than negative or zero). Seven respondents had potential reversals when considering the first two levels (35 cents versus 49 cents), whereas five respondents had potential reversals for the last two levels (49 cents versus 79 cents).

A key question must still be answered: How large does the difference have to be to denote a reversal? Any difference greater than zero would theoretically meet the monotonic relationship.

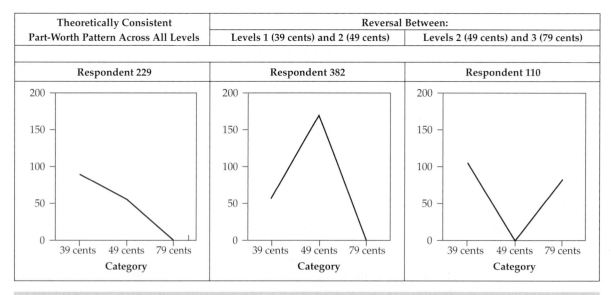

Theoretically Consistent Part-Worth Pattern Across All Levels	Reversal Between:	
	Levels 1 (39 cents) and 2 (49 cents)	Levels 2 (49 cents) and 3 (79 cents)
Respondent 229	Respondent 382	Respondent 110

FIGURE 6-6 Identifying Reversals

TABLE 6-14 Identifying Reversals of the Monotonic Relationship in the Price per Application Attribute

Possible Reversals Between Level 1 (35 cents) and Level 2 (49 cents)		Possible Reversals Between Level 2 (49 cents) and Level 3 (79 cents)	
Respondent	Part-Worth Difference[a]	Respondent	Part-Worth Difference[a]
382	112.68	110	83.33
194	15.87	129	55.56
580	12.82	194	15.87
260	12.66	538	12.82
370	11.90	440	8.77
336	11.49		
514	9.80		

[a] The expected part-worth difference is negative (i.e., a decrease in utility as you go from 35 cents to 49 cents or from 49 cents to 79 cents). Positive values indicate a possible violation of the monotonic relationship.

Subjective and empirical approaches to identifying reversals have been discussed. A researcher should never rely totally on just subjective or empirical approaches, because either approach should act only as a guide to the researcher's judgment in assessing the appropriateness of the part-worths in representing the respondent's preference structure.

Reviewing the potential reversals in Table 6-14, we can see that in each instance one or more respondents have part-worth differences that are substantially higher than the remainder. For example, in the differences between levels 1 and 2, respondent 382 has a difference of 112.68, whereas the next largest difference is 15.87. Likewise, for the differences between levels 2 and 3, respondents 110 and 129 have values much higher (83.33 and 55.56, respectively) than the other respondents. If using a more qualitative approach to examine the distribution of the differences, these three respondents would seem likely to be categorized as having reversals that justify their removal.

A more quantitative approach is to examine statistically the differences. Although no direct statistical test is available, one approach is to calculate the standard error of the differences between levels 1 and 2 and levels 2 and 3 (7.49 and 5.33, respectively) and use them to specify a confidence interval. Using a 99% confidence level, the confidence intervals would be 19.32 (7.49 × 2.58) for the differences between levels 1 and 2 and 13.75 between levels 2 and 3. Applying these results about a difference of zero, we see that the outlying values identified in our visual examination also fall outside the confidence intervals.

Combining these two approaches leads to the identification of three respondents (382, 110, and 129) with reversals in their part-worth estimates. The researcher is now faced with the task of identifying the approach for dealing with these reversals.

Remedies for Reversals and Poor Levels of Goodness-of-Fit. As discussed earlier, the three basic remedies for reversals are to do nothing if reversals are small enough or disaggregate results are the only focus of the analysis, to apply constraints in the estimation process, or to eliminate the respondents. The issue of reversals is distinct, and the ultimate choice for the remedy should be coupled with remedies for respondents with poor levels of estimation or validation fit.

Given the emphasis on the preference structure of respondents, HBAT felt that the only appropriate remedy was elimination of respondents with substantial reversals. Moreover, respondents were also to be eliminated if significantly low levels of estimation or validation fit were found. Three respondents had reversals (110, 129, and 382), and four respondents had low levels of model fit (110, 229, 266, and 372). *Only one respondent failed on both criteria, but all six respondents were eliminated resulting in a sample size of 80 respondents.* Elimination was made to ensure the most representative set of respondents for depicting the preference structures while also maintaining an adequate sample size. *The reduced sample will be used for additional interpretation or further analyses.*

CALCULATING ATTRIBUTE IMPORTANCE   A final approach to examining the preference structure of part-worths is to calculate attribute importance. These values reflect the relative impact each attribute has in the calculation of overall preference (i.e., utility scores). As described earlier, these values are calculated for each respondent and provide another concise basis of comparing the preference structures of respondents.

Table 6-15 compares the derived importance values of each attribute for both the aggregate results and the disaggregate results of three respondents. Although we see a general consistency in the results, each respondent has unique aspects differing from each other and from the aggregate results. The greatest differences are seen for the attribute of *Price per Application,* although substantial variation is also seen on the attributes of *Biodegradability* and *Number of Applications.* Just these limited results show the wide range of part-worth profiles among the respondents and highlight the need for a complete depiction of the preference structures at the disaggregate level as well as the aggregate level.

One extension of conjoint analysis is to define groups of respondents with similar part-worth estimates or importance values of the factors using cluster analysis. These segments may then be profiled and assessed for their unique preference structures and market potential.

Stage 6: Validation of the Results

The final step is to assess the internal and external validity of the conjoint task. As noted earlier, internal validity involves confirmation of the selected composition rule (i.e., additive versus interactive). One approach is to compare alternative models (additive versus interactive) in a pretest study. The second approach is to make sure the levels of model fit are acceptable for each respondent. In general, external validation involves in general the ability of conjoint analysis to predict actual choices and in specific terms the issue of sample representativeness. The validation process with the holdout profiles is the most common approach to assess

TABLE 6-15   Derived Attribute Importance Values for Overall Sample and Three Selected Respondents

	Product Form	Number of Applications	Disinfectant	Biodegradable	Price per Application
Derived Attribute Importance[a]					
Overall Sample[b]					
	15.1	17.6	18.6	9.6	39.1
Select Respondents					
107	14.3	20.4	5.1	13.3	46.9
123	11.4	7.6	9.5	12.4	59.1
135	12.8	12.8	4.2	4.2	66.0

[a] Attribute importance scores sum to 100 across all five attributes for each respondent.
[b] Based on the 80 respondents remaining after elimination of 6 respondents as the remedy for reversals and poor model fit.

external validity; ensuring sample representativeness requires analysis outside the conjoint modeling process.

The high levels of predictive accuracy for both the estimation and holdout profiles across respondents confirm the additive composition rule for this set of respondents. In terms of external validity, the holdout validation process identified four respondents with poor levels of model fit and they were excluded from the analysis. The issue of representativeness of the sample must be addressed based on the research design rather than a specific assessment of the conjoint results. In this situation, HBAT would most likely proceed to a larger-scale project with greater coverage of its customer bases to ensure representativeness. Another consideration is the inclusion of noncustomers, especially if the goal is to understand the entire market, not just HBAT customers.

A Managerial Application: Use of a Choice Simulator

In addition to understanding the aggregate and individual preference structures of the respondents, the part-worth estimates provide a useful approach to representing the preference structure of respondents using other multivariate techniques (e.g., the use of part-worths or attribute importance scores in multiple regression or cluster analysis) or applications. One specific application is the choice simulator, which utilizes the part-worth estimates to make predictions of choice between specified sets of profiles. The respondent can construct a set of profiles to represent any competitive position (i.e., current competitive market or new product entry) and then use the choice simulator to simulate the market and derive market share estimates among the profiles.

The process of running a choice simulation involves three steps: (1) specifying the scenarios, (2) simulating choices, and (3) calculating share of preference. Each of these steps will be discussed in terms of our conjoint example of the industrial cleanser.

STEP 1: SPECIFYING THE SCENARIOS   HBAT also used the conjoint results to simulate choices among three possible products. The products were formulated to identify whether a new value product line might be viable. As such, the new product plus two existing product configurations were developed to represent the existing products. In our example, products 1 and 2 are existing products, whereas product 3 is new:

- **Product 1:** A premixed cleanser in a handy-to-use size (50 applications per container) that is environmentally safe (biodegradable) and still meets all sanitary standards (disinfectant) at only 79 cents per application.
- **Product 2:** An industrial version of product 1 with the same environmental and sanitary features, but in a concentrate form in large containers (200 applications) at the low price of 49 cents per application.
- **Product 3:** A real cleanser value in powder form in economical sizes (200 applications per container) for the lowest feasible price of 35 cents per application.

STEP 2: SIMULATING CHOICES   Once the product configurations were specified, they were submitted to the choice simulator using the results from the remaining 80 respondents. In this process, the part-worths for each respondent were used to calculate the expected utility of each product.

For example, for respondent 107 (see Table 6-10), the utility of product 1 is calculated by taking that respondent's part-worth estimates for the levels of premixed (−.0556), 50 applications per container (.4444), biodegradable (−.5417), disinfectant (−.2083), and 79 cents per application (−2.3889), plus the constant (4.111) for a total utility value of 1.361. Utility values for the other two products were calculated in a similar manner. It should be noted that rescaled utilities could also be used just as easily, because the prediction of choice preferences in the next step focuses on the relative size of the utility values.

Thus, the process derives a set of utility values for each product that is unique to each individual. In this way the preference of each respondent is used to simulate that individual's expected choices when faced with this choice of products. The three products used in the choice simulator are most representative of the differential impact effect among products when similarity among products was minimized.

STEP 3: CALCULATING SHARE OF PREFERENCE The choice simulator then calculated the preference estimates for the products for each respondent. Predictions of the expected market shares were made with two choice models: the maximum utility model and a probabilistic model. The maximum utility model counts the number of times each of the three products had the highest utility across the set of respondents. The probabilistic approach to predicting market shares uses either the BTL or logit model. Both models assess the relative preference of each product and estimate the proportion of times a respondent or the set of respondents will purchase a product.

As seen in Table 6-16, product 1 was preferred (it had the highest predicted preference value) by only 6.88 percent of the respondents. Product 2 was next, preferred by 21.5 percent, and the most preferred was product 3, with 71.88 percent. The fractional percentages are due to tied predictions among products 2 and 3.

As an example of the calculations, the aggregate results can be used. The aggregated predicted preference values for the products were 2.5, 4.9, and 5.9 for products 1, 2, and 3, respectively. The predicted market shares of the aggregate model results using the BTL model are then calculated as follows:

$$\text{Market share}_{\text{product 1}} = 2.5/(2.5 + 4.9 + 5.9) = .188, \text{ or } 18.8\%$$
$$\text{Market share}_{\text{product 2}} = 4.9/(2.5 + 4.9 + 5.9) = .368, \text{ or } 36.8\%$$
$$\text{Market share}_{\text{product 3}} = 5.9/(2.5 + 4.9 + 5.9) = .444, \text{ or } 44.4\%$$

These results are very close to the results derived from using the individual respondent utilities as shown in Table 6-16.

Similar results are obtained using the logit probabilistic model and are shown in Table 6-16 as well. Using the model recommended in situations involving repetitive choices (probability models), as is the case with an industrial cleanser, HBAT has market share estimates indicating an ordering of product 3, product 2, and finally product 1.

It should be remembered that these results represent the entire sample, and the market shares may differ within specific segments of the respondents.

TABLE 6-16 Choice Simulator Results for the Three Product Formulations

| | MARKET SHARE PREDICTIONS | | |
| | | Probabilistic Models | |
Product Formulation	Maximum Utility Model (%)	BTL (%)	Logit (%)
1	6.88	18.00	7.85
2	21.25	36.58	29.09
3	71.88	45.42	63.06

Summary

Conjoint analysis places more emphasis on the ability of the researcher or manager to theorize about the behavior of choice than it does on analytical technique. As such, it should be viewed primarily as exploratory, because many of its results are directly attributable to basic assumptions made during the course of the design and the execution of the study. This chapter helps you to do the following:

Explain the managerial uses of conjoint analysis. Conjoint analysis is a multivariate technique developed specifically to understand how respondents develop preferences for objects (products, services, or ideas). The flexibility of conjoint analysis means it can be used in almost any area in which decisions are studied. Conjoint analysis assumes that any set of objects (e.g., brands, companies) or concepts (e.g., positioning, benefits, images) is evaluated as a bundle of attributes. Having determined the contribution of each factor to the consumer's overall evaluation, the researcher can then (1) define the object or concept with the optimum combination of features, (2) show the relative contributions of each attribute and each level to the overall evaluation of the object, (3) use estimates of purchaser or customer judgments to predict preferences among objects with differing sets of features, (4) isolate groups of potential customers that place differing importance on the features to define high and low potential segments, and (5) identify marketing opportunities by exploring the market potential for combinations of features not currently available. Knowledge of the preference structure for each individual enables almost unlimited flexibility to examine both individual and aggregate reactions to a wide range of product- or service-related issues.

Know the guidelines for selecting the variables to be examined by conjoint analysis. Conjoint analysis employs a variate quite similar in form to what we have seen in other multivariate techniques. The conjoint variate is a linear combination of effects of the independent variables (factors) on a dependent variable. The researcher specifies both the independent variables (factors) and their levels, but the respondent only provides information on the dependent measure. The design of the profiles involves specifying the conjoint variate by selecting the factors and levels to be included in the profiles. When operationalizing factors or levels, the researcher should ensure the measures are both communicable and actionable. Having selected the factors and ensured the measures will be communicable and actionable, the researcher still must address three issues specific to defining factors: the number of factors to

be included, multicollinearity among the factors, and the unique role of price as a factor.

Formulate the experimental plan for a conjoint analysis. For conjoint analysis to explain a respondent's preference structure based only on overall evaluations of a set of profiles, the researcher must make two key decisions regarding the underlying conjoint model: specify the composition rule to be used and select the type of relationships between part-worth estimates. These decisions affect both the design of the profiles and the analysis of respondent evaluations. The composition rule describes how the researcher postulates that the respondent combines the part-worths of the factors to obtain overall worth or utility. It is a critical decision because it defines the basic nature of the preference structure that will be estimated. The most common composition rule is an additive model. The composition rule using interaction effects is similar to the additive form in that it assumes that the consumer sums the part-worths to get an overall total across the set of attributes. It differs in that it allows for certain combinations of levels to be more or less than just their sum. The choice of a composition rule determines the types and number of treatments or profiles the respondent must evaluate, along with the form of estimation method used. The use of one approach over the other has trade-offs. An additive form requires fewer evaluations from the respondent and makes it easier to obtain estimates for the part-worths. However, the interactive form is a more accurate representation because respondents use more complex decision rules in evaluating a product or service.

Understand how to create factorial designs. Having specified the factors and levels, plus the basic model form, the researcher must next make three decisions involving data collection: type of presentation method for the profiles (trade-off, full-profile, or pairwise comparison), type of response variable, and the method of data collection. The overriding objective is to present the attribute combinations (profiles) to respondents in the most realistic and efficient manner possible. In a simple conjoint analysis with a small number of factors and levels, the respondent evaluates all possible profiles in what is known as a *factorial design*. As the number of factors and levels increases, this design becomes impractical. So with the number of choice tasks specified, what is needed is a method for developing a subset of the total profiles that will still provide the information necessary for making accurate and reliable part-worth estimates. The process of selecting a subset of all possible profiles must be done in a manner to preserve the orthogonality (no correlation among levels of an attribute) and balance (each level in a factor appears the same number of times) of the design. A fractional factorial design is the most common method for defining a subset of profiles for evaluation. The process develops a sample of possible profiles, and the number of profiles depends on the type of composition rule assumed to be used by respondents. If the number of factors becomes too large and adaptive conjoint is not acceptable, a bridging design can be employed in which the factors are divided in subsets of appropriate size, with some attributes overlapping between the sets so that each set has a factor(s) in common with other sets of factors. The profiles are then constructed for each subset so that the respondents never see the original number of factors in a single profile.

Explain the impact of choosing rank choice versus ratings as the measure of preference. The measure of preference—rank ordering versus rating (e.g., a 1–10 scale)—also must be selected. Although the trade-off method employs only ranking data, both the pairwise comparison and full-profile methods can evaluate preferences either by obtaining a rating of preference of one profile over the other or just a binary measure of which is preferred. A rank-order preference measure is likely to be more reliable because ranking is easier than rating with a reasonably small number (20 or fewer) of profiles and it provides more flexibility in estimating different types of composition rules. In contrast, rating scales are easily analyzed and administered, even by mail. Still, respondents can be less discriminating in their judgments than when they are rank-ordering. The decision on the type of preference measure to be used must be based on practical as well as conceptual issues. Many researchers favor the rank-order measure because it depicts the underlying choice process inherent in conjoint analysis—choosing among objects. From a practical perspective, however, the effort of ranking large numbers of profiles becomes overwhelming, particularly when the data collection is done in a setting other than personal interview. The ratings measure has the inherent advantage of being easy to administer in any type of data collection context, yet it too has drawbacks. If the respondents are not engaged and involved in the choice task, a ratings measure may provide little differentiation among profiles (e.g., all profiles rated about the same). Moreover, as the choice task becomes more involved with additional profiles, the researcher must be concerned with not only task fatigue, but reliability of the ratings across the profiles.

Assess the relative importance of the predictor variables and each of their levels in affecting consumer judgments. The most common method of interpretation is an examination of the part-worth estimates for each factor in order to determine their magnitude and pattern. Part-worth estimates are typically scaled so the higher the part-worth (either positive or negative) the more impact it has on overall utility. In addition to portraying the impact of each level with the part-worth estimates, conjoint analysis can assess the relative importance of each factor. Because part-worth estimates are typically converted to a common scale, the greatest contribution to overall utility—and hence the most important factor—is the factor with the greatest range (low to high) of part-worths. The importance values of each factor can be converted to percentages summing to 100 percent by dividing each factor's range by the sum of all range values. In evaluating any set of part-worth estimates, the researcher must consider both practical relevance as well as correspondence to any theory-based relationships among levels. In terms of practical relevance, the primary consideration is the degree of differentiation among part-worths within each attribute. Many times an attribute has a theoretically based structure for the relationships between levels. The most common is a monotonic relationship, such that the part-worths of level C should be greater than those of level B, which should in turn be greater than the part-worths of level A. A problem arises when the part-worths do not follow the theorized pattern and violate the assumed monotonic relationship, causing what is referred to as a *reversal*. Reversals can cause serious distortions in the representation of a preference structure.

Apply a choice simulator to conjoint results for the prediction of consumer judgments of new attribute combinations. Conjoint findings reveal the relative importance of the attributes and the impact of specific levels on preference structures. Another primary objective of conjoint analysis is to conduct what-if analyses to predict the share of preferences a profile (real or hypothetical) is likely to capture in various competitive scenarios of interest to management. Choice simulators enable the researcher to simulate any number of competitive scenarios and then estimate how the respondents would react to each scenario. Their real benefit, however, involves the ability of the researcher to specify conditions or relationships among the profiles to more realistically represent market conditions. For example, will all objects compete equally with all others? Does similarity among the objects create differing patterns of preference? Can the unmeasured characteristics of the market be included in the simulation? When using a choice simulator, at least three basic types of effects should be included: (1) differential impact—the impact of any attribute/level is most important when the respondent values that object among the top two objects, indicating its role in actual choice among these objects; (2) differential substitution—the similarity among objects affects choice, with similar objects sharing overall preference (e.g., when choosing whether to ride the bus or take a car, adding buses of differing colors would not increase the chance of taking a bus, but rather the two objects would split the overall chance of taking a bus); and (3) differential enhancement—two highly similar objects of the same basic type can be distinguished by rather small differences on an attribute that is relatively inconsequential when comparing two objects of different types. The final step in conjoint simulation is to predict preference for each individual and then calculate share of preferences for each profile by aggregating the individual choices.

Compare a main effects model and a model with interaction terms and show how to evaluate the validity of one model versus the other. A key benefit of conjoint analysis is the ability to represent many types of relationships in the conjoint variate. A crucial consideration is the type of effects (main effects plus any desired interaction terms) that are to be included, because they require modifications in the research design. Use of interaction terms adds generalizability to the composition rule. The addition of interaction terms does present certain drawbacks in that each interaction term requires an additional part-worth estimate with at least one additional profile for each respondent to evaluate. Unless the researcher knows exactly which interaction terms to estimate, the number of profiles rises dramatically. Moreover, if respondents do not utilize an interactive model, estimating the additional interaction terms in the conjoint variate reduces the statistical efficiency (more part-worth estimates) of the estimation process as well as makes the conjoint task more arduous. Even when used by respondents, interactions predict substantially less variance than the additive effects, most often not exceeding a 5 to 10 percent increase in explained variance. Thus, in many instances, the increased predictive power will be minimal. Interaction terms are most likely to be substantial in cases where the attributes are less tangible, particularly when aesthetic or emotional reactions play a large role. The potential for increased explanation from interaction terms must be balanced with the negative consequences from adding interaction terms.

The interaction term is most effective when the researcher can hypothesize that unexplained portions of utility are associated with only certain levels of an attribute.

Recognize the limitations of traditional conjoint analysis and select the appropriate alternative methodology (e.g., choice-based or adaptive conjoint) when necessary. The full-profile and trade-off methods are unmanageable with more than 10 attributes, yet many conjoint studies need to incorporate 20, 30, or even more attributes. In these cases, some adapted or reduced form of conjoint analysis is used to simplify the data collection effort and still represent a realistic choice decision. The two options include (1) an adaptive/self-explicated conjoint for dealing with a large number of attributes and (2) a choice-based conjoint for providing more realistic choice tasks. In the self-explicated model, the respondent provides a rating of the desirability of each level of an attribute and then rates the relative importance of the attribute overall. With the adaptive (hybrid) model, the self-explicated and part-worth conjoint models are combined. The self-explicated values are used to create a small subset of profiles selected from a fractional factorial design. The profiles are then evaluated in a manner similar to traditional conjoint analysis. The sets of profiles differ among respondents, and although each respondent evaluates only a small number, collectively all profiles are evaluated by a portion of the respondents. To make the conjoint task more realistic, an alternative conjoint methodology, known as *choice-based conjoint,* can be used. It asks the respondent to choose a full profile from a set of alternative profiles known as a *choice set.* This process is much more representative of the actual process of selecting a product from a set of competing products. Moreover, choice-based conjoint provides an option of not choosing any of the presented profiles by including a "No Choice" option in the choice set. Although traditional conjoint assumes respondents' preferences will always be allocated among the set of profiles, the choice-based approach allows for market contraction if all the alternatives in a choice set are unattractive.

To use conjoint analysis the researcher must assess many facets of the decision-making process. Our focus has been on providing a better understanding of the principles of conjoint analysis and how they represent the consumer's choice process. This understanding should enable researchers to avoid misapplication of this relatively new and powerful technique whenever faced with the need to understand choice judgments and preference structures.

Questions

1. Ask three of your classmates to evaluate choice combinations based on the following variables and levels relative to their preferred textbook style for a class and specify the compositional rule you think they will use. Collect information with both the trade-off and full-profile methods.

Factor	Level
Depth	Goes into great depth on each subject
	Introduces each subject in a general overview
Illustrations	Each chapter includes humorous pictures
	Illustrative topics are presented
	Each chapter includes graphics to illustrate the numeric issues
References	General references are included at the end of the textbook
	Each chapter includes specific references for the topics covered

2. How difficult was it for respondents to handle the wordy and slightly abstract concepts they were asked to evaluate? How would you improve on the descriptions of the factors or levels? Which presentation method was easier for the respondents?

3. Using either the simple numerical procedure discussed earlier or a computer program, analyze the data from the experiment in question 1.

4. Design a conjoint analysis experiment with at least four variables and two levels of each variable that is appropriate to a marketing decision. In doing so, define the compositional rule you will use, the experimental design for creating profiles, and the analysis method. Use at least five respondents to support your logic.

5. What are the practical limits of conjoint analysis in terms of variables or types of values for each variable? What types of choice problems are best suited to analysis with conjoint analysis? Which are least well served by conjoint analysis?

6. How would you advise a market researcher to choose among the three types of conjoint methodologies? What are the most important issues to consider, along with each methodology's strengths and weaknesses?

Suggested Readings

A list of suggested readings illustrating issues and applications of multivariate techniques in general is available on the Web at www.pearsonglobaleditions.com/hair or www.mvstats.com.

References

1. Addelman, S. 1962. Orthogonal Main-Effects Plans for Asymmetrical Factorial Experiments. *Technometrics* 4: 21–46.

2. Akaah, I. 1991. Predictive Performance of Self-Explicated, Traditional Conjoint, and Hybrid Conjoint Models under Alternative Data Collection Modes. *Journal of the Academy of Marketing Science* 19: 309–14.

3. Allenby, G. M., N. Arora, and J. L. Ginter. 1995. Incorporating Prior Knowledge into the Analysis of Conjoint Studies. *Journal of Marketing Research* 32 (May): 152–62.

4. Allenby, G. M., N. Arora, and G. L. Ginter. 1998. On the Heterogeneity of Demand. *Journal of Marketing Research* 35 (August): 384–89.

5. Alpert, M. 1971. Definition of Determinant Attributes: A Comparison of Methods. *Journal of Marketing Research* 8(2): 184–91.

6. Andrews, Rick L., Asim Ansari, and Imran S. Currim. 2002. Hierarchical Bayes Versus Finite Mixture Conjoint Analysis Models: A Comparison of Fit, Prediction and Partworth Recovery. *Journal of Marketing Research,* 39 (May): 87–98.

7. Ashok, Kalidas, William R. Dollon, and Sophie Yuan. 2002. Extending Discrete Choice Models to Incorporate Attitudinal and Other Latent Variables. *Journal of Marketing Research* 39 (February): 31–46.

8. Baalbaki, I. B., and N. K. Malhotra. 1995. Standardization Versus Customization in International Marketing: An

Investigation Using Bridging Conjoint Analysis. *Journal of the Academy of Marketing Science* 23(3): 182–94.

9. Bretton-Clark. 1988. *Conjoint Analyzer.* New York: Bretton-Clark.

10. Bretton-Clark. 1988. *Conjoint Designer.* New York: Bretton-Clark.

11. Bretton-Clark. 1988. *Simgraf.* New York: Bretton-Clark.

12. Bretton-Clark. 1988. *Bridger.* New York: Bretton-Clark.

13. Carmone, F. J., Jr., and C. M. Schaffer. 1995. Review of Conjoint Software. *Journal of Marketing Research* 32 (February): 113–20.

14. Carroll, J. D., and P. E. Green. 1995. Psychometric Methods in Marketing Research: Part 1, Conjoint Analysis. *Journal of Marketing Research* 32 (November): 385–91.

15. Chakraborty, Goutam, Dwayne Ball, Gary J. Graeth, and Sunkyu Jun. 2002. The Ability of Ratings and Choice Conjoint to Predict Market Shares: A Monte Carlo Simulation. *Journal of Business Research* 55: 237–49.

16. Chrzan, Keith, and Bryan Orme. 2000. An Overview and Comparison of Design Strategies for Choice-Based Conjoint Analysis. *Sawtooth Software Research Paper Series.* Sequim, WA: Sawtooth Software, Inc.

17. Conner, W. S., and M. Zelen. 1959. *Fractional Factorial Experimental Designs for Factors at Three Levels, Applied Math Series S4.* Washington, DC: National Bureau of Standards.

18. D'Souza, Giles, and Seungoog Weun. 1997. Assessing the Validity of Market Segments Using Conjoint Analysis. *Journal of Managerial Issues* IX (4): 399–418.

19. Darmon, Rene Y., and Dominique Rouzies. 1999. Internal Validity of Conjoint Analysis Under Alternative Measurement Procedures. *Journal of Business Research* 46: 67–81.

20. DeSarbo, Wayne, Venkat Ramaswamy, and Steve H. Cohen. 1995. Market Segmentation with Choice-Based Conjoint Analysis. *Marketing Letters* 6: 137–48.

21. Elrod, T., J. J. Louviere, and K. S. Davey. 1992. An Empirical Comparison of Ratings-Based and Choice-Based Conjoint Models. *Journal of Marketing Research* 29: 368–77.

22. Gelmen, A., J. B. Carlin, H. S. Stern, and D. B. Rubin. 1998. *Bayesian Data Analysis.* Suffolk: Chapman and Hall.

23. Green, P. E. 1984. Hybrid Models for Conjoint Analysis: An Exploratory Review. *Journal of Marketing Research* 21 (May): 155–69.

24. Green, P. E., S. M. Goldberg, and M. Montemayor. 1981. A Hybrid Utility Estimation Model for Conjoint Analysis. *Journal of Marketing* 45 (Winter): 33–41.

25. Green, P. E., and A. M. Kreiger. 1988. Choice Rules and Sensitivity Analysis in Conjoint Simulators. *Journal of the Academy of Marketing Science* 16 (Spring): 114–27.

26. Green, P. E., and A. M. Kreiger. 1991. Segmenting Markets with Conjoint Analysis. *Journal of Marketing* 55 (October): 20–31.

27. Green, P. E., A. M. Kreiger, and M. K. Agarwal. 1991. Adaptive Conjoint Analysis: Some Caveats and Suggestions. *Journal of Marketing Research* 28 (May): 215–22.

28. Green, P. E., and V. Srinivasan. 1978. Conjoint Analysis in Consumer Research: Issues and Outlook. *Journal of Consumer Research* 5 (September): 103–23.

29. Green, P. E., and V. Srinivasan. 1990. Conjoint Analysis in Marketing: New Developments with Implications for Research and Practice. *Journal of Marketing* 54(4): 3–19.

30. Green, P. E., and Y. Wind. 1975. New Way to Measure Consumers' Judgments. *Harvard Business Review* 53 (July–August): 107–17.

31. Gustafsson, Anglers, Andreas Herrmann, and Frank Huber (eds.). 2000. *Conjoint Measurement: Methods and Applications.* Berlin: Springer-Verlag.

32. Haaijer, Rinus, Wagner Kamakura, and Michael Widel. 2001. The "No Choice" Alternative in Conjoint Experiments. *International Journal of Market Research* 43(1): 93–106.

33. Hahn, G. J., and S. S. Shapiro. 1966. *A Catalog and Computer Program for the Design and Analysis of Orthogonal Symmetric and Asymmetric Fractional Factorial Experiments, Report No. 66-C-165.* Schenectady, NY: General Electric Research and Development Center.

34. Holmes, Thomas, Keith Alger, Christian Zinkhan, and Evan Mercer. 1998. The Effect of Response Time on Conjoint Analysis Estimates of Rainforest Protection Values. *Journal of Forest Economics* 4(1): 7–28.

35. Huber, J. 1987. Conjoint Analysis: How We Got Here and Where We Are. In *Proceedings of the Sawtooth Conference on Perceptual Mapping, Conjoint Analysis and Computer Interviewing,* M. Metegrano (ed.). Ketchum, ID: Sawtooth Software, pp. 2–6.

36. Huber, J., and W. Moore. 1979. A Comparison of Alternative Ways to Aggregate Individual Conjoint Analyses. In *Proceedings of the AMA Educator's Conference,* L. Landon (ed.). Chicago: American Marketing Association, pp. 64–68.

37. Huber, Joel, Bryan Orme, and Richard Miller. 1999. Dealing with Product Similarity in Conjoint Simulations. In *Sawtooth Software Conference Proceedings,* M. Metegrano (ed.). Ketchum, ID: Sawtooth Software.

38. Huber, J., D. R. Wittink, J. A. Fielder, and R. L. Miller. 1993. The Effectiveness of Alternative Preference Elicitation Procedures in Predicting Choice. *Journal of Marketing Research* 30 (February): 105–14.

39. Huber, J., D. R. Wittink, R. M. Johnson, and R. Miller. 1992. Learning Effects in Preference Tasks: Choice-Based Versus Standard Conjoint. In *Sawtooth Software Conference Proceedings,* M. Metegrano (ed.). Ketchum, ID: Sawtooth Software, pp. 275–82.

40. Huber, J., and K. Zwerina. 1996. The Importance of Utility Balance in Efficient Choice Designs. *Journal of Marketing Research* 33 (August): 307–17.

41. Intelligent Marketing Systems, Inc. 1993. *CONSURV— Conjoint Analysis Software, Version 3.0.* Edmonton, Alberta: Intelligent Marketing Systems.

42. Intelligent Marketing Systems, Inc. 1993. *NTELOGIT, Version 3.0.* Edmonton, Alberta: Intelligent Marketing Systems.

43. Jaeger, Sara R., Duncan Hedderly, and Halliday J. H. MacFie. 2001. Methodological Issues in Conjoint Analysis: A Case Study. *European Journal of Marketing* 35(11/12): 1217–37.

44. Jedidi, K., R. Kohli, and W. S. DeSarbo. 1996. Consideration Sets in Conjoint Analysis. *Journal of Marketing Research* 33 (August): 364–72.

45. Jedidi, Kamel, and Z. John Zhang. 2002. Augmenting Conjoint Analysis to Estimate Consumer Reservation Price. *Management Science* 48(10): 1350–68.

46. Johnson, R. M. 1975. A Simple Method for Pairwise Monotone Regression. *Psychometrika* 40 (June): 163–68.

47. Johnson, R. M. 1991. Comment on Adaptive Conjoint Analysis: Some Caveats and Suggestions. *Journal of Marketing Research* 28 (May): 223–25.

48. Johnson, Richard M. 1997. Including Holdout Choice Tasks in Conjoint Tasks. *Sawtooth Software Research Paper Series.* Sequim, WA: Sawtooth Software, Inc.

49. Johnson, Richard M. 2000. Monotonicity Constraints in Choice-Based Conjoint with Hierarchical Bayes. *Sawtooth Software Research Paper Series.* Sequim, WA: Sawtooth Software, Inc.

50. Johnson, R. M., and K. A. Olberts. 1991. Using Conjoint Analysis in Pricing Studies: Is One Price Variable Enough? In *Advanced Research Techniques Forum Conference Proceedings.* Beaver Creek, CO: American Marketing Association, pp. 12–18.

51. Johnson, R. M., and B. K. Orme. 1996. How Many Questions Should You Ask in Choice-Based Conjoint Studies? In *Advanced Research Techniques Forum Conference Proceedings.* Beaver Creek, CO: American Marketing Association, pp. 42–49.

52. Krieger, Abba M., Paul E. Green, and U. N. Umesh. 1998. Effect of Level Disaggregation on Conjoint Cross Validations: Some Comparative Findings. *Decision Sciences* 29(4): 1047–58.

53. Kruskal, J. B. 1965. Analysis of Factorial Experiments by Estimating Monotone Transformations of the Data. *Journal of the Royal Statistical Society* B27: 251–63.

54. Kuhfield, Warren F. 1997. Efficient Experimental Designs Using Computerized Searches. *Sawtooth Software Research Paper Series.* Sequim, WA: Sawtooth Software, Inc.

55. Kuhfeld, W. F., R. D. Tobias, and M. Garrath. 1994. Efficient Experimental Designs with Marketing Research Applications. *Journal of Marketing Research* 31 (November): 545–57.

56. Lenk, P. J., W. S. DeSarbo, P. E. Green, and M. R. Young. 1996. Hierarchical Bayes Conjoint Analysis: Recovery of Partworth Heterogeneity from Reduced Experimental Designs. *Marketing Science* 15: 173–91.

57. Loosschilder, G. H., E. Rosbergen, M. Vriens, and D. R. Wittink. 1995. Pictorial Stimuli in Conjoint Analysis to Support Product Styling Decisions. *Journal of the Marketing Research Society* 37(1): 17–34.

58. Louviere, J. J. 1988. *Analyzing Decision Making: Metric Conjoint Analysis.* Sage University Paper Series on Quantitative Applications in the Social Sciences, vol. 67. Beverly Hills, CA: Sage.

59. Louviere, J. J., and G. Woodworth. 1983. Design and Analysis of Simulated Consumer Choice or Allocation Experiments: An Approach Based on Aggregate Data. *Journal of Marketing Research* 20: 350–67.

60. Luce, R. D. 1959. *Individual Choice Behavior: A Theoretical Analysis.* New York: Wiley.

61. Mahajan, V., and J. Wind. 1991. *New Product Models: Practice, Shortcomings and Desired Improvements—Report No. 91–125.* Cambridge, MA: Marketing Science Institute.

62. Marshall, Pablo, and Eric T. Bradlow. 2002. A Unified Approach to Conjoint Analysis Models. *Journal of the American Statistical Association* 97 (September): 674–82.

63. McCullough, Dick. 2002. A User's Guide to Conjoint Analysis. *Marketing Research* 14(2): 19–23.

64. McFadden, D. L. 1974. Conditional Logit Analysis of Qualitative Choice Behavior. In *Frontiers in Econometrics,* P. Zarembka (ed.). New York: Academic Press, pp. 105–42.

65. McLean, R., and V. Anderson. 1984. *Applied Factorial and Fractional Designs.* New York: Marcel Dekker.

66. Molin, Eric J. E., Harmen Oppewal, and Harry J. P. Timmermans. 2001. Analyzing Heterogeneity in Conjoint Estimates of Residential Preferences. *Journal of Housing and the Built Environment* 16: 267–84.

67. Ofek, Elie, and V. Srinivasan. 2002. How Much Does the Market Value an Improvement in a Product Attribute? *Marketing Science* 21(4): 398–411.

68. Oliphant, K., T. C. Eagle, J. J. Louviere, and D. Anderson. 1992. Cross-Task Comparison of Ratings-Based and Choice-Based Conjoint. In *Sawtooth Software Conference Proceedings,* M. Metegrano (ed.). Ketchum, ID: Sawtooth Software, pp. 383–404.

69. Oppewal, H. 1995. A Review of Conjoint Software. *Journal of Retailing and Consumer Services* 2(1): 55–61.

70. Oppewal, H. 1995. A Review of Choice-Based Conjoint Software: CBC and MINT. *Journal of Retailing and Consumer Services* 2(4): 259–64.

71. Orme, Bryan K. 1998. Reducing the Number-of-Attribute-Levels Effect in ACA with Optimal Weighting. *Sawtooth Software Research Paper Series.* Sequim, WA: Sawtooth Software, Inc.

72. Orme, Bryan. 1998. Sample Size Issues for Conjoint Analysis Studies. *Sawtooth Software Research Paper Series.* Sequim, WA: Sawtooth Software, Inc.

73. Orme, Bryan. 1998. Reducing the IIA Problem with a Randomized First Choice Model. *Sawtooth Software*

Research Paper Series. Sequim, WA: Sawtooth Software, Inc.

74. Orme, Bryan. 2003. Which Conjoint Method Should I Choose? *Sawtooth Software Research Paper Series.* Sequim, WA: Sawtooth Software, Inc.

75. Orme, Bryan K., and Gary C. Baker. 2000. Comparing Hierarchical Bayes Draws and Randomized First Choice for Conjoint Simulations. *Sawtooth Software Research Paper Series.* Sequim, WA: Sawtooth Software, Inc.

76. Orme, Bryan K., and Michael A. Heft. 1999. Predicting Actual Sales with CBC: How Capturing Heterogeneity Improves Results. *Sawtooth Software Research Paper Series.* Sequim, WA: Sawtooth Software, Inc.

77. Orme, Bryan K., and Richard Johnson. 1996. Staying Out of Trouble with ACA. *Sawtooth Software Research Paper Series.* Sequim, WA: Sawtooth Software, Inc.

78. Orme, Bryan, and Joel Huber. 2000. Improving the Value of Conjoint Simulations. *Marketing Research* (Winter): 12–20.

79. Orme, Bryan K., and W. Christopher King. 1998. Conducting Full-Profile Conjoint Analysis Over the Internet. *Sawtooth Software Research Paper Series.* Sequim, WA: Sawtooth Software, Inc.

80. Orme, Bryan, and W. Christopher King. 1998. Conducting Full-Profile Conjoint Analysis Over the Internet. *Quirk's Marketing Research Review* (July): #0359.

81. Pinnell, J. 1994. Multistage Conjoint Methods to Measure Price Sensitivity. In *Advanced Research Techniques Forum.* Beaver Creek, CO: American Marketing Association, pp. 65–69.

82. Pullman, Madeline, Kimberly J. Dodson, and William L. Moore, 1999. A Comparison of Conjoint Methods When There Are Many Attributes. *Marketing Letters* 10(2): 1–14.

83. Research Triangle Institute. 1996. *Trade-Off VR.* Research Triangle Park, NC: Research Triangle Institute.

84. Reibstein, D., J.E.G. Bateson, and W. Boulding. 1988. Conjoint Analysis Reliability: Empirical Findings. *Marketing Science* 7 (Summer): 271–86.

85. Sandor, Zsolt, and Michael Wedel. 2002. Profile Construction in Experimental Choice Designs for Mixed Logit Models. *Marketing Science* 21(4): 455–75.

86. SAS Institute, Inc. 1992. *SAS Technical Report R-109: Conjoint Analysis Examples.* Cary, NC: SAS Institute, Inc.

87. Sawtooth Software, Inc. 2002. *ACA 5.0 Technical Paper.* Sequim, WA: Sawtooth Software Inc.

88. Sawtooth Software, Inc. 2001. *Choice-Based Conjoint (CBC) Technical Paper.* Sequim, WA: Sawtooth Software Inc.

89. Sawtooth Software, Inc. 2003. *ACA/Hierarchical Bayes v. 2.0 Technical Paper.* Sequim, WA: Sawtooth Software Inc.

90. Sawtooth Software, Inc. 2003. *Advanced Simulation Module (ASM) for Product Optimization v 1.5 Technical Paper.* Sequim, WA: Sawtooth Software Inc.

91. Sawtooth Software, Inc. 2003. *CBC Hierarchical Bayes Analysis Technical Paper (version 2.0).* Sequim, WA: Sawtooth Software Inc.

92. Sawtooth Software, Inc. 2003. *Conjoint Value Analysis (CVA).* Sequim, WA: Sawtooth Software Inc.

93. Sawtooth Software, Inc. 2003. *HB_Reg v2: Hierarchical Bayes Regression Analysis Technical Paper.* Sequim, WA: Sawtooth Software Inc.

94. Sawtooth Technologies. 1997. *SENSUS, Version 2.0.* Evanston IL: Sawtooth Technologies.

95. Schocker, A. D., and V. Srinivasan. 1977. LINMAP (Version II): A Fortran IV Computer Program for Analyzing Ordinal Preference (Dominance) Judgments Via Linear Programming Techniques for Conjoint Measurement. *Journal of Marketing Research* 14 (February): 101–3.

96. Smith, Scott M. 1989. *PC-MDS: A Multidimensional Statistics Package.* Provo, UT: Brigham Young University Press.

97. SPSS, Inc. 2003. *SPSS Conjoint 12.0.* Chicago: SPSS, Inc.

98. Simmons, Sid, and Mark Esser. 2000. Developing Business Solutions from Conjoint Analysis. In *Conjoint Measurement: Methods and Applications,* Gustafsson, Anglers, Andreas Herrmann, and Frank Huber (eds.). Berlin: Springer-Verlag, pp. 67–96.

99. Srinivasan, V. 1988. A Conjunctive-Compensatory Approach to the Self-Explication of Multiattitudinal Preference. *Decision Sciences* 19 (Spring): 295–305.

100. Srinivasan, V., A. K. Jain, and N. Malhotra. 1983. Improving Predictive Power of Conjoint Analysis by Constrained Parameter Estimation. *Journal of Marketing Research* 20 (November): 433–38.

101. Srinivasan, V., and C. S. Park. 1997. Surprising Robustness of the Self-Explicated Approach to Customer Preference Structure Measurement. *Journal of Marketing Research* 34 (May): 286–91.

102. Steckel, J., W. S. DeSarbo, and V. Mahajan. 1991. On the Creation of Acceptable Conjoint Analysis Experimental Design. *Decision Sciences* 22(2): 435–42.

103. Ter Hofstede, Fenkel, Youngchan Kim, and Michel Wedel. 2002. Bayesian Prediction in Hybrid Conjoint Models. *Journal of Marketing Research* 39 (May): 253–61.

104. Thurstone, L. L. 1927. A Law of Comparative Judgment. *Psychological Review* 34: 273–86.

105. Tumbush, J. J. 1991. Validating Adaptive Conjoint Analysis (ACA) Versus Standard Concept Testing. In *Sawtooth Software Conference Proceedings,* M. Metegrano (ed.). Ketchum, ID: Sawtooth Software, pp. 177–84.

106. van der Lans, I. A., and W. Heiser. 1992. Constrained Part-Worth Estimation in Conjoint Analysis Using the Self-Explicated Utility Model. *International Journal of Research in Marketing* 9: 325–44.

107. van der Lans, I. A., Dick R. Wittink, Joel Huber, and Mareo Vriens. 1992. Within- and Across-Attribute Constraints in ACA and Full-Profile Conjoint Analysis. *Sawtooth Software Research Paper Series.* Sequim, WA: Sawtooth Software, Inc.

108. Vriens, Marco, Harmen Oppewal, and Michel Wedel. 1998. Ratings-Based Versus Choice-Based Latent Class

Conjoint Models: An Empirical Comparison. *Journal of the Marketing Research Society* 40(3): 237–48.

109. Veiens, M., M. Wedel, and T. Wilms. 1996. Metric Conjoint Segmentation Methods: A Monte Carlo Comparison. *Journal of Marketing Research* 33 (February): 73–85.

110. Verlecon, P. W. J., H. N. J. Schifferstein, and Dick R. Wittink. 2002. Range and Number-of-Levels Effects in Derived and Stated Measures of Attribute Importance. *Marketing Letters* 13(1): 41–52.

111. Wedel, Michel, et al. 2002. Discrete and Continuous Representations of Unobserved Heterogeneity in Choice Modeling. *Marketing Letters* 10(3): 219–32.

112. William, Peter, and Dennis Kilroy. 2000. Calibrating Price in ACA: The ACA Price Effect and How to Manage It. *Sawtooth Software Research Paper Series.* Sequim, WA: Sawtooth Software, Inc.

113. Witt, Karlan J., and Steve Bernstein. 1992. Best Practices in Disk-by-Mail Surveys. In *Sawtooth Software Conference Proceedings,* M. Metegrano (ed.). Ketchum, ID: Sawtooth Software.

114. Wittink, D. R., and P. Cattin. 1989. Commercial Use of Conjoint Analysis: An Update. *Journal of Marketing* 53 (July): 91–96.

115. Wittink, Dick R., Joel Huber, John A. Fiedler, and Richard L. Miller. 1991. Attribute Level Effects in Conjoint Revisted: ACA Versus Full-Profile. In *Advanced Research Techniques Forum: Proceedings of the Second Conference,* Rene Mora (ed.). Chicago: American Marketing Association, pp. 51–61.

116. Wittink, D. R., J. Huber, P. Zandan, and R. M. Johnson. 1992. The Number of Levels Effect in Conjoint: Where Does It Come From, and Can It Be Eliminated? In *Sawtooth Software Conference Proceedings,* M. Metegrano (ed.). Ketchum, ID: Sawtooth Software, pp. 355–64.

117. Wittink, D. R., L. Krishnamurthi, and J. B. Nutter. 1982. Comparing Derived Importance Weights Across Attributes. *Journal of Consumer Research* 8 (March): 471–74.

118. Wittink, D. R., L. Krishnamurthi, and D. J. Reibstein. 1990. The Effect of Differences in the Number of Attribute Levels on Conjoint Results. *Marketing Letters* 1(2): 113–29.

119. Wittink, D. R., M. Vriens, and W. Burhenne. 1994. Commercial Use of Conjoint Analysis in Europe: Results and Critical Reflections. *International Journal of Research in Marketing* 11: 41–52.

CHAPTER 7

Multiple Discriminant Analysis and Logistic Regression

LEARNING OBJECTIVES

Upon completing this chapter, you should be able to do the following:

- State the circumstances under which linear discriminant analysis or logistic regression should be used instead of multiple regression.
- Identify the major issues relating to types of variables used and sample size required in the application of discriminant analysis.
- Understand the assumptions underlying discriminant analysis in assessing its appropriateness for a particular problem.
- Describe the two computation approaches for discriminant analysis and the method for assessing overall model fit.
- Explain what a classification matrix is and how to develop one, and describe the ways to evaluate the predictive accuracy of the discriminant function.
- Tell how to identify independent variables with discriminatory power.
- Justify the use of a split-sample approach for validation.
- Understand the strengths and weaknesses of logistic regression compared to discriminant analysis and multiple regression.
- Interpret the results of a logistic regression analysis, with comparisons to both multiple regression and discriminant analysis.

CHAPTER PREVIEW

Multiple regression is undoubtedly the most widely used multivariate dependence technique. The primary basis for the popularity of regression has been its ability to predict and explain metric variables. But what happens when nonmetric dependent variables make multiple regression unsuitable? This chapter introduces two techniques—discriminant analysis and logistic regression—that address the situation of a nonmetric dependent variable. In this type of situation, the researcher is interested in the prediction and explanation of the relationships that affect the category in which an object is located, such as why a person is or is not a customer, or if a firm will succeed or fail. The two major objectives of this chapter are the following:

1. To introduce the underlying nature, philosophy, and conditions of multiple discriminant analysis and logistic regression
2. To demonstrate the application and interpretation of these techniques with an illustrative example

Chapter 1 stated that the basic purpose of discriminant analysis is to estimate the relationship between a single nonmetric (categorical) dependent variable and a set of metric independent variables, in this general form:

$$Y_1 = X_1 + X_2 + X_3 + \cdots + X_n$$
$$\text{(nonmetric)} \qquad \text{(metric)}$$

Multiple discriminant analysis and logistic regression have widespread application in situations in which the primary objective is to identify the group to which an object (e.g., person, firm, or product) belongs. Potential applications include predicting the success or failure of a new product, deciding whether a student should be admitted to graduate school, classifying students as to vocational interests, determining the category of credit risk for a person, or predicting whether a firm will be successful. In each instance, the objects fall into groups, and the objective is to predict and explain the bases for each object's group membership through a set of independent variables selected by the researcher.

KEY TERMS

Before starting the chapter, review the key terms to develop an understanding of the concepts and terminology to be used. Throughout the chapter the key terms appear in **boldface.** Other points of emphasis in the chapter and key term cross-references are *italicized.*

Analysis sample   Group of cases used in estimating the *discriminant function(s)* or the *logistic regression* model. When constructing *classification matrices,* the original sample is divided randomly into two groups, one for model estimation (the analysis sample) and the other for validation (the *holdout sample*).

Box's M   Statistical test for the equality of the covariance matrices of the independent variables across the groups of the dependent variable. If the statistical significance does not exceed the critical level (i.e., nonsignificance), then the equality of the covariance matrices is supported. If the test shows statistical significance, then the groups are deemed different and the assumption is violated.

Categorical variable   See *nonmetric variable.*

Centroid   Mean value for the *discriminant Z scores* of all objects within a particular category or group. For example, a two-group discriminant analysis has two centroids, one for the objects in each of the two groups.

Classification function   Method of classification in which a linear function is defined for each group. Classification is performed by calculating a score for each observation on each group's classification function and then assigning the observation to the group with the highest score. It differs from the calculation of the *discriminant Z score,* which is calculated for each *discriminant function.*

Classification matrix   Means of assessing the predictive ability of the *discriminant function(s)* or *logistic regression* (also called a confusion, assignment, or prediction matrix). Created by cross-tabulating actual group membership with predicted group membership, this matrix consists of numbers on the diagonal representing correct classifications, and off-diagonal numbers representing incorrect classifications.

Cross-validation   Procedure of dividing the sample into two parts: the *analysis sample* used in estimation of the discriminant function(s) or *logistic regression* model, and the *holdout sample* used to validate the results. Cross-validation avoids the overfitting of the discriminant function or logistic regression by allowing its validation on a totally separate sample.

Cutting score   Criterion against which each individual's *discriminant Z score* is compared to determine predicted group membership. When the analysis involves two groups, group prediction is determined by computing a single cutting score. Entities with discriminant Z scores below this score are assigned to one group, whereas those with scores above it are classified in the other group. For three or more groups, multiple discriminant functions are used, with a different cutting score for each function.

Discriminant coefficient   See *discriminant weight.*

Discriminant function   A variate of the independent variables selected for their discriminatory power used in the prediction of group membership. The predicted value of the discriminant function is the *discriminant Z score,* which is calculated for each object (person, firm, or product) in the analysis. It takes the form of the linear equation

$$Z_{jk} = a + W_1 X_{1k} + W_2 X_{2k} + \cdots + W_n X_{nk}$$

where

$$Z_{jk} = \text{discriminant } Z \text{ score of discriminant function } j \text{ for object } k$$
$$a = \text{intercept}$$
$$W_i = \text{discriminant weight for independent variable } i$$
$$X_{ik} = \text{independent variable } i \text{ for object } k$$

Discriminant loadings   Measurement of the simple linear correlation between each independent variable and the *discriminant Z score* for each discriminant function; also called *structure correlations.* Discriminant loadings are calculated whether or not an independent variable is included in the discriminant function(s).

Discriminant weight   Weight whose size relates to the discriminatory power of that independent variable across the groups of the dependent variable. Independent variables with large discriminatory power usually have large weights, and those with little discriminatory power usually have small weights. However, multicollinearity among the independent variables will cause exceptions to this rule. Also called the *discriminant coefficient.*

Discriminant Z score   Score defined by the *discriminant function* for each object in the analysis and usually stated in standardized terms. Also referred to as the *Z score*, it is calculated for each object on each discriminant function and used in conjunction with the *cutting score* to determine predicted group membership. It is different from the *z* score terminology used for standardized variables.

Exponentiated logistic coefficient   Antilog of the *logistic coefficient,* which is used for interpretation purposes in logistic regression. The exponentiated coefficient minus 1.0 equals the percentage change in the *odds.* For example, an exponentiated coefficient of .20 represents a negative 80 percent change in the odds (.20 − 1.0 = −.80) for each unit change in the independent variable (the same as if the odds were multiplied by .20). Thus, a value of 1.0 equates to no change in the odds and values above 1.0 represent increases in the predicted odds.

Fisher's linear discriminant function   See *classification function.*

Hit ratio   Percentage of objects (individuals, respondents, firms, etc.) correctly classified by the *discriminant function.* It is calculated as the number of objects in the diagonal of the *classification matrix* divided by the total number of objects. Also known as the *percentage correctly classified.*

Holdout sample   Group of objects not used to compute the *discriminant function(s)* or *logistic regression* model. This group is then used to validate the discriminant function or logistic regression model with a separate sample of respondents. Also called the *validation sample.*

Likelihood value   Measure used in *logistic regression* to represent the lack of predictive fit. Even though this method does not use the least squares procedure in model estimation, as is done in multiple regression, the likelihood value is similar to the sum of squared error in regression analysis.

Logistic coefficient   Coefficient in the *logistic regression* model that acts as the weighting factor for the independent variables in relation to their discriminatory power. Similar to a regression weight or *discriminant coefficient.*

Logistic curve   An S-shaped curve formed by the *logit transformation* that represents the probability of an event. The S-shaped form is nonlinear because the probability of an event must approach 0 and 1, but never fall outside these limits. Thus, although the midrange involves a linear component, the probabilities as they approach the lower and upper bounds of probability (0 and 1) must flatten out and become asymptotic to these bounds.

Logistic regression   Special form of regression in which the dependent variable is a nonmetric, dichotomous (binary) variable. Although some differences exist, the general manner of interpretation is quite similar to linear regression.

Logit analysis   See *logistic regression.*

Logit transformation   Transformation of the values of the discrete binary dependent variable of *logistic regression* into an S-shaped curve (*logistic curve*) representing the probability of an event. This probability is then used to form the *odds ratio,* which acts as the dependent variable in logistic regression.

Maximum chance criterion   Measure of predictive accuracy in the *classification matrix* that is calculated as the percentage of respondents in the largest group. The rationale is that the best uninformed choice is to classify every observation into the largest group.

Metric variable   Variable with a constant unit of measurement. If a metric variable is scaled from 1 to 9, the difference between 1 and 2 is the same as that between 8 and 9. A more complete discussion of its characteristics and differences from a *nonmetric* or *categorical variable* is found in Chapter 1.

Nonmetric variable   Variable with values that serve merely as a label or means of identification, also referred to as a *categorical,* nominal, binary, qualitative, or taxonomic variable. The number on a football jersey is an example. A more complete discussion of its characteristics and its differences from a *metric variable* is found in Chapter 1.

Odds   The ratio of the probability of an event occurring to the probability of the event not happening, which is used as a measure of the dependent variable in *logistic regression.*

Optimal cutting score   *Discriminant Z score* value that best separates the groups on each discriminant function for classification purposes.

Percentage correctly classified   See *hit ratio.*

Polar extremes approach   Method of constructing a categorical dependent variable from a *metric variable.* First, the metric variable is divided into three categories. Then the extreme categories are used in the discriminant analysis or *logistic regression,* and the middle category is not included in the analysis.

Potency index   Composite measure of the discriminatory power of an independent variable when more than one *discriminant function* is estimated. Based on *discriminant loadings,* it is a relative measure used for comparing the overall discrimination provided by each independent variable across all significant discriminant functions.

Press's *Q* statistic   Measure of the classificatory power of the *discriminant function* when compared with the results expected from a chance model. The calculated value is compared to a critical value based on the chi-square distribution. If the calculated value exceeds the critical value, the classification results are significantly better than would be expected by chance.

Proportional chance criterion   Another criterion for assessing the *hit ratio,* in which the average probability of classification is calculated considering all group sizes.

Pseudo R^2   A value of overall model fit that can be calculated for *logistic regression;* comparable to the R^2 measure used in multiple regression.

Simultaneous estimation   Estimation of the *discriminant function(s)* or the *logistic regression* model in a single step, where weights for all independent variables are calculated simultaneously; contrasts with *stepwise estimation* in which independent variables are entered sequentially according to discriminating power.

Split-sample validation   See *cross-validation.*

Stepwise estimation   Process of estimating the *discriminant function(s)* or *logistic regression* model whereby independent variables are entered sequentially according to the discriminatory power they add to the group membership prediction.

Stretched vector   Scaled vector in which the original vector is scaled to represent the corresponding F ratio. Used to graphically represent the *discriminant loadings* in a combined manner with the group *centroids.*

Structure correlations   See *discriminant loadings.*

Territorial map   Graphical portrayal of the *cutting scores* on a two-dimensional graph. When combined with the plots of individual cases, the dispersion of each group can be viewed and the misclassifications of individual cases identified directly from the map.

Tolerance   Proportion of the variation in the independent variables not explained by the variables already in the model (function). It can be used to protect against multicollinearity. Calculated as $1 - R_i^{2*}$, where R_i^{2*} is the amount of variance of independent variable i explained by all of the other independent variables. A tolerance of 0 means that the independent variable under consideration is a perfect linear combination of independent variables already in the model. A tolerance of 1 means that an independent variable is totally independent of other variables already in the model.

Validation sample   See *holdout sample.*

Variate   Linear combination that represents the weighted sum of two or more independent variables that comprise the *discriminant function.* Also called linear combination or linear compound.

Vector   Representation of the direction and magnitude of a variable's role as portrayed in a graphical interpretation of discriminant analysis results.

Wald statistic   Test used in *logistic regression* for the significance of the *logistic coefficient.* Its interpretation is like the F or t values used for the significance testing of regression coefficients.

Z score   See *discriminant Z score.*

WHAT ARE DISCRIMINANT ANALYSIS AND LOGISTIC REGRESSION?

In attempting to choose an appropriate analytical technique, we sometimes encounter a problem that involves a categorical dependent variable and several metric independent variables. For example, we may wish to distinguish good from bad credit risks. If we had a metric measure of credit risk, then we could use multiple regression. In many instances we don't have the metric measure necessary for multiple regression. Instead, we are only able to ascertain whether someone is in a particular group (e.g., good or bad credit risk).

Discriminant analysis and logistic regression are the appropriate statistical techniques when the dependent variable is a **categorical** (nominal or **nonmetric**) **variable** and the independent variables are **metric variables**. In many cases, the dependent variable consists of two groups or classifications, for example, male versus female or high versus low. In other instances, more than two groups are involved, such as low, medium, and high classifications. Discriminant analysis is capable of handling either two groups or multiple (three or more) groups. When two classifications are involved, the technique is referred to as *two-group discriminant analysis.* When three or more classifications are identified, the technique is referred to as *multiple discriminant analysis (MDA).* **Logistic regression**, also known as **logit analysis**, is limited in its basic form to two groups, although other formulations can handle more groups.

Discriminant Analysis

Discriminant analysis involves deriving a **variate**. The discriminant variate is the linear combination of the two (or more) independent variables that will discriminate best between the objects (persons, firms, etc.) in the groups defined *a priori*. Discrimination is achieved by calculating the variate's weights for each independent variable to maximize the differences between the groups (i.e., the between-group variance relative to the within-group variance). The variate for a discriminant analysis, also known as the **discriminant function**, is derived from an equation much like that seen in multiple regression. It takes the following form:

$$Z_{jk} = a + W_1 X_{1k} + W_2 X_{2k} + \cdots + W_n X_{nk}$$

where

$$
\begin{aligned}
Z_{jk} &= \text{discriminant } Z \text{ score of discriminant function } j \text{ for object } k \\
a &= \text{intercept} \\
W_i &= \text{discriminant weight for independent variable } i \\
X_{ik} &= \text{independent variable } i \text{ for object } k
\end{aligned}
$$

As with the variate in regression or any other multivariate technique we see the discriminant score for each object in the analysis (person, firm, etc.) being a summation of the values obtained by multiplying each independent variable by its discriminant weight. What is unique about discriminant analysis is that more than one discriminant function may be present, resulting in each object possibly having more than one discriminant score. We will discuss what determines the number of discriminant functions later, but here we see that discriminant analysis has both similarities and unique elements when compared to other multivariate techniques.

Discriminant analysis is the appropriate statistical technique for testing the hypothesis that the group means of a set of independent variables for two or more groups are equal. By averaging the discriminant scores for all the individuals within a particular group, we arrive at the group mean. This group mean is referred to as a **centroid**. When the analysis involves two groups, there are two centroids; with three groups, there are three centroids; and so forth. The centroids indicate the most typical location of any member from a particular group, and a comparison of the group centroids shows how far apart the groups are in terms of that discriminant function.

The test for the statistical significance of the discriminant function is a generalized measure of the distance between the group centroids. It is computed by comparing the distributions of the discriminant scores for the groups. If the overlap in the distributions is small, the discriminant function separates the groups well. If the overlap is large, the function is a poor discriminator between the groups. Two distributions of discriminant scores shown in Figure 7-1 further illustrate this concept. The top diagram represents the distributions of discriminant scores for a function that separates the groups well, showing minimal overlap (the shaded area) between the groups. The lower diagram shows the distributions of discriminant scores on a discriminant function that is a relatively poor discriminator between groups A and B. The shaded areas of overlap represent the instances where misclassifying objects from group A into group B, and vice versa, can occur.

Multiple discriminant analysis is unique in one characteristic among the dependence relationships. If the dependent variable consists of more than two groups, discriminant analysis will calculate more than one discriminant function. As a matter of fact, it will calculate $NG - 1$ functions, where NG is the number of groups. Each discriminant function will calculate a separate discriminant Z score. In the case of a three-group dependent variable, each object (respondent, firm, etc.) will have a separate score for discriminant functions one and two, which enables the objects to be plotted in two dimensions, with each dimension representing a discriminant function. Thus, discriminant analysis is not limited to a single variate, as is multiple regression, but creates multiple variates representing dimensions of discrimination among the groups.

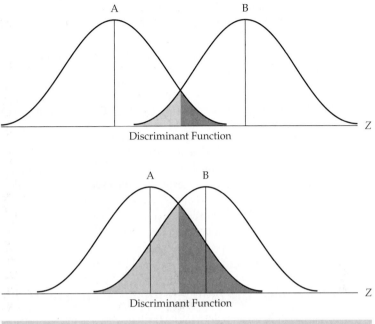

FIGURE 7-1   Univariate Representation of Discriminant *Z* Scores

Logistic Regression

Logistic regression is a specialized form of regression that is formulated to predict and explain a binary (two-group) categorical variable rather than a metric dependent measure. The form of the logistic regression variate is similar to the variate in multiple regression. The variate represents a single multivariate relationship with regression-like coefficients indicating the relative impact of each predictor variable.

The differences between logistic regression and discriminant analysis will become more apparent in our discussion of logistic regression's unique characteristics later in this chapter. Yet many similarities also exist between the two methods. When the basic assumptions of both methods are met, they each give comparable predictive and classificatory results and employ similar diagnostic measures. Logistic regression, however, has the advantage of being less affected than discriminant analysis when the basic assumptions, particularly normality of the variables, are not met. It also can accommodate nonmetric variables through dummy variable coding, just as regression can. Logistic regression is limited, however, to prediction of only a two-group dependent measure. Thus, in cases for which three or more groups form the dependent measure, discriminant analysis is better suited.

ANALOGY WITH REGRESSION AND MANOVA

The application and interpretation of discriminant analysis is much the same as in regression analysis. That is, the discriminant function is a linear combination (variate) of metric measurements for two or more independent variables and is used to describe or predict a single dependent variable. The key difference is that discriminant analysis is appropriate for research problems in which the dependent variable is categorical (nominal or nonmetric), whereas regression is utilized when the dependent variable is metric. As discussed earlier, logistic regression is a variant of regression with many similarities except for the type of dependent variable.

Discriminant analysis is also comparable to "reversing" multivariate analysis of variance (MANOVA), which we discuss in Chapter 8. In discriminant analysis, the single dependent variable is categorical, and the independent variables are metric. The opposite is true of MANOVA, which involves metric dependent variables and categorical independent variable(s). The two techniques both use the same statistical measures of overall model fit as will be seen later in this chapter and Chapter 8.

HYPOTHETICAL EXAMPLE OF DISCRIMINANT ANALYSIS

Discriminant analysis is applicable to any research question with the objective of understanding group membership, whether the groups comprise individuals (e.g., customers versus noncustomers), firms (e.g., profitable versus unprofitable), products (e.g., successful versus unsuccessful), or any other object that can be evaluated on a series of independent variables. To illustrate the basic premises of discriminant analysis, we examine two research settings, one involving two groups (purchasers versus nonpurchasers) and the other three groups (levels of switching behavior). Logistic regression operates in a manner quite comparable to discriminant analysis for two groups. Therefore, we do not specifically illustrate logistic regression here, deferring our discussion until a separate discussion of logistic regression later in this chapter.

A Two-Group Discriminant Analysis: Purchasers Versus Nonpurchasers

Suppose KitchenAid wants to determine whether one of its new products—a new and improved food mixer—will be commercially successful. In carrying out the investigation, KitchenAid is primarily interested in identifying (if possible) those consumers who would purchase the new product versus those who would not. In statistical terminology, KitchenAid would like to minimize the number of errors it would make in predicting which consumers would buy the new food mixer and which would not. To assist in identifying potential purchasers, KitchenAid devised rating scales on three characteristics—durability, performance, and style—to be used by consumers in evaluating the new product. Rather than relying on each scale as a separate measure, KitchenAid hopes that a weighted combination of all three would better predict purchase likelihood of consumers.

The primary objective of discriminant analysis is to develop a weighted combination of the three scales for predicting the likelihood that a consumer will purchase the product. In addition to determining whether consumers who are likely to purchase the new product can be distinguished from those who are not, KitchenAid would also like to know which characteristics of its new product are useful in differentiating likely purchasers from nonpurchasers. That is, evaluations on which of the three characteristics of the new product best separate purchasers from nonpurchasers?

For example, if the response "would purchase" is always associated with a high durability rating and the response "would not purchase" is always associated with a low durability rating, KitchenAid could conclude that the characteristic of durability distinguishes purchasers from nonpurchasers. In contrast, if KitchenAid found that about as many persons with a high rating on style said they would purchase the food mixer as those who said they would not, then style is a characteristic that discriminates poorly between purchasers and nonpurchasers.

IDENTIFYING DISCRIMINATING VARIABLES To identify variables that may be useful in discriminating between groups (i.e., purchasers versus nonpurchasers), emphasis is placed on group differences rather than measures of correlation used in multiple regression.

Table 7-1 lists the ratings of the new mixer on these three characteristics (at a specified price) by a panel of 10 potential purchasers. In rating the food mixer, each panel member is implicitly comparing it with products already on the market. After the product was evaluated, the evaluators were asked to state their buying intentions ("would purchase" or "would not purchase"). Five stated that they would purchase the new mixer and five said they would not.

Examining Table 7-1 identifies several potential discriminating variables. First, a substantial difference separates the mean ratings of X_1 (durability) for the "would purchase" and "would not

TABLE 7-1 KitchenAid Survey Results for the Evaluation of a New Consumer Product

Groups Based on Purchase Intention	Evaluation of New Product*		
	X_1 Durability	X_2 Performance	X_3 Style
Group 1: Would purchase			
Subject 1	8	9	6
Subject 2	6	7	5
Subject 3	10	6	3
Subject 4	9	4	4
Subject 5	4	8	2
Group mean	7.4	6.8	4.0
Group 2: Would not purchase			
Subject 6	5	4	7
Subject 7	3	7	2
Subject 8	4	5	5
Subject 9	2	4	3
Subject 10	2	2	2
Group mean	3.2	4.4	3.8
Difference between group means	4.2	2.4	0.2

*Evaluations are made on a 10-point scale (1 = very poor to 10 = excellent).

purchase" groups (7.4 versus 3.2). As such, durability appears to discriminate well between the two groups and is likely to be an important characteristic to potential purchasers. However, the characteristic of style (X_3) has a much smaller difference of 0.2 between mean ratings (4.0 − 3.8 = 0.2) for the "would purchase" and "would not purchase" groups. Therefore, we would expect this characteristic to be less discriminating in terms of a purchase decision. However, before we can make such statements conclusively, we must examine the distribution of scores for each group. Large standard deviations within one or both groups might make the difference between means nonsignificant and inconsequential in discriminating between the groups.

Because we have only 10 respondents in two groups and three independent variables, we can also look at the data graphically to determine what discriminant analysis is trying to accomplish. Figure 7-2 shows the 10 respondents on each of the three variables. The "would purchase" group is represented by circles and the "would not purchase" group by the squares. Respondent identification numbers are inside the shapes.

- X_1 (Durability) had a substantial difference in mean scores, enabling us to almost perfectly discriminate between the groups using only this variable. If we established the value of 5.5 as our cutoff point to discriminate between the two groups, then we would misclassify only respondent 5, one of the "would purchase" group members. This variable illustrates the discriminatory power in having a large difference in the means for the two groups and a lack of overlap between the distributions of the two groups.
- X_2 (Performance) provides a less clear-cut distinction between the two groups. However, this variable does provide high discrimination for respondent 5, who was misclassified if we used only X_1. In addition, the respondents who would be misclassified using X_2 are well separated on X_1. Thus, X_1 and X_2 might be used quite effectively in combination to predict group membership.
- X_3 (Style) shows little differentiation between the groups. Thus, by forming a variate of only X_1 and X_2, and omitting X_3, a discriminant function may be formed that maximizes the separation of the groups on the discriminant score.

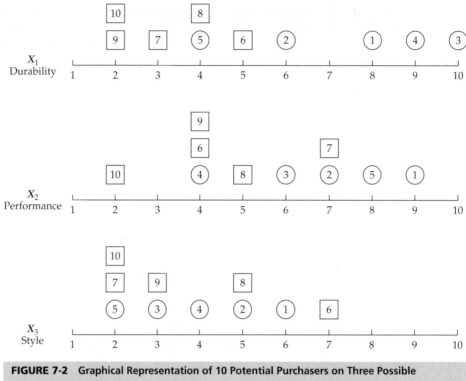

FIGURE 7-2   **Graphical Representation of 10 Potential Purchasers on Three Possible Discriminating Variables**

CALCULATING A DISCRIMINANT FUNCTION   With the three potential discriminating variables identified, attention shifts toward investigation of the possibility of using the discriminating variables in combination to improve upon the discriminating power of any individual variable. To this end, a variate can be formed with two or more discriminating variables to act together in discriminating between the groups.

Table 7-2 contains the results for three different formulations of a discriminant function, each representing different combinations of the three independent variables.

- The first discriminant function contains just X_1, equating the value of X_1 to the discriminant Z score (also implying a weight of 1.0 for X_1 and weights of zero for all other variables). As shown earlier, use of only X_1, the best discriminator, results in the misclassification of subject 5 as shown in Table 7-2, where four out of five subjects in group 1 (all but subject 5) and five of five subjects in group 2 are correctly classified (i.e., lie on the diagonal of the classification matrix). The percentage correctly classified is thus 90 percent (9 out of 10 subjects).

- Because X_2 provides discrimination for subject 5, we can form a second discriminant function by equally combining X_1 and X_2 (i.e., implying weights of 1.0 for X_1 and X_2 and a weight of 0.0 for X_3) to utilize each variable's unique discriminatory powers. Using a cutting score of 11 with this new discriminant function (see Table 7-2) achieves a perfect classification of the two groups (100% correctly classified). Thus, X_1 and X_2 in combination are able to make better predictions of group membership than either variable separately.

- The third discriminant function in Table 7-2 represents the actual estimated discriminant function ($Z = -4.53 + .476X_1 + .359X_2$). Based on a cutting score of 0, this third function also achieves a 100 percent correct classification rate with the maximum separation possible between groups.

TABLE 7-2   Creating Discriminant Functions to Predict Purchasers Versus Nonpurchasers

	Calculated Discriminant Z Scores		
Group	**Function 1:** $Z = X_1$	**Function 2:** $Z = X_1 + X_2$	**Function 3:** $Z = -4.53 + .476X_1 + .359X_2$
Group 1: Would purchase			
Subject 1	8	17	2.51
Subject 2	6	13	.84
Subject 3	10	16	2.38
Subject 4	9	13	1.19
Subject 5	4	12	.25
Group 2: Would not purchase			
Subject 6	5	9	−.71
Subject 7	3	10	−.59
Subject 8	4	9	−.83
Subject 9	2	6	−2.14
Subject 10	2	4	−2.86
Cutting score	5.5	11	0.0

Classification Accuracy:

	Predicted Group		Predicted Group		Predicted Group	
Actual Group	1	2	1	2	1	2
1: Would purchase	4	1	5	0	5	0
2: Would not purchase	0	5	0	5	0	5

As seen in this simple example, discriminant analysis identifies the variables with the greatest differences between the groups and derives a discriminant coefficient that weights each variable to reflect these differences. The result is a discriminant function that best discriminates between the groups based on a combination of the independent variables.

A Geometric Representation of the Two-Group Discriminant Function

A graphical illustration of another two-group analysis will help to further explain the nature of discriminant analysis [7]. Figure 7-3 demonstrates what happens when a two-group discriminant function is computed. Assume we have two groups, A and B, and two measurements, V_1 and V_2, on each member of the two groups. We can plot in a scatter diagram of the association of variable V_1 with variable V_2 for each member of the two groups. In Figure 7-3 the small dots represent the variable measurements for the members of group B and the large dots those for group A. The ellipses drawn around the large and small dots would enclose some prespecified proportion of the points, usually 95 percent or more in each group. If we draw a straight line through the two points at which the ellipses intersect and then project the line to a new Z axis, we can say that the overlap between the univariate distributions A′ and B′ (represented by the shaded area) is smaller than would be obtained by any other line drawn through the ellipses formed by the scatterplots [7].

The important thing to note about Figure 7-3 is that the Z axis expresses the two-variable profiles of groups A and B as single numbers (discriminant scores). By finding a linear combination of the original variables V_1 and V_2, we can project the results as a discriminant function. For example, if the large and small dots are projected onto the new Z axis as discriminant Z scores, the result condenses the information about group differences (shown in the V_1V_2 plot) into a set of points (Z scores) on a single axis, shown by distributions A′ and B′.

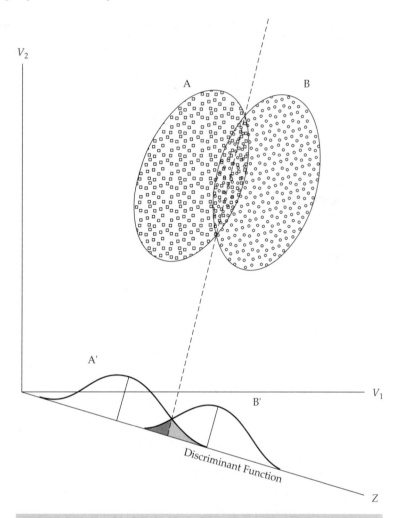

FIGURE 7-3 Graphical Illustration of Two-Group Discriminant Analysis

To summarize, for a given discriminant analysis problem, a linear combination of the independent variables is derived, resulting in a series of discriminant scores for each object in each group. The discriminant scores are computed according to the statistical rule of maximizing the variance between the groups and minimizing the variance within them. If the variance between the groups is large relative to the variance within the groups, we say that the discriminant function separates the groups well.

A Three-Group Example of Discriminant Analysis: Switching Intentions

The two-group example just examined demonstrates the rationale and benefit of combining independent variables into a variate for purposes of discriminating between groups. Discriminant analysis also has another means of discrimination—the estimation and use of multiple variates—in instances of three or more groups. These discriminant functions now become dimensions of discrimination, each dimension separate and distinct from the other. Thus, in addition to improving the explanation of group membership, these additional discriminant functions add insight into the various combinations of independent variables that discriminate between groups.

As an illustration of a three-group application of discriminant analysis, we examine research conducted by HBAT concerning the possibility of a competitor's customers switching suppliers. A small-scale pretest involved interviews of 15 customers of a major competitor. In the course of the interviews, the customers were asked their probability of switching suppliers on a three-category scale. The three possible responses were "definitely switch," "undecided," and "definitely not switch." Customers were assigned to groups 1, 2, or 3, respectively, according to their responses. The customers also rated the competitor on two characteristics: price competitiveness (X_1) and service level (X_2). The research issue is now to determine whether the customers' ratings of the competitor can predict their probability of switching suppliers. Because the dependent variable of switching suppliers was measured as a categorical (nonmetric) variable and the ratings of price and service are metric, discriminant analysis is appropriate.

IDENTIFYING DISCRIMINATING VARIABLES   With three categories of the dependent variable, discriminant analysis can estimate two discriminant functions, each representing a different dimension of discrimination.

Table 7-3 contains the survey results for the 15 customers, 5 in each category of the dependent variable. As we did in the two-group example, we can look at the mean scores for each group to see whether one of the variables discriminates well among all the groups. For X_1, price competitiveness, we see a rather large mean difference between group 1 and group 2 or 3 (2.0 versus 4.6 or 3.8). X_1 may discriminate well between group 1 and group 2 or 3, but is much less effective in discriminating between groups 2 and 3. For X_2, service level, we see that the difference between groups 1 and 2 is very small (2.0 versus 2.2), whereas a large difference exists between group 3 and group 1 or 2 (6.2 versus 2.0 or 2.2). Thus, X_1 distinguishes group 1 from groups 2 and 3, and X_2 distinguishes group 3 from groups 1 and 2. As a result, we see that X_1 and X_2 provide different *dimensions* of discrimination between the groups.

CALCULATING TWO DISCRIMINANT FUNCTIONS   With the potential discriminating variables identified, the next step is to combine them into discriminant functions that will utilize their combined discriminating power for distinguishing between groups.

TABLE 7-3   HBAT Survey Results of Switching Intentions by Potential Customers

Groups Based on Switching Intention	Evaluation of Current Supplier*	
	X_1 Price Competitiveness	X_2 Service Level
Group 1: Definitely switch		
Subject 1	2	2
Subject 2	1	2
Subject 3	3	2
Subject 4	2	1
Subject 5	2	3
Group mean	2.0	2.0
Group 2: Undecided		
Subject 6	4	2
Subject 7	4	3
Subject 8	5	1
Subject 9	5	2
Subject 10	5	3
Group mean	4.6	2.2
Group 3: Definitely not switch		
Subject 11	2	6
Subject 12	3	6
Subject 13	4	6
Subject 14	5	6
Subject 15	5	7
Group mean	3.8	6.2

*Evaluations are made on a 10-point scale (1 = very poor to 10 = excellent).

To illustrate these dimensions graphically, Figure 7-4 portrays the three groups on each of the independent variables separately. Viewing the group members on any one variable, we can see that no variable discriminates well among all the groups. However, if we construct two simple discriminant functions, using just simple weights of 0.0 or 1.0, the results become much clearer. Discriminant function 1 gives X_1 a weight of 1.0, and X_2 a weight of 0.0. Likewise, discriminant function 2 gives X_2 a weight of 1.0, and X_1 a weight of 0.0. The functions can be stated mathematically as

$$\text{Discriminant function 1} = 1.0(X_1) + 0.0(X_2)$$
$$\text{Discriminant function 2} = 0.0(X_1) + 1.0(X_2)$$

These equations show in simple terms how the discriminant analysis procedure estimates weights to maximize discrimination.

With the two functions, we can now calculate two discriminant scores for each respondent. Moreover, the two discriminant functions provide the dimensions of discrimination.

Figure 7-4 also contains a plot of each respondent in a two-dimensional representation. The separation between groups now becomes quite apparent, and each group can be easily distinguished. We can establish values on each dimension that will define regions containing each group (e.g., all members of group 1 are in the region less than 3.5 on dimension 1 and less than 4.5 on dimension 2). Each of the other groups can be similarly defined in terms of the ranges of their discriminant function scores.

In terms of dimensions of discrimination, the first discriminant function, price competitiveness, distinguishes between undecided customers (shown with a square) and those customers who have decided to switch (circles). But price competitiveness does not distinguish those who have decided not to switch (diamonds). Instead, the perception of service level, defining the second discriminant function, predicts whether a customer will decide not to switch versus whether a customer is undecided or determined to switch suppliers. The researcher can present to management the separate impacts of both price competitiveness and service level in making this decision.

The estimation of more than one discriminant function, when possible, provides the researcher with both improved discrimination and additional perspectives on the features and the combinations that best discriminate among the groups. The following sections detail the necessary steps for performing a discriminant analysis, assessing its level of predictive fit, and then interpreting the influence of independent variables in making that prediction.

THE DECISION PROCESS FOR DISCRIMINANT ANALYSIS

The application of discriminant analysis can be viewed from the six-stage model-building perspective introduced in Chapter 1 and portrayed in Figure 7-5 (stages 1–3) and Figure 7-6 (stages 4–6). As with all multivariate applications, setting the objectives is the first step in the analysis. Then the researcher must address specific design issues and make sure the underlying assumptions are met. The analysis proceeds with the derivation of the discriminant function and the determination of whether a statistically significant function can be derived to separate the two (or more) groups. The discriminant results are then assessed for predictive accuracy by developing a classification matrix. Next, interpretation of the discriminant function determines which of the independent variables contributes the most to discriminating between the groups. Finally, the discriminant function should be validated with a holdout sample. Each of these stages is discussed in the following sections. We discuss logistic regression in a separate section after examining the decision process for discriminant analysis. In this way, the similarities and differences between these two techniques can be highlighted.

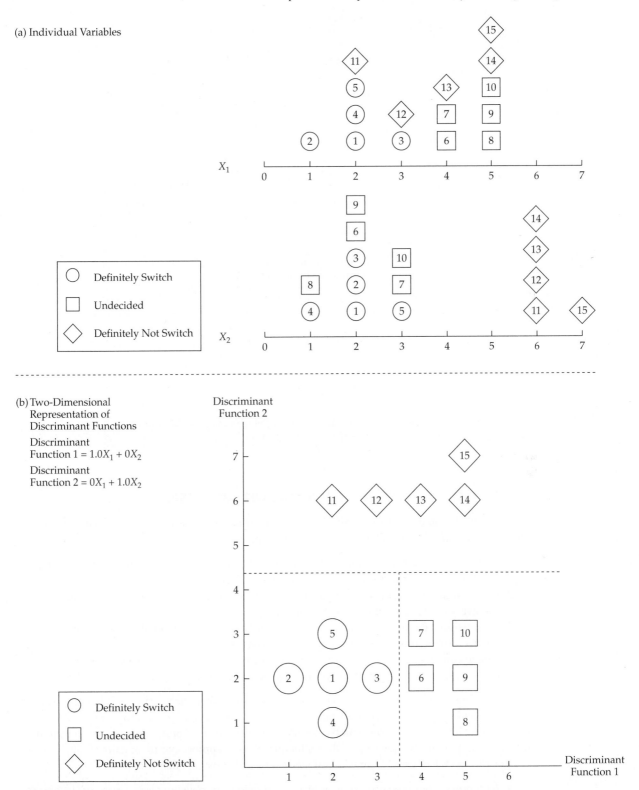

FIGURE 7-4 Graphical Representation of Potential Discriminating Variables for a Three-Group Discriminant Analysis

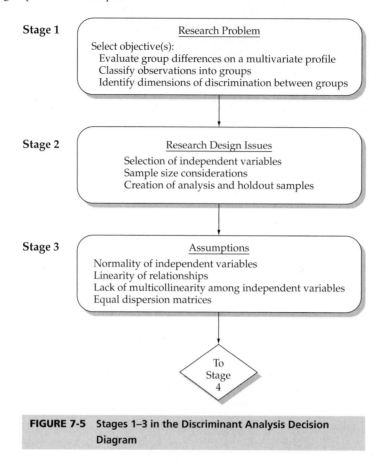

Stage 1

Research Problem

Select objective(s):
    Evaluate group differences on a multivariate profile
    Classify observations into groups
    Identify dimensions of discrimination between groups

Stage 2

Research Design Issues

    Selection of independent variables
    Sample size considerations
    Creation of analysis and holdout samples

Stage 3

Assumptions

    Normality of independent variables
    Linearity of relationships
    Lack of multicollinearity among independent variables
    Equal dispersion matrices

To
Stage
4

FIGURE 7-5   Stages 1–3 in the Discriminant Analysis Decision Diagram

STAGE 1: OBJECTIVES OF DISCRIMINANT ANALYSIS

A review of the objectives for applying discriminant analysis should further clarify its nature. Discriminant analysis can address any of the following research objectives:

1. Determining whether statistically significant differences exist between the average score profiles on a set of variables for two (or more) *a priori* defined groups
2. Determining which of the independent variables most account for the differences in the average score profiles of the two or more groups
3. Establishing the number and composition of the dimensions of discrimination between groups formed from the set of independent variables
4. Establishing procedures for classifying objects (individuals, firms, products, etc.) into groups on the basis of their scores on a set of independent variables

As noted in these objectives, discriminant analysis is useful when the researcher is interested either in understanding group differences or in correctly classifying objects into groups or classes. Discriminant analysis, therefore, can be considered either a type of profile analysis or an analytical predictive technique. In either case, the technique is most appropriate in situations with a single categorical dependent variable and several metrically scaled independent variables.

- As a *profile analysis,* discriminant analysis provides an objective assessment of differences between groups on a set of independent variables. In this situation, discriminant analysis is quite similar to multivariate analysis of variance (see Chapter 8 for a more detailed discussion of multivariate analysis of variance). For understanding group differences, discriminant analysis lends insight into the role of individual variables as well as defining combinations of these variables that represent dimensions of discrimination between groups. These dimensions are the collective effects of several variables that work jointly to distinguish between the groups. The use of sequential estimation methods also allows for identifying subsets of variables with the greatest discriminatory power.

• For *classification purposes,* discriminant analysis provides a basis for classifying not only the sample used to estimate the discriminant function but also any other observations that can have values for all the independent variables. In this way, the discriminant analysis can be used to classify other observations into the defined groups.

STAGE 2: RESEARCH DESIGN FOR DISCRIMINANT ANALYSIS

Successful application of discriminant analysis requires consideration of several issues. These issues include the selection of both dependent and independent variables, the sample size needed for estimation of the discriminant functions, and the division of the sample for validation purposes.

Selecting Dependent and Independent Variables

To apply discriminant analysis, the researcher first must specify which variables are to be independent measures and which variable is to be the dependent measure. Recall that the dependent variable is nonmetric and the independent variables are metric.

THE DEPENDENT VARIABLE   The researcher should focus on the dependent variable first. The number of dependent variable groups (categories) can be two or more, but these groups must be mutually exclusive and exhaustive. In other words, each observation can be placed into only one group. In some cases, the dependent variable may involve two groups (dichotomous), such as good versus bad. In other cases, the dependent variable may involve several groups (multichotomous), such as the occupations of physician, attorney, or professor.

How Many Categories in the Dependent Variable?   Theoretically, discriminant analysis can handle an unlimited number of categories in the dependent variable. As a practical matter, however, the researcher should select a dependent variable and the number of categories based on several considerations:

1. In addition to being mutually exclusive and exhaustive, the dependent variable categories should be distinct and unique on the set of independent variables chosen. Discriminant analysis assumes that each group *should* have a unique profile on the independent variables used and thus develops the discriminant functions to maximally separate the groups based on these variables. Discriminant analysis does not, however, have a means of accommodating or combining categories that are not distinct on the independent variables. If two or more groups have quite similar profiles, discriminant analysis will not be able to uniquely profile each group, resulting in poorer explanation and classification of the groups as a whole. As such, the researcher must select the dependent variables and its categories to reflect differences in the independent variables. An example will help illustrate this issue.

Assume the researcher wishes to identify differences among occupational categories based on a number of demographic characteristics (e.g., income, education, household characteristics). If occupations are represented by a small number of categories (e.g., blue-collar, white-collar, clerical/staff, and professional/upper management), then we would expect unique differences between the groups and that discriminant analysis would be best able to develop discriminant functions that would explain the group differences and successfully classify individuals into their correct category.

If, however, the number of occupational categories was expanded, discriminant analysis may have a harder time identifying differences. For example, assume the professional/upper management category was expanded to the categories of doctors, lawyers, upper management, college professors, and so on. Although this expansion provides a more refined occupational classification, it would be much harder to distinguish between each of these categories on the demographic variables. The results would be poorer performance by discriminant analysis in both explanation and classification.

2. The researcher should also strive, all other things equal, for a smaller rather than larger number of categories in the dependent measure. It may seem more logical to expand the number of categories in search of more unique groupings, but expanding the number of categories presents more complexities in the profiling and classification tasks of discriminant analysis. If discriminant analysis can estimate up to $NG - 1$ (number of groups minus one) discriminant functions, then increasing the number of groups expands the number of possible discriminant functions, increasing the complexity in identifying the underlying dimensions of discrimination reflected by each discriminant function as well as representing the overall effect of each independent variable.

As these two issues suggest, the researcher must always balance the desire to expand the categories for increased uniqueness versus the increased effectiveness in a smaller number of categories. The researcher should try and select a dependent variable with categories that have the maximum differences among all groups while maintaining both conceptual support and managerial relevance.

Converting Metric Variables. The preceding examples of categorical variables were true dichotomies (or multichotomies). In some situations, however, discriminant analysis is appropriate even if the dependent variable is not a true nonmetric (categorical) variable. We may have a dependent variable that is an ordinal or interval measurement that we wish to use as a categorical dependent variable. In such cases, we would have to create a categorical variable, and two approaches are the most commonly used:

- The most common approach is to establish categories using the metric scale. For example, if we had a variable that measured the average number of cola drinks consumed per day, and the individuals responded on a scale from zero to eight or more per day, we could create an artificial trichotomy (three groups) by simply designating those individuals who consumed none, one, or two cola drinks per day as light users, those who consumed three, four, or five per day as medium users, and those who consumed six, seven, eight, or more as heavy users. Such a procedure would create a three-group categorical variable in which the objective would be to discriminate among light, medium, and heavy users of colas. Any number of categorical groups can be developed. Most frequently, the approach would involve creating two, three, or four categories. A larger number of categories could be established if the need arose.

- When three or more categories are created, the possibility arises of examining only the extreme groups in a two-group discriminant analysis. The **polar extremes approach** involves comparing only the extreme two groups and excluding the middle group from the discriminant analysis. For example, the researcher could examine the light and heavy users of cola drinks and exclude the medium users. This approach can be used any time the researcher wishes to examine only the extreme groups. However, the researcher may also want to try this approach when the results of a regression analysis are not as good as anticipated. Such a procedure may be helpful because it is possible that group differences may appear even though regression results are poor. That is, the polar extremes approach with discriminant analysis can reveal differences that are not as prominent in a regression analysis of the full data set [7].

Such manipulation of the data naturally would necessitate caution in interpreting one's findings.

THE INDEPENDENT VARIABLES   After a decision has been made on the dependent variable, the researcher must decide which independent variables to include in the analysis. Independent variables usually are selected in two ways. The first approach involves identifying variables either from previous research or from the theoretical model that is the underlying basis of the research question. The second approach is intuition—utilizing the researcher's knowledge and intuitively selecting variables for which no previous research or theory exists but that logically might be related to predicting the groups for the dependent variable.

In both instances, the most appropriate independent variables are those that differ across at least two of the groups of the dependent variable. Remember that the purpose of any independent variable is to present a unique profile of at least one group as compared to others. Variables that do not differ across the groups are of little use in discriminant analysis.

Sample Size

Discriminant analysis, like the other multivariate techniques, is affected by the size of the sample being analyzed. As discussed in Chapter 1, very small samples have so much sampling error that identification of all but the largest differences is improbable. Moreover, very large sample sizes will make all differences statistically significant, even though these same differences may have little or no managerial relevance. In between these extremes, the researcher must consider the impact of sample sizes on discriminant analysis, both at the overall level and on a group-by-group basis.

OVERALL SAMPLE SIZE   The first consideration involves the overall sample size. Discriminant analysis is quite sensitive to the ratio of sample size to the number of predictor variables. As a result, many studies suggest a ratio of 20 observations for each predictor variable. Although this ratio may be difficult to maintain in practice, the researcher must note that the results become unstable as the sample size decreases relative to the number of independent variables. The minimum size recommended is five observations per independent variable. Note that this ratio applies to all variables considered in the analysis, even if all of the variables considered are not entered into the discriminant function (such as in stepwise estimation).

SAMPLE SIZE PER CATEGORY   In addition to the overall sample size, the researcher also must consider the sample size of each category. At a minimum, the smallest group size of a category must exceed the number of independent variables. As a practical guideline, each category should have at least 20 observations. Even when all categories exceed 20 observations, however, the researcher must also consider the relative sizes of the categories. Wide variations in the groups' size will impact the estimation of the discriminant function and the classification of observations. In the classification stage, larger groups have a disproportionately higher chance of classification. If the group sizes do vary markedly, the researcher may wish to randomly sample from the larger group(s), thereby reducing their size to a level comparable to the smaller group(s). Always remember, however, to maintain an adequate sample size both overall and for each group.

Division of the Sample

One final note about the impact of sample size in discriminant analysis. As will be discussed later in stage 6, the preferred means of validating a discriminant analysis is to divide the sample into two subsamples, one used for estimation of the discriminant function and another for validation purposes. In terms of sample size considerations, it is essential that each subsample be of adequate size to support conclusions from the results. As such, all of the considerations discussed in the previous section apply not only to the total sample, but also to each of the two subsamples (especially the sub-sample used for estimation). No hard-and-fast rules have been established, but it seems logical that the researcher would want at least 100 in the total sample to justify dividing it into the two groups.

CREATING THE SUBSAMPLES   A number of procedures have been suggested for dividing the sample into subsamples. The usual procedure is to divide the total sample of respondents randomly into two subsamples. One of these subsamples, the **analysis sample**, is used to develop the discriminant function. The second, the **holdout sample**, is used to test the discriminant function. This method of validating the function is referred to as the **split-sample validation** or **cross-validation** [1, 5, 9, 18].

No definite guidelines have been established for determining the relative sizes of the analysis and holdout (or validation) subsamples. The most popular approach is to divide the total sample so that one-half of the respondents are placed in the analysis sample and the other half are placed in the holdout sample. However, no hard-and-fast rule has been established, and some researchers prefer a 60–40 or even 75–25 split between the analysis and the holdout groups, depending on the overall sample size.

When selecting the analysis and holdout samples, one usually follows a proportionately stratified sampling procedure. Assume first that the researcher desired a 50–50 split. If the categorical groups for the discriminant analysis are equally represented in the total sample, then the estimation and holdout samples should be of approximately equal size. If the original groups are unequal, the sizes of the estimation and holdout samples should be proportionate to the total sample distribution. For instance, if a sample consists of 50 males and 50 females, the estimation and holdout samples would have 25 males and 25 females. If the sample contained 70 females and 30 males, then the estimation and holdout samples would consist of 35 females and 15 males each.

WHAT IF THE OVERALL SAMPLE IS TOO SMALL?   If the sample size is too small to justify a division into analysis and holdout groups, the researcher has two options. First, develop the function on the entire sample and then use the function to classify the same group used to develop the function. This procedure results in an upward bias in the predictive accuracy of the function, but is certainly better than not testing the function at all. Second, several techniques discussed in stage 6 can perform a type of holdout procedure in which the discriminant function is repeatedly estimated on the sample, each time "holding out" a different observation. In this approach, much smaller samples sizes can be used because the overall sample need not be divided into subsamples.

STAGE 3: ASSUMPTIONS OF DISCRIMINANT ANALYSIS

As with all multivariate techniques, discriminant analysis is based on a number of assumptions. These assumptions relate to both the statistical processes involved in the estimation and classification procedures and issues affecting the interpretation of the results. The following section discusses each type of assumption and the impacts on the proper application of discriminant analysis.

Impacts on Estimation and Classification

The key assumptions for deriving the discriminant function are multivariate normality of the independent variables and unknown (but equal) dispersion and covariance structures (matrices) for the groups as defined by the dependent variable [8, 10]. Although mixed evidence exists concerning the sensitivity of discriminant analysis to violations of these assumptions, the researcher must understand the impacts on the results that can be expected. Moreover, if the assumptions are violated and the potential remedies are not acceptable or do not address the severity of the problem, the researcher should consider alternative methods (e.g., logistic regression, which is described in the next section).

IDENTIFYING ASSUMPTION VIOLATIONS   As discussed in Chapter 2, achieving univariate normality of individual variables will many times suffice to achieve multivariate normality. A number of tests for normality discussed in Chapter 2 are available to the researcher, along with the appropriate remedies, those most often being transformations of the variables.

The issue of equal dispersion of the independent variables (i.e., equivalent covariance matrices) is similar to homoscedasticity between individual variables (also discussed in Chapter 2). The most common test is the **Box's M** test assessing the significance of differences in the matrices between the groups. Here the researcher is looking for a *nonsignificant* probability level, which would indicate that there were not differences between the group covariance matrices. Given the sensitivity of the Box's M test, however, to the size of the covariance matrices and the number of groups in the analysis, researchers should use very conservative levels of significant differences (e.g., .01 rather than .05) when assessing whether differences are present. And as the research design increases in sample size or terms of groups or number of independent variables, even more conservative levels of significance may be considered acceptable. The reader is also referred to the discussion of this issue in Chapter 8 where the implications in MANOVA are discussed.

IMPACT ON ESTIMATION   Data not meeting the multivariate normality assumption can cause problems in the estimation of the discriminant function. Remedies may be possible through transformations of the data to reduce the disparities among the covariance matrices. However, in many instances these remedies are ineffectual. In these situations, the models should be thoroughly validated. If the dependent measure is binary, logistic regression should be used if at all possible.

IMPACT ON CLASSIFICATION   Unequal covariance matrices also negatively affect the classification process. If the sample sizes are small and the covariance matrices are unequal, then the statistical significance of the estimation process is adversely affected. The more likely case is that of unequal covariances among groups of adequate sample size, whereby observations are overclassified into the groups with larger covariance matrices. This effect can be minimized by increasing the sample size and also by using the group-specific covariance matrices for classification purposes, but this approach mandates cross-validation of the discriminant results. Finally, quadratic classification techniques are available in many of the statistical programs if large differences exist between the covariance matrices of the groups and the remedies do not minimize the effect [6, 12, 14].

Impacts on Interpretation

Another characteristic of the data that affects the results is multicollinearity among the independent variables. Multicollinearity, measured in terms of **tolerance**, denotes that two or more independent variables are highly correlated, so that one variable can be highly explained or predicted by the other variable(s) and thus it adds little to the explanatory power of the entire set. This consideration becomes especially critical when stepwise procedures are employed. The researcher, in interpreting the discriminant function, must be aware of the level of multicollinearity and its impact on determining which variables enter the stepwise solution. For a more detailed discussion of multicollinearity and its impact on stepwise solutions, see Chapter 4. The procedures for detecting the presence of multicollinearity are also addressed in Chapter 4.

As with any of the multivariate techniques employing a variate, an implicit assumption is that all relationships are linear. Nonlinear relationships are not reflected in the discriminant function unless specific variable transformations are made to represent nonlinear effects. Finally, outliers can

---

RULES OF THUMB 7-1

Discriminant Analysis Design

- The dependent variable must be nonmetric, representing groups of objects that are expected to differ on the independent variables
- Choose a dependent variable that:
  - Best represents group differences of interest
  - Defines groups that are substantially different
  - Minimizes the number of categories while still meeting the research objectives
- In converting metric variables to a nonmetric scale for use as the dependent variable, consider using extreme groups to maximize the group differences
- Independent variables must identify differences between at least two groups to be of any use in discriminant analysis
- The sample size must be large enough to:
  - Have at least one more observation per group than the number of independent variables, but striving for at least 20 cases per group
  - Strive to maximize the number of observations per variable, with a minimum ratio of five observations per independent variable
  - Have a large enough sample to divide it into estimation and holdout samples, each meeting the above requirements
- Assess the equality of covariance matrices with the Box's M test, but apply a conservative significance level of 0.1 and become even more conservative as the analysis becomes more complex with a larger number of groups and/or independent variables
- Examine the independent variables for univariate normality, because that is most direct remedy for ensuring both multivariate normality and equality of covariance matrices
- Multicollinearity among the independent variables can markedly reduce the estimated impact of independent variables in the derived discriminant function(s), particularly if a stepwise estimation process is used

---

have a substantial impact on the classification accuracy of any discriminant analysis results. The researcher is encouraged to examine all results for the presence of outliers and to eliminate true outliers if needed. For a discussion of some of the techniques for assessing the violations in the basic statistical assumptions or outlier detection, see Chapter 2.

STAGE 4: ESTIMATION OF THE DISCRIMINANT MODEL AND ASSESSING OVERALL FIT

To derive the discriminant function, the researcher must decide on the method of estimation and then determine the number of functions to be retained (see Figure 7-6). With the functions estimated, overall model fit can be assessed in several ways. First, **discriminant Z scores**, also known as the **Z scores**, can be calculated for each object. Comparison of the group means (centroids) on the Z scores provides one measure of discrimination between groups. Predictive accuracy can be measured as the number of observations classified into the correct groups, with a number of criteria available to assess whether the classification process achieves practical or statistical significance. Finally, casewise diagnostics can identify the classification accuracy of each case and its relative impact on the overall model estimation.

Selecting an Estimation Method

The first task in deriving the discriminant function(s) is to choose the estimation method. In making this choice, the researcher must balance the need for control over the estimation process versus a

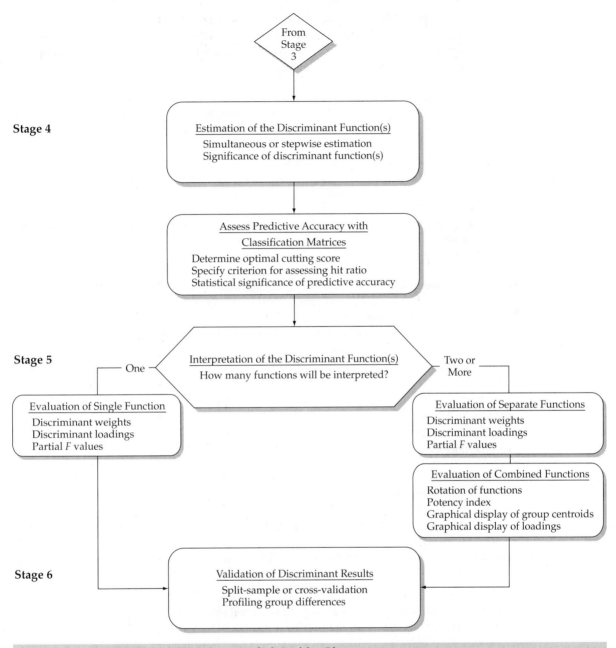

Stage 4

Estimation of the Discriminant Function(s)

Simultaneous or stepwise estimation
Significance of discriminant function(s)

Assess Predictive Accuracy with

Classification Matrices

Determine optimal cutting score
Specify criterion for assessing hit ratio
Statistical significance of predictive accuracy

Stage 5

One — Interpretation of the Discriminant Function(s)   Two or
How many functions will be interpreted?   More

Evaluation of Single Function

Discriminant weights
Discriminant loadings
Partial F values

Evaluation of Separate Functions

Discriminant weights
Discriminant loadings
Partial F values

Evaluation of Combined Functions

Rotation of functions
Potency index
Graphical display of group centroids
Graphical display of loadings

Stage 6

Validation of Discriminant Results

Split-sample or cross-validation
Profiling group differences

FIGURE 7-6   Stages 4–6 in the Discriminant Analysis Decision Diagram

desire for parsimony in the discriminant functions. The two methods available are the simultaneous (direct) method and the stepwise method, each discussed next.

SIMULTANEOUS ESTIMATION   **Simultaneous estimation** involves computing the discriminant function so that all of the independent variables are considered concurrently. Thus, the discriminant function is computed based upon the entire set of independent variables, regardless of the discriminating power of each independent variable. The simultaneous method is appropriate when, for theoretical reasons, the researcher wants to include all the independent variables in the analysis and is not interested in seeing intermediate results based only on the most discriminating variables.

STEPWISE ESTIMATION **Stepwise estimation** is an alternative to the simultaneous approach. It involves entering the independent variables into the discriminant function one at a time on the basis of their discriminating power. The stepwise approach follows a sequential process of adding or deleting variables in the following manner:

1. Choose the single best discriminating variable.
2. Pair the initial variable with each of the other independent variables, one at a time, and select the variable that is best able to improve the discriminating power of the function in combination with the first variable.
3. Select additional variables in a like manner. Note that as additional variables are included, some previously selected variables may be removed if the information they contain about group differences is available in some combination of the other variables included at later stages.
4. Consider the process completed when either all independent variables are included in the function or the excluded variables are judged as not contributing significantly to further discrimination.

The stepwise method is useful when the researcher wants to consider a relatively large number of independent variables for inclusion in the function. By sequentially selecting the next best discriminating variable at each step, variables that are not useful in discriminating between the groups are eliminated and a reduced set of variables is identified. The reduced set typically is almost as good as—and sometimes better than—the complete set of variables.

The researcher should note that stepwise estimation becomes less stable and generalizable as the ratio of sample size to independent variable declines below the recommended level of 20 observations per independent variable. It is particularly important in these instances to validate the results in as many ways as possible.

Statistical Significance

After estimation of the discriminant function(s), the researcher must assess the level of significance for the collective discriminatory power of the discriminant function(s), as well as the significance of each separate discriminant function. Evaluating the overall significance provides the researcher with the information necessary to decide whether to continue on to the interpretation of the analysis or if respecification is necessary. If the overall model is significant, then evaluating the individual functions identifies the function(s) that should be retained and interpreted.

OVERALL SIGNIFICANCE In assessing the statistical significance of the overall model, different statistical criteria are applicable for simultaneous versus stepwise estimation procedures. In both situations, the statistical tests relate to the ability of the discriminant function(s) to derive discriminant Z scores that are significantly different between the groups.

Simultaneous Estimation. When a simultaneous approach is used, the measures of Wilks' lambda, Hotelling's trace, and Pillai's criterion all evaluate the statistical significance of the discriminatory power of the discriminant function(s). Roy's greatest characteristic root evaluates only the first discriminant function. For a more detailed discussion of the advantages and disadvantages of each criterion, see the discussion of significance testing in multivariate analysis of variance in Chapter 8.

Stepwise Estimation. If a stepwise method is used to estimate the discriminant function, the Mahalanobis D^2 and Rao's V measures are most appropriate. Both are measures of generalized distance. The Mahalanobis D^2 procedure is based on generalized squared Euclidean distance that adjusts for unequal variances. The major advantage of this procedure is that it is computed in the original space of the predictor variables rather than as a collapsed version used in other measures. The Mahalanobis D^2 procedure becomes particularly critical as the number of predictor variables increases because it does not result in any reduction in dimensionality. A loss in dimensionality would cause a loss of information because it decreases variability of the independent variables. In general, Mahalanobis D^2 is the preferred procedure when the researcher is interested in the maximal use of available information in a stepwise process.

SIGNIFICANCE OF INDIVIDUAL DISCRIMINANT FUNCTIONS   If the number of groups is three or more, then the researcher must decide not only whether the discrimination between groups overall is statistically significant but also whether each of the estimated discriminant functions is statistically significant. As discussed earlier, discriminant analysis estimates one less discriminant function than there are groups. If three groups are analyzed, then two discriminant functions will be estimated; for four groups, three functions will be estimated; and so on. The computer programs all provide the researcher the information necessary to ascertain the number of functions needed to obtain statistical significance, without including discriminant functions that do not increase the discriminatory power significantly.

The conventional significance criterion of .05 or beyond is often used, yet some researchers extend the required significance level (e.g., .10 or more) based on the trade-off of cost versus the value of the information. If the higher levels of risk for including nonsignificant results (e.g., significance levels > .05) are acceptable, discriminant functions may be retained that are significant at the .2 or even the .3 level.

If one or more functions are deemed not statistically significant, the discriminant model should be reestimated with the number of functions to be derived limited to the number of significant functions. In this manner, the assessment of predictive accuracy and the interpretation of the discriminant functions will be based only on significant functions.

Assessing Overall Model Fit

Once the significant discriminant functions have been identified, attention shifts to ascertaining the overall fit of the retained discriminant function(s). This assessment involves three tasks:

1. Calculating discriminant Z scores for each observation
2. Evaluating group differences on the discriminant Z scores
3. Assessing group membership prediction accuracy

RULES OF THUMB 7-2

Model Estimation and Model Fit

- Although stepwise estimation may seem optimal by selecting the most parsimonious set of maximally discriminating variables, beware of the impact of multicollinearity on the assessment of each variable's discriminatory power
- Overall model fit assesses the statistical significance between groups on the discriminant Z score(s), but it does not assess predictive accuracy
- With more than two groups, do not confine your analysis to only the statistically significant discriminant function(s), but consider if nonsignificant functions (with significance levels of up to .3) add explanatory power

The discriminant Z score is calculated for each discriminant function for every observation in the sample. The discriminant score acts as a concise and simple representation of each discriminant function, simplifying the interpretation process and the assessment of the contribution of independent variables. Groups can be distinguished by their discriminant scores, and, as we will see, the discriminant scores can play an instrumental role in predicting group membership.

CALCULATING DISCRIMINANT Z SCORES With the retained discriminant functions defined, the basis for calculating the discriminant Z scores has been established. As discussed earlier, the discriminant Z score of any discriminant function can be calculated for each observation by the following formula:

$$Z_{jk} = a + W_1 X_{1k} + W_2 X_{2k} + \cdots + W_n X_{nk}$$

where

$$Z_{jk} = \text{discriminant } Z \text{ score of discriminant function } j \text{ for object } k$$
$$a = \text{intercept}$$
$$W_i = \text{discriminant coefficient for independent variable } i$$
$$X_{ik} = \text{independent variable } i \text{ for object } k$$

The discriminant Z score, a metric variable, provides a direct means of comparing observations on each function. Observations with similar Z scores are assumed more alike on the variables constituting this function than those with disparate scores. The discriminant function can be expressed with either standardized or unstandardized weights and values. The standardized version is more useful for interpretation purposes, but the unstandardized version is easier to use in calculating the discriminant Z score.

EVALUATING GROUP DIFFERENCES Once the discriminant Z scores are calculated, the first assessment of overall model fit is to determine the magnitude of differences between the members of each group in terms of the discriminant Z scores. A summary measure of the group differences is a comparison of the group centroids, the average discriminant Z score for all group members. A measure of success of discriminant analysis is its ability to define discriminant function(s) that result in significantly different group centroids. The differences between centroids are measured in terms of Mahalanobis D^2 measure, for which tests are available to determine whether the differences are statistically significant. The researcher should ensure that even with significant discriminant functions, significant differences occur between each of the groups.

Group centroids on each discriminant function can also be plotted to demonstrate the results from a graphical perspective. Plots are usually prepared for the first two or three discriminant functions (assuming they are statistically significant functions). The values for each group show its position in reduced discriminant space (so called because not all of the functions and thus not all of the variance are plotted). The researcher can see the differences between the groups on each function; however, visual inspection does not totally explain what these differences are. Circles can be drawn enclosing the distribution of observations around their respective centroids to clarify group differences further, but this procedure is beyond the scope of this text (see Dillon and Goldstein [4]).

ASSESSING GROUP MEMBERSHIP PREDICTION ACCURACY Given that the dependent variable is nonmetric, it is not possible to use a measure such as R^2, as is done in multiple regression, to assess predictive accuracy. Rather, each observation must be assessed as to whether it was correctly classified. In doing so, several major considerations must be addressed:

- The statistical and practical rationale for developing classification matrices
- Classifying individual cases
- Construction of the classification matrices
- Standards for assessing classification accuracy

Why Classification Matrices Are Developed. The statistical tests for assessing the significance of the discriminant function(s) only assess the degree of difference between the groups based on the discriminant Z scores. They do not indicate how well the function(s) predicts. These statistical tests suffer the same drawbacks as the classical tests of hypotheses. For example, suppose the two groups are deemed significantly different beyond the .01 level. Yet with sufficiently large sample sizes, the group means (centroids) could be virtually identical and still have statistical significance. To determine the predictive ability of a discriminant function, the researcher must construct classification matrices.

The **classification matrix** procedure provides a perspective on practical significance rather than statistical significance. With multiple discriminant analysis, the **percentage correctly classified,** also termed the **hit ratio,** reveals how well the discriminant function classified the objects. With a sufficiently large sample size in discriminant analysis, we could have a statistically significant difference between the two (or more) groups and yet correctly classify only 53 percent (when chance is 50 percent, with equal group sizes) [16]. In such instances, the statistical test would indicate statistical significance, yet the hit ratio would allow for a separate judgment to be made in terms of practical significance. Thus, we must use the classification matrix procedure to assess predictive accuracy beyond just statistical significance.

Classifying Individual Observations. The development of classification matrices requires that each observation be classified into one of the groups of the dependent variable based on the discriminant function(s). The objective is to characterize each observation on the discriminant function(s) and then determine the extent to which observations in each group can be consistently described by the discriminant functions. There are two approaches to classifying observations, one employing the discriminant scores directly and another developing a specific function for classification. Each approach, as well as the importance of determining the role that the sample size for each group plays in the classification process, will be discussed in the following section.

- *Cutting Score Calculation* Using the discriminant functions deemed significant, we can develop classification matrices by calculating the **cutting score** (also called the critical Z value) for each discriminant function. The cutting score is the criterion against which each object's discriminant score is compared to determine into which group the object should be classified. The cutting score represents the dividing point used to classify observations into groups based on their discriminant function score. The calculation of a cutting score between any two groups is based on the two group centroids (group mean of the discriminant scores) and the relative size of the two groups. The group centroids are easily calculated and provided at each stage of the stepwise process.
- *Developing a Classification Function* As noted earlier, using the discriminant function is only one of two possible approaches to classification. The second approach employs a **classification function**, also known as **Fisher's linear discriminant function**. The classification functions, one for each group, are used strictly for classifying observations. In this method of classification, an observation's values for the independent variables are inserted in the classification functions and a classification score for each group is calculated for that observation. The observation is then classified into the group with the highest classification score.

Defining Prior Probabilities. The impact and importance of each group's sample size in the classification process is many times overlooked, yet it is critical in making the appropriate assumptions in the classification process. Do the relative group sizes tell us something about the expected occurrence of each group in the population or are they just an artifact of the data collection process? Here we are concerned about the representativeness of the sample as it relates to representation of the relative sizes of the groups in the actual population, which can be stated as prior probabilities (i.e., the relative proportion of each group to the total sample).

The fundamental question is: Are the relative group sizes representative of the group sizes in the population? The default assumption for most statistical programs is equal prior probabilities; in other words, each group is assumed to have an equal chance of occurring even if the group sizes in the sample are unequal. If the researcher is unsure about whether the observed proportions in the sample are representative of the population proportions, the conservative approach is to employ equal probabilities. In some instances, estimates of the prior probabilities may be available, such as from prior research. Here the default assumption of equal prior probabilities is replaced with values specified by the researcher. In either instance the actual group sizes are replaced based on the specified prior probabilities.

If, however, the sample was conducted randomly and the researcher feels that the group sizes are representative of the population, then the researcher can specify prior probabilities to be based on the estimation sample. Thus, the actual group sizes are assumed representative and used directly in the calculation of the cutting score (see the following discussion). In all instances, however, the researcher must specify how the prior probabilities are to be calculated, which affects the group sizes used in the calculation as illustrated.

For example, consider a holdout sample consisting of 200 observations, with group sizes of 60 and 140 that relate to prior probabilities of 30 percent and 70 percent, respectively. If the sample is assumed representative, then the sample sizes of 60 and 140 are used in calculating the cutting score. If, however, the sample is deemed not representative, the researcher must specify the prior probabilities. If they are specified as equal (50% and 50%), sample sizes of 100 and 100 would be used in the cutting score calculation rather than the actual sample sizes. Specifying other values for the prior probabilities would result in differing sample sizes for the two groups.

- *Calculating the Optimal Cutting Score* The importance of the prior probabilities can be illustrated in the calculation of the "optimal" cutting score, which takes into account the prior probabilities through the use of group sizes. The basic formula for computing the **optimal cutting score** between any two groups is:

$$Z_{CS} = \frac{N_A Z_B + N_B Z_A}{N_A + N_B}$$

where

Z_{CS} = optimal cutting score between groups A and B
N_A = number of observations in group A
N_B = number of observations in group B
Z_A = centroid for group A
Z_B = centroid for group B

With unequal group sizes, the optimal cutting score for a discriminant function is now the weighted average of the group centroids. The cutting score is weighted toward the smaller group, hopefully making for a better classification of the larger group.

If the groups are specified to be of equal size (prior probabilities defined as equal), then the optimum cutting score will be halfway between the two group centroids and becomes simply the average of the two centroids:

$$Z_{CE} = \frac{Z_A + Z_B}{2}$$

where

Z_{CE} = critical cutting score value for equal group sizes
Z_A = centroid for group A
Z_B = centroid for group B

Both of the formulas for calculating the optimal cutting score assume that the distributions are normal and the group dispersion structures are known.

The concept of an optimal cutting score for equal and unequal groups is illustrated in Figures 7-7 and 7-8, respectively. Both the weighted and unweighted cutting scores are shown. It is apparent that if group A is much smaller than group B, the optimal cutting score will be closer to the centroid of group A than to the centroid of group B. Also, if the unweighted cutting score was used, none of the objects in group A would be misclassified, but a substantial portion of those in group B would be misclassified.

Costs of Misclassification. The optimal cutting score also must consider the cost of misclassifying an object into the wrong group. If the costs of misclassifying are approximately equal for all groups, the optimal cutting score will be the one that will misclassify the fewest number of objects across all groups. If the misclassification costs are unequal, the optimum cutting score will be the one that minimizes the costs of misclassification. More sophisticated approaches to determining cutting scores are discussed in Dillon and Goldstein [4] and Huberty et al. [13]. These approaches are based upon a Bayesian statistical model and are appropriate when the costs of misclassification into certain groups are high, when the groups are of grossly different sizes, or when one wants to take advantage of *a priori* knowledge of group membership probabilities.

In practice, when calculating the cutting score, it is not necessary to insert the raw variable measurements for every individual into the discriminant function and to obtain the discriminant score for each person to use in computing the Z_A and Z_B (group A and B centroids). The computer program will provide the discriminant scores as well as the Z_A and Z_B as regular output. When the researcher has the group centroids and sample sizes, the optimal cutting score can be obtained by merely substituting the values into the appropriate formula.

Constructing Classification Matrices. To validate the discriminant function through the use of classification matrices, the sample should be randomly divided into two groups. One of the groups (the analysis sample) is used to compute the discriminant function. The other group (the holdout or validation sample) is retained for use in developing the classification matrix.

The classification of each observation can be accomplished through either of the classification approaches discussed earlier. For the Fisher's approach, an observation is classified into the group

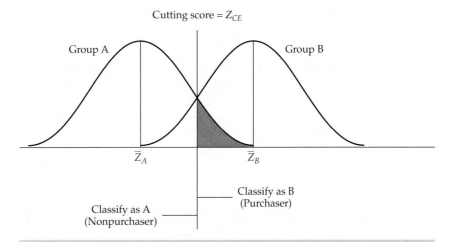

FIGURE 7-7 Optimal Cutting Score with Equal Sample Sizes

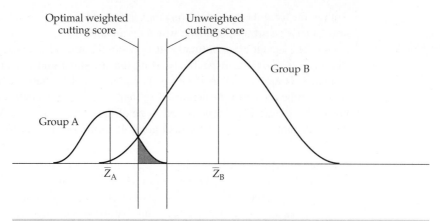

FIGURE 7-8 Optimal Cutting Score with Unequal Sample Sizes

with the largest classification function score. When using the discriminant scores and the optimal cutting score, the procedure is as follows:

Classify an individual into group A if $Z_n < Z_{ct}$

or

Classify an individual into group B if $Z_n > Z_{ct}$

where

$$Z_n = \text{discriminant } Z \text{ score for the } n\text{th individual}$$
$$Z_{ct} = \text{critical cutting score value}$$

The results of the classification procedure are presented in matrix form, as shown in Table 7-4. The entries on the diagonal of the matrix represent the number of individuals correctly classified. The numbers off the diagonal represent the incorrect classifications. The entries under the column labeled "Actual Group Size" represent the number of individuals actually in each of the two groups. The entries at the bottom of the columns represent the number of individuals assigned to the groups by the discriminant function. The percentage correctly classified for each group is shown at the right side of the matrix, and the overall percentage correctly classified, also known as the *hit ratio,* is shown at the bottom.

TABLE 7-4 Classification Matrix for Two-Group Discriminant Analysis

Actual Group	Predicted Group		Actual Group Size	Percentage Correctly Classified
	1	2		
1	22	3	25	88
2	5	20	25	80
Predicted group size	27	23	50	84[a]

[a]Percent correctly classified = (Number correctly classified/Total number of observations) × 100

$$= [(22 + 20)/50] \times 100$$

$$= 84\%$$

In our example, the number of individuals correctly assigned to group 1 is 22, whereas 3 members of group 1 are incorrectly assigned to group 2. Similarly, the number of correct classifications to group 2 is 20, and the number of incorrect assignments to group 1 is 5. Thus, the classification accuracy percentages of the discriminant function for the actual groups 1 and 2 are 88 and 80 percent, respectively. The overall classification accuracy (hit ratio) is 84 percent.

One final topic regarding classification procedures is the *t*-test available to determine the level of significance for the classification accuracy. The formula for a two-group analysis (equal sample size) is

$$t = \frac{p - .5}{\sqrt{\dfrac{.5(1.0 - .5)}{N}}}$$

where

$$p = \text{proportion correctly classified}$$
$$N = \text{sample size}$$

This formula can be adapted for use with more groups and unequal sample sizes.

Establishing Standards of Comparison for the Hit Ratio. As noted earlier, the predictive accuracy of the discriminant function is measured by the hit ratio, which is obtained from the classification matrix. The researcher may ask, What is considered an acceptable level of predictive accuracy for a discriminant function? For example, is 60 percent an acceptable level, or should one expect to obtain 80 to 90 percent predictive accuracy? To answer this question, the researcher must first determine the percentage that could be classified correctly by *chance (without the aid of the discriminant function).*

- *Standards of Comparison for the Hit Ratio for Equal Group Sizes* When the sample sizes of the groups are equal, the determination of the chance classification is rather simple; it is obtained by dividing 1 by the number of groups. The formula is:

$$C_{\text{EQUAL}} = 1 \div \text{ Number of groups}$$

  For instance, for a two-group function the chance probability would be .50; for a three-group function the chance probability would be .33; and so forth.
- *Standards of Comparison for the Hit Ratio for Unequal Group Sizes* The determination of the chance classification for situations in which the group sizes are unequal is somewhat more involved. Should we consider just the largest group, the combined probability of all different size groups, or some other standard? Let us assume that we have a total sample of 200 observations divided into holdout and analysis samples of 100 observations each. In the holdout sample, 75 subjects belong to one group and 25 to the other. We will examine the possible ways in which we can construct a standard for comparison and what each represents.
  - Referred to as the **maximum chance criterion,** we could arbitrarily assign all the subjects to the largest group. The maximum chance criterion should be used when the sole objective of the discriminant analysis is to maximize the percentage correctly classified [16]. It is also the most conservative standard because it will generate the highest standard of comparison. However, situations in which we are concerned only about maximizing the percentage correctly classified are rare. Usually the researcher uses discriminant analysis to correctly identify members of all groups. In cases where the sample sizes are unequal and the researcher wants to classify members of all groups, the discriminant function defies the

odds by classifying a subject in the smaller group(s). The maximum chance criterion does not take this fact into account [16].

In our simple example of a sample with two groups (75 and 25 people each), using this method would set a 75 percent classification accuracy, what would be achieved by classifying everyone into the largest group without the aid of any discriminant function. It could be concluded that unless the discriminant function achieves a classification accuracy higher than 75 percent, it should be disregarded because it has not helped us improve the prediction accuracy we could achieve without using any discriminant analysis at all.

• When group sizes are unequal and the researcher wishes to correctly identify members of all of the groups, not just the largest group, the **proportional chance criterion** is deemed by many to be the most appropriate. The formula for this criterion is

$$C_{PRO} = p^2 + (1 - p)^2$$

where

$$p = \text{proportion of individuals in group 1}$$
$$1 - p = \text{proportion of individuals in group 2}$$

Using the group sizes from our earlier example (75 and 25), we see that the proportional chance criterion would be 62.5 percent [$.75^2 + (1.0 - .75)^2 = .625$] compared with 75 percent. Therefore, in this instance, the actual prediction accuracy of 75 percent may be acceptable because it is above the 62.5 percent proportional chance criterion.

• An issue with either the maximum chance or proportional chance criteria is the sample sizes used for calculating the standard. Do you use the group sizes from the overall sample, the analysis/estimation sample, or the validation/holdout sample? A couple of suggestions:

  • If the sample sizes of the analysis and estimation samples are each deemed sufficiently large (i.e., total sample of 100 with each group having at least 20 cases), derive separate standards for each sample.
  • If the separate samples are not deemed sufficiently large, use the group sizes from the total sample in calculating the standards.
  • Be aware of differing group sizes between samples when using the maximum chance criterion, because it is dependent on the largest group size. This guideline is especially critical when the sample size is small or when group size proportions vary markedly from sample to sample. It is another reason to be cautious in the use of the maximum chance criterion.

These chance model criteria are useful only when computed with holdout samples (split-sample approach). If the individuals used in calculating the discriminant function are the ones being classified, the result will be an upward bias in the prediction accuracy. In such cases, both of these criteria would have to be adjusted upward to account for this bias.

Comparing the Hit Ratio to the Standard. The question of "How high does classification accuracy have to be?" is crucial. If the percentage of correct classifications is significantly larger than would be expected by chance, the researcher can proceed in interpreting the discriminant functions and group profiles. However, if the classification accuracy is no greater than can be expected by chance, whatever differences appear to exist actually merit little or no interpretation; that is, differences in score profiles would provide no meaningful information for identifying group membership.

The question, then, is how high should the classification accuracy be relative to chance? For example, if chance is 50 percent (two-group, equal sample size), does a classification (predictive) accuracy of 60 percent justify moving to the interpretation stage? Ultimately, the decision depends on the cost relative to the value of the information. The cost-versus-value argument offers little assistance to the neophyte data researcher, but the following criterion is suggested: *The classification accuracy should be at least one-fourth greater than that achieved by chance.*

For example, if chance accuracy is 50 percent, the classification accuracy should be 62.5 percent (62.5% = 1.25 × 50%). If chance accuracy is 30 percent, the classification accuracy should be 37.5 percent (37.5% = 1.25 × 30%).

This criterion provides only a rough estimate of the acceptable level of predictive accuracy. The criterion is easy to apply with groups of equal size. With groups of unequal size, an upper limit is reached when the maximum chance model is used to determine chance accuracy. It does not present too great a problem, however, because under most circumstances the maximum chance model would not be used with unequal group sizes.

Overall Versus Group-Specific Hit Ratios. To this point, we focused on evaluating the overall hit ratio across all groups in assessing the predictive accuracy of a discriminant analysis. The researcher also must be concerned with the hit ratio (percent correctly classified) for each separate group. If you focus solely on the overall hit ratio, it is possible that one or more groups, particularly smaller groups, may have unacceptable hit ratios while the overall hit ratio is acceptable. The researcher should evaluate each group's hit ratio and assess whether the discriminant analysis provides adequate levels of predictive accuracy both at the overall level as well as for each group.

Statistically Based Measures of Classification Accuracy Relative to Chance. A statistical test for the discriminatory power of the classification matrix when compared with a chance model is **Press's Q statistic**. This simple measure compares the number of correct classifications with the total sample size and the number of groups. The calculated value is then compared with a critical value (the chi-square value for 1 degree of freedom at the desired confidence level). If it exceeds this critical value, then the classification matrix can be deemed statistically better than chance. The Q statistic is calculated by the following formula:

$$\text{Press's } Q = \frac{[N - (nK)]^2}{N(K - 1)}$$

where

$$N = \text{total sample size}$$
$$n = \text{number of observations correctly classified}$$
$$K = \text{number of groups}$$

For example, in Table 7-4, the Q statistic would be based on a total sample of $N = 50$, $n = 42$ correctly classified observations, and $K = 2$ groups. The calculated statistic would be:

$$\text{Press's } Q = \frac{[50 - (42 \times 2)]^2}{50(2 - 1)} = 23.12$$

The critical value at a significance level of .01 is 6.63. Thus, we would conclude that in the example the predictions were significantly better than chance, which would have a correct classification rate of 50 percent.

This simple test is sensitive to sample size; large samples are more likely to show significance than small sample sizes of the same classification rate.

For example, if the sample size is increased to 100 in the example and the classification rate remains at 84 percent, the Q statistic increases to 46.24. If the sample size increases to 200, but retains the classification rate of 84 percent, the Q statistic increases again to 92.48. But if the sample size was only 20 and the misclassification rate was still 84 percent (17 correct predictions), the Q statistic would be only 9.8. Thus, examine the Q statistic in light of the sample size because increases in sample size will increase the Q statistic even for the same overall classification rate.

One must be careful in drawing conclusions based solely on this statistic, however, because as the sample sizes become larger, a lower classification rate will still be deemed significant.

Casewise Diagnostics

The final means of assessing model fit is to examine the predictive results on a case-by-case basis. Similar to the analysis of residuals in multiple regression, the objective is to understand which observations (1) have been misclassified and (2) are not representative of the remaining group members. Although the classification matrix provides overall classification accuracy, it does not detail the individual case results. Also, even if we can denote which cases are correctly or incorrectly classified, we still need a measure of an observation's similarity to the remainder of the group.

MISCLASSIFICATION OF INDIVIDUAL CASES When analyzing residuals from a multiple regression analysis, an important decision involves setting the level of residual considered substantive and worthy of attention. In discriminant analysis, this issue is somewhat simpler because an observation is either correctly or incorrectly classified. All computer programs provide information that identifies which cases are misclassified and to which group they were misclassified. The researcher can identify not only those cases with classification errors, but a direct representation of the type of misclassification error.

ANALYZING MISCLASSIFIED CASES The purpose of identifying and analyzing the misclassified observations is to identify any characteristics of these observations that could be incorporated into the discriminant analysis for improving predictive accuracy. This analysis may take the form of profiling the misclassified cases on either the independent variables or other variables not included in the model.

Profiling on the Independent Variables. Examining these cases on the independent variables may identify nonlinear trends or other relationships or attributes that led to the misclassification. Several techniques are particularly appropriate in discriminant analysis:

- A graphical representation of the observations is perhaps the simplest yet effective approach for examining the characteristics of observations, especially the misclassified observations. The most common approach is to plot the observations based on their discriminant Z scores and portray the overlap among groups and the misclassified cases. If two or more functions are retained, the optimal cutting points can also be portrayed to give what is known as a **territorial map** depicting the regions corresponding to each group.
- Plotting the individual observations along with the group centroids, as discussed earlier, shows not only the general group characteristics depicted in the centroids, but also the variation in the group members. It is analogous to the areas defined in the three-group example at the beginning of this chapter, in which cutting scores on both functions defined areas corresponding to the classification predictions for each group.
- A direct empirical assessment of the similarity of an observation to the other group members can be made by evaluating the Mahalanobis D^2 distance of the observation to the group centroid. Based on the set of independent variables, observations closer to the centroid have a smaller Mahalanobis D^2 and are assumed more representative of the group than those farther away.
- The empirical measure should be combined with a graphical analysis, however, because although a large Mahalanobis D^2 value does indicate observations that are quite different from

---

RULES OF THUMB 7-3

Assessing Model Fit and Predictive Accuracy

- The classification matrix and hit ratio replace R^2 as the measure of model fit:
  - Assess the hit ratio both overall and by group
  - If the estimation and analysis samples both exceed 100 cases and each group exceeds 20 cases, derive separate standards for each sample; if not, derive a single standard from the overall sample
- Multiple criteria are used for comparison to the hit ratio:
  - The maximum chance criterion for evaluating the hit ratio is the most conservative, giving the highest baseline value to exceed
  - Be cautious in using the maximum chance criterion in situations with overall samples less than 100 and/or group sizes under 20
  - The proportional chance criterion considers all groups in establishing the comparison standard and is the most popular
  - The actual predictive accuracy (hit ratio) should exceed any criterion value by at least 25 percent
- Analyze the missclassified observations both graphically (territorial map) and empirically (Mahalanobis D^2)

---

the group centroids, it does not always indicate misclassification. For example, in a two-group situation, a member of group A may have a large Mahalanobis D^2 distance, indicating it is less representative of the group. However, if that distance is away from the group B centroid, then it would actually increase the chance of correct classification, even though it is less representative of the group. A smaller distance that places an observation between the two centroids would probably have a lower probability of correct classification, even though it is closer to its group centroid than the earlier situation.

Although no prespecified analyses are established, such as found in multiple regression, the researcher is encouraged to evaluate these misclassified cases from several perspectives in attempting to uncover the unique features they hold in comparison to their other group members.

STAGE 5: INTERPRETATION OF THE RESULTS

If the discriminant function is statistically significant and the classification accuracy is acceptable, the researcher should focus on making substantive interpretations of the findings. This process involves examining the discriminant functions to determine the relative importance of each independent variable in discriminating between the groups. Three methods of determining the relative importance have been proposed:

1. Standardized discriminant weights
2. Discriminant loadings (structure correlations)
3. Partial F values

Discriminant Weights

The traditional approach to interpreting discriminant functions examines the sign and magnitude of the standardized **discriminant weight** (also referred to as a **discriminant coefficient**) assigned to each variable in computing the discriminant functions. When the sign is ignored, each weight represents the relative contribution of its associated variable to that function. Independent variables with relatively larger weights contribute more to the discriminating power of the function than do variables with smaller weights. The sign denotes only that the variable makes either a positive or a negative contribution [4].

The interpretation of discriminant weights is analogous to the interpretation of beta weights in regression analysis and is therefore subject to the same criticisms. For example, a small weight may indicate either that its corresponding variable is irrelevant in determining a relationship or that it has been partialed out of the relationship because of a high degree of multicollinearity. Another problem with the use of discriminant weights is that they are subject to considerable instability. These problems suggest caution in using weights to interpret the results of discriminant analysis.

Discriminant Loadings

Discriminant loadings, referred to sometimes as **structure correlations**, are increasingly used as a basis for interpretation because of the deficiencies in utilizing weights. Measuring the simple linear correlation between each independent variable and the discriminant function, the discriminant loadings reflect the variance that the independent variables share with the discriminant function. In that regard they can be interpreted like factor loadings in assessing the relative contribution of each independent variable to the discriminant function. (Chapter 3 further discusses factor-loading interpretation.)

One unique characteristic of loadings is that loadings can be calculated for all variables, whether they were used in the estimation of the discriminant function or not. This aspect is particularly useful when a stepwise estimation procedure is employed and some variables are not included in the discriminant function. Rather than having no way to understand their relative impact, loadings provide a relative effect of every variable on a common measure.

With the loadings, the primary question is: What value must loadings attain to be considered substantive discriminators worthy of note? In either simultaneous or stepwise discriminant analysis, variables that exhibit a loading of ±.40 or higher are considered substantive. With stepwise procedures, this determination is supplemented because the technique prevents nonsignificant variables from entering the function. However, multicollinearity and other factors may preclude a variable from entering the equation, which does not necessarily mean that it does not have a substantial effect.

Discriminant loadings (like weights) may be subject to instability. Loadings are considered relatively more valid than weights as a means of interpreting the discriminating power of independent variables because of their correlational nature. The researcher still must be cautious when using loadings to interpret discriminant functions.

Partial *F* Values

As discussed earlier, two computational approaches—simultaneous and stepwise—can be utilized in deriving discriminant functions. When the stepwise method is selected, an additional means of interpreting the relative discriminating power of the independent variables is available through the use of partial *F* values. It is accomplished by examining the absolute sizes of the significant *F* values and ranking them. Large *F* values indicate greater discriminatory power. In practice, rankings using the *F* values approach are the same as the ranking derived from using discriminant weights, but the *F* values indicate the associated level of significance for each variable.

Interpretation of Two or More Functions

In cases of two or more significant discriminant functions, we are faced with additional problems of interpretation. First, can we simplify the discriminant weights or loadings to facilitate the profiling of each function? Second, how do we represent the impact of each variable across all functions? These problems are found both in measuring the total discriminating effects across functions and in assessing the role of each variable in profiling each function separately. We address these two questions by introducing the concepts of rotation of the functions, the potency index, and stretched vectors representations.

ROTATION OF THE DISCRIMINANT FUNCTIONS   After the discriminant functions are developed, they can be rotated to redistribute the variance. (The concept is more fully explained in Chapter 3.) Basically, rotation preserves the original structure and the reliability of the discriminant solution while making the functions easier to interpret substantively. In most instances, the VARIMAX rotation is employed as the basis for rotation.

POTENCY INDEX   Previously, we discussed using the standardized weights or discriminant loadings as measures of a variable's contribution to a discriminant function. When two or more functions are derived, however, a composite or summary measure is useful in describing the contributions of a variable across all significant functions. The **potency index** is a relative measure among all variables and is indicative of each variable's discriminating power [18]. It includes both the contribution of a variable to a discriminant function (its discriminant loading) and the relative contribution of the function to the overall solution (a relative measure among the functions based on eigenvalues). The composite is simply the sum of the individual potency indices across all significant discriminant functions. Interpretation of the composite measure is limited, however, by the fact that it is useful only in depicting the relative position (such as the rank order) of each variable, and the absolute value has no real meaning. The potency index is calculated by a two-step process:

Step 1:   *Calculate a potency value of each variable for each significant function.* In the first step, the discriminating power of a variable, represented by the squared value of the unrotated discriminant loading, is "weighted" by the relative contribution of the discriminant function to the overall solution. First, the relative eigenvalue measure for each significant discriminant function is calculated simply as:

$$\text{Relative eigenvalue of discriminant function } j = \frac{\text{Eigenvalue of discriminant function } j}{\text{Sum of eigenvalue across all significant funtions}}$$

The potency value of each variable on a discriminant function is then:

$$\text{Potency value of variable } i \text{ on function } j = (\text{Discriminant loading}_{ij})^2 \times \text{Relative eigenvalue of function } j$$

Step 2:   *Calculate a composite potency index across all significant functions.* Once a potency value has been calculated for each function, the composite potency index for each variable is calculated as:

$$\text{Composite potency of variable } i = \text{Sum of potency values of variable } i \text{ across all significant discriminant functions}$$

The potency index now represents the total discriminating effect of the variable across all of the significant discriminant functions. It is only a relative measure, however, and its absolute value has no substantive meaning. An example of calculating the potency index is provided in the three-group example for discriminant analysis.

GRAPHICAL DISPLAY OF DISCRIMINANT SCORES AND LOADINGS   To depict group differences on the predictor variables, the researcher can use two different approaches to graphical display. The territorial map plots the individual cases on the significant discriminant functions to enable the researcher to assess the relative position of each observation based on the discriminant function scores. The second approach is to plot the discriminant loadings to understand the relative grouping and magnitude of each loading on each function. Each approach will be discussed in more detail in the following section.

Territorial Map. The most common graphical method is the territorial map, where each observation is plotted in a graphical display based on the discriminant function Z scores of the observations. For example, assume that a three-group discriminant analysis had two significant discriminant functions. A territorial map is created by plotting each observation's discriminant Z scores for the first discriminant function on the X axis and the scores for the second discriminant function on the Y axis. As such, it provides several perspectives on the analysis:

- Plotting each group's members with differing symbols allows for an easy portrayal of the distinctiveness of each group as well as its overlap with each other group.
- Plotting each group's centroids provides a means for assessing each group member relative to its group centroid. This procedure is particularly useful when assessing whether large Mahalanobis D^2 measures lead to misclassification.
- Lines representing the cutting scores can also be plotted, denoting boundaries depicting the ranges of discriminant scores predicted into each group. Any group's members falling outside these boundaries are misclassified. Denoting the misclassified cases allows for assessing which discriminant function was most responsible for the misclassification as well as the degree to which a case is misclassified.

Vector Plot of Discriminant Loadings. The simplest graphical approach to depicting discriminant loadings is to plot the actual rotated or unrotated loadings on a graph. The preferred approach would be to plot the rotated loadings. Similar to the graphical portrayal of factor loadings (see Chapter 3), this method depicts the degree to which each variable is associated with each discriminant function.

An even more accurate approach, however, involves plotting the loadings and depicting vectors for each loading and group centroid. A **vector** is merely a straight line drawn from the origin (center) of a graph to the coordinates of a particular variable's discriminant loadings or a group centroid. With a **stretched vector** representation, the length of each vector becomes indicative of the relative importance of each variable in discriminating among the groups. The plotting procedure proceeds in three steps:

1. *Selecting variables:* All variables, whether included in the model as significant or not, may be plotted as vectors. In this way, the importance of collinear variables that are not included, such as in a stepwise solution, can still be portrayed.
2. *Stretching the vectors:* Each variable's discriminant loadings are stretched by multiplying the discriminant loading (preferably after rotation) by its respective univariate F value. We note that vectors point toward the groups having the highest mean on the respective predictor and away from the groups having the lowest mean scores.
3. *Plotting the group centroids:* The group centroids are also stretched in this procedure by multiplying them by the approximate F value associated with each discriminant function. If the loadings are stretched, the centroids must be stretched as well to plot them accurately on the same graph. The approximate F values for each discriminant function are obtained by the following formula:

$$F \text{ value }_{\text{Function}_i} = \text{Eigenvalue }_{\text{Function}_i} \left(\frac{N_{\text{Estimation Sample}} - NG}{NG - 1} \right)$$

where

$$N_{\text{Estimation Sample}} = \text{sample size of estimation sample}$$

As an example, assume that the sample of 50 observations was divided into three groups. The multiplier of each eigenvalue would be $(50 - 3) \div (3 - 1) = 23.5$.

When completed, the researcher has a portrayal of the grouping of variables on each discriminant function, the magnitude of the importance of each variable (represented by the length of each vector), and the profile of each group centroid (shown by the proximity to each vector). Although this procedure must be done manually in most instances, it provides a complete portrayal of both discriminant loadings and group centroids. For more details on this procedure, see Dillon and Goldstein [4].

Which Interpretive Method to Use?

Several methods for interpreting the nature of discriminant functions have been discussed, both for single- and multiple-function solutions. Which methods should be used? The loadings approach is more valid than the use of weights and should be utilized whenever possible. The use of univariate and partial *F* values enables the researcher to use several measures and look for some consistency in evaluations of the variables. If two or more functions are estimated, then the researcher can employ several graphical techniques and the potency index, which aid in interpreting the multidimensional solution. The most basic point is that the researcher should employ all available methods to arrive at the most accurate interpretation.

STAGE 6: VALIDATION OF THE RESULTS

The final stage of a discriminant analysis involves validating the discriminant results to provide assurances that the results have external as well as internal validity. *With the propensity of discriminant analysis to inflate the hit ratio if evaluated only on the analysis sample, validation is an essential step.* In addition to validating the hit ratios, the researcher should use group profiling to ensure that the group means are valid indicators of the conceptual model used in selecting the independent variables.

Validation Procedures

Validation is a critical step in any discriminant analysis because many times, especially with smaller samples, the results can lack generalizability (external validity). The most common approach for establishing external validity is the assessment of hit ratios. Validation can occur either with a separate sample (holdout sample) or utilizing a procedure that repeatedly processes the estimation sample. External validity is supported when the hit ratio of the selected approach exceeds the comparison standards that represent the predictive accuracy expected by chance (see earlier discussion).

UTILIZING A HOLDOUT SAMPLE Most often, the validation of the hit ratios is performed by creating a holdout sample, also referred to as the **validation sample**. The purpose of utilizing a holdout sample for validation purposes is to see how well the discriminant function works on a sample of observations not used to derive the discriminant function. This process involves developing a discriminant function with the analysis sample and then applying it to the holdout sample. The justification for dividing the total sample into two groups is that an upward bias will occur in the prediction accuracy of the discriminant function if the individuals used in developing the classification matrix are the same as those used in computing the function; that is, the classification accuracy will be higher than is valid when applied to the estimation sample.

Other researchers have suggested that even greater confidence could be placed in the validity of the discriminant function by following this procedure several times [18]. Instead of randomly dividing the total sample into analysis and holdout groups once, the researcher would randomly divide the total sample into analysis and holdout samples several times, each time testing the validity of the discriminant function through the development of a classification matrix and a hit ratio. Then the several hit ratios would be averaged to obtain a single measure.

CROSS-VALIDATION The cross-validation approach to assessing external validity is performed with multiple subsets of the total sample [2, 4]. The most widely used approach is the jackknife method. Cross-validation is based on the "leave-one-out" principle. The most prevalent use of this method is to estimate $k - 1$ subsamples, eliminating one observation at a time from a sample of k cases. A discriminant function is calculated for each subsample and then the predicted group membership of the eliminated observation is made with the discriminant function estimated on the remaining cases. After all of the group membership predictions have been made, one at a time, a classification matrix is constructed and the hit ratio calculated.

Cross-validation is quite sensitive to small sample sizes. Guidelines suggest that it be used only when the smallest group size is at least three times the number of predictor variables, and most researchers suggest a ratio of 5:1 [13]. However, cross-validation may represent the only possible validation approach in instances where the original sample is too small to divide into analysis and holdout samples, but still exceeds the guidelines already discussed. Cross-validation is also becoming more widely used as major computer programs provide it as a program option.

Profiling Group Differences

Another validation technique is to profile the groups on the independent variables to ensure their correspondence with the conceptual bases used in the original model formulation. After the researcher identifies the independent variables that make the greatest contribution in discriminating between the groups, the next step is to profile the characteristics of the groups based on the group means. This profile enables the researcher to understand the character of each group according to the predictor variables.

For example, referring to the KitchenAid survey data presented in Table 7-1, we see that the mean rating on "durability" for the "would purchase" group is 7.4, whereas the comparable mean rating on "durability" for the "would not purchase" group is 3.2. Thus, a profile of these two groups shows that the "would purchase" group rates the perceived durability of the new product substantially higher than the "would not purchase" group.

Another approach is to profile the groups on a separate set of variables that should mirror the observed group differences. This separate profile provides an assessment of external validity in that the groups vary on both the independent variable(s) and the set of associated variables. This technique is similar in character to the validation of derived clusters described in Chapter 9.

RULES OF THUMB 7-4

Interpreting and Validating Discriminant Functions

- Discriminant loadings are the preferred method to assess the contribution of each variable to a discriminant function because they are:
  - A standardized measure of importance (ranging from 0 to 1)
  - Available for all independent variables whether used in the estimation process or not
  - Unaffected by multicollinearity
- Loadings exceeding ±.40 are considered substantive for interpretation purposes
- In case of more than one discriminant function, be sure to:
  - Use rotated loadings
  - Assess each variable's contribution across all the functions with the potency index
- The discriminant function must be validated either with a holdout sample or one of the "leave-one-out" procedures

---

A TWO-GROUP ILLUSTRATIVE EXAMPLE

To illustrate the application of two-group discriminant analysis, we use variables drawn from the HBAT database introduced in Chapter 1. This example examines each of the six stages of the model-building process to a research problem particularly suited to multiple discriminant analysis.

Stage 1: Objectives of Discriminant Analysis

Recall that one of the customer characteristics obtained by HBAT in its survey was a categorical variable (X_4) indicating the region in which the firm was located: USA/North America or Outside North America. HBAT's management team is interested in any differences in perceptions between those customers located and served by their USA-based salesforce versus those outside the United States who are served mainly by independent distributors. Despite any differences found in terms of sales support issues due to the nature of the salesforce serving each geographic area, the management team is interested to see whether the other areas of operations (product line, pricing, etc.) are viewed differently between these two sets of customers. This inquiry follows the obvious need by management to always strive to better understand their customer, in this instance by focusing on any differences that may occur between geographic areas. If any perceptions of HBAT are found to differ significantly between firms in these two regions, the company would then be able to develop strategies to remedy any perceived deficiencies and develop differentiated strategies to accommodate the differing perceptions.

To do so, discriminant analysis was selected to identify those perceptions of HBAT that best distinguish firms in each geographic region.

Stage 2: Research Design for Discriminant Analysis

The research design stage focuses on three key issues: selecting dependent and independent variables, assessing the adequacy of the sample size for the planned analysis, and dividing the sample for validation purposes.

SELECTION OF DEPENDENT AND INDEPENDENT VARIABLES Discriminant analysis requires a single nonmetric dependent measure and one or more metric independent measures that are affected to provide differentiation between the groups based on the dependent measure.

Because the dependent variable Region (X_4) is a two-group categorical variable, discriminant analysis is the appropriate technique. The survey collected perceptions of HBAT that can now be used to differentiate between the two groups of firms. Discriminant analysis uses as independent variables the 13 perception variables from the database (X_6 to X_{18}) to discriminate between firms in each geographic area.

SAMPLE SIZE Given the relatively small size of the HBAT sample (100 observations), issues of sample size are particularly important, especially the division of the sample into analysis and holdout samples (see discussion in next section).

The sample of 100 observations, when split into analysis and holdout samples of 60 and 40, respectively, barely meets the suggested minimum 5:1 ratio of observations to independent variables (60 observations for 13 potential independent variables) in the analysis sample. Although this ratio would increase to almost 8:1 if the sample were not split, it was deemed more important to validate the results rather than to increase the number of observations in the analysis sample.

The two group sizes of 26 and 34 in the estimation sample also exceed the minimum size of 20 observations per group. Finally, the two groups are comparable enough in size to not adversely impact either the estimation or the classification processes.

DIVISION OF THE SAMPLE Previous discussion emphasized the need for validating the discriminant function by splitting the sample into two parts, one used for estimation and the other validation. Any time a holdout sample is used, the researcher must ensure that the resulting sample sizes are sufficient to support the number of predictors included in the analysis.

The HBAT database has 100 observations; it was decided that a holdout sample of 40 observations would be sufficient for validation purposes. This split would still leave 60 observations for estimation of the discriminant function. Moreover, the relative group sizes in the estimation sample (26 and 34 in the two groups) would allow for estimation without complications due to markedly different group sizes.

It is important to ensure randomness in the selection of the holdout sample so that any ordering of the observations does not affect the processes of estimation and validation. The control cards necessary for both selection of the holdout sample and performance of the two-group discriminant analysis are shown on the text's Web sites at www.pearsonglobaleditions.com/hair or www.mvstats.com.

Stage 3: Assumptions of Discriminant Analysis

The principal assumptions underlying discriminant analysis involve the formation of the variate or discriminant function (normality, linearity, and multicollinearity) and the estimation of the discriminant function (equal variance and covariance matrices). How to examine the independent variables for normality, linearity, and multicollinearity is explained in Chapter 2. For purposes of our illustration of discriminant analysis, these assumptions are met at acceptable levels.

Most statistical programs have one or more statistical tests for the assumption of equal covariance or dispersion matrices addressed in Chapter 2. The most common test is Box's M (for more detail, see Chapter 2).

In this two-group example, the significance of differences in the covariance matrices between the two groups is .011. Even though the significance is less than .05 (in this test the researcher looks for values above the desired significance level), the sensitivity of the test to factors other than just covariance differences (e.g., normality of the variables and increasing sample size) makes this an acceptable level.

No additional remedies are needed before estimation of the discriminant function can be performed.

Stage 4: Estimation of the Discriminant Model and Assessing Overall Fit

The researcher has the choice of two estimation approaches (simultaneous versus stepwise) in determining the independent variables included in the discriminant function. Once the estimation approach is selected, the process determines the composition of the discriminant function subject to the requirement for statistical significance specified by the researcher.

The primary objective of this analysis is to identify the set of independent variables (HBAT perceptions) that maximally differentiates between the two groups of customers. If the set of perception variables was smaller or the objective was simply to determine the discriminating capabilities of the entire set of perception variables, with no regard to the impact of any individual perception, then the simultaneous approach of entering all variables directly into the discriminant function would be employed. But in this case, even with the knowledge of multicollinearity among the perception variables seen in performing factor analysis (see Chapter 3), the stepwise approach is deemed most appropriate. We should note, however, that multicollinearity may impact which variables enter into the discriminant function and thus require particular attention in the interpretation process.

ASSESSING GROUP DIFFERENCES   Let us begin our assessment of the two-group discriminant analysis by examining Table 7-5, which shows the group means for each of the independent variables, based on the 60 observations constituting the analysis sample.

In profiling the two groups, we can first identify five variables with the largest differences in the group means (X_6, X_{11}, X_{12}, X_{13}, and X_{17}). Table 7-5 also shows the Wilks' lambda and univariate ANOVA used to assess the significance between means of the independent variables for the two groups. These tests indicate that the five perception variables are also the only variables with significant univariate differences between the two groups. Finally, the minimum Mahalanobis D^2 values are also given. This value is important because it is the measure used to

TABLE 7-5 Group Descriptive Statistics and Tests of Equality for the Estimation Sample in the Two-Group Discriminant Analysis

Independent Variables	Dependent Variable Group Means: X_4 Region		Test of Equality of Group Means*			Minimum Mahalanobis D^2	
	Group 0: USA/ North America ($n = 26$)	Group 1: Outside North America ($n = 34$)	Wilks' Lambda	F Value	Significance	Minimum D^2	Between Groups
X_6 Product Quality	8.527	7.297	.801	14.387	.000	.976	0 and 1
X_7 E-Commerce Activities	3.388	3.626	.966	2.054	.157	.139	0 and 1
X_8 Technical Support	5.569	5.050	.973	1.598	.211	.108	0 and 1
X_9 Complaint Resolution	5.577	5.253	.986	.849	.361	.058	0 and 1
X_{10} Advertising	3.727	3.979	.987	.775	.382	.053	0 and 1
X_{11} Product Line	6.785	5.274	.695	25.500	.000	1.731	0 and 1
X_{12} Salesforce Image	4.427	5.238	.856	9.733	.003	.661	0 and 1
X_{13} Competitive Pricing	5.600	7.418	.645	31.992	.000	2.171	0 and 1
X_{14} Warranty & Claims	6.050	5.918	.992	.453	.503	.031	0 and 1
X_{15} New Products	4.954	5.276	.990	.600	.442	.041	0 and 1
X_{16} Order & Billing	4.231	4.153	.999	.087	.769	.006	0 and 1
X_{17} Price Flexibility	3.631	4.932	.647	31.699	.000	2.152	0 and 1
X_{18} Delivery Speed	3.873	3.794	.997	.152	.698	.010	0 and 1

*Wilks' lambda (U statistic) and univariate F ratio with 1 and 58 degrees of freedom.

select variables for entry in the stepwise estimation process. Because only two groups are involved, the largest D^2 value also has the most significant difference between groups (note that the same is not necessarily so with three or more groups, where large differences between any two groups may not result in the largest overall differences across all groups, as will be shown in the three-group example).

Examining the group differences leads to identifying five perception variables (X_6, X_{11}, X_{12}, X_{13}, and X_{17}) as the most logical set of candidates for entry into the discriminant analysis. This marked reduction from the larger set of 13 perception variables reinforces the decision to use a stepwise estimation process.

To identify which of these five variables, plus any of the others, best discriminate between the groups, we must estimate the discriminant function.

ESTIMATION OF THE DISCRIMINANT FUNCTION   The stepwise procedure begins with all of the variables excluded from the model and then selects the variable that

1. Shows statistically significant differences across the groups (.05 or less required for entry)
2. Provides the largest Mahalanobis distance (D^2) between the groups

This process continues to include variables in the discriminant function as long as they provide statistically significant additional discrimination between the groups beyond those differences already accounted for by the variables in the discriminant function. This approach is similar to the stepwise process in multiple regression (see Chapter 4), which adds variables with significant increases in the explained variance of the dependent variable. Also, in cases where two or more variables are entered into the model, the variables already in the model are evaluated for possible removal. A variable may be removed if high multicollinearity exists between it and the other included independent variables such that its significance falls below the significance level for removal (.10).

Stepwise Estimation: Adding the First Variable X_{13}. From our review of group differences, we saw that X_{13} had the largest significant difference between groups and the largest Mahalanobis D^2 (see Table 7-5). Thus, X_{13} is entered as the first variable in the stepwise procedure (see Table 7-6). Because only one variable enters in the discriminant model at this time, the significance levels and measures of group differences match those of the univariate tests.

TABLE 7-6 Results from Step 1 of Stepwise Two-Group Discriminant Analysis

Overall Model Fit

	Value	F Value	Degrees of Freedom	Significance
Wilks' Lambda	.645	31.992	1,58	.000

Variable Entered/Removed at Step 1

Variable Entered	Minimum D^2	F Value	F Significance	Between Groups
X_{13} Competitive Pricing	2.171	31.992	.000	0 and 1

Note: At each step, the variable that maximizes the Mahalanobis distance between the two closest groups is entered.

Variables in the Analysis After Step 1

Variable	Tolerance	F to Remove	D^2	Between Groups
X_{13} Competitive Pricing	1.000	31.992		

Variables Not in the Analysis After Step 1

Variable	Tolerance	Minimum Tolerance	F to Enter	Minimum D^2	Between Groups
X_6 Product Quality	.965	.965	4.926	2.699	0 and 1
X_7 E-Commerce Activities	.917	.917	.026	2.174	0 and 1
X_8 Technical Support	.966	.966	.033	2.175	0 and 1
X_9 Complaint Resolution	.844	.844	1.292	2.310	0 and 1
X_{10} Advertising	.992	.992	.088	2.181	0 and 1
X_{11} Product Line	.849	.849	6.076	2.822	0 and 1
X_{12} Salesforce Image	.987	.987	3.949	2.595	0 and 1
X_{14} Warranty & Claims	.918	.918	.617	2.237	0 and 1
X_{15} New Products	1.000	1.000	.455	2.220	0 and 1
X_{16} Order & Billing	.836	.836	3.022	2.495	0 and 1
X_{17} Price Flexibility	1.000	1.000	19.863	4.300	0 and 1
X_{18} Delivery Speed	.910	.910	1.196	2.300	0 and 1

Significance Testing of Group Differences After Step 1[a]

		USA/North America
Outside North America	F	31.992
	Sig.	.000

[a]1, 58 degrees of freedom.

After X_{13} enters the model, the remaining variables are evaluated on the basis of their incremental discriminating ability (group mean differences after the variance associated with X_{13} is removed). Again, variables with significance levels greater than .05 are eliminated from consideration for entry at the next step.

Examining the univariate differences shown in Table 7-5 identifies X_{17} (Price Flexibility) as the variable with the second most significant differences. Yet the stepwise process does not use these univariate results when the discriminant function has one or more variables in the discriminant function. It calculates the D^2 values and statistical significance tests of group differences after the effect of the variable(s) in the models is removed (in this case only X_{13} is in the model).

As shown in the last portion of Table 7-6, three variables (X_6, X_{11}, and X_{17}) clearly met the .05 significance level criteria for consideration at the next stage. X_{17} remains the next best candidate to enter the model because it has the highest Mahalanobis D^2 (4.300) and the largest F to enter value. However, other variables (e.g., X_{11}) have substantial reductions in their significance level and the Mahalanobis D^2 from that shown in Table 7-5 due to the one variable in the model (X_{13}).

Stepwise Estimation: Adding the Second Variable X_{17}. In step 2 (see Table 7-7), X_{17} enters the model as expected. The overall model is significant ($F = 31.129$) and improves in the discrimination between groups as evidenced by the decrease in Wilks' lambda from .645 to .478. Moreover, the discriminating power of both variables included at this point is also statistically significant (F values of 20.113 for X_{13} and 19.863 for X_{17}). With both variables statistically significant, the procedure moves to examining the variables not in the equation for potential candidates for inclusion in the discriminant function based on their incremental discrimination between the groups.

X_{11} is the next variable meeting the requirements for inclusion, but its significance level and discriminating ability has been reduced substantially because of multicollinearity with X_{13} and X_{17} already in the discriminant function. Most noticeable is the marked increase in the Mahalanobis D^2 from the univariate results in which each variable is considered separately. In the case of X_{11} the minimum D^2 value increases from 1.731 (see Table 7-5) to 5.045 (see Table 7-7), indicative of a spreading out and separation of the groups by X_{13} and X_{17} already in the discriminant function. Note that X_{18} is almost identical in remaining discrimination power, but X_{11} will enter in the third step due to its slight advantage.

Stepwise Estimation: Adding a Third Variable X_{11}. Table 7-8 reviews the results of the third step of the stepwise process, where X_{11} does enter the discriminant function. The overall results are still statistically significant and continue to improve in discrimination, as evidenced by the decrease in the Wilks' lambda value (from .478 to .438). Note however that the decrease was much smaller than found when the second variable (X_{17}) was added to the discriminant function. With X_{13}, X_{17}, and X_{11} all statistically significant, the procedure moves to identifying any remaining candidates for inclusion.

As seen in the last portion of Table 7-8, none of the remaining 10 independent variables pass the entry criterion for statistical significance of .05. After X_{11} was entered in the equation, both of the remaining variables that had significant univariate differences across the groups (X_6 and X_{12}) have relatively little additional discriminatory power and do not meet the entry criterion. Thus, the estimation process stops with three variables (X_{13}, X_{17}, and X_{11}) constituting the discriminant function.

Summary of the Stepwise Estimation Process. Table 7-9 provides the overall stepwise discriminant analysis results after all the significant variables are included in the estimation of the discriminant function. This summary table describes the three variables (X_{11}, X_{13}, and X_{17}) that were significant discriminators based on their Wilks' lambda and minimum Mahalanobis D^2 values.

A number of different results are provided addressing both overall model fit and the impact of specific variables.

TABLE 7-7 Results from Step 2 of Stepwise Two-Group Discriminant Analysis

Overall Model Fit

	Value	F Value	Degrees of Freedom	Significance
Wilks' Lambda	.478	31.129	2,57	.000

Variable Entered/Removed at Step 2

		F		Between
Variable Entered	Minimum D^2	Value	Significance	Groups
X_{17} Price Flexibility	4.300	31.129	.000	0 and 1

Note: At each step, the variable that maximizes the Mahalanobis distance between the two closest groups is entered.

Variables in the Analysis After Step 2

Variable	Tolerance	F to Remove	D^2	Between Groups
X_{13} Competitive Pricing	1.000	20.113	2.152	0 and 1
X_{17} Price Flexibility	1.000	19.863	2.171	0 and 1

Variables Not in the Analysis After Step 2

Variable	Tolerance	Minimum Tolerance	F to Enter	Minimum D^2	Between Groups
X_6 Product Quality	.884	.884	.681	4.400	0 and 1
X_7 E-Commerce Activities	.804	.804	2.486	4.665	0 and 1
X_8 Technical Support	.966	.966	.052	4.308	0 and 1
X_9 Complaint Resolution	.610	.610	1.479	4.517	0 and 1
X_{10} Advertising	.901	.901	.881	4.429	0 and 1
X_{11} Product Line	.848	.848	5.068	5.045	0 and 1
X_{12} Salesforce Image	.944	.944	.849	4.425	0 and 1
X_{14} Warranty & Claims	.916	.916	.759	4.411	0 and 1
X_{15} New Products	.986	.986	.017	4.302	0 and 1
X_{16} Order & Billing	.625	.625	.245	4.336	0 and 1
X_{18} Delivery Speed	.519	.519	4.261	4.927	0 and 1

Significance Testing of Group Differences After Step 2[a]

	USA/North America	
Outside North America	F	32.129
	Sig.	.000

[a]2, 57 degrees of freedom.

- The multivariate measures of overall model fit are reported under the heading "Canonical Discriminant Functions." Note that the discriminant function is highly significant (.000) and displays a canonical correlation of .749. We interpret this correlation by squaring it $(.749)^2 =$ .561. Thus, 56.1 percent of the variance in the dependent variable (X_4) can be accounted for (explained) by this model, which includes only three independent variables.
- The standardized discriminant function coefficients are provided, but are less preferred for interpretation purposes than the discriminant loadings. The unstandardized discriminant coefficients are used to calculate the discriminant Z scores that can be used in classification.

TABLE 7-8   Results from Step 3 of Stepwise Two-Group Discriminant Analysis

Overall Model Fit

	Value	F Value	Degrees of Freedom	Significance
Wilks' Lambda	.438	23.923	3,56	.000

Variable Entered/Removed at Step 3

		F		
	Minimum D^2	Value	Significance	Between Groups
X_{11} Product Line	5.045	23.923	.000	0 and 1

Note: At each step, the variable that maximizes the Mahalanobis distance between the two closest groups is entered.

Variables in the Analysis After Step 3

Variable	Tolerance	F to Remove	D^2	Between Groups
X_{13} Competitive Pricing	.849	7.258	4.015	0 and 1
X_{17} Price Flexibility	.999	18.416	2.822	0 and 1
X_{11} Product Line	.848	5.068	4.300	0 and 1

Variables Not in the Analysis After Step 3

Variable	Tolerance	Minimum Tolerance	F to Enter	Minimum D^2	Between Groups
X_6 Product Quality	.802	.769	.019	5.048	0 and 1
X_7 E-Commerce Activities	.801	.791	2.672	5.482	0 and 1
X_8 Technical Support	.961	.832	.004	5.046	0 and 1
X_9 Complaint Resolution	.233	.233	.719	5.163	0 and 1
X_{10} Advertising	.900	.840	.636	5.149	0 and 1
X_{12} Salesforce Image	.931	.829	1.294	5.257	0 and 1
X_{14} Warranty & Claims	.836	.775	2.318	5.424	0 and 1
X_{15} New Products	.981	.844	.076	5.058	0 and 1
X_{16} Order & Billing	.400	.400	1.025	5.213	0 and 1
X_{18} Delivery Speed	.031	.031	.208	5.079	0 and 1

Significance Testing of Group Differences After Step 3[a]

		USA/North America
Outside North America	F	23.923
	Sig.	.000

[a]3, 56 degrees of freedom.

- The discriminant loadings are reported under the heading "Structure Matrix" and are ordered from highest to lowest by the size of the loading. The loadings are discussed later under the interpretation phase (stage 5).
- The classification function coefficients, also known as Fisher's linear discriminant functions, are used in classification and are discussed later.
- Group centroids are also reported, and they represent the mean of the individual discriminant function scores for each group. Group centroids provide a summary measure of the relative position of each group on the discriminant function(s). In this case, Table 7-9 reveals that the group centroid for the firms in USA/North America (group 0) is −1.273, whereas the group

TABLE 7-9 Summary Statistics for Two-Group Discriminant Analysis

Overall Model Fit: Canonical Discriminant Functions

Function	Eigenvalue	Percent of Variance Function %	Cumulative %	Canonical Correlation	Wilks' Lambda	Chi-Square	df	Significance
1	1.282	100	100	.749	.438	46.606	3	.000

Discriminant Function and Classification Function Coefficients

Independent Variables	Discriminant Functions Unstandardized	Standardized	Classification Functions Group 0: USA/North America	Group 1: Outside North America
X_{11} Product Line	−.363	−.417	7.725	6.909
X_{13} Competitive Pricing	.398	.490	6.456	7.349
X_{17} Price Flexibility	.749	.664	4.231	5.912
Constant	−3.752		−52.800	−60.623

Structure Matrix[a]

Independent Variables	Function 1
X_{13} Competitive Pricing	.656
X_{17} Price Flexibility	.653
X_{11} Product Line	−.586
X_7 E-Commerce Activities*	.429
X_6 Product Quality*	−.418
X_{14} Warranty & Claims*	−.329
X_{10} Advertising*	.238
X_9 Complaint Resolution*	−.181
X_{12} Salesforce Image*	.164
X_{16} Order & Billing*	−.149
X_8 Technical Support*	−.136
X_{18} Delivery Speed*	−.060
X_{15} New Products*	.041

*This variable not used in the analysis.

Group Means (Centroids) of Discriminant Functions

X_4 Region	Function 1
USA/North America	−1.273
Outside North America	.973

[a]Pooled within-groups correlations between discriminating variables and standardized canonical discriminant functions variables ordered by absolute size of correlation within function.

centroid for the firms outside North America (group 1) is .973. To show that the overall mean is 0, multiply the number in each group by its centroid and add the result (e.g., $26 \times -1.273 + 34 \times .973 = 0.0$).

The overall model results are acceptable based on statistical and practical significance. However, before proceeding to an interpretation of the results, the researcher needs to assess classification accuracy and examine the casewise results.

ASSESSING CLASSIFICATION ACCURACY With the overall model statistically significant and explaining 56 percent of the variation between the groups (see the preceding discussion and Table 7-9), we move to assessing the predictive accuracy of the discriminant function. In this example, we will illustrate the use of the discriminant scores and the cutting score for classification purposes. In doing so, we must complete three tasks:

1. Calculate the cutting score, the criterion against which each observation's discriminant Z score is judged to determine into which group it should be classified.
2. Classify each observation and develop the classification matrices for both the analysis and the holdout samples.
3. Assess the levels of predictive accuracy from the classification matrices for both statistical and practical significance.

Although examination of the holdout sample and its predictive accuracy is actually performed in the validation stage, the results are discussed now for ease of comparison between estimation and holdout samples.

Calculating the Cutting Score. The researcher must first determine how the prior probabilities of classification are to be determined, either based on the actual group sizes (assuming they are representative of the population) or specified by the researcher, most often specified as equal to be conservative in the classification process.

In this analysis sample of 60 observations, we know that the dependent variable consists of two groups, 26 firms located in the United States and 34 firms outside the United States. If we are not sure whether the population proportions are represented by the sample, then we should employ equal probabilities. However, because our sample of firms is randomly drawn, we can be reasonably sure that this sample does reflect the population proportions. Thus, this discriminant analysis uses the sample proportions to specify the prior probabilities for classification purposes.

Having specified the prior probabilities, the optimum cutting score can be calculated. Because in this situation the groups are assumed representative, the calculation becomes a weighted average of the two group centroids (see Table 7-9 for group centroid values):

$$Z_{CS} = \frac{N_A Z_B + N_B Z_A}{N_A + N_B} = \frac{(26 \times .973) + (34 \times -1.273)}{26 + 34} = -.2997$$

By substitution of the appropriate values in the formula, we can obtain the critical cutting score (assuming equal costs of misclassification) of $Z_{CS} = -.2997$.

Classifying Observations and Constructing the Classification Matrices. Once the cutting score has been calculated, each observation can be classified by comparing its discriminant score to the cutting score.

The procedure for classifying firms with the optimal cutting score is as follows:

- Classify a firm as being in group 0 (United States/North America) if its discriminant score is less than −.2997.
- Classify a firm as being in group 1 (Outside the United States) if its discriminant score is greater than −.2997.

Classification matrices for the observations in both the analysis and the holdout samples were calculated, and the results are shown in Table 7-10. Table 7-11 contains the discriminant scores for each observation as well as the actual and predicted group membership values. Note that cases with a discriminant score less that −.2997 have a predicted group membership value of 0, whereas those greater than −.2997 have a predicted value of 1. The analysis sample, with 86.7 percent prediction accuracy, is slightly higher than the 85.0 percent accuracy of the holdout sample, as anticipated. Moreover, the cross-validated sample achieved a prediction accuracy of 83.3 percent.

TABLE 7-10 Classification Results for Two-Group Discriminant Analysis

Classification Results[a, b, c]

| | | Predicted Group Membership | | |
| | | USA/ North America | Outside North America | Total |
Sample	Actual Group			
Estimation Sample	USA/North America	25 96.2%	1 3.8%	26
	Outside North America	7 20.6%	27 79.4%	34
Cross-validated[d]	USA/North America	24 92.3	2 7.7	26
	Outside North America	8 23.5	26 76.5	34
Holdout Sample	USA/North America	9 69.2	4 30.8	13
	Outside North America	2 7.4	25 92.6	27

[a]86.7% of selected original grouped cases (estimation sample) correctly classified.

[b]85.0% of unselected original grouped cases (validation sample) correctly classified.

[c]83.3% of selected cross-validated grouped cases correctly classified.

[d]Cross-validation is done only for those cases in the analysis (estimation sample). In cross-validation, each case is classified by the functions derived from all cases other than that case.

Evaluating the Achieved Classification Accuracy. Even though all of the measures of classification accuracy are quite high, the evaluation process requires a comparison to the classification accuracy in a series of chance-based measures. These measures reflect the improvement of the discriminant model when compared to classifying individuals without using the discriminant function. Given that the overall sample is 100 observations and group sizes in the holdout/validation sample are less than 20, we will use the overall sample to establish the comparison standards.

The first measure is the proportional chance criterion, which assumes that the costs of misclassification are equal (i.e., we want to identify members of each group equally well). The proportional chance criterion is:

$$C_{PRO} = p^2 + (1 - p)^2$$

where

$$C_{PRO} = \text{proportional chance criterion}$$
$$p = \text{proportion of firms in group 0}$$
$$1 - p = \text{proportion of firms in group 1}$$

The group of customers located within the United States (group 0) constitutes 39.0 percent of the analysis sample (39/100), with the second group representing customers located outside the United States (group 1) forming the remaining 61.0 percent (61/100). The calculated proportional chance value is .524 ($.390^2 + .610^2 = .524$).

The maximum chance criterion is simply the percentage correctly classified if all observations were placed in the group with the greatest probability of occurrence. It reflects our most conservative standard and assumes no difference in cost of misclassification as well.

Because group 1 (customers outside the United States) is the largest group at 61.0 percent of the sample, we would be correct 61.0 percent of the time if we assigned all observations to this group.

TABLE 7-11 Group Predictions for Individual Cases in the Two-Group Discriminant Analysis

Case ID	Actual Group	Discriminant Z Score	Predicted Group	Case ID	Actual Group	Discriminant Z Score	Predicted Group
Analysis Sample							
72	0	−2.10690	0	24	1	−.60937	0
14	0	−2.03496	0	53	1	−.45623	0
31	0	−1.98885	0	32	1	−.36094	0
54	0	−1.98885	0	80	1	−.14687	1
27	0	−1.76053	0	38	1	−.04489	1
29	0	−1.76053	0	60	1	−.04447	1
16	0	−1.71859	0	65	1	.09785	1
61	0	−1.71859	0	35	1	.84464	1
79	0	−1.57916	0	1	1	.98896	1
36	0	−1.57108	0	4	1	1.10834	1
98	0	−1.57108	0	68	1	1.12436	1
58	0	−1.48136	0	44	1	1.34768	1
45	0	−1.33840	0	17	1	1.35578	1
2	0	−1.29645	0	67	1	1.35578	1
52	0	−1.29645	0	33	1	1.42147	1
50	0	−1.24651	0	87	1	1.57544	1
47	0	−1.20903	0	6	1	1.58353	1
88	0	−1.10294	0	46	1	1.60411	1
11	0	−.74943	0	12	1	1.75931	1
56	0	−.73978	0	69	1	1.82233	1
95	0	−.73978	0	86	1	1.82233	1
81	0	−.72876	0	10	1	1.85847	1
5	0	−.60845	0	30	1	1.90062	1
37	0	−.60845	0	15	1	1.91724	1
63	0	−.38398	0	92	1	1.97960	1
43	0	.23553	1	7	1	2.09505	1
3	1	−1.65744	0	20	1	2.22839	1
94	1	−1.57916	0	8	1	2.39938	1
49	1	−1.04667	0	100	1	2.62102	1
64	1	−.67406	0	48	1	2.90178	1
Holdout Sample							
23	0	22.38834	0	25	1	1.47048	1
93	0	−2.03496	0	18	1	1.60411	1
59	0	−1.20903	0	73	1	1.61002	1
85	0	−1.10294	0	21	1	1.69348	1
83	0	−1.03619	0	90	1	1.69715	1
91	0	−.89292	0	97	1	1.70398	1
82	0	−.74943	0	40	1	1.75931	1
76	0	−.72876	0	77	1	1.86055	1
96	0	−.57335	0	28	1	1.97494	1
13	0	.13119	1	71	1	2.22839	1
89	0	.51418	1	19	1	2.28652	1
42	0	.63440	1	57	1	2.31456	1
78	0	.63440	1	9	1	2.36823	1
22	1	−2.73303	0	41	1	2.53652	1
74	1	−1.04667	0	26	1	2.59447	1
51	1	.09785	1	70	1	2.59447	1
62	1	.94702	1	66	1	2.90178	1
75	1	.98896	1	34	1	2.97632	1
99	1	1.13130	1	55	1	2.97632	1
84	1	1.30393	1	39	1	3.21116	1

If we choose the maximum chance criterion as the standard of evaluation, our model should outperform the 61.0 percent level of classification accuracy to be acceptable.

To attempt to ensure practical significance, the achieved classification accuracy must exceed the selected comparison standard by 25 percent. Thus, we must select a comparison standard, calculate the threshold, and compare the achieved hit ratio.

All of the classification accuracy levels (hit ratios) exceed 85 percent, which are substantially higher than the proportional chance criterion of 52.4 percent and the maximum chance criterion of 61.0 percent. All three hit ratios also exceed the suggested threshold of these values (comparison standard plus 25 percent), which in this case are 65.5 percent ($52.4\% \times 1.25 = 65.5\%$) for the proportional chance and 76.3 percent ($61.0\% \times 1.25 = 76.3\%$) for the maximum chance. In all instances (analysis sample, holdout sample, and cross-validation), the levels of classification accuracy are substantially higher than the threshold values, indicating an acceptable level of classification accuracy. Moreover, the hit ratio for individual groups is deemed adequate as well.

The final measure of classification accuracy is Press's Q, which is a statistically based measure comparing the classification accuracy to a random process.

From the earlier discussion, the calculation for the estimation sample is

$$\text{Press's } Q_{\text{estimate sample}} = \frac{[60 - (52 \times 2)]^2}{60(2 - 1)} = 45.07$$

And the calculation for the holdout sample is

$$\text{Press's } Q_{\text{holdout sample}} = \frac{[40 - (34 \times 2)]^2}{40(2 - 1)} = 19.6$$

In both instances, the calculated values exceed the critical value of 6.63. Thus, the classification accuracy for the analysis and, more important, the holdout sample exceeds at a statistically significant level the classification accuracy expected by chance.

CASEWISE DIAGNOSTICS   In addition to examining the overall results, we can examine the individual observations for their predictive accuracy and identify specifically the misclassified cases. In this manner, we can find the specific cases misclassified for each group on both analysis and holdout samples and perform additional analysis profiling for the misclassified cases.

Table 7-11 contains the group predictions for the analysis and holdout samples and enables us to identify the specific cases for each type of misclassification tabulated in the classification matrices (see Table 7-10). For the analysis sample, the seven customers located outside the United States misclassified into the group of customers in the United States can be identified as cases 3, 94, 49, 64, 24, 53, and 32. Likewise, the single customer located in the United States but misclassified is identified as case 43. A similar examination can be performed for the holdout sample.

Once the misclassified cases are identified, further analysis can be performed to understand the reasons for their misclassification. In Table 7-12, the misclassified cases are combined from the analysis and holdout samples and then compared to the correctly classified cases. The attempt is to identify specific differences on the independent variables that might identify either new variables to be added or common characteristics that should be considered.

The five cases (both analysis and holdout samples) misclassified among the United States customers (group 0) show significant differences on two of the three independent variables in the discriminant function (X_{13} and X_{17}) as well as one variable not in the discriminant function (X_6). For that variable not in the discriminant function, the profile of the misclassified cases is not similar to their correct group; thus, it is no help in classification. Likewise, the nine misclassified cases of group 1 (outside the United States) show four significant differences (X_6, X_{11}, X_{13}, and X_{17}), but only X_6 is not in the discriminant function. We can see that here X_6 works against classification accuracy because the misclassified cases are more similar to the incorrect group rather than the correct group.

TABLE 7-12   Profiling Correctly Classified and Misclassified Observations in the Two-Group Discriminant Analysis

Dependent Variable: X_4 Region	Group/Profile Variables	Mean Scores			t Test
		Correctly Classified	Misclassified	Difference	Statistical Significance
USA/North America		($n = 34$)	($n = 5$)		
	X_6 Product Quality	8.612	9.340	−.728	.000[b]
	X_7 E-Commerce Activities	3.382	4.380	−.998	.068[b]
	X_8 Technical Support	5.759	5.280	.479	.487
	X_9 Complaint Resolution	5.356	6.140	−.784	.149
	X_{10} Advertising	3.597	4.700	−1.103	.022
	X_{11} Product Line[a]	6.726	6.540	.186	.345[b]
	X_{12} Salesforce Image	4.459	5.460	−1.001	.018
	X_{13} Competitive Pricing[a]	5.609	8.060	−2.451	.000
	X_{14} Warranty & Claims	6.215	6.060	.155	.677
	X_{15} New Products	5.024	4.420	.604	.391
	X_{16} Order & Billing	4.188	4.540	−.352	.329
	X_{17} Price Flexibility[a]	3.568	4.480	−.912	.000[b]
	X_{18} Delivery Speed	3.826	4.160	−.334	.027[b]
Outside North America		($n = 52$)	($n = 9$)		
	X_6 Product Quality	6.906	9.156	−2.250	.000
	X_7 E-Commerce Activities	3.860	3.289	.571	.159[b]
	X_8 Technical Support	5.085	5.544	−.460	.423
	X_9 Complaint Resolution	5.365	5.822	−.457	.322
	X_{10} Advertising	4.229	3.922	.307	.470
	X_{11} Product Line[a]	4.954	6.833	−1.879	.000
	X_{12} Salesforce Image	5.465	5.467	−.002	.998
	X_{13} Competitive Pricing[a]	7.960	5.833	2.126	.000
	X_{14} Warranty & Claims	5.867	6.400	−.533	.007[b]
	X_{15} New Products	5.194	5.778	−.584	.291
	X_{16} Order & Billing	4.267	4.533	−.266	.481
	X_{17} Price Flexibility[a]	5.458	3.722	1.735	.000
	X_{18} Delivery Speed	3.881	3.989	−.108	.714

Note: Cases from both analysis and validation samples included for total sample of 100.

[a]Variables included in the discriminant function.

[b]t test performed with separate variance estimates rather than pooled estimate because the Levene test detected significant differences in the variations between the two groups.

The findings suggest that the misclassified cases may represent a distinct third group, because they share quite similar profiles across these variables more so than they do with the two existing groups. Management may analyze this group on additional variables or assess whether a geographic pattern among these misclassified cases justifies a new group.

Researchers should examine the patterns in both groups with the objective of understanding the characteristics common to them in an attempt at defining the reasons for misclassification.

Stage 5: Interpretation of the Results

After estimating the discriminant function, the next task is interpretation. This stage involves examining the function to determine the relative importance of each independent variable in discriminating between the groups, interpreting the discriminant function based on the discriminant loadings, and then profiling each group on the pattern of mean values for variables identified as important discriminating variables.

IDENTIFYING IMPORTANT DISCRIMINATING VARIABLES As discussed earlier, discriminant loadings are considered the more appropriate measure of discriminatory power, but we will also consider the discriminant weights for comparative purposes. The discriminant weights, either in unstandardized or standardized form, represent each variable's contribution to the discriminant function. However, as we will discuss, multicollinearity among the independent variables can impact the interpretation using only the weights.

Discriminant loadings are calculated for every independent variable, even for those not included in the discriminant function. Thus, discriminant weights represent the unique impact of each independent variable and are not restricted to only the shared impact due to multicollinearity. Moreover, because they are relatively unaffected by multicollinearity, they more accurately represent each variable's association with the discriminant score.

Table 7-13 contains the entire set of interpretive measures, including unstandardized and standardized discriminant weights, loadings for the discriminant function, Wilks' lambda, and the univariate F ratio. The original 13 independent variables were screened by the stepwise procedure, and three (X_{11}, X_{13}, and X_{17}) are significant enough to be included in the function. For interpretation purposes, we rank the independent variables in terms of their loadings and univariate F values—both indicators of each variable's discriminating power. Signs of the weights or loadings do not affect the rankings; they simply indicate a positive or negative relationship with the dependent variable.

Analyzing Wilks' Lambda and Univariate F. The Wilks' lambda and univariate F values represent the separate or univariate effects of each variable, not considering multicollinearity among the independent variables. Analogous to the bivariate correlations of multiple regression, they indicate each variable's ability to discriminate among the groups, but only separately. To interpret any combination of two or more independent variables requires analysis of the discriminant weights or discriminant loadings as described in the following sections.

TABLE 7-13   Summary of Interpretive Measures for Two-Group Discriminant Analysis

Independent Variables	Discriminant Coefficients		Discriminant Loadings		Wilks' Lambda	Univariate F Ratio		
	Unstandardized	Standardized	Loading	Rank	Value	F Value	Sig.	Rank
X_6 Product Quality	NI	NI	−.418	5	.801	14.387	.000	4
X_7 E-Commerce Activities	NI	NI	.429	4	.966	2.054	.157	6
X_8 Technical Support	NI	NI	−.136	11	.973	1.598	.211	7
X_9 Complaint Resolution	NI	NI	−.181	8	.986	.849	.361	8
X_{10} Advertising	NI	NI	.238	7	.987	.775	.382	9
X_{11} Product Line	−.363	−.417	−.586	3	.695	25.500	.000	3
X_{12} Salesforce Image	NI	NI	.164	9	.856	9.733	.003	5
X_{13} Competitive Pricing	−.398	.490	.656	1	.645	31.992	.000	1
X_{14} Warranty & Claims	NI	NI	−.329	6	.992	.453	.503	11
X_{15} New Products	NI	NI	.041	13	.990	.600	.442	10
X_{16} Order & Billing	NI	NI	−.149	10	.999	.087	.769	13
X_{17} Price Flexibility	.749	.664	.653	2	.647	31.699	.000	2
X_{18} Delivery Speed	NI	NI	−.060	12	.997	.152	.698	12

NI = Not included in estimated discriminant function.

Table 7-13 shows that the variables (X_{11}, X_{13}, and X_{17}) with the three highest F values (and lowest Wilks' lambda values) were also the variables entered into the discriminant function. Two other variables (X_6 and X_{12}) also had significant discriminating effects (i.e., significant group differences), yet were not included by the stepwise process in the discriminant function. This was due to the multicollinearity between these two variables and the three variables included in the discriminant function. These two variables added no incremental discriminating power beyond the variables already in the discriminant function. Interested readers are referred to a more complete discussion of multicollinearity and the stepwise estimation process in Chapter 4. All of the remaining variables had nonsignificant F values and correspondingly high Wilks' lambda values.

Analyzing the Discriminant Weights. The discriminant weights are available in unstandardized and standardized forms. The unstandardized weights (plus the constant) are used to calculate the discriminant score, but can be affected by the scale of the independent variable (just like multiple regression weights). Thus, the standardized weights more truly reflect the impact of each variable on the discriminant function and are more appropriate than unstandardized weights when used for interpretation purposes. If simultaneous estimation is used, multicollinearity among any of the independent variables will impact the estimated weights. However, the impact of multicollinearity can be even greater for the stepwise procedure, because multicollinearity affects not only the weights but may also prevent a variable from even entering the equation.

Table 7-13 provides the standardized weights (coefficients) for the three variables included in the discriminant function. The impact of multicollinearity on the weights can be seen in examining X_{13} and X_{17}. These two variables have essentially equivalent discriminating power when viewed on the Wilks' lambda and univariate F tests. Their discriminant weights, however, reflect a markedly greater impact for X_{17} than X_{13}, which based on the weights is now more comparable to X_{11}. This change in relative importance is due to the collinearity between X_{13} and X_{11}, which reduces the unique effect of X_{13} and thus reduces the discriminant weight as well.

INTERPRETING THE DISCRIMINANT FUNCTION BASED ON DISCRIMINANT LOADINGS The discriminant loadings, in contrast to the discriminant weights, are less affected by multicollinearity and thus more useful for interpretative purposes. Also, because loadings are calculated for all variables, they provide an interpretive measure even for variables not included in the discriminant function. An earlier rule of thumb indicated loadings above ±.40 should be used to identify substantive discriminating variables.

The loadings of the three variables entered in the discriminant function (see Table 7-13) are the three highest and all exceed ±.40, thus warranting inclusion for interpretation purposes. Two additional variables (X_6 and X_7), however, also have loadings above the ±.40 threshold. The inclusion of X_6 is not unexpected, as it was the fourth variable with significant univariate discriminating effect, but was not included in the discriminant function due to multicollinearity (as was shown in Chapter 3, Factor Analysis, where X_6 and X_{13} formed a factor). X_7, however, presents another situation; it did not have a significant univariate effect. The combination of the three variables in the discriminant function created an effect that is associated with X_7, but X_7 does not add any additional discriminating power. In this regard, X_7 can be used to describe the discriminant function for profiling purposes even though it did not enter into the estimation of the discriminant function.

Interpreting the discriminant function and its discrimination between these two groups requires that the researcher consider all five of these variables. To the extent that they characterize or describe the discriminant function, they all represent some component of the function.

The three strongest effects in the discriminant function, which are all generally comparable based on the loading values, are X_{13} (Competitive Pricing), X_{17} (Price Flexibility), and X_{11} (Product Line). X_7 (E-Commerce Activities) and the effect of X_6 (Product Quality) can be added when interpreting the discriminant function. Obviously several different factors are being combined to differentiate between the groups, thus requiring more profiling of the groups to understand the differences.

With the discriminating variables identified and the discriminant function described in terms of those variables with sufficiently high loadings, the researcher then proceeds to profile each group on these variables to understand the differences between them.

PROFILING THE DISCRIMINATING VARIABLES The researcher is interested in interpretations of the individual variables that have statistical and practical significance. Such interpretations are accomplished by first identifying the variables with substantive discriminatory power (see the preceding discussions) and then understanding what the differing group means on each variable indicated.

As described in Chapter 1, higher scores on the independent variables indicate more favorable perceptions of HBAT on that attribute (except for X_{13}, where lower scores are more preferable). Referring back to Table 7-5, we see varied profiles between the two groups on these five variables.

- Group 0 (customers in the USA/North America) has higher perceptions on three variables: X_6 (Product Quality), X_{13} (Competitive Pricing), and X_{11} (Product Line).
- Group 1 (customers outside North America) has higher perceptions on the remaining two variables: X_7 (E-Commerce Activities) and X_{17} (Price Flexibility).

In looking at these two profiles, we can see that the USA/North America customers have much better perceptions of the HBAT products, whereas those customers outside North America feel better about pricing issues and e-commerce. Note that X_6 and X_{13}, both of which have higher perceptions among the USA/North America customers, form the *Product Value* factor developed in Chapter 3. Management should use these results to develop strategies that accentuate these strengths and develop additional strengths to complement them.

The mean profiles also illustrate the interpretation of signs (positive or negative) on the discriminant weights and loadings. The signs reflect the relative mean profile of the two groups. The positive signs, in this example, are associated with variables that have higher scores for group 1. The negative weights and loadings are for those variables with the opposite pattern (i.e., higher values in group 0). Thus, the signs indicate the pattern between the groups.

Stage 6: Validation of the Results

The final stage addresses the internal and external validity of the discriminant function. The primary means of validation is through the use of the holdout sample and the assessment of its predictive accuracy. In this manner, validity is established if the discriminant function performs at an acceptable level in classifying observations that were not used in the estimation process. If the holdout sample is formed from the original sample, then this approach establishes internal validity and an initial indication of external validity. If another separate sample, perhaps from another population or segment of the population, forms the holdout sample, then this addresses more fully the external validity of the discriminant results.

In our example, the holdout sample comes from the original sample. As discussed earlier, the classification accuracy (hit ratios) for both the holdout sample and the cross-validated sample was markedly above the thresholds on all of the measures of predictive accuracy. As such, the analysis does establish internal validity. For purposes of external validity, additional samples should be drawn from relevant populations and the classification accuracy assessed in as many situations as possible.

The researcher is encouraged to extend the validation process through expanded profiling of the groups and the possible use of additional samples to establish external validity. Additional insights from the analysis of misclassified cases may suggest additional variables that could improve even more the discriminant model.

A Managerial Overview

The discriminant analysis of HBAT customers based on geographic location (located within North America or outside) identified a set of perceptual differences that can provide a rather succinct and powerful distinction between the two groups. Several key findings include the following:

- Differences are found in a subset of only five perceptions, allowing for a focus on key variables and not having to deal with the entire set. The variables identified as discriminating between the groups (listed in order of importance) are X_{13} (Competitive Pricing), X_{17} (Price Flexibility), X_{11} (Product Line), X_7 (E-Commerce Activities), and X_6 (Product Quality).
- Results also indicate that firms located in the United States have better perceptions of HBAT than their international counterparts in terms of product value and the product line, whereas the non–North American customers have a more favorable perception of price flexibility and e-commerce activities. These perceptions may result from a better match between USA/North American buyers, whereas the international customers find the pricing policies conducive to their needs.
- The results, which are highly significant, provide the researcher the ability to correctly identify the purchasing strategy used based on these perceptions 85 percent of the time. Their high degree of consistency provides confidence in the development of strategies based on these results.
- Analysis of the misclassified firms revealed a small number of firms that seemed out of place. Identifying these firms may identify associations not addressed by geographic location (e.g., markets served rather than just physical location) or other firm or market characteristics that are associated with geographic location.

Thus, knowing a firm's geographic location provides key insights into their perceptions of HBAT and, more importantly, how the two groups of customers differ so that management can employ a strategy to accentuate the positive perceptions in their dealings with these customers and further solidify their position.

A THREE-GROUP ILLUSTRATIVE EXAMPLE

To illustrate the application of a three-group discriminant analysis, we once again use the HBAT database. In the previous example, we were concerned with discriminating between only two groups, so we were able to develop a single discriminant function and a cutting score to divide the two groups. In the three-group example, it is necessary to develop two separate discriminant functions to distinguish among three groups. The first function separates one group from the other two, and the second separates the remaining two groups. As with the prior example, the six stages of the model-building process are discussed.

Stage 1: Objectives of Discriminant Analysis

HBAT's objective in this research is to determine the relationship between the firms' perceptions of HBAT and the length of time a firm has been a customer with HBAT.

One of the emerging paradigms in marketing is the concept of a customer relationship, based on the establishment of a mutual partnership between firms over repeated transactions. The process of developing a relationship entails the formation of shared goals and values, which should coincide with improved perceptions of HBAT. Thus, the successful formation of a relationship should be seen by improved HBAT perceptions over time. In this analysis, firms are grouped on their tenure as HBAT customers. Hopefully, if HBAT has been successful in establishing relationships with its customers, then perceptions of HBAT will improve with tenure as an HBAT customer.

Stage 2: Research Design for Discriminant Analysis

To test this relationship, a discriminant analysis is performed to establish whether differences in perceptions exist between customer groups based on length of customer relationship. If so, HBAT is then interested in seeing whether the distinguishing profiles support the proposition that HBAT has been successful in improving perceptions among established customers, a necessary step in the formation of customer relationships.

SELECTION OF DEPENDENT AND INDEPENDENT VARIABLES   In addition to the nonmetric (categorical) dependent variables defining the groups of interest, discriminant analysis also requires a set of metric independent variables that are assumed to provide the basis for discrimination or differentiation between the groups.

A three-group discriminant analysis is performed using X_1 (Customer Type) as the dependent variable and the perceptions of HBAT by these firms (X_6 to X_{18}) as the independent variables. Note that X_1 differs from the dependent variable in the two-group example in that it has three categories in which to classify a firm's length of time being an HBAT customer (1 = less than 1 year, 2 = 1 to 5 years, and 3 = more than 5 years).

SAMPLE SIZE AND DIVISION OF THE SAMPLE   Issues regarding sample size are particularly important with discriminant analysis due to the focus on not only overall sample size, but also sample size per group. Coupled with the need for a division of the sample to provide for a validation sample, the researcher must carefully consider the impact of sample division on both samples in terms of the overall sample size and the size of each of the groups.

The HBAT database has a sample size of 100, which again will be split into analysis and holdout samples of 60 and 40 cases, respectively. In the analysis sample, the ratio of cases to independent variables is almost 5:1, the recommended lower threshold. More importantly, in the analysis sample, only one group, with 13 observations, falls below the recommended level of 20 cases per group. Although the group size would exceed 20 if the entire sample were used in the analysis phase, the need for validation dictated the creation of the holdout sample. The three groups are of relatively equal sizes (22, 13, and 25), thus avoiding any need to equalize the group sizes. The analysis proceeds with attention paid to the classification and interpretation of this small group of 13 observations.

Stage 3: Assumptions of Discriminant Analysis

As was the case in the two-group example, the assumptions of normality, linearity, and collinearity of the independent variables have already been discussed at length in Chapter 2. The analyses performed in Chapter 2 indicated that the independent variables met these assumptions at adequate levels to allow for the analysis to continue without additional remedies. The remaining assumption, the equality of the variance/covariance or dispersion matrices, is also addressed in Chapter 2.

Box's M test assesses the similarity of the dispersion matrices of the independent variables among the three groups (categories). The test statistic indicated differences at the .09 significance level. In this case, the differences between groups are nonsignificant and no remedial action is needed. Moreover, no impacts are expected on the estimation or classification processes.

Stage 4: Estimation of the Discriminant Model and Assessing Overall Fit

As in the previous example, we begin our analysis by reviewing the group means and standard deviations to see whether the groups are significantly different on any single variable. With those differences in mind, we then employ a stepwise estimation procedure to derive the discriminant functions and complete the process by assessing classification accuracy both overall and with casewise diagnostics.

ASSESSING GROUP DIFFERENCES   Identifying the most discriminating variables with three or more groups is more problematic than in the two-group situation. For three or more groups, the typical measures of significance for differences across groups (i.e., Wilks' lambda and the F test) only assess the overall differences and do not guarantee that each group is significant from the others. Thus, when examining variables for their overall differences between the groups, be sure to also address individual group differences.

Table 7-14 provides the group means, Wilks' lambda, univariate F ratios (simple ANOVAs), and minimum Mahalanobis D^2 for each independent variable. Review of these measures of discrimination reveals the following:

- On a univariate basis, about one-half (7 of 13) of the variables display significant differences between the group means. The variables with significant differences include X_6, X_9, X_{11}, X_{13}, X_{16}, X_{17}, and X_{18}.
- Although greater statistical significance corresponds to higher overall discrimination (i.e., the most significant variables have the lowest Wilks' lambda values), it does not always correspond to the greatest discrimination between all the groups.
  - Visual inspection of the group means reveal that four of the variables with significant differences (X_{13}, X_{16}, X_{17}, and X_{18}) only differentiate one group versus the other two groups [e.g., X_{18} has significant differences only in the means between group 1 (3.059) versus groups 2 (4.246) and 3 (4.288)]. These variables play a limited role in discriminant analysis because they provide discrimination between only a subset of groups.
  - Three variables (X_6, X_9, and X_{11}) provide some discrimination, in varying degrees, between all three groups simultaneously. One or more of these variables may be used in combination with the four preceding variables to create a variate with maximum discrimination.

TABLE 7-14   Group Descriptive Statistics and Tests of Equality for the Estimation Sample in the Three-Group Discriminant Analysis

Independent Variables	Dependent Variable Group Means: X_1 Customer Type			Test of Equality of Group Means[a]			Minimum Mahalanobis D^2	
	Group 1: Less than 1 Year ($n = 22$)	Group 2: 1 to 5 Years ($n = 13$)	Group 3: More than 5 Years ($n = 25$)	Wilks' Lambda	F Value	Significance	Minimum D^2	Between Groups
X_6 Product Quality	7.118	6.785	9.000	.469	32.311	.000	.121	1 and 2
X_7 E-Commerce Activities	3.514	3.754	3.412	.959	1.221	.303	.025	1 and 3
X_8 Technical Support	4.959	5.615	5.376	.973	.782	.462	.023	2 and 3
X_9 Complaint Resolution	4.064	5.900	6.300	.414	40.292	.000	.205	2 and 3
X_{10} Advertising	3.745	4.277	3.768	.961	1.147	.325	.000	1 and 3
X_{11} Product Line	4.855	5.577	7.056	.467	32.583	.000	.579	1 and 2
X_{12} Salesforce Image	4.673	5.346	4.836	.943	1.708	.190	.024	1 and 3
X_{13} Competitive Pricing	7.345	7.123	5.744	.751	9.432	.000	.027	1 and 2
X_{14} Warranty & Claims	5.705	6.246	6.072	.916	2.619	.082	.057	2 and 3
X_{15} New Products	4.986	5.092	5.292	.992	.216	.807	.004	1 and 2
X_{16} Order & Billing	3.291	4.715	4.700	.532	25.048	.000	.000	2 and 3
X_{17} Price Flexibility	4.018	5.508	4.084	.694	12.551	.000	.005	1 and 3
X_{18} Delivery Speed	3.059	4.246	4.288	.415	40.176	.000	.007	2 and 3

[a]Wilks' lambda (U statistic) and univariate F ratio with 2 and 57 degrees of freedom.

- The Mahalanobis D^2 value provides a measure of the degree of discrimination between groups. For each variable, the minimum Mahalanobis D^2 is the distance between the two closest groups. For example, X_{11} has the highest D^2 value, and it is the variable with the greatest differences between all three groups. Likewise, X_{18}, a variable with little differences between two of the groups, has a small D^2 value. With three or more groups, the minimum Mahalanobis D^2 is important in identifying the variable that provides the greatest difference between the two most similar groups.

All of these measures combine to help identify the sets of variables that form the discriminant functions as described in the next section. When more than one function is created, each function provides discrimination between sets of groups. In the simple example from the beginning of this chapter, one variable discriminated between groups 1 versus 2 and 3, whereas the other discriminated between groups 2 versus 3 and 1. It is one of the primary benefits arising from the use of discriminant analysis.

ESTIMATION OF THE DISCRIMINANT FUNCTION   The stepwise procedure is performed in the same manner as in the two-group example, with all of the variables initially excluded from the model. As noted earlier, the Mahalanobis distance should be used with the stepwise procedure in order to select the variable that has a statistically significant difference across the groups while maximizing the Mahalanobis distance (D_2) between the two closest groups. In this manner, statistically significant variables are selected that maximize the discrimination between the most similar groups at each stage.

This process continues as long as additional variables provide statistically significant discrimination beyond those differences already accounted for by the variables in the discriminant function. A variable may be removed if high multicollinearity with independent variables in the discriminant function causes its significance to fall below the significance level for removal (.10).

Stepwise Estimation: Adding the First Variable, X_{11}. The data in Table 7-14 show that the first variable to enter the stepwise model using the Mahalanobis distance is X_{11} (Product Line) because it meets the criteria for statistically significant differences across the groups and has the largest minimum D^2 value (meaning it has the greatest separation between the most similar groups).

The results of adding X_{11} as the first variable in the stepwise process are shown in Table 7-15. The overall model fit is significant and each of the groups are significantly different, although groups 1 (less than 1 year) and 2 (1 to 5 years) have the smallest difference between them (see bottom section detailing group differences).

With the smallest difference between groups 1 and 2, the discriminant procedure will now select a variable that maximizes that difference while at least maintaining the other differences. If we refer back to Table 7-14, we see that four variables (X_9, X_{16}, X_{17}, and X_{18}) all had significant differences, with substantial differences between groups 1 and 2. Looking in Table 7-15, we see that these four variables have the highest minimum D^2 value, and in each case it is for the difference between groups 2 and 3 (meaning that groups 1 and 2 are not the most similar after adding that variable). Thus, adding any one of these variables would most affect the differences between groups 1 and 2, the pair that was most similar after X_{11} was added in the first step. The procedure will select X_{17} because it will create the greatest distance between groups 2 and 3.

Stepwise Estimation: Adding the Second Variable, X_{17}. Table 7-16 details the second step of the stepwise procedure: adding X_{17} (Price Flexibility) to the discriminant function. The discrimination between groups increased, as reflected in a lower Wilks' lambda value and increase in the minimum D^2 (.467 to .288). The group differences, overall and individual, are still statistically significant. The addition of X_{17} increased the differences between groups 1 and 2 substantially, such that now the two most similar groups are 2 and 3.

TABLE 7-15 Results from Step 1 of Stepwise Three-Group Discriminant Analysis

Overall Model Fit

	Value	F Value	Degrees of Freedom	Significance
Wilks' Lambda	.467	32.583	2,57	.000

Variable Entered/Removed at Step 1

		F		
Variable Entered	Minimum D^2	Value	Significance	Between Groups
X_{11} Product Line	.579	4.729	.000	Less than 1 year and 1 to 5 years

Note: At each step, the variable that maximizes the Mahalanobis distance between the two closest groups is entered.

Variables in the Analysis After Step 1

Variable	Tolerance	F to Remove	D^2	Between Groups
X_{11} Product Line	1.000	32.583	NA	NA

NA = Not applicable.

Variables Not in the Analysis After Step 1

Variable	Tolerance	Minimum Tolerance	F to Enter	Minimum D^2	Between Groups
X_6 Product Quality	1.000	1.000	17.426	.698	Less than 1 year and 1 to 5 years
X_7 E-Commerce Activities	.950	.950	1.171	.892	Less than 1 year and 1 to 5 years
X_8 Technical Support	.959	.959	.733	.649	Less than 1 year and 1 to 5 years
X_9 Complaint Resolution	.847	.847	15.446	2.455	1 to 5 years and more than 5 years
X_{10} Advertising	.998	.998	1.113	.850	Less than 1 year and 1 to 5 years
X_{12} Salesforce Image	.932	.932	3.076	1.328	Less than 1 year and 1 to 5 years
X_{13} Competitive Pricing	.849	.849	.647	.599	Less than 1 year and 1 to 5 years
X_{14} Warranty & Claims	.882	.882	2.299	.839	Less than 1 year and 1 to 5 years
X_{15} New Products	.993	.993	.415	.596	Less than 1 year and 1 to 5 years
X_{16} Order & Billing	.943	.943	12.176	2.590	1 to 5 years and more than 5 years
X_{17} Price Flexibility	.807	.807	17.300	3.322	1 to 5 years and more than 5 years
X_{18} Delivery Speed	.773	.773	19.020	2.988	1 to 5 years and more than 5 years

Significance Testing of Group Differences After Step 1[a]

X_1 – Customer Type		Less than 1 Year	1 to 5 Years
1 to 5 years	F	4.729	
	Sig.	.034	
Over 5 years	F	62.893	20.749
	Sig.	.000	.000

[a]1, 57 degrees of freedom.

Of the variables not in the equation, only X_6 (Product Quality) meets the significance level necessary for consideration. If added, the minimum D^2 will now be between groups 1 and 2.

Stepwise Estimation: Adding the Third and Fourth Variables, X_6 and X_{18}. As noted previously, X_6 becomes the third variable added to the discriminant function. After X_6 is added, only X_{18} exhibits a statistical significance across the groups (*Note:* The details of adding X_6 in step 3 are not shown for space considerations).

TABLE 7-16 **Results from Step 2 of Stepwise Three-Group Discriminant Analysis**

Overall Model Fit

	Value	F Value	Degrees of Freedom	Significance
Wilks' Lambda	.288	24.139	4, 112	.000

Variable Entered/Removed at Step 2

		F		
Variable Entered	Minimum D^2	Value	Significance	Between Groups
X_{17} Price Flexibility	3.322	13.958	.000	1 to 5 years and more than 5 years

Note: At each step, the variable that maximizes the Mahalanobis distance between the two closest groups is entered.

Variables in the Analysis After Step 2

Variable	Tolerance	F to Remove	D^2	Between Groups
X_{11} Product Line	.807	39.405	.005	Less than 1 year and more than 5 years
X_{17} Price Flexibility	.807	17.300	.579	Less than 1 year and 1 to 5 years

Variables Not in the Analysis After Step 2

Variable	Tolerance	Minimum Tolerance	F to Enter	Minimum D^2	Between Groups
X_6 Product Quality	.730	.589	24.444	6.071	Less than 1 year and 1 to 5 years
X_7 E-Commerce Activities	.880	.747	.014	3.327	Less than 1 year and 1 to 5 years
X_8 Technical Support	.949	.791	1.023	3.655	Less than 1 year and 1 to 5 years
X_9 Complaint Resolution	.520	.475	3.932	3.608	Less than 1 year and 1 to 5 years
X_{10} Advertising	.935	.756	.102	3.348	Less than 1 year and 1 to 5 years
X_{12} Salesforce Image	.884	.765	.662	3.342	Less than 1 year and 1 to 5 years
X_{13} Competitive Pricing	.794	.750	.989	3.372	Less than 1 year and 1 to 5 years
X_{14} Warranty & Claims	.868	.750	2.733	4.225	Less than 1 year and 1 to 5 years
X_{15} New Products	.963	.782	.504	3.505	Less than 1 year and 1 to 5 years
X_{16} Order & Billing	.754	.645	2.456	3.323	Less than 1 year and 1 to 5 years
X_{18} Delivery Speed	.067	.067	3.255	3.598	Less than 1 year and 1 to 5 years

Significance Testing of Group Differences After Step 2[a]

X_1 Customer Type		Less than 1 Year	1 to 5 Years
1 to 5 years	F	21.054	
	Sig.	.000	
More than 5 years	F	39.360	13.958
	Sig.	.000	.000

[a]2, 56 degrees of freedom.

The final variable added in step 4 is X_{18} (see Table 7-17), with the discriminant function now including four variables (X_{11}, X_{17}, X_6, and X_{18}). The overall model is significant, with the Wilks' lambda declining to .127. Moreover, significant differences exist between all of the individual groups.

With these four variables in the discriminant function, no other variable exhibits the statistical significance necessary for inclusion and the stepwise procedure is completed in terms of adding variables. The procedure, however, also includes a check on the significance of each variable in order to be retained in the discriminant function. In this case, the "F to Remove" for both X_{11} and

TABLE 7-17   Results from Step 4 of Stepwise Three-Group Discriminant Analysis

Overall Model Fit

	Value	F Value	Degrees of Freedom	Significance
Wilks' Lambda	.127	24.340	8, 108	.000

Variable Entered/Removed at Step 4

Variable Entered	Minimum D^2	F Value	F Significance	Between Groups
X_{18} Delivery Speed	6.920	13.393	.000	Less than 1 year and 1 to 5 years

Note: At each step, the variable that maximizes the Mahalanobis distance between the two closest groups is entered.

Variables in the Analysis After Step 4

Variable	Tolerance	F to Remove	D^2	Between Groups
X_{11} Product Line	.075	.918	6.830	Less than 1 year and 1 to 5 years
X_{17} Price Flexibility	.070	1.735	6.916	Less than 1 year and 1 to 5 years
X_6 Product Quality	.680	27.701	3.598	1 to 5 years and more than 5 years
X_{18} Delivery Speed	.063	5.387	6.071	Less than 1 year and 1 to 5 years

Variables Not in the Analysis After Step 4

Variable	Tolerance	Minimum Tolerance	F to Enter	Minimum D^2	Between Groups
X_7 E-Commerce Activities	.870	.063	.226	6.931	Less than 1 year and 1 to 5 years
X_8 Technical Support	.940	.063	.793	7.164	Less than 1 year and 1 to 5 years
X_9 Complaint Resolution	.453	.058	.292	7.019	Less than 1 year and 1 to 5 years
X_{10} Advertising	.932	.063	.006	6.921	Less than 1 year and 1 to 5 years
X_{12} Salesforce Image	.843	.061	.315	7.031	Less than 1 year and 1 to 5 years
X_{13} Competitive Pricing	.790	.063	.924	7.193	Less than 1 year and 1 to 5 years
X_{14} Warranty & Claims	.843	.063	2.023	7.696	Less than 1 year and 1 to 5 years
X_{15} New Products	.927	.062	.227	7.028	Less than 1 year and 1 to 5 years
X_{16} Order & Billing	.671	.062	1.478	7.210	Less than 1 year and 1 to 5 years

Significance Testing of Group Differences After Step 4[a]

X_1 Customer Type		Less than 1 Year	1 to 5 Years
1 to 5 years	F	13.393	
	Sig.	.000	
More than 5 years	F	56.164	18.477
	Sig.	.000	.000

[a]4, 54 degrees of freedom.

X_{17} is nonsignificant (.918 and 1.735, respectively), indicating that one or both are candidates for removal from the discriminant function.

Stepwise Estimation: Removal of X_{17} and X_{11}. When X_{18} is added to the model in the fourth step (see the preceding discussion), X_{11} had the lowest "*F* to Remove" value (.918), causing the stepwise procedure to eliminate that variable from the discriminant function in step 5 (details of this step 5 are omitted for space considerations). With now three variables in the discriminant function (X_{11}, X_6, and X_{18}), the overall model fit is still statistically significant and the Wilks' lambda increased only slightly to .135. All of the groups are significantly different. No variables reach the level of statistical significance necessary to be added to the discriminant function, and one more variable (X_{11}) has an "F to Remove" value of 2.552, which indicates that it can also be removed from the function.

Table 7-18 contains the details of step 6 of the stepwise procedure where X_{11} is also removed from the discriminant function, with only X_6 and X_{18} as the two variables remaining. Even with the

TABLE 7-18   Results from Step 6 of Stepwise Three-Group Discriminant Analysis

Overall Model Fit

	Value	F Value	Degrees of Freedom	Significance
Wilks' Lambda	.148	44.774	4, 112	.000

Variable Entered/Removed at Step 6

Variable Removed	Minimum D^2	F Value	F Significance	Between Groups
X_{11} Product Line	6.388	25.642	.000	Less than 1 year and 1 to 5 years

Note: At each step, the variable that maximizes the Mahalanobis distance between the two closest groups is entered.

Variables in the Analysis After Step 6

Variable	Tolerance	F to Remove	D^2	Between Groups
X_6 Product Quality	.754	50.494	.007	1 to 5 years and more than 5 years
X_{18} Delivery Speed	.754	60.646	.121	Less than 1 year and 1 to 5 years

Variables Not in the Analysis After Step 6

Variable	Tolerance	Minimum Tolerance	F to Enter	Minimum D^2	Between Groups
X_7 E-Commerce Activities	.954	.728	.177	6.474	Less than 1 year and 1 to 5 years
X_8 Technical Support	.999	.753	.269	6.495	Less than 1 year and 1 to 5 years
X_9 Complaint Resolution	.453	.349	.376	6.490	Less than 1 year and 1 to 5 years
X_{10} Advertising	.954	.742	.128	6.402	Less than 1 year and 1 to 5 years
X_{11} Product Line	.701	.529	2.552	6.916	Less than 1 year and 1 to 5 years
X_{12} Salesforce Image	.957	.730	.641	6.697	Less than 1 year and 1 to 5 years
X_{13} Competitive Pricing	.994	.749	1.440	6.408	Less than 1 year and 1 to 5 years
X_{14} Warranty & Claims	.991	.751	.657	6.694	Less than 1 year and 1 to 5 years
X_{15} New Products	.984	.744	.151	6.428	Less than 1 year and 1 to 5 years
X_{16} Order & Billing	.682	.514	2.397	6.750	Less than 1 year and 1 to 5 years
X_{17} Price Flexibility	.652	.628	3.431	6.830	Less than 1 year and 1 to 5 years

Significance Testing of Group Differences After Step 6[a]

X_1 Customer Type		Less than 1 Year	1 to 5 Years
1 to 5 years	F	25.642	
	Sig.	.000	
More than 5 years	F	110.261	30.756
	Sig.	.000	.000

[a]6, 52 degrees of freedom.

removal of the second variable (X_{11}), the overall model is still significant and the Wilks' lambda is quite small (.148). We should note that this two-variable model of X_6 and X_{18} is an improvement over the first two-variable model of X_{11} and X_{17} formed in step 2 (Wilks' lambda is .148 versus the first model's value of .288 and all of the individual group differences are much greater). With no variables reaching the significance level necessary for addition or removal, the stepwise procedure terminates.

Summary of the Stepwise Estimation Process. The estimated discriminant functions are linear composites similar to a regression line (i.e., they are a linear combination of variables). Just as a regression line is an attempt to explain the maximum amount of variation in its dependent variable, these linear composites attempt to explain the variations or differences in the dependent categorical variable. The first discriminant function is developed to explain (account for) the largest amount of variation (difference) in the discriminant groups. The second discriminant function, which is orthogonal and independent of the first, explains the largest percentage of the remaining (residual) variance after the variance for the first function is removed.

The information provided in Table 7-19 summarizes the steps of the three-group discriminant analysis, with the following results:

- Variables X_6 and X_{18} are the two variables in the final discriminant function, although X_{11} and X_{17} were added in the first two steps and then removed after X_6 and X_{18} were added. The unstandardized and standardized discriminant function coefficients (weights) and the structure matrix of discriminant loadings, unrotated and rotated, are also provided. Rotation of the discriminant loadings facilitates interpretation in the same way that factors were simplified for interpretation by rotation (see Chapter 3 for a more detailed discussion of rotation). We examine the unrotated and rotated loadings more fully in step 5.
- Discrimination increased with the addition of each variable (as evidenced by decreases in Wilks' lambda) even though only two variables remained in the final model. By comparing the final Wilks' lambda for the discriminant analysis (.148) with the Wilks' lambda (.414) for the best result from a single variable, X_9, we see that a marked improvement is made using just two variables in the discriminant functions rather than a single variable.
- The overall goodness-of-fit for the discriminant model is statistically significant and both functions are statistically significant as well. The first function accounts for 91.5 percent of the variance explained by the two functions, with the remaining variance (8.5%) due to the second function. The total amount of variance explained by the first function is $.893^2$, or 79.7 percent. The next function explains $.517^2$, or 26.7 percent, of the remaining variance (20.3%). Therefore, the total variance explained by both functions is 85.1 percent [79.7% + (26.7% × .203)] of the total variation in the dependent variable.

Even though both discriminant functions are statistically significant, the researcher must always ensure that the discriminant functions provide differences among all of the groups. It is possible to have statistically significant functions, but have at least one pair of groups not be statistically different (i.e., not discriminated between). This problem becomes especially prevalent as the number of groups increases or a number of small groups are included in the analysis.

The last section of Table 7-18 provides the significance tests for group differences between each pair of groups (e.g., group 1 versus group 2, group 1 versus group 3, etc.). All pairs of groups show statistically significant differences, denoting that the discriminant functions created separation not only in an overall sense, but for each group as well. We also examine the group centroids graphically in a later section.

ASSESSING CLASSIFICATION ACCURACY Because it is a three-group discriminant analysis model, two discriminant functions are calculated to discriminate among the three groups. Values for each case

TABLE 7-19 Summary Statistics for Three-Group Discriminant Analysis

Overall Model Fit: Canonical Discriminant Functions

| Function | Eigenvalue | Percent of Variance | | Canonical Correlation | Wilks' Lambda | Chi-Square | df | Significance |
		Function %	Cumulative %					
1	3.950	91.5	91.5	.893	.148	107.932	4	.000
2	.365	8.5	100.0	.517	.733	17.569	1	.000

Discriminant Function and Classification Function Coefficients

| Independent Variables | DISCRIMINANT FUNCTION | | | | Classification Functions | | |
| | Unstandardized Discriminant Function | | Standardized Discriminant Function | | | | |
	Function 1	Function 2	Function 1	Function 2	Less than 1 Year	1 to 5 Years	Over 5 Years
X_6 Product Quality	.308	1.159	.969	.622	14.382	15.510	18.753
X_{18} Delivery Speed	2.200	.584	1.021	−.533	25.487	31.185	34.401
(Constant)	−10.832	−11.313			−91.174	−120.351	−159.022

Structure Matrix

| Independent Variables | Unrotated Discriminant Loadings[a] | | Rotated Discriminant Loadings[b] | |
	Function 1	Function 2	Function 1	Function 2
X_9 Complaint Resolution*	.572	−.470	.739	.039
X_{16} Order & Billing	.499	−.263	.546	.143
X_{11} Product Line*	.483	−.256	.529	.137
X_{15} New Products*	.125	−.005	.096	.080
X_8 Technical Support*	.030	−.017	.033	.008
X_6 Product Quality*	.463	.886	−.257	.967
X_{18} Delivery Speed	.540	−.842	.967	−.257
X_{17} Price Flexibility*	.106	−.580	.470	−.356
X_{10} Advertising*	.028	−.213	.165	−.138
X_7 E-Commerce Activities*	−.095	−.193	.061	−.207
X_{12} Salesforce Image*	−.088	−.188	.061	−.198
X_{14} Warranty & Claims*	.030	−.088	.081	.044
X_{13} Competitive Pricing*	−.055	−.059	−.001	−.080

[a]Pooled within-groups correlations between discriminating variables and standardized canonical discriminant functions variables ordered by absolute size of correlation within function.

[b]Pooled within-groups correlations between discriminating variables and rotated standardized canonical discriminant functions.

*This variable is not used in the analysis.

Group Means (Centroids) of Discriminant Functions[c]

X_1 Customer Type	Function 1	
Less than 1 year	−1.911	−1.274
1 to 5 years	.597	−.968
More than 5 years	1.371	1.625

[c]Unstandardized canonical discriminant functions evaluated at group means.

are entered into the discriminant model and linear composites (discriminant Z scores) are calculated. The discriminant functions are based only on the variables included in the discriminant model.

Table 7-19 provides the discriminant weights of both variables (X_6 and X_{18}) and the group means of each group on both functions (lower portion of the table). As we can see by examining the group means, the first function primarily distinguishes group 1 (Less than 1 year) from the other two groups (although a marked difference occurs between groups 2 and 3 as well), while the second function primarily separates group 3 (More than 5 years) from the other two groups. Therefore, the first function provides the most separation between all three groups, but is complemented by the second function, which discriminates best (1 and 2 versus 3) where the first function is weakest.

Assessing Group Membership Prediction Accuracy. The final step of assessing overall model fit is to determine the predictive accuracy level of the discriminant function(s). This determination is accomplished in the same fashion as with the two-group discriminant model, by examination of the classification matrices and the percentage correctly classified (hit ratio) in each sample.

The classification of individual cases can be performed either by the cutoff method described in the two-group case or by using the classification functions (see Table 7-19) where each case is scored on each classification function and classified to the group with the highest score.

Table 7-20 shows that the two discriminant functions in combination achieve a high degree of classification accuracy. The hit ratio for the analysis sample is 86.7 percent. However, the hit ratio for the holdout sample falls to 55.0 percent. These results demonstrate the upward bias that is likely when applied only to the analysis sample and not also to a holdout sample.

TABLE 7-20   Classification Results for Three-Group Discriminant Analysis

Classification Results[a, b, c]

	Actual Group	Less than 1 Year	1 to 5 Years	More than 5 Years	Total
			Predicted Group Membership		
Estimation Sample	Less than 1 year	21	1	0	22
		95.5	4.5	0.0	
	1 to 5 years	2	7	4	13
		15.4	53.8	30.8	
	More than 5 years	0	1	24	25
		0.0	4.0	96.0	
Cross-validated	Less than 1 year	21	1	0	22
		95.5	4.5	0.0	
	1 to 5 years	2	7	4	13
		15.4	53.8	30.8	
	More than 5 years	0	1	24	25
		0.0	4.0	96.0	
Holdout Sample	Less than 1 year	5	3	2	10
		50.0	30.0	20.0	
	1 to 5 years	1	9	12	22
		4.5	40.9	54.5	
	More than 5 years	0	0	8	8
		0.0	0.0	100.0	

[a]86.7% of selected original grouped cases correctly classified.
[b]55.0% of unselected original grouped cases correctly classified.
[c]86.7% of selected cross-validated grouped cases correctly classified.

Both hit ratios must be compared with the maximum chance and the proportional chance criteria to assess their true effectiveness. The cross-validation procedure is discussed in step 6.

- The maximum chance criterion is simply the hit ratio obtained if we assign all the observations to the group with the highest probability of occurrence. In the present sample of 100 observations, 32 were in group 1, 35 in group 2, and 33 in group 3. From this information, we can see that the highest probability would be 35 percent (group 2). The threshold value for the maximum chance ($35\% \times 1.25$) is 43.74 percent.
- The proportional chance criterion is calculated by squaring the proportions of each group, with a calculated value of 33.36 percent ($.32^2 + .35^2 + .33^2 = .334$) and a threshold value of 41.7 percent ($33.4\% \times 1.25 = 41.7\%$).

The hit ratios for the analysis and holdout samples (86.7% and 55.0%, respectively) exceed both threshold values of 43.74 and 41.7 percent. In the estimation sample, all of the individual groups surpass both threshold values. In the holdout sample, however, group 2 has a hit ratio of only 40.9 percent, and it increased to only 53.8 percent in the analysis sample. These results show that group 2 should be the focus of improving classification, possibly by the addition of independent variables or a review of classification of firms in this group to identify the characteristics of this group not represented in the discriminant function.

The final measure of classification accuracy is Press's Q, calculated for both analysis and holdout samples. It tests the statistical significance that the classification accuracy is better than chance.

$$\text{Press's } Q_{\text{Estimation Sample}} = \frac{[60 - (52 \times 3)]^2}{60(3 - 1)} = 76.8$$

And the calculation for the holdout sample is

$$\text{Press's } Q_{\text{Holdout Sample}} = \frac{[40 - (22 \times 3)]^2}{40(3 - 1)} = 8.45$$

Because the critical value at a .01 significance level is 6.63, the discriminant analysis can be described as predicting group membership better than chance.

When completed, we can conclude that the discriminant model is valid and has adequate levels of statistical and practical significance for all groups. The markedly lower values for the holdout sample on all the standards of comparison, however, support the concerns raised earlier about the overall and group-specific hit ratios.

CASEWISE DIAGNOSTICS In addition to the classification tables showing aggregate results, case-specific information is also available detailing the classification of each observation. This information can detail the specifics of the classification process or represent the classification through a territorial map.

Case-Specific Classification Information. A series of case-specific measures is available for identifying the misclassified cases as well as diagnosing the extent of each misclassification. Using this information, patterns among the misclassified may be identified.

Table 7-21 contains additional classification data for each individual case that was misclassified (similar information is also available for all other cases, but was omitted for space considerations). The basic types of classification information include the following:

- *Group membership.* Both the actual and predicted groups are shown to identify each type of misclassification (e.g., actual membership in group 1, but predicted in group 2). In this instance, we see the eight cases misclassified in the analysis sample (verify by adding the off-diagonal values in Table 7-20) and the 18 cases misclassified in the holdout sample.

TABLE 7-21 Misclassified Predictions for Individual Cases in the Three-Group Discriminant Analysis

	GROUP MEMBERSHIP			DISCRIMINANT SCORES		CLASSIFICATION PROBABILITY		
Case ID	(X_1) Actual	Predicted	Mahalanobis Distance to Centroid[a]	Function 1	Function 2	Group 1	Group 2	Group 3
Analysis/Estimation Sample								
10	1	2	.175	.81755	−1.32387	.04173	.93645	.02182
8	2	1	1.747	−.78395	−1.96454	.75064	.24904	.00032
100	2	1	2.820	−.70077	−.11060	.54280	.39170	.06550
1	2	3	2.947	−.07613	.70175	.06527	.28958	.64515
5	2	3	3.217	−.36224	1.16458	.05471	.13646	.80884
37	2	3	3.217	−.36224	1.16458	.05471	.13646	.80884
88	2	3	2.390	.99763	.12476	.00841	.46212	.52947
58	3	2	.727	.30687	−.16637	.07879	.70022	.22099
Holdout/Validation Sample								
25	1	2	1.723	−.18552	−2.02118	.40554	.59341	.00104
77	1	2	.813	.08688	−.22477	.13933	.70042	.16025
97	1	2	1.180	−.41466	−.57343	.42296	.54291	.03412
13	1	3	.576	1.77156	2.26982	.00000	.00184	.99816
96	1	3	3.428	−.26535	.75928	.09917	.27855	.62228
83	2	1	2.940	−1.58531	.40887	.89141	.08200	.02659
23	2	3	.972	.61462	.99288	.00399	.10959	.88641
34	2	3	1.717	.86996	.41413	.00712	.31048	.68240
39	2	3	.694	1.59148	.82119	.00028	.08306	.91667
41	2	3	2.220	.30230	.58670	.02733	.30246	.67021
42	2	3	.210	1.08081	1.97869	.00006	.00665	.99330
55	2	3	1.717	.86996	.41413	.00712	.31048	.68240
57	2	3	6.041	3.54521	.47780	.00000	.04641	.95359
62	2	3	4.088	−.32690	.52743	.17066	.38259	.44675
75	2	3	2.947	−.07613	.70175	.06527	.28958	.64515
78	2	3	.210	1.08081	1.97869	.00006	.00665	.99330
85	2	3	2.390	.99763	.12476	.00841	.46212	.52947
89	2	3	.689	.54850	1.51411	.00119	.03255	.96625

[a]Mahalanobis distance to predicted group centroid.

- *Mahalanobis distance to the predicted group centroid.* Denotes the proximity of these misclassified cases to the predicted group. Some observations, such as case 10, obviously are similar to the observations of the predicted group rather than their actual group. Other observations, such as case 57 (Mahalanobis distance of 6.041), are likely to be outliers in the predicted group as well as the actual group. The territorial map discussed in the next section graphically portrays the position of each observation and assists in interpretation of the distance measures.
- *Discriminant scores.* The discriminant Z score for each case on each discriminant function provides a means of direct comparison between cases as well as a relative positioning versus the group means.
- *Classification probability.* Derived from use of the discriminant classification functions, the probability of membership for each group is given. The probability values enable the researcher to assess the extent of misclassification. For example, two cases, 85 and 89, are the same type of

misclassification (actual group 2, predicted group 3), but quite different in their misclassification when the classification probabilities are viewed. Case 85 represents a marginal misclassification, because the prediction probability for the actual group 2 was .462 and the incorrect predicted group 3 was only slightly higher (.529). This misclassification is in contrast to case 89, where the actual group probability was .032 and the predicted probability for group 3 (the misclassified group) was .966. In both situations of a misclassification, the extent or magnitude varies widely.

The researcher should evaluate the extent of misclassification for each case. Cases that are obvious misclassifications should be selected for additional analysis (profiling, examining additional variables, etc.) discussed in the two-group analysis.

Territorial Map. The analysis of misclassified cases can be supplemented by the graphical examination of the individual observations by plotting them based on their discriminant Z scores.

Figure 7-9 plots each observation based on its two rotated discriminant Z scores with an overlay of the territorial map representing the boundaries of the cutting scores for each function. In viewing each group's dispersion around the group centroid, we can observe several findings:

- Group 3 (More than 5 years) is most concentrated, with little overlap with the other two groups as shown in the classification matrix where only one observation was misclassified (see Table 7-20).
- Group 1 (Less than 1 year) is the least compact, but the range of cases does not overlap to any great degree with the other groups, thus making predictions much better than might be expected for such a diverse group. The only misclassified cases that are substantially different are case 10, which is close to the centroid for group 2, and case 13, which is close to the centroid of group 3. Both of these cases merit further investigation as to their similarities to the other groups.
- Both of these groups are in contrast to group 2 (1 to 5 years), which can be seen to have substantial overlap with group 3 and to a lesser extent with group 1 (Less than 1 year). This overlap results in the lowest levels of classification accuracy in both the analysis and holdout samples.
- The overlap that occurs between groups 2 and 3 in the center and right of the graph suggests the possible existence of a fourth group. Analysis could be undertaken to determine the actual length of time of customers, perhaps with the customers over 1 year divided into three groups instead of two.

The graphical portrayal is useful not only for identifying these misclassified cases that may form a new group, but also in identifying outliers. The preceding discussion identifies possible options for identifying outliers (case 57) as well as the possibility of group redefinition between groups 2 and 3.

Stage 5: Interpretation of Three-Group Discriminant Analysis Results

The next stage of the discriminant analysis involves a series of steps in the interpretation of the discriminant functions.

- Calculate the loadings for each function and review the rotation of the functions for purposes of simplifying interpretation.
- Examine the contributions of the predictor variables: (a) to each function separately (i.e., discriminant loadings), (b) cumulatively across multiple discriminant functions with the potency index, and (c) graphically in a two-dimensional solution to understand the relative position of each group and the interpretation of the relevant variables in determining this position.

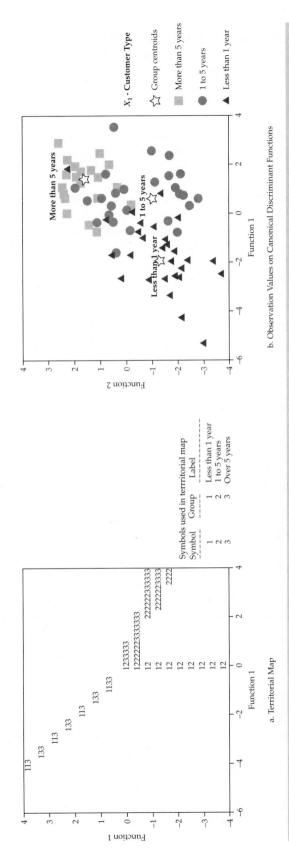

FIGURE 7-9 Territorial Map and Plot of Individual Cases on Discriminant Functions

b. Observation Values on Canonical Discriminant Functions

X_1 - Customer Type

☆ Group centroids

▨ More than 5 years

● 1 to 5 years

▲ Less than 1 year

Symbols used in terrritorial map

Symbol	Group	Label
1	1	Less than 1 year
2	2	1 to 5 years
3	3	Over 5 years

a. Territorial Map

DISCRIMINANT LOADINGS AND THEIR ROTATION   Once the discriminant functions are calculated, they are correlated with all the independent variables, even those not used in the discriminant function, to develop a structure (loadings) matrix. This procedure enables us to see where the discrimination would occur if all the independent variables were included in the model (i.e., if none were excluded by multicollinearity or lack of statistical significance).

Discriminant Loadings.   The unrotated loadings represent the association of each independent variable with each function, even if not included in the discriminant function. Discriminant loadings, similar to factor loadings described in Chapter 3, are the correlations between each independent variable and the discriminant score.

Table 7-19 contains the structure matrix of unrotated discriminant loadings for both discriminant functions. Selecting variables with loadings of .40 or above as descriptive of the functions, we see that function 1 has five variables exceeding .40 (X_9, X_{18}, X_{16}, X_{11}, and X_6), and four variables are descriptive of function 2 (X_6, X_{18}, X_{17}, and X_9). Even though we could use these variables to describe each function, we are faced with the issue that three variables (X_9, X_6, and X_{18}) have double loadings (variables selected as descriptive of both functions). If we were to proceed with the unrotated loadings, each function would share more variables with the other than it would have as unique.

The lack of distinctiveness of the loadings with each variable descriptive of a single function can be addressed by rotation of the structure matrix, just as was done with factor loadings. For a more detailed description of the rotation process, see Chapter 3.

Rotation.   After the discriminant function loadings are calculated, they can be rotated to redistribute the variance (similar to rotation of factors, which is explained in Chapter 3). Basically, rotation preserves the original structure and reliability of the discriminant models while making them easier to interpret substantively.

The rotation of discriminant functions, however, is an option in many software programs. In SPSS, for example, the rotated discriminant function coefficients can be obtained only through the use of command syntax rather than the "pull-down" menus. Examples of using command syntax in SPSS and the specific syntax used for discriminant analysis are provided on the text's Web sites (www.pearsonglobaleditions.com/hair or www.mvstats.com).

In the present application we chose the most widely used procedure of VARIMAX rotation. The rotation affects the function coefficients and discriminant loadings, as well as the calculation of the discriminant Z scores and the group centroids (see Table 7-19). Examining the rotated versus unrotated coefficients or loadings reveals a somewhat more simplified set of results (i.e., loadings tend to separate into high versus low values instead of being midrange). The rotated loadings allow for much more distinct interpretations of each function:

- Function 1 is now described by three variables (X_{18}, X_9, and X_{16}) that comprised the *Postsale Customer Service* factor during factor analysis (see Chapter 3 for more details), plus X_{11} and X_{17}. Thus, customer service, plus product line and price flexibility, are the descriptors of function 1.
- Function 2 shows only one variable, X_6 (Product Quality), which has a loading above .40 for the second function. Although X_{17} has a value just under the threshold ($-.356$), this variable has a higher loading on the first function, which makes it a descriptor of that function. Thus, the second function can be described by the single variable of Product Quality.

With two or more estimated functions, rotation can be a powerful tool that should always be considered to increase the interpretability of the results. In our example, each of the variables entered into the stepwise process was descriptive of one of the discriminant functions. What we must do now is assess the impact of each variable in terms of the overall discriminant analysis (i.e., across both functions).

ASSESSING THE CONTRIBUTION OF PREDICTOR VARIABLES   Having described the discriminant functions in terms of the independent variables—both those used in the discriminant functions and those not included in the functions—we turn our attention to gaining a better understanding of the impact of the functions themselves and then the individual variables.

Impact of the Individual Functions.   The first task is to examine the discriminant functions in terms of how they differentiate between the groups.

We start by examining the group centroids on the two functions as shown in Table 7-19. An easier approach is through viewing the territorial map (Figure 7-9):

- Examining the group centroids and the distribution of cases in each group, we see that function 1 primarily differentiates between group 1 versus groups 2 and 3, whereas function 2 distinguishes between group 3 versus groups 1 and 2.
- The overlap and misclassification of the cases of groups 2 and 3 can be addressed by examining the strength of the discriminant functions and the groups differentiated by each. Looking back to Table 7-19, function 1 was by far the most potent discriminator, and it primarily separated group 1 from the other groups. function 2, which separated group 3 from the others, was much weaker in terms of discriminating power. It is not surprising that the greatest overlap and misclassification would occur between groups 2 and 3, which are differentiated primarily by function 2.

This graphical approach illustrates the differences in the groups due to the discriminant functions but it does not provide a basis for explaining these differences in terms of the independent variables.

To assess the contributions of the individual variables, the researcher has a number of measures to employ—discriminant loadings, univariate F ratios, and the potency index. The techniques involved in the use of discriminant loadings and the univariate F ratios were discussed in the two-group example. We will examine in more detail the potency index, a method of assessing a variable's contribution across multiple discriminant functions.

Potency Index.   The potency index is an additional interpretational technique that is quite useful in situations with more than one discriminant function. Even though it must be calculated "by hand," it is very useful in portraying each individual variable's contribution across all discriminant functions.

The potency index reflects both the loadings of each variable and the relative discriminatory power of each function. The rotated loadings represent the correlation between the independent variable and the discriminant Z score. Thus, the squared loading is the variance in the independent variable associated with the discriminant function. By weighting the explained variance of each function by the relative discriminatory power of the functions and summing across functions, the potency index represents the total discriminating effect of each variable across all discriminant functions.

Table 7-22 provides the details on calculating a potency index for each of the independent variables. Comparing the variables on their potency index reveals the following:

- X_{18} (Delivery Speed) is the independent variable providing the greatest discrimination between the three types of customer groups.
- It is followed in impact by four variables not included in the discriminant function (X_9, X_{16}, X_{11}, and X_{17}).
- The second variable in the discriminant function (X_6) has only the sixth highest potency value.

Why does X_6 have only the sixth highest potency value even though it was one of the two variables included in the discriminant function?

- First, remember that multicollinearity affects stepwise solutions due to redundancy among highly multicollinear variables. X_9 and X_{16} were the two variables highly associated with X_{18}

TABLE 7-22 Calculation of the Potency Indices for the Three-Group Discriminant Analysis

Independent Variables	Discriminant Function 1				Discriminant Function 2				Potency Index
	Loading	Squared Loading	Relative Eigenvalue	Potency Value	Loading	Squared Loading	Relative Eigenvalue	Potency Value	
X_6 Product Quality	-.257	.066	.915	.060	.967	.935	.085	.079	.139
X_7 E-Commerce Activities	.061	.004	.915	.056	-.207	.043	.085	.004	.060
X_8 Technical Support	.033	.001	.915	.001	.008	.000	.085	.000	.001
X_9 Complaint Resolution	.739	.546	.915	.500	.039	.002	.085	.000	.500
X_{10} Advertising	.165	.027	.915	.025	-.138	.019	.085	.002	.027
X_{11} Product Line	.529	.280	.915	.256	.137	.019	.085	.002	.258
X_{12} Salesforce Image	.061	.004	.915	.004	-.198	.039	.085	.003	.007
X_{13} Competitive Pricing	-.001	.000	.915	.000	-.080	.006	.085	.001	.001
X_{14} Warranty & Claims	.081	.007	.915	.006	.044	.002	.085	.000	.006
X_{15} New Products	.096	.009	.915	.008	.080	.006	.085	.001	.009
X_{16} Order & Billing	.546	.298	.915	.273	.143	.020	.085	.002	.275
X_{17} Price Flexibility	.470	.221	.915	.202	-.356	.127	.085	.011	.213
X_{18} Delivery Speed	.967	.935	.915	.855	-.257	.066	.085	.006	.861

Note: The relative eigenvalue of each discriminant function is calculated as the eigenvalue of each function (shown in Table 7-19 as 3.950 and .365 for discriminant functions I and II respectively) divided by the total of the eigenvalues (3.950 + .365 = 4.315).

(forming the Customer Service factor), thus their impact in a univariate sense, reflected in the potency index, was not needed in the discriminant function due to the presence of X_{18}.

- The other two variables, X_{11} and X_{17}, did enter through the stepwise procedure, but were removed once X_6 was added, again due to multicollinearity. Thus, their greater discriminating power is reflected in their potency values even though they too were not needed in the discriminant function once X_6 was added with X_{18} in the discriminant function.
- Finally, X_6, the second variable in the discriminant function, has a low potency value because it is associated with the second discriminant function, which has relatively little discriminating impact when compared to the first function. Thus, although X_6 is a necessary element in discriminating among the three groups, its overall impact is less than those variables associated with the first function.

Remember that potency values can be calculated for all independent variables, even if they are not in the discriminant function(s), because they are based on discriminant loadings. The intent of the potency index is to provide for the interpretation in just such instances where multicollinearity or other factors may have prevented a variable(s) from being included in the discriminant function.

An Overview of the Empirical Measures of Impact. As seen in the prior discussions, the discriminatory power of variables in discriminant analysis is reflected in many different measures, each providing a unique role in the interpretation of the discriminant results. By combining all of these measures in our evaluation of the variables, we can achieve a well-rounded perspective on how each variable fits into the discriminant results.

Table 7-23 presents the three preferred interpretive measures (rotated loadings, univariate F ratio, and potency index) for each of the independent variables. The results support the stepwise analysis, although several cases illustrate the impact of multicollinearity on the procedures and the results.

- Two variables (X_9 and X_{18}) have the greatest individual impact as evidenced by their univariate F values. However, because both are also highly associated (as evidenced by their inclusion on the Customer Service factor in Chapter 3), only one will be included in a stepwise

TABLE 7-23 Summary of Interpretive Measures for Three-Group Discriminant Analysis

| | Rotated Discriminant Function Loadings | | Univariate | |
	Function 1	Function 2	F Ratio	Potency Index
X_6 Product Quality	−.257	.967	32.311	.139
X_7 E-Commerce Activities	.061	−.207	1.221	.060
X_8 Technical Support	.033	.008	.782	.001
X_9 Complaint Resolution	.739	.039	40.292	.500
X_{10} Advertising	.165	−.138	1.147	.027
X_{11} Product Line	.529	.137	32.583	.258
X_{12} Salesforce Image	.061	−.198	1.708	.007
X_{13} Competitive Pricing	−.001	−.080	9.432	.001
X_{14} Warranty & Claims	.081	.044	2.619	.006
X_{15} New Products	.096	.080	.216	.009
X_{16} Order & Billing	.546	.143	25.048	.275
X_{17} Price Flexibility	.470	−.356	12.551	.213
X_{18} Delivery Speed	.967	−.257	40.176	.861

solution. Even though X_9 has a marginally higher univariate F value, the ability of X_{18} to provide better discrimination between all of the groups (as evidenced by its larger minimum Mahalanobis D^2 value described earlier) made it a better candidate for inclusion. Thus, X_9, on an individual basis, has a comparable discriminating power, but X_{18} will be seen to work better in combination with other variables.

- Three additional variables (X_6, X_{11}, and X_{16}) are next highest in impact, but only one, X_6, is retained in the discriminant function. Note that X_{16} is highly correlated with X_{18} (both part of the Customer Service factor) and not included in the discriminant function, while X_{11} did enter the discriminant function, but was one of those variables removed after X_6 was added.
- Finally, two variables (X_{17} and X_{13}) had almost equal univariate effects, but only X_{17} had a substantial association with one of the discriminant functions (a loading of .470 on the first function). The result is that even though X_{17} can be considered descriptive of the first function and considered having an impact in discrimination based in these functions, X_{13} does not have any impact, either in association with these two functions, or in addition once these functions are accounted for.
- All of the remaining variables had low univariate F values and low potency values, indicative of little or no impact in both a univariate and multivariate sense.

Of particular note is the interpretation of the two dimensions of discrimination. This interpretation can be done solely through examination of the loadings, but is complemented by a graphical display of the discriminant loadings, as described in the following section.

Graphical Display of Discriminant Loadings. To depict the differences in terms of the predictor variables, the loadings and the group centroids can be plotted in reduced discriminant space. As noted earlier, the most valid representation is the use of stretched attribute vectors and group centroids.

Table 7-24 shows the calculations for stretching both the discriminant loadings (used for attribute vectors) and the group centroids. The plotting process always involves all the variables included in the model by the stepwise procedure (in our example, X_6 and X_{18}). However, we will also plot the variables not included in the discriminant function if their respective univariate F ratios are significant, which adds X_9, X_{11}, and X_{16} to the reduced discriminant space. This procedure shows the importance of collinear variables that were not included in the final stepwise model, similar to the potency index.

The plots of the stretched attribute vectors for the rotated discriminant loadings are shown in Figure 7-10, which is based on the reduced space coordinates for both the five variables used to describe the discriminant functions and each of the groups (see Table 7-24). The vectors plotted using this procedure point to the groups having the highest mean on the respective independent variable and away from the groups having the lowest mean scores. Thus, interpretation of the plot in Figure 7-10 indicates the following:

- As noted in the territorial map and analysis of group centroids, the first discriminant function distinguishes between group 1 versus groups 2 and 3, and the second discriminant function separates group 3 from groups 1 and 2.
- The correspondence of X_{11}, X_{16}, X_9, and X_{18} with the X axis reflects their association with the first discriminant function, but only X_6 is associated with the second discriminant function. The figure graphically illustrates the rotated loadings for each function and distinguishes the variables descriptive of each function.

Stage 6: Validation of the Discriminant Results

The hit ratios for the cross-classification and holdout matrices can be used to assess the internal and external validity, respectively, of the discriminant analysis. If the hit ratios exceed the threshold values on the comparison standards, then validity is established. As described earlier, the

TABLE 7-24   Calculation of the Stretched Attribute Vectors and Group Centroids in Reduced Discriminant Space

Independent Variables	Rotated Discriminant Function Loadings		Univariate F Ratio	Reduced Space Coordinates	
	Function 1	Function 2		Function 1	Function 2
X_6 Product Quality	−.257	.967	32.311	−8.303	31.244
X_7 E-Commerce Activities[a]	.061	−.207	1.221		
X_8 Technical Support[a]	.033	.008	.782		
X_9 Complaint Resolution	.739	.039	40.292	29.776	1.571
X_{10} Advertising[a]	.165	−.138	1.147		
X_{11} Product Line	.529	.137	32.583	17.236	4.464
X_{12} Salesforce Image[a]	.061	−.198	1.708		
X_{13} Competitive Pricing[a]	−.001	−.080	9.432		
X_{14} Warranty & Claims[a]	.081	.044	2.619		
X_{15} New Products[a]	.096	.080	.216		
X_{16} Order & Billing	.546	.143	25.048	13.676	3.581
X_{17} Price Flexibility[a]	.470	−.356	12.551		
X_{18} Delivery Speed	.967	−.257	40.176	38.850	−10.325

[a]Variables with nonsignificant univariate rations are not plotted in reduced space.

	Group Centroids		Approximate F Value		Reduced Space Coordinates	
	Function 1	Function 2	Function 1	Function 2	Function 1	Function 2
Group 1: Less than 1 year	−1.911	−1.274	66.011	56.954	−126.147	−72.559
Group 2: 1 to 5 years	.597	−.968	66.011	56.954	39.408	−55.131
Group 3: More than 5 years	1.371	1.625	66.011	56.954	90.501	92.550

threshold values are 41.7 percent for the proportional chance criterion and 43.7 percent for the maximum chance criterion. The classification results shown in Table 7-20 provide the following support for validity:

Internal validity is assessed by the cross-classification approach, where the discriminant model is estimated by leaving out one case and then predicting that case with the estimated model. This process is done in turn for each observation, such that an observation never influences the discriminant model that predicts its group classification.

As seen in Table 7-20, the overall hit ratio for the cross-classification approach of 86.7 substantially exceeds both standards, both overall and for each group. However, even though all three groups also have individual hit ratios above the standards, the group 2 hit ratio (53.8) is substantially less than that over the other two groups.

External validity is addressed through the holdout sample, which is a completely separate sample that uses the discriminant functions estimated with the analysis sample for group prediction.

In our example, the holdout sample has an overall hit ratio of 55.0 percent, which exceeds both threshold values, although not to the extent found in the cross-classification approach. Group 2, however, did not exceed either threshold value. When the misclassifications are analyzed, we see that more cases are misclassified into group 3 than correctly classified into

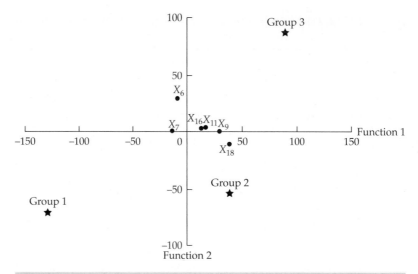

FIGURE 7-10   **Plot of Stretched Attribute Vectors (Variables) in Reduced Discriminant Space**

group 2, which suggests that these misclassified cases be examined for the possibility of a redefinition of groups 2 and 3 to create a new group.

The researcher is also encouraged to extend the validation process through profiling the groups on additional sets of variables or applying the discriminant function to another sample(s) representative of the overall population or segments within the population. Moreover, analysis of the misclassified cases will help establish whether any additional variables are needed or whether the dependent group classifications need revision.

A Managerial Overview

The discriminant analysis aimed at understanding the perceptual differences of customers based on their length of time as an HBAT customer. Hopefully, examining differences in HBAT perceptions based on tenure as a customer will identify perceptions that are critical to the development of a customer relationship, which is typified by those customers of long standing. Three customer groups were formed—less than 1 year, 1 to 5 years, and more than 5 years—and HBAT perceptions were measured on 13 variables. The analysis produced several major findings, both in terms of the types of variables that distinguished between the groups and the patterns of changes over time:

- First, there are two dimensions of discrimination between the three customer groups. The first dimension is typified by higher perceptions of customer service (Complaint Resolution, Delivery Speed, and Order & Billing), along with Product Line and Price Flexibility. In contrast, the second dimension is characterized solely in terms of Product Quality.
- Profiling the three groups on these two dimensions and variables associated with each dimension enables management to understand the perceptual differences among them.
  - Group 1, customers of less than 1 year, generally has the lowest perceptions of HBAT. For the three customer service variables (Complaint Resolution, Order & Billing, and Delivery Speed), these customers are lower than either other group. For Product Quality, Product Line, and Competitive Pricing, this group is comparable to group 2 (customers of 1 to 5 years), but still has lower perceptions than customers of more than 5 years. Only for Price

Flexibility is this group comparable to the oldest customers, and both have lower values than the customers of 1 to 5 years. Overall, the perceptions of these newest customers follow the expected pattern of being lower than other customers, but hopefully improving as they remain customers over time.

- Group 2, customers of between 1 and 5 years, has similarities to both the newest and oldest customers. On the three customer service variables, they are comparable to group 3 (customers of more than 5 years). For Product Quality, Product Line, and Competitive Pricing, their perceptions are more comparable to the newer customers (and lower than the oldest customers). They hold the highest perceptions of the three groups on Price Flexibility.
- Group 3, representing those customer of 5 years or more, holds the most favorable perceptions of HBAT as would be expected. Although they are comparable to the customers of group 2 on the three customer service variables (with both groups greater than group 1), they are significantly higher than customers in the other two groups in terms of Product Quality, Product Line, and Competitive Pricing. Thus, this group represents those customers that have the positive perceptions and have progressed in establishing a customer relationship through the strength of their perceptions.

- Using the three customer groups as indicators in the development of customer relationships, we can identify two stages in which HBAT perceptions change in this development process:
  - *Stage 1*. The first set of perceptions to change is that related to customer service (seen in the differences between groups 1 and 2). This stage reflects the ability of HBAT to positively affect perceptions with service-related operations.
  - *Stage 2*. A longer-run development is needed to foster improvements in more core elements (Product Quality, Product Line, and Competitive Pricing). When these changes occur, the customer hopefully becomes more committed to the relationship, as evidenced by a long tenure with HBAT.
- It should be noted that evidence exists that numerous customers make the transition through stage 2 more quickly than the 5 years as shown by the substantial number of customers who have been customers between 1 and 5 years, yet hold the same perceptions as those long-time customers. Thus, HBAT can expect that certain customers can move through this process possible quite quickly, and further analysis on these customers may identify characteristics that facilitate the development of customer relationships.

Thus, management is presented managerial input for strategic and tactical planning from not only the direct results of the discriminant analysis, but also from the classification errors.

LOGISTIC REGRESSION: REGRESSION WITH A BINARY DEPENDENT VARIABLE

As already discussed, discriminant analysis is appropriate when the dependent variable is nonmetric. However, when the dependent variable has only two groups, logistic regression may be preferred for two reasons:

- Discriminant analysis relies on strictly meeting the assumptions of multivariate normality and equal variance–covariance matrices across groups—assumptions that are not met in many situations. Logistic regression does not face these strict assumptions and is much more robust when these assumptions are not met, making its application appropriate in many situations.
- Even if the assumptions are met, many researchers prefer logistic regression because it is similar to multiple regression. It has straightforward statistical tests, similar approaches to incorporating metric and nonmetric variables and nonlinear effects, and a wide range of diagnostics.

For these and more technical reasons, logistic regression is equivalent to two-group discriminant analysis and may be more suitable in many situations.

Our discussion of logistic regression does not cover each of the six steps of the decision process, but instead highlights the differences and similarities between logistic regression and discriminant analysis or multiple regression. For a complete review of multiple regression see Chapter 4.

Representation of the Binary Dependent Variable

In discriminant analysis, the nonmetric character of a dichotomous dependent variable is accommodated by making predictions of group membership based on discriminant Z scores. It requires the calculation of cutting scores and the assignment of observations to groups.

Logistic regression approaches this task in a manner more similar to that found in multiple regression. Logistic regression represents the two groups of interest as a binary variable with values of 0 and 1. It does not matter which group is assigned the value of 1 versus 0, but this assignment must be noted for the interpretation of the coefficients.

- If the groups represent characteristics (e.g., gender), then either group can be assigned the value of 1 (e.g., females) and the other group the value of 0 (e.g., males). In such a situation, the coefficients would reflect the impact of the independent variable(s) on the likelihood of the person being female (i.e., the group coded as 1).
- If the groups represent outcomes or events (e.g., success or failure, purchase or nonpurchase), the assignment of the group codes impacts interpretation as well. Assume that the group with success is coded as 1, with failure coded as 0. Then, the coefficients represent the impacts on the likelihood of success. Just as easily, the codes could be reversed (code of 1 now denotes failure) and the coefficients represent the forces increasing the likelihood of failure.

Logistic regression differs from multiple regression, however, in being specifically designed to predict the probability of an event occurring (i.e., the probability of an observation being in the group coded 1). Although probability values are metric measures, they are fundamental differences between multiple regression and logistic regression.

USE OF THE LOGISTIC CURVE   Because the binary dependent variable has only the values of 0 and 1, the predicted value (probability) must be bounded to fall within the same range. To define a relationship bounded by 0 and 1, logistic regression uses the **logistic curve** to represent the relationship between the independent and dependent variables (see Figure 7-11). At very low levels of the independent variable, the probability approaches 0, but never reaches 0. Likewise, as the independent variable increases, the predicted values increase up the curve, but then the slope starts decreasing so that at any level of the independent variable, the probability will approach 1.0 but never exceed it. As we saw in our discussions of regression in Chapter 4, the linear models of regression cannot accommodate such a relationship, as it is inherently nonlinear. The linear relationship of regression, even with additional terms of transformations for nonlinear effects, cannot guarantee that the predicted values will remain within the range of 0 and 1.

UNIQUE NATURE OF THE DEPENDENT VARIABLE   The binary nature of the dependent variable (0 or 1) has properties that violate the assumptions of multiple regression. First, the error term of a discrete variable follows the binomial distribution instead of the normal distribution, thus invalidating all statistical testing based on the assumptions of normality. Second, the variance of a dichotomous variable is not constant, creating instances of heteroscedasticity as well. Moreover, neither violation can be remedied through transformations of the dependent or independent variables.

Logistic regression was developed to specifically deal with these issues. Its unique relationship between dependent and independent variables, however, requires a somewhat different

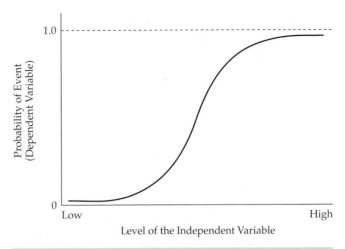

FIGURE 7-11   Form of the Logistic Relationship Between Dependent and Independent Variables

approach in estimating the variate, assessing goodness-of-fit, and interpreting the coefficients when compared to multiple regression.

Sample Size

Logistic regression, like every other multivariate technique, must consider the size of the sample being analyzed. As discussed in Chapter 1, very small samples have so much sampling error that identification of all but the largest differences is improbable. Very large sample sizes increase the statistical power so that any difference, whether practically relevant or not, will be considered statistically significant. Yet most research situations fall somewhere in between these extremes, meaning the researcher must consider the impact of sample sizes on the results, both at the overall level and on a group-by-group basis.

OVERALL SAMPLE SIZE   The first aspect of sample size is the overall sample size needed to adequately support estimation of the logistic model. One factor that distinguishes logistic regression from the other techniques is its use of maximum likelihood (MLE) as the estimation technique. MLE requires larger samples such that, all things being equal, logistic regression will require a larger sample size that multiple regression. For example, Hosmer and Lemeshow recommend sample sizes greater than 400 [4]. Moreover, the researcher should strongly consider dividing the sample into analysis and holdout samples as a means of validating the logistic model (see more detailed discussion in Stage 6). In making this split of the sample, the sample size requirements still hold for both the analysis and holdout samples separately, thus effectively doubling the overall sample size needed based on the model specification (number of parameters estimates, etc.).

SAMPLE SIZE PER CATEGORY OF THE DEPENDENT VARIABLE   The second consideration is that the overall sample size is important, but equally as critical is the sample size per group of the dependent variable. As we discussed for discriminant analysis, there a considerations on the minimum group size as well. The recommended sample size for each group is at least ten observations per estimated parameter. This is much greater than multiple regression which had a minimum of five observations per parameter and that was for the overall sample, not the sample size for each group as seen with logistic regression.

IMPACT OF NONMETRIC INDEPENDENT VARIABLES A final consideration comes into play with the use of nonmetric independent variables. When they are included in the model, they further subdivide the sample into cells created by the combination of dependent and nonmetric independent variables. For example, a simple binary independent variable creates four groups when combined with the binary dependent variable. While it is not necessary for each of these groups to meet the sample size requirements described above, the researcher must still be aware that if any one of these cells has a very small sample size then it is effectively eliminated from the analysis. Moreover, if too many of these cells have zero or very small sample sizes, then the model may have trouble converging and reaching a solution.

Estimating the Logistic Regression Model

Logistic regression has a single variate composed of estimated coefficients for each independent variable—as found in multiple regression. This variate is estimated in a different manner. Logistic regression derives its name from the **logit transformation** used with the dependent variable, creating several differences in the estimation process (as well as the interpretation process discussed in a following section).

TRANSFORMING THE DEPENDENT VARIABLE As shown earlier, the logit model uses the specific form of the logistic curve, which is S-shaped to stay within the range of 0 to 1. To estimate a logistic regression model, this curve of predicted values is fitted to the actual data, just as was done with a linear relationship in multiple regression. However, because the actual data values of the dependent variables can only be either 1 or 0, the process is somewhat different.

Figure 7-12 portrays two hypothetical examples of fitting a logistic relationship to sample data. The actual data represent whether an event either happened or not by assigning values of either 1 or 0 to the outcomes (in this case a 1 is assigned when the event happened, 0 otherwise, but they could have just as easily been reversed). Observations are represented by the dots at either the top or bottom of the graph. These outcomes (happened or not) occur at each value of the independent variable (the X axis). In part (a), the logistic curve cannot fit the data well because a number of values of the independent variable have both outcomes (1 and 0). In this case the independent variable does not distinguish between the two outcomes as shown by the high overlap of the two groups.

However, in part (b), a much more well-defined relationship is based on the independent variable. Lower values of the independent variable correspond to the observations with 0 for the dependent variable, while larger values of the independent variable correspond well with those observations with a value of 1 on the dependent variable. Thus, the logistic curve should be able to fit the data quite well.

But how do we predict group membership from the logistic curve? For each observation, the logistic regression technique predicts a probability value between 0 and 1. Plotting the predicted values for all values of the independent variable generates the curve shown in Figure 7-12. This predicted probability is based on the value(s) of the independent variable(s) and the estimated coefficients. If the predicted probability is greater than .50, then the prediction is that the outcome is 1 (the event happened); otherwise, the outcome is predicted to be 0 (the event did not happen). Let us return to our example and see how it works.

In parts (a) and (b) of Figure 7-12, a value of 6.0 for X (the independent variable) corresponds to a probability of .50. In part (a), we can see that a number of observations of both groups fall on both sides of this value, resulting in a number of misclassifications. The misclassifications are most noticeable for the group with values of 1.0, yet even several observations in the other group (dependent variable = 0.0) are also misclassified. In part (b), we make perfect classification of the two groups when using the probability value of .50 as a cutoff value.

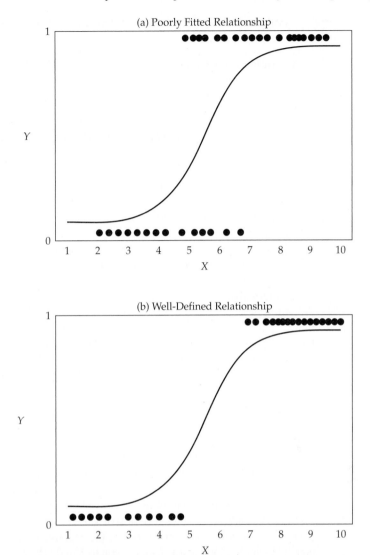

FIGURE 7-12 Examples of Fitting the Logistic Curve to Sample Data

Thus, with an estimated logistic curve, we can estimate the probability for any observation based on its values for the independent variable(s) and then predict group membership using .50 as a cutoff value. Once we have the predicted membership, we can create a classification matrix just as was done for discriminant analysis and assess predictive accuracy.

ESTIMATING THE COEFFICIENTS Where does the curve come from? In multiple regression, we estimate a linear relationship that best fits the data. In logistic regression, we follow the same process of predicting the dependent variable by a variate composed of the **logistic coefficient(s)** and the corresponding independent variable(s). Yet what differs is that in logistic regression the predicted values can never be outside the range of 0 to 1. Although a complete discussion of the conceptual and statistical issues involved in the estimation process is beyond the scope of this text, several excellent sources with complete treatments of these issues are available [3, 15, 17]. We can describe the estimation process in two basic steps as we introduce some common terms and provide a brief overview of the process.

Transforming a Probability into Odds and Logit Values. Just as with multiple regression, logistic regression predicts a metric dependent variable, in this case probability values constrained to the range between 0 and 1. But how can we ensure that estimated values do not fall outside this range? The logistic transformation accomplishes this process in two steps.

- *Restating a Probability as Odds* In their original form, probabilities are not constrained to values between 0 and 1. So, what if we were to restate the probability in a way that the new variable would always fall between 0 and 1? We restate it by expressing a probability as **odds**—the ratio of the probability of the two outcomes or events, $Prob_i \div (1 - Prob_i)$. In this form, any probability value is now stated in a metric variable that can be directly estimated. Any odds value can be converted back into a probability that falls between 0 and 1. We have solved our problem of constraining the predicted values to within 0 and 1 by predicting the odds value and then converting it into a probability.

  Let us use some examples of the probability of success or failure to illustrate how the odds are calculated. If the probability of success is .80, then we also know that the probability of the alternative outcome (i.e., failure) is .20 (.20 = 1.0 − .80). This probability means that the odds of success are 4.0 (.80 ÷ .20), or that success is four times more likely to happen than failure. Conversely, we can state the odds of failure as .25 (.20 ÷ .80), or in other words, failure happens at one-fourth the rate of success. Thus, no matter which outcome we look at (success or failure), we can state the probability as odds.

  As you can probably surmise, a probability of .50 results in odds of 1.0 (both outcomes have an equal chance of occurring). Odds less than 1.0 represent probabilities less than .50 and odds greater than 1.0 correspond to a probability greater than .50. We now have a metric variable that can always be converted back to a probability value within 0 and 1.

- *Calculating the Logit Value* The odds variable solves the problem of making probability estimates between 0 and 1, but we have another problem: How do we keep the odds values from going below 0, which is the lower limit of the odds (there is no upper limit)? The solution is to compute what is termed the *logit value*—calculated by taking the logarithm of the odds. Odds less than 1.0 will have a negative logit value, odds ratios greater than 1.0 will have positive logit values, and the odds ratio of 1.0 (corresponding to a probability of .5) has a logit value of 0. Moreover, no matter how low the negative value gets, it still can be transformed by taking the antilog into an odds value greater than 0. The following shows some typical probability values and the associated odds and log odds values.

Probability	Odds	Log Odds (Logit)
.00	.00	NC
.10	.111	−2.197
.30	.428	−.847
.50	1.000	.000
.70	2.333	.847
.90	9.000	2.197
1.00	NC	NC

NC = Cannot be calculated.

With the logit value, we now have a metric variable that can have both positive and negative values, but that can always be transformed back to a probability value that is between 0 and 1. Note, however, that the logit can never actually reach either 0 or 1. This value now becomes the dependent variable of the logistic regression model.

Model Estimation. Once we understand how to interpret the values of either the odds or logit measures, we can proceed to using them as the dependent measure in our logistic regression. The process of estimating the logistic coefficients is similar to that used in regression, although in this case only two actual values are used for the dependent variable (0 and 1). Moreover, instead of using ordinary least squares as a means of estimating the model, the maximum likelihood method is used.

- *Estimating the Coefficients* The estimated coefficients for the independent variables are estimated using either the logit value or the odds value as the dependent measure. Each of these model formulations is shown here:

$$Logit_i = \ln\left(\frac{prob_{event}}{1 - prob_{event}}\right) = b_0 + b_1 X_1 + \cdots + b_n X_n$$

or

$$Odds_i = \left(\frac{prob_{event}}{1 - prob_{event}}\right) = e^{b_0 + b_1 X_1 + \cdots + b_n X_n}$$

Both model formulations are equivalent, but whichever is chosen affects how the coefficients are estimated. Many software programs provide the logistic coefficients in both forms, so the researcher must understand how to interpret each form. We will discuss interpretation issues in a later section.

This process can accommodate one or more independent variables, and the independent variables can be either metric or nonmetric (binary). As we will see later in our discussion of interpreting the coefficients, both forms of the coefficients reflect both direction and magnitude of the relationship, but are interpreted differently.

- *Using Maximum Likelihood for Estimation* Multiple regression employs the method of least squares, which minimizes the sum of the squared differences between the actual and predicted values of the dependent variable. The nonlinear nature of the logistic transformation requires that another procedure, the maximum likelihood procedure, be used in an iterative manner to find the most likely estimates for the coefficients. Instead of minimizing the squared deviations (least squares), logistic regression maximizes the likelihood that an event will occur. The likelihood value instead of the sum of squares is then used when calculating a measure of overall model fit. Using this alternative estimation technique also requires that we assess model fit in different ways.

Assessing the Goodness-of-Fit of the Estimation Model

The goodness-of-fit for a logistic regression model can be assessed in two ways. One way is to assess model estimation fit using "pseudo" R^2 values, similar to that found in multiple regression. The second approach is to examine predictive accuracy (like the classification matrix in discriminant analysis). The two approaches examine model fit from different perspectives, but should yield similar conclusions.

MODEL ESTIMATION FIT The basic measure of how well the maximum likelihood estimation procedure fits is the **likelihood value,** similar to the sums of squares values used in multiple regression. Logistic regression measures model estimation fit with the value of -2 times the log of the likelihood value, referred to as $-2LL$ or -2 log likelihood. The minimum value for $-2LL$ is 0, which corresponds to a perfect fit (likelihood = 1 and $-2LL$ is then 0). Thus, the lower the $-2LL$ value, the better fitting the model. As will be discussed in the following section, the $-2LL$ value can be used to compare between equations for the change in fit or used to calculate measures comparable to the R^2 measure in multiple regression.

BETWEEN MODEL COMPARISONS   The likelihood value can be compared between equations to assess the difference in predictive fit from one equation to another, with statistical tests for the significance of these differences. The basic approach follows three steps:

1. *Estimate a null model.* The first step is to calculate a null model, which acts as the baseline for making comparisons of improvement in model fit. The most common null model is one without any independent variables, which is similar to calculating the total sum of squares using only the mean in multiple regression. The logic behind this form of null model is that it can act as a baseline against which any model containing independent variables can be compared.
2. *Estimate the proposed model.* This model contains the independent variables to be included in the logistic regression model. Hopefully, model fit will improve from the null model and result in a lower $-2LL$ value. Any number of proposed models can be estimated (e.g., models with one, two, and three independent variables can all be separate proposed models).
3. *Assess –2LL difference.* The final step is to assess the statistical significance of the $-2LL$ value between the two models (null model versus proposed model). If the statistical tests support significant differences, then we can state that the set of independent variable(s) in the proposed model is significant in improving model estimation fit.

In a similar fashion, comparisons also can be made between any two proposed models. In these instances, the $-2LL$ difference reflects the difference in model fit due to the different model specifications. For example, a model with two independent variables may be compared to a model with three independent variables to assess the improvement by adding one independent variable. In these instances, one model is selected to act as the null model and then compared against another model.

For example, assume that we wanted to test the significance of a set of independent variables collectively to see if they improved model fit. The null model would be specified as a model without these variables and the proposed model would include the variables to be evaluated. The difference in $-2LL$ would signify the improvement from the set of independent variables. We could perform similar tests of the differences in $-2LL$ between other pairs of models varying in the number of independent variables included in each model.

The chi-square test and the associated test for statistical significance are used to evaluate the reduction in the log likelihood value. However, these statistical tests are particularly sensitive to sample size (for small samples it is harder to show statistical significance, and vice versa, for large samples). Therefore, researchers must be particularly careful in drawing conclusions solely based on the significance of the chi-square test in logistic regression.

Pseudo R^2 Measures.   In addition to the statistical chi-square tests, several different "R^2-like" measures have been developed and are presented in various statistical programs to represent overall model fit. These pseudo R^2 measures are interpreted in a manner similar to the coefficient of determination in multiple regression. A **pseudo R^2** value can be easily derived for logistic regression similar to the R^2 value in regression analysis [6]. The pseudo R^2 for a logit model (R^2_{LOGIT}) can be calculated as

$$R^2_{\text{LOGIT}} = \frac{-2LL_{\text{null}} - (-2LL_{\text{model}})}{-2LL_{\text{null}}}$$

Just like the multiple regression counterpart, the logit R^2 value ranges from 0.0 to 1.0. As the proposed model increases model fit, the $-2LL$ value decreases. A perfect fit has a $-2LL$ value of 0.0 and a R^2_{LOGIT} of 1.0.

Two other measures are similar in design to the pseudo R^2 value and are generally categorized as pseudo R^2 measures as well. The Cox and Snell R^2 measure operates in the same manner, with higher values indicating greater model fit. However, this measure is limited in that it cannot reach the maximum value of 1, so Nagelkerke proposed a modification that had the range of 0 to 1. Both of these additional measures are interpreted as reflecting the amount of variation accounted for by the logistic model, with 1.0 indicating perfect model fit.

A Comparison to Multiple Regression. In discussing the procedures for assessing model fit in logistic regression, we made several references to similarities with multiple regression in terms of various measures of model fit. In the following table, we show the correspondence between concepts used in multiple regression and their counterparts in logistic regression.

Correspondence of Primary Elements of Model Fit	
Multiple Regression	**Logistic Regression**
Total sum of squares	$-2LL$ of base model
Error sum of squares	$-2LL$ of proposed model
Regression sum of squares	Difference of $-2LL$ for base and proposed models
F test of model fit	Chi-square test of $-2LL$ difference
Coefficient of determination (R^2)	Pseudo R^2 measures

As we can see, the concepts between multiple regression and logistic regression are similar. The basic approaches to testing overall model fit are comparable, with the differences arising from the estimation methods used in the two techniques.

PREDICTIVE ACCURACY   Just as we borrowed the concept of R^2 from regression as a measure of overall model fit, we can look to discriminant analysis for measure of overall predictive accuracy. The two most common approaches are the classification matrix and chi-square-based measures of fit.

Classification Matrix. This classification matrix approach is identical to that used with discriminant analysis, that is, measuring how well group membership is predicted and developing a hit ratio. The case of logistic regression will always include only two groups, but all of the chance-related measures (e.g., maximum chance, proportional chance, or Press's Q) used earlier are applicable here as well.

Chi-Square-Based Measure. Hosmer and Lemeshow [11] developed a classification test where the cases are first divided into approximately 10 equal classes. Then, the number of actual and predicted events is compared in each class with the chi-square statistic. This test provides a comprehensive measure of predictive accuracy that is based not on the likelihood value, but instead on the actual prediction of the dependent variable. The appropriate use of this test requires a sample size of at least 50 cases to ensure that each class has at least five observations and generally even a larger sample because the number of predicted events should never fall below 1. Also, the chi-square statistic is sensitive to sample size, enabling this measure to find small differences statistically significant when the sample size becomes large.

We typically examine as many of these measures of model fit as possible. Hopefully, a convergence of indications from these measures will provide the necessary support for the researcher in evaluating the overall model fit.

Testing for Significance of the Coefficients

Logistic regression tests hypotheses about individual coefficients just as was done in multiple regression. In multiple regression, the statistical test was to see whether the coefficient was significantly different from 0. A coefficient of 0 indicates the coefficient has no impact on the dependent variable. In logistic regression, we also use a statistical test to see whether the logistic coefficient is different from 0. Remember, however, in logistic regression using the logit as the dependent measure, a value of 0 corresponds to the odds of 1.00 or a probability of .50—values that indicate the probability is equal for each group (i.e., again no effect of the independent variable on predicting group membership).

In multiple regression, the *t* value is used to assess the significance of each coefficient. Logistic regression uses a different statistic, the **Wald statistic.** It provides the statistical significance for each estimated coefficient so that hypothesis testing can occur just as it does in multiple regression. If the logistic coefficient is statistically significant, we can interpret it in terms of how it impacts the estimated probability and thus the prediction of group membership.

Interpreting the Coefficients

One of the advantages of logistic regression is that we need to know only whether an event (purchase or not, good credit risk or not, firm failure or success) occurred or not to define a dichotomous value as our dependent variable. When we analyze these data using the logistic transformation, however, the logistic regression and its coefficients take on a somewhat different meaning from those found in regression with a metric dependent variable. Similarly, discriminant loadings from a two-group discriminant analysis are interpreted differently from a logistic coefficient.

From the estimation process described earlier, we know that the coefficients $(B_0, B_1, B_2, \ldots, B_n)$ are actually measures of the change in the ratio of the probabilities (the odds). However, logistic coefficients are difficult to interpret in their original form because they are expressed in terms of logarithms when we use the logit as the dependent measure. Thus, most computer programs also provide an **exponentiated logistic coefficient,** which is just a transformation (antilog) of the original logistic coefficient. In this way, we can use either the original or exponentiated logistic coefficients for interpretation. The two types of logistic coefficient differ in that they reflect the relationship of the independent variable with the two forms of the dependent variable, as shown here:

Logistic Coefficient	Reflects Changes in . . .
Original	Logit (log of the odds)
Exponentiated	Odds

We will discuss in the next section how each form of the coefficient reflects both the direction and magnitude of the independent variable's relationship, but requires differing methods of interpretation.

DIRECTIONALITY OF THE RELATIONSHIP   The direction of the relationship (positive or negative) reflects the changes in the dependent variable associated with changes in the independent variable. A positive relationship means that an increase in the independent variable is associated with an increase in the predicted probability, and vice versa for a negative relationship. We will see that the direction of the relationship is reflected differently for the original and exponentiated logistic coefficients.

Interpreting the Direction of Original Coefficients. The sign of the original coefficients (positive or negative) indicates the direction of the relationship, just as seen in regression coefficients. A positive coefficient increases the probability, whereas a negative value decreases the predicted probability, because the original coefficients are expressed in terms of logit values, where a value of 0.0 equates to an odds value of 1.0 and a probability of .50 Thus, negative numbers relate to odds less than 1.0 and probabilities less than .50.

Interpreting the Direction of Exponentiated Coefficients. Exponentiated coefficients must be interpreted differently because they are the logarithms of the original coefficient. By taking the logarithm, we are actually stating the exponentiated coefficient in terms of odds, which means that exponentiated coefficients will not have negative values. Because the logarithm of 0 (no effect) is 1.0, an exponentiated coefficient of 1.0 actually corresponds to a relationship with no direction. Thus, exponentiated coefficients above 1.0 reflect a positive relationship and values less than 1.0 represent negative relationships.

An Example of Interpretation. Let us look at a simple example to see what we mean in terms of the differences between the two forms of logistic coefficients.

If B_i (the original coefficient) is positive, its transformation (exponentiated coefficient) will be greater than 1, meaning that the odds will increase for any positive change in the independent variable. Thus the model will have a higher predicted probability of occurrence. Likewise, if B_i is negative, the exponentiated coefficient is less than one and the odds will be decreased. A coefficient of zero equates to an exponentiated coefficient value of 1.0, resulting in no change in the odds.

A more detailed discussion of interpretation of coefficients, logistic transformation, and the estimation procedures can be found in numerous texts [11].

MAGNITUDE OF THE RELATIONSHIP OF METRIC INDEPENDENT VARIABLES   To determine how much the probability will change given a one-unit change in the independent variable, the numeric value of the coefficient must be evaluated. Just as in multiple regression, the coefficients for metric and nonmetric variables must be interpreted differently because each reflects different impacts on the dependent variable.

For metric variables, the question is: How much will the estimated probability change for each unit change in the independent variable? In multiple regression, we knew that the regression coefficient was the slope of the linear relationship of the independent and dependent measure. A coefficient of 1.35 indicated that the dependent variable increased by 1.35 units each time that independent variable increased by one unit. In logistic regression, we know that we have a nonlinear relationship bounded between 0 and 1, so the coefficients are likely to be interpreted somewhat differently. Moreover, we have both the original and exponentiated coefficients to consider.

Original Logistic Coefficients. Although most appropriate for determining the direction of the relationship, the original logistic coefficients are less useful in determining the magnitude of the relationship. They reflect the change in the logit (logged odds) value, a unit of measure not particularly understandable in depicting how much the probabilities actually change.

Exponentiated Logistic Coefficients. Exponentiated coefficients directly reflect the magnitude of the change in the odds value. Because they are exponents, they are interpreted slightly differently. Their impact is multiplicative, meaning that the coefficient's effect is not added to the dependent variable (the odds), but multiplied for each unit change in the independent variable. As such, an exponentiated coefficient of 1.0 denotes no change ($1.0 \times$ independent variable = no change). This outcome corresponds to our earlier discussion, where exponentiated coefficients less than 1.0 reflect negative relationships and values above 1.0 denote positive relationships.

An Example of Assessing Magnitude of Change. Perhaps an easier approach to determine the amount of change in probability from these values is as follows:

$$\text{Percentage change in odds} = (\text{Exponentiated coefficient}_i - 1.0) \times 100$$

The following examples illustrate how to calculate the probability change due to a one-unit change in the independent variable for a range of exponentiated coefficients:

	Value				
Exponentiated Coefficient ($e^{b}{}_i$)	.20	.50	1.0	1.5	1.7
Exponentiated Coefficient - 1.0	−.80	−.50	0.0	.50	.70
Percentage change in odds	− 80%	− 50%	0%	50%	70%

If the exponentiated coefficient is .20, a one-unit change in the independent variable will reduce the odds by 80 percent (the same as if the odds were multiplied by .20). Likewise, an exponentiated coefficient of 1.5 denotes a 50 percent increase in the odds ratio.

A researcher who knows the existing odds and wishes to calculate the new odds value for a change in the independent variable can do so directly through the exponentiated coefficient as follows:

$$\text{New odds value} = \text{Old odds value} \times \text{Exponentiated coefficient}$$
$$\times \text{Change in independent variable}$$

Let us use a simple example to illustrate the manner in which the exponentiated coefficient affects the odds value.

Assume that the odds are 1.0 (i.e., 50–50) when the independent variable has a value of 5.5 and the exponentiated coefficient is 2.35. We know that if the exponentiated coefficient is greater than 1.0, then the relationship is positive, but we would like to know how much the odds would change. If we expected that the value of the independent variable would increase 1.5 points to 7.0, we could calculate the following:

$$\text{New odds} = 1.0 \times 2.35 \times (7.0 - 5.5) = 3.525$$

Odds can be translated into probability values by the simple formula of Probability = Odds/(1 + Odds). Thus, the odds of 3.525 translate into a probability of 77.9 percent (3.25/(1 + 3.25) = .779), indicating that increasing the independent variable by 1.5 points will increase the probability from 50 percent to 78 percent, an increase of 28 percent.

The nonlinear nature of the logistic curve is demonstrated, however, when we apply the same increase to the odds again. This time, assume that the independent variable increased another 1.5 points to 8.5. Would we also expect the probability to increase by another 28 percent? It cannot, because that would make the probability greater than 100 percent (78% + 28% = 106%). Thus, the probability increase or decrease slows down so that the curve approaches, but never reaches the two end points (0 and 1). In this example, another increase of 1.5 points creates a new odds value of 12.426, translating into odds of 92.6 percent, an increase of 14 percent. Note that in this case of increasing the probability from 78 percent, the increase in probability for the 1.5 increase in the independent variable is one-half (14%) of what it was for the same increase when the probability was 50 percent.

The researcher may find that exponentiated coefficients are quite useful not only in assessing the impact of an independent variable, but in calculating the magnitude of the effects.

INTERPRETING MAGNITUDE FOR NONMETRIC (DUMMY) INDEPENDENT VARIABLES As we discussed in multiple regression, dummy variables represent a single category of a nonmetric variable (see Chapter 4 for a more detailed discussion of dummy variables). As such, they are not like metric variables that vary across a range of values, but instead take on just the values of 1 or 0, indicating the presence or absence of a characteristic. As we saw in the preceding discussion for metric variables, the exponentiated coefficients are the best means of interpreting the impact of the dummy variable, but are interpreted differently from the metric variables.

Any time a dummy variable is used, it is essential to note the reference or omitted category. In a manner similar to the interpretation in regression, the exponentiated coefficient represents the relative level of the dependent variable for the represented group versus the omitted group. We can state this relationship as follows:

$$\text{Odds}_{\text{represented category}} = \text{Exponentiated coefficient} \times \text{Odds}_{\text{reference category}}$$

Let us use a simple example of two groups to illustrate these points.

If the nonmetric variable is gender, the two possibilities are male and female. The dummy variable can be defined as representing males (i.e., value of 1 if male, 0 if female) or females (i.e., value of 1 if female, 0 if male). Whichever way is chosen, however, determines how the coefficient is interpreted. Let us assume that a 1 is given to females, making the exponentiated coefficient represent the percentage of the odds ratio of females compared to males. If the coefficient is 1.25, then females have 25 percent higher odds than males ($1.25 - 1.0 = .25$). Likewise, if the coefficient is .80, then the odds for females are 20 percent less ($.80 - 1.0 = -.20$) than males.

Calculating Probabilities for a Specific Value of the Independent Variable

In the earlier discussion of the assumed distribution of possible dependent variables, we described an S-shaped or logistic curve. To represent the relationship between the dependent and independent variables, the coefficients must actually represent nonlinear relationships among the dependent and independent variables. Although the transformation process of taking logarithms provides a linearization of the relationship, the researcher must remember that the coefficients actually represent different slopes in the relationship across the values of the independent variable. In this way, the S-shaped distribution can be estimated. If the researcher is interested in the slope of the relationship at various values of the independent variable, the coefficients can be calculated and the relationship assessed [6].

Overview of Interpreting Coefficients

The similarity of the coefficients to those found in multiple regression has been a primary reason for the popularity of logistic regression. As we have seen in the prior discussion, many aspects are quite similar, but the unique nature of the dependent variable (the odds ratio) and the logarithmic form of the variate (necessitating use of the exponentiated coefficients) requires a somewhat different approach to interpretation. The researcher, however, still has the ability to assess the direction and magnitude of each independent variable's impact on the dependent measure and ultimately the classification accuracy of the logistic model.

Summary

The researcher faced with a dichotomous dependent variable need not resort to methods designed to accommodate the limitations of multiple regression nor be forced to employ discriminant

---

RULES OF THUMB 7-5

Logistic Regression

- Logistic regression is the preferred method for two-group (binary) dependent variables due to its robustness, ease of interpretation, and diagnostics
- Model significance tests are made with a chi-square test on the differences in the log likelihood values ($-2LL$) between two models
- Sample size considerations for logistic regression are primarily focused on the size of each group, which should have 10 times the number of estimated model coefficients
- Sample size requirements should be met in both the analysis and the holdout samples
- Coefficients are expressed in two forms: original and exponentiated to assist in interpretation
- Interpretation of the coefficients for direction and magnitude is as follows:
  - Direction can be directly assessed in the original coefficients (positive or negative signs) or indirectly in the exponentiated coefficients (less than 1 are negative, greater than 1 are positive)
  - Magnitude is best assessed by the exponentiated coefficient, with the percentage change in the dependent variable shown by:

Percentage change = (Exponentiated coefficient $-$ 1.0) $\times$ 100

analysis, especially if its statistical assumptions are violated. Logistic regression addresses these problems and provides a method developed to deal directly with this situation in the most efficient manner possible.

AN ILLUSTRATIVE EXAMPLE OF LOGISTIC REGRESSION

Logistic regression is an attractive alternative to discriminant analysis whenever the dependent variable has only two categories. Its advantages over discriminant analysis include the following:

1. Less affected than discriminant analysis by the variance–covariance inequalities across the groups, a basic assumption of discriminant analysis.
2. Handles categorical independent variables easily, whereas in discriminant analysis the use of dummy variables created problems with the variance–covariance equalities.
3. Empirical results parallel those of multiple regression in terms of their interpretation and the casewise diagnostic measures available for examining residuals.

The following example, identical to the two-group discriminant analysis discussed earlier, illustrates these advantages and the similarity of logistic regression to the results obtained from multiple regression. As we will see, even though logistic regression has many advantages as an alternative to discriminant analysis, the researcher must carefully interpret the results due to the unique aspects of how logistic regression handles the prediction of probabilities and group membership.

Stages 1, 2, and 3: Research Objectives, Research Design, and Statistical Assumptions

The issues addressed in the first three stages of the decision process are identical for the two-group discriminant analysis and logistic regression.

The research problem is still to determine whether differences in the perceptions of HBAT (X_6 to X_{18}) exist between the customers in the USA/North America versus those in the rest of the world (X_4). The sample of 100 customers is divided into an analysis sample of 60 observations, with the remaining 40 observations constituting the holdout or validation sample.

We now focus on the results stemming from the use of logistic regression to estimate and understand the differences between these two types of customers.

Stage 4: Estimation of the Logistic Regression Model and Assessing Overall Fit

Before the estimation process begins, it is possible to review the individual variables and assess their univariate results in terms of differentiating between the groups. Given that the objectives of discriminant analysis and logistic regression are the same, we can use the same measures of discrimination for assessing univariate effects as was done for discriminant analysis.

If we revisit our discussion of the differences of the groups on the 13 independent variables (refer to Table 7-5), we recall that five variables ($X_6, X_{11}, X_{12}, X_{13}$, and X_{17}) had statistically significant differences between the two groups. If you refer back to the discussion in the two-group example, we also recall an indication of multicollinearity among these variables, because both X_6 and X_{13} were part of the Product Value factor derived by factor analysis (see Chapter 3). Logistic regression is affected by multicollinearity among the independent variables in a manner similar to discriminant analysis and regression analysis.

Just as in discriminant analysis, these five variables would be the logical candidates for inclusion in the logistic regression variate, because they demonstrate the greatest differences between groups. Logistic regression may include one or more of these variables in the model, as well as even other variables that do not have significant differences at this stage if they work in combination with other variables to significantly improve prediction.

MODEL ESTIMATION   A stepwise logistic regression model is estimated much like multiple regression in that a base model is first estimated to provide a standard for comparison (see earlier discussion for more detail). In multiple regression, the mean is used to set the base model and calculate the total sum of squares. In logistic regression, the same process is used, with the mean used in the estimated model not to calculate the sum of squares, but instead to calculate the log likelihood value. From this model, the partial correlations for each variable can be established and the most discriminating variable chosen in a stepwise model according to the selection criteria.

Estimating the Base Model.   Table 7-25 contains the base model results for the logistic regression analysis based on the 60 observations in the analysis sample. The log likelihood value $(-2LL)$ here is 82.108. The score statistics, a measure of association used in logistic regression, is the measure used for selecting variables in the stepwise procedure. Several criteria can be used to guide entry: greatest reduction in the $-2LL$ value, greatest Wald coefficient, or highest conditional probability. In our example, we employ the criteria of reduction of the log likelihood ratio.

In reviewing the score statistics of variables not in the model at this time, we see that the same five variables with statistically significant differences (X_6, X_{11}, X_{12}, X_{13}, and X_{17}) also are the only variables with significant score statistics in Table 7-25. Because the stepwise procedure selects the variable with the highest score statistic, X_{13} should be the variable added in the first step.

Stepwise Estimation: Adding the First Variable, X_{13}.   As expected, X_{13} was selected for entry in the first step of the estimation process (see Table 7-26). It corresponded to the highest score statistic across all 13 perception variables. The entry of X_{13} into the logistic regression model obtained a reasonable model fit, with pseudo R^2 values ranging from .306 (pseudo R^2) to .459 (Nagelkerke R^2) and the hit ratios of 73.3 percent and 75.0 percent for the analysis and holdout samples, respectively.

TABLE 7-25   Logistic Regression Base Model Results

Overall Model Fit: Goodness-of-Fit Measures

	Value
-2 Log Likelihood $(-2LL)$	82.108

Variables Not in the Equation

Independent Variables	Score Statistic	Significance
X_6 Product Quality	11.925	.001
X_7 E-Commerce Activities	2.052	.152
X_8 Technical Support	1.609	.205
X_9 Complaint Resolution	.866	.352
X_{10} Advertising	.791	.374
X_{11} Product Line	18.323	.000
X_{12} Salesforce Image	8.622	.003
X_{13} Competitive Pricing	21.330	.000
X_{14} Warranty & Claims	.465	.495
X_{15} New Products	.614	.433
X_{16} Order & Billing	.090	.764
X_{17} Price Flexibility	21.204	.000
X_{18} Delivery Speed	.157	.692

TABLE 7-26 Logistic Regression Stepwise Estimation: Adding X_{13} (Competitive Pricing)

Overall Model Fit: Goodness-of-Fit Measures

		CHANGE IN $-2LL$			
		From Base Model		From Prior Step	
	Value	Change	Significance	Change	Significance
-2 Log Likelihood ($-2LL$)	56.971	25.136	.000	25.136	.000
Cox and Snell R^2	.342				
Nagelkerke R^2	.459				
Pseudo R^2	.306				

	Value	Significance
Hosmer and Lemeshow χ^2	17.329	.027

Variables in the Equation

Independent Variable	B	Std. Error	Wald	df	Sig.	Exp(B)
X_{13} Competitive Pricing	1.129	.287	15.471	1	.000	3.092
Constant	-7.008	1.836	14.570	1	.000	.001

B = logistic coefficient, Exp(B) = exponentiated coefficient

Variables Not in the Equation

Independent Variables	Score Statistic	Significance
X_6 Product Quality	4.859	.028
X_7 E-Commerce Activities	.132	.716
X_8 Technical Support	.007	.932
X_9 Complaint Resolution	1.379	.240
X_{10} Advertising	.129	.719
X_{11} Product Line	6.154	.013
X_{12} Salesforce Image	2.745	.098
X_{14} Warranty & Claims	.640	.424
X_{15} New Products	.344	.557
X_{16} Order & Billing	2.529	.112
X_{17} Price Flexibility	13.723	.000
X_{18} Delivery Speed	1.206	.272

Classification Matrix

	Predicted Group Membership[c]					
	ANALYSIS SAMPLE[a]			HOLDOUT SAMPLE[b]		
	X_4 Region			X_4 Region		
Actual Group Membership	USA/ North America	Outside North America	Total	USA/ North America	Outside North America	Total
USA/ North America	19 (73.1)	7	26	4 (30.8)	9	13
Outside North America	9	25 (73.5)	34	1	26 (96.3)	27

[a]73.3% of analysis sample correctly classified.
[b]75.0% of holdout sample correctly classified.
[c]Values in parentheses are percentage correctly classified (hit ratio).

Examination of the results, however, identifies two reasons for considering an additional stage(s) to add variable(s) to the logistic regression model:

- Three variables not in the current logistic model (X_{17}, X_{11}, and X_6) have statistically significant score statistics, indicating that their inclusion would significantly improve the overall model fit.
- The overall hit ratio for the holdout sample is good (75.0%), but one of the groups (USA/North America customers) has an unacceptably low hit ratio of 30.8 percent.

Stepwise Estimation: Adding the Second Variable, X_{17}, the Second Variable. Hopefully one or more steps in the stepwise procedure will result in the inclusion of all independent variables with significant score statistics as well as achieve acceptable hit ratios (overall and group-specific) for both the analysis and holdout samples.

X_{17}, with the highest score statistic after adding X_{13}, was selected for entry at step 2 (Table 7-27). Improvement in all measures of model fit ranged from a decrease in the $-2LL$ value to the various R^2 measures. More important, from a model estimation perspective, none of the variables not in the equation had statistically significant change scores.

Thus, the two-variable logistic model including X_{13} and X_{17} will be the final model to be used for purposes of assessing model fit, prediction accuracy, and interpretation of the coefficients.

ASSESSING OVERALL MODEL FIT   In making an assessment of the overall fit of a logistic regression model, we can draw upon three approaches: statistical measures of overall model fit, pseudo R^2 measures, and classification accuracy as expressed in the hit ratio. Each of these approaches will be examined for the one-variable and two-variable logistic regression models that resulted from the stepwise procedure.

Statistical Measures.   The first statistical measure is the chi-square test for the change in the $-2LL$ value from the base model, which is comparable to the overall F-test in multiple regression. Smaller values of the $-2LL$ measure indicate better model fit, and the statistical test is available for assessing the difference between both the base model and other proposed models (in a stepwise procedure, this test is always based on improvement from the prior step).

- In the single-variable model (see Table 7-26), the $-2LL$ value is reduced from the base model value of 82.108 to 59.971, a decrease of 25.136. This increase in model fit was statistically significant at the .000 level.
- In the two-variable model, the $-2LL$ value decreased further to 39.960, resulting in significant decreases not only from the base model (42.148), but also a significant decrease from the one-variable model (17.011). Both of these improvements in model fit were significant at the .000 level.

The second statistical measure is the Hosmer and Lemeshow measure of overall fit [11]. This statistical test measures the correspondence of the actual and predicted values of the dependent variable. In this case, better model fit is indicated by a smaller difference in the observed and predicted classification.

The Hosmer and Lemeshow test shows significance for the one-variable logistic model (.027 from Table 7-26), indicating that significant differences still remain between actual and expected values. The two-variable model, however, reduces the significance level to .722 (see Table 7-27), a nonsignificant value indicating that the model fit is acceptable.

For the two-variable logistic model, both statistically based measures of overall model fit indicate that the model is acceptable and at a statistically significant level. It is necessary, however, to examine the other measures of overall model fit to assess whether the results reach the necessary levels of practical significance as well.

TABLE 7-27 Logistic Regression Stepwise Estimation: Adding X_{17} (Price Flexibility)

Overall Model Fit: Goodness-of-Fit Measures

		CHANGE IN $-2LL$			
		From Base Model		From Prior Step	
	Value	Change	Significance	Change	Significance
-2 Log Likelihood ($-2LL$)	39.960	42.148	.000	17.011	.000
Cox and Snell R^2	.505				
Nagelkerke R^2	.677				
Pseudo R^2	.513				

	Value	Significance
Hosmer and Lemeshow χ^2	5.326	.722

Variables in the Equation

Independent Variable	B	Std. Error	Wald	df	Sig.	Exp(B)
X_{13} Competitive Pricing	1.079	.357	9.115	1	.003	2.942
X_{17} Price Flexibility	1.844	.639	8.331	1	.004	6.321
Constant	-14.192	3.712	14.614	1	.000	.000

B = logistic coefficient, Exp(B) = exponentiated coefficient

Variables Not in the Equation

Independent Variables	Score Statistic	Significance
X_6 Product Quality	.656	.418
X_7 E-Commerce Activities	3.501	.061
X_8 Technical Support	.006	.937
X_9 Complaint Resolution	.693	.405
X_{10} Advertising	.091	.762
X_{11} Product Line	3.409	.065
X_{12} Salesforce Image	.849	.357
X_{14} Warranty & Claims	2.327	.127
X_{15} New Products	.026	.873
X_{16} Order & Billing	.010	.919
X_{18} Delivery Speed	2.907	.088

Classification Matrix

	Predicted Group Membership[c]					
	ANALYSIS SAMPLE[a]			HOLDOUT SAMPLE[b]		
	X_4 Region			X_4 Region		
Actual Group Membership	USA/ North America	Outside North America	Total	USA/ North America	Outside North America	Total
USA/ North America	25 (96.2)	1	26	9 (69.2)	4	13
Outside North America	6	28 (82.4)	34	2	25 (92.6)	27

[a]88.3% of analysis sample correctly classified.
[b]85.0% of holdout sample correctly classified.
[c]Values in parentheses are percentage correctly classified (hit ratio).

Pseudo R^2 Measures. Three available measures are comparable to the R^2 measure in multiple regression: the Cox and Snell R^2, the Nagelkerke R^2, and a pseudo R^2 measure based on the reduction in the $-2LL$ value.

For the one-variable logistic regression model, these values were .342, .459, and .306, respectively. In combination, they indicate that the one-variable regression model accounts for approximately one-third of the variation in the dependent measure. Although the one-variable model was deemed statistically significant on several overall measures of fit, these R^2 measures are somewhat low for purposes of practical significance.

The two-variable model (see Table 7-27) has R^2 values that are each over .50, indicating that the logistic regression model accounts for at least one-half of the variation between the two groups of customers. One would always want to improve these values, but this level is deemed practically significant in this situation.

The R^2 values of the two-variable model showed substantive improvement over the single-variable model and indicate good model fit when compared to the R^2 values usually found in multiple regression. Coupled with the statistically based measures of model fit, the model is deemed acceptable in terms of both statistical and practical significance.

Classification Accuracy. The third examination of overall model fit will be to assess the classification accuracy of the model in a final measure of practical significance. The classification matrices, identical in nature to those used in discriminant analysis, represent the levels of predictive accuracy achieved by the logistic model. The measure of predictive accuracy used is the hit ratio, the percentage of cases correctly classified. These values will be calculated for both the analysis and holdout samples, and group-specific measures will be examined in addition to the overall measures. Moreover, comparisons can be made, as was done in discriminant analysis, to comparison standards representing the levels of predictive accuracy achieved by chance (see more detailed discussion in the discriminant analysis section).

The comparison standards for the classification matrix hit ratios will be the same as were calculated for the two-group discriminant analysis. The values are 65.5 percent for the proportional chance criterion (the preferred measure) and 76.3 percent for the maximum chance criterion. If you are unfamiliar with the methods of calculating these measures, refer back to the discussion earlier in the chapter regarding assessment of classification measurement accuracy.

- The overall hit ratios for the single-variable logistic model are 73.3 percent and 75.0 percent for the analysis and holdout samples, respectively. Even though the overall hit ratios are greater than the proportional chance criterion and comparable to the maximum chance criterion, a significant problem appears in the holdout sample for the USA/North America customers, where the hit ratio is only 30.8 percent. This level is below both standards and requires that the logistic model be expanded hopefully to the extent that this group-specific hit ratio will exceed the standards.
- The two-variable model shows substantial improvement in both the overall hit ratios as well as the group-specific values. The overall hit ratios increased to 88.3 percent and 85.0 percent for the analysis and holdout samples, respectively. Moreover, the problematic group-specific hit ratio in the holdout sample increases to 69.2 percent, above the standard value for the proportional chance criterion. With these improvements at both the overall and group-specific levels, the two-variable logistic regression model is deemed acceptable in terms of classification accuracy.

Across all three of the basic types of measures of overall model fit, the two-variable model (with X_{13} and X_{17}) demonstrates acceptable levels of both statistical and practical significances. With overall model fit acceptable, we turn our attention to assessing the statistical tests of the logistic coefficients in order to identify the coefficients that have significant relationships affecting group membership.

STATISTICAL SIGNIFICANCE OF THE COEFFICIENTS   The estimated coefficients for the two independent variables and the constant can also be evaluated for statistical significance. The Wald statistic is used to assess significance in a manner similar to the *t* test used in multiple regression.

The logistic coefficients for X_{13} (1.079) and X_{17} (1.844) and the constant (−14.190) are all significant at the .01 level based on the statistical tests of the Wald statistic. No other variables would enter the model and achieve at least a .05 level of significance.

Thus, the individual variables are significant and can be interpreted to identify the relationships affecting the predicted probabilities and subsequently group membership.

CASEWISE DIAGNOSTICS   The analysis of the misclassification of individual observations can provide further insight into possible improvements of the model. Casewise diagnostics such as residuals and measures of influence are available, as well as the profile analysis discussed earlier for discriminant analysis (for a more detailed discussion of measures such as Cook's distance and DFBETA, see the supplement to Chapter 4 available on the Web at www.pearsonglobaleditions.com/hair or www.mvstats.com).

In this case, only 13 cases have been misclassified (7 in the analysis sample and 6 in the holdout sample). Given the high degree of correspondence between these misclassified cases and the misclassified cases analyzed in the two-group discriminant analysis, the profiling process will not be undertaken again (interested readers can refer back to the two-group example). Casewise diagnostics such as residuals and measures of influence are available. Given the low levels of misclassification, however, no further analysis of misclassification is performed.

Stage 5: Interpretation of the Results

The stepwise logistic regression procedure produced a variate quite similar to that of the two-group discriminant analysis, although with one less independent variable. We will examine the logistic coefficients to assess both the direction and impact each variable has on predicted probability and group membership.

INTERPRETING THE LOGISTIC COEFFICIENTS   The final logistic regression model included two variables (X_{13} and X_{17}) with logistic regression coefficients of 1.079 and 1.844, respectively, and a constant of −14.190 (see Table 7-27). Comparing these results to the two-group discriminant analysis reveals almost identical results, as discriminant analysis included three variables in the two-group model—X_{13} and X_{17} along with X_{11}.

Direction of the Relationships.   To assess the direction of the relationship of each variable, we can examine either the original logistic coefficients or the exponentiated coefficients. Let us start with the original coefficients.

If you recall from our earlier discussion, we can interpret the direction of the relationship directly from the sign of the original logistic coefficients. In this case both variables have positive signs, indicating a positive relationship between both independent variables and predicted probability. As the values of either X_{13} or X_{17} increase, the predicted probability will increase, thus increasing the likelihood that a customer will be categorized as residing outside North America.

Turning our attention to the exponentiated coefficients, we should recall that values above 1.0 indicate a positive relationship and below 1.0 indicate a negative relationship. In our case, the values of 2.942 and 6.319 also indicate positive relationships.

Magnitude of the Relationships.   The most direct method of assessing the magnitude of the change in probability due to each independent variable is to examine the exponentiated coefficients. As you recall, the exponentiated coefficients minus one equals the percentage change in odds.

In our case, it means that an increase by one point increases the odds by 194 percent for X_{13} and 531 percent for X_{17}. These numbers can exceed 100-percent change because they are increasing the odds, not the probabilities themselves. The impacts are large because the constant term (-14.190) defines a starting point of almost zero for the probability values. Thus, large increases in the odds are needed to reach larger probability values.

Another approach in understanding how the logistic coefficients define probability is to calculate the predicted probability for any set of values for the independent variables.

For the independent variables X_{13} and X_{17}, let us use the group means for the two groups. In this manner, we can see what the predicted probability would be for a "typical" member of each group.

Table 7-28 shows the calculations for predicting the probability of the two group centroids. First, we calculate the logit value for each group centroid by inserting the group centroid values (e.g., 5.60 and 3.63 for group 0 on X_{13} and X_{17}, respectively) into the logit equation. Remember from Table 7-27 that the estimated weights were 1.079 and 1.844 for X_{13} and X_{17}, respectively with a constant of -14.192. Thus, substitution of the group centroid values into this equation results in logit values of -1.452 (group 0) and 2.909 (group 1). Taking the antilog of the logit values results in odds of .234 and 18.332. Then, the probability of a group is calculated as its odds value over the sum of the odds for both groups. This results in the "typical" member of group 0 having a probability of being incorrectly assigned to group 1 of .189 (.189 = .234/(.234 + 18.332)), whereas the "typical" member of group 1 has a probability of .948 of being correctly assigned to group 1. This example demonstrates that the logistic model does create separation between the two group centroids in terms of predicted probability, resulting in the excellent classification results achieved for both analysis and holdout samples.

The logistic coefficients define positive relationships for both independent variables and provide a means of assessing the impact of a change in either or both variables on the odds and thus the predicted probability. It becomes apparent why many researchers prefer logistic regression to discriminant analysis when comparisons are made on the more useful information available from logistic coefficients versus discriminant loadings.

Stage 6: Validation of the Results

The validation of the logistic regression model is accomplished in this example through the same method used in discriminant analysis: creation of analysis and holdout samples. By examining the hit ratio for the holdout sample, the researcher can assess the external validity and practical significance of the logistic regression model.

TABLE 7-28 Calculating Estimated Probability Values for the Group Centroids of X_4 Region

	X_4 (Region)	
	Group 0: **USA/North America**	**Group 1:** **Outside North America**
Centroid: X_{13}	5.60	7.42
Centroid: X_{17}	3.63	4.93
Logit Value[a]	−1.452	2.909
Odds[b]	.234	18.332
Probability[c]	.189	.948

[a]Calculated as: Logit $= -14.190 + 1.079X_{13} + 1.844X_{17}$
[b]Calculated as: Odds $= e^{\text{Logit}}$
[c]Calculated as: Probability $= \text{Odds}/(1 + \text{Odds})$

For the final two-variable logistic regression model, the hit ratios for both the analysis and holdout samples exceed all of the comparison standards (proportional chance and maximum chance criteria). Moreover, all of the group-specific hit ratios are sufficiently large for acceptance. This aspect is especially important for the holdout sample, which is the primary indicator of external validity.

These outcomes lead to the conclusion that the logistic regression model, as found with the discriminant analysis model as well, demonstrated sufficient external validity for complete acceptance of the results.

A Managerial Overview

Logistic regression presents an alternative to discriminant analysis that may be more comfortable to many researchers due to its similarity to multiple regression. Given its robustness in the face of data conditions that can negatively affect discriminant analysis (e.g., unequal variance–covariance matrices), logistic regression is also the preferred estimation technique in many applications.

When compared to discriminant analysis, logistic regression provides comparable predictive accuracy with a simpler variate that used the same substantive interpretation, only with one fewer variable. From the logistic regression results, the researcher can focus on competitive pricing and price flexibility as the primary differentiating variables between the two groups of customers. The objective in this analysis is not to increase probability (as might be the case of analyzing success versus failure), yet logistic regression still provides a straightforward approach for HBAT to understand the relative impact of each independent variable in creating differences between the two groups of customers.

Summary

The underlying nature, concepts, and approach to multiple discriminant analysis and logistic regression have been presented. Basic guidelines for their application and interpretation were included to clarify further the methodological concepts. This chapter helps you to do the following:

State the circumstances under which linear discriminant analysis or logistic regression should be used instead of multiple regression.   In choosing an appropriate analytical technique, we sometimes encounter a problem that involves a categorical dependent variable and several metric independent variables. Recall that the single dependent variable in regression was measured metrically. Multiple discriminant analysis and logistic regression are the appropriate statistical techniques when the research problem involves a single categorical dependent variable and several metric independent variables. In many cases, the dependent variable consists of two groups or classifications, for example, male versus female, high versus low, or good versus bad. In other instances, more than two groups are involved, such as low, medium, and high classifications. Discriminant

analysis and logistic regression are capable of handling either two groups or multiple (three or more) groups. The results of a discriminant analysis and logistic regression can assist in profiling the intergroup characteristics of the subjects and in assigning them to their appropriate groups.

Identify the major issues relating to types of variables used and sample size required in the application of discriminant analysis.   To apply discriminant analysis, the researcher first must specify which variables are to be independent measures and which variable is to be the dependent measure. The researcher should focus on the dependent variable first. The number of dependent variable groups (categories) can be two or more, but these groups must be mutually exclusive and exhaustive. After a decision has been made on the dependent variable, the researcher must decide which independent variables to include in the analysis. Independent variables are selected in two ways: (1) by identifying variables either from previous research or from the theoretical model underlying the research question and (2) by utilizing the researcher's knowledge and intuition to select variables for which no previous research or theory exists but that logically might be related to predicting the dependent variable groups.

Discriminant analysis, like the other multivariate techniques, is affected by the size of the sample being analyzed. A ratio of 20 observations for each predictor variable is recommended. Because the results become unstable as the sample size decreases relative to the number of independent variables, the minimum size recommended is five observations per independent variable. The sample size of each group also must be considered. At a minimum, the smallest group size of a category must exceed the number of independent variables. As a practical guideline, each category should have at least 20 observations. Even if all categories exceed 20 observations, however, the researcher also must consider the relative sizes of the groups. Wide variations in the sizes of the groups will affect the estimation of the discriminant function and the classification of observations.

Understand the assumptions underlying discriminant analysis in assessing its appropriateness for a particular problem. The assumptions for discriminant analysis relate to both the statistical processes involved in the estimation and classification procedures and issues affecting the interpretation of the results. The key assumptions for deriving the discriminant function are multivariate normality of the independent variables and unknown (but equal) dispersion and covariance structures (matrices) for the groups as defined by the dependent variable. If the assumptions are violated, the researcher should understand the impact on the results that can be expected and consider alternative methods for analysis (e.g., logistic regression).

Describe the two computation approaches for discriminant analysis and the method for assessing overall model fit. The two approaches for discriminant analysis are the simultaneous (direct) method and the stepwise method. Simultaneous estimation involves computing the discriminant function by considering all of the independent variables at the same time. Thus, the discriminant function is computed based upon the entire set of independent variables, regardless of the discriminating power of each independent variable. Stepwise estimation is an alternative to the simultaneous approach. It involves entering the independent variables into the discriminant function one at a time on the basis of their discriminating power. The stepwise approach follows a sequential process of adding or deleting variables to the discriminant function. After the discriminant function(s) is estimated, the researcher must evaluate the significance or fit of the discriminant function(s). When a simultaneous approach is used, Wilks' lambda, Hotelling's trace, and Pillai's criterion all evaluate the statistical significance of the discriminatory power of the discriminant function(s). If a stepwise method is used to estimate the discriminant function, the Mahalanobis D^2 and Rao's V measures are most appropriate to assess fit.

Explain what a classification matrix is and how to develop one, and describe the ways to evaluate the predictive accuracy of the discriminant function. The statistical tests for assessing the significance of the discriminant function(s) only assess the degree of difference between the groups based on the discriminant Z scores, but do not indicate how well the function(s) predicts. To determine the predictive ability of a discriminant function, the researcher must construct classification matrices. The classification matrix procedure provides a perspective on practical significance rather than statistical significance. Before a classification matrix can be constructed, however, the researcher must determine the cutting score for each discriminant function. The cutting score represents the dividing point used to classify observations into each of the groups based on discriminant function score. The calculation of a cutting score between any two groups is based on the two group centroids (group mean of the discriminant scores) and the relative size of the two groups. The results of the classification procedure are presented in matrix form. The entries on the diagonal of the matrix represent the number of individuals correctly classified. The numbers off the diagonal represent the incorrect classifications. The percentage correctly classified, also termed the *hit ratio,* reveals how well the discriminant function predicts the objects. If the costs of misclassifying are approximately equal for all groups, the optimal cutting score will be the one that will misclassify the fewest number of objects across all groups. If the misclassification costs are unequal, the optimum cutting score will be the one that minimizes the costs of misclassification. To evaluate the hit ratio, we must look at chance classification. When the group sizes are equal, determination of chance classification is based on the number of groups. When the group sizes are unequal, calculating chance classification can be done two ways: maximum chance and proportional chance.

Tell how to identify independent variables with discriminatory power. If the discriminant function is statistically significant and the classification accuracy (hit ratio) is acceptable, the researcher should focus on making substantive interpretations of the findings. This process involves determining the relative importance of each independent variable in discriminating between the groups. Three methods of determining the relative importance have been proposed: (1) standardized discriminant weights, (2) discriminant loadings (structure correlations), and (3) partial F values. The traditional approach to interpreting discriminant functions examines the sign and magnitude of the standardized discriminant weight assigned to each variable in computing the discriminant functions. Independent variables with relatively larger weights contribute more to the discriminating power of the function than do variables with smaller weights. The sign denotes whether the variable makes either a positive or a negative contribution. Discriminant loadings are increasingly used as a basis for interpretation because of the deficiencies in utilizing weights. Measuring the simple linear correlation between each independent variable and the discriminant function, the discriminant loadings reflect the variance that the independent variables share with the discriminant function. They can be interpreted like factor loadings in assessing the relative contribution of each independent variable to the discriminant function. When a stepwise estimation method is used, an additional means of interpreting the relative discriminating power of the independent variables is through the use of partial F values, which is accomplished by examining the absolute sizes of the significant F values and ranking them. Large F values indicate greater discriminatory power.

Justify the use of a split-sample approach for validation. The final stage of a discriminant analysis involves validating the discriminant results to provide assurances that the results have external as well as internal validity. In addition to validating the hit ratios, the researcher should use group profiling to ensure that the group means are valid indicators of the conceptual model used in selecting the independent variables. Validation can occur either with a separate sample (holdout sample) or utilizing a procedure that repeatedly processes the estimation sample. Validation of the hit ratios is performed most often by creating a holdout sample, also referred to as the validation sample. The purpose of utilizing a holdout sample for validation purposes is to see how well the discriminant function works on a sample of observations not used to derive the discriminant function. This assessment involves developing a discriminant function with the analysis sample and then applying it to the holdout sample.

Understand the strengths and weaknesses of logistic regression compared to discriminant analysis and multiple regression. Discriminant analysis is appropriate when the dependent variable is nonmetric. If the dependent variable has only two groups, then logistic regression may be preferred for two reasons: First, discriminant analysis relies on strictly meeting the assumptions of multivariate normality and equal variance–covariance matrices across groups—assumptions that are not met in many situations. Logistic regression does not face these strict assumptions and is much more robust when these assumptions are not met, making its application appropriate in many situations. Second, even if the assumptions are met, many researchers prefer logistic regression because it is similar to multiple regression. As such, it has straightforward statistical tests, similar approaches to incorporating metric and nonmetric variables and nonlinear effects, and a wide range of diagnostics. Logistic regression is equivalent to two-group discriminant analysis and may be more suitable in many situations.

Interpret the results of a logistic regression analysis, with comparisons to both multiple regression and discriminant analysis. The goodness-of-fit for a logistic regression model can be assessed in two ways: (1) using pseudo R^2 values, similar to that found in multiple regression, and (2) examining predictive accuracy (i.e., the classification matrix in discriminant analysis). The two approaches examine model fit from different perspectives, but should yield similar conclusions. One of the advantages of logistic regression is that we need to know only whether an event occurred to define a dichotomous value as our dependent variable. When we analyze these data using the logistic transformation, however, the logistic regression and its coefficients take on a somewhat different meaning from those found in regression with a metric dependent variable. Similarly, discriminant loadings in discriminant analysis are interpreted differently from a logistic coefficient. The logistic coefficient reflects both the direction and magnitude of the independent variable's relationship, but requires differing methods of interpretation. The direction of the relationship (positive or negative) reflects the changes in the dependent variable associated with changes in the independent variable. A positive relationship means that an increase in the independent variable is associated with an increase in the predicted probability, and vice versa for a negative relationship. To determine the magnitude of the coefficient, or how much the probability will change given a one-unit change in the independent variable, the numeric value of the coefficient must be evaluated. Just as in multiple regression, the coefficients for metric and nonmetric variables must be interpreted differently because each reflects different impacts on the dependent variable.

Multiple discriminant analysis and logistic regression help us to understand and explain research problems that involve a single categorical dependent variable and several metric independent variables. Both techniques can be used to profile the intergroup characteristics of the subjects and assign them to their appropriate groups. Potential applications of these two techniques to both business and nonbusiness problems are numerous.

Questions

1. How would you differentiate among multiple discriminant analysis, regression analysis, logistic regression analysis, and analysis of variance?
2. When would you employ logistic regression rather than discriminant analysis? What are the advantages and disadvantages of this decision?
3. What criteria could you use in deciding whether to stop a discriminant analysis after estimating the discriminant function(s)? After the interpretation stage?
4. What procedure would you follow in dividing your sample into analysis and holdout groups? How would you change this procedure if your sample consisted of fewer than 100 individuals or objects?
5. How would you determine the optimum cutting score?
6. How would you determine whether the classification accuracy of the discriminant function is sufficiently high relative to chance classification?
7. How does a two-group discriminant analysis differ from a three-group analysis?
8. Why should a researcher stretch the loadings and centroid data in plotting a discriminant analysis solution?
9. How do logistic regression and discriminant analysis each handle the relationship of the dependent and independent variables?
10. What are the differences in estimation and interpretation between logistic regression and discriminant analysis?
11. Explain the concept of odds and why it is used in predicting probability in a logistic regression procedure.

Suggested Readings

A list of suggested readings illustrating issues and applications of discriminant analysis and logistic regression is available on the Web at www.pearsonglobaleditions.com/hair or www.mvstats.com.

References

1. Cohen, J. 1988. *Statistical Power Analysis for the Behavioral Sciences,* 2nd ed. Hillsdale, NJ: Lawrence Erlbaum Associates.
2. Crask, M., and W. Perreault. 1977. Validation of Discriminant Analysis in Marketing Research. *Journal of Marketing Research* 14 (February): 60–68.
3. Demaris, A. 1995. A Tutorial in Logistic Regression. *Journal of Marriage and the Family* 57: 956–68.
4. Dillon, W. R., and M. Goldstein. 1984. *Multivariate Analysis: Methods and Applications.* New York: Wiley.
5. Frank, R. E., W. E. Massey, and D. G. Morrison. 1965. Bias in Multiple Discriminant Analysis. *Journal of Marketing Research* 2(3): 250–58.
6. Gessner, Guy, N. K. Maholtra, W. A. Kamakura, and M. E. Zmijewski. 1988. Estimating Models with Binary Dependent Variables: Some Theoretical and Empirical Observations. *Journal of Business Research* 16(1): 49–65.
7. Green, P. E., D. Tull, and G. Albaum. 1988. *Research for Marketing Decisions.* Upper Saddle River, NJ: Prentice Hall.
8. Green, P. E. 1978. *Analyzing Multivariate Data.* Hinsdale, IL: Holt, Rinehart and Winston.
9. Green, P. E., and J. D. Carroll. 1978. *Mathematical Tools for Applied Multivariate Analysis.* New York: Academic Press.
10. Harris, R. J. 2001. *A Primer of Multivariate Statistics,* 3rd ed. Hillsdale, NJ: Lawrence Erlbaum Associates.
11. Hosmer, D. W., and S. Lemeshow. 2000. *Applied Logistic Regression,* 2nd ed. New York: Wiley.

12. Huberty, C. J. 1984. Issues in the Use and Interpretation of Discriminant Analysis. *Psychological Bulletin* 95: 156–71.

13. Huberty, C. J., J. W. Wisenbaker, and J. C. Smith. 1987. Assessing Predictive Accuracy in Discriminant Analysis. *Multivariate Behavioral Research* 22 (July): 307–29.

14. Johnson, N., and D. Wichern. 2002. *Applied Multivariate Statistical Analysis,* 5th ed. Upper Saddle River, NJ: Prentice Hall.

15. Long, J. S. 1997. *Regression Models for Categorical and Limited Dependent Variables: Analysis and Interpretation.* Thousand Oaks, CA: Sage.

16. Morrison, D. G. 1969. On the Interpretation of Discriminant Analysis. *Journal of Marketing Research* 6(2): 156–63.

17. Pampel, F. C. 2000. *Logistic Regression: A Primer,* Sage University Papers Series on Quantitative Applications in the Social Sciences, # 07–096. Newbury Park, CA: Sage.

18. Perreault, W. D., D. N. Behrman, and G. M. Armstrong. 1979. Alternative Approaches for Interpretation of Multiple Discriminant Analysis in Marketing Research. *Journal of Business Research* 7: 151–73.

CHAPTER 8

ANOVA and MANOVA

LEARNING OBJECTIVES

Upon completing this chapter, you should be able to do the following:

- Explain the difference between the univariate null hypothesis of ANOVA and the multivariate null hypothesis of MANOVA.
- Discuss the advantages of a multivariate approach to significance testing compared to the more traditional univariate approaches.
- State the assumptions for the use of MANOVA.
- Discuss the different types of test statistics that are available for significance testing in MANOVA.
- Describe the purpose of post hoc tests in ANOVA and MANOVA.
- Interpret interaction results when more than one independent variable is used in MANOVA.
- Describe the purpose of multivariate analysis of covariance (MANCOVA).

CHAPTER PREVIEW

Multivariate analysis of variance (MANOVA) is an extension of analysis of variance (ANOVA) to accommodate more than one dependent variable. It is a dependence technique that measures the differences for two or more metric dependent variables based on a set of categorical (nonmetric) variables acting as independent variables. ANOVA and MANOVA can be stated in the following general forms:

Analysis of Variance

$$Y_1 = X_1 + X_2 + X_3 + \cdots + X_n$$

(metric)   (nonmetric)

Multivariate Analysis of Variance

$$Y_1 + Y_2 + Y_3 + \cdots + Y_n = X_1 + X_2 + X_3 + \cdots + X_n$$

(metric)   (nonmetric)

Like ANOVA, MANOVA is concerned with differences between groups (or experimental treatments). ANOVA is termed a *univariate procedure* because we use it to assess group differences on a single metric dependent variable. MANOVA is termed a *multivariate procedure* because we use it to assess group differences across multiple metric dependent variables simultaneously. In MANOVA, each treatment group is observed on two or more dependent variables.

The concept of multivariate analysis of variance was introduced more than 70 years ago by Wilks [26]. However, it was not until the development of appropriate test statistics with tabled distributions and the more recent widespread availability of computer programs to compute these statistics that MANOVA became a practical tool for researchers.

Both ANOVA and MANOVA are particularly useful when used in conjunction with experimental designs; that is, research designs in which the researcher directly controls or manipulates one or more independent variables to determine the effect on the dependent variable(s). ANOVA and MANOVA provide the tools necessary to judge the observed effects (i.e., whether an observed difference is due to a treatment effect or to random sampling variability). However, MANOVA has a role in nonexperimental designs (e.g., survey research) where groups of interest (e.g., gender, purchaser/nonpurchaser) are defined and then the differences on any number of metric variables (e.g., attitudes, satisfaction, purchase rates) are assessed for statistical significance.

KEY TERMS

Before starting the chapter, review the key terms to develop an understanding of the concepts and terminology to be used. Throughout the chapter the key terms appear in **boldface.** Other points of emphasis in the chapter and key term cross-references are *italicized.*

Alpha (α) Significance level associated with the statistical testing of the differences between two or more groups. Typically, small values, such as .05 or .01, are specified to minimize the possibility of making a *Type I error.*

Analysis of variance (ANOVA) Statistical technique used to determine whether samples from two or more groups come from populations with equal means (i.e., Do the group means differ significantly?). Analysis of variance examines one dependent measure, whereas multivariate analysis of variance compares group differences on two or more dependent variables.

A priori test See *planned comparison.*

Beta (β) See *Type II error.*

Blocking factor Characteristic of respondents in the *ANOVA* or MANOVA that is used to reduce within-group variability by becoming an additional *factor* in the analysis. Most often used as a control variable (i.e., a characteristic not included in the analysis but one for which differences are expected or proposed). By including the blocking factor in the analysis, additional groups are formed that are more homogeneous and increase the chance of showing significant differences. As an example, assume that customers were asked about their buying intentions for a product and the independent measure used was age. Prior experience showed that substantial variation in buying intentions for other products of this type was also due to gender. Then gender could be added as a further factor so that each age category was split into male and female groups with greater within-group homogeneity.

Bonferroni inequality Approach for adjusting the selected *alpha* level to control for the overall *Type I error* rate when performing a series of separate tests. The procedure involves calculating a new *critical value* by dividing the proposed *alpha* rate by the number of statistical tests to be performed. For example, if a .05 *significance level* is desired for a series of five separate tests, then a rate of .01 (.05 ÷ 5) is used in each separate test.

Box's M test Statistical test for the equality of the variance–covariance matrices of the dependent variables across the groups. It is especially sensitive to the presence of nonnormal variables. Use of a conservative *significance level* (i.e., .01 or less) is suggested as an adjustment for the sensitivity of the statistic.

Contrast Procedure for investigating specific group differences of interest in conjunction with *ANOVA* and MANOVA (e.g., comparing group mean differences for a specified pair of groups).

Covariates, or covariate analysis Use of regression-like procedures to remove extraneous (nuisance) variation in the dependent variables due to one or more uncontrolled metric independent variables (covariates). The covariates are assumed to be linearly related to the dependent variables. After adjusting for the influence of covariates, a standard *ANOVA* or MANOVA is carried out. This adjustment process (known as ANCOVA or MANCOVA) usually allows for more sensitive tests of treatment effects.

Critical value   Value of a test statistic (*t* test, *F* test) that denotes a specified *significance level*. For example, 1.96 denotes a .05 significance level for the *t* test with large sample sizes.

Discriminant function   Dimension of difference or discrimination between the groups in the MANOVA analysis. The discriminant function is a *variate* of the dependent variables.

Disordinal interaction   Form of *interaction effect* among independent variables that invalidates interpretation of the *main effects* of the treatments. A disordinal interaction is exhibited graphically by plotting the means for each group and having the lines intersect or cross. In this type of interaction the mean differences not only vary, given the unique combinations of independent variable levels, but the relative ordering of groups changes as well.

Effect size   Standardized measure of group differences used in the calculation of statistical *power*. Calculated as the difference in group means divided by the standard deviation, it is then comparable across research studies as a generalized measure of effect (i.e., differences in group means).

Experimental design   Research plan in which the researcher directly manipulates or controls one or more independent variables (see *treatment* or *factor*) and assesses their effect on the dependent variables. Common in the physical sciences, it is gaining in popularity in business and the social sciences. For example, respondents are shown separate advertisements that vary systematically on a characteristic, such as different appeals (emotional versus rational) or types of presentation (color versus black-and-white) and are then asked their attitudes, evaluations, or feelings toward the different advertisements.

Experimentwide error rate   The combined or overall error rate that results from performing multiple *t* tests or *F* tests that are related (e.g., *t* tests among a series of correlated variable pairs or a series of *t* tests among the pairs of categories in a multichotomous variable).

Factor   Nonmetric independent variable, also referred to as a *treatment* or experimental variable.

Factorial design   Design with more than one *factor* (treatment). Factorial designs examine the effects of several factors simultaneously by forming groups based on all possible combinations of the levels (values) of the various treatment variables.

General linear model (GLM)   Generalized estimation procedure based on three components: (1) a *variate* formed by the linear combination of independent variables, (2) a probability distribution specified by the researcher based on the characteristics of the dependent variables, and (3) a *link function* that denotes the connection between the variate and the probability distribution.

Hotelling's T^2   Test to assess the statistical significance of the difference on the means of two or more variables between two groups. It is a special case of MANOVA used with two groups or levels of a treatment variable.

Independence   Critical assumption of *ANOVA* or MANOVA that requires that the dependent measures for each respondent be totally uncorrelated with the responses from other respondents in the sample. A lack of independence severely affects the statistical validity of the analysis unless corrective action is taken.

Interaction effect   In *factorial designs*, the joint effects of two *treatment* variables in addition to the individual *main effects*. It means that the difference between groups on one treatment variable varies depending on the level of the second treatment variable. For example, assume that respondents were classified by income (three levels) and gender (males versus females). A significant interaction would be found when the differences between males and females on the independent variable(s) varied substantially across the three income levels.

Link function   A primary component of the *general linear model (GLM)* that specifies the transformation between the variate of independent variables and the specified probability distribution. In MANOVA (and regression) the identity link is used with a normal distribution, corresponding to our statistical assumptions of normality.

Main effect   In factorial designs, the individual effect of each *treatment* variable on the dependent variable.

Multivariate normal distribution   Generalization of the univariate normal distribution to the case of *p* variables. A multivariate normal distribution of sample groups is a basic assumption required for the validity of the significance tests in MANOVA (see Chapter 2 for more discussion of this topic).

Null hypothesis   Hypothesis with samples that come from populations with equal means (i.e., the group means are equal) for either a dependent variable (univariate test) or a set of dependent variables (multivariate test). The null hypothesis can be accepted or rejected depending on the results of a test of statistical significance.

Ordinal interaction   Acceptable type of *interaction effect* in which the magnitudes of differences between groups vary but the groups' relative positions remain constant. It is graphically represented by plotting mean values and observing nonparallel lines that do not intersect.

Orthogonal   Statistical independence or absence of association. Orthogonal *variates* explain unique variance, with no variance explanation shared between them. Orthogonal *contrasts* are *planned comparisons* that are statistically independent and represent unique comparisons of group means.

Pillai's criterion   Test for multivariate differences similar to *Wilks' lambda*.

Planned comparison   *A priori test* that tests a specific comparison of group mean differences. These tests are performed in conjunction with the tests for *main* and *interaction effects* by using a *contrast*.

Post hoc test   Statistical test of mean differences performed after the statistical tests for *main effects* have been performed. Most often, post hoc tests do not use a single *contrast*, but instead test for differences among all possible combinations of groups. Even though they provide abundant diagnostic information, they do inflate the overall *Type I error* rate by performing multiple statistical tests and thus must use strict confidence levels.

Power   Probability of identifying a treatment effect when it actually exists in the sample. Power is defined as $1 - \beta$ (see *beta*). Power is determined as a function of the statistical significance level (α) set by the researcher for a *Type I error*, the sample size used in the analysis, and the *effect size* being examined.

Repeated measures   Use of two or more responses from a single individual in an *ANOVA* or MANOVA analysis. The purpose of a repeated measures design is to control for individual-level differences that may affect the within-group variance. Repeated measures represent a lack of *independence* that must be accounted for in a special manner in the analysis.

Replication   Repeated administration of an experiment with the intent of validating the results in another sample of respondents.

Roy's greatest characteristic root (gcr)   Statistic for testing the null hypothesis in MANOVA. It tests the first *discriminant function* of the dependent variables for its ability to discern group differences.

Significance level   See *alpha.*

Standard error   Measure of the dispersion of the means or mean differences expected due to sampling variation. The standard error is used in the calculation of the *t statistic*.

Stepdown analysis   Test for the incremental discriminatory power of a dependent variable after the effects of other dependent variables have been taken into account. Similar to stepwise regression or discriminant analysis, this procedure, which relies on a specified order of entry, determines how much an additional dependent variable adds to the explanation of the differences between the groups in the MANOVA analysis.

_t_ statistic   Test statistic that assesses the statistical significance between two groups on a single dependent variable (see *t test*).

_t_ test   Test to assess the statistical significance of the difference between two sample means for a single dependent variable. The *t* test is a special case of *ANOVA* for two groups or levels of a treatment variable.

Treatment   Independent variable (*factor*) that a researcher manipulates to see the effect (if any) on the dependent variables. The treatment variable can have several levels. For example, different intensities of advertising appeals might be manipulated to see the effect on consumer believability.

Type I error   Probability of rejecting the null hypothesis when it should be accepted, that is, concluding that two means are significantly different when in fact they are the same. Small values of *alpha* (e.g., .05 or .01), also denoted as α, lead to rejection of the null hypothesis and acceptance of the alternative hypothesis that population means are not equal.

Type II error   Probability of failing to reject the null hypothesis when it should be rejected, that is, concluding that two means are not significantly different when in fact they are different. Also known as the *beta (β) error.*

U **statistic**   See *Wilks' lambda.*

Variate   Linear combination of variables. In MANOVA, the dependent variables are formed into variates in the discriminant function(s).

Vector   Set of real numbers (e.g., $X_1 \ldots X_n$) that can be written in either columns or rows. Column vectors are considered conventional, and row vectors are considered transposed. Column vectors and row vectors are shown as follows:

$$X = \begin{bmatrix} X_1 \\ X_2 \\ \vdots \\ X_n \end{bmatrix} \qquad X^T = [X_1\ X_2 \cdots X_n]$$

Column vector   Row vector

Wilks' lambda   One of the four principal statistics for testing the null hypothesis in MANOVA. Also referred to as the maximum likelihood criterion or *U statistic.*

MANOVA: EXTENDING UNIVARIATE METHODS FOR ASSESSING GROUP DIFFERENCES

Many times multivariate techniques are extensions of univariate techniques, as in the case for multiple regression, which extended simple regression (with only one independent variable) to a multivariate analysis where two or more independent variables could be used. A similar situation is found in analyzing group differences. These procedures are classified as univariate not because of the number of independent variables, but instead because of the number of dependent variables. In multiple regression, the terms *univariate* and *multivariate* refer to the number of independent variables, but for ANOVA and MANOVA the terminology applies to the use of single or multiple dependent variables. Both of these techniques have long been associated with the analysis of **experimental designs.**

The univariate techniques for analyzing group differences are the *t* **test** (two groups) and **analysis of variance (ANOVA)** for two or more groups. The multivariate equivalent procedures are the Hotelling T^2 and multivariate analysis of variance, respectively. The relationships between the univariate and multivariate procedures are as follows:

	Number of Dependent Variables	
Number of Groups in Independent Variable	One (Univariate)	Two or More (Multivariate)
Two Groups (Specialized Case)	*t* test	Hotelling's T^2
Two or More Groups (Generalized Case)	Analysis of variance (ANOVA)	Multivariate analysis of variance (MANOVA)

The t test and Hotelling's T^2 are portrayed as specialized cases in that they are limited to assessing only two groups (categories) for an independent variable. Both ANOVA and MANOVA can also handle the two group situations as well as address analyses where the independent variables have more than two groups. A review of both the t test and ANOVA are available in the Basic Stats appendix on the Web sites (www.pearsonglobaleditions.com/hair or www.mvstats.com).

Multivariate Procedures for Assessing Group Differences

As statistical inference procedures, both the univariate techniques (t test and ANOVA) and their multivariate extensions (Hotelling's T^2 and MANOVA) are used to assess the statistical significance of differences between groups. In the t test and ANOVA, the **null hypothesis** tested is the equality of a single dependent variable means across groups. In the multivariate techniques, the null hypothesis tested is the equality of **vectors** of means on multiple dependent variables across groups. The distinction between the hypotheses tested in ANOVA and MANOVA is illustrated in Figure 8-1. In the univariate case, a single dependent measure is tested for equality across the groups. In the multivariate case, a variate is tested for equality. The concept of a **variate** has been instrumental in our discussions of the previous multivariate techniques and is covered in detail in Chapter 1.

In MANOVA, the researcher actually has two variates, one for the dependent variables and another for the independent variables. The dependent variable variate is of more interest because the metric-dependent measures can be combined in a linear combination as we have already seen in multiple regression and discriminant analysis. The unique aspect of MANOVA is that the *variate optimally combines the multiple dependent measures into a single value that maximizes the differences across groups.*

ANOVA

$H_0 : \mu_1 = \mu_2 = \ldots \mu_k$

Null hypothesis (H_0) = all the group means are equal, that is, they come from the same population.

MANOVA

$$H_0 : \begin{bmatrix} \mu_{11} \\ \mu_{21} \\ \\ \\ \mu_{p1} \end{bmatrix} = \begin{bmatrix} \mu_{12} \\ \mu_{22} \\ \\ \\ \mu_{p2} \end{bmatrix} = \ldots \ldots = \begin{bmatrix} \mu_{1k} \\ \mu_{2k} \\ \\ \\ \mu_{pk} \end{bmatrix}$$

Null hypothesis (H_0) = all the group mean vectors are equal, that is, they come from the same population.

μ_{pk} = means of variable p, group k

FIGURE 8-1 Null Hypothesis Testing of ANOVA and MANOVA

THE TWO-GROUP CASE: HOTELLING'S T^2   Assume that researchers were interested in both the appeal and purchase intent generated by two advertising messages. If only univariate analyses were used, the researchers would perform separate t tests on the ratings of both the appeal of the messages and the purchase intent generated by the messages. Yet the two measures are interrelated; thus, what is really desired is a test of the differences between the messages on both variables collectively. Here is where **Hotelling's T^2,** a specialized form of MANOVA and a direct extension of the univariate t test, can be used.

Controlling for Type I Error Rate. Hotelling's T^2 provides a statistical test of the variate formed from the dependent variables, which produces the greatest group difference. It also addresses the problem of inflating the **Type I error** rate that arises when making a series of t tests of group means on several dependent measures. It controls this inflation of the Type I error rate by providing a single overall test of group differences across all dependent variables at a specified α level.

How does Hotelling's T^2 achieve these goals? Consider the following equation for a variate of the dependent variables:

$$C = W_1X_1 + W_2X_2 + \cdots + W_nX_n$$

where

$$\begin{aligned}
C &= \text{composite or variate score for a respondent} \\
W_i &= \text{weight for dependent variable } i \\
X_i &= \text{dependent variable } i
\end{aligned}$$

In our example, the ratings of message appeal are combined with the purchase intentions to form the composite. For any set of weights, we could compute composite scores for each respondent and then calculate an ordinary ***t* statistic** for the difference between groups on the composite scores. However, if we can find a set of weights that gives the maximum value for the t statistic for this set of data, these weights would be the same as the discriminant function between the two groups (as shown in Chapter 7). The maximum t statistic that results from the composite scores produced by the discriminant function can be squared to produce the value of Hotelling's T^2 [11].

The computational formula for Hotelling's T^2 represents the results of mathematical derivations used to solve for a maximum t statistic (and, implicitly, the most discriminating linear combination of the dependent variables). It is equivalent to saying that if we can find a discriminant function for the two groups that produces a significant T^2, the two groups are considered different across the mean vectors.

Statistical Testing. How does Hotelling's T^2 provide a test of the hypothesis of no group difference on the vectors of mean scores? Just as the t statistic follows a known distribution under the null hypothesis of no treatment effect on a single dependent variable, Hotelling's T^2 follows a known distribution under the null hypothesis of no treatment effect on any of a set of dependent measures. This distribution turns out to be an F distribution with p and $N_1 + N_2 - 2 - 1$ degrees of freedom after adjustment (where p = the number of dependent variables). To get the **critical value** for Hotelling's T^2, we find the tabled value for F_{crit} at a specified α level and compute T^2_{crit} as follows:

$$T^2_{\text{crit}} = \frac{p(N_1 + N_2 - 2)}{N_1 + N_2 - p - 1} \times F_{\text{crit}}$$

THE K-GROUP CASE: MANOVA   Just as ANOVA was an extension of the t test, MANOVA can be considered an extension of Hotelling's T^2 procedure. We devise dependent variable weights to produce a variate score for each respondent that is maximally different across all of the groups.

Many of the same analysis design issues discussed for ANOVA apply to MANOVA, but the method of statistical testing differs markedly from that of ANOVA.

Analysis Design. All of the issues of analysis design applicable to ANOVA (number of levels per factor, number of factors, etc.) also apply to MANOVA. Moreover, the number of dependent variables and the relationships among these dependent measures raise additional issues that will be discussed later. MANOVA enables the researcher to assess the impact of multiple independent variables on not only the individual dependent variables, but on the dependent variables collectively as well.

Statistical Testing. In the case of two groups, once the variate is formed, the procedures of ANOVA are basically used to identify whether differences exist. With three or more groups (either by having a single independent variable with three levels or by using two or more independent variables), the analysis of group differences becomes more closely allied to discriminant analysis (see Chapter 7). For three or more groups, just as in discriminant analysis, multiple variates of the dependent measures are formed. The first variate, termed a **discriminant function,** specifies a set of weights that maximize the differences between groups, thereby maximizing the F value. The maximum F value itself enables us to compute directly what is called **Roy's greatest characteristic root (gcr)** statistic, which allows for the statistical test of the first discriminant function. The greatest characteristic root statistic can be calculated as [11]:

$$\text{Roy's gcr} = (k - 1)\,F_{\text{max}}/(N - k)$$

To obtain a single test of the hypothesis of no group differences on this first vector of mean scores, we could refer to tables of Roy's gcr distribution. Just as the F statistic follows a known distribution under the null hypothesis of equivalent group means on a single dependent variable, the gcr statistic follows a known distribution under the null hypothesis of equivalent group mean vectors (i.e., group means are equivalent on a set of dependent measures). A comparison of the observed gcr to Roy's gcr_{crit} gives us a basis for rejecting the overall null hypothesis of equivalent group mean vectors.

Any subsequent discriminant functions are **orthogonal;** they maximize the differences among groups based on the remaining variance not explained by the prior function(s). Thus, in many instances, the test for differences between groups involves not just the first variate score but also a set of variate scores that are evaluated simultaneously. In these cases, a set of multivariate tests is available (e.g., Wilks' lambda, Pillai's criterion), each best suited to specific situations of testing these multiple variates.

Differences Between MANOVA and Discriminant Analysis. We noted earlier that in statistical testing MANOVA employs a discriminant function, which is the variate of dependent variables that maximizes the difference between groups. The question may arise: What is the difference between MANOVA and discriminant analysis? In some aspects, MANOVA and discriminant analysis are mirror images. The dependent variables in MANOVA (a set of metric variables) are the independent variables in discriminant analysis, and the single nonmetric dependent variable of discriminant analysis becomes the independent variable in MANOVA. Moreover, both use the same methods in forming the variates and assessing the statistical significance between groups.

The differences, however, center around the objectives of the analyses and the role of the nonmetric variable(s).

- Discriminant analysis employs a single nonmetric variable as the dependent variable. The categories of the dependent variable are assumed as given, and the independent variables are used to form variates that maximally differ between the groups formed by the dependent variable categories.
- MANOVA uses the set of metric variables as dependent variables and the objective becomes finding groups of respondents that exhibit differences on the set of dependent variables.

The groups of respondents are not prespecified; instead, the researcher uses one or more independent variables (nonmetric variables) to form groups. MANOVA, even while forming these groups, still retains the ability to assess the impact of each nonmetric variable separately.

A HYPOTHETICAL ILLUSTRATION OF MANOVA

A simple example can illustrate the benefits of using MANOVA while also illustrating the use of two independent variables to assess differences on two dependent variables.

Assume that HBAT's advertising agency identified two characteristics of HBAT's advertisements (product type being advertised and customer status), which they thought caused differences in how people evaluated the advertisements. They asked the research department to develop and execute a study to assess the impact of these characteristics on advertising evaluations.

Analysis Design

In designing the study, the research team defined the following elements relating to factors used, the dependent variables, and sample size:

- *Factors:* Two factors were defined representing Product Type and Customer Status. For each factor, two levels were also defined: product type (product 1 versus product 2) and customer status (current customer versus ex-customer). In combining these two variables, we get four distinct groups:

	Product Type	
Customer Status	**Product 1**	**Product 2**
Current Customer	Group 1	Group 3
Ex-Customer	Group 2	Group 4

- *Dependent variables:* Evaluation of the HBAT advertisements used two variables (ability to gain attention and persuasiveness) measured with a 10-point scale.
- *Sample:* Respondents were shown advertisements and asked to rate them on the two dependent measures (see Table 8-1).

TABLE 8-1   Hypothetical Example of MANOVA

Customer Type/Product Line	Product 1 $\bar{x}_{attention} = 3.50$ $\bar{x}_{purchase} = 4.50$ $\bar{x}_{total} = 8.00$				Product 2 $\bar{x}_{attention} = 5.50$ $\bar{x}_{purchase} = 5.625$ $\bar{x}_{total} = 11.125$			
	ID	Attention	Purchase	Total	ID	Attention	Purchase	Total
Ex-customer	1	1	3	4	5	3	4	7
$\bar{x}_{attention} = 3.00$	2	2	1	4	6	4	3	7
$\bar{x}_{purchase} = 3.25$	3	2	3	5	7	4	5	9
$\bar{x}_{total} = 6.25$	4	3	2	5	8	5	5	10
Average		2.0	2.25	4.25		4.0	4.25	8.25
Customer	9	4	7	11	13	6	7	13
$\bar{x}_{attention} = 6.00$	10	5	6	11	14	7	8	15
$\bar{x}_{purchase} = 6.875$	11	5	7	12	15	7	7	14
$\bar{x}_{total} = 12.875$	12	6	7	13	16	8	6	14
Average		5.0	6.75	11.75		7.0	7.0	14.0

Values are responses on a 10-point scale (1 = Low, 10 = High).

Differences from Discriminant Analysis

Although MANOVA constructs the variate and analyzes differences in a manner similar to discriminant analysis, the two techniques differ markedly in how the groups are formed and analyzed. Let us use the following example to illustrate these differences:

- With discriminant analysis, we could only examine the differences between the set of four groups, without distinction as to a group's characteristics (product type or customer status). The researcher would be able to determine whether the variate significantly differed only across the groups, but could not assess which characteristics of the groups related to these differences.
- With MANOVA, however, the researcher analyzes the differences in the groups while also assessing whether the differences are due to product type, customer type, or both. Thus, MANOVA focuses the analysis on the composition of the groups based on their characteristics (the independent variables).

MANOVA enables the researcher to propose a more complex research design by using any number of independent nonmetric variables (within limits) to form groups and then look for significant differences in the dependent variable variate associated with specific nonmetric variables.

Forming the Variate and Assessing Differences

With MANOVA we can combine multiple dependent measures into a single variate that will then be assessed for differences across one or more independent variables. Let us see how a variate is formed and used in our example.

Assume for this example that the two dependent measures (attention and purchase) were equally weighted when summed into the variate value (variate total = $\text{score}_{\text{ability to gain attention}} + \text{score}_{\text{persuasiveness}}$). This first step is identical to discriminant analysis and provides a single composite value with the variables weighted to achieve maximum differences among the groups.

With the variate formed, we can now calculate means for each of the four groups as well as the overall means for each level. From Table 8-1 we can see several patterns:

- The four group means for the composite variable total (i.e., 4.25, 8.25, 11.75, and 14.0) vary significantly between each group, being quite different from each other. If we were to use discriminant analysis with these four groups specified as the dependent measure, it would determine that significant differences appeared on the composite variable and also that both dependent variables (attention and purchase) did contribute to the differences. Yet in doing so, we would still have no insight into how the two independent variables contributed to these differences.
- MANOVA, however, goes beyond analyzing only the differences across groups by assessing whether product type and/or customer status created groups with these differences. This determination is accomplished by evaluating the category means (denoted by the symbol ■), which are shown in Figure 8-2 along with the individual group means (the two lines connect the groups—ex-customer and customer—for product 1 and product 2). If we look at product type (ignoring distinctions as to customer status), we can see a mean value of 8.0 for users of product 1 versus a mean value of 11.125 for users of product 2. Likewise, for customer status, ex-customers had a mean value of 6.25 and customers a mean value of 12.875. Visual examination suggests that both category means show significant differences, with the differences for customer type (12.875 – 6.25 = 6.625) greater than that for product (11.125 – 8.00 = 3.125).

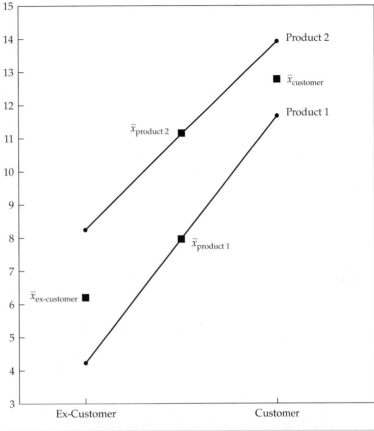

15

14 ● Product 2

13 ■ $\bar{x}_{\text{customer}}$

12 ● Product 1

11 $\bar{x}_{\text{product 2}}$ ■

10

9

8 ■ $\bar{x}_{\text{product 1}}$

7

6 $\bar{x}_{\text{ex-customer}}$ ■

5

4

3

Ex-Customer   Customer

**FIGURE 8-2   Graphical Display of Group Means of Variate (Total)
for Hypothetical Example**

By being able to represent these independent variable category means in the analysis, the MANOVA analysis not only shows that overall differences between the four groups do occur (as was done with discriminant analysis), but also that both customer type and product type contribute significantly to forming these differing groups. Therefore, both characteristics "cause" significant differences, a finding not possible with discriminant analysis.

A DECISION PROCESS FOR MANOVA

The process of performing a multivariate analysis of variance is similar to that found in many other multivariate techniques, so it can be described through the six-stage model-building process described in Chapter 1. The process begins with the specification of research objectives. It then proceeds to a number of design issues facing a multivariate analysis and then an analysis of the assumptions underlying MANOVA. With these issues addressed, the process continues with estimation of the MANOVA model and the assessment of overall model fit. When an acceptable MANOVA model is found, then the results can be interpreted in more detail. The final step involves efforts to validate the results to ensure generalizability to the population. Figure 8-3 (stages 1–3) and Figure 8-4 (stages 4–6, shown later in the text) provide a graphical portrayal of the process, which is discussed in detail in the following sections.

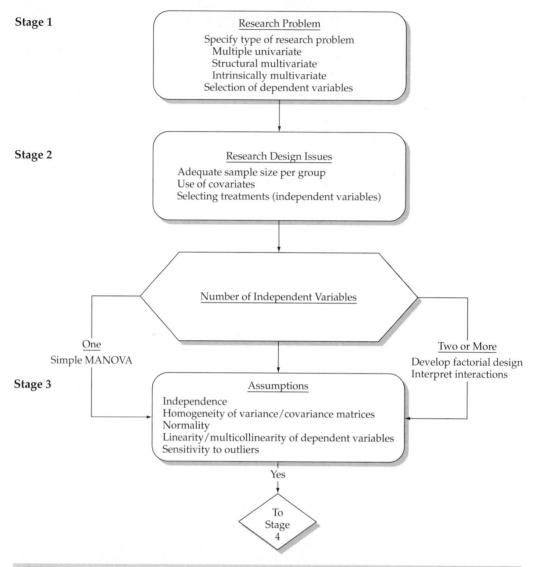

FIGURE 8-3 Stages 1–3 in the Multivariate Analysis of Variance (MANOVA) Decision Diagram

STAGE 1: OBJECTIVES OF MANOVA

The selection of MANOVA is based on the desire to analyze a dependence relationship represented as the differences in a set of dependent measures across a series of groups formed by one or more categorical independent measures. As such, MANOVA represents a powerful analytical tool suitable to a wide array of research questions. Whether used in actual or quasi-experimental situations (i.e., field settings or survey research for which the independent measures are categorical), MANOVA can provide insights into not only the nature and predictive power of the independent measures but also the interrelationships and differences seen in the set of dependent measures.

When Should We Use MANOVA?

With the ability to examine several dependent measures simultaneously, the researcher can gain in several ways from the use of MANOVA. Here we discuss the issues in using MANOVA from the perspectives of controlling statistical accuracy and efficiency while still providing the appropriate forum for testing multivariate questions.

CONTROL OF EXPERIMENTWIDE ERROR RATE The use of separate univariate ANOVAs or *t* tests can create a problem when trying to control the **experimentwide error rate** [12]. For example, assume that we evaluate a series of five dependent variables by separate ANOVAs, each time using .05 as the significance level. Given no real differences in the dependent variables, we would expect to observe a significant effect on any given dependent variable 5 percent of the time. However, across our five separate tests, the probability of a Type I error lies somewhere between 5 percent, if all dependent variables are perfectly correlated, and 23 percent $(1 - .95^5)$, if all dependent variables are uncorrelated. Thus, a series of separate statistical tests leaves us without control of our effective overall or experimentwide Type I error rate. If the researcher desires to maintain control over the experimentwide error rate and at least some degree of correlation is present among the dependent variables, then MANOVA is appropriate.

DIFFERENCES AMONG A COMBINATION OF DEPENDENT VARIABLES A series of univariate ANOVA tests also ignores the possibility that some composite (linear combination) of the dependent variables may provide evidence of an overall group difference that may go undetected by examining each dependent variable separately. Individual tests ignore the correlations among the dependent variables and in the presence of multicollinearity among the dependent variables, MANOVA will be more powerful than the separate univariate tests in several ways:

- MANOVA may detect *combined* differences not found in the univariate tests.
- If multiple variates are formed, then they may provide *dimensions* of differences that can distinguish among the groups better than single variables.
- If the number of dependent variables is kept relatively low (five or fewer), the statistical power of the MANOVA tests equals or exceeds that obtained with a single ANOVA [4].

The considerations involving sample size, number of dependent variables, and statistical power are discussed in a subsequent section.

Types of Multivariate Questions Suitable for MANOVA

The advantages of MANOVA versus a series of univariate ANOVAs extend past the statistical issues in its ability to provide a single method of testing a wide range of differing multivariate questions. Throughout the text, we emphasize the interdependent nature of multivariate analysis. MANOVA has the flexibility to enable the researcher to select the test statistics most appropriate for the question of concern. Hand and Taylor [10] have classified multivariate problems into three categories, each of which employs different aspects of MANOVA in its resolution. These three categories are multiple univariate, structured multivariate, and intrinsically multivariate questions.

MULTIPLE UNIVARIATE QUESTIONS A researcher studying multiple univariate questions identifies a number of separate dependent variables (e.g., age, income, education of consumers) that are to be analyzed separately but needs some control over the experimentwide error rate. In this instance, MANOVA is used to assess whether an overall difference is found between groups, and then the separate univariate tests are employed to address the individual issues for each dependent variable.

STRUCTURED MULTIVARIATE QUESTIONS A researcher dealing with structured multivariate questions gathers two or more dependent measures that have specific relationships among them. A common situation in this category is repeated measures, where multiple responses are gathered from each subject, perhaps over time or in a pretest–posttest exposure to some stimulus, such as an advertisement. Here MANOVA provides a structured method for specifying the comparisons of group differences on a set of dependent measures while maintaining statistical efficiency.

INTRINSICALLY MULTIVARIATE QUESTIONS An intrinsically multivariate question involves a set of dependent measures in which the principal concern is how they differ *as a whole* across the groups. Differences on individual dependent measures are of less interest than their collective effect. One example is the testing of multiple measures of response that should be consistent, such as attitudes, preference, and intention to purchase, all of which relate to differing advertising campaigns. The full power of MANOVA is utilized in this case by assessing not only the overall differences but also the differences among combinations of dependent measures that would not otherwise be apparent. This type of question is served well by MANOVA's ability to detect multivariate differences, even when no single univariate test shows differences.

Selecting the Dependent Measures

In identifying the questions appropriate for MANOVA, it is important also to discuss the development of the research question, specifically the selection of the dependent measures. A common problem encountered with MANOVA is the tendency of researchers to misuse one of its strengths—the ability to handle multiple dependent measures—by including variables without a sound conceptual or theoretical basis. The problem occurs when the results indicate that a subset of the dependent variables has the ability to influence interpretation of the overall differences among groups. If some of the dependent measures with the strong differences are not really appropriate for the research question, then "false" differences may lead the researcher to draw incorrect conclusions about the set as a whole. Thus, the researcher should always scrutinize the dependent measures and form a solid rationale for including them. Any ordering of the variables, such as possible sequential effects, should also be noted. MANOVA provides a special test, stepdown analysis, to assess the statistical differences in a sequential manner, much like the addition of variables to a regression analysis.

In summary, the researcher should assess all aspects of the research question carefully and ensure that MANOVA is applied in the correct and most powerful way. The following sections address many issues that have an impact on the validity and accuracy of MANOVA; however, it is ultimately the responsibility of the researcher to employ the technique properly.

RULES OF THUMB 8-1

Decision Processes for MANOVA

- MANOVA is an extension of ANOVA that examines the effect of one or more nonmetric independent variables on two or more metric dependent variables
- In addition to the ability to analyze multiple dependent variables, MANOVA also has the advantages of:
  - Controlling the experimentwide error rate when some degree of intercorrelation among dependent variables is present
  - Providing more statistical power than ANOVA when the number of dependent variables is five or fewer
- Nonmetric independent variables create groups between which the dependent variables are compared; many times the groups represent experimental variables or "treatment effects"
- Researchers should include only dependent variables that have strong theoretical support

STAGE 2: ISSUES IN THE RESEARCH DESIGN OF MANOVA

MANOVA follows all of the basic design principles of ANOVA, yet in some instances the multivariate nature of the dependent measures requires a unique perspective. In the following section we will review the basic design principles and illustrate those unique issues arising in a MANOVA analysis.

Sample Size Requirements—Overall and by Group

MANOVA, like all of the other multivariate techniques, can be markedly affected by the sample size used. What differs most for MANOVA (and the other techniques assessing group differences such as the *t* test and ANOVA) is that the sample size requirements relate to individual group sizes and not the total sample per se. A number of basic issues arise concerning the sample sizes needed in MANOVA:

- As a bare minimum, the sample in each cell (group) must be greater than the number of dependent variables. Although this concern may seem minor, the inclusion of just a small number of dependent variables (from 5 to 10) in the analysis places a sometimes bothersome constraint on data collection. This problem is particularly prevalent in field experimentation or survey research, where the researcher has less control over the achieved sample.
- As a practical guide, a recommended minimum cell size is 20 observations. Again, remember this quantity is per group, necessitating fairly large overall samples even for fairly simple analyses. In our earlier example of advertising messages, we had only two factors, each with two levels, but this analysis would require 80 observations for an adequate analysis.
- As the number of dependent variables increases, the sample size required to maintain statistical power increases as well. We will defer our discussion of sample size and power until a later section, but as an example, required samples sizes increase by almost 50 percent as the number of dependent variables goes from two to just six.

Researchers should strive to maintain equal or approximately equal sample sizes per group. Although computer programs can easily accommodate unequal group sizes, the objective is to ensure that an adequate sample size is available for all groups. In most instances, the effectiveness of the analysis is dictated by the smallest group sizes, thus always making sample size considerations a primary concern.

Factorial Designs—Two or More Treatments

Many times the researcher wishes to examine the effects of several independent variables or treatments rather than using only a single treatment in either the ANOVA or MANOVA tests. This capability is a primary distinction between MANOVA and discriminant analysis in being able to determine the impact of multiple independent variables in forming the groups with significant group differences. An analysis with two or more treatments (factors) is called a **factorial design.** In general, a design with *n* treatments is called an *n*-way factorial design.

SELECTING TREATMENTS   The most common use of factorial designs involves those research questions that relate two or more nonmetric independent variables to a set of dependent variables. In these instances, the independent variables are specified in the design of the experiment or included in the design of the field experiment or survey questionnaire.

Types of Treatments. As discussed throughout the chapter, a **treatment** or **factor** is a nonmetric independent variable with a defined number of levels (categories). Each level represents a different condition or characteristic that affects the dependent variable(s). In an experiment these treatments and levels are designed by the researcher and administered in the course of the experiment. In field or survey research, they are characteristics of the respondents gathered by researcher and then included in the analysis.

But in some instances, treatments are needed in addition to those in the original analysis design. The most common use of additional treatments is to control for a characteristic that affects the dependent variables but is not part of the study design. In these instances the researcher is aware of conditions (e.g., method of data collection) or characteristics of the respondents (e.g., geographic location, gender, etc.) that potentially create differences in the dependent measures. Even though they are not independent variables of interest to the study, neglecting them ignores potential sources of difference that, left unaccounted for, may obscure some results of interest to the study.

The most direct way to account for such effects is through a **blocking factor,** which is a nonmetric characteristic used post hoc to segment the respondents. The objective is to group the respondents to obtain greater within-group homogeneity and reduce the MS_W source of variance. By doing so, the ability of the statistical tests to identify differences is enhanced.

Assume in our earlier advertising example we discovered that males in general reacted differently than females to the advertisements. If gender is then used as a blocking factor, we can evaluate the effects of the independent variables separately for males and females. Hopefully, this approach will make the effects more apparent than when we assume they both react similarly by not making a distinction on gender. The effects of message type and customer type can now be evaluated for males and females separately, providing a more precise test of their individual effects.

Thus, any nonmetric characteristic can be incorporated directly into the analysis to account for its impact on the dependent measures. However, if the variables you wish to control for are metric, they can be included as covariates, which are discussed in the next section.

Number of Treatments. One of the advantages of multivariate techniques is the use of multiple variables in a single analysis. For MANOVA, this feature relates to the number of dependent variables that can be analyzed concurrently. As already discussed, the number of dependent variables affects the sample sizes required and other issues. But what about the number of treatments (i.e., independent variables)? Although ANOVA and MANOVA can analyze several treatments at the same time, several considerations relate to the number of treatments in an analysis.

- **Number of Cells Formed.** Perhaps the most limiting issue involving multiple treatments involves the number of cells (groups) formed. As discussed in our earlier example, the number of cells is the product of the number of levels for each treatment. For example, if we had two treatments with two levels each and one treatment with four levels, a total of 16 cells ($2 \times 2 \times 4 = 16$) would be formed. Maintaining a sufficient sample size for each cell (assuming 20 respondents per cell) would then require a total sample of 320.

    When applied to survey or field experimentation data, however, increasing the number of cells becomes much more problematic. Because field research is generally not able to administer the survey individually to each cell of the design, the researcher must plan for a large enough overall sample to fill each cell to the required minimum. The proportions of the total sample in each cell most likely vary widely (i.e., some cells would be much more likely to occur than others), especially as the number of cells increases. In such a situation, the researcher must plan for an even larger sample size than the size determined by multiplying the number of cells by the minimum per cell. Let's look back to our earlier example to illustrate this problem.

    Assume that we have a simple two-factor design with two levels for each factor (2×2). If this four-cell design were a controlled experiment, the researcher would be able to randomly assign 20 respondents per cell for an overall sample size of 80. What then if it is a field survey? If it were equally likely that respondents would fall into each cell, then the researcher could get a total sample of 80 and each cell should have a sample of 20. Such tidy proportions and samples rarely happen. What if one cell was thought to represent only 10 percent of the population? If we use a total sample of 80, then this cell would be expected to have a sample of only 8. Thus, if the researcher wanted a sample of 20 even for this small cell, the overall sample would have to be increased to 200.

Unless sophisticated sampling plans are used to ensure the necessary sample per cell, increasing the number of cells (thus the likelihood of unequal population proportions across the cells) will necessitate an even greater sample size than in a controlled experiment. Failure to do so would create situations in which the statistical properties of the analysis could be markedly diminished.

• **Creation of Interaction Effects.** Any time more than one treatment is used, **interaction effects** are created. The interaction term represents the joint effect of two or more treatments. In simple terms, it means that the difference between groups of one treatment depends on the values of another treatment. Let us look at a simple example:

Assume that we have two treatments: region (East versus West) and customer status (customer and noncustomer). First, assume that on the dependent variable (attitude toward HBAT) customers score 15 points higher than noncustomers. However, an interaction of region and customer status would indicate that the amount of the difference between customer and noncustomer depended on the region of the customer. For example, when we separated the two regions, we might see that customers from the East scored 25 points higher than non-customers in the East, while in the West the difference was only 5 points. In both cases the customers scored higher, but the amount of the difference depended on the region. This out-come would be an interaction of the two treatments.

Interaction terms are created for each combination of treatment variables. Two-way interactions are variables taken two at a time. Three-way interactions are combinations of three variables, and so on. The number of treatments determines the number of interaction terms possible. The following chart shows the interactions created for two, three, and four independent variables:

	Interaction Terms		
Treatments	**Two-Way**	**Three-Way**	**Four-Way**
A, B	$A \times B$		
A, B, C	$A \times B$	$A \times B \times C$	
	$A \times C$		
	$B \times C$		
A, B, C, D	$A \times B$	$A \times B \times C$	$A \times B \times C \times D$
	$A \times C$		
	$A \times D$	$A \times B \times D$	
	$B \times C$		
	$B \times D$	$B \times C \times D$	
	$C \times D$		
		$A \times C \times D$	

We will discuss the various types of interaction terms and their interpretation in the following section, but the researcher must be ready to interpret and explain the interaction terms, whether significant or not, depending on the research question.

Obviously, the sample size considerations are of most importance, but the researcher should not overlook the implications of interaction terms. Besides using at least one degree of freedom for each interaction, they present issues of interpretation discussed in stage 4.

Using Covariates—ANCOVA and MANCOVA

We discussed earlier the use of a blocking factor to control for influences on the dependent variable that are not part of the research design yet need to be accounted for in the analysis. It enables the researcher to control for nonmetric variables, but what about metric variables?

One approach would be to convert the metric variable into a nonmetric variable (e.g., median splits, etc.), but this process is generally deemed unsatisfactory because much of the information contained in the metric variable is lost in the conversion. A second approach is to include the metric variables as **covariates.** These variables can extract extraneous influences from the dependent variable, thus increasing the within-group variance (MS_W). The process follows two steps:

1. Procedures similar to linear regression are employed to remove variation in the dependent variable associated with one or more covariates.
2. A conventional analysis is carried out on the adjusted dependent variable. In a simplistic sense, it becomes an analysis of the regression residuals once the effects of the covariate(s) are removed.

When used with ANOVA, the analysis is termed *analysis of covariance* (ANCOVA) and the simple extension of the principles of ANCOVA to multivariate (multiple dependent variables) analysis is termed MANCOVA.

OBJECTIVES OF COVARIANCE ANALYSIS   The objective of the covariate is to eliminate any effects that (1) affect only a portion of the respondents or (2) vary among the respondents. Similar to the uses of a blocking factor, **covariate analysis** can achieve two specific purposes:

1. To eliminate some systematic error outside the control of the researcher that can bias the results
2. To account for differences in the responses due to unique characteristics of the respondents

In experimental settings, most systematic bias can be eliminated by the random assignment of respondents to various treatments. However, in nonexperimental research, such controls are not possible. For example, in testing advertising, effects may differ depending on the time of day or the composition of the audience and their reactions. Moreover, personal differences, such as attitudes or opinions, may affect responses, but the analysis does not include them as a treatment factor. The researcher uses a covariate to take out any differences due to these factors before the effects of the experiment are calculated.

SELECTING COVARIATES   An effective covariate is one that is *highly correlated with the dependent variable(s) but not correlated with the independent variables.* Let us examine why. Variance in the dependent variable forms the basis of our error term.

- If the covariate is correlated with the dependent variable and *not* the independent variable(s), we can explain some of the variance with the covariate (through linear regression), leaving a smaller residual (unexplained) variance in the dependent variable. This residual variance provides a smaller error term (MS_W) for the F statistic and thus a more efficient test of treatment effects. The amount explained by the uncorrelated covariate would not have been explained by the independent variable anyway (because the covariate is not correlated with the independent variable). Thus, the test of the independent variable(s) is more sensitive and powerful.
- However, if the covariate is correlated with the independent variable(s), then the covariate will explain some of the variance that could have been explained by the independent variable and reduce its effects. Because the covariate is extracted first, any variation associated with the covariate is not available for the independent variables.

Thus, it is critical that the researcher ensure that the correlation of the covariates and independent variable(s) is small enough such that the reduction in explanatory power from reducing the variance that could have been explained by the independent variable(s) is less than the decrease in unexplained variance attributable to the covariates.

Number of Covariates. A common question involves how many covariates to add to the analysis. Although the researcher wants to account for as many extraneous effects as possible, too large a number will reduce the statistical efficiency of the procedures. A rule of thumb [13] dictates that the maximum number of covariates is determined as follows:

$$\text{Maximum number of covariates} = (.10 \times \text{Sample size}) - (\text{Number of groups} - 1)$$

For example, for a sample size of 100 respondents and 5 groups, the number of covariates should be less than 6 [$6 = .10 \times 100 - (5 - 1)$]. However, for only two groups, the analysis could include up to nine covariates.

The researcher should always attempt to minimize the number of covariates, while still ensuring that effective covariates are not eliminated, because in many cases, particularly with small sample sizes, they can markedly improve the sensitivity of the statistical tests.

Assumptions for Covariance Analysis. Two requirements for use of an analysis of covariance are the following:

1. The covariates must have some relationship (correlation) with the dependent measures.
2. The covariates must have a homogeneity of regression effect, meaning that the covariate(s) have equal effects on the dependent variable across the groups. In regression terms, it implies equal coefficients for all groups.

Statistical tests are available to assess whether this assumption holds true for each covariate used. If either of these requirements is not met, then the use of covariates is inappropriate.

MANOVA Counterparts of Other ANOVA Designs

Many types of ANOVA designs exist and are discussed in standard experimental design texts [15, 19, 22]. Every ANOVA design has its multivariate counterpart; that is, any ANOVA on a single dependent variable can be extended to MANOVA designs. To illustrate this fact, we would have to discuss each ANOVA design in detail. Clearly, this type of discussion is not possible in a single chapter because entire books are devoted to the subject of ANOVA designs. For more information, the reader is referred to more statistically oriented texts [1, 2, 5, 7, 8, 9, 11, 20, 25].

A Special Case of MANOVA: Repeated Measures

We discussed a number of situations in which we wish to examine differences on several dependent measures. A special situation of this type occurs when the same respondent provides several measures, such as test scores over time, and we wish to examine them to see whether any trend emerges. Without special treatment, however, we would be violating the most important assumption, independence. Special MANOVA models, termed **repeated measures** models, account for this dependence and still ascertain whether any differences occurred across individuals for the set of dependent variables. The within-person perspective is important so that each person is placed on equal footing.

For example, assume we were assessing improvement on test scores over the semester. We must account for the earlier test scores and how they relate to later scores, and we might expect to see different trends for those with low versus high initial scores. Thus, we must match each respondent's scores when performing the analysis. The differences we are interested in become how much each person changes, not necessarily the changes in group means over the semester.

RULES OF THUMB 8-2

Research Design of MANOVA

- Cells (groups) are formed by the combination of independent variables; for example, a three-category nonmetric variable (e.g., low, medium, high) combined with a two-category nonmetric variable (e.g., gender of male versus female) will result in a 3 × 2 design with six cells (groups)
- Sample size per group is a critical design issue:
  - Minimum sample size per group must be greater than the number of dependent variables
  - The recommended minimum cell size is 20 observations per cell (group)
  - Researchers should try to have approximately equal sample sizes per cell (group)
- Covariates and blocking variables are effective ways of controlling for external influences on the dependent variables that are not directly represented in the independent variables
  - An effective covariate is one that is highly correlated with the dependent variable(s) but not correlated with the independent variables
  - The maximum number of covariates in a model should be (.10 × Sample size) − (Number of groups − 1)

We do not address the details of repeated measures models in this text because it is a specialized form of MANOVA. The interested reader is referred to any number of excellent treatments on the subject [1, 2, 5, 7, 8, 9, 11, 20, 25].

STAGE 3: ASSUMPTIONS OF ANOVA AND MANOVA

The univariate test procedures of ANOVA described in this chapter are valid (in a statistical sense) if it is assumed that the dependent variable is normally distributed, the groups are independent in their responses on the dependent variable, and variances are equal for all treatment groups. Some evidence [19, 27], however, indicates that F tests in ANOVA are robust with regard to these assumptions except in extreme cases.

For the multivariate test procedures of MANOVA to be valid, three assumptions must be met:

- Observations must be independent.
- Variance–covariance matrices must be equal for all treatment groups.
- The set of dependent variables must follow a multivariate normal distribution (i.e., any linear combination of the dependent variables must follow a normal distribution) [11].

In addition to the strict statistical assumptions, the researcher must also consider several issues that influence the possible effects—namely, the linearity and multicollinearity of the variate of dependent variables.

Independence

The most basic, yet most serious, violation of an assumption comes from a lack of **independence** among observations, meaning that the responses in each cell (group) are not made independently of responses in any other group. Violations of this assumption can occur as easily in experimental as well as nonexperimental situations. Any number of extraneous and unmeasured effects can affect the results by creating dependence between the groups, but two of the most common violations of independence follow:

- Time-ordered effects (serial correlation) occurring if measures are taken over time, even from different respondents

- Gathering information in group settings, so that a common experience (such as a noisy room or confusing set of instructions) would cause a subset of individuals (those with the common experience) to have answers that are somewhat correlated

Although no tests provide an absolute certainty of detecting all forms of dependence, the researcher should explore all possible effects and correct for them if found. One potential solution is to combine those within the groups and analyze the group's average score instead of the scores of the separate respondents. Another approach is to employ a blocking factor or some form of covariate analysis to account for the dependence. In either case, or when dependence is suspected, the researcher should use a stricter level of significance (.01 or even lower).

Equality of Variance–Covariance Matrices

The second assumption of MANOVA is the equivalence of covariance matrices across the groups. Here we are concerned with substantial differences in the amount of variance of one group versus another for the dependent variables (similar to the problem of heteroscedasticity in multiple regression). In MANOVA, with multiple dependent variables, the interest is in the variance–covariance matrices of the dependent measures for each group.

The variance equivalence test is a very "strict" test because instead of equal variances for a single variable in ANOVA, the MANOVA test examines all elements of the covariance matrix of the dependent variables. For example, for 5 dependent variables, the 5 correlations and 10 covariances are all tested for equality across the groups. As a result, increases in the number of dependent variables and/or the number of cells/groups in the analysis make the test more sensitive to finding differences and thus influence the significance levels used to determine if a violation has occurred.

MANOVA programs conduct the test for equality of covariance matrices—typically the **Box's M test**—and provide significance levels for the test statistic, which indicate the likelihood of differences between the groups. Thus, the researcher is looking for *nonsignificant* differences between the groups, and the observed significance level of the test statistic is considered acceptable if it is less significant than the threshold value for comparison. For example, if a .01 level was considered the threshold level for indicating violations of the assumption, values greater than .01 (e.g., .02) would be considered acceptable because they indicate no differences between groups, whereas values less than .01 (e.g., .001) would be problematic because they indicate that significant differences were present.

Given the sensitivity of the Box's M test to the size of the covariance matrices and the number of groups in the analysis, even simple research designs (four to six groups) with a small number of dependent variables will want to use to use very conservative levels of significant differences (e.g., .01 rather than .05) when assessing whether differences are present. As the design complexity increases, even more conservative levels of significance may be considered acceptable.

The Box's M test is especially sensitive to departures from normality [11, 23]. Thus, one should always check for univariate normality of all dependent measures before performing this test. Fortunately, a violation of this assumption has minimal impact if the groups are of approximately equal size (i.e., Largest group size ÷ Smallest group size < 1.5).

If the group sizes differ more than this amount and the significance levels of Box's M test are not within acceptable levels, then the researcher has several options:

- First, apply one of the many variance-stabilizing transformations available (see Chapter 2 for a discussion of these approaches) and retest to see whether the problem is remedied.
- If the unequal variances persist after transformation and the group sizes differ markedly, the researcher should make adjustments for their effects in the interpretation of the significance levels of both main and interaction effects. First, one must ascertain which group has the largest variance. This determination is easily made either by examining the variance–covariance matrix

or by using the determinant of the variance–covariance matrix, which is provided by all statistical programs. In both measures high values indicate greater variance. Thus,

- If the larger variances are found with the larger group sizes, the alpha level is overstated, meaning that differences should actually be assessed using a somewhat lower value (e.g., use .03 instead of .05).
- If the larger variance is found in the smaller group sizes, then the reverse is true. The power of the test has been reduced, and the researcher should increase the significance level.

In most situations the presence of relatively equal sample sizes among the groups mitigates any violations of this assumption. Thus, it is important to reinforce the importance of analysis design in maintaining equal sample sizes among the groups.

Normality

The last assumption for MANOVA concerns normality of the dependent measures. In the strictest sense, the assumption is that all the variables are multivariate normal. A **multivariate normal distribution** assumes that the joint effect of two variables is normally distributed. Even though this assumption underlies most multivariate techniques, no direct test is available for multivariate normality. Therefore, most researchers test for univariate normality of each variable. Although univariate normality does not guarantee multivariate normality, if all variables meet this requirement, then any departures from multivariate normality are usually inconsequential.

Violations of this assumption have little impact with larger sample sizes, just as is found with ANOVA. Violating this assumption primarily creates problems in applying the Box's M test, but transformations can correct these problems in most situations. For a discussion of transforming variables, refer to Chapter 2. With moderate sample sizes, modest violations can be accommodated as long as the differences are due to skewness and not outliers.

Linearity and Multicollinearity Among the Dependent Variables

Although MANOVA assesses the differences across combinations of dependent measures, it can construct a linear relationship only between the dependent measures (and any covariates, if included). The researcher is again encouraged first to examine the data, this time assessing the presence of any nonlinear relationships. If these exist, then the decision can be made whether they need to be incorporated into the dependent variable set, at the expense of increased complexity but greater representativeness. Chapter 2 addresses such tests.

In addition to the linearity requirement, the dependent variables should not have high multicollinearity (discussed in Chapter 4), which indicates redundant dependent measures and decreases statistical efficiency. We discuss the impact of multicollinearity on the statistical power of the MANOVA in the next section.

Sensitivity to Outliers

In addition to the impact of heteroscedasticity discussed earlier, MANOVA (and ANOVA) are especially sensitive to outliers and their affect on the Type I error. The researcher is strongly encouraged first to examine the data for outliers and eliminate them from the analysis, if at all possible, because their impact will be disproportionate in the overall results.

STAGE 4: ESTIMATION OF THE MANOVA MODEL AND ASSESSING OVERALL FIT

Once the MANOVA analysis is formulated and the assumptions tested for compliance, the assessment of significant differences among the groups formed by the treatment(s) can proceed (see Figure 8-4). Estimation procedures based on the general linear model are becoming more common and the basic

RULES OF THUMB 8-3

MANOVA/ANOVA Assumptions

- For the multivariate test procedures used with MANOVA to be valid:
  - Observations must be independent
  - Variance–covariance matrices must be equal (or comparable) for all treatment groups
  - The dependent variables must have a multivariate normal distribution
  - Multivariate normality is assumed, but many times hard to assess; univariate normality does not guarantee multivariate normality, but if all variables meet the univariate normality requirement, then departures from multivariate normality are inconsequential
- F tests are generally robust if violations of these assumptions are modest

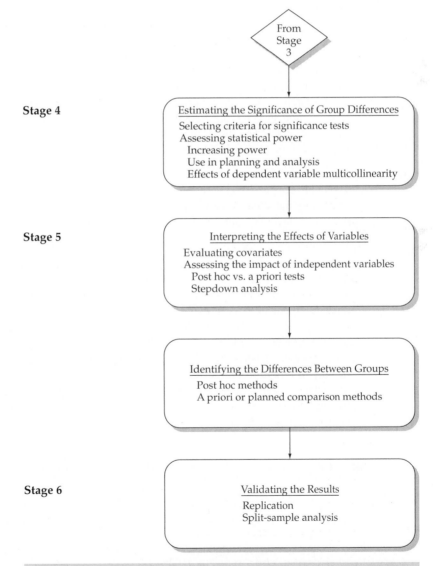

FIGURE 8-4 **Stages 4–6 in the Multivariate Analysis of Variance (MANOVA) Decision Diagram**

issues will be addressed. With the estimated model, the researcher then can assess the differences in means based on the test statistics most appropriate for the study objectives. Moreover, in any situation, but especially as the analysis becomes more complex, the researcher must evaluate the power of the statistical tests to provide the most informed perspective on the results obtained.

Estimation with the General Linear Model

The traditional means of calculating the appropriate test statistics for ANOVA and MANOVA were established more than 70 years ago [26]. In recent years, however, the **general linear model (GLM)** [18, 21] has become a popular means of estimating ANOVA and MANOVA models. The GLM procedure, as the name implies, is a family of models, each composed of three elements:

- *Variate.* The linear combination of independent variables as specified by the researcher. Each independent variable has an estimated weight representing that variable's contribution to the predicted value.
- *Random component.* The probability distribution assumed to underlie the dependent variable(s). Typical probability distributions are the normal, Poisson, binomial, and multinomial distributions. Each distribution is associated with a type of response variable (e.g., continuous variables are associated with a normal distribution, proportions are associated with the binomial distribution, and dichotomous variables are associated with the Poisson distribution). The researcher selects the random component based on the type of response variable.
- *Link function.* Provides the theoretical connection between the variate and the random component to accommodate differing model formulations. The **link function** specifies the type of transformation needed to specify the desired model. The three most common link functions are the identity, logit, and log links.

The GLM approach provides the researcher with a single estimation model within which any number of differing statistical models can be accommodated. Two unique advantages of the GLM approach are its flexibility and simplicity in model design.

- By specifying a specific combination of the random component and link function coupled with a type of variable in the variate, a wide range of multivariate models can be estimated. As shown in Table 8-2, combinations of these components correspond to many of the multivariate techniques already discussed. Thus, a single estimation procedure can be used for a wide range of empirical models.
- The researcher can also vary either the link function or probability distribution to best match the actual properties of the data rather than employing extensive transformations of the data. Two examples illustrate this feature. First, in cases of heteroscedasticity, substitution of the gamma distribution would allow for estimation of the model without transforming the dependent measure. Second, if the variate was assumed to be multiplicative rather than additive, one alternative would be to use a logarithmic transformation of the variate. In a GLM, the variate can remain in the additive formulation with a log link function employed instead.

TABLE 8-2   **Specifying Multivariate Models as GLM Components**

Multivariate Technique	Response (Dependent) Variable	Independent Variable	Link Function	Probability Distribution
Multiple Regression	Metric	Metric	Identity	Normal
Logistic Regression	Nonmetric	Metric	Logit	Binomial
ANOVA/MANOVA	Metric	Nonmetric	Identity	Normal

A more thorough discussion of the GLM procedure and its many variations are available in several texts [6, 14, 18]. Here, we provide this brief introduction to the concept of the GLM because it has become the preferred method of estimation for ANOVA and MANOVA among many researchers and some statistical packages (e.g., SPSS).

Criteria for Significance Testing

In our discussions of the similarity of MANOVA to discriminant analysis we referred to the greatest characteristic root and the first discriminant function, and these terms imply that multiple discriminant functions may act as variates of the dependent variables. The number of functions is defined by the smaller of $(k - 1)$ or p where k is the number of groups and p is the number of dependent variables. Thus, any measure for testing the statistical significance of group differences in MANOVA may need to consider differences across multiple discriminant functions.

STATISTICAL MEASURES   As we first saw in discriminant analysis (Chapter 7), researchers use a number of statistical criteria to apply **significance tests** relating to the differences across dimensions of the dependent variables. The most widely used measures are:

- **Roy's greatest characteristic root (gcr),** as the name implies, measures the differences on only the first discriminant function among the dependent variables. This criterion provides advantages in power and specificity of the test but makes it less useful in certain situations where all dimensions should be considered. Roy's gcr test is most appropriate when the dependent variables are strongly interrelated on a single dimension, but it is also the measure most likely to be severely affected by violations of the assumptions.
- **Wilks' lambda** (also known as the *U* **statistic**) is many times referred to as the *multivariate F* and is commonly used for testing overall significance between groups in a multivariate situation. Unlike Roy's gcr statistic, which is based on the first discriminant function, Wilks' lambda considers all the discriminant functions; that is, it examines whether groups are somehow different without being concerned with whether they differ on at least one linear combination of the dependent variables. Although the distribution of Wilks' lambda is complex, good approximations for significance testing are available by transforming it into an *F* statistic [22].
- **Pillai's criterion** and **Hotelling's** T^2 are two other measures similar to Wilks' lambda because they consider all the characteristic roots and can be approximated by an *F* statistic.

With only two groups, all of the measures are equivalent. Differences occur as the number of discriminant functions increase. Rules of Thumb 8-4 identify the measure(s) best suited to differing situations.

Statistical Power of the Multivariate Tests

In simple terms for MANOVA, **power** is the probability that a statistical test will identify a treatment's effect if it actually exists. Power can also be expressed as one minus the probability of a **Type II error** or **beta (β)** error (i.e., Power $= 1 - \beta$). Statistical power plays a critical role in any MANOVA analysis because it is used both in the planning process (i.e., determining necessary sample size) and as a diagnostic measure of the results, particularly when nonsignificant effects are found. The following sections first examine the impacts on statistical power and then address issues unique to utilizing power analysis in a MANOVA design. The reader is also encouraged to review the discussion of power in Chapter 1.

RULES OF THUMB 8-4

Selecting a Statistical Measure

- The preferred measure is the one that is most immune to violations of the assumptions underlying MANOVA and yet maintains the greatest power
- Each measure is preferred in differing situations:
  - Pillai's criterion or Wilks' lambda is the preferred measure when the basic design considerations (adequate sample size, no violations of assumptions, approximately equal cell sizes) are met
  - Pillai's criterion is considered more robust and should be used if sample size decreases, unequal cell sizes appear, or homogeneity of covariances is violated
  - Roy's gcr is a more powerful test statistic if the researcher is confident that all assumptions are strictly met and the dependent measures are representative of a single dimension of effects
- In a vast majority of the situations, all of the statistical measures provide similar conclusions
- When faced with conflicting conditions, however, statistical measures can be selected that meet the situation faced by the researcher

IMPACTS ON STATISTICAL POWER The level of power for any of the four statistical criteria—Roy's gcr, Wilks' lambda, Hotelling's T^2, or Pillai's criterion—is based on three considerations: the alpha (α) level, the effect size of the treatment, and the sample size of the groups. Each of these considerations is controllable in varying degrees in a MANOVA design and provides the researcher with a number of options in managing the power in order to achieve the desired level of power in the range of .80 or above.

Statistical Significance Level (alpha α). As discussed in Chapter 1, power is inversely related to the **alpha** (α) level selected. Many researchers assume that the significance level is fixed at some level (e.g., .05), but it actually is a judgment by the researcher as to where to place the emphasis of the statistical testing. Many times the other two elements affecting power (effect size and sample size) are already specified or the data have been collected, thus the alpha level becomes the primary tool in defining the power of an analysis.

By setting the alpha level required to denote statistical significance, the researcher is balancing the desire to be strict in what is deemed a significant difference between groups while still not setting the criterion so high that differences cannot be found.

- Increasing alpha (i.e., α becomes more conservative, such as moving from .05 to .01) reduces the chances of accepting differences as significant when they are not really significant. However, doing so decreases power because being more selective in what is considered a statistical difference also increases the difficulty in finding a significant difference.
- Decreasing the alpha level required for statistical significance (e.g., α moves from .05 to .10) is considered many times as being "less statistical" because the researcher is willing to accept smaller group differences as significant. However, in instances where effect sizes or sample sizes are smaller than desired it may be necessary to be less concerned about accepting these false positives and decreasing the alpha level to increase power. One such example is when making multiple comparisons. To control experimentwide error rate, the alpha level is increased for each separate comparison. However, to make several comparisons and still achieve an overall rate of .05 may require strict levels (e.g., .01 or less) for each separate comparison, thus making it hard to find significant differences (i.e., lower power). Here the researcher may increase the overall alpha level to allow a more reasonable alpha level for the separate tests.

The researcher must always be aware of the implications of adjusting the alpha level, because the overriding objective of the analysis is not only avoiding Type I errors but also identifying the treatment effects if they do indeed exist. If the alpha level is set too stringently, then the power may be too low to identify valid results. The researcher should try to maintain an acceptable alpha level with power in the range of .80. For a more detailed discussion of the relationships between Type I and Type II errors and power, see Chapter 1.

Effect Size. How does the researcher increase power once an alpha level is specified? The primary tool at the researcher's disposal is the sample size of the groups. Before we assess the role of sample size, however, we need to understand the impact of **effect size,** which is a standardized measure of group differences, typically expressed as the differences in group means divided by their standard deviation. This formula leads to several generalizations:

- As would be expected, all other things equal, larger effect sizes have more power (i.e., are easier to find) than smaller effect sizes.
- The magnitude of the effect size has a direct impact on the power of the statistical test. For any given sample size, the power of the statistical test will be higher the larger the effect size. Conversely, if a treatment has a small expected effect size, it is going to take a much larger sample size to achieve the same power as a treatment with a large effect size.

Researchers are always hoping to design experiments with large effect sizes. However, when used with field research, researchers must "take what they get" and thus be aware of the possible effect sizes when planning their research as well as when analyzing results.

Sample Size. With the alpha level specified and the effect size identified, the final element affecting power is the sample size. In many instances, this element is the most controllable by the researcher. As discussed before, increased sample size generally reduces sampling error and increases the sensitivity (power) of the test. Other factors discussed earlier (alpha level and effect size) also affect power, and we can draw some generalizations for ANOVA and MANOVA designs:

- In analyses with group sizes of fewer than 30 members, obtaining desired power levels can be quite problematic. If effect sizes are small, then the researcher may be required to decrease alpha (e.g., .05 to .10) to obtain desired power.
- Increasing sample sizes in each group has noticeable effects until group sizes of approximately 150 are reached, and then the increase in power slows markedly.
- Remember that large sample sizes (e.g., 400 or larger) reduce the sampling error component to such a small level that most small differences are regarded as statistically significant. When the sample sizes do become large and statistical significance is indicated, the researcher must examine the power and effect sizes to ensure not only statistical significance but practical significance as well.

Unique Issues with MANOVA. The ability to analyze multiple dependent variables in MANOVA creates additional constraints on the power in a MANOVA analysis. One source [17] of published tables presents power in a number of common situations for which MANOVA is applied. However, we can draw some general conclusions from examining a series of conditions encountered in many research designs. Table 8-3 provides an overview of the sample sizes needed for various levels of analysis complexity. A review of the table leads to several general points.

- Increasing the number of dependent variables requires increased sample sizes to maintain a given level of power. The additional sample size needed is more pronounced for the smaller effect sizes.

TABLE 8-3 Sample Size Requirements per Group for Achieving Statistical Power of .80 in MANOVA

	NUMBER OF GROUPS											
	3				4				5			
	Number of Dependent Variables				Number of Dependent Variables				Number of Dependent Variables			
Effect Size	2	4	6	8	2	4	6	8	2	4	6	8
Very large	13	16	18	21	14	18	21	23	16	21	24	27
Large	26	33	38	42	29	37	44	46	34	44	52	58
Medium	44	56	66	72	50	64	74	84	60	76	90	100
Small	98	125	145	160	115	145	165	185	135	170	200	230

Source: J. Läuter, "Sample Size Requirements for the T^2 Test of MANOVA (Tables for One-Way Classification),"
Biometrical Journal 20 (1978): 389–406.

- For small effect sizes, the researcher must be prepared to engage in a substantial research effort to achieve acceptable levels of power. For example, to achieve the suggested power of .80 when assessing small effect sizes in a four-group design, 115 subjects per group are required if two dependent measures are used. The required sample size increases to 185 per group if eight dependent variables are considered.

As we can see, the advantages of utilizing multiple dependent measures come at a cost in our analysis. As such the researcher must always balance the use of more dependent measures versus the benefits of parsimony in the dependent variable set that occur not only in interpretation but in the statistical tests for group differences as well.

Calculating Power Levels. To calculate power for ANOVA analyses, published sources [3, 24] as well as computer programs are now available. The methods of computing the power of MANOVA, however, are much more limited. Fortunately, most computer programs provide an assessment of power for the significance tests and enable the researcher to determine whether power should play a role in the interpretation of the results.

In terms of published material for planning purposes, little exists for MANOVA because many elements affect the power of a MANOVA analysis. The researcher, however, should utilize the tools available for ANOVA and then make adjustments described to approximate the power of a MANOVA design.

USING POWER IN PLANNING AND ANALYSIS The estimation of power should be used both in planning the analysis and in assessing the results. In the planning stage, the researcher determines the sample size needed to identify the estimated effect size. In many instances, the effect size can be estimated from prior research or reasoned judgments, or even set at a minimum level of practical significance. In each case, the sample size needed to achieve a given level of power with a specified alpha level can be determined.

By assessing the power of the test criteria after analysis is completed, the researcher provides a context for interpreting the results, especially if significant differences are not found. The researcher must first determine whether the achieved power is sufficient (.80 or above). If not, can the analysis be reformulated to provide more power? A possibility includes some form of blocking treatment or covariate analysis that will make the test more efficient by accentuating the effect size. If the power was adequate and statistical significance was not found for a treatment effect, then most likely the effect size for the treatment was too small to be of statistical or practical significance.

THE EFFECTS OF DEPENDENT VARIABLE MULTICOLLINEARITY ON POWER Up to this point we discussed power from a perspective applicable to both ANOVA and MANOVA. In MANOVA, however, the researcher must also consider the effects of multicollinearity of the dependent variables on the power of the statistical tests. The researcher, whether in the planning or analysis stage, must consider the strength and direction of the correlations as well as the effect sizes of the dependent variables. If we classify variables by their effect sizes as strong or weak, then several patterns emerge [4].

- First, if the correlated variable pair is made up of either strong–strong or weak–weak variables, then the greatest power is achieved when the correlation between variables is highly negative. This result suggests that MANOVA is optimized by adding dependent variables with high negative correlations. For example, rather than including two redundant measures of satisfaction, the researcher might replace them with correlated measures of satisfaction and dissatisfaction to increase power.
- When the correlated variable pair is a mixture (strong–weak), then power is maximized when the correlation is high, either positive or negative.
- One exception to this general pattern is the finding that using multiple items to increase reliability results in a net gain of power, even if the items are redundant and positively correlated.

REVIEW OF POWER IN MANOVA One of the most important considerations in a successful MANOVA is the statistical power of the analysis. Even though researchers engaged in experiments have much more control over the three elements that affect power, they must be sure to address the issues raised in the preceding sections or potential problems that reduce power below the desired value of .80 can easily occur. In field research, the researcher is faced not only with less certainty about the effect sizes in the analysis, but also the lack of control of group sizes and the potentially small group sizes that may occur in the sampling process. Thus, issues in the design and execution of field research discussed in stage 2 are critical in a successful analysis as well.

RULES OF THUMB 8-5

MANOVA Estimation

- The four most widely used measures for assessing statistical significance between groups on the independent variables are:
  - Roy's greatest characteristic root
  - Wilks' lambda
  - Pillai's criterion
  - Hotelling's T^2
- In most situations the results/conclusions will be the same across all four measures, but in some unique instances the results will differ between the measures
- Maintaining adequate statistical power is critical:
  - Power in the .80 range for the selected alpha level is acceptable
  - When the effect size is small, the researcher should use larger sample sizes per group to maintain acceptable levels of statistical power
- The general linear model (GLM) is widely used today in testing ANOVA or MANOVA models; GLM is available in most statistical packages such as SPSS and SAS

STAGE 5: INTERPRETATION OF THE MANOVA RESULTS

Once the statistical significance of the treatments has been assessed, the researcher turns attention to examining the results to understand how each treatment affects the dependent measures. In doing so, a series of three steps should be taken:

1. Interpret the effects of covariates, if employed.
2. Assess which dependent variable(s) exhibited differences across the groups of each treatment.
3. Identify whether the groups differ on a single dependent variable or the entire dependent variate.

We first examine the methods by which the significant covariates and dependent variables are identified, and then we address the methods by which differences among individual groups and dependent variables can be measured.

Evaluating Covariates

Covariates can play an important role by including metric variables into a MANOVA or ANOVA design. However, because covariates act as a control measure on the dependent variate, they must be assessed before the treatments are examined. Having met the assumptions for applying covariates, the researcher can interpret the actual effect of the covariates on the dependent variate and their impact on the actual statistical tests of the treatments.

ASSESSING OVERALL IMPACT   The most important role of the covariate(s) is the overall impact in the statistical tests for the treatments. The most direct approach to evaluating these impacts is to run the analysis with and without the covariates. Effective covariates will improve the statistical power of the tests and reduce within-group variance. If the researcher does not see any substantial improvement, then the covariates may be eliminated, because they reduce the degrees of freedom available for the tests of treatment effects. This approach also can identify those instances in which the covariate is too powerful and reduces the variance to such an extent that the treatments are all nonsignificant. Often this situation occurs when a covariate is included that is correlated with one of the independent variables and thus removes this variance, thereby reducing the explanatory power of the independent variable.

INTERPRETING THE COVARIATES   Because MANCOVA and ANCOVA are applications of regression procedures within the analysis of variance method, assessing the impact of the covariates on the dependent variables is quite similar to examining regression equations. If the overall impact is deemed significant, then each covariate can be examined for the strength of the predictive relationship with the dependent measures. If the covariates represent theoretically based effects, then these results provide an objective basis for accepting or rejecting the proposed relationships. In a practical vein, the researcher can examine the impact of the covariates and eliminate those with little or no effect.

Assessing Effects on the Dependent Variate

With the impacts, if any, of the covariates accounted for in the analysis, the next step is to examine the impacts of each treatment (independent variable) on the dependent variables. In doing so, we will first discuss how to assess the differences attributable to each treatment. With the treatment effects established, we will then assess whether those effects are independent in the case of two or more treatments. Finally, we will examine whether the effects of the treatments extend to the entire set of dependent measures or are reflected in only a subset of measures.

MAIN EFFECTS OF THE TREATMENTS   We already discussed the measures available to assess the statistical significance of a treatment. When a significant effect is found, we call it a **main effect,** meaning that significant differences between two or more groups are defined by the treatment. With two levels of the treatment, a significant main effect ensures that the two groups are significantly different. With three or more levels, however, a significant main effect *does not* guarantee that all three groups are significantly different, instead just that at least one significant difference is present between a pair of groups. As we will see in the next section, a wide array of statistical tests is available to assess which groups differ on both the variate and separate dependent variables.

So how do we portray a main effect? A main effect is typically described by the difference between groups on the dependent variables in the analysis. Assume that gender had a significant main effect on a 10-point satisfaction scale. We could then look to the difference in means as a way of describing the impact. If the female group had a mean score of 7.5 and males had an average score of 6.0, we could state that the difference due to gender was 1.5. Thus, all other things equal, females would be expected to score 1.5 points higher than males.

To define a main effect in these terms, however, requires two additional analyses:

1. If the analysis includes more than one treatment, the researcher must examine the interaction terms to see whether they are significant and, if so, whether they allow for an interpretation of the main effects.
2. If a treatment involves more than two levels, then the researcher must perform a series of additional tests between the groups to see which pairs of groups are significantly different.

We will discuss the interpretation of interaction terms in the next section and then examine the types of statistical tests available for assessing group differences when the analysis involves more than two groups.

IMPACTS OF THE INTERACTION TERMS   The interaction term represents the joint effect of two or more treatments. Any time a research design has two or more treatments, the researcher must first examine the interactions before any statement can be made about the main effects. First, we will discuss how to identify significant interactions. Then we will discuss how to classify significant interactions in order to interpret their impact on the main effects of the treatment variables.

Assessing Statistical Significance.   Interaction effects are evaluated with the same criteria as main effects, namely both multivariate and univariate statistical tests and statistical power. Software programs provide a complete set of results for each interaction term in addition to the main effects. All of the criteria discussed earlier apply to evaluating interactions as well as main effects.

Statistical tests that indicate the interaction is nonsignificant denote the independent effects of the treatments. Independence in factorial designs means that the effect of one treatment (i.e., group differences) is the same for each level of the other treatment(s) and that the main effects can be interpreted directly. Here we can describe the differences between groups as constant when considered in combination with the second treatment. We will discuss interpretation of the main effect in a simple example in a later section.

If the interactions are deemed statistically significant, it is critical that the researcher identify the type of interaction (ordinal versus disordinal) because it has direct bearing on the conclusion that can be drawn from the results. As we will see in the next section, interactions can potentially confound any description of the main effects depending on their nature.

Types of Significant Interactions.   The statistical significance of an interaction term is made with the same statistical criteria used to assess the impact of main effects. Upon assessing the significance of the interaction term, the researcher must examine effects of the treatment (i.e., the differences between groups) to determine the type of interaction and the impact of the interaction on the interpretation of the main effect. Significant interactions can be classified into one of two types: ordinal or disordinal interactions.

Ordinal Interactions. When the effects of a treatment are not equal across all levels of another treatment, but the group difference(s) is always the same direction, we term this an **ordinal interaction.** In other words, the group means for one level are always greater/lower than another level of the same treatment no matter how they are combined with the other treatment.

Assume that two treatments (gender and age) are used to examine satisfaction. An ordinal interaction occurs, for example, when females are always more satisfied than males, but the amount of the difference between males and females differs by age group.

When significant interactions are ordinal, the researcher must interpret the interaction term to ensure that its results are acceptable conceptually. Here the researcher must identify where the variation in group differences occurs and how that variation relates to the conceptual model underlying the analysis. If so, then the effects of each treatment must be described in terms of the other treatments it interacts with.

In the preceding example, we can make the general statement that gender does affect satisfaction in that females are always more satisfied than males. However, the researcher cannot state the difference in simple terms as could be done with a simple main effect. Rather the differences on gender must be described for each age category because the male/female differences vary by age.

Disordinal Interactions. When the differences between levels switch, depending on how they are combined with levels from another treatment, this is termed a **disordinal interaction.** Here the effects of one treatment are positive for some levels and negative for other levels of the other treatment.

In our example of examining satisfaction by gender and age, a disordinal interaction occurs when females have higher satisfaction than males in some age categories, but males are more satisfied in other age categories.

If the significant interaction is deemed disordinal, then the main effects of the treatments involved in the interaction cannot be interpreted and the study should be redesigned. This suggestion stems from the fact that with disordinal interactions, the main effects vary not only across treatment levels but also in direction (positive or negative). Thus, the treatments do not represent a consistent effect.

An Example of Interpreting Interactions. Interactions represent the differences between group means when grouped by levels of another treatment variable. Even though we could interpret interactions by viewing a table of values, graphical portrayals are quite effective in identifying the type of interaction between two treatments. The result is a multiple line graph, with levels of one treatment represented on the horizontal axis. Each line then represents one level of the second treatment variable.

Figure 8-5 portrays each type of interaction using the example of interactions between two treatments: cereal shapes and colors. Cereal shapes has three levels (balls, cubes, and stars) as does color (red, blue, and green). The vertical axis represents the mean evaluations (the dependent variable) of each group of respondents across the combinations of treatment levels. The X axis represents the three categories for color (red, blue, and green). The lines connect the category means for each shape across the three colors. For example, in the upper graph the value for red balls is about 4.0, the value for blue balls is about 5.0, and the value increases slightly to about 5.5 for green balls.

How do the graphs identify the type of interaction? As we will discuss, each of the three interactions has a specific pattern:

- *No interaction.* Shown by the parallel lines representing the differences of the various shapes across the levels of color (the same effect would be seen if the differences in color were graphed across the three types of shape). In the case of no interaction, the effects of each treatment (the differences between groups) are constant at each level and the lines are roughly parallel.

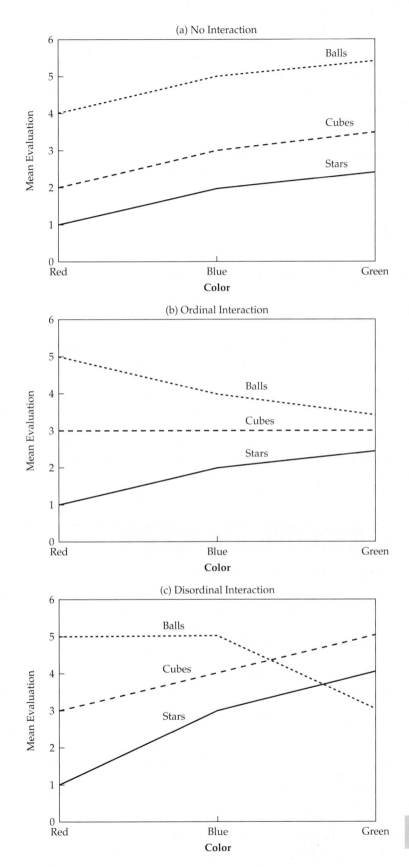

FIGURE 8-5 Interaction Effects in Factorial Designs

- *Ordinal interaction.* The effects of each treatment are not constant and thus the lines are not parallel. The differences for red are large, but they decline slightly for blue cereal and even more for green cereal. Thus, the differences by color vary across the shapes. The relative ordering among levels of shape are the same, however, with stars always highest, followed by the cubes and then the ball shapes.
- *Disordinal interaction.* The differences in color vary not only in magnitude but also in direction. This interaction is shown by lines that are not parallel and that cross between levels. The evaluation of balls is higher than cubes and stars for red and blue, but is evaluated lower than both for the color green.

The graphs complement the statistical significance tests by enabling the researcher to quickly categorize the interaction, especially determining whether significant interactions fall into the ordinal or disordinal categories.

Identifying Differences Between Individual Groups

Although the univariate and multivariate tests of ANOVA and MANOVA enable us to reject the null hypothesis that the groups' means are all equal, they do not pinpoint where the significant differences lie among more than two groups. Multiple *t* tests without any form of adjustment are not appropriate for testing the significance of differences between the means of paired groups because the probability of a Type I error increases with the number of intergroup comparisons made (similar to the problem of using multiple univariate ANOVAs versus MANOVA). Many procedures are available for further investigation of specific group mean differences of interest using different approaches to control Type I error rates across multiple tests.

MULTIPLE UNIVARIATE TESTS ADJUSTING FOR THE EXPERIMENTWIDE ERROR RATE   Many times the simplest approach is to perform a series of univariate tests with some form of manual adjustment by the researcher to account for the experimentwide error rate. Researchers can make these adjustments based on whether the treatments involve two or more levels (groups).

RULES OF THUMB 8-6

Interpreting Covariates and Interaction Effects

- When covariates are involved in a GLM model:
  - Analyze the model both with and without the covariates
  - If the covariates do not improve the statistical power or have no effect on the significance of the treatment effects, then they can be dropped from the final analysis
- Any time two or more independent variables (treatments) are included in the analysis, interactions must be examined before drawing conclusions about main effects for any independent variable
  - If the interactions are not statistically significant, then main effects can be interpreted directly because the difference between treatments is considered constant across combinations of levels
  - If the interaction is statistically significant and the differences are not constant across combinations of levels, then the interaction must be determined to be ordinal or disordinal:
    - Ordinal interactions mean that the direction of differences does not vary by level (e.g., males always less than females) even though the difference between males/females varies by level on the other treatment; in this case, the size of the main effect (e.g., males versus females) should only be described separately for each level of the other treatment
    - Significant disordinal interactions occur when the direction of an observed main effect changes with the level of another treatment (e.g., males greater than females for one level and less than females for another level); disordinal interactions interfere with the interpretation of main effects

---

Two-Group Analyses. Two-group treatments reduce to a series of *t* tests across the specified dependent measures. However, researchers should be aware that as the number of these tests increases, one of the major benefits of the multivariate approach to significance testing—control of the Type I error rate—is negated unless specific adjustments in the T^2 statistic are made that control for the inflation of the Type I error.

If we wish to test the group differences individually for each of the dependent variables, we could use the square root of T^2_{crit} (i.e., T_{crit}) as the critical value needed to establish significance. This procedure would ensure that the probability of any Type I error across all the tests would be held to α (where α is specified in the calculation of T^2_{crit}) [11].

k-Group Analyses. We could make similar tests for *k*-group situations by adjusting the α level by the **Bonferroni inequality,** which adjusts the alpha level for the number of tests being made. The adjusted alpha level used in any separate test is defined as the overall alpha level divided by the number of tests [adjusted α = (overall α)/(number of tests)].

For example, if the overall error rate (α) is .05 and five statistical tests are to be made, then a Bonferroni adjustment would call for a .01 level to be used for each individual test.

STRUCTURED MULTIGROUP TESTS The procedures described in the previous section are best used in simple situations with a few tests being considered. If the researcher wants to systematically examine group differences across specific pairs for one or more dependent measures, more structured statistical tests should be used. In this section we will examine two types of tests:

- **Post hoc tests.** Tests of the dependent variables between all possible pairs of group differences that are tested after the data patterns are established.
- **A priori tests.** Tests planned from a theoretical or practical decision-making viewpoint prior to looking at the data.

The principal distinction between the two types of tests is that the post hoc approach tests all possible combinations, providing a simple means of group comparisons but at the expense of lower power. A priori tests examine only specified comparisons, so that the researcher must explicitly define the comparison to be made, but with a resulting greater level of power. Either method can be used to examine one or more group differences, although the a priori tests also provide the researcher with total control over the types of comparisons made between groups.

Post Hoc Methods. Post hoc methods are widely used because of the ease in which multiple comparisons are made. Among the more common post hoc procedures are (1) the Scheffé method, (2) Tukey's honestly significant difference (HSD) method, (3) Tukey's extension of the Fisher least significant difference (LSD) approach, (4) Duncan's multiple-range test, and (5) the Newman-Keuls test.

Each method identifies which comparisons among groups (e.g., group 1 versus groups 2 and 3) have significant differences. Although they simplify the identification of group differences, these methods all share the problem of having quite low levels of power for any individual test because they examine all possible combinations. These five post hoc or multiple-comparison tests of significance have been contrasted for power [23] and several conclusions can be drawn:

- Scheffé method is the most conservative with respect to Type I error, and the remaining tests are ranked in this order: Tukey HSD, Tukey LSD, Newman-Keuls, and Duncan.
- If the effect sizes are large or the number of groups is small, the post hoc methods may identify the group differences. However, the researcher must recognize the limitations of these methods and employ other methods if more specific comparisons can be identified.

A discussion of the options available with each method is beyond the scope of this chapter. Excellent discussions and explanations of these procedures can be found in other texts [13, 27].

A Priori or Planned Comparisons. The researcher can also make specific comparisons between groups by using a priori tests (also known as **planned comparisons**). This method is similar to the post hoc tests in the statistical methods for making the group comparisons, but differs in design and control by the researcher in three aspects:

- The researcher specifies which group comparisons are to be made versus testing the entire set, as done in the post hoc tests.
- Planned comparisons are more powerful because of the smaller number of comparisons, but more power is of little use if the researcher does not specifically test for correct group comparisons.
- Planned comparisons are most appropriate when conceptual bases can support the specific comparisons to be made. They should not be used in an exploratory manner, however, because they do not have effective controls against inflating the experimentwide Type I error levels.

The researcher specifies the groups to be compared through a **contrast,** which is a combination of group means that represents a specific planned comparison. Contrasts can be stated generally as

$$C = W_1 G_1 + W_2 G_2 + \ldots + W_k G_k$$

where

$$C = \text{contrast value}$$
$$W = \text{weights}$$
$$G = \text{group means}$$

The contrast is formulated by assigning positive and negative weights to specify the groups to be compared while ensuring that the weights sum to 0.

For example, assume we have three group means (G_1, G_2, and G_3). To test for a difference between G_1 and G_2 (and ignoring G_3 for this comparison), the contrast would be:

$$C = (1)G_1 + (-1)G_2 + (0)G_3$$

To test whether the average of G_1 and G_2 differs from G_3, the contrast is:

$$C = (.5)G_1 + (.5)G_2 + (-1)G_3$$

A separate F statistic is computed for each contrast.

In this manner, the researcher can create any comparisons desired and test them directly, but the probability of a Type I error for each a priori comparison is equal to α. Thus, several planned comparisons will inflate the overall Type I error level. All the statistical packages can perform either a priori or post hoc tests for single dependent variables or the variate.

If the researcher wishes to perform comparisons of the entire dependent variate, extensions of these methods are available. After concluding that the group mean vectors are not equivalent, the researcher might be interested in whether any group differences occur on the composite dependent variate. A standard ANOVA F statistic can be calculated and compared to $F_{\text{crit}} = (N - k)\text{gcr}_{\text{crit}}/(k - 1)$, where the value of gcr_{crit} is taken from the gcr distribution with appropriate degrees of freedom. Many software packages have the ability to perform planned comparisons for the dependent variate as well as individual dependent variables.

Assessing Significance for Individual Dependent Variables

Up to this time we have examined only the multivariate tests of significance for the collective set of dependent variables. What about each separate dependent variable? Does a significant difference with a multivariate test ensure that each dependent variable also is significantly different? Or does a nonsignificant effect mean that all dependent variables also have nonsignificant differences? In both instances the answer is no. The result of a multivariate test of differences across the set of dependent measures does not necessarily extend to each variable separately, just collectively. Thus, the researcher should always examine the multivariate results for the extent to which they extend to the individual dependent measures.

UNIVARIATE SIGNIFICANCE TESTS   The first step is to assess which of the dependent variables contribute to the overall differences indicated by the statistical tests. This step is essential because a subset of variables in the set of dependent variables may accentuate the differences, whereas another subset of variables may be nonsignificant or may mask the significant effects of the remainder.

Most statistical packages provide separate univariate significance tests for each dependent measure in addition to the multivariate tests, providing an individual assessment of each variable. The researcher can then determine how each individual dependent variable corresponds to the effects on the variate.

STEPDOWN ANALYSIS   A procedure known as **stepdown analysis** [16, 23] may also be used to assess individually the differences of the dependent variables. This procedure involves computing a univariate F statistic for a dependent variable after eliminating the effects of other dependent variables preceding it in the analysis. The procedure is somewhat similar to stepwise regression, but here we examine whether a particular dependent variable contributes unique (uncorrelated) information on group differences. The stepdown results would be exactly the same as performing a covariate analysis, with the other preceding dependent variables used as the covariates.

A critical assumption of stepdown analysis is that the researcher must know the order in which the dependent variables should be entered, because the interpretations can vary dramatically given different entry orders. If the ordering has theoretical support, then the stepdown test is valid. Variables indicated to be nonsignificant are redundant with the earlier significant variables, and they add no further information concerning differences about the groups. The order of dependent variables may be changed to test whether the effects of variables are either redundant or unique, but the process becomes rather complicated as the number of dependent variables increases.

Both of these analyses are directed toward assisting the researcher in understanding which of the dependent variables contribute to the differences in the dependent variate across the treatment(s).

STAGE 6: VALIDATION OF THE RESULTS

Analysis of variance techniques (ANOVA and MANOVA) were developed in the tradition of experimentation, with **replication** as the primary means of validation. The specificity of experimental treatments allows for a widespread use of the same experiment in multiple populations to assess the

---

RULES OF THUMB 8-7

Interpreting Differences Between Individual Groups

- When the independent variable has more than two groups, two types of procedures can be used to isolate the source of differences:
  - Post hoc tests examine potential statistical differences among all possible combinations of group means; post hoc tests have limited power and thus are best suited to identify large effects
  - Planned comparisons are appropriate when a priori theoretical reasons suggest that certain groups will differ from another group or other groups; Type I error is inflated as the number of planned comparisons increases

generalizability of the results. Although it is a principal tenet of the scientific method, in social science and business research, true experimentation is many times replaced with statistical tests in nonexperimental situations such as survey research. The ability to validate the results in these situations is based on the replicability of the treatments. In many instances, demographic characteristics such as age, gender, income, and the like are used as treatments. These treatments may seem to meet the requirement of comparability, but the researcher must ensure that the additional element of randomized assignment to a cell is also met; however, many times in survey research randomness is not fully achieved.

For example, having age and gender be the independent variables is a common example of the use of ANOVA or MANOVA in survey research. In terms of validation, the researcher must be wary of analyzing multiple populations and comparing results as the sole proof of validity. Because respondents in a simple sense select themselves, the treatments in this case cannot be assigned by the researcher, and thus randomized assignment is impossible.

The researcher should strongly consider the use of covariates to control for other features that might be characteristic of the age or gender groups that could affect the dependent variables but are not included in the analysis.

Another issue is the claim of causation when experimental methods or techniques are employed. The principles of causation are examined in more detail in Chapter 11. For our purposes here, the researcher must remember that in all research settings, including experiments, certain conceptual criteria (e.g., temporal ordering of effects and outcomes) must be established before causation may be supported. The single application of a particular technique used in an experimental setting does not ensure causation.

SUMMARY

We discussed the appropriate applications and important considerations of MANOVA in addressing multivariate analyses with multiple dependent measures. Although considerable benefits stem from its use, MANOVA must be carefully and appropriately applied to the question at hand. When doing so, researchers have at their disposal a technique with flexibility and statistical power. We now illustrate the applications of MANOVA (and its univariate counterpart ANOVA) in a series of examples.

ILLUSTRATION OF A MANOVA ANALYSIS

Multivariate analysis of variance (MANOVA) affords researchers with the ability to assess differences across one or more nonmetric independent variables for a set of metric dependent variables. It provides a means for determining the extent to which groups of respondents (formed by their characteristics on the nonmetric independent variables) differ in terms of the dependent measures. Examining these differences can be done separately or in combination. In the following sections, we will detail the analyses necessary to examine two characteristics (X_1 and X_5) for their impact on a set of purchase outcomes (X_{19}, X_{20}, and X_{21}). First we will analyze each characteristic separately and then both in combination. The reader should note that an expanded version of HBAT (HBAT200 with a sample size of 200) is used in this analysis to allow for the analysis of a two-factor design. This data set is available on the Web at www.pearsonglobaleditions.com/hair or www.mvstats.com.

Recent years have seen increased attention to the area of distribution systems. Fueled by the widespread use of Internet-based systems for channel integration and the cost savings being realized by improved logistical systems, upper management at HBAT is interested in assessing the current state of affairs in their distribution system, which utilizes both indirect (broker-based) and direct channels. In the indirect channel, products are sold to customers by brokers acting as both an external salesforce and even wholesalers in some instances. HBAT also employs a salesforce of its own; salespeople contact and service customers directly from both the corporate office and field offices.

TABLE 8-4 **Group Sizes for a Two-Factor Analysis Using the HBAT Data (100 observations)**

		X_5 Distribution System		
		Indirect Through Broker	Direct to Customer	Total
X_1	Less than 1 year	23	9	32
Customer	1 to 5 years	16	19	35
Type	More than 5 years	18	15	33
	Total	57	43	100

A concern has arisen that changes may be necessary in the distribution system, particularly focusing on the broker system that is perceived to not be performing well, especially in fostering long-term relationships with HBAT. To address these concerns, three questions were posed:

1. What differences are present in customer satisfaction and other purchase outcomes between the two channels in the distribution system?
2. Is HBAT establishing better relationships with its customers over time, as reflected in customer satisfaction and other purchase outcomes?
3. What is the relationship between the distribution system and these relationships with customers in terms of the purchase outcomes?

With the research questions defined, the researcher now turns attention to defining the independent and dependent variables to be used and the ensuing sample size requirements.

To examine these issues, researchers decided to employ MANOVA to examine the effects of X_5 (Distribution System) and X_1 (Customer Type) on three Purchase Outcome measures (X_{19}, Satisfaction; X_{20}, Likelihood of Recommending HBAT; and X_{21}, Likelihood of Future Purchase). Although a sample size of 100 observations would be sufficient for either of the analyses of the individual variables, it would not be appropriate for addressing them in combination. A quick calculation of group sizes for this two-factor analysis (see Table 8-4) identified at least one group with fewer than 10 observations and several more with fewer than 20 observations.

Because these group sizes would not afford the ability to detect medium or small effect sizes with a desired level of statistical power (see Table 8-3), a decision was made to gather additional responses to supplement the 100 observations already available. A second research effort added 100 more observations for a total sample size of 200. This new dataset is named HBAT200 and will be used for the MANOVA analyses that follow. Preliminary analyses indicated that the supplemented data set had the same basic characteristics as the HBAT, thus eliminating the need for additional examination of this new data to determine its basic properties.

EXAMPLE 1: DIFFERENCE BETWEEN TWO INDEPENDENT GROUPS

To introduce the practical benefits of a multivariate analysis of group differences, we begin our discussion with one of the best-known designs: the two-group design in which each respondent is classified based on the levels (groups) of the treatment (independent variable). If this analysis was being performed in an experimental setting, then respondents would be assigned to groups randomly (e.g., depending on whether they see an advertisement or which type of cereal they taste). Many times, however, the groups are formed not by random assignment, but based instead on some characteristic of the respondent (e.g., age, gender, occupation, etc.).

In many research settings, however, it is also unrealistic to assume that a difference between any two experimental groups will be manifested in only a single dependent variable. For example, two advertising messages may not only produce different levels of purchase intent, but also may affect a number of other (potentially correlated) aspects of the response to advertising (e.g., overall product evaluation, message credibility, interest, attention).

Many researchers handle this multiple-criterion situation by repeated application of individual univariate t tests until all the dependent variables have been analyzed. However, this approach has serious deficiencies:

- Inflation of the Type I error rate over multiple t tests
- Inability of paired t tests to detect differences among combinations of the dependent variables not apparent in univariate tests

To overcome these problems, MANOVA can be employed to control the overall Type I error rate while still providing a means of assessing the differences on each dependent variable both collectively and individually.

Stage 1: Objectives of the Analysis

The first step involves identifying the appropriate dependent and independent variables. As discussed earlier, HBAT identified the distribution system as a key element in its customer relationship strategy and first needs to understand the impact of the distribution system on customers.

Research Question. HBAT is committed to strengthening its customer relationship strategy, with one aspect focused on distribution system. Concern has been raised about the differences due to distribution channel system (X_5), which is composed of two channels: direct through HBAT's salesforce or indirect through a broker. Three purchase outcomes (X_{19}, Satisfaction; X_{20}, Likelihood of Recommending HBAT; and X_{21}, Likelihood of Future Purchase) have been identified as the focal issues in evaluating the impacts of the two distribution systems. The task is to identify whether any differences exist between these two systems across all or a subset of these purchase outcomes.

Examining Group Profiles. Table 8-5 provides a summary of the group profiles on each of the purchase outcomes across the two groups (direct versus indirect distribution system). A visual inspection reveals that the direct distribution channel has the higher mean scores for each of the purchase outcomes. The task of MANOVA is to examine these differences and assess the extent to which these differences are significantly different, both individually and collectively.

Stage 2: Research Design of the MANOVA

The principal consideration in the design of the two-group MANOVA is the sample size in each of the cells, which directly affects statistical power. Also, as is the case in most survey research, the cell sizes are unequal, making the statistical tests more sensitive to violations of the assumptions,

TABLE 8-5 **Descriptive Statistics of Purchase Outcome Measures (X_{19}, X_{20}, and X_{21}) for Groups of X_5 (Distribution System)**

	X_5 Distribution System	Mean	Std. Deviation	N
X_{19} Satisfaction	Indirect through broker	6.325	1.033	108
	Direct to customer	7.688	1.049	92
	Total	6.952	1.241	200
X_{20} Likely to Recommend	Indirect through broker	6.488	.986	108
	Direct to customer	7.498	.930	92
	Total	6.953	1.083	200
X_{21} Likely to Purchase	Indirect through broker	7.336	.880	108
	Direct to customer	8.051	.745	92
	Total	7.665	.893	200

especially the test for homogeneity of variance of the dependent variable. Both of these issues must be considered in assessing the research design using X_5.

As discussed earlier, the concern for adequate sample sizes across the entire MANOVA analysis resulted in the addition of 100 additional surveys to the original HBAT survey (see Table 8-4). Based on this larger dataset (HBAT200), 108 firms used the indirect broker system and 92 respondents used the direct system from HBAT.

These group sizes will provide more than adequate statistical power at an 80 percent probability to detect medium effect sizes and almost reach the levels necessary for identifying small effects sizes (see Table 8-3). The result is a research design with relatively balanced group sizes and enough statistical power to identify differences at any managerially significant level.

Stage 3: Assumptions in MANOVA

The most critical assumptions relating to MANOVA are the independence of observations, homoscedasticity across the groups, and normality. Each of these assumptions will be addressed in regards to each of the purchase outcomes. Also of concern is the presence of outliers and their potential influence on the group means for the purchase outcome variables.

Independence of Observations. The independence of the respondents was ensured as much as possible by the random sampling plan. If the study had been done in an experimental setting, the random assignment of individuals would have ensured the necessary independence of observations.

Homoscedasticity. A second critical assumption concerns the homogeneity of the variance–covariance matrices among the two groups. The first analysis assesses the univariate homogeneity of variance across the two groups. As shown in Table 8-6, univariate tests (Levene's test) for

TABLE 8-6   **Multivariate and Univariate Measures for Testing Homoscedasticity of X_5**

Multivariate Test of Homoscedasticity

Box's Test of Equality of Covariance Matrices

Box's M	4.597
F	.753
df1	6
df2	265275.824
Sig.	.607

Univariate Tests of Homoscedasticity

Levene's Test of Equality of Error Variances

Dependent Variable	F	df1	df2	Sig.
X_{19} Satisfaction	.001	1	198	.978
X_{20} Likely to Recommend	.643	1	198	.424
X_{21} Likely to Purchase	2.832	1	198	.094

Test for Correlation Among the Dependent Variables

Bartlett's Test of Sphericity

Likelihood Ratio	.000
Approx. Chi-Square	260.055
df	5
Sig.	.000

all three variables are nonsignificant (i.e., significance greater than .05). The next step is to assess the dependent variables collectively by testing the equality of the entire variance–covariance matrices between the groups. Again, in Table 8-6 the Box's M test for equality of the covariance matrices shows a nonsignificant value (.607), indicating no significant difference between the two groups on the three dependent variables collectively. Thus, the assumption of homoscedasticity is met for each individual variable separately and the three variables collectively.

Correlation and Normality of Dependent Variables. Another test should be made to determine whether the dependent measures are significantly correlated. The most widely used test for this purpose is Bartlett's test for sphericity. It examines the correlations among all dependent variables and assesses whether, collectively, significant intercorrelation exists. In our example, a significant degree of intercorrelation does exist (significance = .000) (see Table 8-6).

The assumption of normality for the dependent variables (X_{19}, X_{20}, and X_{21}) was examined in Chapter 2 and found to be acceptable. This supports the results of testing for the equality of the variance–covariance matrices between groups.

Outliers. The final issue to be addressed is the presence of outliers. A simple approach that identifies extreme points for each group is the use of boxplots (see Figure 8-6). Examining the boxplot for each dependent measure shows few, if any, extreme points across the groups. When we examine these extreme points across the three dependent measures, no observation was an extreme value on all three dependent measures, nor did any observation have a value so extreme that it dictated exclusion. Thus, all 200 observations will be retained for further analysis.

Stage 4: Estimation of the MANOVA Model and Assessing Overall Fit

The next step is to assess whether the two groups exhibit statistically significant differences for the three purchase outcome variables, first collectively and then individually. To conduct the test, we first specify the maximum allowable Type I error rate. In doing so, we accept that 5 times out of 100 we might conclude that the type of distribution channel has an impact on the purchase outcome variables when in fact it did not.

MULTIVARIATE STATISTICAL TESTING AND POWER ANALYSIS Having set the acceptable Type I error rate, we first use the multivariate tests to test the set of dependent variables for differences between the two groups and then perform univariate tests on each purchase outcome. Finally, power levels are assessed.

Multivariate Statistical Testing. Table 8-7 contains the four most commonly used multivariate tests (Pillai's criterion, Wilks' lambda, Hotelling's T^2 and Roy's greatest characteristic root). Each of the four measures indicates that the set of purchase outcomes have a highly significant difference (.000) between the two types of distribution channel. This confirms the group differences seen in Table 8-5 and the boxplots of Figure 8-6.

Univariate Statistical Tests. Although we can show that the set of purchase outcomes differs across the groups, we also need to examine each purchase outcome separately for differences across the two types of distribution channel. Table 8-7 also contains the univariate tests for each individual purchase outcome. As we can see, all of the individual tests are also highly significant (significance = .000), indicating that each variable follows the same pattern of higher purchase outcomes (see Table 8-5) for those served by the direct distribution system (Direct Distribution customers have values of 7.688, 7.498, and 8.051 versus values of 6.325, 6.488, and 7.336 for Indirect Through Broker customers on X_{19}, X_{20}, and X_{21}, respectively).

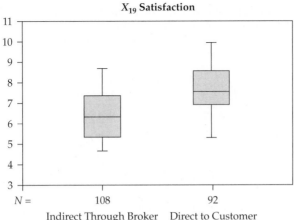

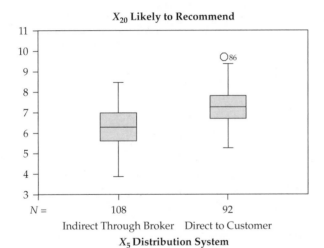

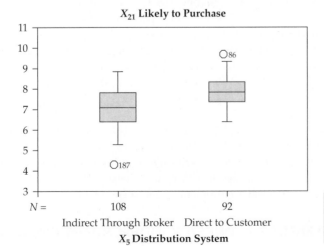

FIGURE 8-6 Boxplots of Purchase Outcome Measures (X_{19}, X_{20}, and X_{21}) for Groups of X_5 (Distribution System)

TABLE 8-7   Multivariate and Univariate Tests for Group Differences in Purchase Outcome Measures (X_{19}, X_{20}, and X_{21}) Across Groups of X_5 (Distribution System)

Multivariate Tests

Statistical Test	Value	F	Hypothesis df	Error df	Sig.	η^2	Observed Power[a]
Pillai's Criterion	.307	28.923	3	196	.000	.307	1.00
Wilks' Lambda	.693	28.923	3	196	.000	.307	1.00
Hotelling's T^2	.443	28.923	3	196	.000	.307	1.00
Roy's greatest characteristic root	.443	28.923	3	196	.000	.307	1.00

[a]Computed using alpha = .05

Univariate Tests (Between-Subjects Effects)

Dependent Variable	Sum of Squares	df	Mean Square	F	Sig.	η^2	Observed Power[a]
X_{19} Satisfaction	92.300[b]	1	92.300	85.304	.000	.301	1.00
X_{20} Likely to Recommend	50.665[c]	1	50.665	54.910	.000	.217	1.00
X_{21} Likely to Purchase	25.396[d]	1	25.396	37.700	.000	.160	1.00

[a]Computed using alpha = .05
[b]R^2 = .301 (Adjusted R^2 = .298)
[c]R^2 = .217 (Adjusted R^2 = .213)
[d]R^2 = .160 (Adjusted R^2 = .156)

Statistical Power. The power for the statistical tests was 1.0, indicating that the sample sizes and the effect size were sufficient to ensure that the significant differences would be detected if they existed beyond the differences due to sampling error.

Stage 5: Interpretation of the Results

The presence of only two groups eliminates the need to perform any type of post hoc tests. The statistical significance of the multivariate and univariate tests indicating group differences on the dependent variate (vector of means) and the individual purchase outcomes leads the researcher to an examination of the results to assess their logical consistency.

As noted earlier, firms using the direct type of distribution system scored significantly higher than those serviced through the broker-based indirect distribution channel. The group means shown in Table 8-5, based on responses to a 10-point scale, indicate that the customers using the direct distribution channel are more satisfied (7.688 − 6.325 = 1.363), more likely to recommend HBAT (7.498 − 6.488 = 1.01), and more likely to purchase in the future (8.051 − 7.336 = .715). These differences are also reflected in the boxplots for the three purchase outcomes in Figure 8-6.

These results confirm that the type of distribution channel does affect customer perceptions in terms of the three purchase outcomes. These statistically significant differences, which are of a sufficient magnitude to denote managerial significance as well, indicate that the direct distribution channel is more effective in creating positive customer perceptions on a wide range of purchase outcomes.

EXAMPLE 2: DIFFERENCE BETWEEN *K* INDEPENDENT GROUPS

The two-group design (example 1) is a special case of the more general *k*-group design. In the general case, each respondent is a member or is randomly assigned to one of *k* levels (groups) of the treatment (independent variable). In a univariate case, a single metric dependent variable is measured, and the null hypothesis is that all group means are equal (i.e., $\mu_1 = \mu_2 = \mu_3 = \cdots = \mu_k$). In the multivariate case, multiple metric dependent variables are measured, and the null hypothesis is that all group vectors of mean scores are equal (i.e., $\nu_1 = \nu_2 = \nu_3 = \cdots \nu_k$), where ν refers to a vector or set of mean scores.

In k-group designs in which multiple dependent variables are measured, many researchers proceed with a series of individual F tests (ANOVAs) until all the dependent variables have been analyzed. As the reader should suspect, this approach suffers from the same deficiencies as a series of t tests across multiple dependent variables; that is, a series of F tests with ANOVA:

- Results in an inflated Type I error rate
- Ignores the possibility that some composite of the dependent variables may provide reliable evidence of overall group differences

In addition, because individual F tests ignore the correlations among the independent variables, they use less than the total information available for assessing overall group differences.

MANOVA again provides a solution to these problems. MANOVA solves the Type I error rate problem by providing a single overall test of group differences at a specified α level. It solves the composite variable problem by implicitly forming and testing the linear combinations of the dependent variables that provide the strongest evidence of overall group differences.

Stage 1: Objectives of the MANOVA

In the prior example, HBAT assessed its performance among customers based on which of the two distribution system channels (X_5) were used. MANOVA was employed due to the desire to examine a set of three purchase outcome variables representing HBAT performance. A second research objective was to determine whether the three purchase outcome variables were affected by the length of their relationship with HBAT (X_1). The null hypothesis HBAT now wishes to test is that the three sample vectors of the mean scores (one vector for each category of customer relationship) are equivalent.

Research Questions. In addition to examining the role of the distribution system, HBAT also has stated a desire to assess whether the differences in the purchase outcomes are attributable solely to the type of distribution channel or whether other nonmetric factors can be identified that show significant differences as well. HBAT specifically selected X_1 (Customer Type) to determine whether the length of HBAT's relationship with the customer has any impact on these purchase outcomes.

Examining Group Profiles. As can be seen in Table 8-8, the mean scores of all three purchase outcome variables increase as the length of the customer relationship increases. The question to be addressed in this analysis is the extent to which these differences as a whole can be considered statistically significant and if those differences extend to each difference between groups. In a second MANOVA analysis, X_1 (Customer Type) is examined for differences in purchase outcomes.

Stage 2: Research Design of MANOVA

As was the situation in the earlier two-group analysis, sample size of the groups is a primary consideration in research design. Even when all instances of the group sizes far exceed the minimum necessary, the researcher should always be concerned with achieving the statistical power needed for the research question at hand.

Analysis of the impact of X_1 now requires that we analyze the sample sizes for the three groups of length of customer relationship (less than 1 year, 1 to 5 years, and more than 5 years). In the HBAT sample, the 200 respondents are almost evenly split across the three groups with sample sizes of 68, 64, and 68 (see Table 8-8). These sample sizes, in conjunction with the three dependent variables, exceed the guidelines shown in Table 8-3 to identify medium effect sizes (suggested sample sizes of 44 to 56), but fall somewhat short of the required sample size (98 to 125) needed to identify small effect sizes with a power of .80. Thus, any nonsignificant results should be examined closely to evaluate whether the effect size has managerial significance, because the low statistical power precluded designating it as statistically significant.

TABLE 8-8 Descriptive Statistics of Purchase Outcome Measures (X_{19}, X_{20}, and X_{21}) for Groups of X_1 (Customer Type)

	X_1 Customer Type	Mean	Std. Deviation	N
X_{19} Satisfaction	Less than 1 year	5.729	.764	68
	1 to 5 years	7.294	.708	64
	More than 5 years	7.853	1.033	68
	Total	6.952	1.241	200
X_{20} Likely to Recommend	Less than 1 year	6.141	.995	68
	1 to 5 years	7.209	.714	64
	More than 5 years	7.522	.976	68
	Total	6.953	1.083	200
X_{21} Likely to Purchase	Less than 1 year	6.962	.760	68
	1 to 5 years	7.883	.643	64
	More than 5 years	8.163	.777	68
	Total	7.665	.893	200

Stage 3: Assumptions in MANOVA

Having already addressed the issues of normality (see Chapter 2) and intercorrelation (Bartlett's test of sphericity in Table 8-6) of the dependent variables in the prior example, the only remaining concern rests in the homoscedasticity of the purchase outcomes across the groups formed by X_1 and identification of any outliers. We first examine this homoscedasticity at the multivariate level (all three purchase outcome variables collectively) and then for each dependent variable separately. The multivariate test for homogeneity of variance of the three purchase outcomes is performed with the Box's M test, while the Levene's test is used to assess each purchase outcome variable separately.

Homoscedasticity. Table 8-9 contains the results of both the multivariate and univariate tests of homoscedasticity. The Box's M test indicates no presence of heteroscedasticity (significance = .069). In the Levene's tests for equality of error variances, two of the purchase outcomes (X_{20} and X_{21})

TABLE 8-9 Multivariate and Univariate Measures for Testing Homoscedasticity of X_1

Multivariate Test of Homoscedasticity

Box's Test of Equality of Covariance Matrices

Box's M	20.363
F	1.659
df1	12
df2	186673.631
Sig.	.069

Univariate Test of Homoscedasticity

Levene's Test of Equality of Error Variances

Dependent Variable	F	df1	df2	Sig.
X_{19} Satisfaction	6.871	2	197	.001
X_{20} Likely to Recommend	2.951	2	197	.055
X_{21} Likely to Purchase	.800	2	197	.451

showed nonsignificant results and confirmed homoscedasticity. In the case of X_{19}, the significance level was .001, indicating the possible existence of heteroscedasticity for this variable. However, given the relatively large sample sizes in each group and the presence of homoscedasticity for the other two purchase outcomes, corrective remedies were not needed for X_{19}.

Outliers. Examination of the boxplot for each purchase outcome variable (see Figure 8-7) reveals a small number of extreme points for each dependent measure (observation 104 for X_{19}; observations 86, 104, 119, and 149 for X_{20}; and observations 104 and 187 for X_{21}). Only one observation had extreme values on all three dependent measures and none of the values were so extreme in any cases as to markedly affect the group values. Thus, no observations were classified as outliers designated for exclusion and all 200 observations were used in this analysis.

Stage 4: Estimation of the MANOVA Model and Assessing Overall Fit

Using MANOVA to examine an independent variable with three or more levels reveals the differences across the levels for the dependent measures with the multivariate and univariate statistical tests illustrated in the earlier example. In these situations, the statistical tests are testing for a significant main effect, meaning that the differences between groups, when viewed collectively, are substantial enough to be deemed statistically significant. It should be noted that statistical significance of the main effect does not ensure that each group is also significantly different from each other group. Rather, separate tests described in the next section can examine which groups do exhibit significant differences.

All three dependent measures show a definite pattern of increasing as the length of the customer relationship increases (see Table 8-8 and Figure 8-7). The first step is to utilize the multivariate tests and assess whether the set of purchase outcomes, which each individually seem to follow a similar increasing pattern as time increases, does vary in a statistically significant manner (i.e., a significant main effect). Table 8-10 contains the four most commonly used multivariate tests and as we see, all four tests indicate a statistically significant difference of the collective set of dependent measures across the three groups.

In addition to the multivariate tests, univariate tests for each dependent measure indicate that all three dependent measures, when considered individually, also have significant main effects. Thus, both collectively and individually, the three purchase outcomes (X_{19}, X_{20}, and X_{21}) do vary at a statistically significant level across the three groups of X_1.

Stage 5: Interpretation of the Results

Interpreting a MANOVA analysis with an independent variable of three or more levels requires a two-step process:

- Examination of the main effect of the independent variable (in this case, X_1) on the three dependent measures
- Identifying the differences between individual groups for each of the dependent measures with either planned comparisons or post hoc tests

The first analysis examines the overall differences across the levels for the dependent measures, whereas the second analysis assesses the differences between individual groups (e.g., group 1 versus group 2, group 2 versus group 3, group 1 versus group 3, etc.) to identify those group comparisons with significant differences.

Assessing the Main Effect of X_5.   All of the multivariate and univariate tests indicated a significant main effect of X_1 (Customer Type) on each individual dependent variable as well as the set of the dependent variables when considered collectively. The significant main effect means that the dependent variable(s) do vary in significant amounts between the three customer groups based on

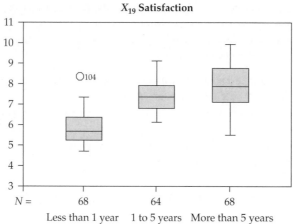

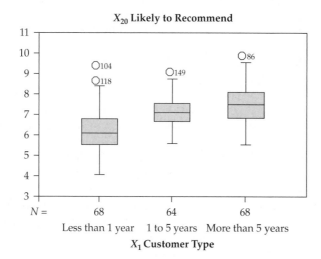

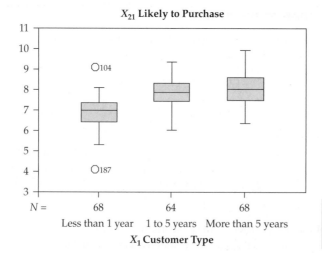

FIGURE 8-7 Boxplots of Purchase Outcome Measures (X_{19}, X_{20}, and X_{21}) for Groups of X_1 (Customer Type)

TABLE 8-10 Multivariate and Univariate Tests for Group Differences in Purchase Outcome Measures (X_{19}, X_{20}, and X_{21}) Across Groups of X_1 (Customer Type)

Multivariate Tests

Statistical Test	Value	F	Hypothesis df	Error df	Sig.	η^2	Observed Power[a]
Pillai's Criterion	.543	24.368	6	392	.000	.272	1.000
Wilks' Lambda	.457	31.103	6	390	.000	.324	1.000
Hotelling's T^2	1.184	38.292	6	388	.000	.372	1.000
Roy's greatest characteristic root	1.183	77.280	3	196	.000	.542	1.000

[a]Computed using alpha = .05

Univariate Tests (Between-Subjects Effects)

Dependent Variable	Type III Sum of Squares	df	Mean Square	F	Sig.	η^2	Observed Power[a]
X_{19} Satisfaction	164.311[b]	2	82.156	113.794	.000	.536	1.00
X_{20} Likely to Recommend	71.043[c]	2	35.521	43.112	.000	.304	1.00
X_{21} Likely to Purchase	53.545[d]	2	26.773	50.121	.000	.337	1.00

[a]Computed using alpha = .05
[b]R^2 = .536 (Adjusted R^2 = .531)
[c]R^2 = .304 (Adjusted R^2 = .297)
[d]R^2 = .337 (Adjusted R^2 = .331)

length of customer relationship. As we can see in Table 8-8 and Figure 8-7 the pattern of purchases increases in each dependent measure as the customer relationship matures. For example, customer satisfaction (X_{19}) is lowest (5.729) for those customers of less than 1 year, increasing (7.294) for those customers between 1 and 5 years until it reaches the highest level (7.853) for those customers of 5 years or more. Similar patterns are seen for the two other dependent measures.

MAKING POST HOC COMPARISONS   As noted earlier, a significant main effect indicates that the total set of group differences (e.g., group 1 versus group 2, etc.) are large enough to be considered statistically significant. It should also be noted that a significant main effect does not guarantee that every one of the group differences is also significant. We may find that a significant main effect is actually due to a single group difference (e.g., group 1 versus group 2) while all of the other comparisons (group 1 versus group 3 and group 2 versus group 3) are not significantly different.

The question becomes: How are these individual group differences assessed while maintaining an acceptable level of overall Type I error rate? This same problem is encountered when considering multiple dependent measures, but in this case in making comparisons for a single dependent variable across multiple groups. This type of question can be tested with one of the a priori procedures. If the contrast is used, a specific comparison is made between two groups (or sets of groups) to see whether they are significantly different. Another approach is to use one of the post hoc procedures that tests all group differences and then identifies those differences that are statistically significant.

Table 8-11 contains three post hoc comparison methods (Tukey HSD, Scheffé, and LSD) applied to all three purchase outcomes across the three groups of X_1. When we examine X_{19} (Satisfaction), we first see that even though the overall main effect is significant, the differences between adjacent groups are not constant. The difference between customers of less than 1 year and those of 1 to 5 years is −1.564 (the minus sign indicates that customers of less than 1 year have the lower value). When we

TABLE 8-11 Post Hoc Comparisons for Individual Group Differences on Purchase Outcome Measures (X_{19}, X_{20}, and X_{21}) Across Groups of X_1 (Customer Type)

Dependent Variable	Groups to Be Compared		Mean Difference Between Groups (I − J)		Statistical Significance of Post Hoc Comparison		
	Group I	Group J	Mean Difference	Standard Error	Tukey HSD	Scheffé	LSD
X_{19} Satisfaction							
	Less than 1 year	1 to 5 years	−1.564	.148	.000	.000	.000
	Less than 1 year	More than 5 years	−2.124	.146	.000	.000	.000
	1 to 5 years	More than 5 years	−.559	.148	.000	.001	.000
X_{20} Likely to Recommend							
	Less than 1 year	1 to 5 years	−1.068	.158	.000	.000	.000
	Less than 1 year	More than 5 years	−1.381	.156	.000	.000	.000
	1 to 5 years	More than 5 years	−.313	.158	.118	.144	.049
X_{21} Likely to Purchase							
	Less than 1 year	1 to 5 years	−.921	.127	.000	.000	.000
	Less than 1 year	More than 5 years	−1.201	.125	.000	.000	.000
	1 to 5 years	More than 5 years	−.280	.127	.071	.091	.029

examine the group difference between customers of 1 to 5 years versus those of more than 5 years, however, the difference is reduced to −.559 (about one-third of the prior difference).

The researcher is thus interested in whether both of these differences are significant, or only significant between the first two groups. When we look to the last three columns in Table 8-11, we can see that all of the separate group differences for X_{19} are significant, indicating that the difference of −.559, even though much smaller than the other group difference, is still statistically significant.

When we examine the post hoc comparisons for the other two purchase outcomes (X_{20} and X_{21}), a different pattern emerges. Again, the differences between the first two groups (less than 1 year and 1 to 5 years) are all statistically significant across all three post hoc tests. Yet when we examine the next comparison (customers of 1 to 5 years versus those of more than 5 years), two of the three tests indicate that the two groups are not different. In these tests, the purchase outcomes of X_{20} and X_{21} for customers of 1 to 5 years are not significantly different from those of more than 5 years. This result is contrary to what was found for satisfaction, in which this difference was significant.

When the independent variable has three or more levels, the researcher must engage in this second level of analysis in addition to the assessment of significant main effects. Here the researcher is not interested in the collective effect of the independent variable, but instead in the differences between specific groups. The tools of either planned comparisons or post hoc methods provide a powerful means of making these tests of group differences while also maintaining the overall Type I error rate.

EXAMPLE 3: A FACTORIAL DESIGN FOR MANOVA WITH TWO INDEPENDENT VARIABLES

In the prior two examples, the MANOVA analyses have been extensions of univariate two- and three-group analyses. In this example, we explore a multivariate factorial design: two independent variables used as treatments to analyze differences of the set of dependent variables. In the course of our discussion, we assess the interactive or joint effects between the two treatments on the dependent variables separately and collectively.

Stage 1: Objectives of the MANOVA

In the previous multivariate research questions, HBAT considered the effect of only a single-treatment variable on the dependent variables. Here the possibility of joint effects among two or more independent variables must also be considered. In this way, the interaction between the independent variables can be assessed along with their main effects.

Research Questions. The first two research questions we examined addressed the impact of two factors—distribution system and duration of customer relationship—on a set of purchase outcomes. In each instance, the factors were shown to have significant impacts (i.e., more favorable purchase outcomes for firms in the direct distribution system or those with longer tenure as an HBAT customer).

Left unresolved is a third question: How do these two factors operate when considered simultaneously? Here we are interested in knowing how the differences between distribution systems hold across the groups based on length of HBAT relationship. We saw that customers in the direct distribution system had significantly greater purchase outcomes (higher satisfaction, etc.), yet are these differences always present for each customer group based on X_1? The following is just a sample of the types of question we can ask when considering the two variables together in a single analysis:

- Is the direct distribution system more effective for newer customers?
- Do the two distribution systems show differences for customers of 5 years or more?
- Is the direct distribution system always preferred over the indirect system across the customer groups of X_1?

By combining both independent variables (X_1 and X_5) into a factorial design, we create six customer groups: the three groups based on length of their relationship with HBAT separated into those groups in each distribution system channel. Known as a 3×2 design, the three levels of X_1 separated for each level of X_5 form a separate group for each customer type within each distribution system channel.

Examining Group Profiles. Table 8-12 provides a profile of each group for the set of purchase outcomes. Many times a quicker and simpler perspective is through a graphical display. One option is to form a line chart, and we will illustrate this when viewing the interaction terms in a later section. We can also utilize boxplots to show not only the differences between group means, but the overlap of the range of values in each group. Figure 8-8 illustrates such a graph for X_{19} (Satisfaction) across the six groups of our factorial design. As we can see, satisfaction increases as the length of relationship with HBAT increases, but the differences between the two distribution systems are not always constant (e.g., they seem much closer for customers of 1 to 5 years).

The purpose of including multiple independent variables into a MANOVA is to assess their effects "contingent on" or "controlling for" the other variables. In this case, we can see how the length of the HBAT relationship changes in any way the more positive perceptions generally seen for the direct distribution system.

Stage 2: Research Design of the MANOVA

Any factorial design of two or more independent variables raises the issue of adequate sample size in the various groups. The researcher must ensure, when creating the factorial design, that each group has sufficient sample size for the following:

1. Meets the minimum requirements of group sizes exceeding the number of dependent variables
2. Provides the statistical power to assess differences deemed practically significant

TABLE 8-12 Descriptive Statistics of Purchase Outcome Measures (X_{19}, X_{20}, and X_{21}) for Groups of X_1 (Customer Type) by X_5 (Distribution System)

Dependent Variable	X_1 Customer Type	X_5 Distribution System	Mean	Std. Deviation	N
X_{19} Satisfaction	Less than 1 year	Indirect through broker	5.462	.499	52
		Direct to customer	6.600	.839	16
		Total	5.729	.764	68
	1 to 5 years	Indirect through broker	7.120	.551	25
		Direct to customer	7.405	.779	39
		Total	7.294	.708	64
	More than 5 years	Indirect through broker	7.132	.803	31
		Direct to customer	8.457	.792	37
		Total	7.853	1.033	68
	Total	Indirect through broker	6.325	1.033	108
		Direct to customer	7.688	1.049	92
		Total	6.952	1.241	200
X_{20} Likely to Recommend	Less than 1 year	Indirect through broker	5.883	.773	52
		Direct to customer	6.981	1.186	16
		Total	6.141	.995	68
	1 to 5 years	Indirect through broker	7.144	.803	25
		Direct to customer	7.251	.659	39
		Total	7.209	.714	64
	More than 5 years	Indirect through broker	6.974	.835	31
		Direct to customer	7.981	.847	37
		Total	7.522	.976	68
	Total	Indirect through broker	6.488	.986	108
		Direct to customer	7.498	.930	92
		Total	6.953	1.083	200
X_{21} Likely to Purchase	Less than 1 year	Indirect through broker	6.763	.702	52
		Direct to customer	7.606	.569	16
		Total	6.962	.760	68
	1 to 5 years	Indirect through broker	7.804	.710	25
		Direct to customer	7.933	.601	39
		Total	7.883	.643	64
	More than 5 years	Indirect through broker	7.919	.648	31
		Direct to customer	8.368	.825	37
		Total	8.163	.777	68
	Total	Indirect through broker	7.336	.880	108
		Direct to customer	8.051	.745	92
		Total	7.665	.893	200

Sample Size Considerations. As noted in the previous section, this analysis is termed a 2×3 design because it includes two levels of X_5 (direct versus indirect distribution) and three levels of X_1 (less than 1 year, 1 to 5 years, and more than 5 years). The issue of sample size per group was such a concern to HBAT researchers that the original HBAT survey of 100 observations was supplemented by 100 additional respondents just for this analysis (see more detailed discussion in the section preceding the examples). Even with the additional respondents, the sample of 200 observations must be split across the six groups, hopefully in a somewhat balanced manner.

The sample sizes per cell are shown in Table 8-12 and can be shown in the following simplified format.

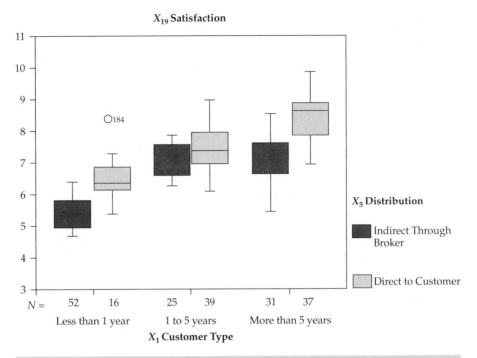

FIGURE 8-8 Boxplot of Purchase Outcome Measure (X_{19}) for Groups of X_5 (Distribution System) by X_1 (Customer Type)

	X_5 Distribution System	
X_1 Customer Type	Indirect	Direct
Less than 1 year	52	16
1 to 5 years	25	39
More than 5 years	31	37

Adequacy of Statistical Power. The sample sizes in all but one of the cells provides enough statistical power to identify at least a large effect size with an 80 percent probability. However, the smaller sample size of 16 for customers of less than 1 year served by the direct distribution channel is of some concern. Thus, we must recognize that unless the effect sizes are substantial, the limited sample sizes in each group, even from this sample of 200 observations, may preclude the identification of significant differences. This issue becomes especially critical when examining nonsignificant difference in that the researcher should determine whether the nonsignificant result is due to insufficient effect size or low statistical power.

Stage 3: Assumptions in MANOVA

As with the prior MANOVA analyses, the assumption of greatest importance is the homogeneity of variance–covariance matrices across the groups. Meeting this assumption allows for direct interpretation of the results without having to consider group sizes, level of covariances in the group, and so forth. Additional statistical assumptions related to the dependent variables (normality and correlation) have already been addressed in the prior examples. A final issue is the presence of outliers and the need for deletion of any observations that may distort the mean values of any group.

TABLE 8-13 Multivariate and Univariate Measures for Testing Homoscedasticity Across Groups of X_1 by X_5

Multivariate Tests for Homoscedasticity

Box's Test of Equality of Covariance Matrices

Box's M	39.721
F	1.263
*df*1	30
*df*2	33214.450
Sig.	.153

Univariate Tests for Homoscedasticity

Levene's Test of Equality of Error Variances

Dependent Variable	*F*	*df*1	*df*2	**Sig.**
X_{19} Satisfaction	2.169	5	194	.059
X_{20} Likely to Recommend	1.808	5	194	.113
X_{21} Likely to Purchase	.990	5	194	.425

Homoscedasticity. For this factorial design, six groups are involved in testing the assumption of homoscedasticity (see Table 8-13). The multivariate test (Box's M) has a nonsignificant value (.153), allowing us to accept the null hypothesis of homogeneity of variance–covariance matrices at the .05 level.

The univariate tests for the three purchase outcome variables separately are also all nonsignificant. With the multivariate and univariate tests showing nonsignificance, the researcher can proceed knowing that the assumption of homoscedasticity has been fully met.

Outliers. The second issue involves examining observations with extreme values and the possible designation of observations as outliers with deletion from the analysis. Interestingly enough, examination of the boxplots for the three purchase outcomes identifies a smaller number of observations with extreme values than found for X_1 by itself. The dependent variable with the most extreme values was X_{21} with only three, whereas the other dependent measures had one and two extreme values. Moreover, no observation had extreme values on more than one dependent measure. As a result, all observations were retained in the analysis.

Stage 4: Estimation of the MANOVA Model and Assessing Overall Fit

The MANOVA model for a factorial design tests not only for the main effects of both independent variables but also their interaction or joint effect on the dependent variables. The first step is to examine the interaction effect and determine whether it is statistically significant. If it is significant, then the researcher must confirm that the interaction effect is ordinal. If it is found to be disordinal, the statistical tests of main effects are not valid. But assuming a significant ordinal or a nonsignificant interaction effect, the main effects can be interpreted directly without adjustment.

ASSESSING THE INTERACTION EFFECT Interaction effects can be identified both graphically and statistically. The most common graphical means is to create line charts depicting pairs of independent variables. As illustrated in Figure 8-4, significant interaction effects are represented by nonparallel lines (with parallel lines denoting no interaction effect). If the lines depart from parallel but never cross in a significant amount, then the interaction is deemed ordinal. If the lines do cross to the degree that in at least one instance the relative ordering of the lines is reversed, then the interaction is deemed disordinal.

Figure 8-9 portrays each dependent variable across the six groups, indicating by the nonparallel pattern that an interaction may exist. As we can see in each graph, the middle level of

X_1 (1 to 5 years with HBAT) has a substantially smaller difference between the two lines (representing the two distribution channels) than the other two levels of X_1. We can confirm this observation by examining the group means from Table 8-12. Using X_{19} (Satisfaction) as an example, we see that the difference between direct and indirect distribution channels is 1.138 for customers of less than 1 year, which is quite similar to the difference between channels (1.325) for customers of greater than 5 years. However, for customers served by HBAT from 1 to 5 years, the difference between customers of the two channels is only (.285). Thus, the differences between the two distribution channels, although found to be significant in earlier examples, can be shown to differ (interact) based on how long the customer has been with

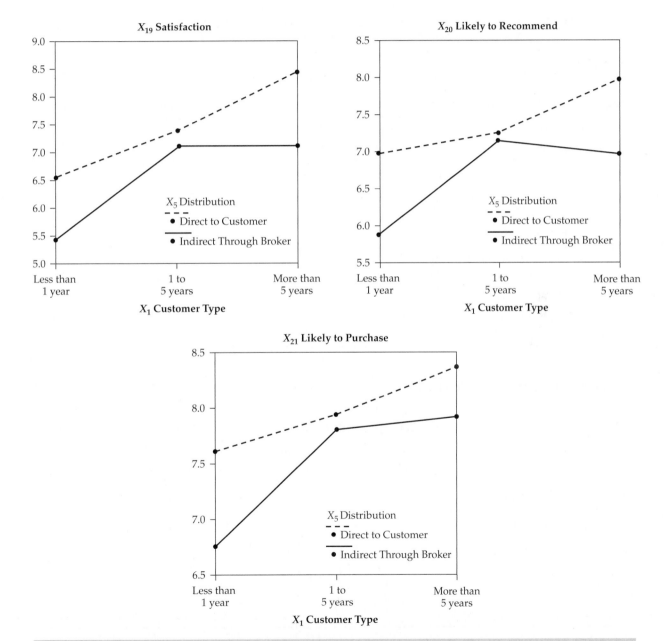

FIGURE 8-9 Graphical Displays of Interaction Effects of Purchase Outcome Measures (X_{19}, X_{20}, and X_{21}) Across Groups of X_5 (Distribution System) by X_1 (Customer Type)

HBAT. The interaction is deemed ordinal because in all instances the direct distribution channel has higher satisfaction scores.

TESTING THE INTERACTION AND MAIN EFFECTS   In addition to the graphical means, interaction effects can also be tested in the same manner as main effects. Thus, the researcher can make a multivariate as well as univariate assessment of the interaction effect with the statistical tests described in earlier examples.

Table 8-14 contains the MANOVA results for testing both the interaction and main effects. Testing for a significant interaction effect proceeds as does any other effect. First, the multivariate

TABLE 8-14   Multivariate and Univariate Tests for Group Differences in Purchase Outcome Measures (X_{19}, X_{20}, and X_{21}) Across Groups of X_1 by X_5

Multivariate Tests

Effect	Statistical Test	Value	F	Hypothesis df	Error df	Sig.	η^2	Observed Power[a]
X_1	Pillai's Criterion	.488	20.770	6	386	.000	.244	1.000
	Wilks' Lambda	.512	25.429	6	384	.000	.284	1.000
	Hotelling's T^2	.952	30.306	6	382	.000	.322	1.000
	Roy's greatest characteristic root	.951	61.211	3	193	.000	.488	1.000
X_5	Pillai's Criterion	.285	25.500	3	192	.000	.285	1.000
	Wilks' Lambda	.715	25.500	3	192	.000	.285	1.000
	Hotelling's T^2	.398	25.500	3	192	.000	.285	1.000
	Roy's greatest characteristic root	.398	25.500	3	192	.000	.285	1.000
$X_1 \times X_5$	Pillai's Criterion	.124	4.256	6	386	.000	.062	.980
	Wilks' Lambda	.878	4.291	6	384	.000	.063	.981
	Hotelling's T^2	.136	4.327	6	382	.000	.064	.982
	Roy's greatest characteristic root	.112	7.194	3	193	.000	.101	.981

[a]Computed using alpha = .05

Univariate Tests (Between-Subjects Effects)

Effect	Dependent Variable	Sum of Squares	df	Mean Square	F	Sig.	η^2	Observed Power[a]
Overall	X_{19} Satisfaction	210.999[b]	5	42.200	85.689	.000	.688	1.000
	X_{20} Likely to Recommend	103.085[c]	5	20.617	30.702	.000	.442	1.000
	X_{21} Likely to Purchase	65.879[d]	5	13.176	27.516	.000	.415	1.000
X_1	X_{19} Satisfaction	89.995	2	44.998	91.370	.000	.485	1.000
	X_{20} Likely to Recommend	32.035	2	16.017	23.852	.000	.197	1.000
	X_{21} Likely to Purchase	26.723	2	13.362	27.904	.000	.223	1.000
X_5	X_{19} Satisfaction	36.544	1	36.544	74.204	.000	.277	1.000
	X_{20} Likely to Recommend	23.692	1	23.692	35.282	.000	.154	1.000
	X_{21} Likely to Purchase	9.762	1	9.762	20.386	.000	.095	.994
$X_1 \times X_5$	X_{19} Satisfaction	9.484	2	4.742	9.628	.000	.090	.980
	X_{20} Likely to Recommend	8.861	2	4.430	6.597	.002	.064	.908
	X_{21} Likely to Purchase	3.454	2	1.727	3.607	.029	.036	.662

[a]Computed using alpha = .05
[b]R^2 = .688 (Adjusted R^2 = .680)
[c]R^2 = .442 (Adjusted R^2 = .427)
[d]R^2 = .415 (Adjusted R^2 = .400)

effects are examined and in this case all four tests show statistical significance. Then, univariate tests for each dependent variable are performed. Again, the interaction effect is also deemed significant for each of the three dependent variables. The statistical tests confirm what was indicated in the graphs: A significant ordinal interaction effect occurs between X_5 and X_1.

ESTIMATING MAIN EFFECTS   If the interaction effect is deemed nonsignificant or even significant and ordinal, then the researcher can proceed to estimating the significance of the main effects for their differences across the groups. In those instances in which a disordinal interaction effect is found, the main effects are confounded by the disordinal interaction and tests for differences should not be performed.

With a significant ordinal interaction, we can proceed to assessing whether both independent variables still have significant main effects when considered simultaneously. Table 8-14 also contains the MANOVA results for the main effects of X_1 and X_5 in addition to the tests for the interaction effect already discussed. As we found when analyzing them separately, both X_1 (Customer Type) and X_5 (Distribution System) have a significant impact (main effect) on the three purchase outcome variables, both as a set and separately, as demonstrated by the multivariate and univariate tests.

The impact of the two independent variables can be compared by examining the relative effect sizes as shown by η^2 (eta squared). The effect sizes for each variable are somewhat higher for X_1 when compared to X_5 on either the multivariate or univariate tests. For example, with the multivariate tests the eta squared values for X_1 range from .244 to .488, but they are lower (all equal to .285) for X_5. Similar patterns can be seen on the univariate tests. This comparison gives an evaluation of practical significance separate from the statistical significance tests. When compared to either independent variable, however, the effect size attributable to the interaction effect is much smaller (e.g., multivariate eta squared values ranging from .062 to .101).

Stage 5: Interpretation of the Results

Interpretation of a factorial design in MANOVA is a combination of judgments drawn from statistical tests and examination of the basic data. The presence of an interaction effect can be assessed statistically, but the resulting conclusions are primarily based on the judgments of the researcher. The researcher must examine the differences for practical significance in addition to statistical significance. If specific comparisons among the groups can be formulated, then planned comparisons can be specified and tested directly in the analysis.

INTERPRETING INTERACTION AND MAIN EFFECTS   Statistical significance may be supported by the multivariate tests, but examining the tests for each dependent variable provides critical insight into the effects seen in the multivariate tests. Moreover, the researcher may employ planned comparisons or even post hoc tests to determine the true nature of differences, particularly when significant interaction terms are found.

With the interaction and main effects found to be statistically significant by both the multivariate and univariate tests, interpretation is still heavily reliant on the patterns of effects shown in the values of the six groups (shown in Table 8-12 and Figure 8-9).

Interaction of X_1 by X_5. The nonparallel lines for each dependent measure notably portray the narrowing of the differences in distribution channels for customers of 1 to 5 years. Although the effects of X_1 and X_5 are still present, we do see some marked differences in these impacts depending on which specific sets of customers we examine. For example, for X_{20} the difference between Direct versus Indirect Distribution customers is 1.098 for customers of less than 1 year, decreases to only .107 for customers of 1 to 5 years, and then increases again to 1.007 for customers of more than 5 years. These substantial differences depending on the Customer Type illustrate the significant interaction effect.

Main Effect of X_1. Its main effect is illustrated for all three purchase outcomes by the upward sloping lines across the three levels of X_1 on the X axis. Here we can see that the effects are consistent with earlier findings in that all three purchase outcomes increase favorably as the length of the relationship with HBAT increases. For example, again examining X_{20}, we see that the overall mean score increases from 6.141 for customers of less than 1 year to 7.209 for customers of 1 to 5 years and finally to 7.522 for customers of more than 5 years.

Main Effect of X_5. The separation of the two lines representing the two distribution channels show us that the direct distribution channel generates more favorable purchase outcomes. Examining Figure 8-9 we see that for each dependent variable, the lines for customers with the direct distribution are greater than those served by the indirect system.

POTENTIAL COVARIATES   The researcher also has an additional tool—adding covariates—to improve in the analysis and interpretation of the independent variables. The role of the covariate is to control for effects outside the scope of the MANOVA analysis that may affect the group differences in some systematic manner (see earlier discussion for more detail). A covariate is most effective when it has correlation with the dependent variables, but is relatively uncorrelated to the independent variables in use. In this way it can account for variance not attributable to the independent variables (due to its low correlation with them), but still reduce the amount of overall variation to be explained (the correlation with the dependent measures).

The HBAT researchers had limited options in choosing covariates for these MANOVA analyses. The only likely candidate was X_{22}, representing the customers' percentage of purchases coming from HBAT. The rationale would be to control for the perceived or actual dependence of firms on HBAT as represented in X_{22}. Firms with more dependence may react quite differently to the variables being considered.

However, X_{22} is a poor candidate for becoming a covariate even though it meets the criterion of being correlated with the dependent variables. Its fatal flaw is the high degree of differences seen on both X_1 and X_5. These differences suggest that the effects of X_1 and X_5 would be severely confounded by the use of X_{22} as a covariate. Thus, no covariates will be utilized in this analysis.

Summary

The results reflected in both the main and interaction effects present convincing evidence that HBAT customers' postpurchase reactions are influenced by the type of distribution system and by the length of the relationship.

The direct distribution system is associated with higher levels of customer satisfaction, as well as likelihood to repurchase and recommend HBAT to others. Similarly, customers with longer relationships also report higher levels of all three dependent variables. The differences between the dependent variables are smallest among those customers who have done business with HBAT for 1 to 5 years.

The use of MANOVA in this process enables the researcher to control the Type I error rate to a far greater extent than if individual comparisons were made on each dependent variable. The interpretations remain valid even after the impact of other dependent variables has been considered. These results confirm the differences found between the effects of the two independent variables.

A MANAGERIAL OVERVIEW OF THE RESULTS

HBAT researchers performed a series of ANOVAs and MANOVAs in an effort to understand how three purchase outcomes (X_{19}, Satisfaction; X_{20}, Likelihood of Recommending HBAT; and X_{21}, Likelihood of Future Purchase) vary across characteristics of the firms involved, such as distribution system (X_5) and customer type (X_1). In our discussion, we focus on the multivariate results as they overlap with the univariate results.

The first MANOVA analysis is direct: Does the type of distribution channel have an effect on the purchase outcomes? In this case the researcher tests whether the sets of mean scores (i.e., the means of the three purchase outcomes) for each distribution group are equivalent. After meeting all assumptions, we find that the results reveal a significant difference in that firms in the direct distribution system had more favorable purchase outcomes when compared to firms served through the broker-based model. Along with the overall results, management also needed to know whether this difference exists not only for the variate but also for the individual variables. Univariate tests revealed significant univariate differences for each purchase outcome as well. The significant multivariate and univariate results indicate to management that the direct distribution system serves customers better as indicated by the more favorable outcome measures. Thus, managers can focus on extending those benefits of the direct system while working on improving the broker-based distribution system.

The next MANOVA follows the same approach, but substitutes a new independent variable, customer type (i.e., the length of time the firm has been a customer), which has three groups (less than 1 year, 1 to 5 years, and more than 5 years). Once again, management focuses on the three outcome measures to assess whether significant differences are found across length of the customer relationship. Both univariate and multivariate tests show differences in the purchase outcome variables across the three groups of customers. Yet one question remains: Is each group different from the other? Group profiles show substantial differences and post hoc tests indicate that for X_{19} (Satisfaction) each customer group is distinct from the other. For the remaining two outcome measures, groups 2 and 3 (customers of 1 to 5 years and customers more than 5 years) are not different from each other, although both are different from customers of less than 1 year. The implication is that for X_{20} and X_{21} the improvements in purchase outcomes are significant in the early years, but do not keep increasing beyond that period. From a managerial perspective, the duration of the customer relationship positively affects the firm's perceptions of purchase outcomes. Even though increases are seen throughout the relationship for the basic satisfaction measure, the only significant increase in the other two outcomes is seen after the first year.

The third example addresses the issue of the combined impact of these two firm characteristics (X_5, distribution system; and X_1, duration of the customer relationship) on the purchase outcomes. The three categories of X_1 are combined with the two categories of X_5 to form six groups. The objective is to establish whether the significant differences seen for each of the two firm characteristics when analyzed separately are also evident when considered simultaneously. The first step is to review the results for significant interactions: Do the purchase outcomes display the same differences between the two types of distribution systems when viewed by duration of the relationship? All three interactions were found to be significant, meaning that the differences between the direct and broker-based systems were not constant across the three groups of customers based on duration of the customer relationship. Examining the results found that the middle group (customers of 1 to 5 years) had markedly smaller differences between the two distribution systems than customers of either shorter or longer relationships. Although this pattern held for all three purchase outcomes and direct systems always were evaluated more favorably (maintaining ordinal interactions), HBAT must realize that the advantages of the direct distribution system are contingent on the length of the customer's relationship. Given these interactions, it was still found that each firm characteristic exhibited significant impacts on the outcome as was found when analyzed separately. Moreover, when considered simultaneously, the impact of each on the purchase outcomes was relatively even.

These results enable HBAT managers to identify the significant effects of these firm characteristics on purchase outcomes, not only individually but also when combined.

Summary

Multivariate analysis of variance (MANOVA) is an extension of analysis of variance (ANOVA) to accommodate more than one dependent variable. It is a dependence technique that measures the differences for two or more metric dependent variables based on a set of categorical (nonmetric) variables acting as independent variables. This chapter helps you to do the following:

Explain the difference between the univariate null hypothesis of ANOVA and the multivariate null hypothesis of MANOVA. Like ANOVA, MANOVA is concerned with differences between groups (or experimental treatments). ANOVA is termed a univariate procedure because we use it to assess group differences on a single metric dependent variable. The null hypothesis is the means of the groups for a single dependent variable are equal (not statistically different). Univariate methods for assessing group differences are the t test (two groups) and analysis of variance (ANOVA) for two or more groups. The t test is widely used because it works with small group sizes and it is quite easy to apply and interpret. But its limitations include: (1) it only accommodates two groups; and (2) it can only assess one independent variable at a time. Although a t test can be performed with ANOVA, the F statistic has the ability to test for differences between more than two groups as well as include more than one independent variable. Also, independent variables are not limited to just two levels, but instead can have as many levels (groups) as desired. MANOVA is considered a multivariate procedure because it is used to assess group differences across multiple metric dependent variables simultaneously. In MANOVA, each treatment group is observed on two or more dependent variables. Thus, the null hypothesis is the vector of means for multiple dependent variables is equal across groups. The multivariate procedures for testing group differences are the Hotelling T^2 and multivariate analysis of variance, respectively.

Discuss the advantages of a multivariate approach to significance testing compared to the more traditional univariate approaches. As statistical inference procedures, both the univariate techniques (t test and ANOVA) and their multivariate extensions (Hotelling's T^2 and MANOVA) are used to assess the statistical significance of differences between groups. In the univariate case, a single dependent measure is tested for equality across the groups. In the multivariate case, a variate is tested for equality. In MANOVA, the researcher actually has two variates, one for the dependent variables and another for the independent variables. The dependent variable variate is of more interest because the metric-dependent measures can be combined in a linear combination, as we have already seen in multiple regression and discriminant analysis. The unique aspect of MANOVA is that the variate optimally combines the multiple dependent measures into a single value that maximizes the differences across groups. To analyze data on multiple groups and variables using univariate methods, the researcher might be tempted to conduct separate t tests for the difference between each pair of means (i.e., group 1 versus group 2; group 1 versus group 3; and group 2 versus group 3). But multiple t tests inflate the overall Type I error rate. ANOVA and MANOVA avoid this Type I error inflation due to making multiple comparisons of treatment groups by determining in a single test whether the entire set of sample means suggests that the samples were drawn from the same general population. That is, both techniques are used to determine the probability that differences in means across several groups are due solely to sampling error.

State the assumptions for the use of MANOVA. The univariate test procedures of ANOVA are valid in a statistical sense if we assume that the dependent variable is normally distributed, the groups are independent in their responses on the dependent variable, and that variances are equal for all treatment groups. There is evidence, however, that F tests in ANOVA are robust with regard to these assumptions except in extreme cases. For the multivariate test procedures of MANOVA to be valid, three assumptions must be met: (1) observations must be independent, (2) variance–covariance matrices must be equal for all treatment groups, and (3) the set of dependent variables must follow a multivariate normal distribution. In addition to these assumptions, the researcher must consider two issues that influence the possible effects—the linearity and multicollinearity of the variate of the dependent variables.

Understand how to interpret MANOVA results. If the treatments result in statistically significant differences in the vector of dependent variable means, the researcher then examines the results to understand how each treatment impacts the dependent measures. Three steps are involved: (1) interpreting the effects of covariates, if included; (2) assessing which dependent variable(s) exhibited differences across the groups of each treatment; and (3) identifying if the groups differ on a single dependent variable or the entire dependent variate. When a significant effect is found, we say that there is a main effect, meaning that there are significant differences between the dependent variables of the two or more groups defined by the treatment. With two levels of the treatment, a significant main effect ensures that the two groups are significantly different. With three or more levels, however, a significant main effect does not guarantee that all three groups are significantly different, instead just that there is at least one significant difference between a pair of groups. If there is more than one treatment in the analysis, the researcher must examine the interaction terms to see if they are significant, and if so, do they allow for an interpretation of the main effects or not. If there are more than two levels for a treatment, then the researcher must perform a series of additional tests between the groups to see which pairs of groups are significantly different.

Describe the purpose of post hoc tests in ANOVA and MANOVA.   Although the univariate and multivariate tests of ANOVA and MANOVA enable us to reject the null hypothesis that the groups' means are all equal, they do not pinpoint where the significant differences lie if there are more than two groups. Multiple *t* tests without any form of adjustment are not appropriate for testing the significance of differences between the means of paired groups because the probability of a Type I error increases with the number of intergroup comparisons made (similar to the problem of using multiple univariate ANOVAs versus MANOVA). If the researcher wants to systematically examine group differences across specific pairs of groups for one or more dependent measures, two types of statistical tests should be used: post hoc and a priori. Post hoc tests examine the dependent variables between all possible pairs of group differences that are tested after the data patterns are established. A priori tests are planned from a theoretical or practical decision-making viewpoint prior to looking at the data. The principal distinction between the two types of tests is that the post hoc approach tests all possible combinations, providing a simple means of group comparisons but at the expense of lower power. A priori tests examine only specified comparisons, so that the researcher must explicitly define the comparison to be made, but with a resulting greater level of power. Either method can be used to examine one or more group differences, although the a priori tests also give the researcher control over the types of comparisons made between groups.

Interpret interaction results when more than one independent variable is used in MANOVA.   The interaction term represents the joint effect of two or more treatments. Any time a research design has two or more treatments, the researcher must first examine the interactions before any

statement can be made about the main effects. Interaction effects are evaluated with the same criteria as main effects. If the statistical tests indicate that the interaction is nonsignificant, this denotes that the effects of the treatments are independent. Independence in factorial designs means that the effect of one treatment (i.e., group differences) is the same for each level of the other treatment(s) and that the main effects can be interpreted directly. If the interactions are deemed statistically significant, it is critical that the researcher identify the type of interaction (ordinal versus disordinal), because this has direct bearing on the conclusion that can be drawn from the results. Ordinal interaction occurs when the effects of a treatment are not equal across all levels of another treatment, but the group difference(s) is always the same direction. Disordinal interaction occurs when the differences between levels "switch" depending on how they are combined with levels from another treatment. Here the effects of one treatment are positive for some levels and negative for other levels of the other treatment.

Describe the purpose of multivariate analysis of covariance (MANCOVA).   Covariates can play an important role by including metric variables into a MANOVA or ANOVA design. However, since covariates act as a "control" measure on the dependent variate, they must be assessed before the treatments are examined. The most important role of the covariate(s) is the overall impact in the statistical tests for the treatments. The most direct approach to evaluating these impacts is to run the analysis with and without the covariates. Effective covariates will improve the statistical power of the tests and reduce within-group variance. If the researcher does not see any substantial improvement, then the covariates may be eliminated, because they reduce the degrees of freedom available for the tests of treatment effects. This approach also can identify those instances in which the covariate is "too powerful" and reduces the variance to such an extent that the treatments are all nonsignificant. Often this occurs when a covariate is included that is correlated with one of the independent variables and thus "removes" this variance, thereby reducing the explanatory power of the independent variable. Because MANCOVA and ANCOVA are applications of regression procedures within the analysis of variance method, assessing the impact of the covariates on the dependent variables is quite similar to examining regression equations. If the overall impact is deemed significant, then each covariate can be examined for the strength of the predictive relationship with the dependent measures. If the covariates represent theoretically based effects, then these results provide an objective basis for accepting or rejecting the proposed relationships. In a practical vein, the researcher can examine the impact of the covariates and eliminate those with little or no effect.

It is often unrealistic to assume that a difference between experimental treatments will be manifested only in a single measured dependent variable. Many researchers handle multiple-criterion situations by repeated application of individual univariate tests until all the dependent variables are analyzed. This approach can seriously inflate Type I error rates, and it ignores the possibility that some composite of the dependent variables may provide the strongest evidence of group differences. MANOVA can solve both of these problems.

Questions

1. What are the differences between MANOVA and discriminant analysis? What situations best suit each multivariate technique?
2. Design a two-way factorial MANOVA experiment. What are the different sources of variance in your experiment? What would a significant interaction tell you?
3. Besides the overall, or global, significance, at least three approaches to follow-up tests include (a) use of Scheffé contrast procedures; (b) stepdown analysis, which is similar to stepwise regression in that each successive F statistic is computed after eliminating the effects of the previous dependent variables; and (c) examination of the discriminant functions. Describe the practical advantages and disadvantages of each of these approaches.
4. How is statistical power affected by statistical and research design decisions? How would you design a study to ensure adequate power?
5. Describe some data analysis situations in which MANOVA and MANCOVA would be appropriate in your areas of interest. What types of uncontrolled variables or covariates might be operating in each of these situations?

Suggested Readings

A list of suggested readings illustrating issues and applications of multivariate techniques in general is available on the Web at www.pearsonglobaleditions.com/hair or www.mvstats.com.

References

1. Anderson, T. W. 2003. *Introduction to Multivariate Statistical Analysis*, 3rd ed. New York: Wiley.
2. Cattell, R. B. (ed.). 1966. *Handbook of Multivariate Experimental Psychology*. Chicago: Rand McNally.
3. Cohen, J. 1988. *Statistical Power Analysis for the Behavioral Sciences*, 2nd ed. Hillsdale, NJ: Lawrence Erlbaum Associates.
4. Cole, D. A., S. E. Maxwell, R. Avery, and E. Salas. 1994. How the Power of MANOVA Can Both Increase and Decrease as a Function of the Intercorrelations Among Dependent Variables. *Psychological Bulletin* 115: 465–74.
5. Cooley, W. W., and P. R. Lohnes. 1971. *Multivariate Data Analysis*. New York: Wiley.
6. Gill, J. 2000. *Generalized Linear Models: A Unified Approach*, Sage University Papers Series on Quantitative Applications in the Social Sciences, #07-134. Thousand Oaks, CA: Sage.
7. Green, P. E. 1978. *Analyzing Multivariate Data*. Hinsdale, IL: Holt, Rinehart and Winston.
8. Green, P. E., and J. Douglas Carroll. 1978. *Mathematical Tools for Applied Multivariate Analysis*. New York: Academic Press.
9. Green, P. E., and D. S. Tull. 1979. *Research for Marketing Decisions,* 3rd ed. Upper Saddle River, NJ: Prentice Hall.
10. Hand, D. J., and C. C. Taylor. 1987. *Multivariate Analysis of Variance and Repeated Measures*. London: Chapman and Hall.
11. Harris, R. J. 2001. *A Primer of Multivariate Statistics,* 3rd ed. Hillsdale, NJ: Lawrence Erlbaum Associates.
12. Hubert, C. J., and J. D. Morris. 1989. Multivariate Analysis Versus Multiple Univariate Analyses. *Psychological Bulletin* 105: 302–8.
13. Huitema, B. 1980. *The Analysis of Covariance and Alternatives*. New York: Wiley.
14. Hutcheson, G., and N. Sofroniou. 1999. *The Multivariate Social Scientist: Introductory Statistics Using Generalized Linear Models*. Thousand Oaks, CA: Sage.
15. Kirk, R. E. 1994. *Experimental Design: Procedures for the Behavioral Sciences,* 3rd ed. Belmont, CA: Wadsworth Publishing.
16. Koslowsky, M., and T. Caspy. 1991. Stepdown Analysis of Variance: A Refinement. *Journal of Organizational Behavior* 12: 555–59.

17. Läuter, J. 1978. Sample Size Requirements for the T^2 Test of MANOVA (Tables for One-Way Classification). *Biometrical Journal* 20: 389–406.
18. McCullagh, P., and J. A. Nelder. 1989. *Generalized Linear Models*, 2nd ed. New York: Chapman and Hall.
19. Meyers, J. L. 1979. *Fundamentals of Experimental Design.* Boston: Allyn & Bacon.
20. Morrison, D. F. 2002. *Multivariate Statistical Methods,* 4th ed. Belmont, CA: Duxbury Press.
21. Nelder, J. A., and R. W. M. Wedderburn. 1972. Generalized Linear Models. *Journal of the Royal Statistical Society, A,* 135: 370–84.
22. Rao, C. R. 1978. *Linear Statistical Inference and Its Application*, 2nd ed. New York: Wiley.
23. Stevens, J. P. 1972. Four Methods of Analyzing Between Variations for the *k*-Group MANOVA Problem. *Multivariate Behavioral Research* 7 (October): 442–54.
24. Stevens, J. P. 1980. Power of the Multivariate Analysis of Variance Tests. *Psychological Bulletin* 88: 728–37.
25. Tatsuoka, M. M. 1988. *Multivariate Analysis: Techniques for Education and Psychological Research,* 2nd ed. New York: Macmillan.
26. Wilks, S. S. 1932. Certain Generalizations in the Analysis of Variance. *Biometrika* 24: 471–94.
27. Winer, B. J., D. R. Brown, and K. M. Michels. 1991. *Statistical Principles in Experimental Design,* 3rd ed. New York: McGraw-Hill.

Analysis Using Interdependence Techniques

OVERVIEW

The dependence methods described in Section II provide the researcher with several methods for assessing relationships between one or more dependent variables and a set of independent variables. Many methods were discussed that accommodated all types (metric and nonmetric) and potentially large numbers of both dependent and independent variables that could be applied to sets of observations. What if the variables or observations are related in ways not captured by the dependence relationships? What if the assessment of interdependence (i.e., structure) is missing? One of the most basic abilities of human beings is to classify and categorize objects and information into simpler schema, such that we can characterize the objects within the groups in total rather than having to deal with each individual object. The objective of the methods in this section is to identify the structure among a defined set of variables, observations, or objects. The identification of structure offers not only simplicity, but also a means of description and even discovery.

Interdependence techniques, however, are focused solely on the definition of structure, assessing interdependence without any associated dependence relationships. None of the interdependence techniques will define structure to optimize or maximize a dependence relationship. It is the researcher's task to first utilize these methods in identifying structure and then to employ it where appropriate. The objectives of dependence relationships are not "mixed" in these interdependence methods—they assess structure for its own sake and no other.

CHAPTERS IN SECTION III

Section III is comprised of only two chapters, which cover two of the three interdependence techniques. The first interdependence technique, factor analysis (Chapter 3), was discussed in Section I on preparing for a multivariate analysis because it provides us with a tool for understanding the relationships among variables, a knowledge fundamental to all of our multivariate analyses. The issues of multicollinearity and model parsimony are reflective of the underlying structure of the variables, and factor analysis provides an objective means of assessing the groupings of variables and the ability to incorporate composite variables reflecting these variable groupings into other multivariate techniques.

It is not only variables that have structure, however. Although we assume independence among the observations and variables in our estimation of relationships, we also know that most populations have subgroups sharing general characteristics. Marketers look for target markets of differentiated groups of homogeneous consumers, strategy researchers look for groups of similar firms to identify common strategic elements, and financial modelers look for stocks with similar fundamentals to create stock portfolios. These and many other situations require techniques that find these groups of similar objects based on a set of characteristics.

This goal is met by cluster analysis, the topic of Chapter 9. Cluster analysis is ideally suited for defining groups of objects with maximal similarity within the groups while also having maximum heterogeneity between the groups—determining the most similar groups that are also most different from each other. As we show, cluster analysis has a rich tradition of application in almost every area of inquiry. Yet, its ability to define groups of similar objects is countered by its rather subjective nature and the instrumental role played by the researcher's judgment in several key decisions. This subjective aspect does not reduce the usefulness of the technique, but it does place a greater burden on the researcher to fully understand the technique and the impact of certain decisions on the ultimate cluster solution.

What if we know only how similar objects are and don't know the source of that similarity or how to best group the objects? This situation is addressed in Chapter 10, Multidimensional Scaling and Correspondence Analysis. Multidimensional scaling is a technique that starts out as a univariate analysis—a single measure of similarity among objects—and infers the dimensionality of the similarities among the objects. It attempts to answer the basic question: Can the objects be grouped in one-, two-, three-, or n-dimensional space in such a way as to adequately represent the similarities among the objects by their proximity? As such, multidimensional scaling is a form of decompositional analysis, somewhat like conjoint analysis (see Chapter 6), but in this case only their similarities are known, not the characteristics of the objects. A special form of multidimensional scaling is correspondence analysis, which analyzes a distinct form of data—cross-tabulated categorical variables. From these data, correspondence analysis is able to portray the relationships between rows and columns (e.g., products and attributes) in a dimensional perspective in which proximity represents similarity.

Cluster analysis, factor analysis, and multidimensional scaling provide the researcher with methods that bring order to the data in the form of structure among the observations or variables. In this way, the researcher can better understand the basic structures of the data, not only facilitating the description of the data, but also providing a foundation for a more refined analysis of the dependence relationships.

CHAPTER 9

Grouping Data with Cluster Analysis

LEARNING OBJECTIVES

Upon completing this chapter, you should be able to do the following:

- Define cluster analysis, its roles, and its limitations.
- Identify the types of research questions addressed by cluster analysis.
- Understand how interobject similarity is measured.
- Understand why different distance measures are sometimes used.
- Understand the differences between hierarchical and nonhierarchical clustering techniques.
- Know how to interpret results from cluster analysis.
- Follow the guidelines for cluster validation.

CHAPTER PREVIEW

Researchers often encounter situations best resolved by defining groups of homogeneous objects, whether they are individuals, firms, or even behaviors. Strategy options based on identifying groups within the population, such as segmentation and target marketing, would not be possible without an objective methodology. This same need is encountered in other areas, ranging from the physical sciences (e.g., creating a biological taxonomy for the classification of various animal groups—insects versus mammals versus reptiles) to the social sciences (e.g., analyzing various psychiatric profiles). In all instances, the researcher is searching for a "natural" structure among the observations based on a multivariate profile.

The most commonly used technique for this purpose is cluster analysis. Cluster analysis groups individuals or objects into clusters so that objects in the same cluster are more similar to one another than they are to objects in other clusters. The attempt is to maximize the homogeneity of objects within the clusters while also maximizing the heterogeneity between the clusters. This chapter explains the nature and purpose of cluster analysis and provides the researcher with an approach for obtaining and using cluster results.

KEY TERMS

Before starting the chapter, review the key terms to develop an understanding of the concepts and terminology used. Throughout the chapter the key terms appear in **boldface.** Other points of emphasis in the chapter and key term cross-references are *italicized.*

Absolute Euclidean distance   See *squared Euclidean distance.*

Agglomerative methods   *Hierarchical procedure* that begins with each *object* or observation in a separate cluster. In each subsequent step, the two clusters that are most similar are combined to build a new aggregate cluster. The process is repeated until all objects are finally combined into a single cluster. This process is the opposite of the *divisive method.*

Average linkage   *Hierarchical* clustering *algorithm* that represents *similarity* as the average distance from all objects in one cluster to all objects in another. This approach tends to combine clusters with small variances.

Centroid method   *Hierarchical* clustering *algorithm* in which *similarity* between clusters is measured as the distance between *cluster centroids.* When two clusters are combined, a new centroid is computed. Thus, cluster centroids migrate, or move, as the clusters are combined.

City-block distance   Method of calculating distances based on the sum of the absolute differences of the coordinates for the *objects.* This method assumes that the variables in the *cluster variate* are uncorrelated and that unit scales are compatible.

Cluster centroid   Average value of the objects contained in the cluster on all the variables in the *cluster variate.*

Cluster seed   Initial value or starting point for a cluster. These values are selected to initiate *nonhierarchical* clustering *procedures,* in which clusters are built around these prespecified points.

Cluster solution   A specific number of clusters selected as representative of the data structure of the sample of *objects.*

Cluster variate   Set of variables or characteristics representing the *objects* to be clustered and used to calculate the *similarity* between objects.

Clustering algorithm   Set of rules or procedures; similar to an equation.

Complete-linkage method   *Hierarchical* clustering *algorithm* in which *interobject similarity* is based on the maximum distance between *objects* in two clusters (the distance between the most dissimilar members of each cluster). At each stage of the *agglomeration,* the two clusters with the smallest maximum distance (most similar) are combined.

Cubic clustering criterion (CCC)   A direct measure of *heterogeneity* in which the highest CCC values indicate the final *cluster solution.*

Dendrogram   Graphical representation (tree graph) of the results of a *hierarchical procedure* in which each *object* is arrayed on one axis, and the other axis portrays the steps in the *hierarchical procedure.* Starting with each object represented as a separate cluster, the dendrogram shows graphically how the clusters are combined at each step of the procedure until all are contained in a single cluster.

Diameter method   See *complete-linkage method.*

Divisive method   *Hierarchical* clustering *algorithm* that begins with all *objects* in a single cluster, which is then divided at each step into two additional clusters that contain the most dissimilar objects. The single cluster is divided into two clusters, then one of these two clusters is split for a total of three clusters. This continues until all observations are in single-member clusters. This method is the opposite of the *agglomerative method.*

Entropy group   Group of *objects* independent of any cluster (i.e., they do not fit into any cluster) that may be considered outliers and possibly eliminated from the cluster analysis.

Euclidean distance   Most commonly used measure of the *similarity* between two *objects.* Essentially, it is a measure of the length of a straight line drawn between two objects when represented graphically.

Farthest-neighbor method   See *complete-linkage method.*

Heterogeneity   A measure of diversity of all observations across all clusters that is used as a general element in *stopping rules.* A large increase in heterogeneity when two clusters are combined indicates that a more natural structure exists when the two clusters are separate.

Hierarchical procedures   Stepwise clustering procedures involving a combination (or division) of the objects into clusters. The two alternative procedures are the *agglomerative* and *divisive methods*. The result is the construction of a hierarchy, or treelike structure (*dendrogram*), depicting the formation of the clusters. Such a procedure produces $N - 1$ cluster solutions, where N is the number of objects. For example, if the agglomerative procedure starts with five objects in separate clusters, it will show how four clusters, then three, then two, and finally one cluster are formed.

Interobject similarity   The correspondence or association of two *objects* based on the variables of the *cluster variate.* Similarity can be measured in two ways. First is a measure of association, with higher positive correlation coefficients representing greater similarity. Second, proximity, or closeness, between each pair of objects can assess similarity. When measures of distance or difference are used, smaller distances or differences represent greater similarity.

K-means   A group of *nonhierarchical clustering algorithms* that work by partitioning observations into a user-specified number of clusters and then iteratively reassigning observations until some numeric goal related to cluster distinctiveness is met.

Mahalanobis distance (D^2)   Standardized form of *Euclidean distance.* Scaling responses in terms of standard deviations standardizes the data with adjustments made for correlations between the variables.

Manhattan distance   See *city-block distance.*

Nearest-neighbor method   See *single-linkage method.*

Nonhierarchical procedures   Procedures that produce only a single cluster solution for a set of *cluster seeds* and a given number of clusters. Instead of using the treelike construction process found in the *hierarchical procedures*, cluster seeds are used to group objects within a prespecified distance of the seeds. Nonhierarchical procedures do not produce results for all possible numbers of clusters as is done with a hierarchical procedure.

Object   Person, product or service, firm, or any other entity that can be evaluated on a number of attributes.

Optimizing procedure   *Nonhierarchical clustering* procedure that allows for the reassignment of *objects* from the originally assigned cluster to another cluster on the basis of an overall optimizing criterion.

Profile diagram   Graphical representation of data that aids in screening for outliers or the interpretation of the final cluster solution. Typically, the variables of the cluster variate or those used for validation are listed along the horizontal axis, and the scale is the vertical axis. Separate lines depict the scores (original or standardized) for individual objects or cluster centroids in a graphic plane.

Response-style effect   Series of systematic responses by a respondent that reflect a bias or consistent pattern. Examples include responding that an object always performs excellently or poorly across all attributes with little or no variation.

Root mean square standard deviation (RMSSTD)   The square root of the variance of the new cluster formed by joining the two clusters across the *cluster variate.* Large increases indicate that the two clusters represent a more natural data structure than when joined.

Row-centering standardization   See *within-case standardization.*

Similarity   See *interobject similarity.*

Single-linkage method   *Hierarchical clustering algorithm* in which *similarity* is defined as the minimum distance between any single *object* in one cluster and any single object in another, which simply means the distance between the closest objects in two clusters. This procedure has the potential for creating less compact, or even chainlike, clusters. It differs from the *complete-linkage method,* which uses the maximum distance between objects in the cluster.

Squared Euclidean distance   Measure of *similarity* that represents the sum of the squared distances without taking the square root (as done to calculate *Euclidean distance*).

Stopping rule   *Clustering algorithm* for determining the final number of clusters to be formed. With no stopping rule inherent in cluster analysis, researchers developed several criteria and guidelines for this determination. Two classes of rules that are applied post hoc and calculated by the researcher are (1) measures of similarity and (2) adapted statistical measures.

Taxonomy   Empirically derived classification of actual *objects* based on one or more characteristics, as typified by the application of cluster analysis or other grouping procedures. This classification can be contrasted to a *typology*.

Typology   Conceptually based classification of objects based on one or more characteristics. A typology does not usually attempt to group actual observations, but instead provides the theoretical foundation for the creation of a *taxonomy,* which groups actual observations.

Ward's method   *Hierarchical clustering algorithm* in which the similarity used to join clusters is calculated as the sum of squares between the two clusters summed over all variables. This method has the tendency to result in clusters of approximately equal size due to its minimization of within-group variation.

Within-case standardization   Method of standardization in which a respondent's responses are not compared to the overall sample but instead to the respondent's own responses. In this process, also known as ipsitizing, the respondents' average responses are used to standardize their own responses.

WHAT IS CLUSTER ANALYSIS?

Cluster analysis is a group of multivariate techniques whose primary purpose is to group objects based on the characteristics they possess. It has been referred to as Q analysis, typology construction, classification analysis, and numerical taxonomy. This variety of names is due to the usage of clustering methods in such diverse disciplines as psychology, biology, sociology, economics, engineering, and business. Although the names differ across disciplines, the methods all have a common dimension: classification according to relationships among the objects being clustered [1, 2, 4, 10, 22, 27]. This common dimension represents the essence of all clustering approaches—the classification of data as suggested by natural groupings of the data themselves. Cluster analysis is comparable to factor analysis (see Chapter 3) in its objective of assessing structure. Cluster analysis differs from factor analysis, however, in that cluster analysis groups objects, whereas factor analysis is primarily concerned with grouping variables. Additionally, factor analysis makes the groupings based on patterns of variation (correlation) in the data whereas cluster analysis makes groupings on the basis of distance (proximity).

Cluster Analysis as a Multivariate Technique

Cluster analysis classifies **objects** (e.g., respondents, products, or other entities), on a set of user selected characteristics. The resulting clusters should exhibit high internal (within-cluster) homogeneity and high external (between-cluster) heterogeneity. Thus, if the classification is successful, the objects within clusters will be close together when plotted geometrically, and different clusters will be far apart.

The concept of the variate is again important in understanding how cluster analysis mathematically produces results. The **cluster variate** represents a mathematical representation of the selected set of variables which compares the objects' similarities.

The variate in cluster analysis is determined quite differently from other multivariate techniques. Cluster analysis is the only multivariate technique that does not estimate the variate empirically but instead uses the variate as specified by the researcher. The focus of cluster analysis is on the comparison of objects based on the variate, not on the estimation of the variate itself. This distinction makes the researcher's definition of the variate a critical step in cluster analysis.

Conceptual Development with Cluster Analysis

Cluster analysis has been used in every research setting imaginable. Ranging from the derivation of taxonomies in biology for grouping all living organisms, to psychological classifications based on personality and other personal traits, to segmentation analyses of markets, cluster analysis applications have focused largely on grouping individuals. However, cluster analysis can classify objects other than individual people, including the market structure, analyses of the similarities and differences among new products, and performance evaluations of firms to identify groupings based on the firms' strategies or strategic orientations.

In many instances, however, the grouping is actually a means to an end in terms of a conceptually defined goal. The more common roles cluster analysis can play in conceptual development include the following:

- *Data reduction:* A researcher may be faced with a large number of observations that are meaningless unless classified into manageable groups. Cluster analysis can perform this data reduction procedure objectively by reducing the information from an entire population or sample to information about specific groups.

    For example, if we can understand the attitudes of a population by identifying the major groups within the population, then we have reduced the data for the entire population into profiles of a number of groups. In this fashion, the researcher provides a more concise, understandable description of the observations, with minimal loss of information.

- *Hypothesis generation:* Cluster analysis is also useful when a researcher wishes to develop hypotheses concerning the nature of the data or to examine previously stated hypotheses.

    For example, a researcher may believe that attitudes toward the consumption of diet versus regular soft drinks could be used to separate soft-drink consumers into logical segments or groups. Cluster analysis can classify soft-drink consumers by their attitudes about diet versus regular soft drinks, and the resulting clusters, if any, can be profiled for demographic similarities and differences.

The large number of applications of cluster analysis in almost every area of inquiry creates not only a wealth of knowledge on its use, but also the need for a better understanding of the technique to minimize its misuse.

Necessity of Conceptual Support in Cluster Analysis

Believe it or not, cluster analysis can be criticized for working too well in the sense that statistical results are produced even when a logical basis for clusters is not apparent. Thus, the researcher should have a strong conceptual basis to deal with issues such as why groups exist in the first place and what variables logically explain why objects end up in the groups that they do. Even if cluster analysis is being used in conceptual development as just mentioned, some conceptual rationale is essential. The following are the most common criticisms that must be addressed by conceptual rather than empirical support:

- *Cluster analysis is descriptive, atheoretical, and noninferential.* Cluster analysis has no statistical basis upon which to draw inferences from a sample to a population, and many contend that it is only an exploratory technique. Nothing guarantees unique solutions, because the cluster membership for any number of solutions is dependent upon many elements of the procedure, and many different solutions can be obtained by varying one or more elements.
- *Cluster analysis will always create clusters, regardless of the actual existence of any structure in the data.* When using cluster analysis, the researcher is making an assumption of some structure among the objects. The researcher should always remember that just because clusters can be found does not validate their existence. Only with strong conceptual support and then validation are the clusters potentially meaningful and relevant.
- *The cluster solution is not generalizable because it is totally dependent upon the variables used as the basis for the similarity measure.* This criticism can be made against any statistical technique, but cluster analysis is generally considered more dependent on the measures used to characterize the objects than other multivariate techniques. With the cluster variate

completely specified by the researcher, the addition of spurious variables or the deletion of relevant variables can have a substantial impact on the resulting solution. As a result, the researcher must be especially cognizant of the variables used in the analysis, ensuring that they have strong conceptual support.

Thus, in any use of cluster analysis the researcher must take particular care in ensuring that strong conceptual support predates the application of the technique. Only with this support in place should the researcher then address each of the specific decisions involved in performing a cluster analysis.

HOW DOES CLUSTER ANALYSIS WORK?

Cluster analysis performs a task innate to all individuals—pattern recognition and grouping. The human ability to process even slight differences in innumerable characteristics is a cognitive process inherent in human beings that is not easily matched with all of our technological advances. Take for example the task of analyzing and grouping human faces. Even from birth, individuals can quickly identify slight differences in facial expressions and group different faces in homogeneous groups while considering hundreds of facial characteristics. Yet we still struggle with facial recognition programs to accomplish the same task. The process of identifying natural groupings is one that can become quite complex rather quickly.

To demonstrate how cluster analysis operates, we examine a simple example that illustrates some of the key issues: measuring similarity, forming clusters, and deciding on the number of clusters that best represent structure. We also briefly discuss the balance of objective and subjective considerations that must be addressed by any researcher.

A Simple Example

The nature of cluster analysis and the basic decisions on the part of the researcher will be illustrated by a simple example involving identification of customer segments in a retail setting.

Suppose a marketing researcher wishes to determine market segments in a community based on patterns of loyalty to brands and stores. A small sample of seven respondents is selected as a pilot test of how cluster analysis is applied. Two measures of loyalty—V_1 (store loyalty) and V_2 (brand loyalty)—were measured for each respondent on a 0–10 scale. The values for each of the seven respondents are shown in Figure 9-1, along with a scatter diagram depicting each observation on the two variables.

The primary objective of cluster analysis is to define the structure of the data by placing the most similar observations into groups. To accomplish this task, we must address three basic questions:

1. *How do we measure similarity?* We require a method of simultaneously comparing observations on the two clustering variables (V_1 and V_2). Several methods are possible, including the correlation between objects or perhaps a measure of their proximity in two-dimensional space such that the distance between observations indicates similarity.
2. *How do we form clusters?* No matter how similarity is measured, the procedure must group those observations that are most similar into a cluster, thereby determining the cluster group membership of each observation for each set of clusters formed.
3. *How many groups do we form?* The final task is to select one set of clusters as the final solution. In doing so, the researcher faces a trade-off: fewer clusters and less homogeneity within clusters versus a larger number of clusters and more within-group homogeneity. Simple structure, in striving toward parsimony, is reflected in as few clusters as possible. Yet as the number of clusters decreases, the heterogeneity within the clusters necessarily increases. Thus, a balance must be made between defining the most basic structure (fewer clusters) that still achieves an acceptable level of heterogeneity between the clusters.

Once we have procedures for addressing each of these issues, we can perform a cluster analysis. We will illustrate the principles underlying each of these issues through our simple example.

Data Values

Clustering Variable	Respondents						
	A	B	C	D	E	F	G
V_1	3	4	4	2	6	7	6
V_2	2	5	7	7	6	7	4

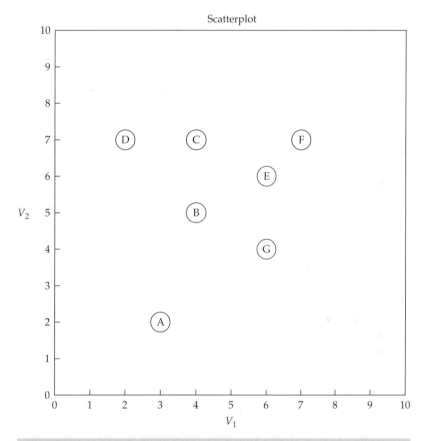

FIGURE 9-1 Data Values and Scatterplot of the Seven Observations Based on the Two Clustering Variables (V_1 and V_2)

MEASURING SIMILARITY   The first task is developing some measure of similarity between each object to be used in the clustering process. **Similarity** represents the degree of correspondence among objects across all of the characteristics used in the analysis. In a way, similarity measures are more descriptively dissimilarity measures in that smaller numbers represent greater similarity and larger numbers represent less similarity.

   Similarity must be determined between each of the seven observations (respondents A–G) to enable each observation to be compared to each other. In this example, similarity will be measured according to the Euclidean (straight-line) distance between each pair of observations (see Table 9-1) based on the two characteristics (V_1 and V_2). In this two-dimensional case (where each characteristic forms one axis of the graph) we can view distance as the proximity of each point to the others. In using distance as the measure of proximity, we must remember that smaller distances indicate greater similarity, such that observations E and F are the most similar (1.414), and A and F are the most dissimilar (6.403).

TABLE 9-1 Proximity Matrix of Euclidean Distances Between Observations

Observation	Observation						
	A	B	C	D	E	F	G
A	—						
B	3.162	—					
C	5.099	2.000	—				
D	5.099	2.828	2.000	—			
E	5.000	2.236	2.236	4.123	—		
F	6.403	3.606	3.000	5.000	1.414	—	
G	3.606	2.236	3.606	5.000	2.000	3.162	—

FORMING CLUSTERS   With similarity measures calculated, we now move to forming clusters based on the similarity measure of each observation. Typically we form a number of cluster solutions (a two-cluster solution, a three-cluster solution, etc.). Once clusters are formed, we then select the final cluster solution from the set of possible solutions. First we will discuss how clusters are formed and then examine the process for selecting a final cluster solution.

Having calculated the similarity measure, we must develop a procedure for forming clusters. As shown later in this chapter, many methods have been proposed, but for our purposes here, we use this simple rule:

> Identify the two most similar (closest) observations not already in the same cluster and combine them.

We apply this rule repeatedly to generate a number of cluster solutions, starting with each observation as its own "cluster" and then combining two clusters at a time until all observations are in a single cluster. This process is termed a **hierarchical procedure** because it moves in a stepwise fashion to form an entire range of cluster solutions. It is also an **agglomerative method** because clusters are formed by combining existing clusters.

Table 9-2 details the steps of the hierarchical agglomerative process, first depicting the initial state with all seven observations in single-member clusters, joining them in an agglomerative process until only one cluster remains. The six-step clustering process is described here:

Step 1: Identify the two closest observations (E and F) and combine them into a cluster, moving from seven to six clusters.

Step 2: Find the next closest pairs of observations. In this case, three pairs have the same distance of 2.000 (E-G, C-D, and B-C). For our purposes, choose the observations E-G. G is a single-member cluster, but E was combined in the prior step with F. So, the cluster formed at this stage now has three members: G, E, and F.

Step 3: Combine the single-member clusters of C and D so that we now have four clusters.

Step 4: Combine B with the two-member cluster C-D that was formed in step 3. At this point, we now have three clusters: cluster 1 (A), cluster 2 (B, C, and D), and cluster 3 (E, F, and G).

Step 5: Combine the two three-member clusters into a single six-member cluster. The next smallest distance is 2.236 for three pairs of observations (E-B, B-G, and C-E). We use only one of these distances, however, as each observation pair contains a member from each of the two existing clusters (B, C, and D versus E, F, and G).

Step 6: Combine observation A with the remaining cluster (six observations) into a single cluster at a distance of 3.162. You will note that distances smaller or equal to 3.162 are not used because they are between members of the same cluster.

TABLE 9-2 **Agglomerative Hierarchical Clustering Process**

| | AGGLOMERATION PROCESS | | CLUSTER SOLUTION | | |
Step	Minimum Distance Between Unclustered Observations[a]	Observation Pair	Cluster Membership	Number of Clusters	Overall Similarity Measure (Average Within-Cluster Distance)
	Initial Solution		**(A) (B) (C) (D) (E) (F) (G)**	**7**	**0**
1	1.414	E-F	(A) (B) (C) (D) (E-F) (G)	6	1.414
2	2.000	E-G	(A) (B) (C) (D) (E-F-G)	5	2.192
3	2.000	C-D	(A) (B) (C-D) (E-F-G)	4	2.144
4	2.000	B-C	(A) (B-C-D) (E-F-G)	3	2.234
5	2.236	B-E	(A) (B-C-D-E-F-G)	2	2.896
6	3.162	A-B	(A-B-C-D-E-F-G)	1	3.420

[a]Euclidean distance between observations.

The hierarchical clustering process can be portrayed graphically in several ways. Figure 9-2 illustrates two such methods. First, because the process is hierarchical, the clustering process can be shown as a series of nested groupings (see Figure 9-2a). This process, however, can represent the proximity of the observations for only two or three clustering variables in the scatterplot or three-dimensional graph. A more common approach is a dendrogram, which represents the clustering process in a treelike graph. The horizontal axis represents the agglomeration coefficient, in this instance the distance used in joining clusters. This approach is particularly useful in identifying outliers, such as observation A. It also depicts the relative size of varying clusters, although it becomes unwieldy when the number of observations increases.

DETERMINING THE NUMBER OF CLUSTERS IN THE FINAL SOLUTION   A hierarchical method results in a number of cluster solutions—in this case starting with a seven-cluster solution and ending in a one-cluster solution. Which solution do we choose? We know that as we move from single-member clusters in the seven-cluster solution, heterogeneity increases. So why not stay at seven clusters, the most homogeneous possible? If all observations are treated as their own unique cluster, no data reduction has taken place and no true segments have been found. The goal is identifying segments by combining observations, but at the same time introducing only small amounts of hetereogeneity.

Measuring Heterogeneity. Any measure of **heterogeneity** of a cluster solution should represent the overall diversity among observations in all clusters. In the initial solution of an agglomerative approach where all observations are in separate clusters, no heterogeneity exists. As observations are combined to form clusters, heterogeneity increases. The measure of heterogeneity thus should start with a value of zero and increase to show the level of heterogeneity as clusters are combined.

In this example, we use a simple measure of heterogeneity: the average of all distances between observations within clusters (see Table 9-2). As already described, the measure should increase as clusters are combined:

- In the initial solution with seven clusters, our overall similarity measure is 0—no observation is paired with another.
- Six clusters: The overall similarity is the distance between the two observations (1.414) joined in step 1.
- Five clusters: Step 2 forms a three-member cluster (E, F, and G), so that the overall similarity measure is the mean of the distances between E and F (1.414), E and G (2.000), and F and G (3.162), for an average of 2.192.

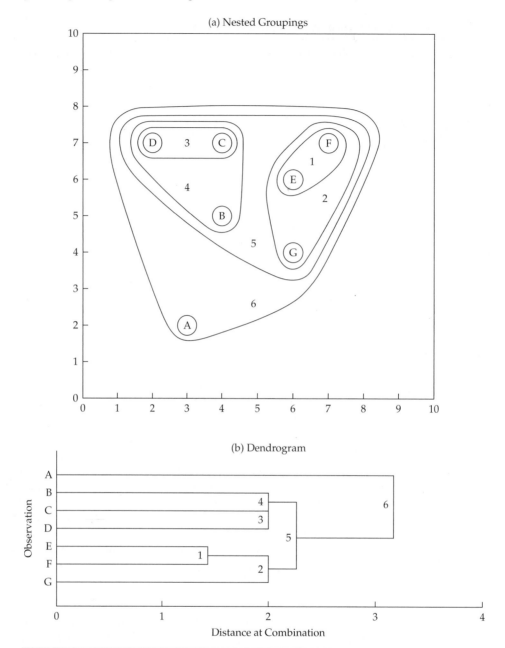

FIGURE 9-2   Graphical Portrayals of the Hierarchical Clustering Process

- Four clusters: In the next step a new two-member cluster is formed with a distance of 2.000, which causes the overall average to fall slightly to 2.144.
- Three, two, and one clusters: The final three steps form new clusters in this manner until a single-cluster solution is formed (step 6), in which the average of all distances in the distance matrix is 3.420.

Selecting a Final Cluster Solution. Now, how do we use this overall measure of similarity to select a **cluster solution?** Remember that we are trying to get the simplest structure possible that still represents homogeneous groupings. If we monitor the heterogeneity measure as the number of clusters decreases, large increases in heterogeneity indicate that two rather dissimilar clusters were joined at that stage.

From Table 9-2, we can see that the overall measure of heterogeneity increases as we combine clusters until we reach the final one-cluster solution. To select a final cluster solution, we examine the changes in the homogeneity measure to identify large increases indicative of merging dissimilar clusters:

- When we first join two observations (step 1) and then again when we make our first three-member cluster (step 2), we see fairly large increases.
- In the next two steps (3 and 4), the overall measure does not change substantially, which indicates that we are forming other clusters with essentially the same heterogeneity of the existing clusters.
- When we get to step 5, which combines the two three-member clusters, we see a large increase. This change indicates that joining these two clusters resulted in a single cluster that was markedly less homogeneous. As a result, we would consider the three-cluster solution of step 4 much better than the two-cluster solution found in step 5.
- We can also see that in step 6 the overall measure again increased markedly, indicating when this single observation was joined at the last step, it substantially changed the cluster homogeneity. Given the rather unique profile of this observation (observation A) compared to the others, it might best be designated as a member of the **entropy group,** those observations that are outliers and independent of the existing clusters.

Thus, when reviewing the range of cluster solutions, the three-cluster solution of step 4 seems the most appropriate for a final cluster solution, with two equally sized clusters and the single outlying observation.

Objective Versus Subjective Considerations

As is probably clear by now, the selection of the final cluster solution requires substantial researcher judgment and is considered by many as too subjective. Even though sophisticated methods have been developed to assist in evaluating the cluster solutions, it still falls to the researcher to make the final decision as to the number of clusters to accept as the final solution. Moreover, decisions on the characteristics to be used, the methods of combining clusters, and even the interpretation of cluster solutions rely as much on the judgment of the researcher as any empirical test.

Even this rather simple example of only two characteristics and seven observations demonstrates the potential complexity of performing a cluster analysis. Researchers in realistic settings are faced with analyses containing many more characteristics with many more observations.

It is thus imperative researchers employ whatever objective support is available and be guided by reasoned judgment, especially in the design and interpretation stages.

CLUSTER ANALYSIS DECISION PROCESS

Cluster analysis, like the other multivariate techniques discussed earlier, can be viewed from the six-stage model-building approach introduced in Chapter 1 (see Figure 9-3 for stages 1–3 and Figure 9-6 for stages 4–6). Starting with research objectives that can be either exploratory or confirmatory, the design of a cluster analysis deals with the following:

- Partitioning the data set to form clusters and selecting a cluster solution
- Interpreting the clusters to understand the characteristics of each cluster and develop a name or label that appropriately defines its nature
- Validating the results of the final cluster solution (i.e., determining its stability and generalizability), along with describing the characteristics of each cluster to explain how they may differ on relevant dimensions such as demographics

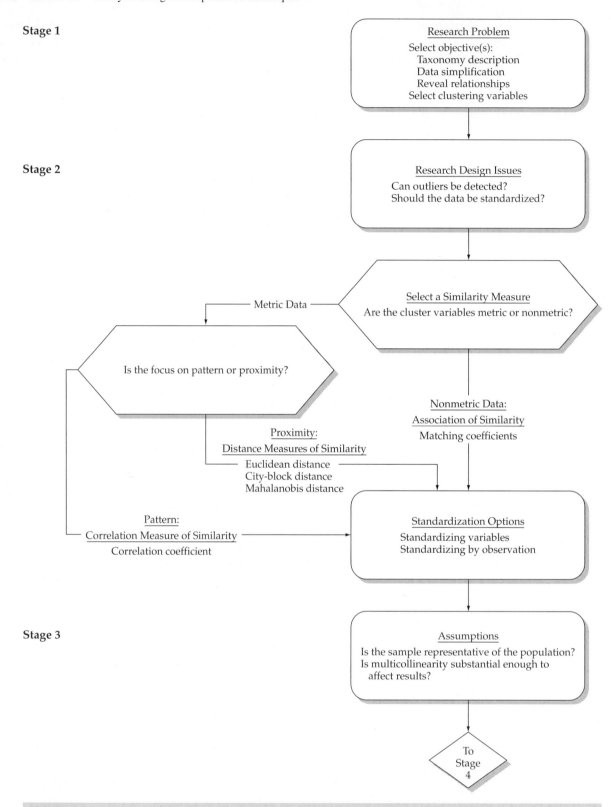

FIGURE 9-3   **Stages 1–3 of the Cluster Analysis Decision Diagram**

The following sections detail all these issues through the six stages of the model-building process.

Stage 1: Objectives of Cluster Analysis

The primary goal of cluster analysis is to partition a set of objects into two or more groups based on the similarity of the objects for a set of specified characteristics (cluster variate). In fulfilling this basic objective, the researcher must address two key issues: the research questions being addressed in this analysis and the variables used to characterize objects in the clustering process. We will discuss each issue in the following section.

RESEARCH QUESTIONS IN CLUSTER ANALYSIS   In forming homogeneous groups, cluster analysis may address any combination of three basic research questions:

1. *Taxonomy description.* The most traditional use of cluster analysis has been for exploratory purposes and the formation of a **taxonomy**—an empirically based classification of objects. As described earlier, cluster analysis has been used in a wide range of applications for its partitioning ability. Cluster analysis can also generate hypotheses related to the structure of the objects. Finally, although viewed principally as an exploratory technique, cluster analysis can be used for confirmatory purposes. In such cases, a proposed **typology** (theoretically based classification) can be compared to that derived from the cluster analysis.

2. *Data simplification.* By defining structure among the observations, cluster analysis also develops a simplified perspective by grouping observations for further analysis. Whereas factor analysis attempts to provide dimensions or structure to variables (see Chapter 3), cluster analysis performs the same task for observations. Thus, instead of viewing all of the observations as unique, they can be viewed as members of clusters and profiled by their general characteristics.

3. *Relationship identification.* With the clusters defined and the underlying structure of the data represented in the clusters, the researcher has a means of revealing relationships among the observations that typically is not possible with the individual observations. Whether analyses such as discriminant analysis are used to empirically identify relationships, or the groups are examined by more qualitative methods, the simplified structure from cluster analysis often identifies relationships or similarities and differences not previously revealed.

SELECTION OF CLUSTERING VARIABLES   The objectives of cluster analysis cannot be separated from the selection of variables used to characterize the objects being clustered. Whether the objective is exploratory or confirmatory, the researcher effectively constrains the possible results by the variables selected for use. The derived clusters reflect the inherent structure of the data and are defined only by the variables. Thus, selecting the variables to be included in the cluster variate must be done with regard to theoretical and conceptual as well as practical considerations.

Conceptual Considerations.   Any application of cluster analysis must have some rationale upon which variables are selected. Whether the rationale is based on an explicit theory, past research, or supposition, the researcher must realize the importance of including only those variables that (1) characterize the objects being clustered and (2) relate specifically to the objectives of the cluster analysis. The cluster analysis technique has no means of differentiating relevant from irrelevant variables and derives the most consistent, yet distinct, groups of objects across all variables. Thus, one should never include variables indiscriminately. Instead, carefully choose the variables with the research objective as the criterion for selection.

Let's use the HBAT data set to provide an example of how to select the appropriate variables for a cluster analysis. First, variables X_1 to X_5 are nonmetric data warehouse classification variables. Thus, they are not appropriate for cluster analysis. Next, let us consider variables X_6 to X_{18}.

These 13 variables are appropriate because they all have a common foundation—they relate to customer's perceptions of the performance of HBAT and they are measured metrically. If we used these perceptions variables for a cluster analysis, the objective would be to see if there are groups of HBAT customers that exhibit distinctively different perceptions of the performance of HBAT between the groups, but similar perceptions within each of the groups. Finally, we need to consider variables X_{19} to X_{23}. These variables would not be considered part of the perceptions cluster variables because they are distinct from variables X_6 to X_{18}. We might consider X_{19} to X_{21} for clustering because they all relate to the construct of customer commitment or loyalty. But, they would be considered by themselves for a cluster solution different from the perceptions variables.

Practical Considerations. Cluster analysis can be affected dramatically by the inclusion of only one or two inappropriate or undifferentiated variables [17]. The researcher is always encouraged to examine the results and to eliminate the variables that are not distinctive (i.e., that do not differ significantly) across the derived clusters. This procedure enables the cluster techniques to maximally define clusters based only on those variables exhibiting differences across the objects.

Stage 2: Research Design in Cluster Analysis

With the objectives defined and variables selected, the researcher must address four questions before starting the partitioning process:

1. Is the sample size adequate?
2. Can outliers be detected and, if so, should they be deleted?
3. How should object similarity be measured?
4. Should the data be standardized?

Many different approaches can be used to answer these questions. However, none of them has been evaluated sufficiently to provide a definitive answer to any of these questions, and unfortunately, many of the approaches provide different results for the same data set. Thus, cluster analysis, along with factor analysis, is as much an art as a science. For this reason, our discussion reviews these issues by providing examples of the most commonly used approaches and an assessment of the practical limitations where possible.

The importance of these issues and the decisions made in later stages becomes apparent when we realize that although cluster analysis is seeking structure in the data, it must actually impose a structure through a selected methodology. Cluster analysis cannot evaluate all the possible partitions because even the relatively small problem of partitioning 25 objects into five nonoverlapping

---

RULES OF THUMB 9-1

Objectives of Cluster Analysis

- Cluster analysis is used for:
  - Taxonomy description: Identifying natural groups within the data
  - Data simplification: The ability to analyze groups of similar observations instead of all individual observations
  - Relationship identification: The simplified structure from cluster analysis portrays relationships not revealed otherwise
- Theoretical, conceptual, and practical considerations must be observed when selecting clustering variables for cluster analysis:
  - Only variables that relate specifically to objectives of the cluster analysis are included
  - Variables selected characterize the individuals (objects) being clustered

clusters involves 2.431×10^{15} possible partitions [2]. Instead, based on the decisions of the researcher, the technique identifies a small subset of possible solutions as "correct." From this viewpoint, the research design issues and the choice of methodologies made by the researcher perhaps have greater impact than with any other multivariate technique.

SAMPLE SIZE   The issue of sample size in cluster analysis does not relate to any statistical inference issues (i.e., statistical power). Instead the sample size must be large enough to provide sufficient representation of small groups within the population and represent the underlying structure. This issue of representation becomes critical in detecting outliers (see next section), with the primary question being: When an outlier is detected, is it a representative of a small but substantive group? Small groups will naturally appear as small numbers of observations, particularly when the sample size is small. For example, when a sample contains only 100 or fewer observations, groups that actually make up 10 percent of the population may be represented by only one or two observations due to the sampling process. In such instances the distinction between outliers and representatives of a small group is much harder to make. Larger samples increase the chance that small groups will be represented by enough cases to make their presence more easily identified.

As a result, the researcher should ensure the sample size is sufficiently large to adequately represent all of the relevant groups of the population. In determining the sample size, the researcher should specify the group sizes necessary for relevance for the research questions being asked. Obviously, if the analysis objectives require identification of small groups within the population, the researcher should strive for larger sample sizes. If the researcher is interested only in larger groups (e.g., major segments for promotional campaigns), however, then the distinction between an outlier and a representative of a small group is less important and they can both be handled in a similar manner.

New programs have also been developed for applications using large sample sizes approaching 1,000 observations or more. SPSS includes a two-step cluster program that has the ability to quickly determine an appropriate number of groups and then classify them using a nonhierarchical routine. This procedure is relatively new, but it may prove useful in applications with large samples where traditional clustering methods are inefficient.

DETECTING OUTLIERS   In its search for structure, we have already discussed how cluster analysis is sensitive to the inclusion of irrelevant variables. But cluster analysis is also sensitive to outliers (objects different from all others). Outliers can represent either:

- Truly aberrant observations that are not representative of the general population
- Representative observations of small or insignificant segments within the population
- An undersampling of actual group(s) in the population that causes poor representation of the group(s) in the sample

In the first case, the outliers distort the actual structure and make the derived clusters unrepresentative of the actual population structure. In the second case, the outlier is removed so that the resulting clusters more accurately represent the relevant segments in the populations. However, in the third case the outliers should be included in the cluster solutions, even if they are underrepresented in the sample, because they represent valid and relevant groups. For this reason, a preliminary screening for outliers is always necessary.

Graphical Approaches.   One of the simplest ways to screen data for outliers is to prepare a graphic **profile diagram,** listing the variables along the horizontal axis and the variable values along the vertical axis. Each point on the graph represents the value of the corresponding variable, and the points are connected to facilitate visual interpretation. Profiles for all objects are then plotted on the graph, a line for each object. Outliers are those respondents that have very different profiles from the more typical respondents. An example of a graphic profile diagram is shown in Figure 9-4.

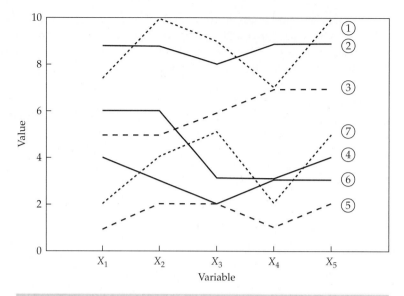

FIGURE 9-4 Profile Diagram

Empirical Approaches. Although quite simple, the graphical procedures become cumbersome with a large number of objects and even more difficult as the number of variables increases. Moreover, detecting outliers must extend beyond a univariate approach, because outliers also may be defined in a multivariate sense as having unique profiles across an entire set of variables that distinguish them from all of the other observations. As a result, an empirical measure is needed to facilitate comparisons across objects. For these instances, the procedures for identifying outliers discussed in Chapter 2 can be applied. The combination of bivariate and multivariate approaches provides a comprehensive set of tools for identifying outliers from many perspectives.

Another approach is to identify outliers through the measures of similarity. The most obvious examples of outliers are single observations that are the most dissimilar to the other observations. Before the analysis, the similarities of all observations can be compared to the overall group centroid (typical respondent). Isolated observations showing great dissimilarity can be dropped. Clustering patterns can also be observed once the cluster program has been run. However, as the number of objects to cluster increases, multiple iterations are needed to identify outliers. Moreover, some of the clustering approaches are quite sensitive to removing just a few cases [14]. Thus, emphasis should be placed on identifying outliers before the analysis begins.

MEASURING SIMILARITY The concept of similarity is fundamental to cluster analysis. **Interobject similarity** is an empirical measure of correspondence, or resemblance, between objects to be clustered. Comparing the two interdependence techniques (factor analysis and cluster analysis) will demonstrate how similarity works to define structure in both instances.

- In our discussion of factor analysis, the correlation matrix between all pairs of variables was used to group variables into factors. The correlation coefficient represented the similarity of each variable to another variable when viewed across all observations. Thus, factor analysis grouped together variables that had high correlations among themselves.
- A comparable process occurs in cluster analysis. Here, the similarity measure is calculated for all pairs of objects, with similarity based on the profile of each observation across the characteristics specified by the researcher. In this way, any object can be compared to any other object through the similarity measure, just as we used correlations between variables in factor analysis. The cluster analysis procedure then proceeds to group similar objects together into clusters.

Interobject similarity can be measured in a variety of ways, but three methods dominate the applications of cluster analysis: correlational measures, distance measures, and association measures. Both the correlational and distance measures require metric data, whereas the association measures are for nonmetric data.

Correlational Measures. The interobject measure of similarity that probably comes to mind first is the correlation coefficient between a pair of objects measured on several variables. In effect, instead of correlating two sets of variables, we invert the data matrix so that the columns represent the objects and the rows represent the variables. Thus, the correlation coefficient between the two columns of numbers is the correlation (or similarity) between the profiles of the two objects. High correlations indicate similarity (the correspondence of patterns across the characteristics) and low correlations denote a lack of it. This procedure is also followed in the application of Q-type factor analysis (see Chapter 3).

The correlation approach is illustrated by using the example of seven observations shown in Figure 9-4. A correlational measure of similarity does not look at the observed mean value, or magnitude, but instead at the patterns of movement seen as one traces the data for each case over the variables measured; in other words, the similarity in the profiles for each case. In Table 9-3, which contains the correlations among these seven observations, we can see two distinct groups. First, cases 1, 5, and 7 all have similar patterns and corresponding high positive correlations. Likewise, cases 2, 4, and 6 also have high positive correlations among themselves but low or negative correlations with the other observations. Case 3 has low or negative correlations with all other cases, thereby perhaps forming a group by itself.

Correlations represent patterns across the variables rather than the magnitudes, which are comparable to a Q-type factor analysis (see Chapter 3). Correlational measures are rarely used because emphasis in most applications of cluster analysis is on the magnitudes of the objects, not the patterns of values.

Distance Measures. Even though correlational measures have an intuitive appeal and are used in many other multivariate techniques, they are not the most commonly used measure of similarity in cluster analysis. Instead, the most commonly used measures of similarity are distance measures. These distance measures represent similarity as the proximity of observations to one another across the variables in the cluster variate. Distance measures are actually a measure of dissimilarity, with larger values denoting lesser similarity. Distance is converted into a similarity measure by using an inverse relationship.

A simple illustration of using distance measures was shown in our hypothetical example (see Figure 9-2), in which clusters of observations were defined based on the proximity of observations to one another when each observation's scores on two variables were plotted graphically. Even though proximity may seem to be a simple concept, several distance measures are available, each with specific characteristics.

- **Euclidean distance** is the most commonly recognized measure of distance, many times referred to as straight-line distance. An example of how Euclidean distance is obtained is shown geometrically in Figure 9-5. Suppose that two points in two dimensions have coordinates (X_1, Y_1) and (X_2, Y_2), respectively. The Euclidean distance between the points is the length of the hypotenuse of a right triangle, as calculated by the formula under the figure. This concept is easily generalized to more than two variables.
- **Squared (or absolute) Euclidean distance** is the sum of the squared differences without taking the square root. The squared Euclidean distance has the advantage of not having to take the square root, which speeds computations markedly. It is the recommended distance measure for the centroid and Ward's methods of clustering.
- **City-block (Manhattan) distance** is not based on Euclidean distance. Instead, it uses the sum of the absolute differences of the variables (i.e., the two sides of a right triangle rather

TABLE 9-3   Calculating Correlational and Distance Measures of Similarity

Original Data

Case	X_1	X_2	X_3	X_4	X_5
1	7	10	9	7	10
2	9	9	8	9	9
3	5	5	6	7	7
4	6	6	3	3	4
5	1	2	2	1	2
6	4	3	2	3	3
7	2	4	5	2	5

Similarity Measure: Correlation

Case	Case						
	1	**2**	**3**	**4**	**5**	**6**	**7**
1	1.00						
2	−.147	1.00					
3	.000	.000	1.00				
4	.087	.516	−.824	1.00			
5	.963	−.408	.000	−.060	1.00		
6	−.466	.791	−.354	.699	−.645	1.00	
7	.891	−.516	.165	−.239	.963	−.699	1.00

Similarity Measure: Euclidean Distance

Case	Case						
	1	**2**	**3**	**4**	**5**	**6**	**7**
1	nc						
2	3.32	nc					
3	6.86	6.63	nc				
4	10.25	10.20	6.00	nc			
5	15.78	16.19	10.10	7.07	nc		
6	13.11	13.00	7.28	3.87	3.87	nc	
7	11.27	12.16	6.32	5.10	4.90	4.36	nc

nc = distances not calculated.

than the hypotenuse). This procedure is the simplest to calculate, but may lead to invalid clusters if the clustering variables are highly correlated [26].

- **Mahalanobis distance (D^2)** is a generalized distance measure that accounts for the correlations among variables in a way that weights each variable equally. It also relies on standardized variables and will be discussed in more detail in a following section. Although desirable in many situations, it is not available as a proximity measure in either SAS or SPSS.

Other distance measures (other forms of differences or the powers applied to the differences) are available in many clustering programs. The researcher is encouraged to explore alternative cluster solutions obtained when using different distance measures in an effort to best represent the

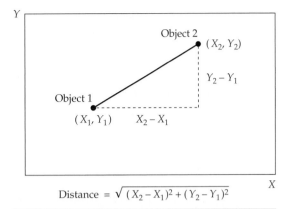

$$\text{Distance} = \sqrt{(X_2 - X_1)^2 + (Y_2 - Y_1)^2}$$

FIGURE 9-5   **An Example of Euclidean Distance Between Two Objects Measured on Two Variables, *X* and *Y***

underlying data patterns. Although these distance measures are said to represent similarity, in a very real sense they better represent dissimilarity, because higher values typically mean relatively less similarity. Greater distance means observations are less similar. Some software packages actually use the term *dissimilarity* because of this fact.

Comparison to Correlational Measures. The difference between correlational and distance measures perhaps can be best illustrated by referring again to Figure 9-4. Distance measures focus on the magnitude of the values and portray as similar the objects that are close together, even if they have different patterns across the variables. In contrast, correlation measures focus on the patterns across the variables and do not consider the magnitude of the differences between objects. Let us look at our seven observations to see how these approaches differ.

Table 9-3 contains the values for the seven observations on the five variables (X_1 to X_5), along with both distance and correlation measures of similarity. Cluster solutions using either similarity measure seem to indicate three clusters, but the membership in each cluster is quite different.

With smaller distances representing greater similarity, we see that cases 1 and 2 form one group (distance of 3.32), and cases 4, 5, 6, and 7 (distances ranging from 3.87 to 7.07) make up another group. The distinctiveness of these two groups from each other is shown in that the smallest distance between the two clusters is 10.20. These two clusters represent observations with higher versus lower values. A third group, consisting of only case 3, differs from the other two groups because it has values that are both low and high.

Using the correlation as the measure of similarity, three clusters also emerge. First, cases 1, 5, and 7 are all highly correlated (.891 to .963), as are cases 2, 4, and 6 (.516 to .791). Moreover, the correlations between clusters are generally close to zero or even negative. Finally, case 3 is again distinct from the other two clusters and forms a single-member cluster.

A correlational measure focuses on patterns rather than the more traditional distance measure and requires a different interpretation of the results by the researcher. Because of this, the researcher will not focus on the actual group centroids on the clustering variables, as is done when distance measures are used. Interpretation depends much more heavily on patterns that become evident in the results.

Which Distance Measure Is Best? In attempting to select a particular distance measure, the researcher should remember the following caveats:

Different distance measures or a change in the scales of the variables may lead to different cluster solutions. Thus, it is advisable to use several measures and compare the results with theoretical or known patterns.

When the variables are correlated (either positively or negatively), the Mahalanobis distance measure is likely to be the most appropriate because it adjusts for correlations and weights all variables equally. Alternatively, the researcher may wish to avoid using highly redundant variables as input to cluster analysis.

Association Measures. Association measures of similarity are used to compare objects whose characteristics are measured only in nonmetric terms (nominal or ordinal measurement). As an example, respondents could answer yes or no on a number of statements. An association measure could assess the degree of agreement or matching between each pair of respondents. The simplest form of association measure would be the percentage of times agreement occurred (both respondents said yes to a question or both said no) across the set of questions.

Extensions of this simple matching coefficient have been developed to accommodate multi-category nominal variables and even ordinal measures. Many computer programs, however, offer only limited support for association measures, and the researcher is forced to first calculate the similarity measures and then input the similarity matrix into the cluster program. Reviews of the various types of association measures can be found in several sources [8, 13, 14, 27].

Selecting a Similarity Measure. Although three different forms of similarity measures are available, the most frequently used and preferred form is the distance measure for several reasons. First, the distance measure best represents the concept of proximity, which is fundamental to cluster analysis. Correlational measures, although having widespread application in other techniques, represent patterns rather than proximity. Second, cluster analysis is typically associated with characteristics measured by metric variables. In some applications, nonmetric characteristics dominate, but most often the characteristics are represented by metric measures making distance again the preferred measure. Thus, in any situation, the researcher is provided measures of similarity that can represent the proximity of objects across a set of metric or nonmetric variables.

STANDARDIZING THE DATA With the similarity measure selected, the researcher must address one more question: Should the data be standardized before similarities are calculated? In answering this question, the researcher must consider that most cluster analyses using distance measures are quite sensitive to differing scales or magnitudes among the variables. In general, variables with larger dispersion (i.e., larger standard deviations) have more impact on the final similarity value.

Clustering variables that are not all of the same scale should be standardized whenever necessary to avoid instances where a variable's influence on the cluster solution is greater than it should be [3]. We will now examine several approaches to standardization available to researchers.

Standardizing the Variables. The most common form of standardization is the conversion of each variable to standard scores (also known as Z scores) by subtracting the mean and dividing by the standard deviation for each variable. This option can be found in all computer programs and many times is even directly included in the cluster analysis procedure. The process converts each raw data score into a standardized value with a mean of 0 and a standard deviation of 1, and in turn, eliminates the bias introduced by the differences in the scales of the several attributes or variables used in the analysis.

There are two primary benefits from standardization. First, it is much easier to compare between variables because they are on the same scale (a mean of 0 and standard deviation of 1). Positive values are above the mean, and negative values are below. The magnitude represents the number of standard deviations the original value is from the mean. Second, no difference occurs in the standardized values when only the scale changes. For example, when we standardize a measure of time duration, the standardized values are the same whether measured in minutes or seconds.

Thus, using standardized variables truly eliminates the effects due to scale differences not only across variables, but for the same variable as well. The need for standardization is minimized when all of the variables are measured on the same response scale (e.g., a series of attitudinal questions), but becomes quite important whenever variables using quite different measurement scales are included in the cluster variate. At times, even when the standarization is not necessary because all variables are measured on the same scale, the researcher may choose to center each variable by subtracting the overall mean for that variable from each observation. The result is a set of variables with a mean of zero but retaining their unique variability. This step simply facilitates interpretation when the variables do not have the same means.

Using a Standardized Distance Measure. A measure of Euclidean distance that directly incorporates a standardization procedure is the Mahalanobis distance (D^2). The Mahalanobis approach not only performs a standardization process on the data by scaling in terms of the standard deviations but it also sums the pooled within-group variance–covariance, which adjusts for correlations among the variables. Highly correlated sets of variables in cluster analysis can implicitly overweight one set of variables in the clustering procedures (see discussion of multicollinearity in stage 3). In short, the Mahalanobis generalized distance procedure computes a distance measure between objects comparable to R^2 in regression analysis. Although many situations are appropriate for use of the Mahalanobis distance, not all programs include it as a measure of similarity. In such cases, the researcher usually selects the squared Euclidean distance.

Standardizing by Observation. Up to now we discussed standardizing only variables. Why might we standardize respondents or cases? Let us take a simple example.

Suppose we collected a number of ratings on a 10-point scale of the importance for several attributes used in purchase decisions for a product. We could apply cluster analysis and obtain clusters, but one distinct possibility is that what we would get are clusters of people who said everything was important, some who said everything had little importance, and perhaps some clusters in between. What we are seeing are patterns of responses specific to an individual. These patterns may reflect a specific way of responding to a set of questions, such as yea-sayers (answer favorably to all questions) or naysayers (answer unfavorably to all questions).

These patterns of yea-sayers and naysayers represent what are termed **response-style effects.** If we want to identify groups according to their response style and even control for these patterns, then the typical standardization through calculating Z scores is not appropriate. What is desired in most instances is the relative importance of one variable to another for each individual. In other words, is attribute 1 more important than the other attributes, and can clusters of respondents be found with similar patterns of importance? In this instance, standardizing by respondent would standardize each question not to the sample's average but instead to that respondent's average score. This **within-case** or **row-centering standardization** can be quite effective in removing response-style effects and is especially suited to many forms of attitudinal data [25]. We should note that this approach is similar to a correlational measure in highlighting the pattern across variables, but the proximity of cases still determines the similarity value.

Should You Standardize? Standardization provides a remedy to a fundamental issue in similarity measures, particularly distance measures, and many recommend its widespread use [11, 13]. However, the researcher should not apply standardization without consideration for its consequences of removing some natural relationship reflected in the scaling of the variables, while others have said it may be appropriate [1]. Some researchers demonstrate that it may not even have noticeable effects [7, 17]. Thus, no single reason tells us to use standardized variables versus unstandardized variables. The decision to standardize should be based on both empirical and conceptual issues reflecting both the research objectives and the empirical qualities of the data. For example, a researcher may wish to consider standardization if clustering variables with different scales or if preliminary analysis shows that the cluster variables display large differences in standard deviations.

Stage 3: Assumptions in Cluster Analysis

Cluster is not a statistical inference technique in which parameters from a sample are assessed as representing a population. Instead, cluster analysis is a method for quantifying the structural characteristics of a set of observations. As such, it has strong mathematical properties but not statistical foundations. The requirements of normality, linearity, and homoscedasticity that were so important in other techniques really have little bearing on cluster analysis. The researcher must focus, however, on two other critical issues: representativeness of the sample and multicollinearity among variables in the cluster variate.

REPRESENTATIVENESS OF THE SAMPLE   Rarely does the researcher have a census of the population to use in the cluster analysis. Usually, a sample of cases is obtained and the clusters are derived in the hope that they represent the structure of the population. The researcher must therefore be confident that the obtained sample is truly representative of the population. As mentioned earlier, outliers may really be only an undersampling of divergent groups that, when discarded, introduce bias in the estimation of structure. The researcher must realize that cluster analysis is only as good as the representativeness of the sample. Therefore, all efforts should be made to ensure that the sample is representative and the results are generalizable to the population of interest.

RULES OF THUMB 9-2

Research Design in Cluster Analysis

- The sample size required is not based on statistical considerations for inference testing, but rather:
  - Sufficient size is needed to ensure representativeness of the population and its underlying structure, particularly small groups within the population
  - Minimum group sizes are based on the relevance of each group to the research question and the confidence needed in characterizing that group
- Similarity measures calculated across the entire set of clustering variables allow for the grouping of observations and their comparison to each other
  - Distance measures are most often used as a measure of similarity, with higher values representing greater dissimilarity (distance between cases), not similarity
  - Less frequently used are correlational measures, where large values do indicate similarity
- Given the sensitivity of some procedures to the similarity measure used, the researcher should employ several distance measures and compare the results from each with other results or theoretical/known patterns
- Outliers can severely distort the representativeness of the results if they appear as structure (clusters) inconsistent with the research objectives
  - They should be removed if the outlier represents:
    - Aberrant observations not representative of the population
    - Observations of small or insignificant segments within the population and of no interest to the research objectives
  - They should be retained if an undersampling/poor representation of relevant groups in the population; the sample should be augmented to ensure representation of these groups
- Outliers can be identified based on the similarity measure by:
  - Finding observations with large distances from all other observations
  - Graphic profile diagrams highlighting outlying cases
  - Their appearance in cluster solutions as single-member or small clusters
- Clustering variables that have scales using widely differing numbers of scale points or that exhibit large differences in standard deviations should be standardized
  - The most common standardization conversion is Z scores
  - If groups are to be identified according to an individual's response style, then within-case or row-centering standardization is appropriate

IMPACT OF MULTICOLLINEARITY   Multicollinearity was an issue in other multivariate techniques because of the difficulty in discerning the true impact of multicollinear variables. In cluster analysis the effect is different because multicollinearity is actually a form of implicit weighting. Let us start with an example that illustrates the effect of multicollinearity.

Suppose that respondents are being clustered on 10 variables, all attitudinal statements concerning a service. When multicollinearity is examined, we see two sets of variables, the first made up of eight statements and the second consisting of the remaining two statements. If our intent is to really cluster the respondents on the dimensions of the service (in this case represented by the two groups of variables), then using the original 10 variables will be quite misleading. Because each variable is weighted equally in cluster analysis, the first dimension will have four times as many chances (eight items compared to two items) to affect the similarity measure. As a result, similarity will be predominantly affected by the first dimension with eight items rather than the second dimension with two items.

Multicollinearity acts as a weighting process not apparent to the observer but affecting the analysis nonetheless. For this reason, the researcher is encouraged to examine the variables used in cluster analysis for substantial multicollinearity and, if found, either reduce the variables to equal numbers in each set or use a distance measure that takes multicollinearity into account. Another possible solution involves factoring the variables prior to clustering and either selecting one cluster variable from each factor or using the resulting factor scores as cluster variables. Recall that principal components or varimax rotated factors are uncorrelated. In this way, the research can take a proactive approach to dealing with multicollinearity.

One last issue is whether to use factor scores in cluster analysis. The debate centers on research showing that the variables that truly discriminate among the underlying groups are not well represented in most factor solutions. Thus, when factor scores are used, it is quite possible that a poor representation of the actual structure of the data will be obtained [23]. The researcher must deal with both multicollinearity and discriminability of the variables to arrive at the best representation of structure.

Stage 4: Deriving Clusters and Assessing Overall Fit

With the clustering variables selected and the similarity matrix calculated, the partitioning process begins (see Figure 9-6). The researcher must:

- Select the partitioning procedure used for forming clusters.
- Make the decision on the number of clusters to be formed.

Both decisions have substantial implications not only on the results that will be obtained but also on the interpretation that can be derived from the results [15]. First, we examine the available partitioning

---

RULES OF THUMB 9-3

Assumptions in Cluster Analysis

- Input variables should be examined for substantial multicollinearity and if present:
  - Reduce the variables to equal numbers in each set of correlated measures, or
  - Use a distance measure that compensates for the correlation, such as Mahalanobis distance
  - Take a proactive approach and include only cluster variables that are not highly correlated

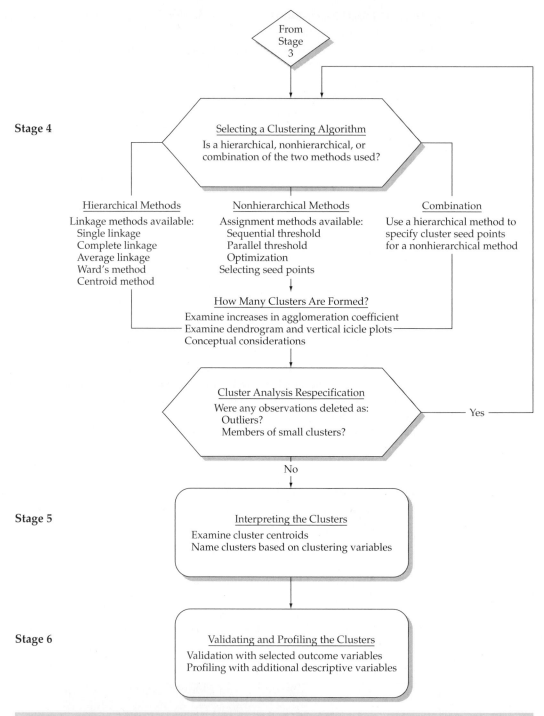

FIGURE 9-6   Stages 4–6 of the Cluster Analysis Decision Diagram

procedures and then discuss the options available for deciding on a cluster solution by defining the number of clusters and membership for each observation. Partitioning procedures work on a simple principle. They seek to maximize the distance between groups while minimizing the differences of in-group members (see Figure 9-7).

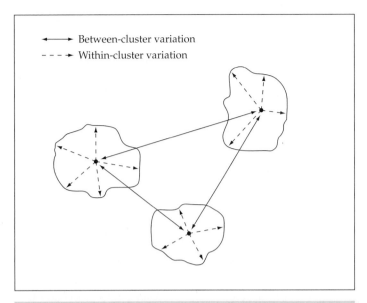

FIGURE 9-7 **Cluster Diagram Showing Between- and Within-Cluster Variation**

HIERARCHICAL CLUSTER PROCEDURES   Hierarchical procedures involve a series of $n - 1$ clustering decisions (where n equals the number of observations) that combine observations into a hierarchy or a treelike structure. The two basic types of hierarchical clustering procedures are agglomerative and divisive. In the **agglomerative methods,** each object or observation starts out as its own cluster and is successively joined, the two most similar clusters at a time until only a single cluster remains. In **divisive methods** all observations start in a single cluster and are successively divided (first into two clusters, then three, and so forth) until each is a single-member cluster. In Figure 9–8, agglomerative methods move from left to right, and divisive methods move from right to left. Because most commonly used computer packages use agglomerative methods, and divisive methods act almost as agglomerative methods in reverse, we focus here on the agglomerative methods.

To understand how a hierarchical procedure works, we will examine the most common form—the agglomerative method—which follows a simple, repetitive process:

1. Start with all observations as their own cluster (i.e., each observation forms a single-member cluster), so that the number of clusters equals the number of observations.
2. Using the similarity measure, combine the two most similar clusters into a new cluster (now containing two observations), thus reducing the number of clusters by one.
3. Repeat the clustering process again, using the similarity measure to combine the two most similar clusters into a new cluster.
4. Continue this process, at each step combining the two most similar clusters into a new cluster. Repeat the process a total of $n - 1$ times until all observations are contained in a single cluster.

Assume that we had 100 observations. We would initially start with 100 separate clusters, each containing a single observation. At the first step, the two most similar clusters would be combined, leaving us with 99 clusters. At the next step, we combine the next two most similar clusters, so that we then have 98 clusters. This process continues until the last step where the final two remaining clusters are combined into a single cluster.

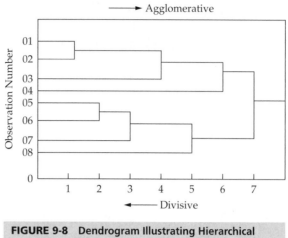

FIGURE 9-8   **Dendrogram Illustrating Hierarchical Clustering**

An important characteristic of hierarchical procedures is that the results at an earlier stage are always nested within the results at a later stage, creating a similarity to a tree. For example, an agglomerative six-cluster solution is obtained by joining two of the clusters found at the seven-cluster stage. Because clusters are formed only by joining existing clusters, any member of a cluster can trace its membership in an unbroken path to its beginning as a single observation. This process is shown in Figure 9-8; the representation is referred to as a **dendrogram** or tree graph, which can be useful, but becomes unwieldy with large applications. The dendrogram is widely available in most clustering software.

Clustering Algorithms. The **clustering algorithm** in a hierarchical procedure defines how similarity is defined between multiple-member clusters in the clustering process. So how do we measure similarity between clusters when one or both clusters have multiple members? Do we select one member to act as a typical member and measure similarity between these members of each cluster, or do we create some composite member to represent the cluster, or even combine the similarities between all members of each cluster? We could employ any of these approaches, or even devise other ways to measure similarity between multiple-member clusters. Among numerous approaches, the five most popular agglomerative algorithms are (1) single-linkage, (2) complete-linkage, (3) average linkage, (4) centroid method, and (5) Ward's method. In our discussions we will use distance as the similarity measure between observations, but other similarity measures could be used just as easily.

- *Single-Linkage* The **single-linkage method** (also called the **nearest-neighbor method**) defines the similarity between clusters as the shortest distance from any object in one cluster to any object in the other. This rule was applied in the example at the beginning of this chapter and enables us to use the original distance matrix between observations without calculating new distance measures. Just find all the distances between observations in the two clusters and select the smallest as the measure of cluster similarity.

  This method is probably the most versatile agglomerative algorithm, because it can define a wide range of clustering patterns (e.g., it can represent clusters that are concentric circles, like rings of a bull's-eye). This flexibility also creates problems, however, when clusters are poorly delineated. In such cases, single-linkage procedures can form long, snakelike chains [15, 20]. Individuals at opposite ends of a chain may be dissimilar, yet still within the same cluster. Many times, the presence of such chains may contrast with the objectives of deriving the most compact clusters. Thus, the researcher must carefully examine the patterns of observations within the clusters to ascertain whether these chains are occurring. It becomes increasingly difficult by simple graphical means as the number of clustering variables

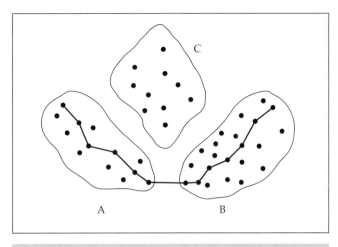

FIGURE 9-9   Example of Single Linkage Joining Dissimilar
Clusters A and B

increases, and requires that the researcher carefully profile the internal homogeneity among
the observations in each cluster.

An example of this arrangement is shown in Figure 9-9. Three clusters (A, B, and C) are
to be joined. The single-linkage algorithm, focusing on only the closest points in each cluster,
would link clusters A and B because of their short distance at the extreme ends of the clusters.
Joining clusters A and B creates a cluster that encircles cluster C. Yet in striving for within-
cluster homogeneity, it would be much better to join cluster C with either A or B. This figure
shows the principal disadvantage of the single-linkage algorithm.

• *Complete-Linkage*   The **complete-linkage method** (also known as the **farthest-neighbor**
or **diameter method**) is comparable to the single-linkage algorithm, except that cluster
similarity is based on maximum distance between observations in each cluster. Similarity
between clusters is the smallest (minimum diameter) sphere that can enclose all observations
in both clusters. This method is called complete-linkage because all objects in a cluster are
linked to each other at some maximum distance. Thus, within-group similarity equals group
diameter.

This technique eliminates the chaining problem identified with single-linkage and has
been found to generate the most compact clustering solutions [3]. Even though it represents
only one aspect of the data (i.e., the farthest distance between members), many researchers
find it the most appropriate for a wide range of clustering applications [12].

Figure 9-10 compares the shortest (single-linkage) and longest (complete-linkage) distances in representing similarity between clusters. Both measures reflect only one aspect of
the data. The use of the single-linkage reflects only a closest single pair of objects, and the
complete-linkage also reflects a single pair, this time the two most extreme.

• *Average Linkage*   The **average linkage** procedure differs from the single-linkage or complete-linkage procedures in that the similarity of any two clusters is the average similarity of
all individuals in one cluster with all individuals in another. This algorithm does not depend
on extreme values (closest or farthest pairs) as do single-linkage or complete-linkage. Instead,
similarity is based on all members of the clusters rather than on a single pair of extreme members and are thus less affected by outliers. Average linkage approaches, as a type of compromise between single- and complete-linkage methods, tend to generate clusters with small
within-cluster variation. They also tend toward the production of clusters with approximately
equal within-group variance.

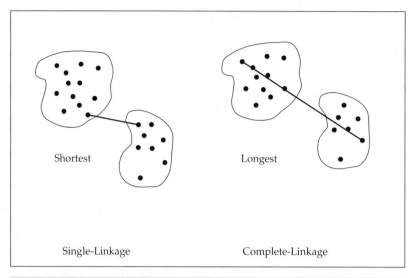

FIGURE 9-10 Comparison of Distance Measures for Single-Linkage and Complete-Linkage

- *Centroid Method* In the **centroid method** the similarity between two clusters is the distance between the cluster centroids. **Cluster centroids** are the mean values of the observations on the variables in the cluster variate. In this method, every time individuals are grouped, a new centroid is computed. Cluster centroids migrate as cluster mergers take place. In other words, a cluster centroid changes every time a new individual or group of individuals is added to an existing cluster.

  These methods are the most popular in the physical and life sciences (e.g., biology) but may produce messy and often confusing results. The confusion occurs because of reversals, that is, instances when the distance between the centroids of one pair may be less than the distance between the centroids of another pair merged at an earlier combination. The advantage of this method, like the average linkage method, is that it is less affected by outliers than are other hierarchical methods.

- *Ward's Method* The **Ward's method** differs from the previous methods in that the similarity between two clusters is not a single measure of similarity, but rather the sum of squares within the clusters summed over all variables. It is quite similar to the simple heterogeneity measure used in the example at the beginning of the chapter to assist in determining the number of clusters. In the Ward's procedure, the selection of which two clusters to combine is based on which combination of clusters minimizes the within-cluster sum of squares across the complete set of disjoint or separate clusters. At each step, the two clusters combined are those that minimize the increase in the total sum of squares across all variables in all clusters.

  This procedure tends to combine clusters with a small number of observations, because the sum of squares is directly related to the number of observations involved. The use of a sum of squares measure makes this method easily distorted by outliers [17]. Moreover, the Ward's method also tends to produce clusters with approximately the same number of observations. If the researcher expects or desires the clustering patterns to reflect somewhat equally sized clusters, then this method is quite appropriate. However, the use of this method also makes it more difficult to identify clusters representing small proportions of the sample.

Overview. Hierarchical clustering procedures are a combination of a repetitive clustering process combined with a clustering algorithm to define the similarity between clusters with multiple members. The process of creating clusters generates a treelike diagram that represents the combinations/divisions of clusters to form the complete range of cluster solutions. Hierarchical procedures generate a complete set of cluster solutions, ranging from all single-member clusters to the one-cluster solution where all observations are in a single cluster. In doing so, the hierarchical procedure provides an excellent framework with which to compare any set of cluster solutions and help in judging how many clusters should be retained.

NONHIERARCHICAL CLUSTERING PROCEDURES   In contrast to hierarchical methods, **nonhierarchical procedures** do not involve the treelike construction process. Instead, they assign objects into clusters once the number of clusters is specified. For example, a six-cluster solution is not just a combination of two clusters from the seven-cluster solution, but is based only on finding the best six-cluster solution. The nonhierarchical cluster software programs usually proceed through two steps:

1. *Specify cluster seeds:* The first task is to identify starting points, known as **cluster seeds,** for each cluster. A cluster seed may be prespecified by the researcher or observations may be selected, usually in a random process.
2. *Assignment:* With the cluster seeds defined, the next step is to assign each observation to one of the cluster seeds based on similarity. Many approaches are available for making this assignment (see later discussion in this section), but the basic objective is to assign each observation to the most similar cluster seed. In some approaches, observations may be reassigned to clusters that are more similar than their original cluster assignment.

We discuss several different approaches for selecting cluster seeds and assigning objects in the next sections.

Selecting Seed Points. Even though the nonhierarchical clustering algorithms discussed in the next section differ in the manner in which they assign observations to the seed points, they all face the same problem: How do we select the cluster seeds? The different approaches can be classified into two basic categories:

1. *Researcher specified.* In this approach, the researcher provides the seed points based on external data. The two most common sources of the seed points are prior research or data from another multivariate analysis. Many times the researcher has knowledge of the cluster profiles being researched. For example, prior research may have defined segment profiles and the task of the cluster analysis is to assign individuals to the most appropriate segment cluster. It is also possible that other multivariate techniques may be used to generate the seed points. One common example is the use of a hierarchical clustering algorithm to establish the number of clusters and then generate seed points from these results (a more detailed description of this approach is contained in the following section). The common element is that the researcher, while knowing the number of clusters to be formed, also has information about the basic character of these clusters.
2. *Sample generated.* The second approach is to generate the cluster seeds from the observations of the sample, either in some systematic fashion or simply through random selection. For example, in the FASTCLUS program in SAS, the first seed is the first observation in the data set with no missing values. The second seed is the next complete observation (no missing data) that is separated from the first seed by a specified minimum distance. The default option is a zero minimum distance. After all seeds are selected, the program assigns each observation to the cluster with the nearest seed. The researcher can specify that the cluster seeds be revised (updated) by calculating seed cluster means each time an observation is assigned. In contrast, the K-means program in SPSS can select the necessary seed points randomly from among the observations. In any of these approaches, the researcher relies on the selection process to choose seed points that reflect natural clusters as starting points for the clustering algorithms. A possible limitation is that replication of the results is difficult if the observations are reordered.

In either approach, the researcher must be aware of the impact of the cluster seed selection process on the final results. All of the clustering algorithms, even those of an optimizing nature (see the following discussion), will generate different cluster solutions depending on the initial cluster seeds. The differences among cluster solutions will hopefully be minimal using different seed points, but they underscore the importance of cluster seed selection and its impact on the final cluster solution.

Nonhierarchical Clustering Algorithms. Several clustering algorithms have been proposed [9]. The most frequently cited are sequential, parallel and optimization. The sequential threshold method starts by selecting one cluster seed and includes all objects within a prespecified distance. A second cluster seed is then selected, and all objects within the prespecified distance of that seed are included. A third seed is then selected, and the process continues as before. The primary disadvantage of this approach is that when an observation is assigned to a cluster, it cannot be reassigned to another cluster, even if that cluster seed is more similar. The parallel threshold method considers all cluster seeds simultaneously and assigns observations within the threshold distance to the nearest seed. The third method, referred to as the optimizing procedure, is similar to the other two nonhierarchical procedures except that it allows for reassignment of observations to a seed other than the one with which it was originally associated. All of these belong to a group of clustering algorithms known as K-means. **K-means** algorithms work by portioning the data into a user-specified number of clusters and then iteratively reassigning observations to clusters until some numerical criterion is met. The criterion specifies a goal related to minimizing the distance of observations from one another in a cluster and maximizing the distance between clusters. K-means is so commonly used that the term is used by some to refer to nonhierarchical cluster analysis in general. For example, in SPSS, the nonhierarchical clustering routine is referred to as K-means. New algorithms are also being developed that combine the hierarchical and nonhierarchical steps in one analysis. This procedure is a part of the later SPSS versions and is considered appropriate for large samples.

An **optimizing procedure** allows for reassignment of observations based on the goal of creating the most distinct clusters. If, in the course of assigning observations, an observation becomes closer to another cluster that is not the cluster to which it is currently assigned, then an optimizing procedure switches the observation to the more similar (closer) cluster.

SHOULD HIERARCHICAL OR NONHIERARCHICAL METHODS BE USED?   A definitive answer to this question cannot be given for two reasons. First, the research problem at hand may suggest one method or the other. Second, what we learn with continued application to a particular context may suggest one method over the other as more suitable for that context. We can examine the strengths and weaknesses of each method to determine which is most appropriate for a unique research setting.

Pros and Cons of Hierarchical Methods. Hierarchical clustering techniques have long been the more popular clustering method, with Ward's method and average linkage probably being the best available [17]. Besides the fact that hierarchical procedures were the first clustering methods developed, they do offer several advantages that lead to their widespread usage:

1. *Simplicity:* Hierarchical techniques, with their development of the treelike structures depicting the clustering process, do afford the researcher with a simple, yet comprehensive, portrayal of the entire range of clustering solutions. In doing so, the researcher can evaluate any of the possible clustering solutions from one analysis.
2. *Measures of similarity:* The widespread use of the hierarchical methods led to an extensive development of similarity measures for almost any type of clustering variables. Hierarchical techniques can be applied to almost any type of research question.

3. *Speed:* Hierarchical methods have the advantage of generating an entire set of clustering solutions (from all separate clusters to one cluster) in an expedient manner. This ability enables the researcher to examine a wide range of alternative clustering solutions, varying measures of similarities and linkage methods, in an efficient manner.

Even though hierarchical techniques have been widely used, they do have several distinct disadvantages that affect any of their cluster solutions:

1. Hierarchical methods can be misleading because undesirable early combinations may persist throughout the analysis and lead to artificial results. Of specific concern is the substantial impact of outliers on hierarchical methods, particularly with the complete-linkage method.
2. To reduce the impact of outliers, the researcher may wish to cluster analyze the data several times, each time deleting problem observations or outliers. The deletion of cases, however, even those not found to be outliers, can many times distort the solution. Thus, the researcher must employ extreme care in the deletion of observations for any reason.
3. Although computations of the clustering process are relatively fast, hierarchical methods are not amenable to analyzing large samples or even large numbers of variables. As sample size increases, the data storage requirements increase dramatically. For example, a sample of 400 cases requires storage of approximately 80,000 similarities, which increases to almost 125,000 for a sample of 500. Even with the computing power of today's personal computers, these data requirements can limit the application in many instances. The researcher may take a random sample of the original observations to reduce sample size but must now question the representativeness of the sample taken from the original sample.

Emergence of Nonhierarchical Methods. Nonhierarchical methods have gained increased acceptability and usage, but any application depends on the ability of the researcher to select the seed points according to some practical, objective, or theoretical basis. In these instances, nonhierarchical methods offer several advantages over hierarchical techniques.

1. The results are less susceptible to outliers in the data, the distance measure used, and the inclusion of irrelevant or inappropriate variables.
2. Nonhierarchical methods can analyze extremely large data sets because they do not require the calculation of similarity matrices among all observations, but instead just the similarity of each observation to the cluster centroids. Even the optimizing algorithms that allow for reassignment of observations between clusters can be readily applied to all sizes of data sets.

Although nonhierarchical methods do have several distinct advantages, several shortcomings can markedly affect their use in many types of applications.

1. The benefits of any nonhierarchical method are realized only with the use of nonrandom (i.e., specified) seed points. Thus, the use of nonhierarchical techniques with random seed points is generally considered inferior to hierarchical techniques.
2. Even a nonrandom starting solution does not guarantee an optimal clustering of observations. In fact, in many instances the researcher will get a different final solution for each set of specified seed points. How is the researcher to select the optimum answer? Only by analysis and validation can the researcher select what is considered the best representation of structure, realizing that many alternatives may be as acceptable.
3. Nonhierarchical methods are also not as efficient when examining a large number of potential cluster solutions. Each cluster solution is a separate analysis, in contrast to the hierarchical techniques that generate all possible cluster solutions in a single analysis. Thus, nonhierarchical techniques are not as well suited to exploring a wide range of solutions based on varying elements such as similarity measures, observations included, and potential seed points.

A Combination of Both Methods. Many researchers recommend a combination approach using both methods. In this way, the advantages of one approach can compensate for the weaknesses or the other [17]. This is accomplished in two steps:

1. First, a hierarchical technique is used to generate a complete set of cluster solutions, establish the applicable cluster solutions (see next section for discussion of this topic), and establish the appropriate number of clusters.
2. After outliers are eliminated, the remaining observations can then be clustered by a nonhierarchical method.

In this way, the advantages of the hierarchical methods are complemented by the ability of the nonhierarchical methods to refine the results by allowing the switching of cluster membership.

SHOULD THE CLUSTER ANALYSIS BE RESPECIFIED? Even before identifying an acceptable cluster analysis solution (see next section), the researcher should examine the fundamental structure represented in the defined clusters. Of particular note are widely disparate cluster sizes or clusters of only one or two observations. Generally, one-member or extremely small clusters are not acceptable given the research objectives, and thus should be eliminated.

Researchers must examine widely varying cluster sizes from a conceptual perspective, comparing the actual results with the expectations formed in the research objectives. More troublesome are single-member clusters, which may be outliers not detected in earlier analyses. If a single-member cluster (or one of small size compared with other clusters) appears, the researcher must decide whether it represents a valid structural component in the sample or should be deleted as unrepresentative. If any observations are deleted, especially when hierarchical solutions are employed, the researcher should rerun the cluster analysis and start the process of defining clusters anew.

HOW MANY CLUSTERS SHOULD BE FORMED? Perhaps the most critical issue for any researcher performing either a hierarchical or nonhierarchical cluster analysis is determining the number of clusters most representative of the sample's data structure [6]. This decision is critical for hierarchical techniques because even though the process generates the complete set of cluster solutions, the researcher must select the cluster solution(s) to represent the data structure (also known as the **stopping rule**). This number is carried forward into nonhierarchical cluster analysis. Cluster results do not always provide unambiguous information as to the best number of clusters. Therefore, researchers commonly use a stopping rule that suggests two or more cluster solutions which can be compared before making the final decision.

Unfortunately, no standard objective selection procedure exists [5, 11]. Because no internal statistical criterion is used for inference, such as the statistical significance tests of other multivariate methods, researchers have developed many criteria for approaching the problem. The principal issues facing any of these stopping rules include the following:

- These ad hoc procedures must be computed by the researcher and often involve fairly complex approaches [1, 18].
- Many of these criteria are specific to a particular software program and are not easily calculated if not provided by the program.
- A natural increase in heterogeneity comes from the reduction in number of clusters. Thus, the researcher must look at the trend in the values of these stopping rules across cluster solutions to identify marked increases. If not, in most instances the two-cluster solution would always be chosen because the value of any stopping rule is normally highest when going from two to one cluster.

Even with the similarities among the stopping rules, they show enough differences to place them into one of two general classes, as described next.

Measures of Heterogeneity Change. One class of stopping rules examines some measure of heterogeneity between clusters at each successive step, with the cluster solution defined when the heterogeneity measure exceeds a specified value or when the successive values between steps makes a sudden jump. **Heterogeneity** refers to how different the observations in a cluster are from each other (i.e., heterogeneity refers to a lack of similarity among group members). A simple example was used at the beginning of the chapter, which looked for large increases in the average within-cluster distance. When a large increase occurs, the researcher selects the prior cluster solution on the logic that its combination caused a substantial increase in heterogeneity. This type of stopping rule has been shown to provide fairly accurate decisions in empirical studies [18], but it is not uncommon for a number of cluster solutions to be identified by these large increases in heterogeneity. It is then the researcher's task to select a final cluster solution from these selected cluster solutions. Various stopping rules follow this general approach.

- *Percentage Changes in Heterogeneity* Probably the simplest and most widespread rule is a simple percentage change in some measure of heterogeneity. A typical example is using the agglomeration coefficient in SPSS, which measures heterogeneity as the distance at which clusters are formed (if a distance measure of similarity is used) or the within-cluster sum of squares if the Ward's method is used. With this measure, the percentage increase in the agglomeration coefficient can be calculated for each cluster solution. Then the researcher selects cluster solutions as a potential final solution when the percentage increase is markedly larger than occurring at other steps.
- *Measures of Variance Change* The **root mean square standard deviation (RMSSTD)** is the square root of the variance of the new cluster formed by joining the two clusters. The variance for the newly formed cluster is calculated as the variance across all clustering variables. Large increases in the RMSSTD suggest the joining of two quite dissimilar clusters, indicating the previous cluster solution (in which the two clusters were separate) was a candidate for selection as the final cluster solution.
- *Statistical Measures of Heterogeneity Change* A series of test statistics attempts to portray the degree of heterogeneity for each new cluster solution (i.e., joining of two clusters). One of the most widely used is a pseudo F statistic, which compares the goodness-of-fit of k clusters

---

RULES OF THUMB 9-4

Deriving Clusters

- Selection of hierarchical or nonhierarchical methods is based on:
  - Hierarchical clustering solutions are preferred when:
    - A wide range of alternative clustering solutions is to be examined
    - The sample size is moderate (under 300–400, not exceeding 1,000) or a sample of the larger data set is acceptable
  - Nonhierarchical clustering methods are preferred when:
    - The number of clusters is known and/or initial seed points can be specified according to some practical, objective, or theoretical basis
    - Outliers cause concern, because nonhierarchical methods generally are less susceptible to outliers
- A combination approach using a hierarchical approach followed by a nonhierarchical approach is often advisable
  - A hierarchical approach is used to select the number of clusters and profile cluster centers that serve as initial cluster seeds in the nonhierarchical procedure
  - A nonhierarchical method then clusters all observations using the seed points to provide more accurate cluster memberships

---

RULES OF THUMB 9-5

Deriving the Final Cluster Solution

- No single objective procedure is available to determine the correct number of clusters; rather the researcher must evaluate alternative cluster solutions on the following considerations to select the optimal solution:
  - Single-member or extremely small clusters are generally not acceptable and should be eliminated
  - For hierarchical methods, ad hoc stopping rules, based on the rate of change in a total heterogeneity measure as the number of clusters increases or decreases, are an indication of the number of clusters
  - All clusters should be significantly different across the set of clustering variables
  - Cluster solutions ultimately must have theoretical validity assessed through external validation

---

to $k - 1$ clusters. Highly significant values indicate that the $k - 1$ cluster solution is more appropriate than the k cluster solution. The researcher should not consider any significant value, but instead look to those values markedly more significant than for other solutions.

Direct Measures of Heterogeneity. A second general class of stopping rules attempts to directly measure heterogeneity of each cluster solution. The most common measure in this class is the **cubic clustering criterion (CCC)** [18] contained in SAS, a measure of the deviation of the clusters from an expected distribution of points formed by a multivariate uniform distribution. Here the researcher selects the cluster solution with the largest value of CCC (i.e., the cluster solution where CCC peaks) [24]. Despite its inclusion in SAS and its advantage of selecting a single-cluster solution, it has been shown to many times generate too many clusters as the final solution [18] and is based on the assumption that the variables are uncorrelated. However, it is a widely used measure and is generally as efficient as any other stopping rule [18].

Summary. Given the number of stopping rules available and the lack of evidence supporting any single stopping rule, it is suggested that a researcher employ several stopping rules and look for a consensus cluster solution(s). Even with a consensus based on empirical measures, however, the researcher should complement the empirical judgment with any conceptualization of theoretical relationships that may suggest a natural number of clusters. One might start this process by specifying some criteria based on practical considerations, such as saying, "My findings will be more manageable and easier to communicate if I use three to six clusters," and then solving for this number of clusters and selecting the best alternative after evaluating all of them. In the final analysis, however, it is probably best to compute a number of different cluster solutions (e.g., two, three, four) and then decide among the alternative solutions by using a priori criteria, practical judgment, common sense, or theoretical foundations. The cluster solutions will be improved by restricting the solution according to conceptual aspects of the problem.

Stage 5: Interpretation of the Clusters

The interpretation stage involves examining each cluster in terms of the cluster variate to name or assign a label accurately describing the nature of the clusters. To clarify this process, let us refer to the example of diet versus regular soft drinks.

Let us assume that an attitude scale was developed that consisted of statements regarding consumption of soft drinks, such as "diet soft drinks taste harsher," "regular soft drinks have a fuller taste," "diet drinks are healthier," and so forth. Further, let us assume that demographic and soft-drink consumption data were also collected.

When starting the interpretation process, one measure frequently used is the cluster's centroid. If the clustering procedure was performed on the raw data, it would be a logical description. If the data were standardized or if the cluster analysis was performed using factor analysis (component factors), the researcher would have to go back to the raw scores for the original variables and develop profile diagrams using these data.

Continuing with our soft-drink example, in this stage we examine the average score profiles on the attitude statements for each group and assign a descriptive label to each cluster. Many times discriminant analysis is applied to generate score profiles, but we must remember that statistically significant differences would not indicate an optimal solution because statistical differences are expected, given the objective of cluster analysis. Examination of the profiles allows for a rich description of each cluster. For example, two of the clusters may have favorable attitudes about diet soft drinks and the third cluster negative attitudes. Moreover, of the two favorable clusters, one may exhibit favorable attitudes toward only diet soft drinks, whereas the other may display favorable attitudes toward both diet and regular soft drinks. From this analytical procedure, one would evaluate each cluster's attitudes and develop substantive interpretations to facilitate labeling each. For example, one cluster might be labeled "health-and calorie-conscious," whereas another might be labeled "get a sugar rush."

The profiling and interpretation of the clusters, however, achieve more than just description and are essential elements in selecting between cluster solutions when the stopping rules indicate more than one appropriate cluster solution.

- They provide a means for assessing the correspondence of the derived clusters to those proposed by prior theory or practical experience. If used in a confirmatory mode, the cluster analysis profiles provide a direct means of assessing the correspondence.
- The cluster profiles also provide a route for making assessments of practical significance. The researcher may require that substantial differences exist on a set of clustering variables and the cluster solution be expanded until such differences arise.

In assessing either correspondence or practical significance, the researcher compares the derived clusters to a preconceived typology. This more subjective judgment by the researcher combines with the empirical judgment of the stopping rules to determine the final cluster solution to represent the data structure of the sample.

Stage 6: Validation and Profiling of the Clusters

Given the somewhat subjective nature of cluster analysis about selecting an optimal cluster solution, the researcher should take great care in validating and ensuring practical significance of the final cluster solution. Although no single method exists to ensure validity and practical significance, several approaches have been proposed to provide some basis for the researcher's assessment.

VALIDATING THE CLUSTER SOLUTION   Validation includes attempts by the researcher to assure that the cluster solution is representative of the general population, and thus is generalizable to other objects and is stable over time.

Cross-Validation. The most direct approach in this regard is to cluster analyze separate samples, comparing the cluster solutions and assessing the correspondence of the results. This approach, however, is often impractical because of time or cost constraints or the unavailability of objects (particularly consumers) for multiple cluster analyses. In these instances, a common approach is to split the sample into two groups. Each cluster is analyzed separately, and the results are then compared. Cross-tabulation also can be used, because the members of any specific cluster in one solution should stay together in a cluster in another solution. Therefore, the cross-tabulation should display obvious patterns of matching cluster membership. Other approaches include (1) a modified form of split sampling whereby cluster centers obtained from one cluster solution are employed to define clusters from the other observations and the results are compared [16], and (2) a direct form of cross-validation [22].

For any of these methods, stability of the cluster results can be assessed by the number of cases assigned to the same cluster across cluster solutions. Generally, very stable solution would be produced with less than 10 percent of observations being assigned to a different cluster. A stable solution would result with between 10 and 20 percent assigned to a different group, and a somewhat stable solution when between 20 and 25 percent of the observations are to a different cluster than the initial one.

Establishing Criterion Validity. The researcher may also attempt to establish some form of criterion or predictive validity. To do so, the researcher selects variable(s) not used to form the clusters but known to vary across the clusters. In our example, we may know from past research that attitudes toward diet soft drinks vary by age. Thus, we can statistically test for the differences in age between those clusters that are favorable to diet soft drinks and those that are not. The variable(s) used to assess predictive validity should have strong theoretical or practical support because they become the benchmark for selecting among the cluster solutions.

PROFILING THE CLUSTER SOLUTION The profiling stage involves describing the characteristics of each cluster to explain how it may differ on relevant dimensions. This process typically involves the use of discriminant analysis. The procedure begins after the clusters are identified. The researcher utilizes data not previously included in the cluster procedure to profile the characteristics of each cluster. These data typically are demographic characteristics, psychographic profiles, consumption patterns, and so forth. Although no theoretical rationale may exist for their difference across the clusters, such as required for predictive validity assessment, they should at least have practical importance. Using discriminant analysis, the researcher compares average score profiles for the clusters. The categorical dependent variable is the previously identified clusters, and the independent variables are the demographics, psychographics, and so on.

From this analysis, assuming statistical significance, the researcher could conclude, for example, that the "health- and calorie-conscious" cluster from our previous diet soft drink example consists of better-educated, higher-income professionals who are moderate consumers of soft drinks.

RULES OF THUMB 9-6

Interpreting, Profiling, and Validating Clusters

- The cluster centroid, a mean profile of the cluster on each clustering variable, is particularly useful in the interpretation stage:
  - Interpretation involves examining the distinguishing characteristics of each cluster's profile and identifying substantial differences between clusters
  - Cluster solutions failing to show substantial variation indicate other cluster solutions should be examined
  - The cluster centroid should also be assessed for correspondence with the researcher's prior expectations based on theory or practical experience
- Validation is essential in cluster analysis because the clusters are descriptive of structure and require additional support for their relevance:
  - Cross-validation empirically validates a cluster solution by creating two subsamples (randomly splitting the sample) and then comparing the two cluster solutions for consistency with respect to number of clusters and the cluster profiles
  - Validation is also achieved by examining differences on variables not included in the cluster analysis but for which a theoretical and relevant reason enables the expectation of variation across the clusters

---

In short, the profile analysis focuses on describing not what directly determines the clusters but rather on the characteristics of the clusters after they are identified. Moreover, the emphasis is on the characteristics that differ significantly across the clusters and those that could predict membership in a particular cluster. Profiling often is an important practical step in clustering procedures, because identifying characteristics like demographics enables segments to be identified or located with easily obtained information.

AN ILLUSTRATIVE EXAMPLE

We will use the HBAT database to illustrate the application of cluster analysis techniques. Customer perceptions of HBAT provide a basis for illustrating one of the most common applications of cluster analysis—the formation of customer segments. In our example, we follow the stages of the model-building process, starting with setting objectives, then addressing research design issues, and finally partitioning respondents into clusters and interpreting and validating the results. The following sections detail these procedures through each of the stages.

Stage 1: Objectives of the Cluster Analysis

The first stage in applying cluster analysis involves determining the objectives to be achieved. Once the objectives have been agreed upon, the HBAT research team must select the clustering variables to be used in the clustering process.

CLUSTERING OBJECTIVES Cluster analysis can achieve any combination of three objectives: taxonomy development, data simplification, and relationship identification. In this situation, HBAT is primarily interested in the segmentation of customers (taxonomy development), although additional uses of the derived segments are possible.

The primary objective is to develop a taxonomy that segments objects (HBAT customers) into groups with similar perceptions. Once identified, strategies with different appeals can be formulated for the separate groups—the requisite basis for market segmentation. Cluster analysis, with its objective of forming homogeneous groups that are as distinct between one another as possible, provides a unique methodology for developing taxonomies with maximal managerial relevance.

In addition to forming a taxonomy that can be used for segmentation, cluster analysis also facilitates data simplification and even identification of relationships. In terms of data simplification, segmentation enables categorization of HBAT customers into segments that define the basic character of group members. In an effective segmentation, customers are viewed not as only individuals, but also as members of relatively homogeneous groups portrayed through their common profiles. Segments also provide an avenue to examine relationships previously not studied. A typical example is the estimation of the impact of customer perceptions on sales for each segment, enabling the researcher to understand what uniquely impacts each segment rather than the sample as a whole. For example, do customers with more favorable perceptions of HBAT also purchase more?

CLUSTERING VARIABLES The research team has decided to look for clusters based on the variables that indicate how HBAT customers rate the firm's performance on several key attributes. These attributes are measured in the database with variables X_6 to X_{18}. The research team knows that multicollinearity can be an issue in using cluster analysis. Thus, having already analyzed the data using factor analysis, they decide to use variables that are not strongly correlated to one another. To do so, they select a single variable to represent each of the four factors (see an earlier chapter) plus X_{15}, which was eliminated from the factor analysis by the MSA test because it did not share enough variance with the other variables. The variables included in the cluster analysis are ratings of HBAT on various firm attributes:

X_6 — product quality

X_8 — technical support

X_{12} — sales force

X_{15} — new product development

X_{18} — delivery speed

Table 9-4 displays the descriptive statistics for these variables.

Stage 2: Research Design of the Cluster Analysis

In preparing for a cluster analysis, the researcher must address four issues in research design: detecting outliers, determining the similarity measure to be used, deciding the sample size, and standardizing the variable and/or objects. Each of these issues plays an essential role in defining the nature and character of the resulting cluster solutions.

DETECTING OUTLIERS   The first issue is to identify any outliers in the sample before partitioning begins. Univariate procedures discussed in Chapter 2 did not identify any potential candidates for designation as outliers. Multivariate procedures are used because cluster analysis includes all of the selected clustering variables in identifying clusters that exhibit similarity within the groups. Outliers are observations that are different or dissimilar. In our example, we will refer to the term *dissimilarity,* because larger distances mean less similar observations. Moreover, because we are looking for outliers, the focus is squarely on finding observations that are potentially quite different than the others.

We begin looking for outliers by determining the average dissimilarity for each observation. The average dissimilarity is determined using distance measures for each observation. There is no single best way to identify the most dissimilar objects. Most statistical software will compute a matrix of pairwise proximity measures showing, for example, the Euclidean distance from each observation to every other observation. In SPSS, you can do this from the correlation dropdown menu or as an option under the hierarchical clustering routine. Observations with relatively large pair-wise proximities (large differences between) become outlier candidates. This method has one big disadvantage. Determining which observations have the largest average distances can prove difficult when the number of observations exceeds 20 or so. In the HBAT example, the 100 observations would produce a proximity, or, more precisely, a dissimilarity matrix that is 100 rows by 100 columns. Thus, a smaller summary analysis of how different each observation is from an average respondent makes this process much easier.

TABLE 9-4   Descriptive Statistics for Cluster Variables

	N	Minimum	Maximum	Mean	Std. Deviation
X_6, Product Quality	100	5	10	7.81	1.40
X_8, Technical Support	100	1.3	8.5	5.37	1.53
X_{12}, Salesforce Image	100	2.9	8.2	5.12	1.07
X_{15}, New Products	100	1.7	9.5	5.15	1.49
X_{18}, Delivery Speed	100	1.6	5.5	3.89	0.73

TABLE 9-5 Largest Dissimilarity Values for Identifying Potential Outliers

Observation	Differences from Mean for Each Observation:					Squared Differences from Mean					Dissimilarity
	X_6	X_8	X_{12}	X_{15}	X_{18}	X_6	X_8	X_{12}	X_{15}	X_{18}	
87	−2.81	−4.07	−0.22	2.45	−0.79	7.90	16.52	0.05	6.00	0.62	5.58
6	−1.31	−2.27	−1.42	4.35	−0.59	1.72	5.13	2.02	18.92	0.34	5.30
90	−2.31	2.34	3.08	−0.25	1.01	5.34	5.45	9.47	0.06	1.03	4.62
53	1.59	−0.57	−0.52	4.05	0.71	2.53	0.32	0.27	16.40	0.51	4.48
44	−2.71	1.23	2.68	0.05	0.61	7.34	1.53	7.17	0.00	0.38	4.05
41	0.49	−2.07	0.08	−3.45	0.01	0.24	4.26	0.01	11.90	0.00	4.05
72	−1.11	−2.37	−0.62	−2.65	−0.79	1.23	5.59	0.39	7.02	0.62	3.85
31	−0.91	3.14	−0.42	−1.85	−0.59	0.83	9.83	0.18	3.42	0.34	3.82
22	1.79	1.43	2.68	1.35	0.41	3.20	2.06	7.17	1.82	0.17	3.80
88	−0.11	2.64	−0.82	2.55	0.41	0.01	6.94	0.68	6.50	0.17	3.78

Table 9-5 lists the 10 observations with the highest average dissimilarities. Here is how these values can be computed. (Most standard statistical packages do not include a function that produces this table; program syntax provides an option, but a spreadsheet can be used to do these simple computations and then sort the observations by dissimilarity. The book's Web sites [see www.pearsonglobaleditions.com/hair or www.mvstats.com] provides an example spreadsheet that performs this analysis.) Recall from earlier in the chapter that the Euclidean distance measure is easily generalized to more than two observations or dimensions. As a result, we can use a variation of this approach to find the observations that have relatively high dissimilarity. A typical respondent can be thought of as one that responds with the central tendency on each variable (assuming the data follow a conventional distribution). In other words, an average respondent provides responses that match the mean of any variable. Thus, we can compute the average distance of each observation from the typical respondent to determine if any observations are very dissimilar from the others. The following paragraph describes this process.

Using a process like that illustrated in Figure 9-5, generalized to five variables, we can begin computing a dissimilarity measure by taking each individual's score on each cluster variable and subtracting the mean for that variable from the observation. In other words, if a respondent checked a 7.9 on X_6—Product Quality—we would take that 7.9 and subtract the variable mean (7.81) from it to yield a value of 0.09. The resulting mean-centered values for each respondent on each variable provide the basis for this computation, which eventually produces a dissimilarity for each observation from the average or typical respondent. The process can be illustrated in a step-by-step fashion using HBAT respondent 100 as an example. Here is how we arrive at an average dissimilarity of 1.28 for observation 100:

	Observed Values for Observation 100	Less	Variable Means (see Table 9-4)		Difference	Squared Difference
X_6	7.9	−	7.81	=	0.09	0.0081
X_8	4.4	−	5.37	=	−0.97	0.9409
X_{12}	4.8	−	5.12	=	−0.32	0.1024
X_{15}	5.8	−	5.15	=	0.65	0.4225
X_{18}	3.5	−	3.89	=	−0.39	0.1521
			Total Differences Squared			1.63
			Square Root of Total			1.28

- Compute the difference between the observed value for a clustering variable and the variable mean. This process is illustrated under the observed values and variable means headings, respectively, with the result shown under the differences heading. The process is repeated for each cluster variable. In this example, the clustering variables are X_6, X_8, X_{12}, X_{15}, and X_{18}. Each row in the illustration shows how the difference is taken for each individual variable.
- The differences are squared, as in the equation for Euclidean distances and as is typically done in computing measures of variation to avoid the problem of having negative and positive differences cancel each other out. Thus, the individual squared differences for each variable show how different observation 100 is on that particular variable. Here, observation 100 shows the greatest dissimilarity on variable X_8.
- The squared differences for each variable are summed to get a total for this observation across all clustering variables. This represents the squared distance from the typical or average respondent.
- Next, the square root of that sum is taken to create an estimate of the dissimilarity of this observation based on distance from the typical respondent profile.
- The process can be repeated for each observation. Observations with the highest dissimilarities have the potential to be outliers.

Researchers looking for outliers do not focus on the absolute value of dissimilarity. Rather, researchers are simply looking for any values that are relatively large compared to the others. Thus, sorting the observations from the most to the least dissimilar can be convenient. Here, we will look at the 10 percent of observations that are most dissimilar. In examining the 10 most dissimilar observations listed in Table 9-5, two observations—87 and 6—display relatively large values. Each has a dissimilarity of over 5 (5.58 and 5.30, respectively). In contrast, the next largest dissimilarity is for observation 90 (4.62), which is only slightly larger than the fourth highest proximity of 4.57 for observation 53. These two observations—87 and 6—stand out over the others as having relatively high average distances. At this point, we will not eliminate any observations but will continue to watch for other potential signs that observations 87 and/or 6 may truly be outliers. Generally, researchers are more comfortable deleting observations as outliers when multiple pieces of evidence are present.

DEFINING SIMILARITY   The next issue involves the choice of a similarity measure to be used as input to the hierarchical clustering algorithm. The researcher does not have to actually perform these computations separately, but rather only needs to specify which approach will be used by the cluster program. Correlational measures are not used, because when segments are identified with cluster analysis we should consider the magnitude of the perceptions (favorable versus unfavorable) as well as the pattern. Correlational measures only consider the patterns of the responses, not the absolute values. Cluster analysis objectives are therefore best accomplished with a distance measure of similarity.

Given that the five clustering variables are metric, squared Euclidean distance is chosen as the similarity measure. Either squared Euclidean distance or Euclidean distance is typically the default similarity measure in statistical packages. Multicollinearity has been addressed by selecting variables that are not highly correlated with each other based on the previous factor analysis.

SAMPLE SIZE   The third issue relates to the adequacy of the sample of 100 observations. This issue is not a statistical (inferential) issue. Instead, it relates to the ability of the sample to identify managerially useful segments. That is, segments with a large enough sample size to be meaningful. In our example, the HBAT research team believes segments that represent at least 10 percent of the total sample size will be meaningful. Smaller segments are considered too small to justify development of segment-specific marketing programs. Thus, in our example with a sample of 100 observations, we consider segments consisting of 10 or more observations as meaningful. But initial clusters obtained through hierarchical clustering with as few as five observations may be retained because the cluster size will likely change when observations are reassigned in nonhierarchical clustering.

STANDARDIZATION The final issue involves the type of standardization that may be used. It is not useful to apply within-case standardization because the magnitude of the perceptions is an important element of the segmentation objectives. But the issue of standardizing by variable still remains.

All of the clustering variables are measured on the same scale (0 to 10), so there is no need to standardize because of differences in the scale of the variables. There are, however, other considerations, such as the standard deviation of the variables. For example, as shown in Table 9-4, the variables generally have similar amounts of dispersion, with the possible exception of X_{18} (a small standard deviation of 0.73 compared to all others above one). These differences could affect the clustering results, and standardization would eliminate that possibility. But with only one variable exhibiting a difference, we choose not to standardize in the HBAT example.

Another consideration is the means of the variables used in the cluster analysis. For example, the means of the variables in the HBAT example vary somewhat, ranging from less than four for X_{18} (3.89) to nearly eight for X_6 (7.81). This does not suggest standardization, but it may make it more difficult to interpret each cluster's meaning. In some situations, the mean-centered values will facilitate interpretation of clusters. Using mean centering does not affect the cluster results, but it often makes it easier to compare the mean values on each variable for each cluster.

Table 9-4 displays the means for the variables across all 100 observations. Mean-centered values for each variable can be obtained by subtracting the mean from each observation. The HBAT researcher performs this task by using the software's compute function and entering the following instruction for each variable:

$$X_{6C} = X_6 - \text{Mean}(X_6)$$

Where X_{6C} is the name given by the researcher to represent the mean centered values for X_6, X_6 is the variable itself, and Mean(X_6) represents the mean value for that variable (7.81 in this case). This process is repeated for each cluster variable. Mean-centered variables retain the same information as the raw variables because the standard deviations are the same for each clustering variable and its mean-centered counterpart. The mean for each mean-centered variable is zero, however, meaning that each mean-centered variable has this common reference point. This is simply a way of recoding the variables to have a common mean. The common mean may make it easier to interpret the cluster profiles.

Stage 3: Assumptions in Cluster Analysis

In meeting the assumptions of cluster analysis, the researcher is not interested in the statistical qualities of the data (e.g., normality, linearity, etc.) but instead is focused primarily on issues of research design. The two basic issues to be addressed are sample representativeness and multicollinearity among the clustering variables.

SAMPLE REPRESENTATIVENESS A key requirement for using cluster analysis to meet any of the objectives discussed in stage 1 is that the sample is representative of the population of interest. Whether developing a taxonomy, looking for relationships, or simplifying data, cluster analysis results are not generalizable from the sample unless representativeness is established. The researcher must not overlook this key question, because cluster analysis has no way to determine if the research design ensures a representative sample.

The sample of 100 HBAT customers was obtained through a random selection process from among the entire customer base. All issues concerned with data collection were addressed adequately to ensure that the sample is representative of the HBAT customer base. Thus, we can extend the sample findings to the population of HBAT customers.

MULTICOLLINEARITY   If there is multicollinearity among the clustering variables, the concern is that the set of clustering variables is assumed to be independent, but may actually be correlated. This may become problematic if several variables in the set of cluster variables are highly correlated and others are relatively uncorrelated. In such a situation, the correlated variables influence the cluster solution much more so than the several uncorrelated variables.

As indicated earlier, the HBAT research team minimized any effects of multicollinearity through the variable selection process. That is, they chose cluster variables based on the findings of the previous factor analysis.

Employing Hierarchical and Nonhierarchical Methods

In applying cluster analysis to the sample of 100 HBAT customers, the research team decided to use both hierarchical and nonhierarchical methods in combination. To do so, they followed a two-step process:

1. The *first step* was the partitioning stage. A hierarchical procedure was used to identify a preliminary set of cluster solutions as a basis for determining the appropriate number of clusters.
2. The *second step* used nonhierarchical procedures to "fine-tune" the results and then profile and validate the final cluster solution. The hierarchical and nonhierarchical procedures from SPSS are used in this analysis, although comparable results would be obtained with most other clustering programs.

Step 1: Hierarchical Cluster Analysis (Stage 4)

In this step, we utilize the hierarchical clustering procedure's advantage of quickly examining a wide range of cluster solutions to identify a set of preliminary cluster solutions. This range of solutions is then analyzed by nonhierarchical clustering procedures to determine the final cluster solution. Our emphasis in the hierarchical analysis, therefore, is on stage 4 (the actual clustering process). The profiling and validation stages (stages 5 and 6) are then undertaken in step 2—the nonhierarchical process. In the course of performing the hierarchical cluster analysis, the researcher must perform a series of tasks:

1. Select the clustering algorithm.
2. Generate the cluster results and check for single member or other inappropriate clusters.
3. Select the preliminary cluster solution(s) by applying the stopping rule(s).
4. Profile the clustering variables to identify the most appropriate cluster solutions.

In doing so, the researcher must address methodological issues as well as consider managerial and clustering objectives to derive the most representative cluster solution for the sample. In the following sections, we will discuss both types of issues as we address the tasks listed above.

SELECTING A CLUSTERING ALGORITHM   Before actually applying the cluster analysis procedure, we must first ask the following question: Which clustering algorithm should we use? Combined with the similarity measure chosen (squared Euclidean distance), the clustering algorithm provides the means of representing the similarity between clusters with multiple members. The Ward's method was used because of its tendency to generate clusters that are homogeneous and relatively equal in size.

INITIAL CLUSTER RESULTS   With the similarity measure and clustering algorithm defined, the HBAT research team can now apply the hierarchical clustering procedure. The results must be reviewed for the range of cluster solutions selected. This process enables us to identify any clusters that may need to be deleted due to small size or other reasons (outliers, unrepresentative, etc.). After review, any identified clusters or data are deleted and the cluster analysis is run again with the reduced dataset.

Table 9-6 shows a portion of the agglomeration schedule produced by the hierarchical cluster results. The agglomeration schedule can be useful in identifying any unusual clusters or observations that are resisting joining others within a cluster. Key information in the agglomeration schedule includes the information about when a cluster appears and the agglomeration coefficients.

The schedule provides information for each of the stages of the clustering process. A stage is when one or more individual observations combines with another observation to form a cluster. For example, in stage 1 observations 3 and 94 combine to create the first cluster. The hierarchical process concludes (stage 99) when the observations making up cluster 1 combine with the observations from cluster 6 to form a single huge cluster (all 100 observations).

There is other information in the table that helps us to understand the clustering process. Looking down the schedule to stage 75 and then across that row, we first see that observation 6 is combining with observation 87 to form a cluster. By looking further to the right across the table, under the columns labeled Stage Cluster First Appears, we see that neither of these observations has joined a cluster until this stage. The zero in the columns labeled Cluster 1 and Cluster 2 (under the stage cluster first appears) indicates that before this stage neither observations 6 or 87 were members of any cluster. Recall in our earlier discussion of outliers in the HBAT data that these two observations were indicated as relatively different from the other 98, and possibly outliers. At this point, with this additional information, the researcher decides observations 6 and 87 should be removed as outliers. Thus, the cluster analysis must be run again (respecified) after observations 6 and 87 are removed. Observation 84 also joins a cluster late in the game (stage 76), but we decide to keep it because it did not display as high an average distance from the other observations.

TABLE 9-6   Partial Agglomeration Schedule for Initial HBAT Hierarchical Cluster Solution

Agglomeration Schedule

	Cluster Combined			Stage Cluster First Appears		
Stage	Cluster 1	Cluster 2	Coefficients	Cluster 1	Cluster 2	Next Stage
1	3	94	.080	0	0	18
2	75	96	.180	0	0	62
.	.	.	.	.	.	.
.	.	.	.	.	.	.
18	3	38	6.065	1	0	67
.	.	.	.	.	.	.
.	.	.	.	.	.	.
74	2	98	120.542	59	0	92
75	6	87	125.83	0	0	89
76	32	84	131.506	54	0	86
77	3	50	137.566	67	62	83
.	.	.	.	.	.	.
.	.	.	.	.	.	.
98	6	11	659.781	97	95	99
99	1	6	812.825	96	98	0

Note: Stages 3–17, 19–73, and 78–97 have been omitted from the table.

Changes in the agglomeration coefficients can be used, as with the proximity measures, much like eigenvalues from scree plots, to help identify the appropriate number of clusters. We discuss this process later in the chapter.

RESPECIFIED CLUSTER RESULTS   The deletion of two observations requires that the cluster analysis be performed again on the remaining 98 observations. We now discuss the findings of the new cluster solution, including examining cluster sizes and the clustering criteria.

EVALUATING CLUSTER SIZES   The process proceeds as before, with the respecified cluster results also examined for inappropriate cluster sizes. Clusters that are below the size considered managerially significant are candidates for deletion. It is expected that outliers have already been identified before respecification. However, the researcher may consider additional single member or extremely small clusters as outliers as well and may want to remove them from further analysis.

Researchers should be cautioned, however, against "getting into a loop" by continually deleting small clusters and then respecifying the cluster analysis. Judgment must be used in accepting a small cluster and retaining it in the analysis at some point. Otherwise, you may find that the process will start to delete small, but representative, segments. The most problematic small clusters to retain are those that do not combine until very late in the process. But small clusters that are merged in the higher ranges of cluster solutions may be retained. Further, when applying a two-step cluster approach, hierarchical followed by nonhierarchical, final cluster sizes are uncertain until the second step. For several reasons, therefore, care should be taken in deleting small clusters even if they contain slightly fewer observations than would be considered managerially useful.

The HBAT research team chose Ward's algorithm as the clustering method. This reduces the chance of finding small clusters, because it tends to produce clusters of equal size.

Clustering Schedule and Agglomeration Coefficient.   All of the individual clusters in our solutions meet the minimum cluster size criteria. We therefore proceed to further examine the actual clustering process through the clustering schedule and agglomeration coefficient. The clustering schedule produced by SPSS was shown in Table 9-6. The five elements describing each clustering stage are:

- *Stage*: Recall that the stage is the step in the clustering process where the two most similar clusters are combined. For a hierarchical process, there are always $N-1$ stages, where N is the number of observations being clustered. Thus, in the previous HBAT cluster we initially had a sample size of 100 and there were 99 stages.
- *Clusters Combined*: Information detailing which two clusters are combined at each stage.
- *Agglomeration Coefficient*: Measures the increase in heterogeneity (reduction in within cluster similarity) that occurs when two clusters are combined. For most hierarchical methods, the agglomeration coefficient is the distance between the two closest observations in the clusters being combined.
- *Stage Cluster First Appears*: Identifies the prior stage at which each cluster being combined was involved. Values of zero indicate the observation is still a single member cluster. That is, the observation has never been combined before that stage.
- *Next Stage in Which New Cluster Appears*: Denotes the next stage at which the new cluster is combined with another cluster.

Let's return to the agglomeration schedule shown in Table 9-6. We will examine a few stages in detail to illustrate the information contained in the schedule and to demonstrate how this information can help determine an appropriate number of clusters to extract from the solution.

- Stage 1: Clusters 3 and 94 are the first to join with a coefficient (which is a within-subjects sum of squares when using Ward's algorithm) of only 0.080. In this case, both clusters are in reality individual observations. We know this because of the zeros in the columns labeled "stage" in which the cluster first appears (columns on right side of table). This cluster will again be seen in stage 18 when observation (now cluster) 3 joins observation 38. Notice the "1" in the column under "Stage Cluster First Appears" for cluster 1. This refers back to the fact that the last time cluster 1 was joined with anything was in stage 1. The 67 in the next stage column means the next time this cluster will be joined with something else is in Stage 67.
- Stage 99: Clusters 1 and 6 are joined. The result is that all 100 observations are now in a single large cluster and this stage has an agglomeration coefficient of 812.8. This is much larger than the coefficient in stage 1.

The information from the clustering schedule provides an overview of the clustering process, enabling the researcher to follow any single observation or cluster throughout the process. It also provides diagnostic information, such as the ability to quickly identify single-member clusters (i.e., a zero in the "Stage Cluster First Appears" column) and the agglomeration coefficient.

Dendrogram. The dendrogram provides a graphical portrayal of the clustering process. The treelike structure of the dendrogram depicts each stage of the clustering process. Typically, the graph is scaled, so that closer distances between combinations indicate greater homogeneity.

The dendrogram is a visual display of the information from the agglomeration schedule. Results from the revised hierarchical cluster analysis, after deleting observations 6 and 87, produced a new clustering schedule and a new dendrogram. Some may find it easier to visually analyze the information in this graphical form rather than in the agglomeration schedule. A portion of the new agglomeration schedule ($N = 98$) is shown in Table 9-7. The dendrogram is not reproduced here, but generally displays the same pattern as shown in the agglomeration schedule.

Determining the Preliminary Cluster Solution(s). Up to this point we have detailed the aspects of the clustering process. But we still have not addressed the fundamental question: What is the final cluster solution? We should note that in most situations a single final solution will not be identified in hierarchical analysis. Rather, a set of preliminary cluster solutions is identified. These cluster solutions form the basis for the nonhierarchical analysis from which a final cluster solution is selected. Even though a final cluster solution is not identified at this stage, the researcher must make a critical decision as to how many clusters will be used in the nonhierarchical analysis. Hopefully, the decision will be relatively clear and the potential number of solutions can be determined. However, researchers routinely decide that a small number of cluster solutions should be analyzed using nonhierarchical procedures. Further analysis of multiple clusters may help resolve the question of how many clusters are most appropriate.

Applying the Stopping Rule. How many clusters should we have? Because the data involve profiles of HBAT customers and our interest is in identifying customer types or profiles that may form the basis for differing strategies, a manageable number of segments from a strategic and tactical perspective would be more than two but no more than six or seven.

- *Percentage Changes in Heterogeneity* The stopping rule we will apply is based on assessing the changes in heterogeneity between cluster solutions. The basic rationale is that when large increases in heterogeneity occur in moving from one stage to the next, the researcher selects the prior cluster solution because the new combination is joining quite different clusters.

  The agglomeration coefficient is particularly useful for this stopping rule. Small coefficients indicate that fairly homogeneous clusters are being merged, whereas joining two

TABLE 9-7   Agglomeration Schedule for the Reduced HBAT Cluster Sample

Stage	Cluster 1	Combined with Cluster:	Coefficient	Number of Clusters After Combining	Differences	Proportionate Increase in Heterogeneity to Next Stage	Stopping Rule
90	1	2	297.81	8	28.65	9.6%	HBAT not interested in this many clusters.
91	22	27	326.46	7	39.11	12.0%	Increase is larger than the previous stage, arguing against combination.
92	1	5	365.56	6	41.82	11.4%	Increase is relatively small, favoring combination to five clusters.
93	7	10	407.38	5	58.01	14.2%	Increase is larger than the previous stage, favoring five to four clusters.
94	1	4	465.39	4	70.86	15.2%	Increase is relatively large, favoring four clusters over three and suggests a possible stopping point.
95	7	22	536.24	3	77.55	14.5%	Increase is relatively large and favors a three-cluster solution over a two-cluster solution.
96	7	9	613.79	2	138.71	22.6%	Increase from two to one is relatively large (the increase from two to one is normally large).
97	1	7	752.50	1		—	One-cluster solution not meaningful.

very different clusters results in a large coefficient. Because each combination of clusters results in increased heterogeneity, we focus on large percentage changes in the coefficient, similar to the scree test in factor analysis, to identify cluster combination stages that are markedly different. The only caveat is that this approach, although a fairly accurate algorithm, has the tendency to indicate too few clusters.

Look at the bottom portion of the agglomeration (Stages 90 through 97) shown in Table 9-7. This will be helpful in determining an appropriate number of clusters to examine further . . . since no more than 7 clusters are desired. The first four columns contain information similar to that in Table 9-6. However, three additional columns have been added to facilitate an understanding of this solution. The fifth column states the number of clusters that exist upon completing this stage. A sixth column shows the differences in the agglomeration coefficients between a particular stage and the next combination. In other words, it shows how much smaller the coefficient is compared to the next stage. Recall that these coefficients also indicate how much heterogeneity exists in the cluster solution. Thus, the differences indicate how much the heterogeneity increases when you move from one stage to the next. Interpretation of the cluster number stopping rule is facilitated by calculating the percentage change in the clustering coefficients for these final stages. Two points worth remembering in interpreting these solutions are:

- By their nature, the size of the agglomeration coefficient gets larger toward the end of the cluster solution.
- Applying a stopping rule is not an exact science.

We now look for relatively large increases in the agglomeration coefficients. The agglomeration coefficient shows rather large increases in going from stages 94 to 95 (465.38 versus 536.24), stages 95 to 96 (536.24 versus 613.79), and stages 96 to 97 (613.79 versus 752.50). The percentage increases show the proportionate increase associated with the combination of clusters from one stage to the next.

The average proportionate increase for all stages shown (90 to 97) is 14.2 percent and serves as a rough guide in determining what a large increase is. Stage 97 results from condensing a two-cluster solution to a one-cluster solution. That is, all 98 observations are in a single cluster in stage 97. Combining two clusters into one yields a proportionate increase of 22.6 percent ((752.50 – 613.79)/613.79 = .226). Although this is the largest increase, cluster solutions almost always show a large increase for this stage. A two-cluster solution also may represent limited value in meeting many research objectives. Researchers must avoid the temptation to say the two-cluster solution is the best, because it involves the largest change in heterogeneity. A two-cluster solution must be supported by strong theoretical reasoning. Thus, we will not select a two-cluster solution.

Now let us look at the other stages shown in Table 9-7. The movements between stages 93, 94, and 95 also are associated with relatively large increases in heterogeneity (14.2%, 15.2%, and 14.5%, respectively). What does this mean for the cluster solution? Let us take a look at the largest increase (moving from stage 94 to 95). The proportionate increase moving from the four-cluster to the three-cluster solution is 15.2 percent ((536.24 − 465.39)/465.39). This means that the cluster solution associated with four clusters is associated with proportionately less heterogeneity than is the three-cluster solution. More homogeneous clusters are a good cluster characteristic. Thus, we will focus on a four-cluster solution because the largest percent increase (other than stage 97 to 96) would occur if we used the results of this stage. However, the researcher also should note that a quite large increase occurs in the coefficient when moving from a five-cluster to a four-cluster solution. Using more of a scree plot logic, an argument can be made that this would be a stopping point. Also, the second largest increase occurs in moving from a three-cluster to a two-cluster solution (14.5%). Therefore, the three-cluster and five-cluster solutions are plausible candidates to compare with a four-cluster solution, particularly if the four-cluster solution proves difficult to interpret or has otherwise undesirable characteristics. Some may prefer a graphical depiction of the percent change, such as that shown in Figure 9-11.

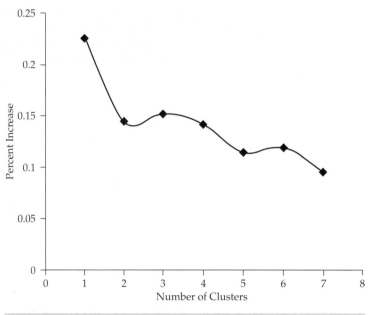

FIGURE 9-11   **Percent Change in Heterogeneity**

Profiling the Clustering Variables. Before proceeding to the nonhierarchical analysis, we will profile the clustering variables for the four-cluster solution to confirm that the differences between clusters are distinctive and significant in light of the research question and to define the characteristics of the clusters. The profiling information is shown in Figure 9-12.

Let us examine the distinctiveness first. At the far right side of the figure are the F statistics from one-way ANOVAs that examine whether there are statistically significant differences between the four clusters on each of the five clustering variables. The independent variable is cluster membership (which of the four clusters each of the 98 observations were placed in by the clustering process), and the dependent variables are the five clustering variables. The results show there are significant differences between the clusters on all five variables. The significant F statistics provide initial evidence that each of the four clusters is distinctive.

Now we examine the means of the five cluster variables. This stage in the profiling process is based on interpretation of both the mean values and the mean-centered values. Cluster 1 contains 49 observations and has a relatively lower mean on X_{15} (New Products) than the other three clusters. The means of the other three clusters are somewhat above average. Cluster 2 contains 18 observations and is best characterized by two variables: a very low mean on X_8 (Technical Support) and the highest score on X_{15} (New Products). Cluster 3 has 14 observations and is best characterized by a relatively low score on X_6 (product quality). Cluster 4 has 17 observations and is characterized by a relatively low score on X_{12} (Salesforce Image). These results indicate that each of the four clusters exhibit somewhat distinctive characteristics. Moreover, no clusters contain less than 10 percent of observations. Therefore, all clusters are retained, because this preliminary assessment is sufficiently favorable to indicate moving on to nonhierarchical clustering. The cluster sizes will change in the nonhierarchical analysis and observations will be reassigned. As a result, the final meanings of the four clusters will be determined in the nonhierarchical analysis.

Step 2: Nonhierarchical Cluster Analysis (Stages 4, 5, and 6)

The hierarchical clustering method facilitated a comprehensive evaluation of a wide range of cluster solutions. These solutions were impacted, however, by a common characteristic—once observations are joined in a cluster, they are never separated (reassigned) in the clustering process. In the hierarchical

Means from Hierarchical Cluster Analysis

Variable	Mean Values Cluster Number:				Mean-Centered Values Cluster Number:				F	Sig
	1	**2**	**3**	**4**	**1**	**2**	**3**	**4**	*F*	*Sig*
X_6 Product Quality	8.21	8.04	5.97	8.18	0.40	0.23	−1.84	0.37	14.56	0.000
X_8 Technical Support	5.37	4.04	6.16	6.47	0.00	−1.33	0.78	1.09	12.64	0.000
X_{12} Salesforce Image	4.91	5.69	6.12	4.42	−0.02	0.57	1.00	−0.72	11.80	0.005
X_{15} New Products	3.97	6.63	5.51	6.28	−1.18	1.45	0.36	1.13	62.74	0.000
X_{18} Delivery Speed	3.83	4.14	4.37	3.45	−0.06	0.25	0.48	−0.44	5.49	0.002
Cluster Sample Sizes	49	18	14	17	49	18	14	17		

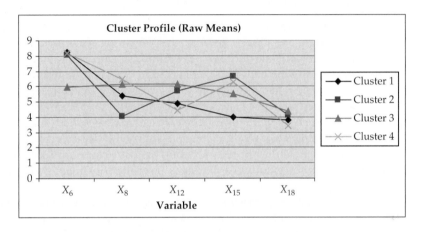

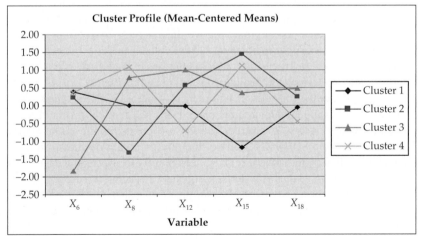

FIGURE 9-12 Profile of Four Clusters from Hierarchical Cluster Analysis

clustering process, we selected the algorithm (Ward's) that minimized the impact of this process. But nonhierarchical clustering methods have the advantage of being able to better "optimize" cluster solutions by reassigning observations until maximum homogeneity (similarity) within clusters is achieved.

This second step in the clustering process uses results of the hierarchical process to execute nonhierarchical clustering. Specifically, the number of clusters is determined from the hierarchical results. Nonhierarchical procedures then develop "optimal" cluster solutions. The cluster solutions are then compared in terms of criterion validity as well as applicability to the research question to select a single solution as the final cluster solution.

STAGE 4: DERIVING CLUSTERS AND ASSESSING OVERALL FIT   The primary objective of the second step is using nonhierarchical techniques to adjust, or "fine-tune," the results from the hierarchical procedures. In performing a nonhierarchical cluster, researchers must make two decisions:

1. How will seed points for the clusters be generated?
2. What clustering algorithm will be used?

The following discussion addresses both of these points by demonstrating how to use the hierarchical results to improve the nonhierarchical procedure.

Specifying Cluster Seed Points.   The first task in nonhierarchical cluster analysis is to select the method for specifying cluster seeds. The cluster seeds are the initial starting point for each cluster. From there, the clustering algorithm assigns observations to each seed and forms clusters. There are two methods for selecting cluster seed points: generation of the sample by the cluster software (i.e., random selection) and specification by the researcher. Sample-generated methods sometimes produce clusters that are difficult to replicate across samples and are not based on theoretical support. In contrast, researcher-specified cluster seeds require some conceptual or empirical basis for selecting the seed points. Researcher-specified methods reduce problems with replicability of cluster solutions. But choosing the best seed points can be difficult, and with some software packages inserting researcher determined cluster seeds is complicated.

The most common approach to identify researcher-specified cluster seeds is the hierarchical solution. This involves either selecting a single observation from each cluster to represent the cluster or, more commonly, to use the cluster centroids as the seed points. Note that deriving the cluster centroids typically requires additional analysis to (a) select the cluster solution(s) to be used in the nonhierarchical analysis and (b) to derive the centroids by profiling each cluster solution. These profiles are not typically generated in the hierarchical analysis, because doing so requires a tremendous effort, as $N - 1$ cluster solutions (97 solutions in the HBAT example) are generated, and deriving a profile for each would be time-consuming and inefficient.

Recall that the four-cluster hierarchical solution was selected as the one that would be further analyzed using nonhierarchical procedures. All five clustering variables will be used in the nonhierarchical analysis. Thus, the cluster seed points require initial values on each variable for each cluster. For our HBAT example, the research team decides to use the random initial seed points identified by the software. These seed points are affected by the ordering of the observations in the data file. To evaluate cluster solution stability, some researchers reorganize the data (change the order of the observations) and rerun the cluster analysis. If the cluster solutions change substantially, which indicates they are highly unstable, researcher-specified seed points may be need to be used.

Selecting a Clustering Algorithm.   The researcher must now select the clustering algorithm to be used in forming clusters. A primary benefit of nonhierarchical cluster methods is the ability to develop cluster solutions later in the process that are not based on clusters formed earlier. This is because observations assigned to a particular cluster early in the clustering process can later be reassigned (moved from one cluster to another) to another cluster formed later in the process. This is in contrast to hierarchical methods, where cluster solutions formed later in the clustering process are directly based on combining two clusters formed earlier in the process. For this reason, nonhierarchical methods are generally preferred, when possible, for their "fine-tuning" of an existing cluster solution from a hierarchical process.

For the HBAT example, we selected the optimizing algorithm in SPSS that allows for reassignment of observations among clusters until a minimum level of heterogeneity is reached. Using this algorithm, observations are initially grouped to the closest cluster seed. When all observations are assigned, each observation is evaluated to see if it is still in the closest cluster. If it is not, it is reassigned to a closer cluster. The process continues until the heterogeneity within clusters cannot be increased by further movement (reassignment) of observations between clusters.

Forming Clusters. With the cluster seeds and clustering algorithm specified, the clustering process can begin. To execute the nonhierarchical cluster, we specify the number of clusters as four, based on the results of the hierarchical cluster solution. Using the optimizing algorithm, the process continues to reassign observations until reassignment will not improve within-cluster homogeneity.

Results from the nonhierarchical four-cluster solution are shown in Figure 9-13. There are two notable differences between the hierarchical and nonhierarchical results:

- **Cluster sizes.** The nonhierarchical solution, perhaps due to the ability to reassign observations between clusters, has a more even dispersion of observations among the clusters. As an example, nonhierarchical analysis resulted in cluster sizes of 25, 29, 17, and 27, compared to clusters of 49, 18, 14, and 17 in the hierarchical analysis.
- **Significance of clustering variable differences.** Another fundamental difference between the two cluster solutions is the ability of the nonhierarchical process to delineate clusters that are usually more distinctive than the hierarchical cluster solution. Figure 9-13 includes ANOVA results showing the differences in variable means across four clusters. Given that the five clustering variables were used to produce the clusters, the results should be statistically significant. The F-values indicate that the means of four of the five variables are significantly different. Only the means of X_{18} (delivery speed) are not significantly different across groups. In fact, three of the five clustering variables have very large F-values (X_6, X_8, and X_{15}). Thus, the nonhierarchical results suggest that the cluster solution is adequately discriminating observations, with the exception of X_{18}, delivery speed.

The nonhierarchical clustering process produced a four-cluster solution based on the software-generated seed points. Further analysis in terms of profiling the solutions and assessing their criterion validity will provide the elements needed to select a final cluster solution.

STAGE 5: INTERPRETATION OF THE CLUSTERS The HBAT research team interprets the meanings of the clusters by analyzing the pattern of cluster means and mean-centered values shown in Figure 9-13, which are plotted in the profile diagram in the lower portion of the figure. Interpretation begins by looking for extreme values associated with each cluster. In other words, variable means that are the highest or lowest compared to other clusters are useful in this process.

- Cluster 1 has 25 observations and is most distinguished by a relatively low mean for new products (X_{15}). The means for the other variables (except product quality) also are relatively low. Thus, this cluster represents a market segment characterized by the belief that HBAT does not perform well in general, particularly in offering new products, and the overall lower means suggest this segment is not a likely target for new product introductions.
- Cluster 2 has 29 observations and is most distinguished by relatively higher means on technical support (X_8) and on product quality (X_6). Therefore, HBAT interprets this market segment as believing that HBAT provides strong support for its high-quality products. HBAT believes this is a favorable segment for other products and services. This is the largest cluster.
- Cluster 3 has 17 observations and is distinguished by a relatively higher mean for new products (X_{15}). In contrast, cluster 3 has a relatively lower mean for technical support (X_8). Thus, this segment suggests that HBAT offers new and innovative products, but that its support of these new products is not very good. It should be noted that the new product mean of 6.75, although the highest for any cluster, is still only moderate overall, and HBAT could improve in this area even with this cluster. Cluster 3 is somewhat the opposite of cluster 2.

Variable	Mean Values Cluster Number:				Mean-Centered Values Cluster Number:				F	Sig
	1	**2**	**3**	**4**	**1**	**2**	**3**	**4**		
X_6 Product Quality	8.25	8.91	8.18	6.14	0.44	1.10	0.37	−1.67	55.06	0.000
X_8 Technical Support	4.40	6.80	3.92	5.86	−0.97	1.43	−1.44	0.50	45.56	0.000
X_{12} Salesforce Image	4.70	4.89	5.49	5.59	−0.42	−0.23	0.37	0.47	4.56	0.005
X_{15} New Products	3.83	5.25	6.75	5.01	−1.32	0.10	1.60	−0.14	25.56	0.000
X_{18} Delivery Speed	3.72	3.89	4.05	3.98	−0.17	0.01	0.17	0.10	0.88	0.46
Cluster Sample Sizes	25	29	17	27	25	29	17	27		

Cluster Means from Actual Data Plotted

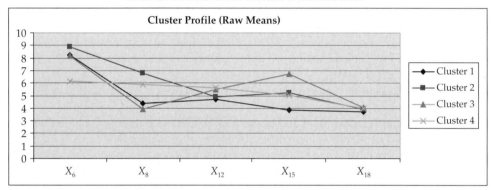

Cluster Means from Mean-Centered Data Plotted

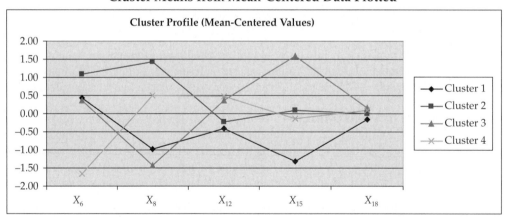

FIGURE 9-13   Means from K-Means Cluster Solution

- Cluster 4 has 27 observations and is distinguished by the lowest mean of all clusters on product quality (X_6). All of the means of variable X_6 in the other clusters are much higher. Moreover, the means on the other clustering variables are relatively average for this cluster, with the exception of a somewhat higher mean on salesforce image (X_{12}) and technical support (X_6). Thus, the segment is characterized as one that indicates that HBAT's products are below average in quality, but the salesforce and technical support are slightly better than average. This is the second largest cluster (X_6).

STAGE 6: VALIDATION AND PROFILING THE CLUSTERS   In this final stage, the processes of validation and profiling are critical due to the exploratory and often atheoretical basis for the cluster analysis. Researchers should perform tests to confirm the validity of the cluster solution while also ensuring the cluster solution has practical significance. Researchers who minimize or skip this step risk accepting a cluster solution that is specific only to the sample and has limited generalizability, or even little use beyond its mere description of the data on the clustering variables.

Cluster Stability. At this point, researchers often assess the stability of the cluster solution. Given that the software chose the initial seed points, factors such as the ordering of the cases in the data can affect cluster membership. To do so, the researcher can sort the observations in a different order and then perform the cluster analysis once again (with the new starting point selected by the software, but with the same number of clusters specified). The cluster solutions can then be compared to see if the same clusters are identified. A cross-classification of cluster membership between solutions should reveal mostly matches between the two solutions. In other words, the observations that cluster together in one analysis should for the most part cluster together in the subsequent cluster solution. This is best illustrated by an example.

The stability of the HBAT four-group nonhierarchical cluster solution is examined. The researcher first sorts the observations into a different order. To do so, the researcher selects a variable from the data set and uses the SPSS sort function to change the order of the observations. In this case, the observations were sorted by customer type (X_1), ranging from those with the least time doing business with HBAT to those with the most time doing business with HBAT. The K-means algorithm is once again used to place observations into one of four clusters. Following the clustering routine, a cross-classification is performed (much like a confusion matrix in discriminant analysis), using the cluster membership variable from the first K-means solution as one variable and the cluster membership variable from the second K-means solution as the other variable. The results are shown in Table 9-8.

Most observations are grouped with the same observations they clustered with in the first K-means solution. Although cluster 1 in the first K-means becomes cluster 4 in the second K-means analysis (as indicated by the 24 in the first row, fourth column of the cross-classification), all but one of the observations ends up clustering together. The one observation not staying together is now in cluster 3. For cluster 2, 8 of the 29 observations end up not clustering together. All cluster 3 observations stay together. Cluster 4, which is now cluster 1, retains 22 of the 27 original members. Perfect cross-validation would appear if only one cell in each row or column of the cross-classification contained a value. Thus, all but 14 observations have retained the same cluster membership across solutions—a result that supports the validity of a four-cluster solution. In other words, the four-cluster solution appears relatively stable with only fourteen percent of the cases switching clusters between solutions. Additional cluster analyses conducted based on sorting the data in a different way could be conducted to further examine the stability of the data.

TABLE 9-8   Cross-Classification to Assess Cluster Stability

Cluster Number from First K-Means	Cluster Number from Second K-Means				Total
	1	2	3	4	
1	0	0	1	24	25
2	2	21	0	6	29
3	0	0	17	0	17
4	22	0	5	0	27
Total	24	21	23	30	98

Assessing Criterion Validity. To assess predictive validity, we focus on variables that have a theoretically based relationship to the clustering variables but were not included in the cluster solution. Given this relationship, we should see significant differences in these variables across the clusters. If significant differences do exist on these variables, we can draw the conclusion that the clusters depict groups that have predictive validity.

For this purpose, we consider four outcome measures from the HBAT dataset:

- X_{19} — Satisfaction
- X_{20} — Likelihood to Recommend
- X_{21} — Likelihood to Purchase
- X_{22} — Purchase Level

A MANOVA model (see Chapter 8) was estimated using the four criterion validity variables as the dependent variables and cluster membership as an independent variable. MANOVA was selected because the dependent variables are known to correlate with each other. Table 9-9 displays the results. First, the overall MANOVA model is significant ($F = 2.23$, $P = .01$), providing initial support for the idea that these variables can be predicted by knowing to which segment an HBAT customer belongs. The individual univariate F-statistics are also significant, further verifying this finding.

The results demonstrate, therefore, that the cluster solution can predict other key outcomes, which provides evidence of criterion validity. For example, cluster 2, which HBAT believed was receptive to more business based on its cluster profile (described above), displays the highest scores on each of these key outcome variables. Thus, HBAT will likely find the cluster solution useful in predicting other key outcomes and forming appropriate strategies.

Profiling the Final Cluster Solution. The final task is to profile the clusters on a set of additional variables not included in the clustering variate or used to assess predictive validity. The importance of identifying unique profiles on these additional variables is in assessing both the

TABLE 9-9 Multivariate *F* Results Assessing Cluster Solution Criterion Validity

Variable	Cluster Number	Cluster Mean	Multivariate F^*	Univariate F^*	Sig.
			2.23		0.01
X_{19} Satisfaction	1	6.76		5.98	0.001
	2	7.44			
	3	7.39			
	4	6.34			
X_{20} Likely to Recommend	1	6.89		3.06	0.032
	2	7.46			
	3	7.14			
	4	6.68			
X_{21} Likely to Purchase	1	7.74		3.53	0.018
	2	8.09			
	3	7.83			
	4	7.33			
X_{22} Purchase Level	1	58.70		6.21	0.001
	2	62.17			
	3	60.92			
	4	53.17			

*Multivariate F has 12,241 degrees of freedom and univariate Fs each have 3,94 degrees of freedom.

practical significance and the theoretical basis of the identified clusters. In assessing practical significance, researchers often require that the clusters exhibit differences on a set of additional variables.

In this example, five characteristics of the HBAT customers are available. These include X_1 (Customer Type), X_2 (Industry Type), X_3 (Firm Size), X_4 (Region), and X_5 (Distribution System). Each of these variables is nonmetric, similar to the variable representing cluster membership for each observation. Thus, cross-classification is used to test the relationships.

Results of the cross-classification are provided in Table 9-10. Significant chi-square values are observed for three of the five profile variables. Several patterns are evident. For example, cluster 4 consists almost entirely of customers from outside of the USA/North America (26 out of 27). In contrast, cluster 2 consists predominantly of customers from the USA/North America. From these variables, distinctive profiles can be developed for each cluster. These profiles support the distinctiveness of the clusters on variables not used in the analysis at any prior point.

A successful segmentation analysis not only requires the identification of homogeneous groups (clusters), but also that the homogeneous groups are identifiable (uniquely described by other variables). When cluster analysis is used to verify a typology or other proposed grouping of objects, associated variables—either antecedents or outcomes—typically are profiled to ensure correspondence of the identified clusters within a larger theoretical model.

EXAMINING AN ALTERNATIVE CLUSTER SOLUTION: STAGES 4, 5, AND 6 The four-cluster solution was examined first because it had the largest reduction in the agglomeration schedule error coefficient (other than the two-group solution—see Table 9-7). The HBAT management team then considered looking at both the five-cluster and three-cluster solutions. After reflection, management suggested that a smaller number of clusters would mean fewer market segments to develop separate strategies, and the result would likely be lower costs to execute the strategies. Moreover, the three-cluster solution not only had fewer clusters, but also exhibited the second largest increase in heterogeneity from three clusters to two, indicating that three clusters are

TABLE 9-10 Results of Cross-Classification of Clusters on X_1, X_2, X_3, X_4, and X_5

		Number of Cases Per Cluster				
Customer Characteristics		**1**	**2**	**3**	**4**	**Total**
X_1 Customer Type	Less than 1 year	8	5	5	12	30
	1 to 5 years	8	6	6	15	35
	more than 5 years	9	18	6	0	33
	Total ($\chi^2 = 24.4, p < .001$)	25	29	17	27	98
X_2 Industry Type	Magazine industry	8	21	10	12	51
	Newsprint industry	17	8	7	15	47
	Total ($\chi^2 = 10.1, p < .05$)	25	29	17	27	98
X_3 Firm Size	Small (0 to 499)	11	19	7	10	47
	Large (500+)	14	10	10	17	51
	Total ($\chi^2 = 5.4, p > .1$)	25	29	17	27	98
X_4 Region	USA/North America	14	14	8	1	39
	Outside North America	11	15	9	26	59
	Total ($\chi^2 = 28.3, p < .001$)	25	29	17	27	98
X_5 Distribution System	Indirect through broker	13	14	8	20	55
	Direct to customer	12	15	9	7	13
	Total ($\chi^2 = 5.2, p > .1$)	25	29	17	27	98

TABLE 9-11   Means from K-Means Three-Cluster Solution

Variable	Mean Values Cluster Number:			Mean-Centered Values Cluster Number:			F	Sig
	1	2	3	1	2	3		
X_6 Product Quality	8.4	6.1	8.7	0.58	−1.70	0.91	79.78	0.000
X_8 Technical Support	5.3	5.5	5.5	−0.03	0.12	.16	0.16	0.851
X_{12} Salesforce Image	4.8	5.7	5.1	−0.31	0.61	−0.04	6.90	0.002
X_{15} New Products	4.0	5.3	6.7	−1.18	0.11	1.57	89.76	0.000
X_{18} Delivery Speed	3.7	4.1	4.0	−0.14	0.19	0.09	1.98	0.144
Cluster sample sizes	44	27	27	44	27	27		

substantially more distinct than two. As a result, the research team decided to examine the nonhierarchical three-cluster solution.

The results of the three cluster solution are shown in Table 9-11. Cluster 1 has 44 customers, whereas clusters 2 and 3 each have 27 customers. Significant differences exist between the three clusters on three variables—X_6, X_{12}, and X_{15}—so the solution is discriminating between the three customer groups.

To interpret the clusters, we examine both the means and the mean-centered values. HBAT is perceived very unfavorably by cluster 1. Three of the variables (X_{12}, X_{15}, and X_{18}) are rated very poorly, whereas X_8 is only average (5.3). Only X_6 (Product Quality) is rated favorably (8.4). Thus, HBAT is definitely doing poorly with cluster 1 and needs improvement. Cluster 2 views HBAT more favorably than does cluster 1 with one big exception. HBAT performs slightly above average on four of the five variables (X_8, X_{12}, X_{15}, and X_{18}) according to cluster 2. The score on X_{12} (5.7) is clearly the highest among all clusters. However, its rating on X_6 is 6.1. This is by far the lowest rating on this variable across the three clusters. HBAT is therefore overall viewed slightly more favorably by cluster 2 than cluster 1, but has an issue with perceived product quality relative to the other groups. Cluster 3 customers view HBAT relatively favorably. Indeed, HBAT performs quite high on X_6 (Product Quality) and the highest of all customer segments on X_{15} (New Products). Thus, HBAT may consider maintaining an emphasis on newness and innovativeness among customers in this group.

Criterion validity for the three-cluster solution was examined using the same approach as with the four-cluster solution. Variables X_{19}, X_{20}, X_{21}, and X_{22} were submitted to a MANOVA analysis as dependent variables, and the independent variable was cluster membership. The overall F-statistic for the MANOVA, as well as the univariate F-statistics, were all significant, thus providing evidence of criterion validity.

The final task is to profile the three clusters so management can determine the characteristics of each cluster and target them with different strategies. Results of the cross-classification are provided in Table 9-12. As was determined earlier, significant chi-square values are observed for three of the five profile variables. Several patterns are evident. For example, cluster 2 consists entirely of customers from outside of the USA/North America (27 out of 27). In contrast, clusters 1 and 3 are rather evenly split between customers from the USA/North America. Other differences indicate that cluster 2 customers are not among the customers who have been with HBAT the longest. These profiles also support the distinctiveness of the clusters on variables not used in the analysis at any prior point. Moreover, these findings can be used to develop different strategies for each customer cluster.

The question remains: Which cluster solution is best? Each solution, including the five-cluster solution, which was not discussed, has strengths and weaknesses. The four-cluster and five-cluster solutions provide more differentiation between the customers, and each cluster represents a smaller and more homogeneous set of customers. In contrast, the three-cluster

TABLE 9-12 Cross-Classifications from Three-Cluster Solution

Customer Characteristics		Number of Cases Per Customer			
		1	2	3	Total
X_1 Customer Type	Less than 1 year	13	10	7	30
	1 to 5 years	12	17	6	35
	More than 5 years	19	0	14	33
	Total ($\chi^2 = 29.2$; $p < .001$)	44	27	27	98
X_2 Industry Type	Magazine industry	20	14	17	51
	Newsprint industry	24	13	10	47
	Total ($\chi^2 = 2.1$; $p = .36$)	44	27	27	98
X_3 Firm Size	Small (0 to 499)	22	10	15	47
	Large (500+)	22	17	12	51
	Total ($\chi^2 = 2.0$; $p = .37$)	44	27	27	98
X_4 Region	USA/North America	25	0	14	39
	Outside North America	19	27	13	59
	Total ($\chi^2 = 34.2$, $p < .001$)	44	27	27	98
X_5 Distribution System	Indirect through broker	20	21	14	55
	Direct to customer	24	6	13	43
	Total ($\chi^2 = 7.8$, $p < .05$)	44	27	27	98

solution is more parsimonious and likely easier and less costly for HBAT management to implement. So, ultimately, the question of which is best is not by determined by statistical results alone.

A Managerial Overview of the Clustering Process. The cluster analyses (hierarchical and nonhierarchical) were successful in performing a market segmentation of HBAT customers. The process not only created homogeneous groupings of customers based on their perceptions of HBAT, but also found that these clusters met the tests of predictive validity and distinctiveness on additional sets of variables, which are all necessary for achieving practical significance. The segments represent quite different customer perspectives of HBAT, varying in both the types of variables that are viewed most positively as well as the magnitude of the perceptions.

One issue that can always be questioned is the selection of a "final" cluster solution. In this example, both the three-cluster and four-cluster solutions exhibited distinctiveness and strong relationships to the relevant outcomes. Moreover, the solutions provide a basic, but useful, delineation of customers that vary in perceptions, buying behavior, and demographic profile. The selection of the best cluster solution needs to involve all interested parties; in this example, the research team and management both have to provide input so a consensus can be reached.

Summary

Cluster analysis can be a very useful data-reduction technique. But its application is more an art than a science, and the technique can easily be abused or misapplied. Different similarity measures and different algorithms can and do affect the results. If the researcher proceeds cautiously, however, cluster analysis can be an invaluable tool in identifying latent patterns by suggesting useful groupings (clusters) of objects that are not discernible through other multivariate techniques. This chapter helps you to do the following:

Define cluster analysis, its roles, and its limitations. Cluster analysis is a group of multivariate methods whose primary purpose is to group objects based on the characteristics they possess. Cluster analysis classifies objects (e.g., respondents, products, or other entities) so that each object is very similar to others in the cluster based on a set of selected characteristics. The resulting clusters of objects should exhibit high internal (within-cluster) homogeneity and high external (between-cluster) heterogeneity. If the process is successful, the objects within clusters will be close together when plotted geometrically, and different clusters will be far apart. Among the more common roles cluster analysis plays are: (1) data reduction of the type common when a researcher is faced with a large number of observations that can be meaningfully classified into groups or segments and (2) hypothesis generation where cluster analysis is used to develop hypotheses concerning the nature of the data or to examine previously stated hypotheses. The most common criticisms, and therefore limitations, of cluster analysis are: (1) it is descriptive, atheoretical, and noninferential; (2) it will always create clusters, regardless of the "true" existence of any structure in the data; and (3) cluster solutions are not generalizable, because they are totally dependent upon the variables used as the basis for the similarity measure.

Identify types of research questions addressed by cluster analysis.   In forming homogeneous groups, cluster analysis can address any combination of three basic research questions: (1) taxonomy description (the most traditional use of cluster analysis has been for exploratory purposes and the formation of a taxonomy—an empirically based classification of objects); (2) data simplification (by defining structure among the observations, cluster analysis also develops a simplified perspective by grouping observations for further analysis); and (3) relationship identification (with the clusters defined and the underlying structure of the data represented in the clusters, the researcher has a means of revealing relationships among the observations that typically is not possible with the individual observations).

Understand how interobject similarity is measured. Interobject similarity can be measured in a variety of ways. Three methods dominate applications of cluster analysis: correlational measures, distance measures, and association measures. Each method represents a particular perspective on similarity, dependent on both its objectives and type of data. Both the correlational and distance measures require metric data, whereas the association measures are for nonmetric data. Correlational measures are rarely used, because the emphasis in most applications of cluster analysis is on the objects' magnitudes, not the patterns of values. Distance measures are the most commonly used measures of similarity in cluster analysis. The distance measures

represent similarity, because the proximity of observations to one another across the variables in the cluster variate. However, the term *dissimilarity* may be more appropriate for distance measures, because higher values represent more dissimilarity, not more similarity.

Understand why different similarity measures are sometimes used.   Euclidean distance is the most commonly recognized measure of distance, and is many times referred to as *straight-line* distance. This concept is easily generalized to more than two variables. Squared (or absolute) Euclidean distance is the sum of the squared differences without taking the square root. City-block (Manhattan) distance is not based on Euclidean distance. Instead, it uses the sum of the absolute differences of the variables (i.e., the two sides of a right triangle rather than the hypotenuse). This procedure is the simplest to calculate, but may lead to invalid clusters if the clustering variables are highly correlated. Researchers sometimes run multiple cluster solutions using different distance measures to compare the results. Alternatively, correlation measures can be used to represent similarity when the researcher is more interested in patterns than in profiles based on similarity of cluster members.

Understand the differences between hierarchical and nonhierarchical clustering techniques.   A wide range of partitioning procedures has been proposed for cluster analysis. The two most widely used procedures are hierarchical versus nonhierarchical. Hierarchical procedures involve a series of $n - 1$ clustering decisions (where n equals the number of observations) that combine observations into a hierarchy or treelike structure. In contrast to hierarchical methods, nonhierarchical procedures do not involve the treelike construction process. Instead, they assign objects into clusters once the number of clusters to be formed is specified. For example, if a six-cluster solution is specified, then the resulting clusters are not just a combination of two clusters from the seven-cluster solution, but are based only on finding the best six-cluster solution. The process has two steps: (1) identifying starting points, known as cluster seeds, for each cluster, and (2) assigning each observation to one of the cluster seeds based on similarity within the group. Nonhierarchical cluster methods are often referred to as K-means.

Know how to interpret results from cluster analysis. Results from hierarchical cluster analysis are interpreted differently from nonhierarchical cluster analysis. Perhaps the most crucial issue for any researcher is determining the number of clusters. The researcher must select the cluster solution(s) that will best represent the data by applying a stopping rule. In hierarchical cluster analysis, the agglomeration schedule becomes crucial in determining this number. The researcher also must analyze clusters for distinctiveness and the possibility of outliers, which would be identified by very small cluster sizes or by observations that join clusters late in the agglomeration schedule. Hierarchical cluster results, including the number of clusters and possibly the cluster seed points, are input to a K-means approach. Here, interpretation focuses closely on profiling the clusters on the clustering variables and on other variables. The profiling stage involves describing the characteristics of each cluster to explain how they may differ on relevant dimensions. The procedure begins after the clusters are identified and typically involves the use of discriminant analysis or ANOVA. Clusters should be distinct and consistent with theory and can be explained.

Follow the guidelines for cluster validation. The most direct approach to validation is to cluster analyze separate samples, comparing the cluster solutions and assessing the correspondence of the results. The researcher also can attempt to establish some form of criterion or predictive validity. To do so, variable(s) not used to form the clusters but known to vary across the clusters are selected and compared. The comparisons can usually be performed with either MANOVA/ANOVA or a cross-classification table. The cluster solution should also be examined for stability. Nonhierarchical approaches are particularly susceptible to unstable results because of the different processes for selecting initial seed values as inputs. Thus, the data should be sorted into several different orders and the cluster solution redone with the clusters from the different solutions examined against one another for consistency.

Questions

1. What are the basic stages in the application of cluster analysis?
2. What is the purpose of cluster analysis, and when should it be used instead of factor analysis?
3. What should the researcher consider when selecting a similarity measure to use in cluster analysis?
4. How does the researcher know whether to use hierarchical or nonhierarchical cluster techniques? Under which conditions would each approach be used?
5. How does a researcher decide the number of clusters to have in a solution?
6. What is the difference between the interpretation stage and the profiling and validation stages?
7. How can researchers use graphical portrayals of the cluster procedure?

Suggested Readings

A list of suggested readings illustrating issues and applications of multivariate techniques in general is available on the Web at www.pearsonglobaleditions.com/hair or www.mvstats.com.

References

1. Aldenderfer, Mark S., and Roger K. Blashfield. 1984. Cluster Analysis. Thousand Oaks, CA: Sage.
2. Anderberg, M. 1973. *Cluster Analysis for Applications.* New York: Academic Press.
3. Baeza-Yates, R. A. 1992. Introduction to Data Structures and Algorithms Related to Information Retrieval. In *Information Retrieval: Data Structures and Algorithms*, W. B. Frakes and R. Baeza-Yates (eds.). Upper Saddle River, NJ: Prentice Hall, pp. 13–27.
4. Bailey, Kenneth D. 1994. *Typologies and Taxonomies: An Introduction to Classification Techniques.* Thousand Oaks, CA: Sage.
5. Bock, H. H. 1985. On Some Significance Tests in Cluster Analysis. *Communication in Statistics* 3: 1–27.
6. Dubes, R. C. 1987. How Many Clusters Are Best—An Experiment. *Pattern Recognition* 20 (November): 645–63.
7. Edelbrock, C. 1979. Comparing the Accuracy of Hierarchical Clustering Algorithms: The Problem of Classifying Everybody. *Multivariate Behavioral Research* 14: 367–84.
8. Everitt, B., S. Landau, and M. Leese. 2001. *Cluster Analysis,* 4th ed. New York: Arnold Publishers.
9. Green, P. E. 1978. *Analyzing Multivariate Data.* Hinsdale, IL: Holt, Rinehart and Winston.

10. Green, P. E., and J. Douglas Carroll. 1978. *Mathematical Tools for Applied Multivariate Analysis*. New York: Academic Press.

11. Hartigan, J. A. 1985. Statistical Theory in Clustering. *Journal of Classification* 2: 63–76.

12. Jain, A. K., and R. C. Dubes. 1988. *Algorithms for Clustering Data*. Upper Saddle River, NJ: Prentice Hall.

13. Jain, A.K., M. N. Murty, and P. J. Flynn. 1999. Data Clustering: A Review. *ACM Computing Surveys* 31(3): 264–323.

14. Jardine, N., and R. Sibson. 1975. *Mathematical Taxonomy*. New York: Wiley.

15. Ketchen, D.J., and C. L. Shook. 1996. The Application of Cluster Analysis in Strategic Management Research: An Analysis and Critique. *Strategic Management Journal* 17: 441–58.

16. McIntyre, R. M., and R. K. Blashfield. 1980. A Nearest-Centroid Technique for Evaluating the Minimum-Variance Clustering Procedure. *Multivariate Behavioral Research* 15: 225–38.

17. Milligan, G. 1980. An Examination of the Effect of Six Types of Error Perturbation on Fifteen Clustering Algorithms. *Psychometrika* 45 (September): 325–42.

18. Milligan, Glenn W., and Martha C. Cooper. 1985. An Examination of Procedures for Determining the Number of Clusters in a Data Set. *Psychometrika* 50(2): 159–79.

19. Morrison, D. 1967. Measurement Problems in Cluster Analysis. *Management Science* 13 (August): 775–80.

20. Nagy, G. 1968. State of the Art in Pattern Recognition. *Proceedings of the IEEE* 56: 836–62.

21. Overall, J. 1964. Note on Multivariate Methods for Profile Analysis. *Psychological Bulletin* 61(3): 195–98.

22. Punj, G., and D. Stewart. 1983. Cluster Analysis in Marketing Research: Review and Suggestions for Application. *Journal of Marketing Research* 20 (May): 134–48.

23. Rohlf, F. J. 1970. Adaptive Hierarchical Clustering Schemes. *Systematic Zoology* 19: 58.

24. Sarle, W. S. 1983. *Cubic Clustering Criterion, SAS Technical Report A-108*. Cary, NC: SAS Institute, Inc.

25. Schaninger, C. M., and W. C. Bass. 1986. Removing Response-Style Effects in Attribute-Determinance Ratings to Identify Market Segments. *Journal of Business Research* 14: 237–52.

26. Shephard, R. 1966. Metric Structures in Ordinal Data. *Journal of Mathematical Psychology* 3: 287–315.

27. Sneath, P. H. A., and R. R. Sokal. 1973. *Numerical Taxonomy*. San Francisco: Freeman Press.

CHAPTER 10

MDS and Correspondence Analysis

LEARNING OBJECTIVES

Upon completing this chapter, you should be able to do the following:

- Define multidimensional scaling and describe how it is performed.
- Understand the differences between similarity data and preference data.
- Select between a decompositional or compositional approach.
- Determine the comparability and number of objects.
- Understand how to create a perceptual map.
- Explain correspondence analysis as a method of perceptual mapping.

CHAPTER PREVIEW

Multidimensional scaling (MDS) refers to a series of techniques that help the researcher identify key dimensions underlying respondents' evaluations of objects and then position these objects in this dimensional space. For example, multidimensional scaling is often used in marketing to identify key dimensions underlying customer evaluations of products, services, or companies. Other common applications include the comparison of physical qualities (e.g., food tastes or various smells), perceptions of political candidates or issues, and even the assessment of cultural differences between distinct groups. Multidimensional scaling techniques can infer the underlying dimensions using only a series of similarity or preference judgments about the objects provided by respondents. Once the data are in hand, multidimensional scaling can help determine the number and relative importance of the dimensions respondents use when evaluating objects, as well as how the objects are related perceptually on these dimensions, most often portrayed graphically.

Correspondence analysis (CA) is a related technique with similar objectives. CA infers the underlying dimensions that are evaluated as well as the positioning of objects, yet follows a quite different approach. First, instead of using overall evaluations of similarity or preference concerning the objects, each object is evaluated (in nonmetric terms) on a series of attributes. Then, with this information CA develops the dimensions of comparison between objects and places each object in this dimensional space to allow for comparisons among both objects and attributes simultaneously.

KEY TERMS

Before starting the chapter, review the key terms to develop an understanding of the concepts and terminology used. Throughout the chapter the key terms appear in **boldface.** Other points of emphasis in the chapter and key term cross-references are *italicized.*

Aggregate analysis   Approach to MDS in which a *perceptual map* is generated for a group of respondents' evaluations of *objects.* This composite perceptual map may be created by a computer program or by the researcher to find a few "average" or representative subjects.

Compositional method   An approach to perceptual mapping that derives overall *similarity* or *preference* evaluations from evaluations of separate attributes by each respondent. With compositional methods separate attribute evaluations are combined (composed) into an overall evaluation. The most common examples of compositional methods are the techniques of factor analysis and discriminant analysis.

Confusion data   Procedure to obtain respondents' perceptions of *similarities data.* Respondents indicate the similarities between pairs of stimuli. The pairing (or confusing) of one stimulus with another is taken to indicate similarity. Also known as *subjective clustering.*

Contingency table   Cross-tabulation of two nonmetric or categorical variables in which the entries are the frequencies of responses that fall into each cell of the matrix. For example, if three brands were rated on four attributes, the brand-by-attribute contingency table would be a three-row by four-column table. The entries would be the number of times a brand (e.g., Coke) was rated as having an attribute (e.g., sweet taste).

Correspondence analysis (CA)   *Compositional approach* to perceptual mapping that is based on categories of a *contingency table.* Most applications involve a set of *objects* and attributes, with the results portraying both objects and attributes in a common *perceptual map.* To derive a multidimensional map, you must have a minimum of three attributes and three objects.

Cross-tabulation table   See *contingency table.*

Decompositional method   Perceptual mapping method associated with MDS techniques in which the respondent provides only an overall evaluation of *similarity* or *preference* between *objects.* This set of overall evaluations is then decomposed into a set of dimensions that best represent the objects' differences.

Degenerate solution   MDS solution that is invalid because of (1) inconsistencies in the data or (2) too few objects compared with the number of dimensions specified by the researcher. Even though the computer program may indicate a valid solution, the researcher should disregard the degenerate solution and examine the data for the cause. This type of solution is typically portrayed as a circular pattern with illogical results.

Derived measures   Procedure to obtain respondents' perceptions of *similarities data.* Derived similarities are typically based on a series of scores given to stimuli by respondents, which are then combined in some manner. The semantic differential scale is frequently used to elicit such scores.

Dimensions   Features of an *object.* A particular object can be thought of as possessing both *perceived/subjective* dimensions (e.g., expensive, fragile) and *objective* dimensions (e.g., color, price, features).

Disaggregate analysis   Approach to MDS in which the researcher generates *perceptual maps* on a respondent-by-respondent basis. The results may be difficult to generalize across respondents. Therefore, the researcher may attempt to create fewer maps by some process of *aggregate analysis,* in which the results of respondents are combined.

Disparities   Differences in the computer-generated distances representing *similarity* and the distances provided by the respondent.

Ideal point   Point on a perceptual map that represents the most preferred combination of perceived attributes (according to the respondents). A major assumption is that the position of the ideal point (relative to the other objects on the perceptual map) would define relative *preference* such that objects farther from the ideal point should be preferred less.

Importance–performance grid   Two-dimensional approach for assisting the researcher in labeling dimensions. The vertical axis is the respondents' perceptions of the importance (e.g., as measured on a scale of "extremely important" to "not at all important"). The horizontal axis is performance (e.g., as measured on a scale of "excellent performance" to "poor performance") for each brand or product/service on various attributes. Each object is represented by its values on importance and performance.

Index of fit   Squared correlation index (R^2) that may be interpreted as indicating the proportion of variance of the *disparities* (optimally scaled data) that can be accounted for by the MDS procedure. It measures how well the raw data fit the MDS model. This index is an alternative to the *stress measure* for determining the number of dimensions. Similar to measures of covariance in other multivariate techniques, measures of .60 or greater are considered acceptable.

Inertia   A relative measure of chi-square used in correspondence analysis. The total inertia of a cross-tabulation table is calculated as the total chi-square divided by the total frequency count (sum of either rows or columns). Inertia can then be calculated for any row or column category to represent its contribution to the total.

Initial dimensionality   A starting point in selecting the best spatial configuration for data. Before beginning an MDS procedure, the researcher must specify how many *dimensions* or features are represented in the data.

Mass   A relative measure of frequency used in *correspondence analysis* to describe the size of any single cell or category in a *cross-tabulation*. It is defined as the value (cell or category total) divided by the total frequency count, making it the percentage of the total frequency represented by the value. As such, the total mass across rows, columns, or all cell entries is 1.0.

Multiple correspondence analysis   Form of *correspondence analysis* that involves three or more categorical variables related in a common perceptual space.

Object   Any stimulus that can be compared and evaluated by the respondent, including tangible entities (product or physical object), actions (service), sensory perceptions (smell, taste, sights), or even thoughts (ideas, slogans).

Objective dimension   Physical or tangible characteristics of an *object* that have an objective basis of comparison. For example, a product has size, shape, color, weight, and so on.

Perceived dimension   A respondent's subjective attachment of features to an *object* represent its intangible characteristics. Examples include "quality," "expensive," and "good-looking." These perceived dimensions are unique to the individual respondent and may bear little correspondence to actual *objective dimensions.*

Perceptual map   Visual representation of a respondent's perceptions of *objects* on two or more *dimensions.* Usually this map has opposite levels of dimensions on the ends of the *X* and *Y* axes, such as "sweet" to "sour" on the ends of the *X* axis and "high-priced" to "low-priced" on the ends of the *Y* axis. Each object then has a spatial position on the perceptual map that reflects the relative *similarity* or *preference* to other objects with regard to the dimensions of the perceptual map.

Preference   Implies that *objects* are judged by the respondent in terms of dominance relationships; that is, the stimuli are ordered in preference with respect to some property. Direct ranking, paired comparisons, and preference scales are frequently used to determine respondent preferences.

Preference data   Data used to determine the *preference* among *objects.* Can be contrasted to *similarities data,* which denotes the similarity among objects, but has no "good–bad" distinction as seen in preference data.

Projections   Points defined by perpendicular lines from an object to a *vector.* Projections are used in determining the *preference* order with vector representations.

Similarities data   Data used to determine which *objects* are the most similar to each other and which are the most dissimilar. Implicit in similarities measurement is the ability to compare all pairs of objects. Three procedures to obtain similarities data are paired comparison of objects, *confusion data*, and *derived measures*.

Similarity   See *similarities data.*

Similarity scale   Arbitrary scale, for example, from −5 to +5, that enables the representation of an ordered relationship between objects from the most similar (closest) to the least similar (farthest apart). This type of scale is appropriate only for representing a single dimension.

Spatial map   See *perceptual map.*

Stress measure   Proportion of the variance of the *disparities* (optimally scaled data) not accounted for by the MDS model. This type of measurement varies according to the type of program and the data being analyzed. The stress measure helps to determine the appropriate number of *dimensions* to include in the model.

Stretched   Transformation of an MDS solution to make the *dimensions* or individual elements reflect the relative weight of preference.

Subjective clustering   See *confusion data.*

Subjective dimension   See *perceived dimension.*

Subjective evaluation   Method of determining how many *dimensions* are represented in the MDS model. The researcher makes a subjective inspection of the spatial maps and asks whether the configuration looks reasonable. The objective is to obtain the best fit with the least number of dimensions.

Unfolding   Representation of an individual respondent's *preferences* within a common (aggregate) stimulus space derived for all respondents as a whole. The individual's preferences are "unfolded" and portrayed as the best possible representation within the aggregate analysis.

Vector   Method of portraying an ideal point or attribute in a perceptual map. Involves the use of *projections* to determine an *object's* order on the vector.

WHAT IS MULTIDIMENSIONAL SCALING?

Multidimensional scaling (MDS), also known as perceptual mapping, is a procedure that enables a researcher to determine the perceived relative image of a set of objects (firms, products, ideas, or other items associated with commonly held perceptions). The purpose of MDS is to transform consumer judgments of overall similarity or **preference** (e.g., preference for stores or brands) into distances represented in multidimensional space.

Comparing Objects

Multidimensional scaling is based on the comparison of **objects** (e.g., product, service, person, aroma). MDS differs from other multivariate methods in that it uses only a single, overall measure of similarity or preference. To perform a multidimensional scaling analysis, the researcher performs three basic steps:

1. Gather measures of similarity or preference across the entire set of objects to be analyzed.
2. Use MDS techniques to estimate the relative position of each object in multidimensional space.
3. Identify and interpret the axes of the dimensional space in terms of perceptual and/or objective attributes.

Assume that objects A and B are judged by respondents to be the most similar compared with all other possible pairs of objects (AC, BC, AD, and so on). MDS techniques will position objects A and B so that the distance between them in multidimensional space is smaller than the distance between any other two pairs of objects. The resulting **perceptual map,** also known as a **spatial map,** shows the relative positioning of all objects, as shown in Figure 10-1.

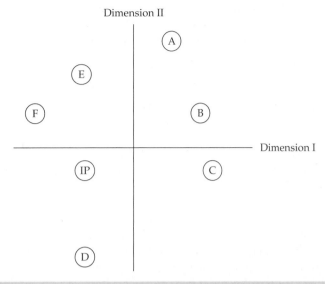

FIGURE 10-1 Illustration of a Multidimensional Map of Perceptions of Six Industrial Suppliers (A to F) and the Ideal Point (IP)

Dimensions: The Basis for Comparison

What is the basis for the relative position of each object? Why were A and B more similar than any other pair of objects (e.g., A and D)? What do the axes of the multidimensional space represent? Before we attempt to answer any of these questions, we first must recognize that any object can be thought of as having **dimensions** that represent an individual's perceptions of attributes or combinations of attributes. These dimensions may represent a single attribute/perception or idea, or they may be a composite of any number of attributes (e.g., reputation).

OBJECTIVE VERSUS SUBJECTIVE DIMENSIONS   When characterizing an object, it is also important to remember that individuals may use different types of measures in making these judgments. For example, one measure is an **objective dimension** that has quantifiable (physical or observable) attributes. Another type of measure is a **perceived dimension** (also known as a **subjective dimension**), in which individuals evaluate the objects based on perceptions. In this case, the perceived dimension is an interpretation by the individual that may or may not be based on objective dimensions.

For example, management may view their product (a lawn mower) as having two color options (red and green), a two-horsepower motor, and a 24-inch blade, which are the objective dimensions. However, customers may (or may not) focus on these attributes. Customers may focus on a perceived dimension, such as the mower being expensive-looking or fragile.

Two objects may have the same physical characteristics (objective dimensions) but be viewed differently because the objects are perceived to differ in quality (a perceived dimension) by many customers. Thus, the following two differences between objective and perceptual dimensions are important:

- *Individual differences:* The dimensions perceived by customers may not coincide with (or may not even include) the objective dimensions assumed by the researcher. We expect that each individual may have different perceived dimensions, but the researcher must also accept that the objective dimensions may also vary substantially. Individuals may consider different sets of objective characteristics as well as vary the importance they attach to each dimension.

- *Interdependence:* The evaluations of the dimensions (even if the perceived dimensions are the same as the objective dimensions) may not be independent and may not agree. Both perceived and objective dimensions may interact with one another to create unexpected evaluations. For example, one soft drink may be judged sweeter than another because the first has a fruitier aroma, although both contain the same amount of sweetener.

RELATING OBJECTIVE AND SUBJECTIVE DIMENSIONS   The challenge to the researcher is to understand how both the perceived dimensions and objective dimensions relate to the axes of the multidimensional space used in the perceptual map, if possible. It is similar to the interpretation of the variate in many other multivariate techniques (e.g., "labeling" the factors from factor analysis), but differs in that the researcher never directly uses any attribute ratings (e.g., ratings of quality, attractiveness, etc.) when obtaining the similarity ratings among objects. Instead, the researcher collects only similarity or preference.

Using only the overall measures (similarity or preference) requires that the researcher first understand the correspondence between perceptual and objective dimensions with the axes of the perceptual map. Then, additional analysis can identify which attributes predict the position of each object in both perceptual and objective space.

A note of caution must be raised, however, concerning the interpretation of dimensions. Because this process is as much an art as a science, the researcher must resist the temptation to allow personal perception to affect the qualitative dimensionality of the perceived dimensions. Given the level of researcher input, caution must be taken to be as objective as possible in this critical, yet still rudimentary, area.

A SIMPLIFIED LOOK AT HOW MDS WORKS

To facilitate a better understanding of the basic procedures in multidimensional scaling, we first present a simple example to illustrate the basic concepts underlying MDS and the procedure by which it transforms similarity judgments into the corresponding spatial positions. We will follow the three basic steps described previously.

Gathering Similarity Judgments

The first step is to obtain similarity judgments from one or more respondents. Here we will ask respondents for a single measure of similarity for each pair of objects.

Market researchers are interested in understanding consumers' perceptions of six candy bars that are currently on the market. Instead of trying to gather information about consumer's evaluations of the candy bars on a number of attributes, the researchers instead gather only perceptions of overall similarities or dissimilarities. The data are typically gathered by having respondents give simple global responses to statements such as these:

- Rate the similarity of products A and B on a 10-point scale.
- Product A is more similar to B than to C.
- I like product A better than product B.

Creating a Perceptual Map

From these simple responses, a perceptual map can be drawn that best portrays the overall pattern of **similarities** among the six candy bars. We illustrate the process of creating a perceptual map with the data from a single respondent, although this process could also be applied to multiple respondents or to the aggregate responses of a group of consumers.

The data are gathered by first creating a set of 15 unique pairs of the 6 candy bars ($6 \times 5 \div 2 = 15$ pairs). After tasting the candy bars, the respondent is asked to rank the 15 candy bar pairs, where a rank of 1 is assigned to the pair of candy bars that is most similar and a rank of 15 indicates the pair that is least alike. The results (rank orders) for all pairs of candy bars for one respondent are shown in Table 10-1. This respondent thought that candy bars D and E were the most similar, candy bars A and B were the next most similar, and so forth until candy bars E and F were the least similar.

If we want to illustrate the similarity among candy bars graphically, a first attempt would be to draw a single **similarity scale** and fit all the candy bars to it. In this one-dimensional portrayal of similarity, distance represents similarity. Thus, objects closer together on the scale are more similar and those farther away less similar. The objective is to position the candy bars on the scale so that the rank orders are best represented (rank order of 1 is closest, rank order of 2 is next closest, and so forth).

Let us try to see how we would place some of the objects. Positioning two or three candy bars is fairly simple. The first real test comes with four objects. We choose candy bars A, B, C, and D. Table 10-1 shows that the rank order of the pairs is as follows: AB < AD < BD < CD < BC < AC (each pair of letters refers to the distance [similarity] between the pair). From these values, we must place the four candy bars on a single scale so that the most similar (AB) are the closest and the least similar (AC) are the farthest apart. Figure 10-2a contains a one-dimensional perceptual map that matches the orders of pairs. If the person judging the similarity between the candy bars had been thinking of a simple rule of similarity that involved only one attribute (dimension), such as amount of chocolate, then all the pairs could be placed on a single scale that reproduces the similarity values.

Although a one-dimensional map can accommodate four objects, the task becomes increasingly difficult as the number of objects increases. The interested reader is encouraged to attempt this task with six objects. When a single dimension is employed with the six objects, the actual ordering varies substantially from the respondent's original rank orders.

Because one-dimensional scaling does not fit the data well, a two-dimensional solution should be attempted. It allows for another scale (dimension) to be used in configuring the six candy bars.

The procedure is quite tedious to attempt by hand. The two-dimensional solution produced by an MDS program is shown in Figure 10-2b. This configuration matches the rank orders of Table 10-1 exactly, supporting the notion that the respondent most probably used two dimensions in evaluating the candy bars. The conjecture that at least two attributes (dimensions) were considered is based on the inability to represent the respondent's perceptions in one dimension. However, we are still not aware of what attributes the respondent used in this evaluation.

Interpreting the Axes

Although we have no information as to what these dimensions are, we may be able to look at the relative positions of the candy bars and infer what attribute(s) the dimensions represent.

For example, suppose that candy bars A, B, and F were a form of combination bar (e.g., chocolate and peanuts, chocolate and peanut butter), and C, D, and E were strictly chocolate bars.

TABLE 10-1   Similarity Data (Rank Orders) for Candy Bar Pairs

Candy Bar	A	B	C	D	E	F
A	—	2	13	4	3	8
B		—	12	6	5	7
C			—	9	10	11
D				—	1	14
E					—	15
F						—

Note: Lower values indicate greater similarity, with 1 the most similar pair and 15 the least similar pair.

(a) One-Dimensional Perceptual Map of Four Observations

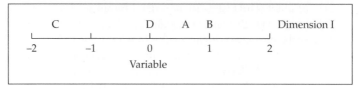

(b) Two-Dimensional Perceptual Map of Six Observations

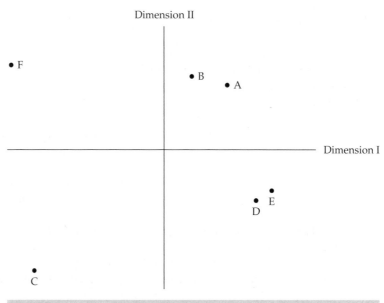

FIGURE 10-2   One- and Two-Dimensional Perceptual Maps

We could then infer that the horizontal dimension represents the type of candy bar (chocolate versus combination). When we look at the position of the candy bars on the vertical dimension, other attributes may emerge as the descriptors of that dimension as well.

MDS enables researchers to understand the similarity between objects (e.g., candy bars) by asking only for overall similarity perceptions. The procedure may also assist in determining which attributes actually enter into the perceptions of similarity. Although we do not directly incorporate the attribute evaluations into the MDS procedure, we can use them in subsequent analyses to assist in interpreting the dimensions and the impacts each attribute has on the relative positions of the candy bars.

COMPARING MDS TO OTHER INTERDEPENDENCE TECHNIQUES

Multidimensional scaling can be compared to the other interdependence techniques such as factor and cluster analysis based on its approach to defining structure:

- *Factor analysis:* Defines structure by grouping variables into variates that represent underlying dimensions in the original set of variables. Variables that highly correlate are grouped together.
- *Cluster analysis:* Defines structure by grouping objects according to their profile on a set of variables (the cluster variate) in which objects in close proximity to each other are grouped together.

MDS differs from factor and cluster analysis in two key aspects: (1) a solution can be obtained for each individual and (2) it does not use a variate.

Individual as the Unit of Analysis

In MDS, each respondent provides evaluations of all objects being considered, so that a solution can be obtained for each individual that is not possible in cluster analysis or factor analysis. As such, the focus is not on the objects themselves but instead on how the individual perceives the objects. The structure being defined is the perceptual dimensions of comparison for the individual(s). Once the perceptual dimensions are defined, the relative comparisons among objects can also be made.

Lack of a Variate

Multidimensional scaling, unlike the other multivariate techniques, does not use a variate. Instead, the variables that make up the variate (i.e., the perceptual dimensions of comparison) are inferred from global measures of similarity among the objects. In a simple analogy, it is like providing the dependent variable (similarity among objects) and figuring out what the independent variables (perceptual dimensions) must be. MDS has the advantage of reducing the influence of the researcher by not requiring the specification of the variables to be used in comparing objects, as was required in cluster analysis. It also has the disadvantage that the researcher is not really sure what variables the respondent is using to make the comparisons.

A DECISION FRAMEWORK FOR PERCEPTUAL MAPPING

Perceptual mapping encompasses a wide range of possible methods, including MDS, and all these techniques can be viewed through the model-building process introduced in Chapter 1. These steps represent a decision framework, depicted in Figure 10-3 (stages 1–3) and Figure 10-5 (stages 4–6, see page 585) within which all perceptual mapping techniques can be applied and the results evaluated.

STAGE 1: OBJECTIVES OF MDS

Perceptual mapping, and MDS in particular, is most appropriate for achieving two objectives:

1. An exploratory technique to identify unrecognized dimensions affecting behavior
2. A means of obtaining comparative evaluations of objects when the specific bases of comparison are unknown or undefined

In MDS, it is not necessary to specify the attributes of comparison for the respondent. All that is required is to specify the objects and make sure that the objects share a common basis of comparison. This flexibility makes MDS particularly suited to image and positioning studies in which the dimensions of evaluation may be too global or too emotional and affective to be measured by conventional scales. MDS methods combine the positioning of objects and subjects in a single overall map, making the relative positions of objects and consumers for segmentation analysis much more direct.

Key Decisions in Setting Objectives

The strength of perceptual mapping is its ability to infer dimensions without the need for defined attributes. The flexibility and inferential nature of MDS places a greater responsibility on the researcher to correctly define the analysis. The conceptual as well as practical considerations essential for MDS to achieve its best results are addressed through three key decisions:

1. Selecting the objects that will be evaluated
2. Deciding whether similarities or preferences are to be analyzed
3. Choosing whether the analysis will be performed at the group or individual level

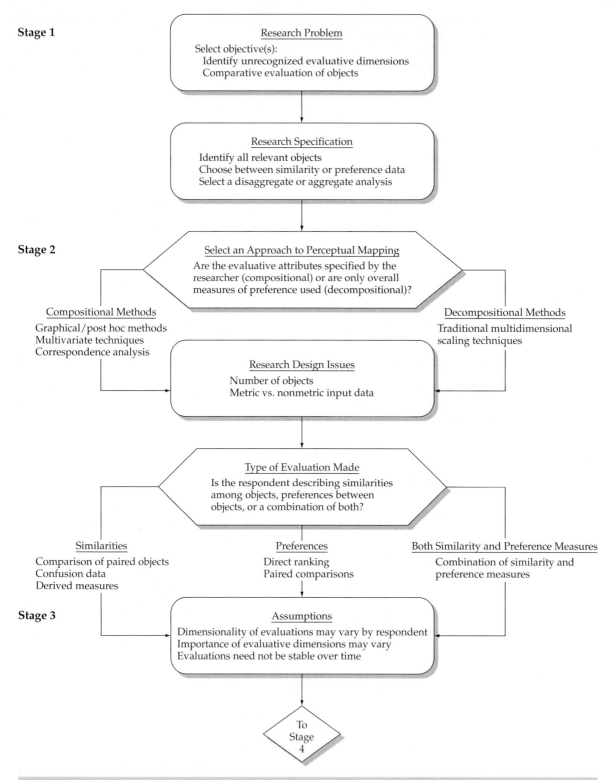

FIGURE 10-3 Stages 1–3 in the Multidimensional Scaling Decision Diagram

IDENTIFICATION OF ALL RELEVANT OBJECTS TO BE EVALUATED   The most basic, but important, issue in perceptual mapping is the defining of the objects to be evaluated. The researcher must ensure that all relevant firms, products, services, or other objects are included, because perceptual mapping is a technique of relative positioning. Relevancy is determined by the research questions to be addressed.

For example, a study of soft drinks must include both sugar-based and sugar-free soft drinks, unless the research question explicitly excludes one type or another. Likewise, a study of soft drinks would not include fruit juices.

The perceptual maps resulting from any of the methods can be greatly influenced by either the omission of objects or the inclusion of inappropriate ones [8, 21]. If irrelevant or noncomparable objects are included, the researcher is forcing the technique not only to infer the perceptual dimensions that distinguish among comparable objects but also to infer those dimensions that distinguish among noncomparable objects as well. This task is beyond the scope of MDS and results in a solution that addresses neither question well. Likewise, omitting a relevant object may prevent the true depiction of the perceptual dimensions.

SIMILARITIES VERSUS PREFERENCE DATA   Having selected the objects for study, the researcher must next select the basis of evaluation: similarity versus preference. To this point, we discussed perceptual mapping and MDS mainly in terms of similarity judgments. In providing **similarities data,** the respondent does not apply any "good–bad" aspects of evaluation in the comparison. The good–bad assessment is done, however, within **preference data,** which assumes that differing combinations of perceived attributes are valued more highly than other combinations.

Both bases of comparison can be used to develop perceptual maps, but with differing interpretations:

- Similarity-based perceptual maps represent attribute similarities and perceptual dimensions of comparison but do not reflect any direct insight into the determinants of choice.
- Preference-based perceptual maps do reflect preferred choices but may not correspond in any way to the similarity-based positions, because respondents may base their choices on entirely different dimensions or criteria from those on which they base comparisons.

Without any optimal base for evaluation, the decision between similarities and preference data must be made with the ultimate research question in mind, because they are fundamentally different in what they represent.

AGGREGATE VERSUS DISAGGREGATE ANALYSIS   In considering similarities or preference data, we are taking respondent's perceptions of stimuli and creating representations (perceptual maps) of stimulus proximity in t-dimensional space (where the number of dimensions t is less than the number of stimuli). At issue, however, is the level of analysis (individual or group) at which the data is analyzed. Each approach has both strengths and weaknesses.

Disaggregate Analysis. One of the distinctive characteristics of MDS techniques is their ability to estimate solutions for each respondent, a method known as a **disaggregate analysis.** Here the researcher generates perceptual maps on a subject-by-subject basis (producing as many maps as subjects). The advantage is the representation of the unique elements of each respondent's perceptions. The primary disadvantage is that the researcher must identify the common dimensions of the perceptual maps across multiple respondents.

Aggregate Analysis. MDS techniques can also combine respondents and create a single perceptual map through an **aggregate analysis.** The aggregation may take place either before or after scaling the subjects' data. Three basic approaches to this type of analysis are aggregating before the MDS analysis, aggregate individual results, and INDSCAL.

- *Aggregating Before the MDS Analysis* The simplest approach is for the researcher to find the average evaluations for all respondents and then obtain a single solution for the group of respondents as a whole. It is also the most typical type of aggregate analysis. To identify subgroups of similar respondents and their unique perceptual maps, the researcher may cluster analyze the subjects' responses to find a few average or representative subjects and then develop perceptual maps for the cluster's average respondent.

- *Aggregate Individual Results* Alternatively, the researcher may develop maps for each individual and cluster the maps according to the coordinates of the stimuli on the maps. It is recommended, however, that the previous approach of finding average evaluations be used rather than clustering the individual perceptual maps because minor rotations of essentially the same map can cause problems in creating reasonable clusters by the second approach.

- *INDSCAL: A Combination Approach* A specialized form of aggregate analysis is available with INDSCAL (individual differences scaling) [4] and its variants, which have characteristics of both disaggregate and aggregate analyses. An INDSCAL analysis assumes that all individuals share a common or group space (an aggregate solution) but that the respondents individually weight the dimensions, including zero weights when totally ignoring a dimension. The analysis proceeds in two steps:

1. As a first step, INDSCAL derives the perceptual space shared by all individuals, just as do other aggregate solutions.
2. However, individuals are also portrayed in a special group space map where each respondent's position is determined by the respondent's weights for each dimension. Respondents positioned closely together employ similar combinations of the dimensions from the common group space. Moreover, the distance of the individual from the origin is an approximate measure of the proportion of variance for that subject accounted for by the solution. Thus, a position farther from the origin indicates better fit. Being at the origin means "no fit" because all weights are zeros. If two or more subjects or groups of subjects are at the origin, separate group spaces need to be configured for each of them.

As an example, let us assume we derived a two-dimensional aggregate solution (see step 1). INDSCAL would also derive weights for each dimension, which would allow for each respondent to be portrayed in a two-dimensional graph (see Figure 10-4). For respondent A, almost all of the solution was oriented around dimension I, whereas the opposite is seen for respondent C. Respondents B and D have a balance between the two dimensions.

We also determine the fit for each respondent given the respondent's distance from the origin. Respondents A, B, and C are relatively similar in their distance from the origin, indicating comparable fit. Respondent D, however, has a substantially lower level of fit given its close proximity to the origin.

In an INDSCAL analysis, the researcher is presented with not only an overall representation of the perceptual map but also the degree to which each respondent is represented by the overall perceptual map. These results for each respondent can then be used to group respondents and even identify different perceptual maps in subsequent analyses.

Selecting a Disaggregate Versus Aggregate Analysis. The choice of aggregate or disaggregate analysis is based on the study objectives. If the focus is on an understanding of the overall evaluations of objects and the dimensions employed in those evaluations, an aggregate analysis is the most suitable. However, if the objective is to understand variation among individuals, particularly as a prelude to segmentation analysis, then a disaggregate approach is the most helpful.

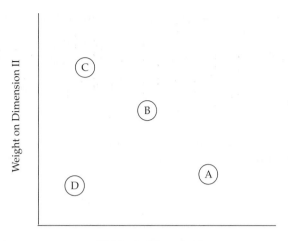

FIGURE 10-4   Respondents' Weights in an INDSCAL Disaggregate Analysis

---

RULES OF THUMB 10-1

Objectives of MDS

- MDS is an exploratory technique well suited for:
  - Identifying unrecognized dimensions used by respondents in making comparisons between objects (brands, products, stores, etc.)
  - Providing an objective basis for comparison between objects based on these dimensions
  - Identifying specific attributes that may correspond to these dimensions
- An MDS solution requires identification of all relevant objects (e.g., all competing brands within a product category) that set the boundaries for the research question
- Respondents provide one or both types of perceptions:
  - Perceptual distances where they indicate how similar/dissimilar objects are to each other, or
  - "Good–bad" assessments of competing objects (*preference* comparisons) that assist in identifying combinations of attributes that are valued highly
- MDS can be performed at the individual or group level:
  - Disaggregate (individual) analysis:
    - Allows for construction of perceptual maps on a respondent-by-respondent basis
    - Assesses variation among individuals
    - Provides a basis for segmentation analysis
  - Aggregate (group) analysis:
    - Creates perceptual maps of one or more groups
    - Helps understand overall evaluations of objects and/or dimensions employed in those evaluations
    - Should be found by using the average evaluations of all respondents within a group

STAGE 2: RESEARCH DESIGN OF MDS

Although MDS looks quite simple computationally, the results, as with other multivariate techniques, are heavily influenced by a number of key issues that must be resolved before the research can proceed. We cover four of the major issues, ranging from discussions of research design (selecting the approach and objects or stimuli for study) to specific methodological concerns (metric versus nonmetric methods) and data collection methods.

Selection of Either a Decompositional (Attribute-Free) or Compositional (Attribute-Based) Approach

Perceptual mapping techniques can be classified into one of two types based on the nature of the responses obtained from the individual concerning the object:

- The **decompositional method** measures only the overall impression or evaluation of an object and then attempts to derive spatial positions in multidimensional space that reflect these perceptions. This technique is typically associated with MDS.
- The **compositional method** is an alternative approach that employs several of the multivariate techniques already discussed that are used in forming an impression or evaluation based on a combination of specific attributes.

Each approach has advantages and disadvantages that we address in the following sections. Our discussion here centers on the distinctions between the two approaches and then we focus primarily on the decompositional methods in the remainder of the chapter.

DECOMPOSITIONAL OR ATTRIBUTE-FREE APPROACH Commonly associated with the techniques of MDS, decompositional methods rely on global or overall measures of similarity from which the perceptual maps and relative positioning of objects are formed. Because of the relatively simple task presented to the respondent, decompositional methods have two distinct advantages:

1. They require only that respondents give their overall perceptions of objects. Respondents do not have to detail the attributes or the importance of each attribute used in the evaluation.
2. Because each respondent gives a full assessment of similarities among all objects, perceptual maps can be developed for individual respondents or aggregated to form a composite map.

Decompositional methods have disadvantages as well, primarily related to the inferences required of the researcher to evaluate the perceptual maps:

1. The researcher has no objective basis provided by the respondent on which to identify the basic dimensions used to evaluate the objects (i.e., the correspondence of perceptual and objective dimensions). In many instances, the usefulness to managers of attribute-free studies is restricted because the studies provide little guidance for specific action. For example, the inability to develop a direct link between actions by the firm (the objective dimension) and market positions of their products (the perceptual dimension) many times diminishes the value of perceptual mapping.
2. Little guidance, other than generalized guidelines or a priori beliefs, is available to determine both the dimensionality of the perceptual map and the representativeness of the solution. Although some overall measures of fit are available, they are nonstatistical, and thus decisions about the final solution involve substantial researcher judgment.

Characterized by the generalized category of MDS techniques, a wide range of possible decompositional techniques is available. Selection of a specific method requires decisions regarding the nature of the respondent's input (rating versus ranking), whether similarities or preferences are obtained, and whether individual or composite perceptual maps are derived. Among the most common multidimensional scaling programs are KYST, MDSCAL, PREFMAP, MDPREF, INDSCAL, ALSCAL, MINISSA, POLYCON, and MULTISCALE. Detailed descriptions of the programs and sources for obtaining them are available [7, 24, 25, 26].

COMPOSITIONAL OR ATTRIBUTE-BASED APPROACH   Compositional methods include some of the more traditional multivariate techniques (e.g., discriminant analysis or factor analysis), as well as methods specifically designed for perceptual mapping, such as correspondence analysis. A principle common to all of these methods, however, is the assessment of similarity in which a defined set of attributes is considered in developing the similarity between objects. The various techniques included in the set of compositional methods can be grouped into one of three basic groups:

1. *Graphical or post hoc approaches.* Included in this set are analyses such as semantic differential plots or **importance–performance grids,** which rely on researcher judgment and univariate or bivariate representations of the objects.
2. *Conventional multivariate statistical techniques.* These techniques, especially factor analysis and discriminant analysis, are particularly useful in developing a dimensional structure among numerous attributes and then representing objects on these dimensions.
3. *Specialized perceptual mapping methods.* Notable in this class is correspondence analysis, developed specifically to provide perceptual mapping with only qualitative or nominally scaled data as input.

Compositional methods in general have two distinct advantages stemming from their defined attributes used in comparison:

- First is the explicit description of the dimensions of perceptual space. Because the respondent provides detailed evaluations across numerous attributes for each object, the evaluative criteria represented by the dimensions of the solution are much easier to ascertain.
- Also, these methods provide a direct way of representing both attributes and objects on a single map, with several methods providing the additional positioning of respondent groups. This information offers unique managerial insight into the competitive marketplace.

Yet the explicit description of the dimensions of comparison also has disadvantages:

- The similarity between objects is limited to only the attributes rated by the respondents. Omitting salient attributes eliminates the opportunity for the respondent to incorporate them, which would be available if a single overall measure were provided.
- The researcher must assume some method of combining these attributes to represent overall similarity, and the chosen method may not represent the respondents' thinking.
- The data collection effort is substantial, especially as the number of choice objects increases.
- Results are not typically available for the individual respondent.

Even though compositional models follow the concept of a variate depicted in many of the other multivariate techniques we have discussed in other sections of the text, they represent a distinctly different approach, with advantages and disadvantages, when compared to the decompositional methods. It is a choice that the researcher must make based on the research objectives of each particular study.

SELECTING BETWEEN COMPOSITIONAL AND DECOMPOSITIONAL TECHNIQUES   Perceptual mapping can be performed with both compositional and decompositional techniques, but each technique has specific advantages and disadvantages that must be considered in view of the research objectives:

- If perceptual mapping is undertaken in the spirit of either of the two basic objectives discussed earlier (see stage 1), the decompositional or attribute-free approaches are the most appropriate.
- If, however, the research objectives shift to the portrayal among objects on a defined set of attributes, then the compositional techniques become the preferred alternative.

Our discussion of compositional methods in past chapters illustrated their uses and application along with their strengths and weaknesses. The researcher must always remember the alternatives that are available in the event the objectives of the research change. Thus, we focus here on the decompositional approaches, followed by a discussion of correspondence analysis, a widely used compositional technique particularly suited to perceptual mapping. As such, we also consider as synonymous the terms *perceptual mapping* and *multidimensional scaling* unless necessary distinctions are made.

Objects: Their Number and Selection

Before undertaking any perceptual mapping study, the researcher must address two key questions dealing with the objects being evaluated. These questions deal with issues concerning the basic task (i.e., ensuring the comparability of the objects) as well as the complexity of the analysis (i.e., the number of objects being evaluated).

SELECTING OBJECTS   The key question when selecting objects is: Are the objects really comparable? An implicit assumption in perceptual mapping is that common characteristics, either objective or perceived, are present and may be used by the respondent for evaluations. Therefore, it is essential that the objects being compared have some set of underlying attributes that characterize each object and form the basis for comparison by the respondent. It is not possible for the researcher to force the respondent to make comparisons by creating pairs of noncomparable objects. Even if responses are given in such a forced situation, their usefulness is questionable.

THE NUMBER OF OBJECTS   A second issue concerns the number of objects to be evaluated. In deciding how many objects to include, the researcher must balance two desires: a smaller number of objects to ease the effort on the part of the respondent versus the required number of objects to obtain a stable multidimensional solution. These opposing considerations each impose limits on the analysis:

- A suggested guideline for stable solutions is to have more than four times as many objects as dimensions desired [10]. Thus, at least five objects are required for a one-dimensional perceptual map, nine objects for a two-dimensional solution, and so on.
- When using the method of evaluating pairs of objects for similarity, the respondent must make 36 comparisons of the 9 objects—a substantial task. A three-dimensional solution suggests at least 13 objects be evaluated, necessitating the evaluation of 78 pairs of objects.

Therefore, a trade-off must be made between the dimensionality accommodated by the objects (and the implied number of underlying dimensions that can be identified) and the effort required on the part of the respondent.

The number of objects also affects the determination of an acceptable level of fit. Many times the estimate of fit is inflated when fewer than the suggested number of objects are present for a given dimensionality. Similar to the overfitting problem we found in regression, falling below the recommended guidelines of at least four objects per dimension greatly increases the chances of a misleading solution.

For example, an empirical study demonstrated that when seven objects are fit to three dimensions with random similarity values, acceptable stress levels and apparently valid perceptual maps are generated more than 50 percent of the time. If the seven objects with random similarities were fit to four dimensions, the stress values decreased to zero, indicating perfect fit, in half the cases [19]. Yet in both instances, no real pattern of similarity emerged among the objects.

Thus, we must be aware of the risks associated with violating the guidelines for the number of objects per dimension and the impact this has on both the measures of fit and the validity of the resulting perceptual maps.

Nonmetric Versus Metric Methods

The original MDS programs were truly nonmetric, meaning that they required only nonmetric input, but they also provided only nonmetric (rank-order) output. The nonmetric output, however, limited the interpretability of the perceptual map. Therefore, all MDS programs used today produce metric output. The metric multidimensional positions can be rotated about the origin, the origin can be changed by adding a constant, the axes can be flipped (reflection), or the entire solution can be uniformly stretched or compressed, all without changing the relative positions of the objects.

Because all programs today produce metric output, the differences in the approaches are based only on the input measures of similarity.

- Nonmetric methods, distinguished by the nonmetric input typically generated by rank-ordering pairs of objects, are more flexible in that they do not assume any specific type of relationship between the calculated distance and the similarity measure. However, because nonmetric methods contain less information for creating the perceptual map, they are more likely to result in degenerate or suboptimal solutions. This problem arises when wide variations occur in the perceptual maps between respondents or the perceptions between objects are not distinct or well defined.
- Metric methods assume that input as well as output is metric. This assumption enables us to strengthen the relationship between the final output dimensionality and the input data. Rather than assuming that only the ordered relationships are preserved in the input data, we can assume that the output preserves the interval and ratio qualities of the input data. Even though the assumptions underlying metric programs are more difficult to support conceptually in many cases, the results of nonmetric and metric procedures applied to the same data are often similar.

Thus, selection of the input data type must consider both the research situation (variations of perceptions among respondents and distinctiveness of objects) and the preferred mode of data collection.

Collection of Similarity or Preference Data

As already noted, the primary distinction among MDS programs is the type of input data (metric versus nonmetric) and whether the data represent similarities or preferences. Here we address issues associated with making similarity-based and preference judgments. For many of the data collection methods, either metric (ratings) or nonmetric (rankings) data may be collected. In some instances, however, the responses are limited to only one type of data.

SIMILARITIES DATA   When collecting similarities data, the researcher is trying to determine which items are the most similar to each other and which are the most dissimilar. The terms of dissimilarities and similarities often are used interchangeably to represent measurement of the differences between objects. Implicit in similarity measurement is the ability to compare all pairs of objects.

If, for example, all pairs of objects of the set A, B, C (i.e., AB, AC, BC) are rank-ordered, then all pairs of objects can also be compared. Assume that the pairs were ranked AB = 1, AC = 2, and BC = 3 (where 1 denotes most similar). Clearly, pair AB is more similar than pair AC, pair AB is more similar than pair BC, and pair AC is more similar than pair BC.

Several procedures are commonly used to obtain respondents' perceptions of the similarities among stimuli. Each procedure is based on the notion that the relative differences between any pair of stimuli must be measured so that the researcher can determine whether the pair is more or less similar to any other pair. We discuss three procedures commonly used to obtain respondents' perceptions of similarities: comparison of paired objects, confusion data, and derived measures.

Comparison of Paired Objects. By far the most widely used method of obtaining similarity judgments is that of paired objects, in which the respondent is asked simply to rank or rate the similarity of all pairs of objects. If we have stimuli A, B, C, D, and E, we could rank pairs AB, AC, AD, AE, BC, BD, BE, CD, CE, and DE from most similar to least similar.

If, for example, pair AB is given the rank of 1, we would assume that the respondent sees that pair as containing the two stimuli that are the most similar, in contrast to all other pairs (see example in preceding section).

This procedure would provide a nonmetric measure of similarity. Metric measures of similarity would involve a rating of similarity (e.g., from 1 "very similar" to 10 "not at all similar"). Either form (metric or nonmetric) can be used in most MDS programs.

Confusion Data. Measuring similarity by pairing (or confusing) stimulus *I* with stimulus *J* is known as **confusion data.** Also known as **subjective clustering,** a typical procedure for gathering these data when the number of objects is large is as follows:

- Place the objects whose similarity is to be measured on small cards, either descriptively or with pictures.
- The respondent is asked to sort the cards into stacks so that all the cards in a stack represent similar candy bars. Some researchers tell the respondents to sort into a fixed number of stacks; others say to sort into as many stacks as the respondent wants.
- The data from each respondent is then aggregated into a similarities matrix similar to a **cross-tabulation table.** Each cell contains the number of times each pair of objects was included in the same stack. These data then indicate which products appeared together most often and are therefore considered the most similar.

Collecting data in this manner allows only for the calculation of aggregate similarity, because the responses from all individuals are combined to obtain the similarities matrix.

Derived Measures. Similarity based on scores given to stimuli by respondents are known as **derived measures.** The researcher defines the dimensions (attributes) and the respondent rates each object on each dimension. From these ratings, the similarity of each object is calculated by such methods as a correlation between objects or some form of index of agreement.

For example, subjects are asked to evaluate three stimuli (cherry, strawberry, and lemon-lime soda) on a number of attributes (diet versus nondiet, sweet versus tart, and light tasting versus heavy tasting) using semantic differential scales. The responses would be evaluated for each respondent (e.g., correlation or index of agreement) to create similarity measures between the objects.

Three important assumptions underlie this approach:

1. The researcher selects the appropriate dimensions to measure.
2. The scales can be weighted (either equally or unequally) to achieve the similarities data for a subject or group of subjects.
3. All individuals have the same weights.

Of the three procedures discussed, the derived measure is the least desirable in meeting the spirit of MDS—that the evaluation of objects be made with minimal influence by the researcher.

PREFERENCES DATA Preference implies that stimuli should be judged in terms of dominance relationships; that is, the stimuli are ordered in terms of the preference for some property. It enables the researcher to make direct statements of which is the more preferred object (e.g., brand A is preferred over brand C). The two most common procedures for obtaining preference data are direct ranking and paired comparisons.

Direct Ranking. Each respondent ranks the objects from most preferred to least preferred. This method of gathering nonmetric similarity data is popular because it is easy to administer for a small to moderate number of objects. It is quite similar in concept to the subjective clustering procedure discussed earlier, only in this case each object must be given a unique rank (no ties).

Paired Comparisons. A respondent is presented with all possible pairs and asked to indicate which member of each pair is preferred. Then, overall preference is based on the total number of times each object was the preferred member of the paired comparison. In this way, the researcher gathers explicit data for each comparison. This approach covers all possible combinations and is much more detailed than just the direct rankings. The principal drawback to this method is the large number of tasks involved with even a relatively small number of objects. For example, 10 objects result in 45 paired comparisons, which are too many tasks for most research situations. Note that paired comparisons are also used in collecting similarity data, as noted in the example at the beginning of the chapter, but there the pairs of objects are ranked or rated as to the degree of similarity between the two objects in the pair.

PREFERENCE DATA VERSUS SIMILARITY DATA Both similarity and preference data provide a basis for constructing a perceptual map that can depict the relative positions of the objects across the perceived (inferred) dimensions. Selecting between the two approaches lies in the objectives to be achieved:

- Similarity-based perceptual maps are best suited to understanding the attributes/dimensions that describe the objects. In this approach, the focus is on characterizing the nature of each object and its composition relative to the other objects.
- Preference data enable the researcher to view the location of objects on a perceptual map for which distance implies differences in preference. This procedure is useful because an individual's perception of objects in a preference context may be different from that in a similarity context. That is, a particular dimension may be useful in describing the similarities between two objects but may not be consequential in determining preference.

The differing bases for comparison in the two approaches many times result in quite different perceptual maps. Two objects could be perceived as different in a similarity-based map but be similar in a preference-based spatial map, resulting in two quite different maps. For example, two different brands of candy bars could be far apart in a similarity-based map but, with equivalent preference, be positioned close to each other on a preference map. The researcher must choose the map that best matches the objectives of the analysis.

RULES OF THUMB 10-2

Research Design of MDS

- Perceptual maps can be generated through decompositional or compositional approaches:
  - Decompositional approaches are the traditional and most common MDS method, requiring only overall comparisons of similarity between objects
  - Compositional approaches are used when the research objectives involve comparing objects on a defined set of attributes
- The number of objects to be evaluated is a trade-off between:
  - A small number of objects to facilitate the respondents' task
  - Four times as many objects as dimensions desired (i.e., 5 objects for one dimension, 9 objects for two dimensions, etc.) to obtain a stable solution

---

STAGE 3: ASSUMPTIONS OF MDS ANALYSIS

Multidimensional scaling has no restraining assumptions on the methodology, type of data, or form of the relationships among the variables, but MDS does require that the researcher accept three tenets concerning perception:

1. *Variation in dimensionality.* Respondents may vary in the dimensionality they use to form their perceptions of an object (although it is thought that most people judge in terms of a limited number of characteristics or dimensions). For example, some might evaluate a car in terms of its horsepower and appearance, whereas others do not consider these factors at all but instead assess it in terms of cost and interior comfort.
2. *Variation in importance.* Respondents need not attach the same level of importance to a dimension, even if all respondents perceive this dimension. For example, two respondents perceive a soft drink in terms of its level of carbonation, but one may consider this dimension unimportant whereas the other may consider it very important.
3. *Variation over time.* Judgments of a stimulus in terms of either dimensions or levels of importance are likely to change over time. In other words, one may not expect respondents to maintain the same perceptions for long periods of time.

In spite of the differences we can expect between individuals, MDS methods can represent perceptions spatially so that all of these differences across individuals are accommodated. This capability enables MDS techniques to not only help a researcher understand each separate individual but also to identify the shared perceptions and evaluative dimensions within the sample of respondents.

STAGE 4: DERIVING THE MDS SOLUTION AND ASSESSING OVERALL FIT

Today the basic MDS programs available in the major statistical programs can accommodate the differing types of input data and types of spatial representations, as well as the varying interpretational alternatives. Our objective here is to provide an overview of MDS to enable a ready understanding of the differences among these programs. However, as with other multivariate techniques, development in both application and knowledge is continual. Thus, we refer the user interested in specific program applications to other texts devoted solely to multidimensional scaling [10, 11, 17, 19, 24].

Determining an Object's Position in the Perceptual Map

The first task of stage 4 involves the positioning of objects to best reflect the similarity evaluations provided by the respondents (see Figure 10-5). Here the MDS techniques determine the optimal locations for each object in a specified dimensionality. The solutions for each dimensionality (two dimensions, three dimensions, etc.) are then compared to select a final solution defining the number of dimensions and each object's relative position on those dimensions.

CREATING THE PERCEPTUAL MAP MDS programs follow a common three-step process for determining the optimal positions in a selected dimensionality:

1. Select an initial configuration of stimuli (S_k) at a desired **initial dimensionality** (t). Various options for obtaining the initial configuration are available. The two most widely used are configurations either applied by the researcher based on previous data or generated by selecting pseudorandom points from an approximately normal multivariate distribution.
2. Compute the distances between the stimuli points and compare the relationships (observed versus derived) with a measure of fit. Once a configuration is found, the interpoint distances between stimuli (d_{ij}) in the starting configurations are compared with distance measures ($\hat{d}_{ij}$)

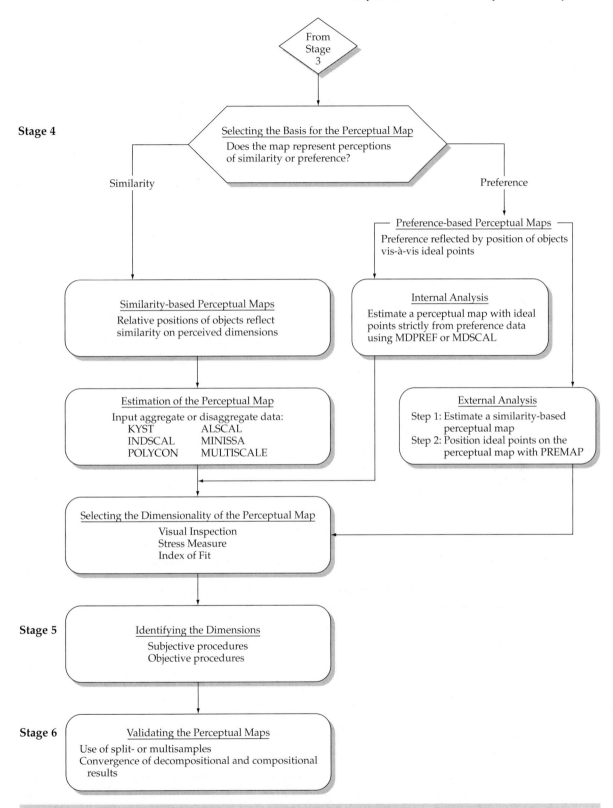

FIGURE 10-5   Stages 4–6 in the Multidimensional Scaling Decision Diagram

derived from the similarity judgments (s_{ij}). The two distance measures are then compared by a measure of fit, typically a measure of stress. (Fit measures are discussed in a later section.)

3. If the measure of fit does not meet a selected predefined stopping value, find a new configuration for which the measure of fit is further minimized. The program determines the directions in which the best improvement in fit can be obtained and then moves the points in the configuration in those directions in small increments.

The need for a computer program versus hand calculations becomes apparent as the number of objects and dimensions increases. Let us look at a typical MDS analysis and see what is actually involved.

With 10 products to be evaluated, each respondent must rank all 45 possible pairs of objects from most similar (1) to least similar (45). With these rank orders, we proceed to attempt to define the dimensionality and positions of each object.

1. First, assume that we are starting with a two-dimensional solution. Although we could define any number of dimensions, it is easiest to visualize the process in a simple two-dimensional situation.
2. Place 10 points (representing the 10 products) randomly on a sheet of graph paper (representing the two dimensions) and then measure the distances between every pair of points (45 distances).
3. Calculate the goodness-of-fit of the solution by measuring the rank-order agreement between the Euclidean (straight-line) distances of the plotted objects and the original 45 ranks.
4. If the straight-line distances do not agree with the original ranks, move the 10 points and try again. Continue to move the objects until you get a satisfactory fit among the distances between all objects and the ranks indicating similarity.
5. You can then position the 10 objects in three-dimensional space and follow the same process. If the fit of actual distances and similarity ranks is better, then the three-dimensional solution may be more appropriate.

As you can see, the process quickly becomes intractable with increases in the number of objects and dimensions. Computers execute the calculations and allow for a more accurate and detailed solution. The program calculates the best solution across any number of dimensions, thus providing a basis to compare the various solutions.

The primary criterion in all instances for finding the best representation of the data is preservation of the ordered relationship between the original rank data and the derived distances between points. Any measure of fit (e.g., stress) is simply a measure of how well (or poorly) the ranks based on the distances on the map agree with the ranks given by the respondents.

AVOIDING DEGENERATE SOLUTIONS   In evaluating a perceptual map, the researcher should always be aware of **degenerate solutions.** Degenerate solutions are derived perceptual maps that are not accurate representations of the similarity responses. Most often they are caused by inconsistencies in the data or an inability of the MDS program to reach a stable solution. They are characterized most often by either a circular pattern, in which all objects are shown to be equally similar, or a clustered solution, in which the objects are grouped at two ends of a single dimension. In both cases, MDS is unable to differentiate among the objects for some reason. The researcher should then reexamine the research design to see where the inconsistencies occur.

Selecting the Dimensionality of the Perceptual Map

As seen in the previous section, MDS defines the optimal perceptual map in a number of solutions of varying dimensionality. With these solutions in hand, the objective of the next step is the selection of a spatial configuration (perceptual map) in a specific number of dimensions. The determination of how many dimensions are actually represented in the data is generally reached through one of three approaches: subjective evaluation, scree plots of the stress measures, or an overall index of fit.

SUBJECTIVE EVALUATION   The spatial map is a good starting point for the evaluation. The number of maps necessary for interpretation depends on the number of dimensions. A map is produced for each combination of dimensions. One objective of the researcher should be to obtain the best fit with the smallest possible number of dimensions. Interpretation of solutions derived in more than three dimensions is extremely difficult and usually is not worth the improvement in fit. The researcher typically makes a **subjective evaluation** of the perceptual maps and determines whether the configuration looks reasonable. This evaluation is important because at a later stage the dimensions will need to be interpreted and explained.

STRESS MEASURES   A second approach is to use a **stress measure,** which indicates the proportion of the variance of the **disparities** (differences in distances between objects on the perceptual map and the similarity judgments of the respondents) not accounted for by the MDS model. This measurement varies according to the type of program and the data being analyzed. Kruskal's [18] stress is the most commonly used measure for determining a model's goodness-of-fit. It is defined by:

$$\text{Stress} = \sqrt{\frac{(d_{ij} - \hat{d}_{ij})^2}{(d_{ij} - \overline{d}_{ij})^2}}$$

where

$$\overline{d} = \text{average distance } (\Sigma d_{ij}/n) \text{ on the map}$$
$$\hat{d}_{ij} = \text{derived distance from the perceptual map}$$
$$d_{ij} = \text{original distance based on similarity judgements}$$

The stress value becomes smaller as the derived ($\hat{d}_{ij}$) approaches the original d_{ij}. Stress is minimized when the objects are placed in a configuration so that the distances between the objects best match the original distances.

   A problem found in using stress, however, is analogous to that of R^2 in multiple regression in that stress always improves with increased dimensions. (Remember that R^2 always increases with additional variables.) A trade-off must then be made between the fit of the solution and the number of dimensions. As was done for the extraction of factors in factor analysis, we can plot the stress value against the number of dimensions to determine the best number of dimensions to utilize in the analysis [19].

   For example, in the scree plot in Figure 10-6, the elbow indicates substantial improvement in the goodness-of-fit when the number of dimensions is increased from 1 to 2. Therefore, the best fit is obtained with a relatively low number of dimensions (two).

INDEX OF FIT   A squared correlation index is sometimes used as an **index of fit.** It can be interpreted as indicating the proportion of variance of the disparities accounted for by the MDS procedure. In other words, it is a measure of how well the raw data fit the MDS model. The R^2 measure in multidimensional scaling represents essentially the same measure of variance as it does in other multivariate techniques. Therefore, it is possible to use similar measurement criteria. That is, measures of .60 or better are considered acceptable. Of course, the higher the R^2, the better the fit.

Incorporating Preferences into MDS

Up to this point, we have concentrated on developing perceptual maps based on similarity judgments. However, perceptual maps can also be derived from preferences. The objective is to determine the preferred mix of characteristics for a set of stimuli that predicts preference, given a set configuration of objects [9, 10]. In doing so, a joint space is developed portraying both the objects (stimuli) and the subjects (ideal points). A critical assumption is the homogeneity of perception across individuals for

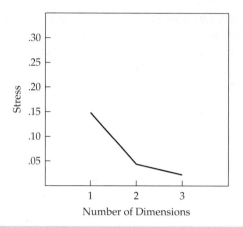

Figure 10-6   Use of a Scree Plot to Determine the Appropriate Dimensionality

the set of objects. This homogeneity enables all differences to be attributed to preferences, not perceptual differences.

IDEAL POINTS   The term **ideal point** has been misunderstood or misleading at times. We can assume that if we locate (on the derived perceptual map) the point that represents the most preferred combination of perceived attributes, we have identified the position of an ideal object. Equally, we can assume that the position of this ideal point (relative to the other products on the derived perceptual map) defines relative preferences so that products farther from the ideal should be less preferred. Thus, an ideal point is positioned so that the distance from the ideal conveys changes in preference.

Consider, for example, Figure 10-7. When preference data on the six candy bars (A to F) were obtained from a respondent, their ideal point (indicated by the dot) was positioned so that increasing the distance from it indicated declining preference. Based on this perceptual map, this respondent's preference order is C, F, D, E, A, B. To imply that the ideal candy bar is exactly at that point or even

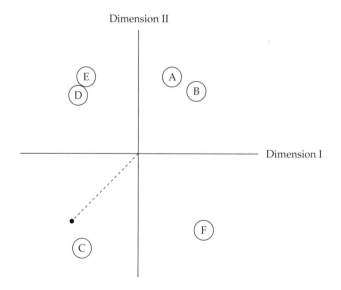

• Indicates respondent's ideal point

Figure 10-7   A Respondent's Ideal Point Within the Perceptual Map

beyond (in the direction shown by the dashed line from the origin) can be misleading. The ideal point simply defines the ordered preference relationship (most preferred to least preferred) among the set of six candy bars for that respondent.

Although ideal points individually may not offer much insight, clusters of them can be useful in defining segments. Many respondents with ideal points in the same general area represent potential market segments of persons with similar preferences, as indicated in Figure 10-8.

DETERMINING IDEAL POINTS   Two approaches generally work to determine ideal points: explicit and implicit estimation. The primary difference between the two approaches is the type of evaluative response requested of the respondent. We discuss each approach in the following sections.

Explicit Estimation. Explicit estimation proceeds from the direct responses of subjects, typically asking the subject to rate a hypothetical ideal on the same attributes on which the other stimuli are rated. Alternatively, the respondent is asked to include among the stimuli used to gather similarities data a hypothetical ideal stimulus (e.g., brand, image).

When asking respondents to conceptualize an ideal of anything, we typically run into problems. Often the respondent conceptualizes the ideal at the extremes of the explicit ratings used or as being similar to the most preferred product from among those with which the respondent has had experience. Also, the respondent must think in terms not of similarities but of preferences, which is often difficult with relatively unknown objects. Often these perceptual problems lead the researcher to use implicit ideal point estimation.

Implicit Estimation. Several procedures for implicitly positioning ideal points are described in the following section. The basic assumption underlying most procedures is that derived measures of ideal points' spatial positions are maximally consistent with individual respondents' preferences. Srinivasan and Schocker [27] assume that the ideal point for all pairs of stimuli is determined so that it violates with least harm the constraint that it be closer to the most preferred in each pair than it is to the least preferred.

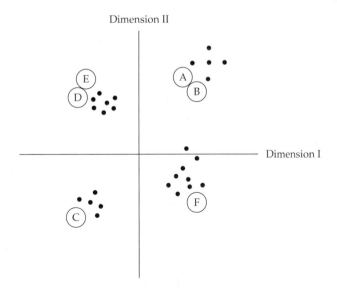

• Indicates a single respondent's ideal point

FIGURE 10-8 Incorporating Multiple Ideal Points in the Perceptual Map

Summary. In summary, ideal point estimation can be approached in many ways, and no single method has been demonstrated as best in all cases. The choice depends on the researcher's skills and the MDS procedure selected.

IMPLICIT POSITIONING OF THE IDEAL POINT   Implicit positioning of the ideal point from preference data can be accomplished through either an internal or an external analysis:

- Internal analysis of preference data refers to the development of a spatial map shared by both stimuli and subject points (or vectors) solely from the preference data.
- External analysis of preference uses a prespecified configuration of objects and then attempts to place the ideal points within this perceptual map.

Each approach has advantages and disadvantages, which are discussed in the following sections.

Internal Analysis.   Internal analyses must make certain assumptions in deriving a combined perceptual map of stimuli and ideal points. The objects' positions are calculated based on **unfolding** preference data for each individual. The results reflect perceptual dimensions that are **stretched** and weighted to predict preference. One characteristic of internal estimation methods is that they typically employ a vector representation of the ideal point (see the following section for a discussion of vector versus point representations), whereas external models can estimate either vector or point representations.

As an example of this approach, MDPREF [5] or MDSCAL [18], two of the more widely used programs, enable the user to find configurations of stimuli and ideal points. In doing so, the researcher must assume the following:

1. No difference between respondents
2. Separate configurations for each respondent
3. A single configuration with individual ideal points

By gathering preference data, the researcher can represent both stimuli and respondents on a single perceptual map.

External Analysis.   External analysis of preference data refers to fitting ideal points (based on preference data) to stimulus space developed from similarities data obtained from the same subjects. For example, we might scale similarities data individually, examine the individual maps for commonality of perception, and then scale the preference data for any group identified in this fashion. If this procedure is followed, the researcher must gather both preference and similarities data to achieve external analysis.

PREFMAP [6] was developed solely to perform external analysis of preference data. Because the similarity matrix defines the objects in the perceptual map, the researcher can now define both attribute descriptors (assuming that the perceptual space is the same as the evaluative dimensions) and ideal points for individuals. PREFMAP provides estimates for a number of different types of ideal points, each based on different assumptions as to the nature of preferences (e.g., vector versus point representations, or equal versus differential dimension weights).

Choosing Between Internal and External Analysis.   It is generally accepted [10, 11, 24] that external analysis is clearly preferable to internal analysis. This conclusion is based on computational difficulties with internal analysis procedures and on the confounding of differences in preference with differences in perception. In addition, the saliencies of perceived dimensions may change as one moves from perceptual space (Are the stimuli similar or dissimilar?) to evaluative space (Which stimulus is preferred?).

We discuss the procedure of external estimation in our example of perceptual mapping with MDS at the end of this chapter.

POINT VERSUS VECTOR REPRESENTATIONS The discussion of perceptual mapping of preference data emphasized an ideal point that portrays the relationship of an individual's preference ordering for a set of stimuli. The previous section discussed the issues relating to the type of data and analysis used in estimating and placing the ideal point. The remaining issue focuses on the manner in which the other objects in the perceptual map relate to the ideal point to reflect preference. The two approaches (point versus vector representation) are discussed next.

Point Representation. The most easily understood method of portraying the ideal point is to use the straight-line (Euclidean) distance measure of preference ordering from the ideal point to all the points representing the objects. We are assuming that the direction of distance from the ideal point is not critical, only the relative distance.

An example is shown in Figure 10-9. Here, the ideal point as positioned indicates that the most preferred object is E, followed by C, then B, D, and finally A. The ordering of preference is directly related to the distance from the ideal point.

Vector Representation. The ideal point can also be shown as a **vector.** To calculate the preferences in this approach, perpendicular lines (also known as **projections**) are drawn from the objects to the vector. Preference increases in the direction the vector is pointing. The preferences can be read directly from the order of the projections.

Figure 10-10 illustrates the vector approach for two subjects with the same set of stimuli positions. For subject 1, the vector has the direction of lower preference in the bottom left-hand corner to higher preference in the upper right-hand corner. When the projection for each object is made, the preference order (highest to lowest) is A, B, C, E, and D. However, the same objects have a quite different preference order for subject 2. For subject 2 the preference order ranges from the most preferred, E, to the least preferred, C. In this manner, a separate vector can represent each subject.

The vector approach does not provide a single ideal point, but it is assumed that the ideal point is at an infinite distance outward on the vector.

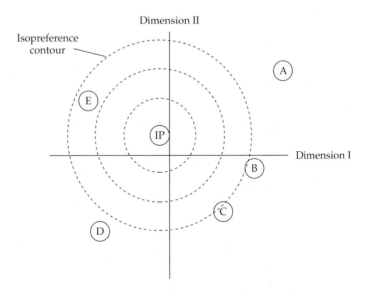

Preference order (highest to lowest): E > C > B > D > A

(A) Object   (IP) Ideal Point

FIGURE 10-9 **Point Representation of an Ideal Point**

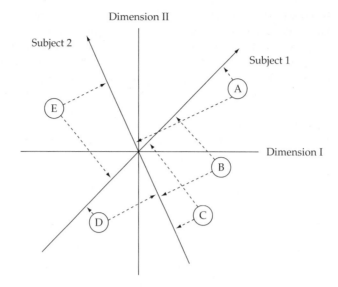

Preference order (highest to lowest): Subject 1: A > B > C > E > D

Subject 2: E > A > D > B > C

FIGURE 10-10   **Vector Representations of Two Ideal Points: Subjects 1 and 2**

Although either the point or vector representations can indicate what combinations of attributes are more preferred, these observations are often not borne out by further experimentation. For example, Raymond [23] cites an example in which the conclusion was drawn that people would prefer brownies on the basis of degree of moistness and chocolate content. When the food technicians applied this result in the laboratory, they found that their brownies made to the experimental specification became chocolate milk. One cannot always assume that the relationships found are independent or linear, or that they hold over time, as noted previously. However, MDS is a beginning in understanding perceptions and choice that will expand considerably as applications extend our knowledge of both the methodology and human perception.

STAGE 5: INTERPRETING THE MDS RESULTS

Once the perceptual map is obtained, the two approaches—compositional and decompositional—again diverge in their interpretation of the results. The differences in interpretation are based on the amount of information directly provided in the analysis (e.g., the attributes incorporated in the compositional analysis versus their absence in the decompositional analysis) and the generalizability of the results to the actual decision-making process.

- For *compositional methods,* the perceptual map can be directly interpreted with the attributes incorporated in the analysis. The solution, however, must be validated against other measures of perception, because the positions are totally defined by the attributes specified by the researcher. For example, discriminant analysis results might be applied to a new set of objects or respondents, assessing the ability to differentiate with these new observations.
- For *decompositional methods,* the most important issue is the description of the perceptual dimensions and their correspondence to attributes. Evaluations of similarity or preference are done without regard to attributes, thus avoiding a specification error issue. The risk, however,

is that the perceptual dimensions are not correctly translated in that the dimensions used in the evaluations are not reflected by the attributes chosen for their interpretation. Descriptive techniques to label the dimensions, as well as to integrate preferences (for objects and attributes) with the similarity judgments, are discussed later. Again, in line with their objectives, the decompositional methods provide an initial look into perceptions from which more formalized perspectives may emerge.

Because other chapters in this text deal with many of the compositional techniques, the remainder of this chapter focuses on decompositional methods, primarily the various techniques used in multidimensional scaling. A notable exception is the discussion of a compositional approach—correspondence analysis—that, to a degree, bridges the gap between the two approaches in its flexibility and methods of interpretation.

Identifying the Dimensions

As discussed in Chapter 3 relating to interpreting factors in factor analysis, identifying underlying dimensions is often a difficult task. Multidimensional scaling techniques have no built-in procedure for labeling the dimensions. The researcher, having developed the maps with a selected dimensionality, can adopt several procedures, either subjective or objective.

SUBJECTIVE PROCEDURES Interpretation must always include some element of researcher or respondent judgment, and in many cases this level of judgment proves adequate for the questions at hand. A quite simple, yet effective, method is labeling (by visual inspection) the dimensions of the perceptual map by the respondent. Respondents may be asked to interpret the dimensionality subjectively by inspecting the maps, or a set of experts may evaluate and identify the dimensions. Although it makes no attempt to quantitatively link the dimensions to attributes, this approach may be the best available if the dimensions are believed to be highly intangible, or affective or emotional in content, so that adequate descriptors cannot be devised.

In a similar manner, the researcher may describe the dimensions in terms of known (objective) characteristics. In this way, the correspondence is made between objective and perceptual dimensions directly, although these relationships are not a result of respondent feedback but of the researcher's judgment.

OBJECTIVE PROCEDURES As a complement to the subjective procedures, a number of more formalized methods have been developed. The most widely used method, PROFIT (PROperty FITting) [3], collects attribute ratings for each object and then finds the best correspondence of each attribute to the derived perceptual space. The attempt is to identify the determinant attributes in the similarity judgments made by individuals. Measures of fit are given for each attribute, as well as their correspondence with the dimensions. The researcher can then determine which attributes best describe the perceptual positions and are illustrative of the dimensions. The need for correspondence between the attributes and the defined dimensions diminishes with the use of metric output, because the dimensions can be rotated freely without any changes in interpretation.

SELECTING BETWEEN SUBJECTIVE AND OBJECTIVE PROCEDURES For either subjective or objective procedures, the researcher must remember that although a dimension can represent a single attribute, it usually does not. A more common procedure is to collect data on several attributes, associate them either subjectively or empirically with the dimensions where applicable, and determine labels for each dimension using multiple attributes, similar to factor analysis. Many researchers suggest that using attribute data to help label the dimensions is the best alternative. The problem, however, is that the researcher may not include all the important attributes in the study. Thus, the researcher can never be totally assured that the labels represent all relevant attributes.

Both subjective and objective procedures illustrate the difficulty of labeling the axes. This task is essential because the dimensional labels are required for further interpretation and use of the results. The researcher must select the type of procedure that best suits both the objectives of the research and the available information. Thus, the researcher must plan for the derivation of the dimensional labels as well as the estimation of the perceptual map.

STAGE 6: VALIDATING THE MDS RESULTS

Validation in MDS is as important as in any other multivariate technique. Because of the highly inferential nature of MDS, this effort should be directed toward ensuring the generalizability of the results both across objects and to the population. As will be seen in the following discussion, MDS presents particularly problematic issues in validation, both from a substantive as well as methodological perspective.

Issues in Validation

Any MDS solution must deal with two specific issues that complicate efforts to validate the results:

- The only output of MDS that can be used for comparative purposes involves the relative positions of the objects. Thus, although the positions can be compared, the underlying dimensions have no basis for comparison. If the positions vary, the researcher cannot determine whether the objects are viewed differently, the perceptual dimensions vary, or both.
- Systematic methods of comparison have not been developed and integrated into the statistical programs. The researcher is left to improvise with procedures that may address general concerns but are not specific to MDS results.

As a result, researchers must strive in their validation efforts to maintain comparability between solutions while providing an empirical basis for comparison.

Approaches to Validation

Any approach to validation attempts to assess generalizability (e.g., similarity across different samples) while maintaining comparability for comparison purposes. The issues discussed in the previous section, however, make these requirements difficult for any MDS solution. Several approaches for validation that meet each requirement to some degree are discussed next.

SPLIT-SAMPLE ANALYSIS The most direct validation approach is a split- or multisample comparison, in which either the original sample is divided or a new sample is collected. In either instance, the researcher must then find a means of comparing the results. Most often the comparison between results is done visually or with a simple correlation of coordinates. Some matching programs are available, such as FMATCH [25], but the researcher still must determine how many of the disparities are due to differences in object perceptions, differing dimensions, or both.

COMPARISON OF DECOMPOSITIONAL VERSUS COMPOSITIONAL SOLUTIONS Another approach is to obtain a convergence of MDS results by applying both decompositional and compositional methods to the same sample. The decompositional method(s) could be applied first, along with interpretation of the dimensions to identify key attributes. Then one or more compositional methods, particularly correspondence analysis, could be applied to confirm the results. The researcher must realize that this is not true validation of the results in terms of generalizability, but does confirm the interpretation of the dimension. From this point, validation efforts with other samples and other objects could be undertaken to demonstrate generalizability to other samples.

RULES OF THUMB 10-3

Deriving and Validating an MDS Solution

- Stress measures (lower values are better) represent an MDS solution's fit
- Researchers can identify a degenerate MDS solution that is generally problematic by looking for:
  - A circular pattern of objects suggesting that all objects are equally similar, or
  - A multiclustered solution in which objects are grouped at two ends of a single continuum
- The appropriate number of dimensions for a perceptual map is based on:
  - A subjective judgment as to whether a solution with a given dimensionality is reasonable
  - Use of a scree plot to identify the elbow where a substantial improvement in fit occurs
  - Use of R^2 as an index of fit; measures of .6 or higher are considered acceptable
- External analysis, such as is performed by PREFMAP, is considered preferable in generating ideal points relative to internal analysis
- The most direct validation method is a split-sample approach
  - Multiple solutions are generated by either splitting the original sample or collecting new data
  - Validity is indicated when the multiple solutions match

OVERVIEW OF MULTIDIMENSIONAL SCALING

Multidimensional scaling represents a distinctive approach to multivariate analysis when compared to other methods in this text. Whereas other techniques are concerned with accurately specifying the attributes comprising independent and/or dependent variables, multidimensional scaling takes a quite different approach. It gathers only global or holistic measures of similarity or preference and then empirically infers the dimensions (both character and number) that reflect the best explanation of an individual's responses either separately or collectively. In this approach, the variate used in many other techniques becomes the perceptual dimensions inferred from the analysis. As such, the researcher does not have to be concerned with such issues as specification error, multicollinearity, or statistical characteristics of the variables. The challenge to the researcher, however, is to interpret the variate; without a valid interpretation the primary objectives of MDS are compromised.

The application of MDS is appropriate when the objective is more oriented toward understanding overall preferences or perceptions rather than detailed perspectives involving individual attributes. One technique, however, combines the specificity of attribute-level analysis within MDS-like solutions. That method, correspondence analysis, is discussed in the following section where the similarities and differences with traditional MDS techniques are highlighted.

CORRESPONDENCE ANALYSIS

Up to this point we have discussed the traditional decompositional approaches to MDS, but what about compositional techniques? In the past, compositional approaches relied on traditional multivariate techniques such as discriminant and factor analysis. Recent developments, however, combine aspects of both methods and MDS to form potent new tools for perceptual mapping.

Distinguishing Characteristics

Correspondence analysis (CA) is an increasingly popular interdependence technique for dimensional reduction and perceptual mapping [1, 2, 12, 14, 20]. It is also known as optimal scaling or scoring, reciprocal averaging or homogeneity analysis. When compared to the MDS techniques described in the earlier portion of this chapter, correspondence analysis has three distinguishing characteristics:

1. It is a compositional technique, rather than a decompositional approach, because the perceptual map is based on the association between objects and a set of descriptive characteristics or attributes specified by the researcher.
2. Its most direct application is in portraying the correspondence of categories of variables, particularly those measured in nominal measurement scales. This correspondence is then the basis for developing perceptual maps.
3. The unique benefits of CA lie in its abilities for representing rows and columns, for example, brands and attributes, in joint space.

Differences from Other Multivariate Techniques

Among the compositional techniques, factor analysis is the most similar by defining composite dimensions (factors) of the variables (e.g., attributes) and then plotting objects (e.g., products) on their scores on each dimension. In discriminant analysis, products can be distinguished by their profiles across a set of variables and plotted in a dimensional space as well. Correspondence analysis extends beyond either of these two compositional techniques:

- CA can be used with nominal data (e.g., frequency counts of preference for objects across a set of attributes) rather than metric ratings of each object on each object. This capability enables CA to be used in many situations in which the more traditional multivariate techniques are inappropriate.
- CA creates perceptual maps in a single step, where variables and objects are simultaneously plotted in the perceptual map based directly on the association of variables and objects. The relationships between objects and variables are the explicit objective of CA.

We first examine a simple example of CA to gain some perspective on its basic principles. Then we discuss each of the six stages of the decision-making process introduced in Chapter 1. The emphasis is on those unique elements of CA as compared to the decompositional methods of MDS discussed earlier.

A Simple Example of CA

Let us examine a simple situation as an introduction to CA. In its most basic form, CA examines the relationships between categories of nominal data in a contingency table, the cross-tabulation of two categorical (nonmetric) variables. Perhaps the most common form of contingency table would be cross-tabulating objects and attributes (e.g., most distinctive attributes for each product or product sales by demographic category). CA can be applied to any contingency table and portray a perceptual map relating the categories of each nonmetric variable in a single perceptual map.

Our example involves product sales across a single demographic variable (age). The cross-tabulated data (see Table 10-2) portray the sales figures for products A, B, and C broken down by three age categories (young adults, who are 18 to 35 years old; middle age, who are 36 to 55 years old; and mature individuals, who are 56 or older).

UTILIZING CROSS-TABULATED DATA   What can we learn from the cross-tabulated data? First, we can look at the column and row totals to identify the ordering of categories (highest to lowest). But more important, we can view the relative sizes of each cell of the contingency table, which reflects the amount of each variable for each object. Comparing the cells may identify patterns reflecting associations among certain objects and attributes.

TABLE 10-2   Cross-Tabulated Data Detailing Product Sales by Age Category

Age Category	Product Sales			
	A	B	C	Total
Young Adults (18–35 years old)	20	20	20	60
Middle Age (36–55 years old)	40	10	40	90
Mature Individuals (56 + years old)	20	10	40	70
Total	80	40	100	220

Viewing Table 10-2, we see that product sales do vary across products (product C has the highest total sales of 100 units; product B the lowest sales, 40 units) and age groups (middle-age group buys 90 units; young-adult group, 40 units). To identify any pattern to the sales, we need to be able to state that a certain group (e.g., young adults) buys more or less of a certain product. But to do this we need to have some way to define what the "expected" sales would be so we can say that the actual sales amount is more or less. For example, the middle-age group purchases 40 units of both product A and C. Do we then assess that this group has equal preference for the two products, or should the fact that product C was generally a more popular product, with 100 units of total sales, than was product A, with only 80 total units, have some impact on how we judge the unit sales of the two products among this age group?

To identify these patterns of "more or less" for any cell in the table, we need two more elements that help in quantifying the amount of "more or less" as well as in portraying it graphically. The first is a means of standardizing the cell counts to make them comparable and then to develop a means of portraying the values of each cell.

Standardizing Frequency Counts. First is a standardized measure of the cell counts that simultaneously considers the differences in row and column totals. We can directly compare the cells when all the row and column totals are equal, which is rarely the case. Instead, the rows and column totals are usually unequal. In this case, we need a measure that compares each cell value to an expected value that reflects the specific row and column totals of that cell.

In our product sales example, let us examine the sales to the young-adults group. As we can see, this group purchased equals amounts of each product (20 units each). But is this what we would expect? How can we tell if they actually prefer one product over another by purchasing it more when compared to all the other age groups? A simple way would be to calculate what we expect product sales to be in direct proportion to overall product sales across all the groups. Product A had 36 percent of total sales (36% = 80 ÷ 220), whereas product B had 18 percent and product C had 45 percent. We can then calculate an "expected" sales amount for each product by applying these percentages to the 60 unit total sales by the young-adults group. This would give us an expected sales of just over 21 units for product A (36% × 60 units), while we would expect only about 11 units for product B and about 27 units for product C. We can now see that this group purchases slightly less than expected for both product A (21.8 units expected versus an actual of 20 units) and product C (27 units expected sales versus actual sales of 20 units), but purchases substantially more than expected for product B (actual sales of 20 units when expected sales are only 11 units).

This illustrates a simple way to calculate the amount of the differences (i.e., the amount that purchases are "more or less" than expected), but we still need a way to "standardize" the differences across age groups and products. As we will discuss later, the chi-square value (a variant of the process we just described) will be used as the standardized measure of comparison.

Portraying Each Cell. Once we have a standardized measure of the differences between expected and actual values for each cell (representing a distinct combination of rows and columns), we then need a method for portraying each cell in a single perceptual map. The task is to portray all of the associations between the rows and columns (age groups and products in our example). We do this by assigning each category of the rows and columns a separate symbol. Then, when the standardized values are higher than expected for a cell (a specific row/column combination) we would expect that the symbols for that row and column would be located closer together, and the symbols for cells with standardized values much lower than expected would be more widely separated. The challenge is to develop a perceptual map that best portrays all of the associations represented by the cells of the contingency table.

In our product example, we would start with six symbols (three symbols for the age groups and three symbols for the products). We would then use the standardized measures of difference discussed earlier to help position the symbols to represent the associations between age groups and products. Going back to the young-adult age group, we would expect the symbol for product B to be located closer to the young-adult symbol than either product A or C, with C the farthest removed from the young-adult symbol because it had the largest difference of actual sales being lower than expected sales. This process is repeated for all the age group/product combinations and then the perceptual map is drawn so that the symbol positions best reflect the values of the standardized measure across all of the age group/product combinations.

In the following sections, we discuss how CA calculates a standardized measure of association based on the cell counts of the contingency and then the process whereby these associations are converted into a perceptual map.

CALCULATING A MEASURE OF ASSOCIATION OR SIMILARITY Correspondence analysis uses one of the most basic statistical concepts—chi-square—to standardize the cell frequency values of the contingency table and form the basis for association or similarity. Chi-square is a standardized measure of actual cell frequencies compared to expected cell frequencies. In cross-tabulated data, each cell contains the values for a specific row/column combination (e.g., sales of a specific product in a specific age group). Thus, the chi-square value represents a measure of association between the row and column categories. Higher levels of association, just like higher levels of similarity, should be represented as closer together in the perceptual map than those with lower levels of association. Readers wishing to review the process of calculating chi-square values can refer to the Basic Stats appendix available on the text's Web sites (www.pearsonglobaleditions.com/hair or www.mvstats.com).

The chi-square value can be easily transformed into a similarity measure. The process of calculating the chi-square (squaring the difference) removes the direction of the similarity. To restore the directionality, we use the sign of the original difference, but reverse it to make it more intuitive. This must be done since the difference in the chi-square calculation is defined as expected minus actual values. This makes negative differences represent those situations that we denote as greater association—where actual counts exceed expected counts. In order to make the similarity measures like those used in MDS (i.e., positive/larger values are greater association and negative/smaller values are less association), we reverse the sign of the original difference and apply it to the chi-square value. The result is a measure that acts just like the similarity measures used in earlier MDS examples. Negative values indicate less association (similarity), and positive values indicate greater association.

Table 10-3 illustrates the calculation of the chi-square value and its transformation into a similarity measure. For each cell, the actual and expected values are given along with the difference. Then the chi-square value is shown along with the transformed (signed) value. For example, the top row shows the actual sales for the young adults across the three products (20 units each) as well as the expected sales (21.82, 10.91, and 27.27 units, respectively). The differences for products A and C are positive, meaning that expected sales were greater than actual sales (a negative association), whereas the difference was positive for product B. The chi-square values act as the standardized measure of difference. The final value in each cell is the signed chi-square value, where positive values represent greater similarity between the age group/product combination and negative values are lower similarity.

TABLE 10-3 Calculating Chi-Square As Similarity Values for Cross-Tabulated Data

Age Category	Product Sales			Total
	A	B	C	
Young Adults				
Actual Sales	20	20	20	60
Expected Sales[a]	21.82	10.91	27.27	60
Difference[b]	1.82	−9.09	7.27	—
Chi-Square Value[c]	.15	7.58	1.94	
Signed Chi-Square Value[d]	−.15	7.58	−1.94	
Middle Age				
Actual Sales	40	10	40	90
Expected Sales	32.73	16.36	40.91	90
Difference	−7.27	6.36	.91	—
Chi-Square Value	1.62	2.47	.02	
Signed Chi-Square Value	1.62	−2.47	−.02	
Mature Individuals				
Actual Sales	20	10	40	70
Expected Sales	25.45	12.73	31.82	70
Difference	5.45	2.73	−8.18	—
Chi-Square Value	1.17	.58	2.10	
Signed Chi-Square Value	−1.17	−.58	2.10	
Total				
Sales	80	40	100	220
Expected Sales	80	40	100	220
Difference	—	—	—	—

[a]Expected sales = (Row total × Column total) ÷ Overall total

    Example: Cell$_{\text{Young Adults, Product A}}$ = (60 × 80) ÷ 220 = 21.82

[b]Difference = Expected sales − Actual sales

    Example: Cell$_{\text{Young Adults, Product A}}$ = 21.82 − 20.00 = 1.82

[c]Chi-square value = $\dfrac{\text{Difference}^2}{\text{Expected sales}}$

    Example: Cell$_{\text{Young Adults, Product A}}$ = 1.82^2 ÷ 21.82 = .15

[d]Signed chi-square value is the chi-square value with a reversed sign of the difference between expected and actual sales.

CREATING THE PERCEPTUAL MAP The similarity (signed chi-square) values provide a standardized measure of association, much like the similarity judgments used in earlier MDS methods. With this association/similarity measure, CA creates a perceptual map by using the standardized measure to estimate orthogonal dimensions upon which the categories can be placed to best account for the strength of association represented by the chi-square distances.

As done in MDS techniques, we consider first a lower-dimensional solution (e.g., one or two dimensions) and then expand the number of dimensions and continue until we reach the maximum number of dimensions. In CA the maximum number of dimensions is one less than the smaller of the number of rows or columns.

Looking back at the signed chi-square values in Table 10-3, which act as a similarity measure, the perceptual map should place certain combinations with positive values closer together on the perceptual map (e.g., young adults close to product B, middle age close to product A, and mature individuals close to product C). Moreover, certain combinations should be farther apart given their negative values (young adults farther from product C, middle age farther from product B, and mature individuals farther from product A).

In our example, we can only have two dimensions (the smaller of the number of rows or columns minus one, or $3 - 1 = 2$), so the two-dimensional perceptual map is shown in Figure 10-11. Corresponding to the similarity values detailed above, the positions of the age groups and products represent the positive and negative associations between age group and products very well. The researcher can examine the perceptual map to understand the product preferences among age groups based on their sales patterns. In contrast to MDS, however, we can relate age groups with product positions. We now have an additional tool that can relate different characteristics (in this case, products and the age characteristics of their buyers) in a single perceptual map. Just as easily we could have related products to attributes, benefits, or any other characteristic. We should note, however, that in CA the positions of objects are now dependent on the characters they are associated with.

In contrast to MDS, however, with CA we now can relate age groups with product positions. We can see that we now have an additional tool that can relate different characteristics (in this case, products and the age characteristics of their buyers) in a single perceptual map. Just as easily we could have related products to attributes, benefits or any other characteristic. We should note, however, that in CA the positions of objects are now dependent on the characters to which they are associated.

A Decision Framework for Correspondence Analysis

Correspondence analysis and the issues associated with a successful analysis can be viewed through the model-building process introduced in Chapter 1. In the following sections we examine the unique issues associated with correspondence analysis compared with MDS methods across the six stages of the decision process.

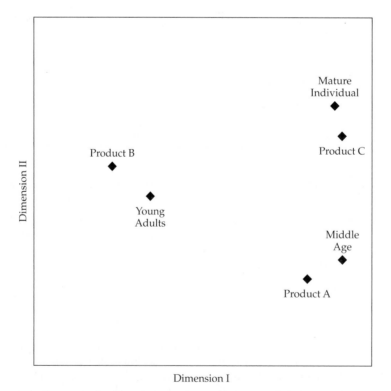

FIGURE 10-11 **Perceptual Map from Correspondence Analysis**

Stage 1: Objectives of CA

Researchers are constantly faced with the need to quantify the qualitative data found in nominal variables. CA differs from other MDS techniques in its ability to accommodate both nonmetric data and nonlinear relationships. It performs dimensional reduction similar to multidimensional scaling and a type of perceptual mapping, in which categories are represented in the multidimensional space. Proximity indicates the level of association among row or column categories. CA can address either of two basic objectives:

1. *Association among only row or column categories.* CA can be used to examine the association among the categories of just a row or column. A typical use is the examination of the categories of a scale, such as the Likert scale (five categories from "strongly agree" to "strongly disagree") or other qualitative scales (e.g., excellent, good, poor, bad). The categories can be compared to see whether two can be combined (i.e., they are in close proximity on the map) or whether they provide discrimination (i.e., they are located separately in the perceptual space).
2. *Association between both row and column categories.* In this application, interest lies is portraying the association between categories of the rows and columns, such as our example of product sales by age group. This use is most similar to the previous example of MDS and has propelled CA into more widespread use across many research areas.

The researcher must determine the specific objectives of the analysis because certain decisions are based on which type of objective is chosen. CA provides a multivariate representation of interdependence for nonmetric data not possible with other methods. With a compositional method, the researcher must be aware that the results are based on the descriptive characteristics (e.g., object attributes or respondent characteristics) of the objects included in the analysis. This contrasts with the decompositional MDS procedures described earlier in which only the overall measure of similarity between objects is needed.

Stage 2: Research Design of CA

Correspondence analysis requires only a rectangular data matrix (cross-tabulation) of nonnegative entries. The most common type of input matrix is a contingency table with specific categories defining the rows and columns. In creating the table, several issues emerge, relating to the nature of the variables and categories comprising the rows and columns:

1. The rows and columns do not have predefined meanings (i.e., attributes do not always have to be rows, and so on) but instead represent the responses to one or more categorical variables. The categories in both rows and columns, however, must have specific meaning for interpretation purposes.
2. The categories for a row or column need not be a single variable but can represent any set of relationships. A prime example is the "pick any" method [15, 16], in which respondents are given a set of objects and characteristics. A common application is where a set of objects (e.g., products) are rated on a set of characteristics (e.g., attributes). Here the rows can be the individual products and each attribute is a separate column. The respondents then indicate which objects, if any, are described by each characteristic. The respondent may choose any number of objects for each characteristic and the cross-tabulation table is the total number of times each object was described by each characteristic.
3. The cross-tabulation may occur for more than two variables in a multiway matrix form. In these cases, **multiple correspondence analysis** is employed. In a procedure quite similar to two-way analysis, the additional variables are fitted so that all the categories are placed in the same multidimensional space.

The generalized nature of the types of relationships that can be portrayed in the contingency table makes CA a widely applicable technique. Its increased usage in recent years is a direct result of the continued development of approaches for using this format for analyzing new types of relationships.

Stage 3: Assumptions of CA

Correspondence analysis shares with the more traditional MDS techniques a relative freedom from assumptions. The use of strictly nonmetric data in its simplest form (cross-tabulated data) represents linear and nonlinear relationships equally well. The lack of assumptions, however, must not cause the researcher to neglect the efforts to ensure the comparability of objects and, because it is a compositional technique, the completeness of the attributes used.

Stage 4: Deriving CA Results and Assessing Overall Fit

With a cross-tabulation table, the frequencies for any row–column combination of categories are related to other combinations based on the marginal frequencies. As depicted in our simple example earlier, correspondence analysis uses this basic relationship in three steps to create a perceptual map:

1. Calculate a conditional expectation (the expected cell count) that represents the similarity or association between row and column categories.
2. Once obtained, compute the differences between the expected and actual cell counts and convert them into a standardized measure (chi-square). Using these results as a distance metric makes them comparable to the input matrices used in the MDS approaches discussed earlier.
3. Using estimation techniques similar to MDS to convert the measure of similarity between categories (i.e., the signed chi-square), a series of dimensional solutions (one-dimensional, two-dimensional, etc.) is created where possible. The dimensions simultaneously relate the rows and columns in a single joint plot. The result is a representation of categories of rows and/or columns (e.g., brands and attributes) in the same plot.

DETERMINING IMPACT OF INDIVIDUAL CELLS   It should be noted that two specific terms, developed in correspondence analysis, describe the properties of the frequency values and their relative contribution to the analysis.

- The first term is **mass,** which is first defined for any single entry in the cross-tabulation table as the percentage of the total represented by that entry. It is calculated as the value of any single entry divided by N (the total for the table, which equals the sum of either the rows or columns). Thus, the sum of all table entries (cells) equals 1.0. We can also calculate the mass of any row or column category by summing across all entries. This result represents the contribution of any row or column category to the total mass.
- The second measure is **inertia,** which is defined as the total chi-square divided by N (the total of the frequency counts). In this way we have a relative measure of chi-square that can be related to any frequency count.

With these similarities to MDS comes a similar set of problems, focused on two primary issues in assessing overall fit: assessing the relative importance of the dimensions and then identifying the appropriate number of dimensions. Each of these issues is discussed in the following section.

ASSESSING THE NUMBER OF DIMENSIONS   Eigenvalues, also known as singular values, are derived for each dimension and indicate the relative contribution of each dimension in explaining the variance in the categories. Similar to factor analysis, we can determine the amount of explained variance both for individual dimensions and the solution as a whole. Some programs, such as those of SPSS, calculate the inertia for a dimension, which also measures explained variation and is directly related to the eigenvalue.

DETERMINING THE NUMBER OF DIMENSIONS   The maximum number of dimensions that can be estimated is one less than the smaller of the number of rows or columns. For example, with six columns and eight rows, the maximum number of dimensions would be five, which is six (the number of columns) minus one.

The researcher selects the number of dimensions based on the overall level of explained variance desired and the incremental explanation gained by adding another dimension. In assessing dimensionality, the researcher is faced with trade-offs much like other MDS solutions or even factor analysis (Chapter 3):

- Each dimension added to the solution increases the explained variance of the solution, but at a decreasing amount (i.e., the first dimension explains the most variance, the second dimension the second greatest, etc.).
- Adding dimensions increases the complexity of the interpretation process; perceptual maps of greater than three dimensions become increasing complex to analyze.

The researcher must balance the desire for increased explained variance versus the more complex solution that may affect interpretation. A rule of thumb is that dimensions with inertia (eigenvalues) greater than .2 should be included in the analysis.

MODEL ESTIMATION   A number of computer programs are available to perform correspondence analysis. Among the more popular programs are ANACOR and HOMALS, available with SPSS; PROC CORRESP in SAS; CORRESP from NewMDSX [25]; CORRAN and CORRESP from PC-MDS [25]; and MAPWISE [22]. A large number of specialized applications have emerged in specific disciplines such as ecology, geology, and many of the social sciences.

Stage 5: Interpretation of the Results

Once the dimensionality has been established, the researcher is faced with two tasks: interpreting the dimensions to understand the basis for the association among categories and assessing the degree of association between categories, either within a row/column or between rows and columns. In doing so, the researcher gains an understanding of the underlying dimensions upon which the perceptual map is based along with the derived association of any specific set of categories.

DEFINING THE CHARACTER OF THE DIMENSIONS   If the researcher is interested in defining the character of one or more dimensions in terms of the row or column categories, descriptive measures in each software program indicate the association of each category with a specific dimension. For example, in SPSS, the inertia measure (used to assess the degree of explained variance) is decomposed across the dimensions. Similar in character to factor loadings, these values represent the extent of association for each category individually with each dimension. The researcher can then name each dimension in terms of the categories most associated with that dimension.

In addition to representing the association of each category with each dimension, the inertia values can be totaled across dimensions in a collective measure. In doing so, we gain an empirical measure of the degree to which each category is represented across all dimensions. In concept, this measure is similar to the communality measure of factor analysis (see Chapter 3).

ASSESSING ASSOCIATION AMONG CATEGORIES   The second task in interpretation is to identify a category's association with other categories, which can be done visually or through empirical measures. No matter which approach is used, the researcher must first select the types of comparison to be made and then the appropriate normalization for the selected comparison. The two types of comparison are:

1. *Between categories of the same row or column.* Here the focus is on only rows or columns, such as when examining the categories of a scale to see whether they can be combined. These types of comparisons can be made directly from any correspondence analysis.

2. *Between rows and columns.* An attempt to relate the association between a row category and a column category. This most common type of comparison relates categories across dimensions (e.g., in our earlier example, product sales most associated with age categories). At this time, however, some debate centers on the appropriateness of comparing between row and column categories. In a strict sense, distances between points representing categories can only be made within a row or a column. It is deemed inappropriate to directly compare a row and column category. It is appropriate to make generalizations regarding the dimensions and each category's position on those dimensions. So the relative positioning of row and column categories can be defined within those dimensions, but should not be directly compared.

Some computer programs provide for a normalization procedure to allow for this direct comparison. If only a row or column normalization is available, alternative procedures are proposed to make all categories comparable [2, 22], yet disagreement still remains as to their success [13]. In the cases for which direct comparisons are not possible, the general correspondence still holds and specific patterns can be distinguished.

Research objectives may focus on either evaluation of the dimensions or comparison of the categories, and the researcher is encouraged to make both interpretations as they reinforce each other. For example, comparing row versus column categories can always be complemented with an understanding of the nature of the dimensions to provide a more comprehensive perspective on the positioning of categories rather than just the specific comparisons. Likewise, evaluating the specific category comparisons can provide specificity to the interpretation of the dimensions.

Stage 6: Validation of the Results

The compositional nature of correspondence analysis provides more specificity for the researcher with which to validate the results. In doing so, the researcher should strive to assess two key questions concerning generalizability of two elements:

- *Sample.* As with all MDS techniques, an emphasis must be made to ensure generalizability through split- or multisample analyses.
- *Objects.* The generalizability of the objects (represented individually and as a set by the categories) must also be established. The sensitivity of the results to the addition or deletion of a category can be evaluated. The goal is to assess whether the analysis is dependent on only a few objects and/or attributes.

In either instance, the researcher must understand the true meaning of the results in terms of the categories being analyzed. The inferential nature of correspondence analysis, like other MDS methods, requires strict confidence in the representativeness and generalizability of the sample of respondents and the objects (categories) being analyzed.

Overview of Correspondence Analysis

Correspondence analysis presents the researcher with a number of advantages, ranging from the generalized nature of the input data to development of unique perceptual maps:

- The simple cross-tabulation of multiple categorical variables, such as product attributes versus brands, can be represented in a perceptual space. This approach enables the researcher either to analyze existing responses or to gather responses at the least restrictive measurement type, the nominal or categorical level. For example, the respondent need rate only yes or no for a set of objects on a number of attributes. These responses can then be aggregated in a cross-tabulation table and analyzed. Other techniques, such as factor analysis, require interval ratings of each attribute for each object.

- CA portrays not only the relationships between the rows and columns, but also the relationships between the categories of either the rows or the columns. For example, if the columns were attributes, multiple attributes in close proximity would all have similar profiles across products, forming a group of attributes quite similar to a factor from principal components analysis.
- CA can provide a joint display of row and column categories in the same dimensionality. Certain program modifications allow for interpoint comparisons in which relative proximity is directly related to higher association among separate points [1, 22]. When these comparisons are possible, they enable row and column categories to be examined simultaneously. An analysis of this type would enable the researcher to identify groups of products characterized by attributes in close proximity.

With the advantages of CA, however, come a number of disadvantages or limitations.

- The technique is descriptive and not at all appropriate for hypothesis testing. If the quantitative relationship of categories is desired, methods such as log-linear models are suggested. CA is best suited for exploratory data analysis.
- CA, as is the case with many dimensionality-reducing methods, has no method for conclusively determining the appropriate number of dimensions. As with similar methods, the researcher must balance interpretability versus parsimony of the data representation.
- The technique is quite sensitive to outliers, in terms of either rows or columns (e.g., attributes or brands). Also, for purposes of generalizability, the problem of omitted objects or attributes is critical.

Overall, correspondence analysis provides a valuable analytical tool for a type of data (nonmetric) normally not the focal point of multivariate techniques. Correspondence analysis also provides the researcher with a complementary compositional technique to MDS for addressing issues where direct comparison of objects and attributes is preferable.

ILLUSTRATIONS OF MDS AND CORRESPONDENCE ANALYSIS

To demonstrate the use of MDS techniques, we examine data gathered in a series of interviews with 18 company representatives from a cross-section of potential customers. In the course of the

---

RULES OF THUMB 10-4

Correspondence Analysis

- Correspondence analysis (CA) is best suited for exploratory research and is not appropriate for hypothesis testing
- CA is a form of compositional technique that requires specification of both objects and attributes to be compared
- Correspondence analysis is sensitive to outliers, which should be eliminated prior to using the technique
- The number of dimensions to be retained in the solution is based on:
  - Dimensions with inertia (eigenvalues) greater than .2
  - Enough dimensions to meet the research objectives (usually two or three)
- Dimensions can be "named" based on the decomposition of inertia measures across a dimension:
  - These values show the extent of association for each category individually with each dimension
  - They can be used for description much like loadings in factor analysis

perceptual mapping analysis, we apply both decompositional (MDS) and compositional (correspondence analysis) methods. The discussion will proceed in four sections:

1. Examination of the initial three stages of the model-building process (Research Objectives, Research Design, and Assumptions), which are common to both methods
2. Discussion of the next two stages (Model Estimation and Interpretation) for the decompositional methods of MDS
3. Discussion of the same two stages for the compositional method (Correspondence Analysis) applied to the same sample of respondents
4. A look at the sixth stage of the model-building process (Validation) through comparison of the results from both types of methods

The application of both compositional and decompositional techniques enables the researcher to gain unique insights from each technique while also establishing a basis of comparison between each method. Data sets for both techniques are available on the text's Web sites at www.pearsonglobaleditions.com/hair or www.mvstats.com.

STAGE 1: OBJECTIVES OF PERCEPTUAL MAPPING

A common purpose of research dealing with perceptual mapping is to explore a firm's image and competitiveness. This exploration includes addressing the perceptions of a set of firms in the market, as well as investigating attributes underlying those firm positions. Information regarding preferences among potential customers can be added to the analysis if desired. In this example, HBAT employs perceptual mapping techniques in a two-phase plan:

1. Identification of the position of HBAT in a perceptual map of major competitors in the market, with an understanding of the dimension comparisons used by potential customers.
2. Analysis of those market positions to identify the relevant attributes that contribute to HBAT's position as well as those of the competitors.

Particular interest is focused on assessing the dimensions of evaluation that may be too subjective or affective in composition to be measured by conventional scales. Moreover, the intent is to create a single overall perceptual map by combining the positioning of objects and subjects and making the relative positions of objects and consumers for segmentation analysis much more direct.

In achieving these objectives, the researcher must address three fundamental issues that dictate the basic character of the results: objects to be considered for comparison, the use of similarity and/or preference data, and the use of disaggregate or aggregate analysis. Each of these issues will be addressed in the following discussion.

Identifying Objects for Inclusion

A critical decision for any perceptual mapping analysis is the selection of the objects to be compared. Given that judgments are made based on the similarity of one object to another, the inclusion or exclusion of objects can have a marked impact. For example, excluding a firm with distinguishing characteristics unique to other firms may help reveal firm-to-firm comparisons or even dimensions not otherwise detected. Likewise, the exclusion of distinctive or otherwise relevant firms may affect the results in a similar manner.

In our example, the objects of study are HBAT and its nine major competitors. To understand the perceptions of these competing firms, mid-level executives of firms representing potential customers are surveyed on their perceptions of HBAT and the competing firms. The resulting perceptual maps hopefully portray HBAT's positioning in the marketplace.

Basing the Analysis on Similarity or Preference Data

The choice of similarity or preference data depends on the basic objectives of the analysis. Similarity data provide the most direct comparison of objects based on their attributes, whereas preference data allow for a direct assessment of respondent sentiment toward an object. It is possible through use of multiple techniques to combine the two types of data if both are gathered.

For this analysis, emphasis will be placed on gathering similarity for use in both the MDS and correspondence analysis techniques. Preference data will be used in supplementary analyses to assess preference ordering among objects.

Using a Disaggregate or Aggregate Analysis

The final decision involves whether to use aggregate or disaggregate analyses individually or in common. Aggregate analyses provide for an overall perspective on the entire sample in a single analysis, with perceptual maps representing the composite perceptions of all respondents. Disaggregate analyses allow for an individualized analysis, where all respondents can be analyzed separately and even portrayed with their own personal perceptual map. It is also possible to combine these two types of analysis such that results for individuals are displayed in conjunction with the aggregate results.

In this HBAT example, most analysis will be presented at the aggregate level, although in certain instances the disaggregate results will be examined to provide diagnostic information about the consistency of the individual results. The aggregate results more closely match the research objectives, which are an overall portrayal of HBAT versus the major competitors. If subsequent research were to focus more on segmentation or targeting issues that directly involve individuals, then disaggregate analyses would be more appropriate.

STAGE 2: RESEARCH DESIGN OF THE PERCEPTUAL MAPPING STUDY

With the objectives defined for the perceptual mapping analyses, HBAT researchers must next address a set of decisions focusing on research design issues that define the methods used and the specific firms to be studied. By doing so, they also define the types of data that need to be collected to perform the desired analyses. Each of these issues is discussed in the following section.

Selecting Decompositional or Compositional Methods

The choice between decompositional (attribute-free) or compositional (attribute-based) methods revolves around the level of specificity the researcher desires. In the decompositional approach, the respondent provides only overall perceptions or evaluations in order to provide the most direct measure of similarity. However, the researcher is left with little objective evidence of how these perceptions are formed or upon what basis they are made. In contrast, the compositional approach provides some points of references (e.g., attributes) when assessing similarities, but then we must be aware of the potential problems when relevant attributes are omitted.

In this example, a combination of decompositional and compositional techniques is employed. First, a traditional MDS technique using overall measures of similarity provides a perceptual map of HBAT and its nine competitors. Then a compositional method (correspondence analysis) is used as a complementary approach in perceptual mapping, contributing its ability to simultaneously portray firms and attributes in a single perceptual map. Supplementary analyses incorporating preference data into the perceptual maps as well as methods of describing the dimensions of the perceptual map in terms of firm attributes are provided as supplementary analyses available on the text's Web sites (www.pearsonglobaleditions.com/hair or www.mvstats.com).

Selecting Firms for Analysis

In selecting firms for analysis, the researcher must address two issues. First, are all of the firms comparable and relevant for the objectives of this study? Second, is the number of firms included enough to portray the dimensionality desired? The design of the research to address each issue is discussed here.

This study includes nine competitors, plus HBAT, representing all of the major firms in this industry and collectively having more than 85 percent of total sales. Moreover, they are considered representative of all of the potential segments existing in the market. All of the remaining firms not included in the analysis are considered secondary competitors to one or more of the firms already included.

By including 10 firms, researchers can be reasonably certain that perceptual maps of at least two dimensions can be identified and portrayed. Although the inclusion of this many firms results in a somewhat extensive evaluation task on the part of respondents, it was deemed necessary to include this set of firms to enable researchers to generate a multidimensional framework representative of the entire industry structure.

Nonmetric Versus Metric Methods

The choice between nonmetric and metric methods is based jointly on the type of analyses to be performed (e.g., compositional versus decompositional) as well as the actual programs to be used. In some instances the requirements for specific programs (e.g., correspondence analysis) dictate the approach, but in most cases both options are available.

In the HBAT study, both metric and nonmetric methods are used. The multidimensional scaling analyses are performed exclusively with metric data (similarities, preferences, and attribute ratings). The correspondence analysis performs a nonmetric analysis using data in the form of cross-tabulated frequency scores.

Collecting Data for MDS

The primary decision in constructing the perceptual map is whether to utilize either similarities or preferences. To make this decision, the researcher must understand the research objectives: Does the analysis focus on understanding how the objects compare on the antecedents of choice (i.e., similarities based on attributes of the objects) or on the outcomes of choice (i.e., preferences)? In selecting one approach, the analyst must then infer the other through additional analysis. For example, if similarities are chosen as the input data, the researcher is still unsure as to what choices would be made in any type of decision. Likewise, if preferences are analyzed, the researcher has no direct basis for understanding the determinants of choice unless additional analysis is performed.

The HBAT study is composed of in-depth interviews with 18 mid-level management personnel from different firms. From the research objectives, the primary goal is to understand the similarities of firms based on firms' attributes. Thus, focus is placed on similarity data for use in the multidimensional scaling analysis and nonmetric attribute ratings for the correspondence analysis. In the course of the interviews, however, additional types of data were collected for use in supplementary MDS analyses, including attribute ratings of firms and preferences for each firm.

SIMILARITY DATA   The starting point for data collection for the MDS analysis was obtaining the perceptions of the respondents concerning the similarity or dissimilarity of HBAT and nine competing firms in the market.

Similarity judgments were made with the comparison-of-paired-objects approach. The 45 pairs of firms $[(10 \times 9)/2]$ were presented to the respondents, who indicated how similar each was on a 9-point scale, with 1 being "not at all similar" and 9 being "very similar." The results are tabulated for each respondent in a lower triangular matrix (see the HBAT_MDS data set for specific example). Note that the use of larger values indicating greater similarity was done to ease the burden on the respondents. The values will need to be "reversed" during analysis because increasing values for the similarity ratings indicate greater similarity, the opposite of a distance measure of similarity that is used in the MDS techniques.

ATTRIBUTE RATINGS   In addition to the similarity judgments, ratings of each firm on a series of attributes were obtained to provide some objective means of describing the dimensions identified in the perceptual maps. These ratings would be used in the supplementary MDS techniques to assist in interpreting the dimensions of the perceptual map.

For the metric ratings used in MDS, each firm was rated on a set of attributes using a 6-point scale, ranging from low (1) to high (6). The ratings indicate the extent to which each attribute is descriptive of a firm. The attributes include 8 of the 10 attributes identified as composing the four factors in Chapter 3: X_6, Product Quality; X_8, Technical Support; X_{10}, Advertising; X_{12}, Salesforce Image; X_{13}, Competitive Pricing; X_{14}, Warranty & Claims; X_{16}, Order & Billing; and X_{18}, Delivery Speed. Two of the attributes from the original set of 10 were eliminated in this analysis. First, X_7, relating to E-Commerce, was not used because about one-half of the firms did not have an e-commerce presence. Also, X_9, Complaint Resolution, which is largely experience-based, was also omitted because evaluation by noncustomers was difficult for the respondents.

Preference Evaluations.   The final type of data assessed the preferences of each respondent in a specific choice context. These data are to be used in conjunction with the perceptual maps derived in multidimensional scaling to provide insight into the correspondence of similarity and preference judgments through a set of supplementary analyses.

Respondents ranked the firms in order of preference for a "typical" buying situation. Although choice criteria may vary given the type of buying context (e.g., straight rebuy versus new purchase), respondents were asked to provide an overall preference for each firm. The respondents indicated preferences with a simple ordinal ranking of firms, where they identified most preferred (rank order = 1), the next most preferred (rank order = 2), and so on until all 10 firms were ranked.

Collecting Data for Correspondence Analysis

As discussed earlier, correspondence analysis employs a somewhat different type of data (i.e., nonmetric data in a cross-tabulation table) from the other techniques discussed in this chapter. The frequency counts in each cell represent the degree of association between the row and column categories. One advantage of correspondence analysis is that it is quite flexible in what can be represented in the rows and columns. For example, in the simple example discussed earlier in this chapter, the rows represented age groups and the columns represented products. The cell counts were sales for each age group/product combination. But this approach could have just as easily substituted a set of attributes for the age groups, with each cell representing the number of times respondents selected that attribute as representative of that firm. This approach corresponds to the "pick any" approach. Note that in most applications of correspondence analysis the respondents provide a nonmetric response (e.g., Yes/No), which is then put in the cross-tabulation table versus a rating on a scale.

For correspondence analysis, nonmetric ratings were gathered using the same eight attributes discussed earlier in the MDS data collection description. In this task, each respondent was asked to indicate his or her perceptions of firms by checking Yes for the firms best characterized by each attribute. Respondents could pick any number of firms for each attribute, ranging from none to all of the firms. The HBAT_CORRESP_INDIV data set shows the individual responses for each individual, with a 1 indicating a YES response and a 0 indicating a NO response. The individual responses were then compiled into a single cross-tabulation table indicating the number of times each attribute/firm combination was mentioned across the 18 respondents. HBAT_CORRESP contains the cross-tabulated data (similar in structure to the cross-tabulated sales data by product used in the simple example earlier in the chapter) that is input into the correspondence analysis program.

STAGE 3: ASSUMPTIONS IN PERCEPTUAL MAPPING

The assumptions of MDS and CA deal primarily with the comparability and representativeness of the objects being evaluated and the respondents. The techniques themselves place few limitations on the data, but their success is based on several characteristics of the data.

With regard to the sample, the sampling plan emphasized obtaining a representative sample of HBAT customers. Moreover, care was taken to obtain respondents of comparable position and market knowledge. Because HBAT and the other firms serve a fairly distinct market, all the firms evaluated in the perceptual mapping should be known, ensuring that positioning discrepancies can be attributed to perceptual differences among respondents.

MULTIDIMENSIONAL SCALING: STAGES 4 AND 5

Having specified the 10 firms to be included in the image study, HBAT's management specified that both decompositional (MDS) and compositional (CA) approaches were to be employed in constructing the perceptual maps. We first discuss a series of decompositional techniques. Then we examine a compositional approach to perceptual mapping in the following section.

Stage 4: Deriving MDS Results and Assessing Overall Fit

The process of developing a perceptual map can vary markedly in terms of the types of input data and associated analyses performed. In this section, we first discuss the process of developing a perceptual map based on similarity judgments. Then, with the perceptual map established, we discuss the supplementary analyses for incorporating preference judgments into the existing perceptual map.

DEVELOPING AND ANALYZING THE PERCEPTUAL MAP   The estimation of the perceptual map starts with the type of input data and the model estimation (aggregate versus disaggregate) chosen. Methods such as the INDSCAL [4] approach are flexible in that they can produce aggregate results (i.e., a single perceptual map across all respondents) but also provide information on the individual respondents relating to the consistency across respondents.

The INDSCAL method of multidimensional scaling in SPSS was used to develop both a composite, or aggregate, perceptual map as well as the measures of the differences between respondents in their perception. The 45 similarity judgments from the 18 respondents were input as separate matrices (see the HBAT_MDS data set), but mean scores across respondents were calculated to illustrate the general pattern of similarities (see Table 10-4). The bottom rows of the table detail the firms with the highest and lowest similarity values for each firm. These relationships are illustrative of the basic patterns that should be identified in the resulting map (i.e., highly similar firms should be located closer together and dissimilar firms farther apart). For example, HBAT is most similar to firm A and least similar to firms C and G.

Establishing the Appropriate Dimensionality.   The first analysis of the MDS results is to determine the appropriate dimensionality and portray the results in a perceptual map. To do so, the researcher should consider both the indices of fit at each dimensionality and the researcher's ability to interpret the solution.

TABLE 10-4   Mean Similarity Ratings for HBAT and Nine Competing Firms

| | Firm | | | | | | | | | |
Firm	HBAT	A	B	C	D	E	F	G	H	I
HBAT	0.00									
A	6.61	0.00								
B	5.94	5.39	0.00							
C	2.33	2.61	3.44	0.00						
D	2.56	2.56	4.11	6.94	0.00					
E	4.06	2.39	2.17	4.06	2.39	0.00				
F	2.50	3.50	4.00	2.22	2.17	4.06	0.00			
G	2.33	2.39	3.72	2.67	2.61	3.67	2.28	0.00		
H	2.44	4.94	6.61	2.50	7.06	5.61	2.83	2.56	0.00	
I	6.17	6.94	2.83	2.50	2.50	3.50	6.94	2.44	2.39	0.00

Maximum and Minimum Similarity Ratings

	HBAT	A	B	C	D	E	F	G	H	I
Maximum similarity	A, I	I	H	D	H	H	I	B	D	A
Minimum similarity	C, G	E, G	E	F	F	B	C	F	I	H

Note: Similarity ratings are on a 9-point scale (1 = Not at All Similar, 9 = Very Similar).

Table 10-5 shows the indices of fit for solutions of two to five dimensions (a one-dimensional solution was not considered a viable alternative for 10 firms). As the table shows, substantial improvement in the stress measure occurs when moving from two to three dimensions, after which the improvement diminishes somewhat and remains consistent as we increase in the number of dimensions. Balancing this improvement in fit against the increasing difficulty of interpretation, the two- or three-dimensional solutions seem the most appropriate. For purposes of illustration, the two-dimensional solution is selected for further analyses, but the methods we discuss here could just as easily be applied to the three-dimensional solution. The researcher is encouraged to explore other solutions to assess whether any substantively different conclusions would be reached based on the dimensionality selected.

Creating the Perceptual Map. With the dimensionality established at two dimensions, the next step is to position each object (firm) in the perceptual map. Remember that the basis for the map (in this case similarity) defines how objects can be compared.

TABLE 10-5   Assessing Overall Model Fit and Determining the Appropriate Dimensionality

Dimensionality of the Solution	Average Fit Measures[a]			
	Stress[b]	Percent Change	R^{2c}	Percent Change
5	.20068	—	.6303	—
4	.21363	6.4	.5557	11.8
3	.23655	10.7	.5007	9.9
2	.30043	27.0	.3932	21.5

[a]Average across 18 individual solutions.
[b]Kruskal's stress formula.
[c]Proportion of original similarity ratings accounted for by scaled data (distances) from the perceptual map.

The two-dimensional aggregate perceptual map is shown in Figure 10-12. To see how the similarity values are represented, let us examine some of the relationships between HBAT and other firms. In Table 10-4, we saw that HBAT is most similar to firm A and least similar to firms C and G. As we view the perceptual map, we can see those relationships depicted—HBAT is closest to firm A and farthest away firms C and G. Similar comparisons for other highly similar pairs of firms (E and G, D and H, and F and I) show that they are closely positioned in the perceptual map as well.

Differences can also be distinguished between firms based on the dimensions of the perceptual map. For example, HBAT differs from firms E and G primarily on dimension II, whereas dimension I differentiates HBAT most clearly from firms C, D, and H in one direction and firms F and I in another direction. All of these differences are reflected in their relative positions in the perceptual map, and similar comparisons can be made among all sets of firms. To understand the sources of these differences, however, the researcher must interpret the dimensions.

To finalize the perceptual map, the researcher should examine the results to identify any potential outliers and check the assumption of homogeneity of respondents. These two additional analyses provide the researcher with some assessment of the results at the respondent level. This will allow for a better understanding of how the perceptions of individuals are combined into the aggregate results, as well as enable a means of identifying the individuals that are candidates for possible deletion due to their inconsistency with the remaining sample.

Assessing Potential Outliers. In the process of selecting the appropriate dimensionality, an overall measure of fit (stress) was examined. However, this measure does not depict in any way the fit of the solution to individual comparisons. Such an analysis can be done visually through a scatterplot of actual distances (scaled similarity values) versus fitted distances from the perceptual map. Each point represents a single similarity judgment between two objects, with poor fit depicted as outlying points in the graph. Outliers are a set of similarity judgments that reflect

FIGURE 10-12 Perceptual Map of HBAT and Major Competitors

consistently poor fit for an object or individual respondent. If a consistent set of objects or individuals is identified as outliers, they can be considered for deletion.

Measures of model fit are also provided for each individual so that the researcher has a quantitative perspective on each respondent. Lower values of stress indicate better fit, as do higher R^2 values. Although there are no absolute standards in assessing these measures, they do provide a sound comparative measure between respondents, such that individuals with relatively poor model fit can be identified and potentially eliminated from the analysis.

Figure 10-13 represents the scatterplot of similarity values versus the derived distances from the MDS program. Each point represents a separate similarity value. Points falling closer to the 45-degree diagonal are those with the best "fit" in the perceptual map, whereas points farther from the diagonal exhibit greater differences between the actual and derived similarity portrayed in the perceptual map. In examining the most outlying points in this analysis, we find that they are not associated with specific individuals to the extent that deletion of that individual would improve the overall solution.

Table 10-6 also provides measures of fit (stress and R^2) for each individual. For example, respondent 7 has low stress and a corresponding high R^2, indicating good overall model fit of the overall model to that respondent's perceptions of similarity. Likewise, respondent 9 has a relatively high stress value and a low R^2 value. This would be a potential candidate to see if the overall solution could be improved by eliminating a single respondent.

Testing the Assumption of Homogeneity of Respondents. In addition to developing the composite perceptual map, INDSCAL also provides the means for assessing one of the assumptions of MDS, the homogeneity of respondents' perceptions. Weights are calculated for each respondent indicating the correspondence of the respondent's own perceptual space and the aggregate perceptual map. These weights provide a measure of comparison among respondents because respondents with similar weights have similar individual perceptual maps.

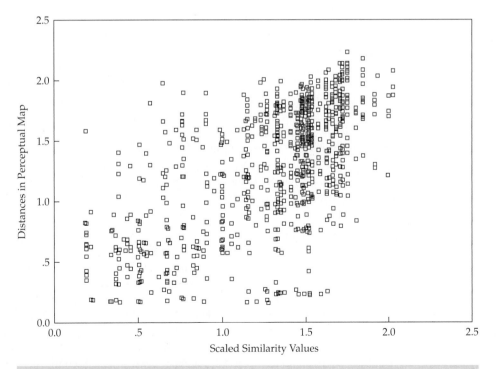

FIGURE 10-13   Scatterplot of Linear Fit

TABLE 10-6   Measures of Individual Differences in Perceptual Mapping: Respondent-Specific Measures of Fit and Dimension Weights

Subject	Measures of Fit		Dimension Weights	
	Stress[b]	R^{2c}	Dimension I	Dimension II
1	.358	.274	.386	.353
2	.297	.353	.432	.408
3	.302	.378	.395	.472
4	.237	.588	.572	.510
5	.308	.308	.409	.375
6	.282	.450	.488	.461
7	.247	.547	.546	.499
8	.302	.332	.444	.367
9	.320	.271	.354	.382
10	.280	.535	.523	.511
11	.299	.341	.397	.429
12	.301	.343	.448	.378
13	.292	.455	.497	.456
14	.302	.328	.427	.381
15	.290	.371	.435	.426
16	.311	.327	.418	.390
17	.281	.433	.472	.458
18	.370	.443	.525	.409
Average[a]	.300	.393		

[a]Average across 18 individual solutions.
[b]Kruskal's stress formula.
[c]Proportion of original similarity ratings accounted for by scaled data (distances) from the perceptual map.

Table 10-6 contains the weights and measures of fit for each respondent, and Figure 10-14 is a graphical portrayal of the individual respondents based on their weights. Examination of the weights (Table 10-6) and Figure 10-14 reveals that the respondents are quite homogeneous in their perceptions, because the weights show few substantive differences on either dimension, and no distinctive clusters of individuals emerge. In Figure 10-14, all of the individual weights fall roughly on a straight line, indicating a consistent weight between dimensions I and II.

The distance of each individual weight from the origin also indicates its level of fit with the solution. Better fits are shown by farther distances from the origin. Thus, respondents 4, 7, and 10 have the highest fit, and respondents 1 and 9 have the lowest fit. Combined with the earlier discussion of stress and R^2 values, no individual emerges as a definite candidate for elimination due to a poor fit in the two-dimensional solution.

SUPPLEMENTARY ANALYSES: INCORPORATING PREFERENCES IN THE PERCEPTUAL MAP   Up to this point, we have dealt only with judgments of firms based on similarity, but many times we may wish to extend the analysis to the decision-making process and to understand the respondents' preferences for the objects. To do so requires additional analyses that attempt to correspond preferences with the similarity-based perceptual maps. Additional programs, such as PREFMAP, can be used to perform this analysis. Although not available in SPSS, preference mapping can be accomplished by more specialized MDS programs such as NewMDSX [25] and the Marketing Engineering package by Lilien and Rangaswamy [16]. Supplementary analyses using PREFMAP are provided on the text's Web sites (www.pearsonglobaleditions.com/hair or www.mvstats.com) where the preferences of s set of respondents are incorporated into the similarity-based perceptual map.

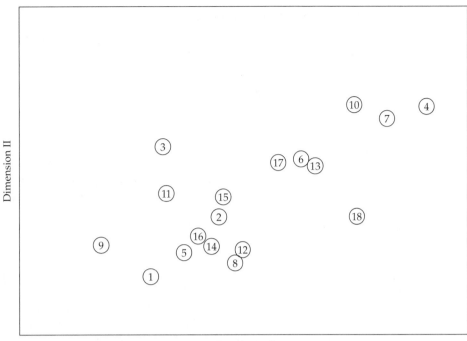

FIGURE 10-14   Individual Subject Weights

Stage 5: Interpretation of the Results

Once the perceptual map is established, we can begin the process of interpretation. Because the MDS procedure uses only the overall similarity judgments, HBAT also gathered ratings of each firm on a series of eight attributes descriptive of typical strategies followed in this industry. The ratings for each firm were then averaged across the respondents for a single overall rating used in describing each firm.

As described in stage 2, the eight attributes included were X_6, Product Quality; X_8, Technical Support; X_{10}, Advertising; X_{12}, Salesforce Image; X_{13}, Competitive Pricing; X_{14}, Warranty & Claims; X_{16}, Order & Billing; and X_{18}, Delivery Speed. These attributes represent the individual variables composing the four factors developed in Chapter 3, exclusive of X_7, relating to E-Commerce, and X_9, Complaint Resolution. The mean scores for each firm are shown in Table 10-7.

A SUBJECTIVE APPROACH TO INTERPRETATION   The researcher can undertake several subjective approaches to interpretation. First, the firms can be profiled in terms of their attribute ratings with distinguishing attributes identified for each firm. In this manner, each firm is characterized on a set of attributes, with the researcher then relating the attributes to the association among firms if possible. Interpreting the dimensions is more complicated in that the researcher must relate the positions of firms to the dimensions in terms of their characteristics. In both approaches, however, the researcher relies upon personal judgment to identify the distinguishing characteristics and then relate them to both firm positions and the resulting interpretation of the dimensions.

These approaches are best suited for use in situations where the objects and basic characteristics are well established. Then the researcher uses general knowledge of existing relationships among attributes and objects to assist in the interpretation. In situations where the researcher must develop these relationships and associations from the analysis itself, the objective approaches described in the following section are recommended because they provide a systematic method of identifying the basic issues involved in the interpretation of objects and dimensions.

TABLE 10-7   Interpreting the Perceptual Map

Variables	AVERAGE ATTRIBUTE RATINGS BY FIRM									
	Firm									
	HBAT	A	B	C	D	E	F	G	H	I
X_6 Product Quality	5.33	3.72	6.33	5.56	6.39	4.72	5.28	5.22	7.33	5.11
X_8 Technical Support	4.17	1.56	6.06	8.22	7.72	4.28	3.89	6.33	7.72	5.06
X_{10} Advertising	4.00	1.83	6.33	7.67	6.00	5.78	5.50	6.11	7.50	4.17
X_{12} Salesforce Image	6.94	7.17	7.67	3.22	4.78	5.11	6.56	1.61	8.78	3.17
X_{13} Competitive Pricing	6.94	5.67	3.39	3.67	3.67	6.94	6.44	7.22	4.94	6.11
X_{14} Warranty & Claims	5.11	1.22	5.78	7.89	6.56	3.83	4.28	6.94	8.67	4.72
X_{16} Order & Billing	5.16	3.47	6.41	5.88	6.06	4.94	5.29	4.82	8.35	4.65
X_{18} Delivery Speed	4.00	3.39	7.33	6.11	7.50	4.22	7.17	4.33	8.22	5.56

Perceptions of HBAT and the nine other firms were gathered for eight attributes (see earlier discussion), with profiles of each firm presented in Table 10-7. These perceptions can be used in a subjective manner to attempt to identify the unique attributes of each firm, as well as to understand the characteristics represented by each dimension. For example, the firms with the highest perceptions on X_8 are C, D, and H, all located on the left side of the perceptual map, whereas firms with lower perceptions are generally on the right side of the perceptual map. This would indicate that dimension I (the X-axis) could be characterized in terms of X_8. Similar analysis of X_{12} indicates firms with higher values located toward the top of the perceptual map, whereas lower scores are toward the bottom. Thus, dimension II would be partially characterized by X_{12}. Although this approach is limited, it does provide some reasonable basis for describing the characteristics of each of the dimensions.

SUPPLEMENTARY ANALYSIS: OBJECTIVE APPROACHES TO INTERPRETATION   To provide an objective means of interpretation, additional programs, such as PROFIT [3], can be used to match the ratings to the firm positions in the perceptual map with attribute ratings for each object. The objective is to identify the determinant attributes in the similarity judgments made by individuals to determine which attributes best describe the perceptual positions of the firms and the dimensions. As was seen in the case of incorporating respondent preferences, a number of specialized MDS programs can assist in the interpretation of the dimensions of the perceptual map (NewMDSx [25] and the Marketing Engineering program [16]). The interested reader is referred to the text's Web sites (www.pearsonglobaleditions.com/hair or www.mvstats.com) where additional analyses employ PROFIT to assist in interpretation of the dimensions.

Overview of the Decompositional Results

The decompositional methods employed in this image study illustrate the inherent trade-off and resulting advantages and disadvantages of attribute-free multidimensional scaling techniques:

- *Advantage:* The use of overall similarity judgments provides a perceptual map based on only the relevant criteria chosen by each respondent. The respondent can make these judgments based on any set of criteria deemed relevant on a single measure of overall similarity.

versus

- *Disadvantage:* The use of an attribute-free technique gives rise, however, to the notable difficulty of interpreting the perceptual map in terms of specific attributes. The researcher is required to infer the bases for comparison among objects without direct confirmation from the respondent.

The researcher using these methods must examine the research objectives and decide whether the benefits accrued from perceptual maps developed through the attribute-free approaches outweigh the limitations imposed in interpretation. We can examine the results from the HBAT analysis to assess the trade-offs, benefits, and costs. In terms of perceptions, HBAT is most associated with firm A and somewhat with firms B and I. Some competitive groupings (e.g., F and I, E and G) must also be considered. None of the firms are so markedly distinct as to be considered an outlier. HBAT can be considered average on several attributes (X_6, X_{16}, and X_{18}), but has lower scores on several attributes (X_8, X_{10}, and X_{14}) countered by a high score on attribute X_{12}.

These results provide HBAT insight into not only its perceptions, but also the perceptions of the other major competitors in the market. Remember, however, that the researcher is not assured of understanding what attributes were actually used in the judgment, just that these attributes may be descriptive of the objects.

CORRESPONDENCE ANALYSIS: STAGES 4 AND 5

An alternative to attribute-free perceptual mapping is correspondence analysis (CA), a compositional method based on nonmetric measures (frequency counts) between objects and/or attributes. In this attribute-based method, the perceptual map is a joint space, showing both attributes and firms in a single representation. Moreover, the positions of firms are relative not only to the other firms included in the analysis but also to the attributes selected as well.

Stage 4: Estimating a Correspondence Analysis

The data preparation and estimation procedure for correspondence analysis is similar in some regards to the multidimensional scaling process discussed earlier, with some notable exceptions. In the following sections we discuss the method of data collection used in the HBAT study and then the issues involved in calculating similarity and determining the dimensionality of the solution. The correspondence analysis procedure in SPSS was used to perform this analysis, although a number of more specialized programs also include correspondence analysis (21, 24, 25).

DATA COLLECTION AND PREPARATION   A unique characteristic of correspondence analysis is the use of nonmetric data to portray relationships between categories (objects or attributes). A common approach to data presentation is the use of a cross-tabulation matrix relating the attributes (represented as rows) to the ratings of objects/firms (the columns). The values represent the number of times each firm is rated as being characterized by that attribute. Thus, higher frequencies indicate a stronger association between that object and the attribute in question.

In the HBAT study, binary firm ratings were gathered for each firm on each of the eight attributes (i.e., a yes–no rating of each firm on each attribute). The individual entries in the cross-tabulation table are the number of times a firm is rated as possessing a specific attribute. Respondents could choose any number of attributes as characterizing each firm. Table 10-8 first presents the responses for a single individual in the HBAT study and then provides the complete cross-tabulation table for all respondents. The HBAT_CORRESP_INDIV data set contains the responses for the individuals, and the HBAT_CORRESP data set has the final cross-tabulation data set used in the analysis.

CALCULATING THE SIMILARITY MEASURE   Correspondence analysis is based on a transformation of the chi-square value into a metric measure of distance, which acts as the similarity measure. The chi-square value is calculated as the actual frequency of occurrence minus the expected frequency. Thus, a negative value indicates, in this case, that a firm was rated less often than would be expected. The expected value for a cell (any firm–attribute combination in the cross-tabulation table) is based on how often the firm was rated on other attributes and how often other firms were

TABLE 10-8 Individual Respondent Ratings and Cross-Tabulated Frequency Data of Attribute Descriptors for HBAT and Nine Competing Firms

RESPONDENT # 1

Variables	Firm									
	HBAT	**A**	**B**	**C**	**D**	**E**	**F**	**G**	**H**	**I**
X_6 Product Quality	1	0	1	1	0	1	1	0	1	0
X_8 Technical Support	0	1	0	0	1	1	1	0	1	0
X_{10} Advertising	1	1	1	1	0	1	1	1	1	0
X_{12} Salesforce Image	1	0	0	1	1	0	0	1	0	0
X_{13} Competitive Pricing	1	0	0	0	0	1	1	1	1	0
X_{14} Warranty & Claims	0	1	1	0	1	0	1	0	0	1
X_{16} Order & Billing	1	1	1	0	1	1	1	0	1	1
X_{18} Delivery Speed	1	1	0	1	1	1	1	1	0	0

OVERALL TOTAL

Variables	Firm									
	HBAT	**A**	**B**	**C**	**D**	**E**	**F**	**G**	**H**	**I**
X_6 Product Quality	6	6	14	10	11	8	7	4	14	4
X_8 Technical Support	15	18	9	2	3	15	16	7	8	8
X_{10} Advertising	15	16	15	11	11	14	16	12	14	14
X_{12} Salesforce Image	4	3	1	13	9	6	3	18	2	10
X_{13} Competitive Pricing	15	14	6	4	4	15	14	13	7	13
X_{14} Warranty & Claims	7	18	13	4	9	16	14	5	4	16
X_{16} Order & Billing	14	14	10	11	11	14	12	13	10	14
X_{18} Delivery Speed	16	13	8	13	9	17	15	16	6	12

rated on that attribute. (In statistical terms, the expected value is based on the row [attribute] and column [firm] marginal probabilities.)

Table 10-9 contains the transformed (metric) chi-square distances for each cell of cross-tabulation from Table 10-8. High positive values indicate a strong degree of correspondence between the attribute and firm, and negative values have an opposite interpretation. For example, the high values for HBAT and firms A and F with the technical support attribute (X_8) indicate that they should be located close together on the perceptual map if at all possible. Likewise, the high negative values for firms C and D on the same variable would indicate that their position should be far from the attribute's location.

TABLE 10-9 Measures of Similarity in Correspondence Analysis: Chi-Square Distances

Variables	Firm									
	HBAT	**A**	**B**	**C**	**D**	**E**	**F**	**G**	**H**	**I**
X_6 Product Quality	−1.02	−1.28	2.37	1.27	1.71	−.73	−.83	−1.59	2.99	−1.66
X_8 Technical Support	1.24	1.69	−.01	−2.14	−1.76	.72	1.32	−1.07	.10	−.85
X_{10} Advertising	.02	−.13	.76	−.01	.04	−.73	.07	−.60	1.07	−.20
X_{12} Salesforce Image	−1.27	−1.83	−2.08	3.19	1.53	−.86	−1.73	4.07	−1.42	.97
X_{13} Competitive Pricing	1.08	.40	−1.10	−1.52	−1.48	.57	.59	.65	−.36	.53
X_{14} Warranty & Claims	−1.32	−1.49	1.15	−1.54	.23	.81	.55	−1.80	−1.44	1.39
X_{16} Order & Billing	.19	−.19	−.30	.37	.42	−.30	−.54	.08	.20	.23
X_{18} Delivery Speed	.68	−.51	−.95	.95	−.27	.40	.20	.86	−1.15	−.37

DETERMINING THE DIMENSIONALITY OF THE SOLUTION Correspondence analysis tries to satisfy all of these relationships simultaneously by producing dimensions representing the chi-square distances. To determine the dimensionality of the solution, the researcher examines the cumulative percentage of variation explained, much as in factor analysis, and determines the appropriate dimensionality. The researcher balances the desire for increased explanation in adding additional dimensions versus interpretability by creating more complexity with each dimension added.

Table 10-10 contains the eigenvalues and cumulative and explained percentages of variation for each dimension up to the maximum of seven. A two-dimensional solution in this situation explains 86 percent of the variation, whereas increasing to a three-dimensional solution adds only an additional 10 percent. In comparing the additional variance explained in relation to the increased complexity in interpreting the results, a two-dimensional solution was deemed adequate for further analysis.

Stage 5: Interpreting CA Results

With the number of dimensions defined, the researcher must proceed with an interpretation of the derived perceptual map. In doing so, at least three issues must be addressed: positioning of row and/or column categories, characterization of the dimensions, and assessing the goodness-of-fit of individual categories. Each will be discussed in the following sections.

RELATIVE POSITIONING OF CATEGORIES The first task is to assess the relative positions of the categories for the rows and columns. In doing so, the researcher can assess the association between categories in terms of their proximity in the perceptual map. Note that the comparison should only be between categories within the same row or column.

The perceptual map shows the relative proximities of both firms and attributes (see Figure 10-15). If we focus on the firms first, we see that the pattern of firm groups is similar to that found in the MDS results. Firms A, E, F, and I, plus HBAT form one group; firms C and D and firms H and B form two other similar groups. However, the relative proximities of the members in each group differ somewhat from the MDS solution. Also, firm G is more isolated and distinct, and firms F and E are now seen as more similar to HBAT.

In terms of attributes, several patterns emerge. First, X_6 and X_{13}, the two variables negatively related, do appear at opposite extremes of the perceptual map. Moreover, variables shown to have high association (e.g., forming factors) also fall in close proximity (X_{16} and X_{18}, X_8 and X_{14}). Perhaps a more appropriate perspective is an attribute's contribution to each dimension, as discussed in the next section.

INTERPRETING THE DIMENSIONS It may be helpful to interpret the dimensions if row or column normalizations are used. For these purposes, the inertia (explained variation) of each dimension can be attributed among categories for rows and columns.

TABLE 10-10 Determining the Appropriate Dimensionality in Correspondence Analysis

Dimension	Eigenvalue (Singular Value)	Inertia (Normalized Chi-Square)	Percentage Explained	Cumulative Percentage
1	.27666	.07654	53.1	53.1
2	.21866	.04781	33.2	86.3
3	.12366	.01529	10.6	96.9
4	.05155	.00266	1.8	98.8
5	.02838	.00081	.6	99.3
6	.02400	.00058	.4	99.7
7	.01951	.00038	.3	100.0

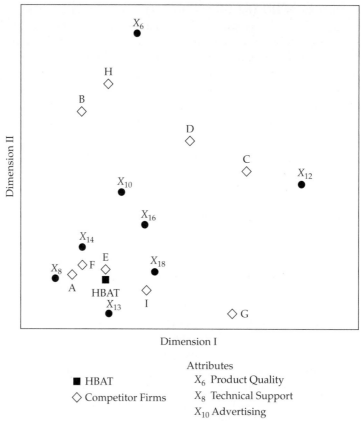

FIGURE 10-15 Perceptual Mapping with Compositional Methods: Correspondence Analysis

Table 10-11 provides the contributions of both sets of categories to each dimension. For the attributes, we can see that X_{12} (Salesforce Image) is the primary contributor to dimension I, and X_8 (Technical Support) is a secondary contributor. Note that these two attributes are extreme in terms of their location on dimension I (i.e., highest or lowest values on dimension I). Between these two attributes, 86 percent of dimension I is accounted for. A similar pattern follows for dimension II, for which X_6 (Product Quality) is the primary contributor followed by X_{13} (Competitive Pricing), which when combined account for 83 percent of the inertia of dimension II.

If we shift our focus to the 10 firms, we see a somewhat more balanced situation, with three firms (A, C, and G) contributing above the average of 10 percent. For the second dimension, four firms (B, D, G, and H) have contributions above average.

Although the comparisons in this example are between both sets of categories and not restricted to a single set of categories (either row or column), these measures of contribution demonstrate the ability to interpret the dimension when desired.

ASSESSING FIT FOR CATEGORIES One final measure provides an assessment of fit for each category. Comparable to squared factor loadings in factor analysis (see Chapter 3 for a more detailed

TABLE 10-11 Interpreting the Dimensions and Their Correspondence to Firms and Attributes

Object	Coordinates		Contribution to Inertia[a]		Explanation by Dimension (Fit)[b]		
	I	II	I	II	I	II	Total
Attribute							
X_6 Product Quality	.044	1.235	.001	.689	.002	.989	.991
X_8 Technical Support	−.676	−.285	.196	.044	.789	.111	.901
X_{10} Advertising	−.081	.245	.004	.045	.093	.678	.772
X_{12} Salesforce Image	1.506	0.298	.665	.033	.961	.030	.991
X_{13} Competitive Pricing	−.202	−.502	.018	.142	.138	.677	.816
X_{14} Warranty & Claims	−.440	−.099	.087	.006	.358	.014	.372
X_{16} Order & Billing	.115	.046	.007	.001	.469	.058	.527
X_{18} Delivery Speed	.204	−.245	.022	.040	.289	.330	.619
Firm							
HBAT	−.247	−.293	.024	.042	.206	.228	.433
A	−.537	−.271	.125	.040	.772	.156	.928
B	−.444	.740	.063	.224	.294	.648	.942
C	1.017	.371	.299	.050	.882	.093	.975
D	.510	.556	.074	.111	.445	.418	.863
E	−.237	−.235	.025	.031	.456	.356	.812
F	−.441	−.209	.080	.023	.810	.144	.954
G	.884	−.511	.292	.123	.762	.201	.963
H	−.206	.909	.012	.289	.049	.748	.797
I	.123	−.367	.006	.066	.055	.390	.446

[a]Proportion of dimension's inertia attributable to each category.
[b]Proportion of category variation accounted for by dimension.

discussion), these values represent the amount of variation in the category accounted for by the dimension. The total value represents the total amount of variation across all dimensions, with the maximum possible being 100 percent.

Table 10-11 also contains fit values for each category on each dimension. As we can see, the total fit values range from a high of 99.1 for X_6 (Product Quality) and X_{12} (Salesforce Image) to a low of .372 for X_{14} (Warranty & Claims). Among the attributes, only X_{14} has a value below 50 percent and only two firms (HBAT and firm I) fall below this value. Even though these values are somewhat low, they still represent a substantial enough explanation to retain them in the analysis and deem the analysis of sufficient practical significance.

Overview of CA

These and other comparisons highlight the differences between MDS and CA methods and their results. CA results provide a means for directly comparing the similarity or dissimilarity of firms and the associated attributes, whereas MDS allows only for the comparison of firms. However, the CA solution is conditioned on the set of attributes included. It assumes that all attributes are appropriate for all firms and that the same dimensionality applies to each firm. Thus, the resulting perceptual map should always be viewed only in the context of both the firms and attributes included in the analysis.

Correspondence analysis is a quite flexible technique applicable to a wide range of issues and situations. The advantages of the joint plot of attributes and objects must always be weighed against the inherent interdependencies that exist and the potentially biasing effects of a single inappropriate attribute or firm or, perhaps more important, the omitted attribute of a firm. Yet CA still provides a powerful tool for gaining managerial insight into the relative position of firms and the attributes associated with those positions.

STAGE 6: VALIDATION OF THE RESULTS

Perhaps the strongest internal validation of this analysis is to assess the convergence between the results from the separate decompositional and compositional techniques. Each technique employs different types of consumer responses, but the resulting perceptual maps are representations of the same perceptual space and should correspond. If the correspondence is high, the researcher can be assured that results reflect the problem as depicted. The researcher should note that this type of convergence does not address the generalizability of the results to other objects or samples of the population.

Comparison of the decompositional and compositional methods, shown in Figures 10-12 and 10-15, can take two approaches: examining the relative positioning of objects and interpreting the axes. Let us start by examining the positioning of firms. When Figures 10-12 and 10-15 are rotated to obtain the same perspective, they show quite similar patterns of firms reflecting two groups: firms B, H, D, and C versus firms E, F, G, and I. Even though the relative distances among firms do vary between the two perceptual maps, we still see HBAT associated strongly with firms A and I in each perceptual map. CA produces more distinction between the firms, but its objective is to define firm positions as a result of differences; thus, it will generate more distinctiveness in its perceptual maps.

The interpretation of axes and distinguishing characteristics also shows similar patterns in the two perceptual maps. For the decompositional method shown in Figure 10-12, we noted in the earlier discussion that by rotating the axes we would obtain a clearer interpretation. If we rotate the axes, then dimension I becomes associated with customer service and product value (X_6, X_{13}, X_{16}, and X_{18}), whereas dimension II reflects marketing and technical support (X_8, X_{10}, and X_{12}). The remaining attributes are not associated strongly with either axis.

To make a comparison with correspondence analysis (Figure 10-15), we must first reorient the axes. As we can see, the dimensions flip between the two analyses. The firm groupings remain essentially the same, but they are in different positions on the perceptual map. In CA, the dimensions reflect somewhat the same elements, with the highest loadings being X_{18} (Delivery Speed) on dimension I and X_{12} (Salesforce Image) on dimension II. This compares quite favorably with the decompositional results except that the other attributes are somewhat more diffused on the dimensions.

Overall, although some differences do exist, owing to the characteristics of each approach, the convergence of the two results does provide some internal validity to the perceptual maps. Perceptual differences may exist for a few attributes, but the overall patterns of firm positions and evaluative dimensions are supported by both approaches. The disparity of the price flexibility attribute illustrates the differences in the two approaches.

The researcher has two complementary tools in the understanding of consumer perceptions. The decompositional method determines position based on overall judgments, with attributes applied only as an attempt to explain the positions. The compositional method positions firms according to the selected set of attributes, thus creating positions based on the attributes. Moreover, each attribute is weighted equally, thus potentially distorting the map with irrelevant attributes. These differences do not make either approach better or optimal but instead must be understood by the researcher to ensure selection of the method most suited to the research objectives.

A MANAGERIAL OVERVIEW OF MDS RESULTS

Perceptual mapping is a unique technique providing overall comparisons not readily possible with any other multivariate method. As such, its results present a wide range of perspectives for managerial use. The most common application of the perceptual maps is for the assessment of image for any firm or group of firms. As a strategic variable, image can be quite important as an overall indicator of market presence or position.

In this study, we found that HBAT is most closely associated with firms A and I, and most dissimilar from firms C, E, and G. Thus, when serving the same product markets, HBAT can identify those firms considered similar to or different from its image. With the results based not on any set of specific attributes, but instead on respondents' overall judgments, they present the benefit of not being subject to a researcher's subjective judgments as to attributes to include or how to weight the individual attributes, in keeping with the true spirit of assessing image. However, MDS technologies are less useful in guiding strategy because they are less helpful in prescribing how to change image. The global responses that are advantageous for comparison now work against us in explanation.

Although MDS techniques can augment the explanation of the perceptual maps, they must be viewed as supplemental and expected to show greater inconsistencies than if integral to the process. Thus, additional research may assist in helping to explain the relative positions.

To this end, CA results are a compromise approach in trying to portray perceptual maps from a compositional perspective. The comparison of CA results to those from the classical MDS solution reveals a number of consistencies, but some discrepancies as well.

Comparing the two solutions identifies some general patterns of associations between firms (such as HBAT and firms A and I) and between groups of attributes. HBAT management can use these results not only as a guide to overall policy, but also as a framework for further investigation with other multivariate techniques into more specific research questions.

The researcher should note that neither technique has the absolute answer, but that each can be used to capitalize on their relative benefits. When used in this manner, the expected differences in the two approaches can actually provide unique and complementary insights into the research question.

Summary

Multidimensional scaling is a set of procedures that may be used to display the relationships tapped by data representing similarity or preference. It has been used successfully (1) to illustrate market segments based on preference judgments, (2) to determine which products are more competitive with each other (i.e., are more similar), and (3) to deduce what criteria people use when judging objects (e.g., products, companies, advertisements). This chapter helps you to do the following:

Define multidimensional scaling and describe how it is performed. Multidimensional scaling (MDS), also known as perceptual mapping, is a procedure that enables a researcher to determine the perceived relative image of a set of objects (firms, products, ideas, or other items associated with commonly held perceptions). The purpose of MDS is to transform consumer judgments of overall similarity or preference (e.g., preference for stores or brands) into distances represented in multidimensional space. To perform a multidimensional scaling analysis, the researcher performs three basic steps: (1) gathers measures of similarity or preference across the entire set of objects to be analyzed, (2) uses MDS

techniques to estimate the relative position of each object in multidimensional space, and (3) identifies and interprets the axes of the dimensional space in terms of perceptual and/or objective attributes. The perceptual map, also known as a spatial map, shows the relative positioning of all objects.

Understand the differences between similarity data and preference data. After choosing objects for study, the researcher must next select the basis of evaluation: similarity versus preference. In providing similarities data, the respondents do not apply any "good–bad" aspects of evaluation in the comparison, but with preference data good–bad assessments are done. In short, preference data assumes that differing combinations of perceived attributes are valued more highly than other combinations. Both bases of comparison can be used to develop perceptual maps, but with differing interpretations: (1) similarity-based perceptual maps represent attribute similarities and perceptual dimensions of comparison but do not reflect any direct insight into the determinants of choice and (2) preference-based perceptual maps do reflect preferred choices but may not correspond in any way to the similarity-based positions, because respondents may base their choices on entirely different dimensions or criteria from those on which they base comparisons. With no optimal base for evaluation, the decision between similarities and preference data must be made with the ultimate research question in mind because they are fundamentally different in what they represent.

Select between a decompositional or compositional approach. Perceptual mapping techniques can be classified into one of two types based on the nature of the responses obtained from the individual concerning the object: (1) the decompositional method measures only the overall impression or evaluation of an object and then attempts to derive spatial positions in multidimensional space that reflect these perceptions (It uses either similarity or preferences data and is the approach typically associated with MDS.) and (2) the compositional method, which employs several of the multivariate techniques already discussed that are used in forming an impression or evaluation based on a combination of specific attributes. Perceptual mapping can be performed with both compositional and decompositional techniques, but each technique has specific advantages and disadvantages that must be considered in view of the research objectives. If perceptual mapping is undertaken either as an exploratory technique to identify unrecognized dimensions or as a means of obtaining comparative evaluations of objects when the specific bases of comparison are unknown or undefined, the decompositional or attribute-free approaches are the most appropriate. In contrast, if the research objectives include the portrayal among objects on a defined set of attributes, then the compositional techniques are the preferred alternative.

Determine the comparability and number of objects. Before undertaking any perceptual mapping study, the researcher must address two key questions dealing with the objects being evaluated. These questions deal with ensuring the comparability of the objects as well as selecting the number of objects to be evaluated. The first question when selecting objects is: Are the objects really comparable? An implicit assumption in perceptual mapping is that of common characteristics, either objective or perceived, used by the respondent for evaluations. Thus, it is essential that the objects being compared have some set of underlying attributes that characterize each object and form the basis for comparison by the respondent. It is not possible for the researcher to force the respondent to make comparisons by creating pairs of noncomparable objects. A second question concerns the number of objects to be evaluated. In deciding how many objects to include, the researcher must balance two desires: a smaller number of objects to ease the effort on the part of the respondent versus the required number of objects to obtain a stable multidimensional solution. Often a trade-off must be made between the number of underlying dimensions that can be identified and the effort required on the part of the respondent to evaluate them.

Understand how to create a perceptual map. Three steps are involved in creating a perceptual map based on the optimal positions of the objects. The first step is to select an initial configuration of stimuli at a desired initial dimensionality. The two most widely used approaches for obtaining the initial configuration are either to base it on previous data or generate one by selecting pseudorandom points from an approximately normal multivariate distribution. The second step is to compute the distances between the stimuli points and compare the relationships (observed versus derived) with a measure of fit. Once a configuration is found, the interpoint distances between stimuli in the starting configurations are compared with distance measures derived from the similarity judgments. The two distance measures are then compared by a measure of fit, typically a measure of stress. The third step is necessary if the measure of fit does not meet a selected predefined stopping value. In such cases, you find a new configuration for which the measure of fit is further minimized. The software determines the directions in which the best improvement in fit can be obtained and then moves the points in the configuration in those directions in small increments.

Explain correspondence analysis as a method of perceptual mapping. Correspondence analysis (CA) is an interdependence technique that has become increasingly popular for dimensional reduction and perceptual mapping. Correspondence analysis has three distinguishing characteristics: (1) it is a compositional technique, rather than a decompositional approach, because the perceptual map is based on the association between objects and a set of descriptive characteristics or attributes specified by the researcher; (2) its most direct application is portraying the correspondence of categories of variables, particularly those measured in nominal measurement scales, which is then used as the basis for developing perceptual maps; and (3) the unique benefits of CA lie in its abilities for representing rows and columns, for example, brands and attributes, in joint space. Overall, correspondence analysis provides a valuable analytical tool for a type of data (nonmetric) that often is not the focal point of multivariate techniques. Correspondence analysis also provides the researcher with a complementary compositional technique to MDS for addressing issues where direct comparison of objects and attributes is preferable.

MDS can reveal relationships that appear to be obscured when one examines only the numbers resulting from a study. Visual perceptual maps emphasize the relationships between the stimuli under study. One must be cautious when using this technique. Misuse is common. The researcher should become familiar with the technique before using it and should view the output as only the first step in the determination of perceptual information.

Questions

1. How does MDS differ from other interdependence techniques (cluster analysis and factor analysis)?
2. What is the difference between preference data and similarities data, and what impact does it have on the results of MDS procedures?
3. How are ideal points used in MDS procedures?
4. How do metric and nonmetric MDS procedures differ?
5. How can the researcher determine when the optimal MDS solution has been obtained?
6. How does the researcher identify the dimensions in MDS? Compare this procedure with the procedure for factor analysis.
7. Compare and contrast CA and MDS techniques.
8. Describe how correspondence, or association, is derived from a contingency table.
9. Describe the methods for interpretation of categories (row or column) in CA. Can categories always be directly compared based on proximity in the perceptual map?

Suggested Readings

A list of suggested readings illustrating issues and applications of multivariate techniques in general is available on the Web at www.pearsonglobaleditions.com/hair or www.mvstats.com.

References

1. Carroll, J. Douglas, Paul E. Green, and Catherine M. Schaffer. 1986. Interpoint Distance Comparisons in Correspondence Analysis. *Journal of Marketing Research* 23 (August): 271–80.
2. Carroll, J. Douglas, Paul E. Green, and Catherine M. Schaffer. 1987. Comparing Interpoint Distances in Correspondence Analysis: A Clarification. *Journal of Marketing Research* 24 (November): 445–50.
3. Chang, J. J., and J. Douglas Carroll. 1968. How to Use PROFIT, a Computer Program for Property Fitting by Optimizing Nonlinear and Linear Correlation. Unpublished paper, Bell Laboratories, Murray Hill, NJ.
4. Chang, J. J., and J. Douglas Carroll. 1969. How to Use INDSCAL, a Computer Program for Canonical Decomposition of *n*-Way Tables and Individual Differences in Multidimensional Scaling. Unpublished paper, Bell Laboratories, Murray Hill, NJ.
5. Chang, J. J., and J. Douglas Carroll. 1969. How to Use MDPREF, a Computer Program for Multidimensional Analysis of Preference Data. Unpublished paper, Bell Laboratories, Murray Hill, NJ.
6. Chang, J. J., and J. Douglas Carroll. 1972. How to Use PREFMAP and PREFMAP2—Programs Which Relate Preference Data to Multidimensional Scaling Solution. Unpublished paper, Bell Laboratories, Murray Hill, NJ.
7. Lilien, G.A., and A. Rangaswamy. Decision Pro. Inc. 2008. *Marketing Engineering for Excel.* State College, PA.
8. Green, P. E. 1975. On the Robustness of Multidimensional Scaling Techniques. *Journal of Marketing Research* 12 (February): 73–81.
9. Green, P. E., and F. Carmone. 1969. Multidimensional Scaling: An Introduction and Comparison of Nonmetric Unfolding Techniques. *Journal of Marketing Research* 7 (August): 33–41.
10. Green, P. E., F. Carmone, and Scott M. Smith. 1989. *Multidimensional Scaling: Concept and Applications.* Boston: Allyn & Bacon.
11. Green, P. E., and Vithala Rao. 1972. *Applied Multidimensional Scaling.* New York: Holt, Rinehart and Winston.
12. Greenacre, Michael J. 1984. *Theory and Applications of Correspondence Analyses.* London: Academic Press.
13. Greenacre, Michael J. 1989. The Carroll-Green-Schaffer Scaling in Correspondence Analysis: A Theoretical and Empirical Appraisal. *Journal of Marketing Research* 26 (August): 358–65.

14. Hoffman, Donna L., and George R. Franke. 1986. Correspondence Analysis: Graphical Representation of Categorical Data in Marketing Research. *Journal of Marketing Research* 23 (August): 213–27.

15. Holbrook, Morris B., William L. Moore, and Russell S. Winer. 1982. Constructing Joint Spaces from Pick-Any Data: A New Tool for Consumer Analysis. *Journal of Consumer Research* 9 (June): 99–105.

16. Levine, Joel H. 1979. Joint-Space Analysis of "Pick-Any" Data: Analysis of Choices from an Unconstrained Set of Alternatives. *Psychometrika* 44 (March): 85–92.

17. Lingoes, James C. 1972. *Geometric Representations of Relational Data.* Ann Arbor, MI: Mathesis Press.

18. Kruskal, Joseph B., and Frank J. Carmone. 1967. How to Use MDSCAL, Version 5-M, and Other Useful Information. Unpublished paper, Bell Laboratories, Murray Hill, NJ.

19. Kruskal, Joseph B., and Myron Wish. 1978. *Multidimensional Scaling.* Sage University Paper Series on Quantitative Applications in the Social Sciences, 07–011, Beverly Hills, CA: Sage.

20. Lebart, Ludovic, Alain Morineau, and Kenneth M. Warwick. 1984. *Multivariate Descriptive Statistical Analysis: Correspondence Analysis and Related Techniques for Large Matrices.* New York: Wiley.

21. Maholtra, Naresh. 1987. Validity and Structural Reliability of Multidimensional Scaling. *Journal of Marketing Research* 24 (May): 164–73.

22. Market ACTION Research Software, Inc. 1989. *MAPWISE: Perceptual Mapping Software.* Peoria, IL: Business Technology Center, Bradley University.

23. Raymond, Charles. 1974. *The Art of Using Science in Marketing.* New York: Harper & Row.

24. Schiffman, Susan S., M. Lance Reynolds, and Forrest W. Young. 1981. *Introduction to Multidimensional Scaling.* New York: Academic Press.

25. Smith, Scott M. 1989. *PC–MDS: A Multidimensional Statistics Package.* Provo, UT: Brigham Young University.

26. Sigma Essex Research and Consultancy. 2007. *NewMDSX.* Argyll, United Kingdom.

27. Srinivasan, V., and A. D. Schocker. 1973. Linear Programming Techniques for Multidimensional Analysis of Preferences. *Psychometrika* 38 (September): 337–69.

Structural Equations Modeling

OVERVIEW

This section provides a simple and concise introduction to a cutting-edge technique in multivariate analysis—structural equation modeling (SEM)—that has grown the most in popularity over the past 20 years. The ability to simultaneously estimate multiple dependence relationships (similar to multiple regression equations) while also incorporating multiple measures for each concept (i.e., akin to factor analysis) has been embraced across almost every academic discipline. This section offers an extended introduction to this technique and is intended to provide the reader with a general understanding of the procedure, the knowledge of when it can be applied, and the ability to apply this technique to basic problems.

CHAPTERS IN SECTION IV

Section IV contains two chapters. Chapter 11 provides an overview of structural equation modeling (SEM), a procedure for accommodating measurement error directly in the estimation of a series of dependence relationships. It is the best multivariate procedure for testing both the construct validity and theoretical relationships among a set of concepts represented by multiple measured variables. Previous to the introduction of SEM, this process would require the application of several different statistical tools and the result would be a less satisfying examination. We do not wish to underestimate the effort involved, but no researcher should avoid SEM solely for this reason because the principles of factor analysis and multiple regression form a basis for understanding SEM.

Following the basic overview, Chapter 12 discusses a number of applications of SEM and is divided into three parts. Part1 is devoted to confirmatory factor analysis, which extends the ideas presented earlier when we discussed exploratory factor analysis. Now, however, the researcher must take a more active role in developing and specifying a theory that will determine how many factors should exist among a set of variables and how these variables relate to, or load on, the smaller number of factors. A test of how well this theory fits the data is provided and enables the reader to thoroughly examine construct validity among this set of measures.

Part 2 is devoted to the testing of theoretical relationships between the factors represented by multiple variables. The goal here is to test the structure of relationships among the factors. Therefore, it is conceptually similar to conducting regression analysis using a set of summated rating scales, each summated rating scale representing a factor that can be recovered with factor analysis. Using SEM, the researcher can assess the strength of relationships between any two factors more accurately because SEM will correct the relationship for measurement error. Furthermore, an overall test of fit is provided that will enable the researcher to assess the validity of a prespecified set of hypotheses, each representing a regression-like relationship between factors.

Finally, Part 3 addresses several advanced topics in SEM, notably higher-order confirmatory factor analysis, testing relationships across groups, evaluating moderated and mediating relationships, plus an examination of the principal estimation approaches (e.g., maximum likelihood versus partial least squares). These issues extend the range of conceptual questions that SEM can address while maintaining the underlying foundation of measurement theory.

CHAPTER 11

SEM: An Introduction

LEARNING OBJECTIVES

Upon completing this chapter, you should be able to do the following:

- Understand the distinguishing characteristics of SEM.

- Distinguish between variables and constructs.

- Understand structural equation modeling and how it can be thought of as a combination of familiar multivariate techniques.

- Know the basic conditions for causality and how SEM can help establish a cause-and-effect relationship.

- Explain the types of relationships involved in SEM.

- Understand that the objective of SEM is to explain covariance and how it translates into the fit of a model.

- Know how to represent a SEM model visually with a path diagram.

- List the six stages of structural equation modeling and understand the role of theory in the process.

CHAPTER PREVIEW

One of the primary objectives of multivariate techniques is to expand the researcher's explanatory ability and statistical efficiency. Multiple regression, factor analysis, multivariate analysis of variance, discriminant analysis, and the other techniques discussed in previous chapters all provide the researcher with powerful tools for addressing a wide range of managerial and theoretical questions. They also all share one common limitation: Each technique can examine only a single relationship at a time. Even the techniques allowing for multiple dependent variables, such as multivariate analysis of variance and canonical analysis, still represent only a single relationship between the dependent and independent variables.

All too often, however, the researcher is faced with a set of interrelated questions. For example, what variables determine a store's image? How does that image combine with other variables to affect purchase decisions and satisfaction at the store? How does satisfaction with the store result in long-term loyalty to it? This series of issues has both managerial and theoretical importance. Yet none of the multivariate techniques we examined thus far enable us to address all these questions with one comprehensive technique. In other words, these techniques do not enable us to test the researcher's entire theory with a technique that considers all possible information. For this reason, we now examine the technique of structural equation modeling (SEM), an extension of several multivariate techniques we already studied, most notably factor analysis and multiple regression analysis.

As briefly described in Chapter 1, structural equation modeling can examine a series of dependence relationships simultaneously. It is particularly useful in testing theories that contain multiple equations involving dependence relationships. In other words, if we believe that image creates satisfaction, and then satisfaction creates loyalty, then satisfaction is both a dependent and an independent variable in the same theory. So, a hypothesized dependent variable becomes an independent variable in a subsequent dependence relationship. None of the previous techniques enable us to assess both measurement properties and test the key theoretical relationships in one technique. SEM will help address these types of questions.

KEY TERMS

Before starting the chapter, review the key terms to develop an understanding of the concepts and terminology used. Throughout the chapter the key terms appear in **boldface.** Other points of emphasis in the chapter and key term cross-references are *italicized.*

Absolute fit indices   Measures of overall *goodness-of-fit* for both the *structural* and *measurement models.* This type of measure does not make any comparison to a specified *null model* (*incremental fit* measure) or adjust for the number of parameters in the estimated model (*parsimonious fit* measure).

All-available approach   Method for handling missing data that computes values based on all available valid observations. Also known as pairwise deletion.

Badness-of-fit   An alternative perspective on *goodness-of-fit* in which larger values represent poorer fit. Examples include the root mean square error of approximation or the standardized root mean square residual.

Causal inference   *Dependence relationship* of two or more variables in which the researcher clearly specifies that one or more variables causes or brings about an outcome represented by at least one other variable. It must meet the requirements for *causation.*

Causation   Principle by which cause and effect are established between two variables. It requires a sufficient degree of association (covariance) between the two variables, that one variable occurs before the other (that one variable is clearly the outcome of the other), and that no other reasonable causes for the outcome are present. Although in its strictest terms causation is rarely found, in practice strong theoretical support can make empirical estimation of causation possible.

Chi-square (χ^2)   Statistical measure of difference used to compare the *observed* and *estimated covariance matrices.* It is the only measure that has a direct statistical test as to its significance, and it forms the basis for many other *goodness-of-fit* measures.

Chi-square (χ^2) difference statistic ($\Delta\chi^2$)   *Competing, nested* SEM models can be compared using this statistic, which is the simple difference between each model's χ^2 statistic. It has *degrees of freedom* equal to the difference in the models' degrees of freedom.

Communality   Total amount of variance a *measured variable* has in common with the *constructs* upon which it loads. Good measurement practice suggests that each measured variable should load on only one construct. Thus, it can be thought of as the variance explained in a measured variable by the construct. In CFA, it is referred to as the squared multiple correlation for a measured variable. Also see *variance extracted* in the next chapter.

Competing models strategy   Modeling strategy that compares the proposed model with a number of alternative models in an attempt to demonstrate that no better-fitting model exists. This approach is particularly relevant in *structural equation modeling* because a model can be shown only to have acceptable fit, but acceptable fit alone does not guarantee that another model will not fit better or equally well.

Complete case approach   Approach for handling missing data that computes values based on data from only complete cases; that is, cases with no missing data. Also known as listwise deletion.

Confirmatory analysis   Use of a multivariate technique to test (confirm) a prespecified relationship. For example, suppose we hypothesize that only two variables should be predictors of a dependent variable. If we empirically test for the significance of these two predictors and the nonsignificance of all others, this test is a confirmatory analysis. It is the opposite of *exploratory analysis.*

Confirmatory modeling strategy   Strategy that statistically assesses a single model for its fit to the observed data. This approach is actually less rigorous than the *competing models strategy* because it does not consider alternative models that might fit better or equally well than the proposed model.

Construct   Unobservable or *latent* concept that the researcher can define in conceptual terms but cannot be directly measured (e.g., the respondent cannot articulate a single response that will totally and perfectly provide a measure of the concept) or measured without error (see *measurement error*). A construct can be defined in varying degrees of specificity, ranging from quite narrow concepts to more complex or abstract concepts, such as intelligence or emotions. No matter what its level of specificity, however, a construct cannot be measured directly and perfectly but must be approximately measured by multiple *indicators.*

Construct validity   Extent to which a set of *measured variables* actually represent the theoretical *latent construct* they are designed to measure. Discussed in detail in Chapter 12.

Degrees of freedom (*df*)   The number of bits of information available to estimate the sampling distribution of the data after all model parameters have been estimated. In SEM models, degrees of freedom are the number of nonredundant covariances/correlations (moments) in the input matrix minus the number of estimated coefficients. The researcher attempts to maximize the degrees of freedom available while still obtaining the best-fitting model. Each estimated coefficient "uses up" a degree of freedom. A model can never estimate more coefficients than the number of nonredundant correlations or covariances, meaning that zero is the lower bound for the degrees of freedom for any model.

Dependence relationship   A regression type of relationship represented by a one-headed arrow flowing from an independent variable or *construct* to a dependent variable or construct. Typical dependence relationships in SEM connect constructs to measured variables and predictor *(exogenous)* constructs to outcome *(endogenous)* constructs.

Endogenous constructs   *Latent,* multi-item equivalent to dependent variables. An endogenous construct is represented by a *variate* of dependent variables. In terms of a *path diagram,* one or more arrows lead into the *endogenous* construct.

Equivalent models   SEM models involving the same observed covariance matrix with the same fit and *degrees of freedom* (*nested models*) but that differ in one or more paths. The number of equivalent models expands quickly as model complexity increases and demonstrates alternative explanations that fit just as well as the proposed model.

Estimated covariance matrix   Covariance matrix composed of the predicted covariances between all *indicator* variables involved in a SEM based on the equations that represent the hypothesized model. Typically abbreviated with Σ_k.

Exogenous constructs   *Latent,* multi-item equivalent of independent variables. They are *constructs* determined by factors outside of the model.

Exploratory analysis   Analysis defining possible relationships in only the most general form and then allowing the multivariate technique to reveal relationship(s). The opposite of *confirmatory analysis,* the researcher is not looking to confirm any relationships specified prior to the analysis, but instead lets the method and the data define the nature of the relationships. An example is stepwise multiple regression, in which the method adds predictor variables until some criterion is met.

Fit   See *goodness-of-fit.*

Fixed parameter   Parameter that has a value specified by the researcher. Most often the value is specified as zero, indicating no relationship, although in some instances an actual value (e.g., 1.0 or such) can be specified.

Free parameter   Parameter estimated by the structural equation program to represent the strength of a specified relationship. These parameters may occur in the *measurement model* (most often denoting loadings of *indicators* to *constructs*) as well as the *structural model* (relationships among constructs).

Goodness-of-fit (GOF)   Measure indicating how well a specified model reproduces the observed covariance matrix among the *indicator* variables.

Imputation   Process of estimating the missing data of an observation based on valid values of the other variables. The objective is to employ known relationships that can be identified in the valid values of the sample to assist in representing or even estimating the replacements for missing values. See also *all-available, complete case,* and *model-based approaches* for missing data.

Incremental fit indices   Group of *goodness-of-fit* indices that assesses how well a specified model fits relative to some alternative baseline model. Most commonly, the baseline model is a *null model* specifying that all *measured variables* are unrelated to each other. Complements the other two types of goodness-of-fit measures, the *absolute fit* and *parsimonious fit* measures.

Indicator   Observed value (also called a *measured* or *manifest variable*) used as a measure of a *latent construct* that cannot be measured directly. The researcher must specify which indicators are associated with each latent construct.

Latent construct   *Operationalization of a construct* in *structural equation modeling.* A latent construct cannot be measured directly but can be represented or measured by one or more variables (*indicators*). In combination, the answers to these questions give a reasonably accurate measure of the latent construct (attitude) for an individual.

Latent factor   See *latent construct.*

Latent variable   See *latent construct.*

LISREL   Most widely used SEM program. The name is derived from LInear Structural RELations.

LISREL notation   A commonly used method of expressing SEM models and results as a series of matrices used by LISREL. The matrices, such as lambda, beta, and gamma, represent specific components in a SEM model. Although the notation is specific to the LISREL program, the widespread use of LISREL has popularized the terminology when describing models and results.

Manifest variable   See *measured variable.*

Maximum likelihood estimation (MLE)   Estimation method commonly employed in structural equation models. An alternative to ordinary least squares used in multiple regression, MLE is a procedure that iteratively improves parameter estimates to minimize a specified fit function.

Measured variable   Observed (measured) value for a specific item or question, obtained either from respondents in response to questions (as in a questionnaire) or from some type of observation. Measured variables are used as the *indicators* of *latent constructs.* Same as *manifest variable.*

Measurement error   Degree to which the variables we can measure do not perfectly describe the *latent construct(s)* of interest. Sources of measurement error can range from simple data entry errors to definition of *constructs* (e.g., abstract concepts such as patriotism or loyalty that mean many things to different people) that are not perfectly defined by any set of *measured variables.* For all practical purposes, all constructs have some measurement error, even with the best *indicator variables.* However, the researcher's objective is to minimize the amount of measurement error. SEM can take measurement error into account in order to provide more accurate estimates of the relationships between constructs.

Measurement model   A SEM *model* that (1) specifies the *indicators* for each *construct* and (2) enables an assessment of *construct validity.* The first of the two major steps in a complete *structural model* analysis, it is discussed in more detail in Chapter 12.

Measurement relationship   *Dependence relationship* between *indicators* or *measured variables* and their associated *construct(s)*. A common specification depicts the construct "causing" or giving rise to the indicators, thus the arrows point from the construct to the indicators. An alternative specification reverses the relationship. These alternative specifications will be discussed in more detail in Chapter 12.

Missing at random (MAR)   Classification of missing data applicable when missing values of Y depend on X, but not on Y. When missing data are MAR, observed data for Y are a truly random sample for the X values in the sample, but not a random sample of all Y values, due to missing values of X.

Missing completely at random (MCAR)   Classification of missing data applicable when missing values of Y are not dependent on X. When missing data are MCAR, observed values of Y are a truly random sample of all Y values, with no underlying process that lends bias to the observed data.

Model   Representation and operationalization of a theory. A conventional model in SEM terminology consists of two parts. The first part is the *measurement model.* It represents the theory showing how *measured variables* come together to represent *constructs.* The second part is the *structural model,* which shows how constructs are associated with each other, often with multiple *dependence relationships.* The model can be formalized in a *path diagram.*

Model development strategy   Structural modeling strategy incorporating *model respecification* as a theoretically driven method of improving a tentatively specified *model.* It enables exploration of alternative model formulations that may be supported by theory. A basic model framework is proposed and the purpose of the modeling effort is to improve this framework through modifications of the *structural* and/or *measurement models.* The modified model would be validated with new data. It does not correspond to an *exploratory approach* in which *model respecifications* are made atheoretically.

Model respecification   Modification of an existing *model* with estimated parameters to correct for inappropriate parameters encountered in the estimation process or to create a *competing model* for comparison.

Model-based approach   Replacement approach for missing data in which values for missing data are estimated based on all nonmissing data for a given respondent. Most widely used methods are maximum likelihood estimation (ML) of missing values and EM, which involves maximum likelihood estimation of the means and covariances given missing data.

Multicollinearity   Extent to which a *construct* can be explained by the other constructs in the analysis. As multicollinearity increases, it complicates the interpretation of relationships because it is more difficult to ascertain the effect of any single construct owing to their interrelationships.

Nested model   A *model* is nested within another model if it contains the same number of *constructs* and can be formed from the other model by altering the relationships. The most common form of nested model occurs when a single relationship is added to or deleted from another model. Thus, the model with fewer estimated relationships is nested within the more general model.

Null model   Baseline or comparison standard used in *incremental fit indices.* The null model is hypothesized to be the simplest *model* that can be theoretically justified.

Observed sample covariance matrix   Typical input matrix for SEM estimation composed of the observed variances and covariances for each *measured variable.* Typically abbreviated with a bold, capital letter S (**S**).

Operationalizing a construct   Key process in the *measurement model;* it involves determining the *measured variables* that will represent a *construct* and the way in which they will be measured.

Parsimony fit indices   Measures of overall *goodness-of-fit* representing the degree of model fit per estimated coefficient. This measure attempts to correct for any overfitting of the *model* and evaluates the parsimony of the model compared to the goodness-of-fit. These measures complement the other two types of goodness-of-fit measures, the *absolute fit* and *incremental fit* measures.

Path analysis   General term for an approach that employs simple bivariate correlations to estimate relationships in a SEM *model*. Path analysis seeks to determine the strength of the paths shown in *path diagrams*.

Path diagram   A visual representation of a *model* and the complete set of relationships among the model's *constructs*. *Dependence relationships* are depicted by straight arrows, with the arrow emanating from the predictor variable and the arrowhead pointing to the dependent construct or variable. Curved arrows represent correlations between constructs or *indicators,* but no *causation* is implied.

Reliability   Measure of the degree to which a set of *indicators* of a *latent construct* is internally consistent in their measurements. The indicators of highly reliable *constructs* are highly interrelated, indicating that they all seem to measure the same thing. Individual item reliability can be computed as 1.0 minus the *measurement error.* Note that high reliability does not guarantee that a construct is representing what it is supposed to represent. It is a necessary but not sufficient condition for validity.

Residual   The difference between the actual and estimated value for any relationship. In SEM analyses, residuals are the differences between the *observed* and *estimated covariance matrices.*

Spurious relationship   A relationship that is false or misleading. A common occurrence in which a relationship can be spurious is when an omitted construct variable explains both the cause and effect (i.e., relationship between original *constructs* becomes nonsignificant upon adding omitted construct).

Structural equation modeling (SEM)   Multivariate technique combining aspects of factor analysis and multiple regression that enables the researcher to simultaneously examine a series of interrelated *dependence relationships* among the *measured variables* and *latent constructs* (*variates*) as well as between several latent constructs.

Structural model   Set of one or more *dependence relationships* linking the hypothesized model's *constructs*. The structural model is most useful in representing the interrelationships of variables between *constructs*.

Structural relationship   *Dependence relationship* (regression type) specified between any two *latent constructs*. Structural relationships are represented with a single-headed arrow and suggest that one *construct* is dependent upon another. *Exogenous constructs* cannot be dependent on another construct. *Endogenous constructs* can be dependent on either exogenous or endogenous constructs (see Chapter 12 for more detail).

Theory   A systematic set of relationships providing a consistent and comprehensive explanation of phenomena. In practice, a theory is a researcher's attempt to specify the entire set of *dependence relationships* explaining a particular set of outcomes. A theory may be based on ideas generated from one or more of three principal sources: (1) prior empirical research; (2) past experiences and observations of actual behavior, attitudes, or other phenomena; and (3) other theories that provide a perspective for analysis.

Variate   A linear combination of *measured variables* that represents a *latent construct.*

WHAT IS STRUCTURAL EQUATION MODELING?

Structural equation modeling (SEM) is a family of statistical models that seek to explain the relationships among multiple variables. In doing so, it examines the *structure* of interrelationships expressed in a series of equations, similar to a series of multiple regression equations. These equations depict all of the relationships among **constructs** (the dependent and independent variables) involved in the analysis. Constructs are unobservable or **latent factors** represented by multiple variables (much like variables representing a factor in factor analysis). So far each multivariate technique has been classified either as an interdependence or dependence technique. SEM can be thought of as a unique combination of both types of techniques because SEM's foundation lies in two familiar multivariate techniques: factor analysis and multiple regression analysis.

SEM is known by many names: covariance structure analysis, latent variable analysis, and sometimes it is even referred to by the name of the specialized software package used (e.g., a LISREL or AMOS model). Although SEM models can be tested in different ways, all structural equation models are distinguished by three characteristics:

1. Estimation of multiple and interrelated dependence relationships
2. An ability to represent unobserved concepts in these relationships and account for measurement error in the estimation process
3. Defining a model to explain the entire set of relationships

Estimation of Multiple Interrelated Dependence Relationships

The most obvious difference between SEM and other multivariate techniques is the use of separate relationships for each of a set of dependent variables. In simple terms, SEM estimates a series of separate, but interdependent, multiple regression equations simultaneously by specifying the **structural model** used by the statistical program. First, the researcher draws upon theory, prior experience, and the research objectives to distinguish which independent variables predict each dependent variable. Dependent variables in one relationship can become independent variables in subsequent relationships, giving rise to the interdependent nature of the structural model. Moreover, many of the same variables affect each of the dependent variables, but with differing effects. The structural model expresses these **dependence relationships** among independent and dependent variables, even when a dependent variable becomes an independent variable in other relationships.

The proposed relationships are then translated into a series of structural equations (similar to regression equations) for each dependent variable. This feature sets SEM apart from techniques discussed previously that accommodate multiple dependent variables—multivariate analysis of variance and canonical correlation—in that they allow only a single relationship between dependent and independent variables.

Incorporating Latent Variables Not Measured Directly

SEM also has the ability to incorporate latent variables into the analysis. A **latent construct** (also termed a **latent variable**) is a hypothesized and unobserved concept that can be represented by observable or measurable variables. It is measured indirectly by examining consistency among multiple **measured variables,** sometimes referred to as **manifest variables,** or **indicators,** which are gathered through various data collection methods (e.g., surveys, tests, observational methods).

THE BENEFITS OF USING LATENT CONSTRUCTS   Yet why would we want to use a latent variable that we cannot measure directly instead of the exact measures the respondents provided? Although it may sound like a nonsensical or "black box" approach, it has both theoretical and practical justification. First, we can better represent theoretical concepts by using multiple measures of a concept to reduce the measurement error of that concept. Second, it improves the statistical estimation of the relationships between concepts by accounting for the measurement error in the concepts.

Representing Theoretical Concepts. As we introduced in Chapter 3 when discussing exploratory factor analysis, most concepts we wish to examine require multiple measures (e.g., items) for adequate representation. From a theoretical perspective, most concepts are relatively complex (e.g., patriotism, consumer confidence, or even satisfaction) and have many meanings and/or dimensions. With concepts such as these, the researcher tries to design the best questions to measure the concept knowing that individuals may interpret any single question somewhat differently. The intent is for the collective set of questions to represent the concept better than any single item [13].

Moreover, the researcher must also be aware of measurement error that occurs with any form of measurement. Although we may be able to minimize it with physical concepts such as time (e.g., measurement with atomic clocks), any more theoretical or abstract concept is necessarily subject to measurement error. In its most basic form, measurement error is due to inaccurate responses. But, more important, it occurs when respondents may be somewhat unsure about how to respond or may interpret the questions in a way that is different from what the researcher intended. Finally, it can result from a natural degree of inconsistency on the part of the respondent when we use multiple perspectives or items to measure the same concept. All of these situations give rise to measurement error. If we know the magnitude of the problem, we can incorporate the extent of the measurement error into the statistical estimation and improve our dependence model.

How do we represent theoretical concepts and then quantify the amount of measurement error? SEM provides the **measurement model,** which specifies the rules of correspondence between measured and latent variables (constructs). The measurement model enables the researcher to use any number of variables for a single independent or dependent construct. Once the constructs are defined, then the model can be used to assess the extent of measurement error (known as *reliability*).

As an example, let us consider the following situation in developing a measurement model for HBAT. HBAT would like to determine which factors may influence the job satisfaction of its employees. The dependent (outcome) variable is job satisfaction, and the two independent variables are how they feel about their supervisor and how they like their work environment. Each of these three variables can be defined as a latent construct. Each latent construct would be measured with several indicator variables. For example, how employees feel about their supervisor might be measured by the following three indicator variables: (1) My supervisor recognizes my potential; (2) My supervisor helps me resolve problems at work; and (3) My supervisor understands the challenges of balancing work and home demands. The researcher identifies the specific indicator variables associated with each construct, typically based on a combination of previous similar studies and the situation at hand. When SEM is applied the researcher can assess the contribution of each indicator variable in representing its associated construct and measure how well the combined set of indicator variables represents the construct (reliability and validity). This is the measurement assessment component of SEM. After the constructs have met the required measurement standards, the relationships between the constructs can be estimated. This is the structural assessment component of SEM.

Improving Statistical Estimation. In all the multivariate techniques to this point, we have assumed we can overlook the measurement error in our variables. As just discussed, we know from both practical and theoretical perspectives that we cannot perfectly measure a concept and that some degree of **measurement error** is always present. For example, when asking about something as straightforward as household income, we know some people will answer incorrectly, either overstating or understating the amount or not knowing it precisely. The answers provided have some measurement error and thus affect the estimate of the true structural coefficient.

Reliability is a measure of the degree to which a set of *indicators* of a *latent construct* is internally consistent based on how highly interrelated the indicators are with each other. In other words, it represents the extent to which the indicators all measure the same thing. Reliability does not guarantee, however, that the measures indicate only one thing. We will discuss this more in the next chapter. Generally though, reliability is inversely related to measurement error. As reliability goes up, the relationships between a construct and the indicators are greater, meaning that the construct explains more of the variance in each indicator. Just as we learned in multiple regression, as more of an outcome is explained (in this case a measured indicator of a construct), the amount of error decreases. In this way, high reliability is associated with lower measurement error.

Statistical theory tells us that a regression coefficient is actually composed of two elements: the *true* structural coefficient between the dependent and independent variable and the reliability of the predictor variable. The impact of measurement error (and the corresponding reliability) can be shown from an expression of the regression coefficient as

$$\beta_{y \cdot x} = \beta_s \times p_x$$

where $\beta_{y \cdot x}$ is the observed regression coefficient, β_s is the true structural coefficient, and ρ_x is the reliability of the predictor variable. What SEM does is make an estimate of the true structural coefficient (β_s) rather than the observed regression coefficient. This is a critical point because *unless the reliability is 100 percent (i.e., no measurement error), the observed correlation (and resulting regression coefficient) will always understate the true relationship.* So SEM "corrects for" or "accounts for" the amount of measurement error in the variables (latent constructs) and estimates what the relationship would be if there was no measurement error. These are the estimates of the causal relationships in the structural model between constructs.

Given this, the relationships we can estimate through regression models will always be weaker in the presence of measurement error (this makes sense when we think about it because error can only detract from the true relationship). The equation means that relationships estimated with typical multivariate procedures will understate the actual or true relationship because reliability can only take on values between 0 (meaning no reliability) and 1 (meaning 100% reliability). So, if one knows the reliability of measures, and the observed regression coefficient, the true regression relationship can be found (true regression relationship = observed regression coefficient ÷ reliability). SEM offers the big advantage of automatically applying this correction. So, the relationships are corrected and should be more accurate than those found with simpler approaches. Because the SEM relationship coefficients are corrected in this fashion, they will tend to be larger than coefficients obtained when multiple regression is used.

Although reliability is important, high reliability does not guarantee that a construct is measured accurately. That conclusion involves an assessment of validity, which will be discussed in the next chapter. Reliability is a necessary but not sufficient condition for validity.

DISTINGUISHING EXOGENOUS VERSUS ENDOGENOUS LATENT CONSTRUCTS   Recall that in multiple regression, multiple discriminant analysis, and MANOVA, it was important to distinguish between independent and dependent variables. Likewise, in SEM a similar distinction must be made. However, because we are now generally predicting latent constructs with other latent constructs, a different terminology is used.

Exogenous constructs are the latent, multi-item equivalent of independent variables. As such, they use a **variate** of measures to represent the construct, which acts as an independent variable in the model. They are determined by factors outside of the model (i.e., they are not explained by any other construct or variable in the model), thus the term *independent*. SEM models are often depicted by a visual diagram, so it is useful to know how to spot an exogenous construct. Given that it is independent of any other construct in the model, visually an exogenous construct does not have any paths (one-headed arrows) from any other construct or variable going into it. We discuss the issues in constructing the visual diagram in the next section.

Endogenous constructs are the latent, multi-item equivalent to dependent variables (i.e., a variate of individual dependent variables). These constructs are theoretically determined by factors within the model. Thus, they are dependent on other constructs, and this dependence is represented visually by a path to an endogenous construct from an exogenous construct (or from another endogenous construct, as we will see later).

Defining a Model

A **model** is a representation of a theory. **Theory** can be thought of as a systematic set of relationships providing a consistent and comprehensive explanation of phenomena. From this definition, we see that theory is not the exclusive domain of academia, but can be rooted in experience and practice obtained by observation of real-world behavior. A conventional model in SEM terminology consists of really two models—the measurement model (representing how measured variables come together to represent constructs) and the structural model (showing how constructs are associated with each other). Chapter 12 addresses the first part of SEM, or the measurement model, in Part 1, whereas Part 2 examines the second part of SEM, or the structural model.

IMPORTANCE OF THEORY   A model should not be developed without some underlying theory. Theory is often a primary objective of academic research, but practitioners may develop or propose a set of relationships that are as complex and interrelated as any academically based theory. Thus, researchers from both academia and industry can benefit from the unique analytical tools provided by SEM. We will discuss in a later section specific issues in establishing a theoretical base for your SEM model, particularly as it relates to establishing causality. In all instances SEM analyses should be dictated first and foremost by a strong theoretical base.

A VISUAL PORTRAYAL OF THE MODEL   A complete SEM model consisting of measurement and structural models can be quite complex. Although all of the relationships can be expressed in path analysis notation (see materials available at www.pearsonglobaleditions.com/hair or www.mvstats.com for more detail), many researchers find it more convenient to portray a model in a visual form, known as a **path diagram.** This visual portrayal of the relationships employs specific conventions both for the constructs and measured variables as well as the relationships between them.

Depicting the Constructs Involved in a Structural Equations Model.   Latent constructs are related to measured variables with a **measurement relationship.** This is a type of dependence relationship (depicted by a straight arrow) between measured variables and constructs. In a typical SEM the arrow is drawn from the latent constructs to the variables that are associated with the constructs. These variables are referred to as *indicators* because no single variable can completely represent a construct, but it can be used as an indication of the construct. The researcher must justify the theoretical basis of the indicators because SEM only examines the empirical character-istics of the variables. An alternative specification where the arrows point from the indictors toward the construct will be discussed later in this chapter and in more detail in Chapter 12. We will also discuss how to assess the quality of the indicators of the constructs in a SEM model. Here, we focus on the basic principles in constructing a diagram of a measurement model:

- Constructs typically are represented by ovals or circles, and measured variables are represented by squares or rectangles.
- To assist in distinguishing the indicators for endogenous versus exogenous constructs, meas-ured variables (indicators) for exogenous constructs are usually referred to as X variables, whereas endogenous construct indicators are usually referred to as Y variables.
- The X and/or Y measured variables are associated with their respective construct(s) by a straight arrow from the construct(s) to the measured variable.

Figure 11-1a illustrates the measurement relationship between a construct and one of its measured variables. Because constructs will likely be indicated by multiple measured variables, the more common depiction is as in Figure 11-1b. Remember that the indicators are labeled as either X or Y depending on whether they are associated with an exogenous or endogenous construct, respectively.

Depicting Structural Relationships.   A structural model involves specifying **structural relationships** between latent constructs. Specifying a relationship generally means that we either specify that a relationship exists or that it does not exist. If it exists, an arrow is drawn; if no relation-ship is expected, then no arrow is drawn. On some occasions, specification may also mean that a

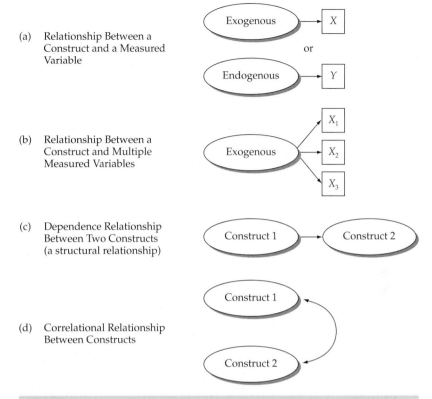

FIGURE 11-1 Common Types of Theoretical Relationships in an SEM Model

specific value is specified for a relationship. Two types of relationships are possible among constructs: dependence relationships and correlational (covariance) relationships.

As has been discussed, measurement relationships are one form of dependence relationships between constructs to variables. The second form is a dependence relationship between constructs. Here, the arrows point from the antecedent (independent variable) to the subsequent effect or outcome (dependent variable). This relationship is depicted in Figure 11-1c. In a later section we discuss issues involved in specifying causation, which is a special form of dependence relationship.

Specification of dependence relationships also determines whether a construct is considered exogenous or endogenous. Recall that an endogenous construct acts like a dependent variable, and any construct with a dependence path (arrow) pointing to it is considered endogenous. An exogenous construct has only correlational relationships with other constructs and acts as an independent variable in structural relationships (i.e., no dependence paths coming into the construct).

In many instances, the researcher wishes to specify a simple correlation between exogenous constructs. The researcher believes the constructs are correlated, but does not assume that one construct is dependent upon another. This relationship is depicted by a two-headed arrow connection as shown in Figure 11-1d. An exogenous construct cannot share this type of relationship with an endogenous construct. Only a dependence relationship can exist between exogenous and endogenous constructs.

Combining Measurement and Structural Relationships. Figure 11-2 illustrates a simple SEM model incorporating both the measurement and structural relationships of two constructs with four indicators each. In Figure 11-2a, there is a correlational relationship between the two constructs, indicated by the curved arrow. The indicators (four on each construct) are labeled X_1 to X_8.

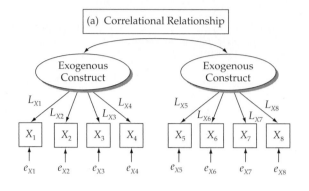

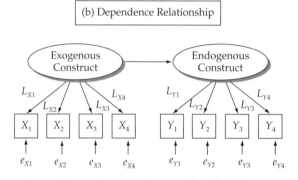

FIGURE 11-2 Visual Representation Measurement and Structural Model Relationships in a Simple SEM Model

Figure 11-2b depicts a dependence relationship between the exogenous and endogenous constructs. The two constructs retain their same indicators, but two changes distinguish it from the correlational relationship. First, the indicators of the exogenous constructs are denoted by X_1 to X_4, whereas the endogenous indicators are Y_1 to Y_4. The measured variables themselves did not change at all, just their designation in the model. Second, the single dependence relationship between the exogenous construct and the endogenous construct is depicted by the straight arrow between the constructs that replaces the curved arrow.

The researcher determines whether constructs are exogenous or endogenous based on the theory being tested. Constructs retain the same indicators, but they can differ based on their role in the model. A single SEM model most likely will contain both dependence and correlational relationships.

HOW WELL DOES THE MODEL FIT?   It is important to remember that in contrast to regression analysis or other dependence techniques, which seek to explain relationships in a single equation, SEM's statistical goal is to test a set of relationships representing multiple equations. Therefore, measures of fit or predictive accuracy for other techniques (i.e., R^2 for multiple regression, classification accuracy in discriminant analysis, or statistical significance in MANOVA) are not well suited for SEM. What is needed is a measure of fit or predictive accuracy that reflects the overall model, not any single relationship. The researcher must "accept or reject" the entire model, determining if the overall model fit is acceptable before examining any specific relationships.

Because the focus is on the entire model, SEM uses a series of measures that depict how well a researcher's theory explains the input data—the observed covariance matrix among measured variables. Model fit is determined by the correspondence between the observed covariance matrix and an estimated covariance matrix that results from the proposed model. If the proposed model properly estimates all of the substantive relationships between constructs and the measurement model adequately defines the constructs, then it should be possible to estimate a covariance matrix between measured variables that closely matches reality as depicted in the observed covariance matrix. We will discuss the process of estimating a covariance matrix from the proposed model along with a number of measures of fit in much greater detail in later sections of this chapter as well as in Chapter 12.

SEM AND OTHER MULTIVARIATE TECHNIQUES

SEM is a multivariate technique based on variates in both the measurement and structural models. In the measurement models, each set of indicators for a construct acts collectively (as a variate) to define the construct. In the structural model, constructs are related to one another in correlational and dependence relationships. SEM is most appropriate when the researcher has multiple constructs, each represented by several measured variables, and these constructs are distinguished based on whether they are exogenous or endogenous. In this sense, SEM shows similarity to other multivariate dependence techniques such as MANOVA and multiple regression analysis. Moreover, the measurement model looks similar in form and function to factor analysis. We will discuss the similarities of SEM to both dependence and interdependence techniques in the following sections.

Similarity to Dependence Techniques

An obvious similarity of SEM is to multiple regression, one of the most widely used dependence techniques. Relationships for each endogenous construct can be written in a form similar to a regression equation. The endogenous construct is the dependent variable, and the independent variables are the constructs with arrows pointing to the endogenous construct. One principal difference in SEM is that a construct that acts as an independent variable in one relationship can be the dependent variable in another relationship. SEM then allows for all of the relationships/equations to be estimated simultaneously.

SEM can also be used to represent other dependence techniques. Variations of the standard SEM models can be used to represent nonmetric, categorical variables, and even a MANOVA model can be examined using SEM. It enables the researcher to take advantage of SEM's ability to accommodate measurement error, for example, within a MANOVA context.

Similarity to Interdependence Techniques

At first glance, the measurement model, associating measured variables with constructs, seems identical to factor analysis where variables have loadings on factors (see Chapter 3 for a more detailed discussion). Despite a great deal of similarity, such as the interpretation of the strength of the relationship of each variable to the construct (known as a *loading* in factor analysis), one difference is critical. Factor analysis of this type is basically an **exploratory analysis** technique that searches for structure among variables by defining factors in terms of sets of variables. As a result, every variable has a loading on every factor.

SEM is the opposite of an exploratory technique. It requires that the researcher specify which variables are associated with each construct, and then loadings are estimated only where variables are associated with constructs (typically there are no cross-loadings). The distinction is not so much with interpretation as with implementation. Exploratory factor analysis requires little to no specification on the part of the researcher. In contrast, SEM requires complete specification of the measurement model.

The advantages of using multiple measures for a construct, discussed earlier and in Chapter 3, are realized through the measurement model in SEM. In this way the estimation procedures for the structural model can include a direct correction for measurement error as discussed earlier. By doing so, the relationships between constructs are estimated more accurately.

The Emergence of SEM

SEM is a relatively new analytical tool, but its roots extend back to the first half of the twentieth century. SEM's development originated with the desires of genetics and economics researchers to be able to establish causal relationships between variables [8, 19, 55]. The mathematical complexity of SEM limited its application until computers and software became widely available. They enabled the two multivariate procedures of factor analysis and multiple regression to be combined. During the late 1960s and early 1970s, the work of Jöreskog and Sörbom led to simultaneous maximum likelihood estimation of the relationships between constructs and measured indicator variables, as well as among latent constructs. This work culminated in the SEM program **LISREL** [27, 28, 29, 30]. It was not the first software to perform SEM or path analysis, but it was the first to gain widespread usage.

SEM's growth remained relatively slow during the 1970s and 1980s, in large part due to its perceived complexity. By 1994, however, more than 150 SEM articles were published in the academic social science literature. That number increased to more than 300 by 2000, and today SEM is "the dominant multivariate technique," followed by cluster analysis and MANOVA [23].

THE ROLE OF THEORY IN STRUCTURAL EQUATION MODELING

SEM should never be attempted without a strong theoretical basis for specification of both the measurement and structural models. The following sections address some fundamental roles played by theory in SEM: (1) specifying relationships that define the model; (2) establishing causation, particularly when using cross-sectional data; and (3) the development of a modeling strategy.

Specifying Relationships

Although theory can be important in all multivariate procedures, it is particularly important for SEM because it is considered a **confirmatory analysis.** That is, it is useful for testing and potentially confirming theory. Theory is needed to specify relationships in both measurement and structural models, modifications to the proposed relationships, and many other aspects of estimating a model.

From a practical perspective, a theory-based approach to SEM is necessary because all relationships must be specified by the researcher before the SEM model can be estimated. With other multivariate techniques the researcher may have been able to specify a basic model and allow default values in the statistical programs to "fill in" the remaining estimation issues. This option of using default values is not possible with SEM. In addition, any model modifications must be made through specific actions by the researcher. Thus, when we stress the need for theoretical justification, we are emphasizing that SEM is a confirmatory method guided more by theory than by empirical results.

The depiction of a set of relationships in a path diagram typically involves a combination of dependence and correlational relationships among exogenous and endogenous constructs. The researcher can specify any combination of relationships that have theoretical support for the research questions at hand. The following examples illustrate how relationships can involve both dependence and correlational elements as well as accommodate interrelated relationships.

Figure 11-3 shows three examples of relationships depicted by path diagrams, along with the corresponding equations. Figure 11-3a shows a simple three-construct model. Both X_1 and X_2 are exogenous constructs related to the endogenous construct Y_1, and the curved arrow between X_1 and X_2 shows the effects of intercorrelation (multicollinearity) on the prediction. We can show this relationship with a single equation, much as we did in our discussion of multiple regression.

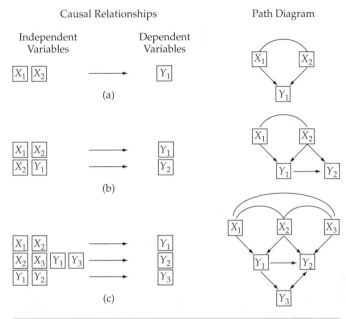

Causal Relationships

Path Diagram

FIGURE 11-3 **Representing Dependence and Correlational Relationships Through Path Diagrams**

In Figure 11-3b we add a second endogenous construct—Y_2. Now, in addition to the model and equation shown in Figure 11-3a, we add a second equation showing the relationship between X_2 and Y_1 with Y_2. Here we can see the unique role played by SEM when more than one relationship shares constructs. We want to know the effects of X_1 on Y_1, the effects of X_2 on Y_1, and simultaneously the effects of X_2 and Y_1 on Y_2. If we did not estimate them in a consistent manner, we would not be assured of representing their true and separate effects. For example, such a technique is needed to show the effects of X_2 on both Y_1 and Y_2.

The relationships become even more intertwined in Figure 11-3c, with three dependent constructs, each related to the others as well as to the independent constructs. A reciprocal relationship (two-headed, straight arrow) even occurs between Y_2 and Y_3. This relationship is shown in the equations by Y_2 appearing as a predictor of Y_3 and Y_3 appearing as a predictor of Y_2. It is not possible to express all the relationships in either Figure 11-3b or 11-3c in a single equation. Separate equations are required for each dependent construct. The need for a method that can estimate all the equations simultaneously is met by SEM.

These examples are just a preview as to the types of relationships that can be portrayed and then empirically examined through SEM. Given the ability for the models to become complex quite easily, it is even more important to use theory as a guiding factor to specification of both the measurement and structural models. Later in this chapter as well as in Chapter 12 we will discuss the criteria by which the researcher can specify SEM models in more detail.

Establishing Causation

Perhaps the strongest type of theoretical inference a researcher can draw is a causal inference, which involves proposing that a dependence relationship actually is based on **causation.** A **causal inference** involves a hypothesized cause-and-effect relationship. If we understand the causal sequence between variables, then we can explain how some cause determines a given effect. In practical terms, the effect can be at least partially managed with some degree of certainty.

So, dependence relationships can sometimes be theoretically hypothesized as causal. However, simply thinking that a dependence relationship is causal doesn't make it so. As such we use the term *cause* with great care in SEM.

Let us consider HBAT's interest in job satisfaction as an example. If feeling positive about a supervisor can be proven to result in (cause) increased job satisfaction, then we know that higher job satisfaction can be achieved by improving how employees feel about their supervisor. Thus, company policies and training can focus on improving supervision approaches. If supervision is indeed causally related as hypothesized, then the resulting improvements will increase employee job satisfaction.

Causal research designs traditionally involve an experiment with some controlled manipulation (e.g., a categorical independent variable as found in MANOVA or ANOVA). SEM models are typically used, however, in nonexperimental situations in which the exogenous constructs are not experimentally controlled variables. This limits the researcher's ability to draw causal inferences and SEM alone cannot establish causality. It can, however, treat dependence relationships as causal if four types of evidence (covariation, sequence, nonspurious covariation, and theoretical support) are reflected in the SEM model [26, 45].

COVARIATION   Because causality means that a change in a cause brings about a corresponding change in an effect, systematic covariance (correlation) between the cause and effect is necessary, but not sufficient, to establish causality. Just as is done in multiple regression by estimating the statistical significance of coefficients of independent variables that affect the dependent variable, SEM can determine systematic and statistically significant covariation between constructs. Thus, statistically significant estimated paths in the structural model (i.e., relationships between constructs) provide evidence of covariation. Structural relationships between constructs are typically the paths for which causal inferences are most often hypothesized.

SEQUENCE   A second requirement for causation is the temporal sequence of events. Let us use our earlier example as an illustration.

If improvements in supervision result in increased job satisfaction, then the changes in supervision cannot occur after the change in job satisfaction. If we picture many dominos standing in a row, and the first one is knocked down by a small ball, it may cause all of the other dominos to fall. In other words, the ball hitting the first domino causes the other dominos to fall. If the ball is the cause of this effect, the ball must hit the first domino before the others fall down. If the others have fallen down before the ball strikes the first domino, then the ball cannot have caused them to fall. Thus, sequence in causation means that improvements in supervision must occur before job satisfaction increases if the relationship between the two variables is causal.

SEM cannot provide this type of evidence without a research design that involves either an experiment or longitudinal data. An experiment can provide this evidence because the researcher maintains control of the causal variable through manipulations. Thus, the research first manipulates a variable and then observes the effect. Longitudinal data can provide this evidence because they enable us to account for the time period in which events occur. A great deal of social science research relies on cross-sectional surveys. Measuring all of the variables at the same point in time does not provide a way of accounting for the time sequence. Thus, theory must be used to argue that the sequence of effects is from one construct to another.

NONSPURIOUS COVARIANCE   A **spurious relationship** is one that is false or misleading. Any relationship is considered spurious when another event not included in the analysis actually explains both the cause and effect. Simply put, the size and nature of the relationship between a cause and the relevant effect should not be affected by including other constructs (or variables) in a model. Many anecdotes describe what can happen with spurious correlation.

For example, a significant correlation between ice cream consumption and the likelihood of drowning can be empirically verified. However, is it safe to say that eating ice cream causes drowning? If we account for some other potential cause (e.g., temperature is associated with increased ice cream consumption and more swimming), we would find no real relationship between ice cream consumption and drowning. Thus we cannot say with any certainty that ice cream consumption causes the likelihood of drowning even though they are significantly correlated.

The Impact of Collinearity. Because a causal inference is supported when we can show that some third construct does not affect the relationship between the cause and effect, a lack of collinearity among the predictors (see Chapter 4 on multicollinearity) is desirable. When collinearity is not present, the researcher comes closest to reproducing the conditions that are present in an experimental design. These conditions include orthogonal, or uncorrelated, experimental predictor variables.

Unfortunately, most structural models involve multiple predictor constructs that exhibit **multicollinearity** with both other predictors and the construct. In these cases, making a causal inference is less certain. Therefore, in SEM models involving cross-sectional survey research, causal evidence is found when (1) the relationship between a cause and an effect remains constant when other predictor constructs are included in the model and (2) when the effect construct's error is independent of the causal construct [45, 51].

Testing for Spurious Relationships. Figure 11-4 shows an example of testing for a nonspurious relationship with two SEM models. The first model specifies the proposed structural relationship between two constructs. The second model incorporates the Alternative Cause construct as an additional predictor variable. If the estimated relationship between constructs found in the first model remains unchanged when the additional predictor is added (the second model), then the relationship is deemed nonspurious. However, if the structural relationship becomes nonsignificant in the second model because of the addition of the other predictors, then the relationship must be considered spurious. We should note that more than one additional construct may be added and that the structural relationships must remain significant no matter how many constructs are added.

In our employee job satisfaction example, we proposed that supervisor perceptions influenced satisfaction (Figure 11-4a). One could argue, however, that how employees feel about their supervisor

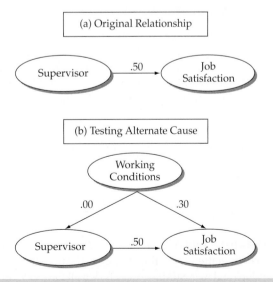

FIGURE 11-4 Testing for a Nonspurious Relationship Between Constructs

does not really determine their level of satisfaction with their job. An alternative explanation, for instance, is that good working conditions act as an alternative cause for both improved supervision and higher job satisfaction (Figure 11-4b). If the working conditions construct is measured along with the other constructs, and a relationship is specified between working conditions and both supervision and job satisfaction, then a SEM model can determine whether a relationship between the constructs exists. In our example, the estimated coefficient remains unchanged (.50), indicating that the relationship between supervisor and job satisfaction is nonspurious. If the estimated coefficient had become nonsignificant when working conditions is added to the model, then we would consider the relationship between supervisor and job satisfaction to be spurious.

THEORETICAL SUPPORT   The final condition for causality is theoretical support, or a compelling rationale to support a cause-and-effect relationship. This condition emphasizes the fact that simply testing a SEM model and analyzing its results cannot establish causality. Theoretical support becomes especially important with cross-sectional data. A SEM model can demonstrate relationships between any constructs that are correlated with one another (e.g., ice cream consumption and drowning statistics). But unless theory can be used to establish a causal ordering and a rationale for the observed covariance, the relationships remain simple association and should not be attributed with any further causal power.

Do employees' feelings about their supervisors cause job satisfaction? A theoretical justification for causation may exist in that as employees spend more time with their supervisors they become more familiar with their supervision approaches, which increases their understanding and reactions to the supervisor, and based on these experiences they become more satisfied with their job situation. Thus, a case can be made that more favorable feelings about supervisors cause increased job satisfaction.

Although SEM is often referred to as *causal modeling,* causal inferences are only possible when evidence is consistent with the four conditions for causality already mentioned. SEM can provide evidence of systematic covariation and can help in demonstrating that a relationship is not spurious. If data are longitudinal, SEM can also help establish the sequence of relationships. However, it is up to the researcher to establish theoretical support. Thus, SEM is helpful in establishing a causal inference, but it cannot do it alone.

Developing a Modeling Strategy

One of the most important concepts a researcher must learn regarding multivariate techniques is that no single correct way exists to apply them. In some instances, relationships are strictly specified, and the objective is a confirmation of the relationship. At other times, the relationships are loosely recognized, and the objective is the discovery of relationships. At each extreme as well as points in between, the researcher must apply the multivariate technique in accordance with the research objectives.

The application of SEM follows this same tenet. Its flexibility provides researchers with a powerful analytical tool appropriate for many research objectives that serve as guidelines in a modeling strategy. The use of the term *strategy* is designed to denote a plan of action toward a specific outcome. For our purposes, we define three distinct strategies in the application of SEM: confirmatory modeling strategy, competing models strategy, and model development strategy.

CONFIRMATORY MODELING STRATEGY   The most direct application of structural equation modeling is a **confirmatory modeling strategy.** The researcher specifies a single model composed of a set of relationships and uses SEM to assess how well the model fits the data. Here the researcher is saying, "It either works or it doesn't." It is quite the opposite of exploratory techniques such as stepwise regression. If the proposed model has acceptable fit, the researcher has found support for that model. But as we will discuss later, that model is just one of several different models having equally acceptable model fits. Perhaps a more insightful test can be achieved by comparing alternative models.

COMPETING MODELS STRATEGY   A **competing models strategy** is based on comparing the estimated model with alternative models through overall model comparisons. The strongest test of a proposed model is to identify and test competing models that represent truly different, but highly plausible, hypothesized structural relationships. When comparing these models, the researcher comes much closer to a test of competing theories, which is much stronger than a test of a single model in isolation.

Equivalent models provide a second perspective on developing a set of comparative models. It has been shown that for any proposed structural equation model, at least one other model exists with the same number of parameters but with different relationships portrayed that fits at least as well as the proposed model. As a general rule of thumb, the more complex a model, the more equivalent models exist. Thus, one should not draw conclusions on the basis of empirical results alone.

MODEL DEVELOPMENT STRATEGY   The **model development strategy** differs from the prior two strategies in that, although a basic model framework is proposed, the purpose of the modeling effort is to improve this framework through modifications of the structural or measurement models. In many applications, theory can provide only a starting point for development of a theoretically justified model that can be empirically supported. Thus, the researcher must employ SEM not just to test the model empirically but also to provide insights into its respecification.

One note of caution must be made. The researcher must be careful not to employ this strategy to the extent that the final model has acceptable fit but cannot be generalized to other samples or populations. Moreover, **model respecification** must always be done with theoretical support rather than just empirical justification. Models developed in this fashion should be verified with an independent sample.

A SIMPLE EXAMPLE OF SEM

The following example illustrates how SEM works with multiple relationships, estimating many equations at once, even when they are interrelated and the dependent variable in one equation is an independent variable in another equation(s). This capability enables the researcher to model complex relationships in a way that is not possible with any of the other multivariate techniques discussed in this text.

We should note, however, that our example will not depict one of SEM's other strengths— the ability to employ multiple measures (the measurement model) to represent a construct in a manner similar to factor analysis. For simplicity, each construct in the following example is treated as a single variable.

Chapter 12 discusses measurement theory and confirmatory factor analysis and will illustrate multiple item measurement in detail. For now, we focus only on the basic principles of model construction and estimating multiple relationships.

The Research Question

Theory must be the foundation of even the simplest of models, because variables could always be linked to one another in multiple ways. Most would be complete nonsense. Theory should make the model plausible. The emphasis on representing dependence relationships necessitates that the researcher carefully detail not only the number of constructs involved, but the expected relationships among those constructs. With these constructs in hand, models and the estimation of relationships can proceed.

To demonstrate how theory can be used to develop a model to test with SEM, let us use our example of employee job satisfaction, but expand it by adding a couple more constructs. Two key research questions are: (1) what factors influence job satisfaction and (2) is job satisfaction related to an employee's likelihood of looking for another job (i.e., quitting their present job)? More specifically, HBAT believes that favorable perceptions of supervision, coworkers, and working conditions will increase job satisfaction, which in turn will decrease the likelihood of searching for another job.

From their experiences, HBAT managers developed a series of relationships they believe explain the process:

- Improved supervision leads to higher job satisfaction.
- Better work environment leads to higher job satisfaction.
- More favorable perceptions of coworkers lead to higher job satisfaction.
- Higher job satisfaction leads to lower likelihood of job search.

These four relationships form the basis of how HBAT management believes they can reduce the likelihood of employees searching for another job. Management would like to reduce job searching activities because the cost of finding, hiring, and training new employees is very high.

The research team could use multiple regression, but that technique could only test part of this model at a time because regression is used to examine relationships between multiple independent variables and a single dependent variable. Because the following model has more than a single dependent variable the research team must use another technique that can examine relationships with more than a single dependent variable.

Setting Up the Structural Equation Model for Path Analysis

Once a series of relationships is specified, the researcher is able to identify the model in a form suitable for analysis. Constructs are identified as either being exogenous or endogenous. Then, to demonstrate the relationships easily, they are portrayed visually in a path diagram, where straight arrows depict the impact of one construct on another. If causal hypotheses are inferred, the arrows representing dependence relationships point from the cause to the subsequent effect.

The relationships identified by HBAT management include five constructs: perceptions of Supervision, Work Environment, and Coworkers, along with Job Satisfaction and Job Search. An initial step is to identify which constructs are considered exogenous and which are endogenous. Remember that exogenous constructs are similar to independent variables, whereas endogenous are the equivalent of dependent variables.

The Supervision, Work Environment, and Coworker constructs are identified as exogenous variables because they are similar to independent variables in the model. Similarly, Job Search is clearly an endogenous variable because it is represented as a dependent variable. But what about Job Satisfaction? It is dependent on the Supervision, Work Environment, and Coworker constructs, but it is also an independent variable because it influences the Job Search construct. This is one of the unique and clearly beneficial characteristics of SEM—it can examine relationships (models) in which a construct operates as both an independent and dependent variable. From our model of the relationships, therefore, we can identify the following types of constructs:

Exogenous Constructs	Endogenous Constructs
Supervision	Job Satisfaction
Work Environment	Job Search
Coworkers	

With the constructs specified as either exogenous or endogenous, the relationships can now be represented in a path diagram as shown in Figure 11-5.

Note that one type of relationship presented in Figure 11-5 was not expressed by the HBAT research team: the correlations among the exogenous constructs. These relationships are typically added in SEM when the researcher feels the exogenous constructs have some degree of association

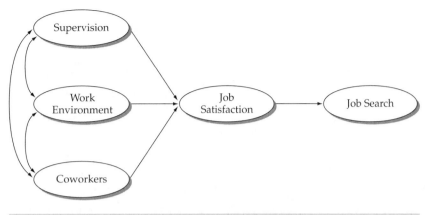

FIGURE 11-5 Path Diagram of a Simple Structural Model

that gives rise to their interrelationships. Typically, this is the result of additional variables not included in the model that impact the exogenous variables (i.e., a common cause). In the case of exogenous variables, it is directly comparable to representing the multicollinearity discussed in multiple regression (see Chapter 4 for more detail). We have added these correlational relationships in our theoretical model because we expect the separate elements of managing HBAT employees (Supervision, Work Environment, and Coworkers) will be coordinated and based on consistent planning and execution. Moreover, including the interconstruct correlations between the exogenous variables often makes the estimates for the dependent relationships more reliable. We will discuss other reasons for adding this type of relationship in the following chapter. The research team can now collect data on the five constructs as a basis for evaluating the proposed theoretical model.

The Basics of SEM Estimation and Assessment

With the relationships and path diagram specified, researchers can now put them in a format suitable for analysis in SEM, estimate the strength of the relationships, and assess how well the data actually fit the model. In the example we illustrate the basic procedures in each of these steps as we investigate the issues raised by employees' work environment and their job satisfaction and desire to engage in job search.

OBSERVED COVARIANCE MATRIX SEM differs from other multivariate techniques in that it is a covariance structure analysis technique rather than a variance analysis technique. As a result, SEM focuses on explaining covariation among the variables measured, or the **observed sample covariance matrix.** Although it may not always be obvious to the user, SEM programs can compute solutions using either a covariance matrix or a correlation matrix as input.

You may wonder if it makes a difference whether we use a covariance matrix instead of a correlation matrix such as that used in multiple regression. We discuss the advantages of a covariance matrix later in this chapter (stage 3 of the decision process), but we should remember that correlation is just a special case of covariance. A correlation matrix is simply the covariance matrix when standardized variables are used (i.e., the standardized covariance matrix). Only values below the diagonal are unique and of particular interest when the focus is on correlations. The key at this point is to realize that the observed covariance matrix can simply be computed from sample observations just as we did in computing a correlation matrix. It is not estimated nor is it dependent on a model imposed by a researcher.

Let us revisit our example and see how the researchers would proceed after the model is defined.

To understand how data are input into SEM, think of the covariance matrix among the five variables. The observed covariance matrix would contain 25 values. The five diagonal values would represent the variance of each variable with 10 unique covariance terms. Because the covariance matrix is symmetric, the 10 unique terms would be repeated above and below the diagonal. As a result, the number of unique values in the matrix is the five diagonal values (variances) plus the 10 unique off-diagonals (covariances) for a total of 15.

For example, suppose the sample involves individuals interviewed using a mall-intercept technique. The resulting covariance matrix is composed of the following values, with each construct simply abbreviated as S for Supervision, WE for Work Environment, CW for Coworkers, SAT for Job Satisfaction, and SRCH for Job Search (as in Figure 11-5). The matrix of unduplicated values would be as follows:

	S	WE	CW	SAT	SRCH
Observed	Var (S)				
Covariance	Cov (S, WE)	Var (WE)			
	Cov (S, CW)	Cov (WE, CW)	Var (CW)		
	Cov (S, SAT)	Cov (WE, SAT)	Cov (CW, SAT)	Var (SAT)	
	Cov (S, SRCH)	Cov (WE, SRCH)	Cov (CW, SRCH)	Cov (SAT, SRCH)	Var (SRCH)

Actual values for this example are shown in Table 11-1a, the observed covariance matrix.

ESTIMATING AND INTERPRETING RELATIONSHIPS   Prior to the widespread use of SEM programs, researchers found solutions for multiple equation models using a process known as **path analysis.**

TABLE 11-1   Observed, Estimated, and Residual Covariance Matrices

a) Observed Covariance Matrix

	Supervision	Work Environment	Coworkers	Job Satisfaction	Job Search
Observed Covariance Matrix: (S)	Var (SP)	—	—	—	—
	.20	Var (WE)	—	—	—
	.20	.15	Var (CW)	—	—
	.20	.30	.50	Var (JS)	—
	−.05	.25	.40	.50	Var (JS)

b) Estimated Covariance Matrix

	Supervision	Work Environment	Coworkers	Job Satisfaction	Job Search
Estimated Covariance Matrix: (Σ)	—	—	—	—	—
	.20	—	—	—	—
	.20	.15	—	—	—
	.20	.30	.50	—	—
	.10	.15	.25	.50	—

c) Residuals: Observed Minus Estimated Covariances

	Supervision	Work Environment	Coworkers	Job Satisfaction	Job Search
Supervision	—	—	—	—	—
Work Environment	.00	—	—	—	—
Co-workers	.00	.00	—	—	—
Job Satisfaction	.00	.00	.00	—	—
Job Search	−.15	.10	.15	.00	—

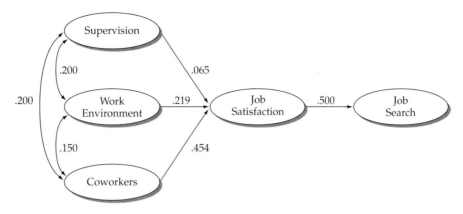

FIGURE 11-6   An Estimated Structural Equation Model of Job Search

Path analysis uses bivariate correlations to estimate the relationships in a system of structural equations. This process estimates the strength of each structural relationship (a straight or curved arrow) in a path diagram. The actual mathematical procedure is briefly described in Appendix 11A.

Path analysis procedures provide estimates for each relationship (arrow) in the model shown in Figure 11-6. These estimates are interpreted like regression coefficients if two separate equations are used—one to predict Job Satisfaction and a second to predict Job Search. But SEM does not keep each equation separate and all estimates of relationships in both equations are computed at the same time using the information from all equations that make up the model. SEM also provides estimates of the correlational relationships between the exogenous constructs, which may be useful in our interpretation of the results as well as directly influencing our assessment of the validity of the exogenous constructs.

With estimates for each path, an interpretation can be made of each relationship represented in the model. When statistical inference tests are applied, the researcher can assess the probability that the estimates are significant (i.e., not equal to zero). Moreover, the estimates can be used like regression coefficients to make estimates of the values of any construct in the model.

The relationships (paths) in the model shown in Figure 11-6 represent the research questions posed by the HBAT research team. When we look at the first three relationships (i.e., impact of Supervision, Work Environment, and Coworkers on Job Satisfaction), we can see that the estimated coefficients are .065, .219, and .454, respectively. The sizes of these coefficients indicate that Coworkers has the biggest impact on job satisfaction, whereas Work Environment is somewhat less and supervision has the smallest impact. Moreover, Job Satisfaction has a substantial impact on Job Search (.50) and provides evidence of that relationship as well.

Now recall from Chapter 4 that regression coefficients can be used to compute predicted values for dependent variables. Those values were referred to as $\hat{y}$. Thus, for any particular values of the independent variables, an estimated value for the outcome can be obtained. In this case where we treat constructs as variables, they would represent predicted values for endogenous constructs, or the outcome. The difference between the actual observed value for the outcome and $\hat{y}$ is error. SEM also can provide estimated values for exogenous constructs when multiple variables are used to indicate the construct. This process will become clearer in the next chapter. Realize that several potential relationships between constructs have no path drawn, which means that the researcher does not expect a direct relationship between these constructs. For instance, no arrows are drawn between Supervision and Job Search, Work Environment and Job Search, or Coworkers and Job Search, which affects the equations for the predicted values.

In our model, if we take any observed values for Supervision, Work Environment, and Coworkers, we can estimate a value for job satisfaction using this equation:

$$\hat{y}_{JobSatisfaction} = .065 \text{ (Supervision)} + .219 \text{ (Work Environment)} + .454 \text{ (Coworkers)}$$

Similarly, predicted values for Job Search can be obtained:

$$\hat{y}_{JobSearch} = .50 \text{ (Job Satisfaction)}$$

This would represent a multiple-equation prediction because Job Satisfaction is also endogenous. Substituting the equation for Job Satisfaction into the equation for Job Search, we get:

$$\hat{y}_{JobSearch} = .50[.065 \text{ (Supervision)} + .219 \text{ (Work Environment)} + .454 \text{ (Coworkers)}]$$

This illustrates, therefore, how path estimates in Figure 11-6 can be used to calculate estimated values for Job Satisfaction and Job Search.

ASSESSING MODEL FIT WITH THE ESTIMATED COVARIANCE MATRIX   The last step in a SEM analysis involves calculating an **estimated covariance matrix** and then assessing the degree of fit to the observed covariance model. The estimated covariance matrix is derived from the path estimates of the model. With these estimates we can calculate all of the covariances that were in the observed covariance matrix using the principles of path analysis "in reverse." Then by comparing the two matrices SEM can test the fit of a model. Models that produce an estimated covariance matrix that is within sampling variation of the observed covariance matrix are generally thought of as good models and would be said to fit well.

The process of calculating an estimated covariance first identifies all of the direct and indirect paths that relate to a specific covariance or correlation. Then, the coefficients are used to calculate the value of each path, which are then totaled to get the estimated value for each covariance/correlation.

Let us look at one relationship (Work Environment and Job Satisfaction) to illustrate what happens. They involve both direct and indirect paths:

Direct path:

$$\text{Work Environment} \rightarrow \text{Job Satisfaction} = .219$$

Indirect paths:

$$\text{Work Environment} \rightarrow \text{Supervision} \rightarrow \text{Job Satisfaction} = .200 \times .065 = .013$$
$$\text{Work Environment} \rightarrow \text{Coworkers} \rightarrow \text{Job Satisfaction} = .150 \times .454 = .068$$

Total:

$$\text{Direct} + \text{Indirect} = .219 + .013 + .068 = .300$$

Thus, the estimated covariance between Work Environment and Job Satisfaction is .300, the sum of both the direct and indirect paths. Similarly, we can think of the estimated covariance matrix as the covariances derived from the estimates of all the variables ($\hat{y}$). The complete estimated covariance matrix is shown in Table 11-1b.

This example illustrates how the researcher impacts the estimated covariances (and ultimately model fit) by the paths specified in the model. In our example, if the Work Environment was not correlated with the Supervision or Coworkers constructs, then the estimated covariance would be different. Therefore the researcher should note that each path added or deleted in the model ultimately controls how well the observed covariances can be predicted. The identification of direct and indirect paths for each covariance is addressed in more detail in the Basic Stats appendix available on the text's Web site (www.pearsonglobaleditions.com/hair or www.mvstats.com).

The last issue in assessing fit is the concept of a residual. With SEM, a residual is the difference between any observed and estimated covariance. Thus, when we compare the observed and actual covariance matrices, any differences we detect are the residuals. The distinction with other multivariate techniques, especially multiple regression, is important. In those techniques, residuals reflected the errors in predicting individual observations. In SEM, individual observations are not the focus of the analysis. When a SEM program refers to residuals, it refers to the difference between the estimated and observed covariances for any pair of indicators.

RULES OF THUMB 11-1

Structural Equation Modeling Introduction

- No model should be developed for use with SEM without some underlying theory, which is needed to develop:
  - Measurement model specification
  - Structural model specification
- Models can be represented visually with a path diagram
  - Dependence relationships are represented with single-headed directional arrows
  - Correlational (covariance) relationships are represented with two-headed arrows
- Causal are the strongest type of inference made in applying multivariate statistics; therefore, they can be supported only when precise conditions for causality exist:
  - Covariance between the cause and effect
  - The cause must occur before the effect
  - Nonspurious association must exist between the cause and effect
  - Theoretical support exists for the relationship between the cause and effect
- Models developed with a model development strategy must be cross-validated with an independent sample

The matrix of residuals (the differences between the observed and estimated covariance matrices, $|\mathbf{S} - \Sigma_k|$), becomes the key driver in assessing the fit of a SEM model. If the estimated covariance matrix is sufficiently close to the observed covariance matrix (the residuals are small), then the model and its relationships are supported. If the reader is familiar with cross-tabulation, it should be no surprise that a χ^2 statistic can be computed based on the difference between the two matrices. Later, we will use this statistic as the basic indicator of the goodness-of-fit of a theoretical model.

In comparing the observed and estimated covariance matrices, some covariances are exactly predicted and some differences are found. For example, if you look in the first column of numbers in both matrices, you will see that the relationships between Supervision and Work Environment, and Coworkers and Job Satisfaction, are all predicted exactly. That is, they are all .20 in both the observed and estimated matrices. For other relationships, such as the relationship between Coworkers and Job Satisfaction, the estimated covariance (.25) is noticeably different from the observed covariance (.40).

The result is the residuals matrix (Table 11-1c). As we have noted, there are three residuals that are not zero. Specifically, only the residuals for the relationships between the three exogenous constructs and Job Search are not zero (−.15, .10, and .15). These findings indicate that the SEM model does not perfectly explain the covariance between these constructs, and it could suggest that the researcher's theory is inadequate. But we need additional information before rejecting the proposed theory.

With these simple rules, the entire model can now be estimated. Notice that dependent variables in one relationship can easily be independent variables in another relationship (as with Job Satisfaction). No matter how large the path diagram gets or how many relationships are included, path analysis provides a way to analyze the set of relationships.

We should note that the researcher does not have to do all of the calculations in path analysis, because they are handled by the computer software. The researcher needs to understand the principles underlying SEM so that the implications of adding or deleting paths or other model modifications can be understood. The next chapter explains how these procedures are implemented in testing measurement and structural theories, respectively.

SIX STAGES IN STRUCTURAL EQUATION MODELING

SEM has become a popular multivariate approach in a relatively short period of time. Researchers are attracted to SEM because it provides a conceptually appealing way to test theory. If a researcher can express a theory in terms of relationships among measured variables

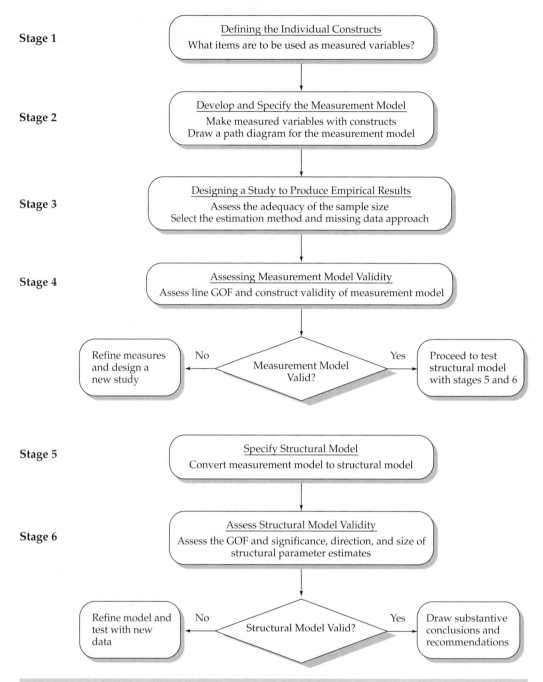

FIGURE 11-7   Six-Stage Process for Structural Equation Modeling

and latent constructs (variates), then SEM will assess how well the theory *fits* reality as represented by data.

This section continues the discussion of SEM by describing a six-stage decision process (see Figure 11-7). This process varies slightly from that introduced in Chapter 1 in order to reflect the unique terminology and procedures of SEM. The six stages are as follows:

Stage 1: Defining individual constructs

Stage 2: Developing the overall measurement model

Stage 3: Designing a study to produce empirical results

Stage 4: Assessing the measurement model validity

Stage 5: Specifying the structural model

Stage 6: Assessing structural model validity

The remainder of this chapter provides a brief overview and introduction of these six stages, which will also be discussed in greater detail in the next chapter. Rather than include an HBAT example as an illustration of the technique here in this chapter, it will be introduced in the next chapter. Many SEM analyses involve testing both measurement theory (how the constructs are represented) and structural theory (how the constructs relate to each other). Chapter 12 covers the first four stages of SEM in Part 1 and the remaining two stages in Part 2.

STAGE 1: DEFINING INDIVIDUAL CONSTRUCTS

A good measurement theory is a necessary condition to obtain useful results from SEM. Hypotheses tests involving the structural relationships among constructs will be no more reliable or valid than is the measurement model in explaining how these constructs are constructed. Researchers often have a number of established scales to choose from, each a slight variant from the others. But in other situations the researcher is faced with the lack of an established scale and must develop a new scale or substantially modify an existing scale to the new context. In each case, how the researcher selects the items to measure each construct sets the foundation for the entire remainder of the SEM analysis. The researcher must invest significant time and effort early in the research process to make sure the measurement quality will enable valid conclusions to be drawn.

Operationalizing the Construct

The process begins with a good theoretical definition of the constructs involved. This definition provides the basis for selecting or designing individual indicator items. A researcher **operationalizes a construct** by selecting its measurement scale items and scale type. In survey research, operationalizing a construct results in a series of scaled indicator items in a common format such as a Likert scale or a semantic differential scale. The definitions and items are derived from two common approaches.

SCALES FROM PRIOR RESEARCH   In many instances, constructs can be defined and operationalized as they were in previous research studies. Researchers may do a literature search on the individual constructs and identify scales that previously performed well. Most research today uses prior scales published in academic studies. As we discussed in Chapter 3, compendiums of prior scales are available in numerous disciplines.

NEW SCALE DEVELOPMENT   Construct measures can be developed. This development is appropriate when a researcher is studying something that does not have a rich history of previous research. The general process for developing scale items can be long and detailed. The essentials of this process are highlighted in the next chapter, but the reader is referred elsewhere for a more thorough discussion [13, 10, 42].

Pretesting

Generally, when measures are either developed for a study or are taken from various sources, some type of pretest should be performed. The pretest should use respondents similar to those from the population to be studied so as to screen items for appropriateness. Pretesting is particularly important when scales are applied in specific contexts (e.g., purchase situations, industries, or other instances where specificity is paramount) or in contexts outside their normal use. Empirical testing of the pretest results is done in a manner identical to the final model analysis (see discussion on stage 4 later in this chapter). Items that do not behave statistically as expected may need to be refined or deleted to avoid these issues when the final model is analyzed.

STAGE 2: DEVELOPING AND SPECIFYING THE MEASUREMENT MODEL

With the scale items specified, the research must now specify the measurement model. In this stage, each latent construct to be included in the model is identified and the measured indicator variables (items) are assigned to latent constructs. Although this identification and assignment can be represented by equations, it is simpler to represent this process with a diagram. Figure 11-8 represents a simple two-construct measurement model, with four indicators associated with each construct and a correlational relationship between constructs.

SEM Notation

A key element in the path diagram is the labeling notation for indicators, constructs, and relationships between them. Each software program utilizes a somewhat unique approach, although a standard convention has been associated with LISREL and is simply referred to as **LISREL notation.** Although widely used, LISREL notation is uniquely tied to the program's use of matrix notation, and thus become unwieldy for those with no experience with LISREL. For purposes of this text we will simplify our notation to be as generalizable as possible among all the software programs. Given the widespread use of LISREL notation, however, we have developed a reference guide to LISREL notation (see Appendix 11B) along with a "conversion" between this notation and the LISREL notation for interested readers that are available at the text's Web site (www.pearsonglobaleditions.com/hair or www.mvstats.com).

Table 11-2 lists the notation used in this text for the measurement and structural models. As discussed earlier, the three types of relationships are measurement relationships between indicators/items and constructs, structural relationships between constructs, and correlational relationships between constructs. There are also two types of error terms, one related to individual indicators and the other to endogenous constructs.

Specification of the complete measurement model uses (1) measurement relationships for the items and constructs, (2) correlational relationships among the constructs, and (3) error terms for the items.

A basic measurement model can be illustrated as shown in Figure 11-8. The model has a total of 17 estimated parameters. The 17 parameters include eight loading estimates, eight error estimates, and one between-construct correlation estimate. The estimate for each arrow linking a

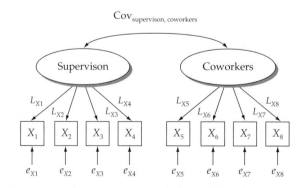

FIGURE 11-8 Visual Representation of a Measurement Model

TABLE 11-2   Notation for Measurement and Structural Models

Element	Symbol	Notation	Example
Type of Indicator			
Exogenous	X	X_{number}	X_1
Endogenous	Y	Y_{number}	Y_1
Type of Relationship			
Measurement (Loading)	L	L_{item}	L_{X1}
Structural (Path coefficient)	P	$P_{outcome, predictor}$	$P_{Job\ Sat,\ Sup}$
Correlational between Constructs	Cov	$Cov_{construct1,\ construct2}$	$Cov_{Sup,\ WE}$
Error Terms			
Indicators	e	e_{item}	e_{X1}
Constructs	E	$E_{construct}$	$E_{Job\ Search}$

construct to a measured variable is an estimate of a variable's loading—the degree to which that item is related to the construct. This stage of SEM can be thought of as assigning individual variables to constructs. Visually, it answers the question, "Where should arrows be drawn linking constructs to variables?"

There are a number of possible paths that were not specified. For example, no paths suggest correlations between indicator variables or loadings of indicators on more than one construct (cross-loadings). In the estimation process these unspecified loadings (there are 19 in total) are set (fixed) at zero, meaning that they will not be estimated.

Creating the Measurement Model

Specification of the measurement model can be a straightforward process, but a number of issues still must be addressed. Chapter 12 provides more detailed discussion of each issue. The types of questions are listed here:

1. Can we empirically support the validity and unidimensionality of the constructs? Essential points must be engaged in establishing the theoretical basis of the constructs and measures.
2. How many indicators should be used for each construct? What is the minimum number of indicators? Is there a maximum? What are the trade-offs for increasing or decreasing the number of indicators?
3. Should the measures be considered as portraying the construct (meaning that they describe the construct) or as explaining the construct (such that we combine indicators into an index)? Each approach brings with it differing interpretations of what the construct represents.

The researcher, even with well-established scales, must still confirm the validity and unidimensionality in this specific context. In any scale development effort, issues as to number of indicators and type of construct specification must be addressed. Researchers should always ensure that these issues are examined, because any unresolved problems at this stage can affect the entire analysis, often in unseen ways.

STAGE 3: DESIGNING A STUDY TO PRODUCE EMPIRICAL RESULTS

With the basic model specified in terms of constructs and measured variables/indicators, the researcher must turn attention to issues involved with research design and estimation. Our discussion will focus on issues related to both research design and model estimation. In the area of research design, we will discuss (1) the type of data to be analyzed, either covariances or correlations; (2) the impact and remedies for missing data; and (3) the impact of sample size. In terms of model estimation, we will address model structure, the various estimation techniques available; and the current computer software being used.

Issues in Research Design

As with any other multivariate technique, SEM requires careful consideration of factors affecting the research design necessary for a successful SEM analysis. SEM can be estimated with either covariances or correlations. Thus, the researcher must choose the appropriate type of data matrix for the research question being addressed. And even though the statistical issues of SEM estimation are discussed in the next section, here it is important to note that sample size and missing data can have a profound effect on the results no matter what method is used.

METRIC VERSUS NONMETRIC DATA   The observed or measured variables have traditionally been restricted to metric data (interval or ordinal). This type of data is directly amenable to the calculation of covariances among items as discussed earlier. Advances in the software programs, however, now allow for the use of many nonmetric data types (censored, binary, ordinal, or nominal). Differing types of variables can even be used as items for the same latent construct. The researcher must be careful to specify the type of data being used for each measured variable so that the appropriate measure of association can be calculated.

COVARIANCE VERSUS CORRELATION   Researchers conducting SEM analyses in the past debated over the use of a covariance versus correlation matrix as input. SEM was originally developed using covariance matrices (hence, it is referred to with the common name of *analysis of covariance structures*). Many researchers advocated the use of correlations as a simpler form of analysis that was easier to interpret. The issue had practical significance because for many years the input matrices were computed using a statistical routine outside the SEM program and then the matrix of correlations or covariances was used as input for the analysis. Today most SEM programs can compute a model solution directly from raw data without the researcher computing a correlation or covariance matrix separately. But researchers must still choose between correlations versus covariance based on interpretive and statistical issues.

Interpretation.   The key advantage of correlational input for SEM lies in the fact that the default parameter estimates are standardized, meaning not scale dependent. All estimated values must fall within the range –1.0 to +1.0, making identification of inappropriate estimates easier than with covariances, which have no defined range. However, it is simple to produce these results from covariance input by requesting a standardized solution. As such, correlations hold no real advantage over the standardized results obtained using covariances.

Statistical Impact.   The primary advantages of using covariances arise from statistical considerations. First, the use of correlations as input can at times lead to errors in standard error computations [12]. In addition, any time hypotheses concern questions related to the scale or magnitude of values (e.g., comparing means), then covariances must be used because this information is not retained using correlations. Finally, any comparisons between samples require that covariances be used as input. Thus, covariances have distinct advantages in terms of their statistical properties versus correlations.

Choosing Between Covariances and Correlations.   In comparing the use of correlations versus covariances, we recommend using covariances whenever possible. Software makes the selection of one type versus another just a matter of selecting the type of data being computed. Covariance matrices provide the researcher with far more flexibility due to the relatively greater information content they contain.

MISSING DATA   Just as with other multivariate procedures, the researcher must make several important decisions regarding missing data. Two questions must be answered concerning missing data to suitably address any problems it may create:

1. Is the missing data sufficient and nonrandom so as to cause problems in estimation or interpretation?
2. If missing data must be remedied, what is the best approach?

We will discuss the issues that relate specifically to SEM and missing data in the following section. The reader is also referred back to Chapter 2, where a more complete discussion is provided on the methods of assessing the extent and pattern of missing data and then the approaches to remedy missing data if needed.

Extent and Pattern of Missing Data.   Most notably, missing data must always be addressed if the missing data are in a nonrandom pattern or more than 10 percent of the data items are missing. Missing data are considered **missing completely at random (MCAR)** if the pattern of missing data for a variable does not depend on any other variable in the data set or on the values of the variable itself [48]. If the pattern of missing data for a variable is related to other variables, but not related to its own values, then it is considered to be **missing at random (MAR)**. Again, Chapter 2 provides much more detailed discussion on the procedures used in assessing the extent and pattern of missing data.

Missing Data Remedies.   Four basic methods are available for solving the missing data problem: the **complete case approach** (known as *listwise* deletion, whereby the respondent is eliminated if missing data on any variable); the **all-available approach** (known as *pairwise* deletion, whereby all nonmissing data are used); **imputation** techniques (e.g., mean substitution); and **model-based approaches.**

Traditionally, listwise deletion has been considered most appropriate for SEM. More recently, pairwise deletion, which allows the use of more data, has been applied. Both of these procedures can produce problems [1]. Model-based approaches extend past the simpler imputation approaches in that missing data are imputed (replaced) based on all available data for a given respondent. The two most common approaches are the (1) maximum likelihood estimation of the missing values (ML) and (2) the EM approach. Discussion of these imputation methods is beyond our scope but is available in a number of sources [15]. Each of the methods, however, creates issues in the resulting data matrix. See Chapter 2 for a more detailed description. Further, although approaches for dealing with missing data exist, researchers should apply caution whenever the missing data exceeds 10 percent, because conclusions about fit become increasingly suspect [49].

SEM programs have also introduced the approach wherein the model is estimated directly from the available data, making allowances for the missing data during the estimation process. Known as a full information likelihood approach (FIML) [2, 18], it eliminates the need to remedy the missing data before estimation with one of the approaches just described. But there are two considerations in using this approach. First, it can only be used when the original data set is available (i.e., cannot be used with only a covariance matrix as input). Second, in many cases it provides only a smaller subset of fit statistics, which may be inadequate for complete model evaluation.

Another consideration in selecting an approach is specification of the sample size. Both the all-available (pairwise) and model-based approaches complicate the specification of sample size, because they potentially utilize differing sample sizes for each covariance term. SEM researchers have investigated the varying effects of setting the overall sample size (N) at the full sample size (the largest number of observations), the average sample size, and the minimum sample size (the smallest N associated with any sample covariance). These results generally suggest that inserting the minimum sample size leads to the fewest problems with convergence, fit bias, and parameter estimate bias [11, 15].

Selecting a Missing Data Approach. What is the best approach for handling missing data for SEM in general? We should first note that if missing data are random, less than 10 percent of observations, and the factor loadings are relatively high (.7 or greater), then any of the approaches are appropriate [14]. When missing data is more problematic than this, the first decision facing the researcher is whether to remedy the missing data problem before the estimation process.

Table 11-3 summarizes the strengths and weaknesses of each approach. When applying a remedy for missing data before estimation, the complete case approach (listwise deletion) becomes particularly problematic when samples and factor loadings are small. Conversely, the advantages of the model-based approaches become particularly apparent as sample sizes and factor loadings become generally smaller and/or the amount of missing data becomes larger. The all-available approach (pairwise deletion) is recommended when sample sizes exceed 250 and the total amount

TABLE 11-3   Some Advantages and Disadvantages of Different Missing Data Procedures

Method	Advantages	Disadvantages
Complete case (listwise)	• χ^2 shows little bias under most conditions. • Effective sample size is known. • Easy to implement using any program.	• Increases the likelihood of nonconvergence (SEM program cannot find a solution) unless factor loadings are high (> .6) and sample sizes are large (> 250). • Increased likelihood of factor loading bias. • Increased likelihood of bias in estimates of relationships among factors.
All-available (pairwise)	• Fewer problems with convergence. • Factor loading estimates relatively free of bias. • Easy to implement using any program.	• χ^2 is biased upward when amount of missing data exceeds 10%, factor loadings are high, and sample size is high. • Effective sample size is uncertain. • Not as well known.
Model-based (ML/EM)	• Fewer problems with convergence. • χ^2 shows little bias under most conditions. • Least bias under conditions of random missing data.	• Not available on older SEM programs. • Effective sample size is uncertain for EM.
Full information maximum likelihood (FIML)	• Remedy directly in estimation process. • In most situations has less bias than other methods.	• Researcher has no control over how missing data remedied. • No knowledge how missing data impacts estimates. • Typically only a subset of fit indices available.

Note: See Enders and Bandalos (2001) and Enders and Peugh (2004) for more detail. ML/EM have been combined based on the negligible differences between the results for the two (Enders and Peugh, 2004).

of missing data involved among the measured variables is below 10 percent. With this approach, the sample size (N) should be set at the minimum (smallest) sample size available for any two covariances. The all-available approach has many good properties, but the user should be aware of the potential inflation of fit statistics when a modest or large amount of data are missing and factor loadings are large.

All of these approaches, however, allow the researcher to explicitly remedy the missing data before estimation and understand the implications of whichever approach is taken. If the decision is to estimate the model without any remedy for missing data, then the FIML approach is the best alternative. Here the estimation process is based on incomplete data, which may or may not have implications for the results. In the absence of serious missing data problems, all of the approaches produce comparable results. It thus becomes as much a choice by the researcher as to whether to remedy the missing data beforehand or to do so in the estimation process. In the authors' experiences, it is most often best to resolve missing data issues before estimating the model.

SAMPLE SIZE   In general, SEM requires a larger sample relative to other multivariate approaches. Some of the statistical algorithms used by SEM programs are unreliable with small samples. Sample size, as in any other statistical method, provides a basis for the estimation of sampling error. As a starting place in discussing sample sizes for SEM, the reader can review the sample size discussions required for exploratory factor analysis (Chapter 3). Given that larger samples are usually more time consuming and expensive to obtain, the critical question in SEM involves how large a sample is needed to produce trustworthy results.

Opinions regarding minimum sample sizes have varied [34, 35]. Proposed guidelines vary with analysis procedures and model characteristics. Five considerations affecting the required sample size for SEM include the following: (1) multivariate normality of the data, (2) estimation technique, (3) model complexity, (4) amount of missing data, and (5) average error variance among the reflective indicators. Each of these considerations is discussed in the following paragraphs.

Multivariate Normality. As data deviate more from the assumption of multivariate normality, then the ratio of respondents to parameters needs to increase. A generally accepted ratio to minimize problems with deviations from normality is 15 respondents for each parameter estimated in the model. Although some estimation procedures are specifically designed to deal with nonnormal data, the researcher is always encouraged to provide sufficient sample size to allow for the sampling error's impact to be minimized, especially for nonnormal data [54].

Estimation Technique. The most common SEM estimation procedure is **maximum likelihood estimation (MLE).** Simulation studies suggest that under ideal conditions, MLE provides valid and stable results with sample sizes as small as 50. As one moves away from conditions with very strong measurement and no missing data, minimum sample sizes to ensure stable MLE solutions increase when confronted with sampling error [34]. Given less than ideal conditions, one study recommends a sample size of 200 to provide a sound basis for estimation. But it should be noted that as the sample size becomes large (>400), the method becomes more sensitive and almost any difference is detected, making goodness-of-fit measures suggest poor fit [52]. As a result, sample sizes in the range of 100 to 400 are suggested subject to the other considerations discussed next.

Model Complexity. Simpler models can be tested with smaller samples. In the simplest sense, more measured or indicator variables require larger samples. However, models can be complex in other ways that all require larger sample sizes:

- More constructs that require more parameters to be estimated.
- Constructs having fewer than three measured/indicator variables.
- Multigroup analyses requiring an adequate sample for each group.

The role of sample size is to produce more information and greater stability. Once a researcher has exceeded the absolute minimum size (one more observation than the number of observed covariances), larger samples mean less variability and increased stability in the solutions. Thus, model complexity leads to the need for larger samples.

Missing Data. Missing data complicate the testing of SEM models and the use of SEM in general because in most approaches to remedying missing data, the sample size is reduced to some extent from the original number of cases. Depending on the missing data approach taken and the extent of missing data anticipated and even the types of issues being addressed, which may include higher levels of missing data, the researcher should plan for an increase in sample size to offset any problems of missing data.

Average Error Variance of Indicators. Recent research indicates the concept of **communality** (see Chapter 3 for more details) is a more relevant way to approach the sample size issue. Communalities represent the average amount of variation among the measured/indicator variables explained by the measurement model. The communality of an item can be directly calculated as the square of the standardized construct loadings (see Chapter 12). Studies show that larger sample sizes are required as communalities become smaller (i.e., the unobserved constructs are not explaining as much variance in the measured items). Models containing multiple constructs with communalities less than .5 (i.e., standardized loading estimates less than .7) also require larger sizes for convergence and model stability [14]. The problem is exaggerated when models have constructs with only one or two items.

Summary on Sample Size. As SEM matures and additional research is undertaken on key research design issues, previous guidelines such as "always maximize your sample size" and "sample sizes of 300 are required" are no longer appropriate. It is still true that larger samples generally produce more stable solutions that are more likely to be replicable, but it has been shown that sample size decisions must be made based on a set of factors.

Based on the discussion of sample size, the following suggestions for minimum sample sizes are offered based on the model complexity and basic measurement model characteristics:

- Minimum sample size—100: Models containing five or fewer constructs, each with more than three items (observed variables), and with high item communalities (.6 or higher).
- Minimum sample size—150: Models with seven or fewer constructs, modest communalities (.5), and no underidentified constructs.
- Minimum sample size—300: Models with seven or fewer constructs, lower communalities (below .45), and/or multiple underidentified (fewer than three items) constructs.
- Minimum sample size—500: Models with large numbers of constructs, some with lower communalities, and/or having fewer than three measured items.

In addition to these characteristics of the model being estimated, sample size should be increased in the following circumstances: (1) data deviates from multivariate normality, (2) sample-intensive estimation techniques (e.g., ADF) are used, or (3) missing data exceeds 10 percent. Also, remember that group analysis requires that each group meet the sample size requirements. Finally, the researcher must remember that the sample size issue goes beyond being able to estimate a model. The sample size, just as with any other statistical inference, must be adequate to represent the population of interest, and this may often be the researcher's overriding concern.

Issues in Model Estimation

In addition to the more general research design issues discussed in the prior section, SEM analysis has several unique issues as well. These issues relate to the model structure, estimation technique used, and computer program selected for the analysis.

MODEL STRUCTURE   Among the most important steps in setting up a SEM analysis is determining and communicating the theoretical model structure to the program. Path diagrams like those used in prior examples can be useful for this purpose. Knowing the theoretical model structure, the researcher can then specify the model parameters to be estimated. These models often include common SEM abbreviations denoting the type of relationship or variable referred to. As discussed earlier, LISREL notation is widely used as a notational form. A guide to LISREL notation is available at the text's Web site (www.pearsonglobaleditions.com/hair or www.mvstats.com).

As we have mentioned many times, the researcher is responsible for specifying both the measurement and structural models. This process varies between different users and software programs, ranging from users relying on syntax and matrix notation in LISREL to those using a completely graphical interface in AMOS. But no matter what approach is taken, each approach is performing the same function—specifying which model parameters to estimate. For every possible parameter, the researcher must decide if it is to be free or fixed. A **free parameter** is one to be estimated in the model, whereas a **fixed parameter** is one in which the value is specified by the researcher. Most often a fixed parameter is set to a value of zero, indicating that no relationship is estimated. SEM requires that each possible parameter be specified as estimated or not. Yet no matter which software is used, the researcher must be able to specify the complete SEM model in terms of each parameter to be estimated.

ESTIMATION TECHNIQUE   Once the model is specified, researchers must choose the estimation method: Which mathematical algorithm will be used to identify estimates for each free parameter? Several options are available for obtaining a SEM solution.

Early attempts at structural equation model estimation were performed with ordinary least squares (OLS) regression. These efforts were quickly supplanted by maximum likelihood estimation (MLE), which is more efficient and unbiased when the assumption of multivariate normality is met. MLE is a flexible approach to parameter estimation in which the "most likely" parameter values to achieve the best model fit are found. The potential sensitivity of MLE to nonnormality, however, created a need for alternative estimation techniques. Methods such as weighted least squares (WLS), generalized least squares (GLS), and asymptotically distribution free (ADF) estimation became available [21]. The ADF technique has received particular attention due to its insensitivity to nonnormality of the data, but its requirement of rather large sample sizes limits its use.

All of the alternative estimation techniques have become more widely available as the computing power of the personal computer has increased, making them feasible for typical problems. MLE continues to be the most widely used approach and is the default in most SEM programs. In fact, it has proven fairly robust to violations of the normality assumption. Researchers compared MLE with other techniques, and it produced reliable results under many circumstances [43, 44, 49].

COMPUTER PROGRAMS   Several readily available statistical programs are convenient for performing SEM. Traditionally, the most widely used program is LISREL (LInear Structural RELations) [9, 30]. LISREL is a flexible program that can be applied in numerous situations (i.e., cross-sectional, experimental, quasi-experimental, and longitudinal studies) and has become almost synonymous with structural equation modeling. EQS (actually an abbreviation for *equations*) is another widely available program that also can perform regression, factor analysis, and test structural models [6]. AMOS (Analysis of Moment Structures) [3] is a program that gained popularity because in addition to being a module in SPSS, it also was among the first SEM programs to use a graphical interface for all functions so that a researcher never has to use syntax commands or computer code. Mplus is a flexible modeling program with multiple techniques that is particularly useful in complex applications involving multiple units of analysis in the same model [41]. Finally, CALIS is a SEM program traditionally available within SAS [20].

Ultimately, the selection of a SEM program is based on researcher preference and availability. Some programs are actually becoming more similar as they evolve. AMOS, EQS, and LISREL are

RULES OF THUMB 11-2

SEM Stages 1–3

- When a model has scales borrowed from various sources reporting other research, a pretest using respondents similar to those from the population to be studied is recommended to screen items for appropriateness
- Pairwise deletion of missing cases (all-available approach) is a good alternative for handling missing data when the amount of missing data is less than 10 percent and the sample size is 250 or more
  - As sample sizes become small or when missing data exceed 10 percent, one of the imputation methods for missing data becomes a good alternative for handling missing data
  - When the amount of missing data becomes very high (15% or more), SEM may not be appropriate
- Covariance matrices provide the researcher with far more flexibility due to the relatively greater information content they contain and are the recommended form of input to SEM models
- The minimum sample size for a particular SEM model depends on several factors, including the model complexity and the communalities (average variance extracted among items) in each factor:
  - SEM models containing five or fewer constructs, each with more than three items (observed variables), and with high item communalities (.6 or higher), can be adequately estimated with samples as small as 100 to 150
  - When the number of factors is greater than six, some of which have fewer than three measured items as indicators, and multiple low communalities are present, sample size requirements may exceed 500
- No matter the modeling approach, the sample size must be sufficient to allow the model to run, but, more important, it must adequately represent the population of interest

all available with point-and-click interface availability and the ability to specify and modify the model through an interactive path diagram. The principal difference is the notation used in specifying the measurement and structural models. LISREL uses Greek letters to represent latent factors, error terms, and parameter estimates, and Latin letters (x and y) to represent observed variables. Known as LISREL notation, it provides a convenient shorthand for describing models and simplifies discussions considerably once one is familiar with the notation. The other programs rely less on Greek abbreviations and rely on a different approach to both model specification and reporting of results. All of these programs are available in versions that can easily be run on virtually any PC. For most standard applications, these programs should produce similar substantive results. An appendix available at the text's Web site (www.pearsonglobaleditions.com/hair or www.mvstats.com) provides examples of the commands needed for several of these programs.

STAGE 4: ASSESSING MEASUREMENT MODEL VALIDITY

With the measurement model specified, sufficient data collected, and the key decisions such as the estimation technique already made, the researcher comes to the most fundamental event in SEM testing: "Is the measurement model valid?" Measurement model validity depends on (1) establishing acceptable levels of goodness-of-fit for the measurement model and (2) finding specific evidence of **construct validity.** Because we are focusing on the structural model in this simple example, we will defer the investigation of construct validity until discussed in greater detail in Chapter 12. The following discussion will focus on assessing goodness-of-fit of the overall model.

      Goodness-of-fit (GOF) indicates how well the specified model reproduces the observed covariance matrix among the indicator items (i.e., the similarity of the observed and estimated covariance matrices). Ever since the first GOF measure was developed, researchers have strived to refine and develop new measures that reflect various facets of the model's ability to represent the data. As such, a number of alternative GOF measures are available to the researcher. Each GOF measure is unique, but the measures are classed into three general groups: absolute measures, incremental measures, and parsimony fit measures. In the following sections, we first review some basic elements underlying all GOF measures, followed by discussions of each class of GOF measures. Readers interested in more detailed and statistically based discussions are referred to Appendix 11C.

The Basics of Goodness-of-Fit

Once a specified model is estimated, model fit compares the theory to reality by assessing the similarity of the estimated covariance matrix (theory) to reality (the observed covariance matrix). If a researcher's theory were perfect, the observed and estimated covariance matrices would be the same. The values of any GOF measure result from a mathematical comparison of these two matrices. The closer the values of these two matrices are to each other, the better the model is said to **fit.**

We start by examining **chi-square** (χ^2) because it is the fundamental measure of differences between the observed and estimated covariance matrices. Then the discussion focuses on calculating degrees of freedom and finally on how statistical inference is affected by sample size and the impetus that provides for alternative GOF measures.

CHI-SQUARE (χ^2) GOF   The difference in the observed and estimated covariance matrices (termed **S** and Σ_k, respectively) is the key value in assessing the GOF of any SEM model. The chi-square (χ^2) test is the only statistical test of the difference between matrices in SEM and is represented mathematically by the following equation:

$$\chi^2 = (N-1) \text{ (Observed sample covariance matrix } - \text{ SEM estimated covariance matrix)}$$

or

$$\chi^2 = (N-1)(\mathbf{S} - \Sigma_k)$$

N is the overall sample size. It should be noted that even if the differences in covariance matrices (i.e., residuals) remained constant, the χ^2 value increases as sample size increases. Likewise, the estimated covariance matrix is influenced by how many parameters are specified (i.e., free) in the model (the k in Σ_k), so the model degrees of freedom also influence the χ^2 GOF test.

DEGREES OF FREEDOM (DF)   As with other statistical procedures, **degrees of freedom** represent the amount of mathematical information available to estimate model parameters. Let us start by reviewing how it is calculated. The number of degrees of freedom for a SEM model is determined by

$$df = \frac{1}{2}[(p)(p+1)] - k$$

where p is the total number of observed variables and k is the number of estimated (free) parameters. Subtracting the number of estimated parameters from the total amount of available mathematical information is similar to other multivariate methods. But the fundamental difference in SEM comes in the first part of the calculation—$1/2[(p)(p+1)]$—which represents the number of covariance terms below the diagonal plus the variances on the diagonal. It is not derived at all from sample size as we saw in other multivariate techniques (e.g., in regression, df is the sample size minus number of estimated coefficients). Thus degrees of freedom in SEM are based on the size of the covariance matrix, which comes from the number of indicators in the model. An important implication is that the researcher does not affect degrees of freedom through sample size, but we will see later how sample size does influence the use of chi-square as a GOF measure.

STATISTICAL SIGNIFICANCE OF χ^2   The implied null hypothesis of SEM is that the observed sample and SEM estimated covariance matrices are equal, meaning that the model fits perfectly. The χ^2 value increases as differences (residuals) are found when comparing the two matrices. With the χ^2 test, we then assess the statistical probability that the observed sample and SEM estimated covariance matrices are actually equal in a given population. This probability is the traditional p-value associated with parametric statistical tests.

An important difference between SEM and other multivariate techniques also occurs in this statistical test for GOF. For other techniques, we typically looked for a smaller p-value (less than .05) to show that a significant relationship existed. But with the χ^2 GOF test in SEM, we make inferences in a way that is in some ways exactly opposite. When we find a p-value for the χ^2 test to be small (statistically significant), it indicates that the two covariance matrices are statistically different and indicates problems with the fit. So in SEM we look for a relatively small χ^2 value (and corresponding large p-value), indicating no statistically significant difference between the two matrices, to support the idea that a proposed theory fits reality. Relatively small χ^2 values support the proposed theoretical model being tested.

We should note that the chi-square can also be used when comparing models because the difference in chi-square between two models can be tested for statistical significance. Thus, if the researcher is expecting differences between models (e.g., differences in two models estimated for males and females), large χ^2 values would lend support that the models are different.

Chi-square (χ^2) is the fundamental statistical measure in SEM to quantify the differences between the covariance matrices. When used as a GOF measure, the comparison is between observed and predicted covariance matrices. Yet the actual assessment of GOF with a χ^2 value alone is complicated by several factors as discussed in the next section. To provide alternative perspectives on model fit, researchers developed a number of alternative goodness-of-fit measures. The discussions that follow present the role of chi-square as well as the alternative measures.

Absolute Fit Indices

Absolute fit indices are a direct measure of how well the model specified by the researcher reproduces the observed data [31]. As such, they provide the most basic assessment of how well a researcher's theory fits the sample data. They do not explicitly compare the GOF of a specified model to any other model. Rather, each model is evaluated independently of other possible models.

χ^2 STATISTIC   The most fundamental absolute fit index is the χ^2 statistic. It is the only statistically based SEM fit measure [9] and is essentially the same as the χ^2 statistic used in cross-classification analysis between two nonmetric measures. The one crucial distinction, however, is that when used as a GOF measure the researcher is looking for no differences between matrices (i.e., low χ^2 values) to support the model as representative of the data.

The χ^2 GOF statistic has two mathematical properties that are problematic in its use as a GOF measure. First, recall that the χ^2 statistic is a mathematical function of the sample size (N) and the difference between the observed and estimated covariance matrices. As N increases so does the χ^2 value, even if the differences between matrices are identical. Second, although perhaps not as obvious, the χ^2 statistic also is likely to be greater when the number of observed variables increases. Thus, all other things equal, just adding indicators to a model will cause the χ^2 values to increase and make it more difficult to achieve model fit.

Although the χ^2 test provides a test of statistical significance, these mathematical properties present trade-offs for the researcher. Although larger sample sizes are often desirable, just the increase in sample size itself will make it more difficult for those models to achieve a statistically insignificant GOF. Moreover, as more indicators are added to the model, because of either more constructs or better measurement of constructs, this will make it more difficult in using chi-square to assess model fit. One could argue that if more variables are needed to represent reality, then they should reflect a better fit, not a worse fit, as long as they produce valid measures. Thus, in some ways the mathematical properties of the χ^2 GOF test punish a model for things that should not be detrimental to its overall validity.

For this reason, the χ^2 GOF test is often not used as the sole GOF measure. Researchers have developed many alternative measures of fit to correct for the bias against large samples and increased model complexity. Several of these GOF indices are presented next. However, the χ^2 issues also impact many of these additional indices, particularly some of the absolute fit indices. This said, the χ^2 value for a model does summarize the fit of a model quite well, and with experience the researcher can make educated judgments about models based on this result. In sum, the statistical test or resulting *p*-value is less meaningful as sample sizes become large or the number of observed variables becomes large.

GOODNESS-OF-FIT INDEX (GFI)   The GFI was an early attempt to produce a fit statistic that was less sensitive to sample size. Even though *N* is not included in the formula, this statistic is still sensitive to sample size due to the effect of *N* on sampling distributions [36]. No statistical test is associated with the GFI, only guidelines to fit [53]. The possible range of GFI values is 0 to 1, with higher values indicating better fit. In the past, GFI values of greater than .90 typically were considered good. Others argue that .95 should be used [24]. Recent development of other fit indices has led to a decline in usage.

ROOT MEAN SQUARE ERROR OF APPROXIMATION (RMSEA)   One of the most widely used measures that attempts to correct for the tendency of the χ^2 GOF test statistic to reject models with a large sample or a large number of observed variables is the root mean square error of approximation (RMSEA). Thus, it better represents how well a model fits a population, not just a sample used for estimation [25]. It explicitly tries to correct for both model complexity and sample size by including each in its computation. Lower RMSEA values indicate better fit.

The question of what is a "good" RMSEA value is debatable. Although previous research had sometimes pointed to a cutoff value of .05 or .08, more recent research points to the fact that drawing an absolute cutoff for RMSEA is inadvisable [17]. An empirical examination of several measures found that the RMSEA was best suited for use in a confirmatory or competing models strategy as samples become larger [47]. Large samples can be considered as consisting of more than 500 respondents. One key advantage to RMSEA is that a confidence interval can be constructed giving the range of RMSEA values for a given level of confidence. Thus, it enables us to report that the RMSEA is between 0.03 and 0.08, for example, with 95% confidence.

ROOT MEAN SQUARE RESIDUAL (RMR) AND STANDARDIZED ROOT MEAN RESIDUAL (SRMR)   As discussed earlier, the error in prediction for each covariance term creates a residual. When covariances are used as input, the residual is stated in terms of covariances, which makes them difficult to interpret because they are impacted by the scale of the indicators. However, standardized residuals (SR) are directly comparable. The average SR value is zero, meaning that both positive and negative residuals can occur. Thus, a predicted covariance lower than the observed value results in a positive residual, whereas a predicted covariance larger than observed results in a negative residual. A common rule is to carefully scrutinize any standardized residual exceeding |4.0| (below −4.0 or above 4.0). Individual SRs enable a researcher to spot potential problems with a measurement model.

Standardized residuals are deviations of individual covariance terms and do not reflect overall model fit. What is needed is an "overall" residual value; two measures have emerged in this regard. First is the root mean square residual (RMR), which is the square root of the mean of these squared residuals; that is, it is an average of the residuals. Yet the RMR has the same problem as residuals in that it is related to the scale of the covariances. An alternative statistic is the standardized root mean residual (SRMR). This standardized value of RMR (i.e., the average standardized residual) is useful for comparing fit across models. Although no statistical threshold level can be established, the researcher can assess the practical significance of the magnitude of the SRMR in light of the research objectives and the observed or actual covariances or correlations [4]. Lower RMR and SRMR values represent better fit and higher values represent worse fits, which puts the RMR, SRMR, and RMSEA into a category of indices sometimes known as **badness-of-fit** measures in which high values are indicative of poor fit. A rule of thumb is that an SRMR over .1 suggests a problem with fit, although there are conditions that make the SRMR inappropriate that are discussed in a later section.

NORMED CHI-SQUARE This GOF measure is a simple ratio of χ^2 to the degrees of freedom for a model. Generally, χ^2:df ratios on the order of 3:1 or less are associated with better-fitting models, except in circumstances with larger samples (greater than 750) or other extenuating circumstances, such as a high degree of model complexity. It is widely used because if it is not provided directly by the software program, it is easily calculated from the model results.

OTHER ABSOLUTE INDICES Most SEM programs today provide the user with many different fit indices. In the preceding discussion, we focused more closely on those that are most widely used. But this is by no means an exhaustive list. For more information, the reader can refer to an extended discussion of these measures on the text's Web sites (www.pearsonglobaleditions.com/hair or www.mvstats.com) as well as the documentation associated with the specific SEM program used.

Incremental Fit Indices

Incremental fit indices differ from absolute fit indices in that they assess how well the estimated model fits relative to some alternative baseline model. The most common baseline model is referred to as a **null model,** one that assumes all observed variables are uncorrelated. It implies that no model specification could possibly improve the model because it contains no multi-item factors (see Chapter 3) or relationships between them. This class of fit indices represents the improvement in fit by the specification of related multi-item constructs.

Most SEM programs provide multiple incremental fit indices as standard output. Different programs provide different fit statistics, however, so you may not find all of these in a particular SEM output. Also, they are sometimes referred to as *comparative fit indices* for obvious reasons. The following are some of the most widely used incremental fit measures, but the TLI and CFI are the most widely reported.

NORMED FIT INDEX (NFI) The normed fit index (NFI) is one of the original incremental fit indices. It is a ratio of the difference in the χ^2 value for the fitted model and a null model divided by the χ^2 value for the null model. It ranges between 0 and 1, and a model with perfect fit would produce an NFI of 1. One disadvantage is models that are more complex will necessarily have higher index values and artificially inflate the estimate of model fit. As a result, it is used less today in relation to either of the following incremental fit measures.

TUCKER-LEWIS INDEX (TLI) The Tucker-Lewis index (TLI) is conceptually similar to the NFI, but it varies in that it is actually a comparison of the normed chi-square values for the null and specified model, which to some degree takes into account model complexity. However, the TLI is not normed, and thus its values can fall below 0 or above 1. Typically though, models with good fit have values that approach 1, and a model with a higher value suggests a better fit than a model with a lower value.

COMPARATIVE FIT INDEX (CFI) The comparative fit index (CFI) is an incremental fit index that is an improved version of the normed fit index (NFI) [5, 7, 25]. The CFI is normed so that values range between 0 and 1, with higher values indicating better fit. Because the CFI has many desirable properties, including its relative, but not complete, insensitivity to model complexity, it is among the most widely used indices. CFI values above .90 are *usually* associated with a model that fits well.

RELATIVE NONCENTRALITY INDEX (RNI)   The RNI also compares the observed fit resulting from testing a specified model to that of a null model. Like the other incremental fit indices, higher values represent better fit and the possible values generally range between 0 and 1. RNIs lower than .90 are *usually* not associated with good fit.

Parsimony Fit Indices

The third group of indices is designed specifically to provide information about which model among a set of competing models is best, considering its fit relative to its complexity. A **parsimony fit** measure is improved either by a better fit or by a simpler model. In this case, a simpler model is one with fewer estimated parameters paths. The parsimony ratio is the basis for these measures and is calculated as the ratio of degrees of freedom used by a model to the total degrees of freedom available [37].

Parsimony fit indices are conceptually similar to the notion of an adjusted R^2 (discussed in Chapter 4) in the sense that they relate model fit to model complexity. More complex models are expected to fit the data better, so fit measures must be relative to model complexity before comparisons between models can be made. The indices are not useful in assessing the fit of a single model, but are quite useful in comparing the fit of two models, one more complex than the other.

The use of parsimony fit indices remains somewhat controversial. Some researchers argue that a comparison of competing models' incremental fit indices provides similar evidence and that we can take parsimony into account further in some other way. It is clear to say that a parsimony index can provide useful information in evaluating competing models, but that it should not be relied upon alone. In theory, parsimony indices are a good idea. In practice, they tend to favor more parsimonious models to a large extent. When used, the PNFI is the most widely applied parsimony fit index.

ADJUSTED GOODNESS OF FIT INDEX (AGFI)   An adjusted goodness-of-fit index (AGFI) tries to take into account differing degrees of model complexity. It does so by adjusting GFI by a ratio of the degrees of freedom used in a model to the total degrees of freedom available. The AGFI penalizes more complex models and favors those with a minimum number of free paths. AGFI values are typically lower than GFI values in proportion to model complexity. No statistical test is associated with AGFI, only guidelines to fit [53]. As with the GFI, however, the AGFI is less frequently used in favor of the other indices that are not as affected by sample size and model complexity.

PARSIMONY NORMED FIT INDEX (PNFI)   The parsimony normed fit index (PNFI) adjusts the normed fit index (NFI) by multiplying it times the PR [40]. Relatively high values represent relatively better fit, so it can be used in the same way as the NFI. The PNFI takes on some of the added characteristics of incremental fit indices relative to absolute fit indices in addition to favoring less complex models. Once again, the values of the PNFI are meant to be used in comparing one model to another with the highest PNFI value being most supported with respect to the criteria captured by this index.

Problems Associated with Using Fit Indices

Ultimately, fit indices are used to establish the acceptability of any SEM model. Probably no SEM topic is more debated than what constitutes an adequate or good fit. The expanding collection of fit indices and the lack of consistent guidelines can tempt the researcher to "pick and choose" an index that provides the best fit evidence in one specific analysis and a different index in another analysis. The researcher is faced with two basic questions in selecting a measure of model fit:

1. What are the best fit indices to objectively reflect a model's fit?
2. What are objective cutoff values suggesting good model fit for a given fit index?

Unfortunately, the answer to both questions is neither simple nor straightforward. Some researchers equate the search for answers to these questions with the "mythical Golden Fleece, the search for the fountain of youth, and the quest for absolute truth and beauty" [38]. Indeed, many problems are associated with the pursuit of good fit. Following is a brief summary of the major issues found in various fit indices.

PROBLEMS WITH THE χ^2 TEST   Perhaps the most clear and convincing evidence that a model's fit is adequate would be a χ^2 value with a p-value indicating no significant difference between the observed and estimated covariance matrices. For example, if a researcher used an error rate of 5 percent, then a p-value greater than .05 would suggest that the researcher's model capably reproduced the observed variables' covariance matrix—a "good" model fit.

However, as we have discussed, so many factors impact the χ^2 significance test that practically any result can be questioned. Does a nonsignificant χ^2 value always enable a researcher to say "Case closed, we have good fit"? Not quite! Very simple models with small samples have a bias toward a nonsignificant χ^2 even though they do not meet other standards of validity or appropriateness. Likewise, there are inherent penalties in the χ^2 for larger sample sizes and larger numbers of indicator variables [5]. The result is that typical models are more complex and have sample sizes that make the χ^2 significance test less useful as a GOF measure that always separates good from poor models. Thus, no matter what the χ^2 result, the researcher should always complement it with other GOF indices, but just as important, the χ^2 value itself and the model degrees of freedom should always be reported [22, 49].

CUTOFF VALUES FOR FIT INDICES: THE MAGIC .90, OR IS THAT .95?   Although we know we need to complement the χ^2 with additional fit indices, one question still remains no matter what index is chosen: What is the appropriate cutoff value for that index? For most of the incremental fit statistics, accepting models producing values of .90 became standard practice in the early 1990s. However, the case was made that .90 was too low and could lead to false models being accepted, and by the end of the decade .95 had become the standard for indices such as the TLI and CFI [25]. In general, .95 somehow became the magic number indicating good-fitting models.

However, research has challenged the use of a single cutoff value for GOF indices, finding instead that a series of additional factors can affect the index values associated with acceptable fit. First, research using simulated data (for which the actual fit is known) provides counterarguments to these cutoff values and does not support .90 as a generally acceptable rule of thumb [25]. It demonstrates that at times even an incremental goodness-of-fit index above .90 would still be associated with a severely misspecified model. This suggests that cutoff values should be set higher than .90. Second, research continues to support the notion that model complexity unduly affects GOF indices, even with something as simple as just more indicators per construct [31]. Finally, the true underlying distribution of data can influence fit indices [16]. In particular, as data become less appropriate for the particular estimation technique selected, the ability of fit indices to accurately reflect misspecification can vary. This issue seems to affect incremental fit indices more than absolute fit indices.

What has become clear is that no single "magic" value always distinguishes good models from bad ones. GOF must be interpreted in light of the characteristics of the research. It is interesting to compare these issues in SEM to the general lack of concern for establishing a magic R^2 number in multiple regression. If a magic minimum R^2 value of .5 had ever been imposed, it would be just an arbitrary limit that would exclude potentially meaningful research. So, we should be cautious in adopting one size fits all standards. It is simply not practical to apply a single set of cutoff rules that apply for all SEM models of any type.

Unacceptable Model Specification to Achieve Fit

Researchers sometimes test theory and sometimes pursue a good fit. The desire to achieve good fit should never compromise the theory being tested. Yet in practice, the pursuit of increasing model fit can lead to several poor practices in model specification [31, 33, 39]. In each of the following instances, a researcher may be able to increase fit, but only in a manner that compromises the theory test. Although each of these actions may be required in very specific instances, they should be avoided whenever possible because each has the potential to unduly limit the ability of SEM to provide a true test of a model. Further, researchers learn not only from theory that is confirmed, but from the areas where theoretical expectations are not confirmed [22].

One area of poor practices involves the number of items per construct. A common mistake is to reduce the number of items per construct to only two or three. While doing so may improve model fit by reducing the total number of indicators and even improve the reliability of the construct, it very likely diminishes its theoretical domain and ultimately its validity. The concept of multiple measures was to include as wide a range as possible of items that could measure the construct, not limit it to a very small subset of these items. An even more extreme action is to use a single item to represent a construct, necessitating an arbitrary specification of measurement error. Here the researcher circumvents the objective of the measurement model by providing the values for the indicator. Single items should only be used when the construct truly is and can be measured by a single item (e.g., a binary variable such as purchase/no purchase, succeed/fail or yes/no). Finally, a test of a measurement model should be performed with the full set of items. The parceling of items, where the full set of indicator variables (e.g., 15 indicators for a construct) is parceled into a small number of composite indicators (e.g., three composites of five items each), can reduce model complexity but may obscure the qualities of individual items. Thus, if parceling of items is performed, it should be employed after the entire set has been evaluated.

Another poor practice is to assess measurement model fit through a separate analysis for each construct instead of one analysis for the entire model. This is an inappropriate use of the GOF indices, which are designed for testing the entire model, not a single construct at a time. The result is not only an incomplete test of the overall model, but a bias toward confirming models because it is easier for single constructs to each meet the fit indices than it is for the entire set to achieve acceptable fit. Moreover, tests of discriminant validity and potential item cross-loadings (see Chapter 12) are impossible unless all of the constructs are tested collectively.

Finally, most model fit indices can be improved by reducing the sample size. This approach, while improving model fit, obviously runs counter to the need for use of as large a sample as possible or feasible to ensure representativeness and generalizability. Moreover, it increases the chances for encountering statistical problems with model convergence, less accurate parameter estimates, and lower statistical power.

Awareness of the problems resulting from these actions does not mean that one of them might not be necessary to address a particular model specification, or to be helpful diagnostically in building a model. Still, improvement in fit is not an appropriate justification for any of these steps. Always remember that these procedures can interfere with the overall test of a measurement model, and thus the measurement theory remains untested until all measured variables are included in a single test.

Guidelines for Establishing Acceptable and Unacceptable Fit

A simple rule for index values that distinguishes good models from poor ones across all situations cannot be offered. It cannot be overemphasized that these are *guides for usage, not rules that guarantee a correct model.* Thus, no specific value on any index can separate models into acceptable and unacceptable fits. However, several general guidelines used together can assist in determining the acceptability of fit for a given model.

USE MULTIPLE INDICES OF DIFFERING TYPES Typically, using three to four fit indices provides adequate evidence of model fit. Current research suggests a fairly common set of indices perform adequately across a wide range of situations and the researcher need not report all GOF indices because they are often redundant. However, the researcher should report at least one incremental index and one absolute index, in addition to the χ^2 value and the associated degrees of freedom because using a single GOF index, even with a relatively high cutoff value, is no better than simply using the χ^2 GOF test alone [38]. Thus, reporting the χ^2 value and degrees of freedom, the CFI or TLI, and the RMSEA will usually provide sufficient unique information to evaluate a model. The SRMR can replace RMSEA to also represent badness of fit, whereas the others represent goodness of fit. When comparing models of varying complexity, the researcher may also wish to add the PNFI.

ADJUST THE INDEX CUTOFF VALUES BASED ON MODEL CHARACTERISTICS Table 11-4 provides some guidelines for using fit indices in different situations. The guidelines are based primarily on simulation research that considers different sample sizes, model complexity, and degrees of error in model specification to examine how accurately various fit indices perform [25, 38]. One key point across the results is that *simpler models* and *smaller samples* should be subject to *more strict evaluation*

TABLE 11-4 **Characteristics of Different Fit Indices Demonstrating Goodness-of-Fit Across Different Model Situations**

No. of Stat. vars. (*m*)	N < 250			N > 250		
	m ≤ 12	**12 < m < 30**	**m ≥ 30**	**m < 12**	**12 < m < 30**	**m ≥ 30**
χ^2	Insignificant p-values expected	Significant p-values even with good fit	Significant p-values expected	Insignificant p-values even with good fit	Significant p-values expected	Significant p-values expected
CFI or TLI	.97 or better	.95 or better	Above .92	.95 or better	Above .92	Above .90
RNI	May not diagnose misspecification well	.95 or better	Above .92	.95 or better, not used with N > 1,000	Above .92, not used with N > 1,000	Above .90, not used with N > 1,000
SRMR	Biased upward, use other indices	.08 or less (with CFI of .95 or higher)	Less than .09 (with CFI above .92)	Biased upward; use other indices	.08 or less (with CFI above .92)	.08 or less (with CFI above .92)
RMSEA	Values < .08 with CFI = .97 or higher	Values < .08 with CFI of .95 or higher	Values < .08 with CFI above .92	Values < .07 with CFI of .97 or higher	Values < .07 with CFI of .92 or higher	Values < .07 with CFI of .90 or higher

Note: m = number of observed variables; *N* applies to number of observations per group when applying CFA to multiple groups at the same time.

than are more *complex models* with *larger samples*. Likewise, more *complex models* with *smaller samples* may require *somewhat less strict* criteria for evaluation with the multiple fit indices [50].

For example, based on a sample of 100 respondents and a four-construct model with only 12 total indicator variables, evidence of good fit would include an insignificant χ^2 value, a CFI of at least .97, and a RMSEA of .08 or lower. It is extremely unrealistic, however, to apply the same criteria to an eight-construct model with 50 indicator variables tested with a sample of 2,000 respondents.

It is important to remember that Table 11-4 is provided more to give the researcher an idea of how fit indices can be used than to suggest absolute rules for standards separating good and bad fit. Moreover, it is worth repeating that even a model with a good fit must still meet the other criteria for validity discussed in the next chapter.

COMPARE MODELS WHENEVER POSSIBLE   Although it is difficult to determine absolutely when a model is good or bad, it is much easier to determine that one model is better than another. The indices in Table 11-4 perform well in distinguishing the relative superiority of one model compared to another. A CFI of .95, for instance, indicates truly better fit than a similar model with a CFI of .85. A more in-depth discussion of competing models is described in stage 6.

THE PURSUIT OF BETTER FIT AT THE EXPENSE OF TESTING A TRUE MODEL IS NOT A GOOD TRADE-OFF   Many model specifications can influence model fit, so the researcher should be sure that all model specifications should be done to best approximate the theory to be tested rather than hopefully increase model fit.

STAGE 5: SPECIFYING THE STRUCTURAL MODEL

Specifying the measurement model (i.e., assigning indicator variables to the constructs they should represent) is a critical step in developing a SEM model. This activity is accomplished in stage 2. Stage 5 involves specifying the structural model by assigning relationships from one construct to another based on the proposed theoretical model. Structural model specification focuses on using the dependence relationship type from Figure 11-1c to represent structural hypotheses of the researcher's model. In other words, what dependence relationships exist among constructs? Each hypothesis represents a specific relationship that must be specified.

We return to the Job Search model from earlier in the chapter. We can specify the full measurement model as shown in Figure 11-9, where there were no structural relationships among the constructs. All constructs were considered exogenous and correlated. This is also known as a confirmatory factor analysis (CFA) model.

In specifying a structural model, the researcher now carefully selects what are believed to be the key factors that influence Job Search. From their experience and judgment, the HBAT research team believes there is a strong reason to suspect that perceptions of supervision, work environment, and coworkers affect job satisfaction, which in turn affects job search. Based on theory the research team proposes the following structural relationships:

H_1:	Supervision perceptions are positively related to Job Satisfaction.
H_2:	Work Environment perceptions are positively related to customer share.
H_3:	Coworkers perceptions are positively related to customer share.
H_4:	Job Satisfaction is negatively related to Job Search.

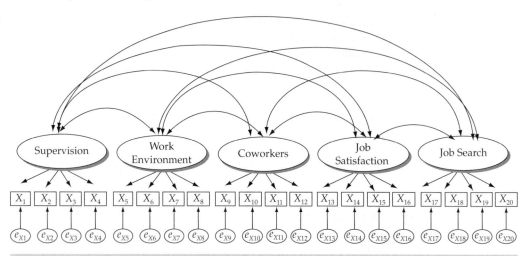

FIGURE 11-9   **A Path Diagram Showing Hypothesized Measurement Model Specification (CFA Model)**

These structural relationships are shown in Figure 11-10. H_1 is specified with the arrow connecting supervision and job satisfaction. In a similar manner, H_2, H_3, and H_4 are specified. The single-headed arrows showing the dependence relationship between constructs represents the structural part of the model. The constructs display the specified measurement structure (links to indicator variables) that would have already been tested in the confirmatory factor analysis stage.

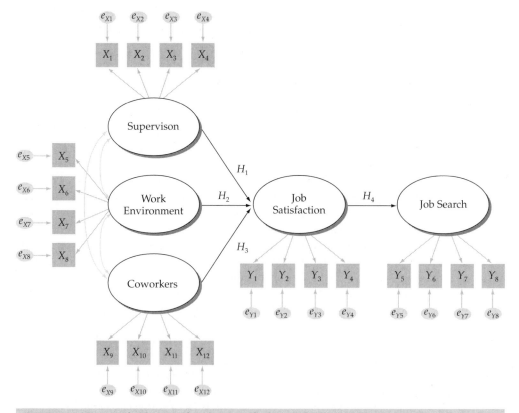

FIGURE 11-10   **A Path Diagram Showing Specified Hypothesized Structural Relationships and Measurement Specification**

Any relationships among exogenous constructs are accounted for with correlational relationships (curved, two-headed arrows). Thus, the three relationships among the two exogenous constructs are specified just as they were in the measurement model.

Another way to view the structural model is that "constraints" can be added to the measurement model. That is, specific structural paths replace the correlations between constructs for each hypothesized relationship. With the exception of correlational relationships among exogenous constructs, no path is drawn between two constructs unless a direct, dependence relationship is hypothesized. Thus, all relationships not shown in the structural model are "constrained" to be equal to zero.

Although the focus in this stage is on the structural model, estimation of the SEM model requires that the measurement specifications be included as well. In this way, the path diagram represents both the measurement and structural part of SEM in one overall model. Thus, the path diagram in Figure 11-10 shows not only the complete set of constructs and indicators in the measurement model, but also imposes the structural relationships among constructs. The model is now ready for estimation. This becomes the test of the overall theory, including both the measurement relationships of indicators to constructs as well as the hypothesized structural relationships among constructs.

STAGE 6: ASSESSING THE STRUCTURAL MODEL VALIDITY

The final stage involves efforts to test the validity of the structural model and its corresponding hypothesized theoretical relationships (e.g., H_1-H_4 in our simple example). Realize that if the measurement model has not survived its tests of reliability and validity in stage 4, stages 5 and 6 cannot be performed. We would have reached a stopping point and must achieve acceptable results in assessing the measurement model before proceeding. If you do not achieve acceptable fit for the measurement model, model fit will not improve when the structural relationships are specified. Only when the measurement model is validated and achieves acceptable model fit can we turn our attention to a test of the structural relationships.

Two key differences arise in testing the fit of a structural model relative to a measurement model. First, even though acceptable overall model fit must be established, alternative or competing models are encouraged to support a model's superiority. Second, particular emphasis is placed on the estimated parameters for the structural relationships, because they provide direct empirical evidence relating to the hypothesized relationships depicted in the structural model.

Structural Model GOF

The process of establishing the structural model's validity follows the general guidelines outlined in stage 4. The observed data are still represented by the observed sample covariance matrix. It does not and should not change. However, a new SEM estimated covariance matrix is computed, and it is different from that for the measurement model. This difference is a result of the structural relationships in the structural model. Remember that the measurement model assumes all constructs are correlated with one another (correlational relationships). Yet in a structural model, the relationships between some constructs are assumed to be zero. Therefore, for almost all conventional SEM models the χ^2 GOF for the measurement model will be less than the χ^2 GOF for the structural model.

The overall fit can be assessed using the same criteria as the measurement model: using the χ^2 value for the structural model and at least one absolute index and one incremental index. These measures establish the validity of the structural model, but comparisons between the overall fit should also be made with the measurement model. Generally, the closer the structural model GOF comes to the measurement model, the better the structural model fit because the measurement model fit provides an upper bound to the GOF of a conventional structural model.

Competitive Fit

Earlier a competing models assessment was discussed as one approach to SEM. The primary objective is to ensure that the proposed model not only has acceptable model fit, but that it performs better than some alternative model. If not, then the alternative theoretical model is supported. Comparing models can be accomplished by assessing differences in incremental or parsimony fit indices along with differences in χ^2 GOF values for each model.

COMPARING NESTED MODELS A powerful test of alternative models is to compare models of similar complexity, yet representing varying theoretical relationships. A common approach is through **nested models,** where a model is nested within another model if it contains the same number of variables and can be formed from the other model by altering the relationships, such as either adding or deleting paths. Generally, competing nested SEM models are compared based on a **chi-square (χ^2) difference statistic ($\Delta\chi^2$).** The χ^2 value from some baseline model (B) is subtracted from the χ^2 value of a lesser constrained, alternative nested model (A). Similarly, the difference in degrees of freedom is found, with one less degree of freedom for each additional path that is estimated. The following equation is used for computation:

$$\Delta\chi^2_{\Delta df} = \chi^2_{df(B)} - \chi^2_{df(A)}$$
$$\Delta df = df(B) - df(A)$$

Because the difference of two χ^2 values is itself χ^2 distributed, we can test for statistical significance given a $\Delta\chi^2$ difference value and the difference in degrees of freedom (Δdf). For example, for a model with one degree of freedom difference ($\Delta df = 1$, meaning one additional path in model A) a $\Delta\chi^2$ of 3.84 or better would be significant at the .05 level. The researcher would conclude that the model with one additional path provides a better fit based on the significant reduction in the χ^2 GOF. Nested models can also be formed by deleting a path(s), with the same process followed in calculating differences in χ^2 and degrees of freedom.

An example of a nested model in Figure 11-10 might be the addition of a structural path from the Supervision construct directly to the Job Search construct. This added path would reduce the degrees of freedom by one. The new model would be re-estimated and the $\Delta\chi^2$ calculated. If it is larger than 3.84, then the researcher would conclude that the alternative model was a significantly better fit. Before the path is added, however, there must be theoretical support for the new relationship.

Comparison to the Measurement Model

Users should know how to perform this chi-square difference test because the SEM program is not likely to provide exactly the test needed in every case. One useful comparison of models is between the CFA and the structural model fit. The structural model is composed of theoretical networks of relationships among constructs. In a conventional CFA (as will be detailed in the next chapter), all constructs are assumed to be related to all other constructs. As discussed earlier, a structural model will generally specify fewer relationships among constructs because not every construct will be hypothesized to have a direct relationship with every other construct. In this sense, a structural model is more constrained than a measurement model because more relationships are fixed to zero and not allowed to be estimated. A way to think of this is that a structural model is formed from a measurement model by adding constraints. Adding a constraint cannot reduce the chi-square value. At best, if a relationship between constructs truly is zero, and the researcher constrains that relationship to zero by not specifying it in a structural model, the actual chi-square value will be unchanged by adding the constraint. When the two constructs truly are related, then adding the constraint will increase the actual chi-square value. Conversely, relaxing a constraint by including a relationship in the model should reduce the chi-square value or keep it the same.

As just described, adding or deleting paths (i.e., adding a path means a constraint has been relaxed and deleting a path means a constraint has been added) changes the degrees of freedom accordingly. Adding one constraint means the chi-square difference test will have one degree of freedom, adding two means the test will have two degrees of freedom, and so forth. When a measurement model and a structural model have approximately the same chi-square value, this means that the constraints added to form the structural model have not significantly added to the χ^2 value.

The overall χ^2 GOF for the Job Search example measurement model can be compared to the overall χ^2 GOF for the example's structural model shown in Figure 11-10. A $\Delta\chi^2$ test can be used to compare these two models. The test would have $\Delta df = 3$ because three relationships that would be estimated in a CFA (measurement model test) are constrained to zero (i.e., not modeled) in the structural model. Specifically, no direct relationship from any exogenous construct (Supervision, Work Environment, or Coworkers) to the rightmost endogenous construct (Job Search) is modeled in this researcher's theory. If the $\Delta\chi^2$ test with three degrees of freedom is insignificant, it would mean that constraining the measurement model (which includes all interconstruct covariances) by not allowing these three direct relationships did not significantly worsen fit. An insignificant $\Delta\chi^2$ test between a measurement model and a structural model would generally provide supporting evidence for the proposed theoretical model.

EQUIVALENT MODELS   Good fit statistics do not prove that a theory is the best way to explain the observed sample covariance matrix. As described earlier, any number of equivalent models exists for any proposed model and they will produce the same estimated covariance matrix. Therefore, any given model, even with good fit, is only one potential explanation. Other model specifications can fit equally well. This means that good empirical fit does not prove that a given model is the "only" true structure. Favorable fit statistics are highly desirable, but many alternative models can provide an equivalent fit [46].

This issue further reinforces the need for building measurement models based on solid theory. More complex models may have a quite large number of equivalent models. Yet it is quite possible that many or all make little sense given the conceptual nature of the constructs involved. Thus, in the end empirical results provide some evidence of validity, but the researcher must provide theoretical evidence that is equally important in validating a model.

Testing Structural Relationships

Good model fit alone is insufficient to support a proposed structural theory. The researcher also must examine the individual parameter estimates that represent each specific hypothesis. A theoretical model is considered valid to the extent that the parameter estimates are:

1. *Statistically significant and in the predicted direction.* That is, they are greater than zero for a positive relationship and less than zero for a negative relationship.
2. *Nontrivial.* This characteristic should be checked using the completely standardized loading estimates. The guideline here is the same as in other multivariate techniques.

Therefore, the structural model shown in Figure 11-10 is considered acceptable only when it demonstrates acceptable model fit *and* the path estimates representing each of the four hypotheses are significant and in the predicted direction. The researcher also can examine the variance-explained estimates for the endogenous constructs, analogous to the analysis of R^2 performed in multiple regression. More detail will be provided about procedures used in this stage in Chapter 12, including discussions on diagnostic measures for both the measurement and structural models.

RULES OF THUMB 11-3

SEM Stages 4–6

- As models become more complex, the likelihood of alternative models with equivalent fit increases
- Multiple fit indices should be used to assess a model's goodness-of-fit and should include:
  - The χ^2 value and the associated *df*
  - One absolute fit index (i.e., GFI, RMSEA, or SRMR)
  - One incremental fit index (i.e., CFI or TLI)
  - One goodness-of-fit index (GFI, CFI, TLI, etc.)
  - One badness-of-fit index (RMSEA, SRMR, etc.)
- No single "magic" value for the fit indices separates good from poor models, and it is not practical to apply a single set of cutoff rules to all measurement models and for that matter to all SEM models of any type
- The quality of fit depends heavily on model characteristics including sample size and model complexity:
  - Simple models with small samples should be held to strict fit standards, even an insignificant *p*-value for a simple model may not be meaningful
  - More complex models with larger samples should not be held to the same strict standards, and so when samples are large and the model contains a large number of measured variables and parameter estimates, cutoff values of .95 on key GOF measures are unrealistic

---

Summary

Several key learning objectives were provided for this chapter. These learning objectives together provide a basic overview of SEM. The basic overview should enable a better understanding of the more specific illustrations that follow in the next chapter.

Understand the distinguishing characteristics of SEM. SEM is a flexible approach to examining how things are related to each other. So, SEM applications can appear quite different. However, three key characteristics of SEM are (1) the estimation of multiple and interrelated dependence relationships, (2) an ability to represent unobserved concepts in these relationships and correct for measurement error in the estimation process, and (3) a focus on explaining the covariance among the measured items.

Distinguish between variables and constructs. The models typically tested using SEM involve both a measurement model and a structural model. Most of the multivariate approaches discussed in the previous chapters focused on analyzing variables directly. Variables are the actual items that are measured using a survey, observation, or some other measurement device. Variables are considered observable in the

sense that we can obtain a direct measure of them. Constructs are unobservable or latent factors that are represented by a variate that consists of multiple variables. Simply put, multiple variables come together mathematically to represent a construct. Constructs can be exogenous or endogenous. Exogenous constructs are the latent, multi-item equivalent of independent variables. They are constructs that are determined by factors outside of the model. Endogenous constructs are the latent, multi-item equivalent of dependent variables.

Understand structural equation modeling and how it can be thought of as a combination of familiar multivariate techniques. SEM can be thought of as a combination of factor analysis and multiple regression analysis. The measurement model part is similar to factor analysis in that it also demonstrates how measured variables load on a smaller number of factors (i.e., constructs). Several different regression analogies apply, but key among them is the fact that key outcome or endogenous constructs are predicted using multiple other constructs in the same way that independent variables predicted dependent variables in multiple regression.

Know the basic conditions for causality and how SEM can help establish a cause-and-effect relationship. Theory can be defined as a systematic set of relationships providing a consistent and comprehensive explanation of a phenomenon. SEM has become the most prominent multivariate tool for testing behavioral theory. SEM's history grew from the desire to test causal models. Theoretically, four conditions must be present to establish causality: (1) covariation, (2) temporal sequence, (3) nonspurious association, and (4) theoretical support. SEM can establish evidence of covariation through the tests of relationships represented by a model. SEM cannot, as a rule, demonstrate that cause occurred before the effect, because cross-sectional data are most often used in SEM. SEM models using longitudinal data can help demonstrate temporal sequentiality. Evidence of nonspurious association between a cause and effect can be supplied, at least in part, by SEM. If the addition of other alternative causes does not eliminate the relationship between the cause and effect, then the causal inference becomes stronger. Finally, theoretical support can only be supplied through reason. Empirical findings alone cannot render a relationship sensible. Thus, SEM can be useful in establishing causality, but simply using SEM on any given data does not mean that causal inferences can be established.

Explain the basic types of relationships involved in SEM. The four key theoretical relationship types in a SEM model are described in Figure 11-1, which also shows the conventional graphical representation of each type. The first shows relationships between latent constructs and measured variables. Latent constructs are represented with ovals and measured variables are represented with rectangles. The second shows simple covariation or correlation between constructs. It does not imply any causal sequence and does not distinguish between exogenous and endogenous constructs. These first two relationship types are fundamental in forming a measurement model. The third relationship type shows how an exogenous construct is related to an endogenous construct and can represent a causal inference in which the exogenous construct is a cause and the endogenous construct is an effect. The fourth relationship type shows how one endogenous construct is related to another. It can also represent a causal sequence from one endogenous construct to another.

Understand that the objective of SEM is to explain covariance and how it translates into the fit of a model. SEM is sometimes known as covariance structure analysis. The algorithms that perform SEM estimation have the goal of explaining the observed covariance matrix of variables, **S,** using an estimated covariance matrix, Σ_k, calculated using the regression equations that represent the researcher's model. In other words, SEM is looking for a set of parameter estimates producing estimated covariance values that most closely match observed covariance values. The closer these values come, the better the model's fit. Fit indicates how well a specified model reproduces the covariance matrix among the measured items. The basic SEM fit statistic is the χ^2 statistic. However, its sensitivity to sample size and model complexity brought about the development of many other fit indices. Fit is best assessed using multiple fit indices. It is also important to realize that no magic values determine when a model is proved best on fit. Rather, the model context must be taken into account in assessing fit. Simple models with small samples should be held to different standards than more complex models tested with larger samples.

Know how to represent a model visually using a path diagram. The entire set of relationships that make up a SEM model can be represented visually using a path diagram. Each type of relationship is conventionally represented with a different type of arrow and abbreviated with a different character. Figure 11-10 depicts a path diagram showing both a measurement and a structural model. The inner portion represents the structural model. The outer portion represents the measurement model.

List the six stages of structural equation modeling and understand the role of theory in the process. Figure 11-7 lists the six stages in the SEM process. It begins with choosing the variables that will be measured. It concludes with assessing the overall structural model fit. It should also be emphasized that theory plays a key role in each step of the process. The goal of a SEM is to provide a test of a theory. Thus, without theory, a true SEM test cannot be conducted.

As mentioned previously, this chapter does not include an extended HBAT example. Rather, a new HBAT example will be introduced in the next chapter. That chapter will illustrate the complete use of SEM to test relationships that will help HBAT make key managerial decisions.

Questions

1. What is the difference between a latent construct and a measured variable?
2. What are the distinguishing characteristics of SEM?
3. Describe conceptually how the estimated covariance matrix in a SEM analysis (Σ_k) can be computed. Why do we compare it to **S**?
4. How is structural equation modeling similar to the other multivariate techniques discussed in the earlier chapters?
5. What is a theory? How is a theory represented in a SEM framework?
6. What is a spurious correlation? How might it be revealed using SEM?
7. What is fit?
8. What is the difference between an absolute and a relative fit index?
9. How does sample size affect structural equation modeling?
10. Why are no magic values available to distinguish good fit from poor fit across all situations?
11. Draw a path diagram with two exogenous constructs and one endogenous construct. The exogenous constructs are each measured by five items and the endogenous construct is measured by four items. Both exogenous constructs are expected to be related negatively to the endogenous construct.

Suggested Readings

A list of suggested readings illustrating issues and applications of multivariate techniques in general is available on the Web at www.pearsonglobaleditions.com/hair or www.mvstats.com.

ESTIMATING RELATIONSHIPS USING PATH ANALYSIS

What was the purpose of developing the path diagram? Path diagrams are the basis for path analysis, the procedure for empirical estimation of the strength of each relationship (path) depicted in the path diagram. Path analysis calculates the strength of the relationships using only a correlation or covariance matrix as input. We will describe the basic process in the following section, using a simple example to illustrate how the estimates are actually computed.

IDENTIFYING PATHS

The first step is to identify all relationships that connect any two constructs. Path analysis enables us to decompose the simple (bivariate) correlation between any two variables into the sum of the compound paths connecting these points. The number and types of compound paths between any two variables are strictly a function of the model proposed by the researcher.

A compound path is a path along the arrows of a path diagram that follow three rules:

1. After going forward on an arrow, the path cannot go backward again; but the path can go backward as many times as necessary before going forward.
2. The path cannot go through the same variable more than once.
3. The path can include only one curved arrow (correlated variable pair).

When applying these rules, each path or arrow represents a path. If only one arrow links two constructs (path analysis can also be conducted with variables), then the relationship between those two is equal to the parameter estimate between those two constructs. For now, this relationship can be called a direct relationship. If there are multiple arrows linking one construct to another as in X → Y → Z, then the effect of X on Z is equal to the product of the parameter estimates for each arrow and is termed an indirect relatonship. This concept may seem quite complicated but an example makes it easy to follow:

Figure 11A.1 portrays a simple model with two exogenous constructs (X_1 and X_2) causally related to the endogenous construct (Y_1). The correlational path A is X_1 correlated with X_2, path B is the effect of X_1 predicting Y_1, and path C shows the effect of X_2 predicting Y_1. The value for Y_1 can be stated simply with a regression-like equation:

$$Y_1 = b_1 X_1 + b_2 X_2$$

We can now identify the direct and indirect paths in our model. For ease in referring to the paths, the causal paths are labeled A, B, and C.

Direct Paths	Indirect Paths
A = X_1 to X_2	
B = X_1 to Y_1	AC = X_1 to Y_1
C = X_2 to Y_1	AB = X_2 to Y_1

ESTIMATING THE RELATIONSHIP

With the direct and indirect paths now defined, we can represent the correlation between each construct as the sum of the direct and indirect paths.

The three unique correlations among the constructs can be shown to be composed of direct and indirect paths as follows:

$$\text{Corr}_{X1X2} = A$$
$$\text{Corr}_{X1Y1} = B + AC$$
$$\text{Corr}_{X2Y1} = C + AB$$

First, the correlation of X_1 and X_2 is simply equal to A. The correlation of X_1 and Y_1 ($\text{Corr}_{X1,Y1}$) can be represented as two paths: B and AC. The symbol B represents the direct path from X_1 to Y_1, and the other path (a compound path) follows the curved arrow from X_1 to X_2 and then to Y_1. Likewise, the correlation of X_2 and Y_1 can be shown to be composed of two causal paths: C and AB.

Once all the correlations are defined in terms of paths, the values of the observed correlations can be substituted and the equations solved for each separate path. The paths then represent either the causal relationships between constructs (similar to a regression coefficient) or correlational estimates.

Using the correlations as shown in Figure 11A.1, we can solve the equations for each correlation (see

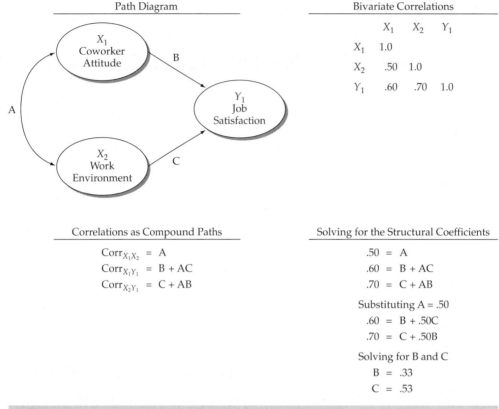

Path Diagram

Bivariate Correlations

	X_1	X_2	Y_1
X_1	1.0		
X_2	.50	1.0	
Y_1	.60	.70	1.0

Correlations as Compound Paths

$$\text{Corr}_{X_1 X_2} = A$$
$$\text{Corr}_{X_1 Y_1} = B + AC$$
$$\text{Corr}_{X_2 Y_1} = C + AB$$

Solving for the Structural Coefficients

$$.50 = A$$
$$.60 = B + AC$$
$$.70 = C + AB$$

Substituting A = .50

$$.60 = B + .50C$$
$$.70 = C + .50B$$

Solving for B and C

$$B = .33$$
$$C = .53$$

FIGURE 11A-1   Calculating Structural Coefficients with Path Analysis

Figure 11A-1) and estimate the causal relationships represented by the coefficients b_1 and b_2.

We know that A equals .50, so we can substitute this value into the other equations. By solving these two equations, we get values of B(b_1) = .33 and C(b_2) = .53. The actual calculations are shown in Figure 11A-1. This approach enables path analysis to solve for any causal relationship based only on the correlations among the constructs and the specified causal model.

As you can see from this simple example, if we change the path model in some way, the causal relationships will change as well. Such a change provides the basis for modifying the model to achieve better fit, if theoretically justified.

With these simple rules, the larger model can now be modeled simultaneously, using correlations or covariances as the input data. We should note that when used in a larger model, we can solve for any number of interrelated equations. Thus, dependent variables in one relationship can easily be independent variables in another relationship. No matter how large the path diagram gets or how many relationships are included, path analysis provides a way to analyze the set of relationships.

SEM ABBREVIATIONS

The following guide will aid in the pronunciation and understanding of common SEM abbreviations. SEM terminology often is abbreviated with a combination of Greek characters and Roman characters to help distinguish different parts of a SEM model.

Symbol	Pronunciation	Meaning
ξ	xi (KSI or KZI)	A construct associated with measured X variables
λ_x	lambda "x"	A path representing the factor loading between a latent construct and a measured x variable
λ_y	lambda "y"	A path representing the factor loading between a latent construct and a measured y variable
Λ	capital lambda	A way of referring to a set of loading estimates represented in a matrix where rows represent measured variables and columns represent latent constructs
η	eta ("eight-ta")	A construct associated with measured Y variables
ϕ	phi (fi)	A path represented by an arced two- headed arrow representing the covariation between one ξ and another ξ
Φ	capital phi	A way of referring to the covariance or correlation matrix between a set of ξ constructs
γ	gamma	A path representing a causal relationship (regression coefficient) from a ξ to an η
Γ	capital gamma	A way of referring to the entire set of γ relationships for a given model
β	beta ("bay-ta")	A path representing a causal relationship (regression coefficient) from one η construct to another η construct
B	capital beta	A way of referring to the entire set of β relationships for a given model
δ	delta	The error term associated with an estimated, measured x variable
θ_δ	theta ("they-ta") delta	A way of referring to the residual variances and covariances associated with the x estimates; the error variance items are the diagonal
ε	epsilon	The error term associated with an estimated, measured y variable
θ_ε	theta-epsilon	A way of referring to the residual variances and covariances associated with the y estimates; the error variance items ar the diagonal
ζ	zeta ("zay-ta")	A way of capturing the covariation between η construct errors
τ	tau (rhymes with "now")	The intercept terms for estimating a measured variable
κ	kappa	The intercept terms for estimating a latent construct
χ^2	chi (ki)-squared	The likelihood ratio

DETAIL ON SELECTED GOF INDICES

The chapter describes how researchers developed many different fit indices that represent the GOF of a SEM model in different ways. Here, a bit more detail is provided about some of the key indices in an effort to provide a better understanding of just what information is contained in each.

GOODNESS-OF-FIT INDEX (GFI)

If we think of F_k as the minimum fit function after a SEM model has been estimated using k degrees of freedom $(\mathbf{S} - \Sigma_k)$, and we think of F_0 as the fit function that would result if all parameters were zero (everything is unrelated to each other; no theoretical relationships), then we can define the GFI simply as:

$$\text{GFI} = 1 - \frac{F_k}{F_0}$$

A model that fits well produces a ratio of F_k/F_0 that is quite small. Conversely, a model that does not fit well produces F_k/F_0 that is relatively large because F_k would not differ much from F_0. This ratio works something like the ratio of SSE/SST discussed in Chapter 4. In the extreme, if a model failed to explain any true covariance between measured variables, F_k/F_0 would be 1, meaning the GFI would be 0.

ROOT MEAN SQUARED ERROR OF APPROXIMATION (RMSEA)

Computation of RMSEA is rather straightforward and provided here to demonstrate how statistics try to correct for the problems of using the χ^2 statistic alone.

$$\text{RMSEA} = \sqrt{\frac{(\chi^2 - df_k)}{(N-1)}}$$

Note that the df are subtracted from the numerator in an effort to capture model complexity. The sample size is used in the denominator to take it into account. To avoid negative RMSEA values, the numerator is set to 0 if df_k exceeds χ^2.

COMPARATIVE FIT INDEX (CFI)

The general computational form of the CFI is:

$$\text{CFI} = 1 - \frac{(\chi_k^2 - df_k)}{(\chi_N^2 - df_N)}$$

Here, the subscript k represents values associated with the researcher's specified model or theory, that is, the resulting fit with k degrees of freedom. The subscript N denotes values associated with the statistical null model. Additionally, the equation is normed to values between 0 and 1—with higher values indicating better fit—by substituting an appropriate value (i.e., 0) if a χ^2 value is less than the corresponding degrees of freedom.

TUCKER-LEWIS INDEX (TLI)

The equation for the TLI is provided here for comparison purposes:

$$\text{TLI} = \frac{\left[\left(\dfrac{\chi_N^2}{df_N}\right) - \left(\dfrac{\chi_k^2}{df_k}\right)\right]}{\left[\left(\dfrac{\chi_N^2}{df_N}\right) - 1\right]}$$

Once again, N and k refer to the null and specified models, respectively. The TLI is not normed and thus its values can fall below 0 or above 1. It produces values similar to the CFI in most situations.

PARSIMONY RATIO (PR)

The parsimony ratio (PR) forms the basis for parsimony GOF measures [31]:

$$\text{PR} = \frac{df_k}{df_t}$$

As can be seen by the formula, it is the ratio of degrees of freedom used by a model to the total degrees of freedom available. Thus, other indices are adjusted by PR to form parsimony fit indices. Although parsimony fit indices can be useful, they tend to strongly favor the more parsimonious measures. These measures have existed for quite some time but are still not widely applied.

References

1. Allison, P. D. 2003. Missing Data Techniques for Structural Equations Models. *Journal of Abnormal Psychology* 112 (November): 545–56.

2. Arbuckle, J. L. (1996). Full information estimation in the presence of incomplete data. In G. A. Marcoulides and R. E. Schumacker (Eds.). *Advanced Structural Equation Modeling.* Mahwah, NJ: Lawrence Erlbaum Publishers.

3. Arbuckle, J. L. (2006). *Amos 7.0 User's Guide.* Chicago, IL: SPSS Inc.

4. Bagozzi, R. P., and Y. Yi. 1988. On the Use of Structural Equation Models in Experimental Designs. *Journal of Marketing Research* 26 (August): 271–84.

5. Bentler, P. M. 1990. Comparative Fit Indexes in Structural Models. *Psychological Bulletin* 107: 238–46.

6. Bentler, P. M. 1992. *EQS: Structural Equations Program Manual.* Los Angeles: BMDP Statistical Software.

7. Bentler, P. M., and D. G. Bonnett. 1980. Significance Tests and Goodness of Fit in the Analysis of Covariance Structures. *Psychological Bulletin* 88: 588–606.

8. Blalock, H. M. 1962. Four-Variable Causal Models and Partial Correlations. *American Journal of Sociology* 68: 182–94.

9. Byrne, B. 1998. *Structural Equation Modeling with LISREL, PRELIS and SIMPLIS: Basic Concepts, Applications and Programming.* Mahwah, NJ: Lawrence Erlbaum Associates.

10. Churchill, G. A. 1979. A Paradigm for Developing Better Measures of Marketing Constructs. *Journal of Marketing Research* 16 (February): 64–73.

11. Collins, L. M., J. L. Schafer, and C. M. Kam. 2001. A Comparison of Inclusive and Restrictive Strategies in Modern Missing-Data Procedures. *Psychological Methods* 6: 352–70.

12. Cudeck, R. 1989. Analysis of Correlation Matrices Using Covariance Structure Models. *Psychological Bulletin* 105: 317–27.

13. DeVellis, Robert. 1991. *Scale Development: Theories and Applications.* Thousand Oaks, CA: Sage.

14. Enders, C. K., and D. L. Bandalos. 2001. The Relative Performance of Full Information Maximum Likelihood Estimation for Missing Data in Structural Equation Models. *Structural Equation Modeling* 8(3): 430–59.

15. Enders, C. K., and J. L. Peugh. 2004. Using an EM Covariance Matrix to Estimate Structural Equation Models with Missing Data: Choosing an Adjusted Sample Size to Improve the Accuracy of Inferences. *Structural Equations Modeling* 11(1): 1–19.

16. Fan, X., B. Thompson, and L. Wang. 1999. Effects of Sample Size, Estimation Methods, and Model Specification on Structural Equation Modeling Fit Indexes. *Structural Equation Modeling* 6: 56–83.

17. Feinian, C., P. J. Curran, K. A. Bollen, J. Kirby, and P. Paxton (2008), "An Empirical Evaluation of the Use of Fixed Cutoff Points in RMSEA Test Statistic in Structural Equation Models," *Sociological Methods & Research,* 36 (4), 462–94.

18. Graham, J. W. (2003), "Adding Missing-Data-Relevant Variables to FIML-Based Structural Equation Models," *Structural Equation Modeling,* 10 (1), 80–100.

19. Habelmo, T. 1943. The Statistical Implications of a System of Simultaneous Equations. *Econometrica* 11: 1–12.

20. Hartmann, W. M. (1992), *The CALIS Procedure Extended User's Guide.* Cary, N.C.: SAS Institute.

21. Hayduk, L. A. 1996. *LISREL Issues, Debates and Strategies.* Baltimore: Johns Hopkins University Press.

22. Hayduck, L., G. Cummings, K. Boadu, H. P. Robinson, and S. Boulianne (2007), "Testing! Testing! One, Two, Three—Testing Theory in Structural Equation Models!" *Personality and Individual Differences,* 42: 841–50.

23. Hershberger, S. L. 2003. The Growth of Structural Equation Modeling: 1994–2001. *Structural Equation Modeling* 10(1): 35–46.

24. Hoelter, J. W. 1983. The Analysis of Covariance Structures: Goodness-of-Fit Indices. *Sociological Methods and Research* 11: 324–44.

25. Hu, L., and P. M. Bentler. 1999. Covariance Structure Analysis: Conventional Criteria Versus New Alternatives. *Structural Equations Modeling* 6(1): 1–55.

26. Hunt, S. D. 2002. *Foundations of Marketing Theory: Toward a General Theory of Marketing.* Armonk, NY: M.E. Sharpe.

27. Jöreskog, K. G. 1970. A General Method for Analysis of Covariance Structures. *Biometrika* 57: 239–51.

28. Jöreskog, K. G. 1981. Basic Issues in the Application of LISREL. *Data* 1: 1–6.

29. Jöreskog, K. G., and D. Sörbom. 1976. *LISREL III: Estimation of Linear Structural Equation Systems by Maximum Likelihood Methods.* Chicago: National Educational Resources, Inc.

30. Jöreskog, K. G., and D. Sörbom (1997). *LISREL 8: A Guide to the Program and Applications.* Chicago, IL: SPSS Inc.

31. Kenny, D. A., and D. B. McCoach. 2003. Effect of the Number of Variables on Measures of Fit in Structural Equations Modeling. *Structural Equations Modeling* 10(3): 333–51.

32. Kline, R. B. (1998). Software Programs for Structural Equation Modeling: AMOS, EQS, and LISREL." *Journal of Psychoeducational Assessment* 16: 302–323.

33. Little, T. D., W. A. Cunningham, G. Shahar, and K. F. Widaman. 2002. To Parcel or Not to Parcel: Exploring the Question, Weighing the Merits. *Structural Equation Modeling* 9: 151–73.

34. MacCallum, R. C. 2003. Working with Imperfect Models. *Multivariate Behavioral Research* 38(1): 113–39.

35. MacCallum, R. C., K. F. Widaman, K. J. Preacher, and S. Hong. 2001. Sample Size in Factor Analysis: The Role of Model Error. *Multivariate Behavioral Research* 36(4): 611–37.

36. Maiti, S. S., and B. N. Mukherjee. 1991. Two New Goodness-of-Fit Indices for Covariance Matrices with Linear Structure. *British Journal of Mathematical and Statistical Psychology* 44: 153–80.

37. Marsh, H. W., and J. Balla. 1994. Goodness-of-Fit in CFA: The Effects of Sample Size and Model Parsimony. *Quality & Quantity* 28 (May): 185–217.

38. Marsh, H. W., K. T. Hau, and Z. Wen. 2004. In Search of Golden Rules: Comment on Hypothesis Testing Approaches to Setting Cutoff Values for Fit Indexes and Dangers in Overgeneralizing Hu and Bentler's (1999) Findings. *Structural Equation Modeling* 11(3): 320–41.

39. Marsh, H. W., K. T. Hau, J. R. Balla, and D. Grayson. 1988. Is More Ever Too Much? The Number of Indicators per Factors in Confirmatory Factor Analysis. *Multivariate Behavioral Research* 33: 181–222.

40. Mulaik, S. A., L. R. James, J. Val Alstine, N. Bennett, S. Lind, and C. D. Stilwell. 1989. Evaluation of Goodness-of-Fit Indices for Structural Equations Models. *Psychological Bulletin* 105 (March): 430–45.

41. Muthén, L. K., and B. O. Muthén. 2007. *Mplus User's Guide.* Fifth Edition. Los Angeles, CA: Muthén & Muthén.

42. Netemeyer, R. G., W. O. Bearden, and S. Sharma. 2003. *Scaling Procedures: Issues and Applications.* Thousand Oaks, CA: Sage.

43. Olsson, U. H., T. Foss, and E. Breivik. 2004. Two Equivalent Discrepancy Functions for Maximum Likelihood Estimation: Do Their Test Statistics Follow a Noncentral CM-square Distribution Under Model Misspecification? *Sociological Methods & Research* 32 (May): 453–510.

44. Olsson, U. H., T. Foss, S. V. Troye, and R. D. Howell. 2000. The Performance of ML, GLS and WLS Estimation in Structural Equation Modeling Under Conditions of Misspecification and Nonnormality. *Structural Equations Modeling* 7: 557–95.

45. Pearl, J. 1998. Graphs, Causality and Structural Equation Models. *Sociological Methods & Research* 27 (November): 226–84.

46. Raykov, T., and G. A. Marcoulides. 2001. Can There Be Infinitely Many Models Equivalent to a Given Covariance Structure Model? *Structural Equation Modeling* 8(1); 142–49.

47. Rigdon, E. E. 1996. CFI Versus RMSEA: A Comparison of Two Fit Indices for Structural Equation Modeling. *Structural Equation Modeling* 3(4): 369–79.

48. Rubin, D. B. 1976. Inference and Missing Data. *Psychometrica* 63: 581–92.

49. Savalei, V. (2008). Is the ML Chi-Square Ever Robust to Nonnormality? A Cautionary Note with Missing Data *Structural Equation Modeling* 15 (1): 1–22.

50. Sharma, S., S. S. Mukherjee, A. Kumar, and W. R. Dillon. 2005. A Simulation Study to Investigate the Use of Cutoff Values for Assessing Model Fit in Covariance Structure Models. *Journal of Business Research* 58 (July): 935–43.

51. Sobel, M. E. 1998. Causal Inferences in Statistical Models of the Process of Socioeconomic Achievement. *Sociological Methods & Research* 27 (November): 318–48.

52. Tanaka, J. 1993. Multifaceted Conceptions of Fit in Structural Equation Models. In K. A. Bollen and J. S. Long (eds.), *Testing Structural Equation Models.* Newbury Park, CA: Sage.

53. Tanaka, J. S., and G. J. Huba. 1985. A Fit-Index for Covariance Structure Models Under Arbitrary GLS Estimation. *British Journal of Mathematics and Statistics* 42: 233–39.

54. Wang, L. L., X. Fan, and V. L. Wilson. 1996. Effects of Nonnormal Data on Parameter Estimates for a Model with Latent and Manifest Variables: An Empirical Study. *Structural Equation Modeling* 3(3): 228–47.

55. Wright, S. 1921. Correlation and Causation. *Journal of Agricultural Research* 20: 557–85.

Applications of SEM

LEARNING OBJECTIVES

Upon completing this chapter, you should be able to do the following:

- Distinguish between exploratory factor analysis and confirmatory factor analysis (CFA).
- Assess the construct validity of a measurement model.
- Know how to represent a measurement model using a path diagram.
- Understand the basic principles of statistical identification and know some of the primary causes of CFA identification problems.
- Understand the concept of model fit as it applies to measurement models and be able to assess the fit of a confirmatory factor analysis model.
- Distinguish a measurement model from a structural model.
- Describe the similarities between SEM and other multivariate techniques.
- Depict a model with dependence relationships using a path diagram.
- Test a structural model using SEM.
- Diagnose problems with the SEM results.
- Understand the differences between reflective and formative constructs.
- Be able to specify formative constructs in SEM models.
- Identify the situation in which higher-order factor analysis models are appropriate.
- Know how SEM can be used to compare results between groups.
- Perform an invariance measurement analysis using multigroup methods.
- Understand the concepts of statistical mediation and moderation.
- Appreciate the differences between SEM and PLS approaches.

CHAPTER PREVIEW

The previous chapter introduced the basics of structural equation modeling. It described the two basic submodels in a conventional structural equation model—the measurement and structural models. This chapter addresses each submodel and an additional set of issues in three parts. Part 1 focuses on confirmatory factor analysis (CFA) and how confirmatory processes can test a proposed measurement theory. The measurement theory can be represented with a model that shows how measured variables come together to represent constructs. CFA enables us to test how well the measured variables represent the constructs. The key advantage is that the researcher can analytically test a conceptually grounded theory explaining how different measured items represent important psychological, sociological, or business measures. When CFA results are combined with construct validity tests, researchers can obtain a better understanding of the quality of their measures.

Part 2 of the chapter focuses on the second model: testing the theoretical or structural model where the primary focus shifts to the relationships between latent constructs. With SEM we examine relationships between latent constructs much as we examined the relationships between independent and dependent variables in multiple regression analysis (Chapter 4). Even though we saw that summated factors representing theoretical constructs could be entered as variables in regression models, regression models treated variables and constructs identically. That is, multiple regression did not take into account any of the measurement properties that go along with forming a multiple-item construct when estimating the relationship. SEM provides a better way of empirically examining a theoretical model by involving both the measurement model and the structural model in one analysis. In other words, it takes information about measurement into account in testing the structural model.

Part 3 of the chapter extends the discussion to several more advanced topics faced by many researchers today. We first examine the current debate regarding the use of formative rather than reflective measurement theory. As will be discussed, there are not just estimation issues involved in the use of formative constructs, but there are also questions involving their appropriateness in structural equation modeling. We then discuss the applicability and use of higher-order factor models. In these instances, the latent constructs we estimate with measured variables now act as "indicators" of a higher-order latent construct. So in a sense we represent a latent construct with other latent constructs. We then focus on multigroup models, a form of SEM analysis comparing the same model across groups of respondents. A specific type of multigroup analysis, measurement invariance testing, is discussed in detail regarding its investigation into the measurement model's equivalence across groups. We shift focus to the structural model and initially deal with two new types of relationships—mediation and moderation. Each is discussed in terms of not only how it is estimated in a SEM model, but the underlying logic of each type of relationship so the researcher can be sure to select the appropriate type when needed. The topic discusses an alternative method of estimating path models—PLS, or partial least squares. This approach provides an alternative to SEM when issues arise in the research methodology such as measurement difficulties or sample size limitations. We conclude with an assessment of the strengths and weaknesses of the PLS methodology so that researchers can select the method that will work best in their situation.

In each section of the chapter we begin with a brief overview of the general topic area and then focus on the six stages of theoretical model testing introduced in Chapter 11. We employ the HBAT_SEM data set to illustrate the issues in each topic area.

KEY TERMS

Before beginning this chapter, review the key terms to develop an understanding of the concepts and terminology used. Throughout the chapter the key terms appear in **boldface.** Other points of emphasis in the chapter and key term cross-references are *italicized*.

Average variance extracted (AVE)   A summary measure of convergence among a set of items representing a latent construct. It is the average percentage of variation explained (*variance extracted*) among the items of a construct.

Between-construct error covariance   Covariance between two error terms of measured variables indicating different constructs.

Between-group constraints   Fixing a relationship to be equivalent across two or more group models. A single estimate is made for all groups rather than a unique estimate in each group.

Causal model   *Structural model* that infers that relationships have a sequential ordering in which a change in one brings about a change in another.

Chi-square difference($\Delta\chi^2$)   Measure for assessing the statistical significance of the difference in overall model fit between two models based on the chi-square values of each model. A nonsignificant value indicates that the two models provide the same level of model fit and can be considered equivalent in terms of explanation. The degrees of freedom for the chi-square is the difference in the number of estimated parameters in the two models.

Communality   See *variance extracted.*

Complete mediation   See *full mediation.*

Configural invariance   Exists when an acceptable fit is obtained from a *multisample CFA* model that simultaneously estimates a factor solution for all groups with each group configured with the same structure (same pattern of free and fixed parameters). Also see *totally free multiple group model.*

Congeneric measurement model   *Measurement model* consisting of several *unidimensional* constructs with all cross-loadings and *between-* and *within-construct error covariances* appropriately fixed at zero.

Constant methods bias   Covariance among measured variables is influenced by the data collection method (e.g., same collection method, questionnaire format, or even scale type).

Constraints   Fixing a potential relationship in a SEM model to some specified value (even if fixed to zero) rather than allowing the value to be estimated (free). Also see *between-group constraints.*

Construct reliability (CR)   Measure of reliability and internal consistency of the measured variables representing a latent construct. Must be established before *construct validity* can be assessed.

Construct validity   Extent to which a set of measured variables actually represents the theoretical latent construct those variables are designed to measure.

Convergent validity   Extent to which indicators of a specific construct converge or share a high proportion of variance in common.

Cross-validation   Attempt to reproduce the results found in one sample using data from a different sample, usually drawn from the same population.

Direct effect   Relationship linking two constructs with a single arrow between the two.

Discriminant validity   Extent to which a construct is truly distinct from other constructs.

Error term invariance   Occurs when the error variance terms for each measured variable are equal across the groups being studied. Achieving error term invariance denotes equal reliabilities for the constructs across groups.

Face validity   Extent to which the content of the items is consistent with the construct definition, based solely on the researcher's judgment.

Factor covariance invariance   An intermediate stage of the *measurement invariance* process where the covariances between constructs are the same across groups.

Factor variance invariance   An intermediate stage of the *measurement invariance* process after *metric* and *scalar invariance* where the variances of the constructs are tested for invariance across the groups. It is necessary to compare standardized parameter estimates (e.g., correlations, standardized loading estimates) across groups.

Feedback loop   Relationship when a construct is seen as both a predictor and an outcome of another single construct. Feedback loops can involve either direct or indirect relationships. Also called a *nonrecursive relationship.*

First-order factor model   Covariances between measured variables explained with a single latent factor layer. See also *second-order factor model,* which has two layers of latent factors.

Formative measurement theory   Theory based on the assumptions that (1) the measured variables cause the construct and (2) the error in measurement is an inability to fully explain the construct. The construct is not latent in this case. See also *reflective measurement theory.*

Full invariance   Achieved when the *chi-square difference* test is nonsignificant for the complete set of constraints when testing for *measurement invariance* in *MCFA*. For example, when comparing loading estimates (*metric invariance*), full invariance is supported with a nonsignificant difference between models where all loading estimates have *between-group constraints*. This contrasts to *partial invariance,* where only a subset of loading estimates is constrained.

Full mediation   Relationship between a predictor and an outcome variable becomes nonsignificant after a mediator is entered as an additional predictor.

Heywood case   Factor solution that produces an error variance estimate of less than zero (a negative error variance). SEM programs will usually generate an improper solution when a Heywood case(s) is present.

Identification   Whether enough information exists to identify a solution for a set of structural equations. An identification problem leads to an inability of the proposed model to generate unique estimates and can prevent the SEM program from producing results. The three possible conditions of identification are *overidentified, just-identified,* and *underidentified.*

Indirect effect   Sequence of relationships with at least one intervening construct involved. That is, a sequence of two or more *direct effects* represented visually by multiple arrows between constructs.

Interpretational confounding   Measurement estimates for one construct are significantly affected by relationships other than those among the specific measures. It is indicated when loading estimates vary substantially from one SEM model to another model that is the same except for the change in specification of one or more relationships.

Just-identified   SEM model containing just enough degrees of freedom to estimate all free parameters. Just-identified models have perfect fit by definition, meaning a fit assessment is not meaningful.

Measurement equivalence   See *measurement invariance.*

Measurement invariance   *Measurement theory* condition in which the measures forming a measurement model have the same meaning and are used in the same way by different groups of respondents. Tested through a series of increasingly rigorous *MCFA* models where *between-group constraints* restrict different elements of the measurement model (e.g., *metric invariance* tests for equivalence of the factor loading estimates).

Measurement model   Specification of the *measurement theory* that shows how constructs are *operationalized* by sets of measured variables. The specification is similar to an EFA by factor analysis, but differs in that the number of factors and the items loading on each factor must be known and specified before the analysis can be conducted.

Measurement theory   Series of relationships that suggest how measured variables represent a construct not measured directly (latent). A measurement theory can be represented by a series of regression-like equations mathematically relating a factor (construct) to the measured variables.

Mediating effect   Effect of a third variable/construct intervening between two other related constructs.

Metric invariance   An important stage in the *measurement invariance* process that assesses the extent to which factor loading estimates are equivalent across groups. Metric invariance provides support that respondents use the rating scales similarly across groups so the differences between values can be compared directly.

Moderating effect   Effect of a third variable or construct changing the relationship between two related variables/constructs. That is, the relationship between two variables changes based on the level/amount of a moderator. For example, if a relationship changes significantly when measured for males versus females, then gender moderates the relationship.

Modification index   Amount the overall model χ^2 value would be reduced by freeing any single particular path that is not currently estimated.

Multiple group analysis   A form of SEM analysis where two or more samples of respondents are compared using similar models. *Between-group constraints* are used to assess the similarities between groups on any model parameter(s).

Multisample confirmatory factor analysis (MCFA)   A form of *multiple group analysis* where multiple CFA models, one for each group of respondents, are estimated and then measures of fit calculated for all of the models collectively.

Nomological validity   Test of validity that examines whether the correlations between the constructs in the *measurement theory* make sense. The construct correlations can be useful in this assessment.

Nonrecursive model   Structural model containing *feedback loops.*

Nuisance factor   An external effect to the SEM model that may impact the results in some fashion and thus needs to be accounted for. Examples can be types of questions used in the questionnaire or differing conditions at various times of data collection. For an example, see *constant methods bias.*

Operationalization   Manner in which a construct can be represented. With CFA, a set of measured variables is used to represent a construct.

Order condition   Requirement that the degrees of freedom for a model be greater than zero; that is, the number of unique covariance and variance terms less the number of free parameter estimates must be positive.

Overidentified model   Model that has more unique covariance and variance terms than parameters to be estimated. It has positive degrees of freedom. This is the preferred type of identification for a SEM model.

Parameter   Numerical representation of some characteristic of a population. In SEM, relationships are the characteristic of interest for which the modeling procedures will generate estimates. Parameters are numerical characteristics of the SEM relationships, comparable to regression coefficients in multiple regression.

Partial invariance   When the *chi-square difference* indicates that only a subset of possible *between-groups constraints* (at least two per construct) are nonsignificant when testing *measurement invariance* in *MCFA.* For example, when comparing loading estimates (*metric invariance*), partial invariance is supported with a nonsignificant difference between models where at least two loading estimates per construct have *between-group constraints.* This contrasts to *full invariance,* where all loading estimates are constrained.

Partial least squares (PLS)   Alternative estimation approach to SEM. The constructs are represented as composites based on factor analysis results, with no attempt to re-create covariances among measured items.

Partial mediation   Effect when a relationship between a predictor and an outcome is reduced but remains significant when a mediator is also entered as an additional predictor.

Path estimate   See *structural parameter estimate.*

Post hoc analysis   After-the-fact tests of relationships for which no hypothesis was theorized. In other words, a path is tested where the original theory did not indicate a path.

Rank condition   Requirement that each individual parameter estimated be uniquely, algebraically defined. If you think of a set of equations that could define any dependent variable, the rank condition is violated if any two equations are mathematical duplicates.

Recursive models   Structural models in which paths between constructs all proceed only from the antecedent construct to the consequences (outcome construct). No construct is both a cause and an effect of any other single construct.

Reflective measurement theory   Theory based on the assumptions that (1) latent constructs cause the measured variables and (2) the measurement error results in an inability to fully explain these measures. It is the typical representation for a latent construct. See also *formative measurement theory.*

Residuals   Individual differences between observed covariance terms and the estimated covariance terms.

Saturated structural model   *Recursive* SEM *model* specifying the same number of direct *structural relationships* as the number of possible construct correlations in the CF2A. The fit statistics for a saturated theoretical model should be the same as those obtained for the CFA model.

Scalar invariance   A stage in the *measurement invariance* process following *metric invariance* that assesses the extent to which intercepts (means) of the measured variables are equivalent across groups in a *MCFA*. Scalar invariance supports valid comparison of the latent construct means between groups.

Second-order factor model   *Measurement theory* involving two "layers" of latent constructs. These models introduce a second-order latent factor(s) that causes multiple *first-order latent factors,* which, in turn, cause the measured variables (x).

Specification search   Empirical trial-and-error approach that may lead to sequential changes in the model based on key model diagnostics.

Squared multiple correlations   Values representing the extent to which a measured variable's variance is explained by a latent factor. It is similar to the idea of communality from EFA.

Standardized residuals   *Residuals* divided by the standard error of the residual. Used as a diagnostic measure of model fit.

Structural model   Set of one or more dependence relationships linking the hypothesized model's constructs (i.e., the *structural theory*). The structural model is most useful in representing the interrelationships of variables between constructs.

Structural relationship   Dependence relationship (regression type) specified between any two latent constructs. Structural relationships are represented with a single-headed arrow and suggest that one construct is dependent upon another. Exogenous constructs cannot be dependent on another construct. Endogenous constructs can be dependent on either exogenous or endogenous constructs.

Structural parameter estimate   SEM equivalent of a regression coefficient that measures the linear relationship between a predictor construct and an outcome construct. Also called a *path estimate.*

Structural theory   Conceptual representation of the relationships between constructs.

Tau-equivalence   Assumption that a *measurement model* is *congeneric* and that all factor loadings are equal.

TF   See *totally free multiple group model.*

Three-indicator rule   Assumes a *congeneric measurement model* in which all constructs have at least three indicators; therefore the model is *identified.*

Totally free multiple group model (TF)   Model that uses the same structure (pattern of fixed and free parameters) on all groups in a *multiple group analysis.*

Two-step SEM process   Approach to SEM in which the measurement model fit and construct validity are first assessed using CFA and then the *structural model* is tested, including an assessment of the significance of relationships. The structural model is tested only after adequate measurement and construct validity are established.

Underidentified model   Model with more parameters to be estimated than there are item variance and covariances. The term *unidentified* is used in the same way as *underidentified.*

Unidentified model   See *underidentified model.*

Unidimensional measures   Set of measured variables (indicators) with only one underlying latent construct. That is, the indicator variables load on only one construct.

Unit of analysis   Unit or level to which results apply. In business research, it often deals with the choice of testing relationships between individuals' (people) perceptions versus between organizations.

Variance extracted   Total amount of variance a measured variable has in common with the constructs upon which it loads. Good measurement practice suggests that each measured variable should load on only one construct. So, it can be thought of as the variance explained in a measured variable by the construct. Also referred to as a *communality*.

Within-construct error covariance   Covariance between two error terms of measured variables that are indicators of the same construct.

PART 1: CONFIRMATORY FACTOR ANALYSIS

This chapter begins by providing a description of confirmatory factor analysis (CFA). CFA is a way of testing how well measured variables represent a smaller number of constructs. The chapter illustrates this process by showing how CFA is similar to other multivariate techniques. Then, a simple example is provided. A few key aspects of CFA are discussed prior to describing the CFA stages in more detail and demonstrating CFA with an extended illustration.

CFA and Exploratory Factor Analysis

Chapter 3 described procedures for conducting exploratory factor analysis (EFA). EFA *explores* the data and provides the researcher with information about how many factors are needed to best represent the data. With EFA, all measured variables are related to *every* factor by a factor loading estimate. Simple structure results when each measured variable loads highly on only one factor and has smaller loadings on other factors (i.e., loadings less than .4).

The distinctive feature of EFA is that the factors are derived from statistical results, not from theory. This means the researcher runs the software and lets the underlying pattern of the data determine the factor structure. Thus, EFA is conducted without knowing how many factors really exist (if any) or which variables belong with which constructs. When EFA is applied, the researcher uses established guidelines to determine which variables load on a particular factor and how many factors are appropriate. The factors that emerge can only be named *after* the factor analysis is performed. In this respect, CFA and EFA are not the same. Note that in this chapter the terms *factor* and *construct* are used interchangeably.

In Chapter 3, EFA was conducted on 13 variables from the HBAT data set. Based on the eigenvalues and the pattern of loadings, a four-factor solution was deemed most appropriate. The four factors were named based on the variables loading highly on each factor. Using this process the factors were named (1) Customer Service, (2) Marketing, (3) Technical Support, and (4) Product Value (see Chapter 3 for more details).

CFA is similar to EFA in some respects, but philosophically it is quite different. With CFA, the researcher must specify both the number of factors that exist for a set of variables and which factor each variable will load on *before* results can be computed. Thus, the statistical technique does not assign variables to factors. Instead the researcher makes this assignment based on the theory being tested before any results can be obtained. Moreover, a variable is assigned to only a single factor (construct) and cross-loadings (loading on more than a single factor) are assigned. CFA is then applied to test the extent to which a researcher's *a-priori, theoretical* pattern of factor loadings on prespecified constructs (variables loading on specific constructs) represents the actual data. Thus, instead of allowing the statistical method to determine the number of factors and loadings as in EFA, CFA statistics tell us how well our theoretical specification of the factors matches reality (the actual data). In a sense, CFA is a tool that enables us to either "confirm" or "reject" our preconceived theory.

CFA is used to provide a confirmatory test of our measurement theory. A **measurement theory** specifies how measured variables logically and systematically represent constructs involved in a theoretical model. In other words, measurement theory specifies a series of relationships that suggest how measured variables represent a latent construct that is not measured directly. The measurement theory may then be combined with a structural theory to fully specify a SEM model.

Measurement theory requires that a construct first be defined. Therefore, unlike EFA, with CFA a researcher uses measurement theory to specify *a priori* the number of factors as well as which variables load on those factors. This specification is often referred to as the way the conceptual constructs in a **measurement model** are **operationalized.** CFA cannot be conducted properly without a measurement theory. In EFA, theory is not needed to derive factors nor is the ability to define constructs ahead of time.

A Simple Example of CFA and SEM

We are now going to illustrate a simple CFA using two constructs from the example first introduced in Chapter 11. We will explain how the measurement theory is represented in a path diagram.

Consider a situation where a researcher is interested in studying factors that impact employee job satisfaction. After reviewing the relevant theory, the researcher concludes that two factors have the largest impact Supervisor Support and Work Environment. The measured variables for both factors are evaluated using a 7-point, agree–disagree Likert scale.

The construct Supervisor Support can be defined as what workers think about the management capabilities of their immediate supervisor. It can be represented by the following four items:

- My supervisor recognizes my potential.
- My supervisor helps me resolve problems at work.
- My supervisor understands the challenges of balancing work and home demands.
- My supervisor supports me when I have a problem.

The construct Work Environment can be defined as the aspects of the environment where people work that impact their productivity. It can be represented by the following four measured variables:

- Supervisors and workers have similar values and ideas about what this organization should be doing.
- My organization provides the equipment needed to perform my job well.
- The temperature of my office and other working areas is comfortable.
- The physical arrangement of work areas at my organization helps me to manage my time on the job well.

A Visual Diagram

Measurement theories often are represented using visual diagrams called *path diagrams.* The path diagram shows the linkages between specific measured variables and their associated constructs, along with the relationships among constructs. "Paths" from the latent construct to the measured items (loadings) are based on the measurement theory. When CFA is applied, only the loadings theoretically linking a measured item to its corresponding latent factor are calculated. All others are assumed to be equal to zero. This highlights a key difference between EFA versus CFA in that EFA produces a loading for every variable on every factor, but with CFA there are no cross-loadings.

In CFA, we must specify five elements: the latent constructs, the measured variables, the item loadings on specific constructs, the relationships among constructs, and the error terms for each indicator. First, latent constructs are drawn as ellipses, and the measured variables are represented by rectangles. Because there are only correlational relationships (depicted by two-headed curved arrows) among constructs in a CFA, all constructs are considered exogenous. This also means that indicator variables are denoted by X (e.g., $X_1, X_2, \ldots$). The relationships between the latent constructs and the respective measured variables (called *factor loadings* as in EFA) are represented by arrows from the construct to the measured variable. Finally, each measured indicator variable has an error term (shown as an *e* in our diagram), which is the extent to which the latent factor does not explain the measured variable.

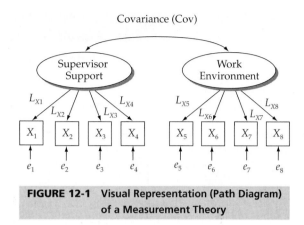

Covariance (Cov)

FIGURE 12-1 **Visual Representation (Path Diagram) of a Measurement Theory**

Figure 12-1 provides a complete specification of the CFA model. There are two latent constructs Supervisor Support and Work Environment. The X_1–X_8 represent the measured indicator variables and the L_{x1}–L_{x8} are the relationships between the latent constructs and the respective measured items (i.e., factor loadings). As you can see, the four items measuring Supervisor Support are linked to that latent construct as are the other four items to the Work Environment construct. The curved arrow between the two constructs denotes a correlational relationship between them. Finally, e_1–e_8 represent the errors associated with each measured item.

All SEM software uses a somewhat unique notation to denote the various elements of a SEM model. Some programs (such as LISREL) rely on a very formal notation based on matrix notation, each represented by Greek characters (e.g., lambda, beta, and gamma). In this text we have developed a simplified notation compatible with all notations. Chapter 11 describes the notation in greater detail. The text's Web sites (www.pearsonglobaleditions.com/hair or www.mvstats.com) also has an overview of this notation as well as a pronunciation guide for Greek characters often used as SEM notations (See Appendix 12B). It also includes visual diagrams with the appropriate notation indicated.

SEM STAGES FOR TESTING MEASUREMENT THEORY VALIDATION WITH CFA

A measurement theory is used to specify how sets of measured items represent a set of constructs. The key relationships link constructs to variables (factor loading estimates) and constructs to each other (construct correlations). With estimates of these relationships, the researcher can make an

---

RULES OF THUMB 12-1

Construct Validity

- Standardized loading estimates should be .5 or higher, and ideally .7 or higher
- AVE should be .5 or greater to suggest adequate convergent validity
- AVE estimates for two factors also should be greater than the square of the correlation between the two factors to provide evidence of discriminant validity
- Construct reliability should be .7 or higher to indicate adequate convergence or internal consistency

---

RULES OF THUMB 12-2

Defining Individual Constructs

- All constructs must display adequate construct validity, whether they are new scales or scales taken from previous research; even previously established scales should be carefully checked for content validity
- Content validity should be of primary importance and judged both qualitatively (e.g., experts opinion) and empirically (e.g., unidimensionality and convergent validity).
- A pretest should be used to purify measures prior to confirmatory testing

---

empirical examination of the proposed measurement theory. A six-stage SEM process was introduced in the previous chapter. Stages 1–4 will be discussed in more detail here because they involve examining measurement theory. Stages 5 and 6, which address the structural theory linking constructs theoretically to each other, will be discussed in the next part of the chapter.

STAGE 1: DEFINING INDIVIDUAL CONSTRUCTS

The process begins by listing the constructs that will comprise the measurement model. If the researcher has experience with measuring one of these constructs before, then perhaps some scale that was previously used can be applied again. If not, numerous compilations of validated scales for a wide range of constructs are available [9, 68]. When a previously applied scale is not available, the researcher may have to develop a scale as described in Chapter 11. The process of designing a new construct measure involves a number of steps through which the researcher translates the theoretical definition of the construct into a set of specific measured variables. As such, it is essential that the researcher consider not only the operational requirements (e.g., number of items, dimensionality) but also establish the construct validity of the newly designed scale. Although designing a new construct measure may provide a greater degree of specificity, the researcher must also consider the amount of time and effort required in the scale development and validation process [59].

STAGE 2: DEVELOPING THE OVERALL MEASUREMENT MODEL

In this step, the researcher must carefully consider how all of the individual constructs will come together to form an overall measurement model. Several key issues should be highlighted at this point.

Unidimensionality

Unidimensionality was first introduced in Chapter 3. **Unidimensional measures** mean that a set of measured variables (indicators) can be explained by only one underlying construct. Unidimensionality becomes critically important when more than two constructs are involved. In such a situation, each measured variable is hypothesized to relate to only a single construct. All cross-loadings are hypothesized to be zero when unidimensional constructs exist.

Figure 12-1 hypothesizes two unidimensional constructs because no measured item is determined by more than one construct (has more than one arrow from a latent construct to it). In other words, all cross-loadings are fixed at zero.

One type of relationship among variables that impacts unidimensionality is when researchers allow a single measured variable to be caused by more than one construct. This situation is represented in the path model by arrows from a single construct pointing toward indicator variables associated with separate constructs. Remember the researcher is seeking a model that produces a good fit. When one frees another path in a model to be estimated, the value of the estimated path can only make the model more accurate. That is, the difference between the estimated and observed covariance matrices ($\Sigma_k - S$) is reduced unless the two variables are completely uncorrelated. Therefore, the χ^2 statistic will almost always be reduced by freeing additional paths.

Figure 12-2 is similar to the original model with the exception that several additional relationships are hypothesized. In contrast to the original measurement model, this one is not hypothesized to be unidimensional. Additional relationships are hypothesized between X_3, a measured variable, and the latent construct Work Environment, and between X_5 and the latent construct Supervisor Support. These relationships are represented by $L_{X3,WE}$ and $L_{X5,SUP}$, respectively. This means that Supervisor Support indicator variable X_3 and Work Environment indicator variable X_5 are each hypothesized as loading on both of the latent constructs.

As a rule, even if the addition of these paths leads to a significantly better fit, the researcher should not free (hypothesize) cross-loadings. Why? Because the existence of significant cross-loadings is evidence of a lack of construct validity. When a significant cross-loading is found to exist, any potential improvement in fit is artificial in the sense that it is obtained with the admission of a corresponding lack of construct validity.

Another form of relationships between variables is the covariance among error terms of two measured variables. Two types of covariance between error terms include covariance among error terms of items indicating the same construct, referred to as **within-construct error covariance.** The second type is covariance between two error terms of items indicating different constructs, referred to as **between-construct error covariance.**

Figure 12-2 also shows covariance (correlation) among some of the error terms. For example, the diagram shows covariance between the error terms of measured variables X_1 and X_2 (i.e., within-construct error covariance). It also indicates covariance between two error terms of indicator variables loading on different constructs—e_{x4} and e_{x7}. Here the covariance between e_{X4} and e_{X7} is an example of between-construct error covariance between measured indicator variables.

You also should not run CFA models that include covariances between error terms or cross-loadings. Allowing these paths to be estimated (freeing them) will reduce the χ^2, but at the same time seriously question the construct validity of the construct. Although more focus is typically on the issues raised with between-construct correlations and their impact on the structural model, within-construct error covariance terms also are threats to construct validity [35]. Significant between-construct error covariances suggest that the two items associated with these error terms are more highly related to each other than the original measurement model predicts. Evidence that a significant cross-loading exists also shows a lack of discriminant validity. So again, although these

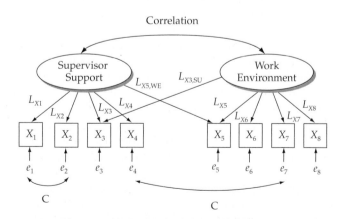

FIGURE 12-2   Measurement Model with Hypothesized Cross-Loadings and Correlated Error Variance

paths can be freed (covariance permitted) and improve the model fit, doing so violates the assumptions of good measurement.

Therefore, we recommend that you not free either type of path in most CFA applications. There are relatively rare and specific situations where researchers may free these paths as a way of explaining some specific measurement issue not represented by standard factor loadings. For more information on this topic, the reader is referred to other sources [5].

Congeneric Measurement Model

SEM terminology often states that a measurement model is *constrained* by the model hypotheses. The **constraints** refer specifically to the set of fixed **parameter** estimates. One type of common constraint is a measurement model hypothesized to consist of several unidimensional constructs with all cross-loadings constrained to zero. In addition, when a measurement model also hypothesizes no covariance between or within construct error variances, meaning they are all fixed at zero, the measurement model is said to be *congeneric*. **Congeneric measurement models** are considered to be sufficiently constrained to represent good measurement properties [21]. A congeneric measurement model that meets these requirements is hypothesized to have construct validity and is consistent with good measurement practice.

Items per Construct

Researchers are faced with somewhat of a dilemma in deciding how many indicators are needed per construct. On the one hand, researchers prefer many indicators in an attempt to fully represent a construct and maximize reliability. On the other hand, parsimony encourages researchers to use the smallest number of indicators to adequately represent a construct.

More items (measured variables or indicators) are not necessarily better. Even though more items do produce higher reliability estimates and generalizability [6], more items also require larger sample sizes and can make it difficult to produce truly unidimensional factors. As researchers increase the number of scale items (indicators) representing a single construct (factor), they may include a subset of items that inadvertently focuses on some specific aspect of a problem and create a subfactor. This problem becomes particularly prevalent when the content of the items has not been carefully screened ahead of time.

In practice, you can find SEM conducted with only a single item representing some factors. However, good practice dictates a minimum of three items per factor, preferably four, to not only provide minimum coverage of the construct's theoretical domain, but also to provide adequate identification for the construct as discussed next. Assessing construct validity of single item measures is problematic. When single items are included, they typically do not represent latent constructs.

ITEMS PER CONSTRUCT AND IDENTIFICATION   A brief introduction to the concept of statistical identification is provided here to clarify why at least three or four items per construct are recommended. We discuss the issue of identification within an overall SEM model in more detail later. In general, **identification** deals with whether enough information exists to *identify* a solution to a set of structural equations. As we saw earlier in our example, information is provided by the sample covariance matrix. In a CFA or SEM model, one parameter can be estimated for each unique variance and covariance in the observed covariance matrix. Thus, the covariance matrix provides the degrees of freedom used to estimate parameters just as the number of respondents provided degrees of freedom in regression.

If there are p measured items, then we can calculate the number of unique variances/covariances as $1/2[p(p+1)]$. For example, if there are six items, then there are 15 unique variances/covariances ($15 = 1/2(5 \times 6)$). One degree of freedom is then lost or used for each parameter estimated. As discussed in the following section, this indicates the level of identification.

Models and even constructs can be characterized by their degree of identification, which is defined by the degrees of freedom of a model after all the parameters to be estimated are specified.

We will discuss the three levels of identification in terms of construct identification at this time and then discuss overall model identification at a later point.

Underidentified. An **underidentified model** (also termed **unidentified**) has more parameters to be estimated than unique indicator variable variances and covariances in the observed variance/covariance matrix. For instance, the measurement model for a single construct with only two measured items as shown in Figure 12-3 is *underidentified.* The covariance matrix would be 2 by 2, consisting of one unique covariance and the variances of the two variables. Thus, there are three unique values. A measurement model of this construct would require, however, that two factor loadings (L_{x1} and L_{x2}) and two error variances (e_1 and e_2) be estimated. Thus, a unique solution cannot be found.

Just-Identified. Using the same logic, the three-item indicator model in Figure 12-3 is **just-identified.** This means there are just enough degrees of freedom to estimate all free parameters. All of the information is used, which means that the CFA analysis will reproduce the sample covariance matrix identically. Because of this, just-identified models have perfect fit. To help understand this, you can use the equation for degrees of freedom given in Chapter 11 and you will see that the resulting degrees of freedom for a three-item factor would be zero:

$$[3(3+1)/2] - 6 = 0$$

In SEM terminology, a model with zero degrees of freedom is referred to as *saturated.* The resulting χ^2 goodness of fit statistic also is zero. Just-identified models do not test a theory because their fit is determined by the circumstance. As a result, they are not of interest to researchers testing theories. Figure 12-3 illustrates the logic of both underidentified and just-identified single construct models. As noted in the figure, the number of unique variances/covariances is either exceeded by the number of estimated parameters (i.e., underidentified model) or equal to the number of estimated parameters (i.e., just-identified model).

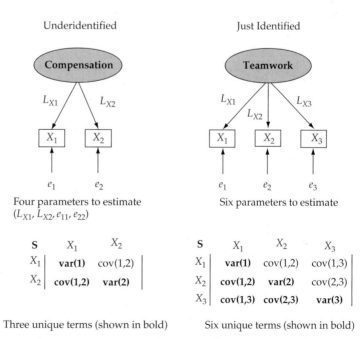

FIGURE 12-3 **Underidentified and Just-Identified CFA Models**

Overidentified. Overidentified models have more unique covariance and variance terms than parameters to be estimated. Thus, for any given measurement model a solution can be found with positive degrees of freedom and a corresponding chi-square goodness-of-fit value. A four-item, unidimensional measurement model produces an overidentified construct for which a fit value can be computed [41]. Increasing the number of measured items only strengthens the overidentified condition. Thus, the objective when applying CFA and SEM is to have an overidentified model and constructs.

Figure 12-4 illustrates an overidentification situation. It shows CFA results testing a unidimensional Empowerment construct measured using the following four items: (1) This organization allows me to do things I find personally satisfying, (2) This organization provides an opportunity for me to excel in my job, (3) I am encouraged to make suggestions about how this organization can be more effective, and (4) This organization encourages people to solve problems by themselves. The items (X_1–X_4) measure how much employees perceive they are empowered on their job.

A SEM program was used to calculate the results. The sample size was 800 respondents. By counting the number of items in the covariance matrix, we can see that there are a total of 10 unique covariance matrix values ($10 = 1/2(4 \times 5)$). We can also count the number of measurement parameters that are free to be estimated. There are four loading estimates ($L_{x1}, L_{x2}, L_{x3}, L_{x4}$) and four error variances (e_1, e_2, e_3, e_4) for a total of eight. Thus, the resulting model has two degrees of freedom ($10 - 8$). The overidentified model produces a χ^2 of 14.9 with two degrees of freedom.

But consider what would happen if only the first three items were used to measure Empowerment. There would be only six items in the covariance matrix and exactly six item parameters to be estimated (three loadings and three error variances). The model would be just-identified. Finally, if only two items—X_1 and X_2—were used (as in Figure 12-3), four parameter estimates (two loadings and two error variances) would be needed but there would be only three items in the covariance matrix. Therefore, the construct would be underidentified.

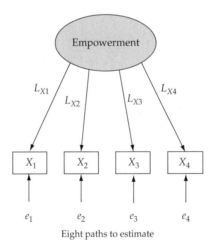

Symmetric Covariance Matrix

	X_1	X_2	X_3	X_4
X_1	2.01			
X_2	1.43	2.01		
X_3	1.31	1.56	2.24	
X_4	1.36	1.54	1.57	2.00

10 unique variance–covariance terms

Model Fit:
$\chi^2 = 14.9$
$df = 2$
$p = .001$
CFI = .99

Eight paths to estimate

Measured Items

X_1 = This organization allows me to do things I find personally satisfying.
X_2 = This organization provides an opportunity for me to excel in my job.
X_3 = I am encouraged to make suggestions about how this organization can be more effective.
X_4 = This organization encourages people to solve problems by themselves.

Loading Estimates	Error Variance Estimates
$L_{X1} = 0.78$	$e_1 = 0.39$
$L_{X2} = 0.89$	$e_2 = 0.21$
$L_{X3} = 0.83$	$e_3 = 0.31$
$L_{X4} = 0.87$	$e_4 = 0.24$

FIGURE 12-4   Four-Item Single Construct Model Is Overidentified

Note that even though a unidimensional two-item construct CFA is underidentified on its own, if it is integrated into a CFA model with other constructs the overall model may be overidentified. The same identification rules still apply to the overall model as described earlier. But the extra degrees of freedom from some of the other constructs can provide the necessary degrees of freedom to identify the overall model.

Does this mean the researcher should not worry about the identification of individual constructs, only the overall model? The answer is NO! Even though the overall model is identified, this does not mean the underlying problems with two-item and single-item measures disappear when we integrate them into a larger model. Strictly speaking, unidimensionality of constructs with fewer than four item indicators cannot be determined separately [1]. The dimensionality of any construct with only one or two items can only be established relative to other constructs. Constructs measured with one or two items also increase the likelihood of problems with interpretational confounding [19]. In the authors' experience, one- and two-item measures are associated with a higher likelihood of estimation problems encountered in later stages of the SEM process, including problems with convergence (identifying an appropriate mathematical solution).

In summary, when specifying the number of indicators per construct, the following is recommended:

- Use four indicators whenever possible.
- Having three indicators per construct is acceptable, particularly when other constructs have more than three.
- Constructs with fewer than three indicators should be avoided.

SINGLE-ITEM CONSTRUCTS   The exception to using multiple items to represent a construct comes when concepts can be adequately represented with a single item. Some concepts are very simple and lack nuance and complexity that accompanies the majority of psychological constructs. In other words, if there is little argument over the meaning of a term and that term is distinct and very easily understood, a single item can be sufficient. In marketing, some behavioral outcomes such as sales can be captured with a single item. Some behaviors are directly observable (purchase/no purchase, fail/succeed, etc.). Some would argue that the concept of liking (How much do you like this store?) is a very simple and easily understood concept that does not require multiple items (see Berkovitz and Rossiter [11] for a perspective on single-item measures). In rare instances, only the summated scale values are available, not the individual scale items themselves. Although single items can adequately represent some phenomena, operationally, they can be difficult to validate. When in doubt, and when multiple items are truly available, using multiple items is the safest approach. We will also address how to operationalize a single-item measure in a SEM model later in this chapter.

Reflective Versus Formative Constructs

The issue of causality affects measurement theory. Behavioral researchers typically study latent factors thought to cause measured variables. At times, however, the causality may be reversed. The contrasting direction of causality leads to different measurement approaches—reflective versus formative measurement models. Until now, our discussion of CFA assumed a reflective measurement theory. A **reflective measurement theory** is based on the idea that latent constructs cause the measured variables and that the error results in an inability to fully explain these measured variables. Thus, the arrows are drawn from latent constructs to measured variables. As such, reflective measures are consistent with classical test theory [60].

In our earlier example of employee job satisfaction, the construct Job Search is believed to cause specific measured indicators, such as how often you search for another job, telling friends about looking for another job, and your frequency of looking at job listings on the Web.

RULES OF THUMB 12-3

Developing the Overall Measurement Model

- In standard CFA applications testing a measurement theory, within- and between-error covariance terms should be fixed at zero and not be estimated
- In standard CFA applications testing a measurement theory, all measured variables should be free to load only on one construct
- Latent constructs should be indicated by at least three measured variables, preferably four or more; in other words, latent factors should be statistically identified
- Formative factors are not latent and are not validated as are conventional reflective factors; as such, they present greater difficulties with statistical identification

---

In contrast, a **formative measurement theory** is modeled based on the assumption that the measured variables cause the construct. The error in formative measurement models, therefore, is an inability of the measured variables to fully explain the construct. A key assumption is that formative constructs are not considered latent. Instead, they are viewed as indices where each indicator is a cause of the construct. A typical example would be a social class index [29]. Social class often is viewed as a composite of one's educational level, occupational prestige, and income (or sometimes wealth). Social class does not cause these indicators as in the reflective case. Rather, any one formative indicator is considered a partial cause of the index.

Formative constructs are becoming more widely used today, but they should be approached with a degree of caution. Besides issues related to specification in a SEM model (which are quite different than reflective constructs), formative constructs have unique qualities in terms of their conceptual and practical meaning. For example, construct validity must be assessed from a different perspective when using formative measures. In Part 3 of this chapter we examine issues related to formative indicators in more detail.

STAGE 3: DESIGNING A STUDY TO PRODUCE EMPIRICAL RESULTS

The third stage involves designing a study that will produce confirmatory results. In other words, the researcher's measurement theory will be tested. Here, all the standard rules and procedures that produce valid descriptive research apply [38]. If all goes well with the measurement model (CFA), the same sample will be used to test the structural model (SEM). We should note that the initial data analysis procedures described in Chapter 2 should first be performed to identify any problems in the data, including issues such as data input errors. After conducting these preliminary analyses, the researcher must make some key decisions on designing the CFA model.

Measurement Scales in CFA

CFA models typically contain reflective indicators measured with an ordinal or better measurement scale. Indicators with ordinal responses of at least four response categories can be treated as interval, or at least as if the variables are continuous. All the indicators for a construct need not be of the same scale type, nor do different scale values have to be normalized (mathematically transformed to a common scale range) prior to using SEM. Sometimes, however, combining scales with different ranges can require significantly longer computational time. Normalization can make interpreting coefficients and response values easier, so it is sometimes done prior to estimating the model. Thus, there are few restrictions on the type of data that can be used in SEM analyses, making typical survey research data suitable for a CFA model using SEM.

To illustrate the possibilities of using different scales when applying CFA, let us consider the job satisfaction example introduced earlier. The Job Satisfaction measure consists of four indicator variables. Each indicator variable could be measured with a different number of potential scale values (7 points, 10 points, 100 points, etc.). Although the researcher could transform the number of scale points to a common scale before estimating the model (e.g., all 7 points), it is not necessary to do so. CFA is capable of analyzing multiple indicator variables using a different number of scale points.

SEM and Sampling

Issues related to sample size and SEM in general were addressed in Chapter 11. But many times CFA requires the use of multiple samples. Testing measurement theory generally requires multiple studies and/or samples. An initial sample can be examined with EFA and the results used for further purification. Then an additional sample(s) should be drawn to perform the CFA. Even after CFA results are obtained, however, evidence of model stability and generalizability can only come from performing the analysis on additional samples and contexts.

Specifying the Model

CFA, not EFA, should be used to test the measurement model. As noted earlier, a critical distinction between CFA and EFA is the ability of the researcher to use CFA to perform an exact test of the measurement theory by specifying the correspondence between indicators and constructs. EFA provides insight into the structure of the items and may be helpful in proposing the measurement model, but it does not test a theory. So it should not be the only technique used to validate a measurement theory. As we have discussed, in CFA the researcher specifies (frees for estimation) the indicators associated with each construct as well as the correlations between constructs. In addition, the researcher does not specify cross-loadings, which fixes the loadings at zero.

One unique feature in specifying the indicators for each construct is the process of "setting the scale" of a latent factor. Because it is unobserved, a latent construct has no metric scale, meaning no range of values. This must be done for both exogenous and endogenous constructs in one of two ways:

- Fix one of the factor loadings on each construct to a specific value (1 is typically used)

  or

- Fix the value of the variance of the construct (again, 1 makes a good value as it transforms the results into a standardized form by having each element on the diagonal of the estimated covariance matrix among constructs equalling one)

SEM software programs will insert "1s" often in appropriate places to set a scale for the latent constructs. For example, the AMOS software automatically fixes one of the factor loading estimates for each construct drawn. However, the researcher should get in the habit of making sure that appropriate values are fixed. As one works with models, variables with fixed values can get changed or deleted, leading to constructs without the proper constraints needed to set a scale. Likewise, the researcher should check that multiple values are not constrained to 1 for the purpose of establishing the scale of a construct.

"Setting the scale" of a construct does not determine the true nature of the relationship. In earlier examples, we had discussed that there were two estimated parameters for each reflective indicator—the loading and the error term. If you fix one of the factor loadings on a construct (i.e., don't estimate the loading parameter), the error variance is still estimated with one parameter. From this, the actual extent of relationship is computed so that when the final standardized values for parameter estimates are examined, the loading estimate is not likely to be 1.0. When the scale is set by fixing the variance of a construct to 1, the actual relationships reflected in loadings remain unchanged but are simply shown as if they were already standardized (based on unit variance). Also, when a CFA is conducted properly, the method of setting the scale does not influence the total number of free (estimated) parameters.

The researcher may also wish to place additional constraints on a CFA model. For instance, it is sometimes useful to set two or more parameters as equal or to set a specific parameter to a specific value. Information about imposing additional constraints can be found in the documentation for the SEM program of choice. We will discuss some procedures (e.g., testing for measurement invariance in Part 3 of this chapter) that employ these types of constraints on the estimated parameters.

Issues in Identification

Once the measurement model is specified, the researcher must revisit the issues relating to identification. Although in the earlier discussion we were concerned about the identification of constructs, here we are concerned about the overall model. As before, overidentification is the desired state for CFA and SEM models in general. Even though comparing the degrees of freedom to the number of parameters to be estimated seems simple, in practice, establishing the identification of a model can be complicated and frustrating. This complexity is partly due to the fact that a wide variety of problems and data idiosyncrasies can manifest themselves in error messages, suggesting a lack of convergence or a lack of identification.

During the estimation process, the most likely cause of the computer program "blowing up" or producing meaningless results is a problem with statistical identification. As SEM models become more complex, however, ensuring that a model is identified can be problematic [15]. Once an identification problem is diagnosed, remedies must still be applied.

Avoiding Identification Problems

Several guidelines can help determine the identification status of a SEM model [65] and assist the researcher in avoiding identification problems. The order and rank conditions for identification are the two most basic rules [13], but they can be supplemented by basic rules in construct specification.

Meeting the Order and Rank Conditions. The order and rank conditions are the required mathematical properties for identification. The **order condition** refers to the requirement discussed earlier that the degrees of freedom for a model be greater than zero. That is, the number of unique covariance and variance terms less the number of free parameter estimates must be positive. The degrees of freedom for the overall model are always provided in the program output.

In contrast, the **rank condition** can be difficult to verify and a detailed discussion would require a working knowledge of linear algebra. In general terms it is the requirement that each parameter be estimated by a unique relationship (equation). As such, diagnosing a violation of the rank condition can be quite difficult. This is a problem encountered more often in the structural model relationships, particularly when nonrecursive or "feedback" relationships are specified (see Part 2 of this chapter for more discussion). But in CFA models it can still occur in the presence of cross-loading of items and/or in correlated error terms. Although we discouraged use of either cross-loadings or correlated errors on the basis of construct validity concerns, if they are used the researcher should be aware of the identification problems that may result.

Three-Indicator Rule. Given the difficulty in establishing the rank condition, researchers turn to more general guidelines. These guidelines include the **three-indicator rule.** It is satisfied when all factors in a congeneric model have at least three significant indicators. A two-indicator rule also states that a congeneric factor model with two significant items per factor will be identified as long as each factor also has a significant relationship with some other factor. One-item factors cause the most problems with identification. As noted above, this rule applies to congeneric models—those models without cross-loadings or correlated error terms. Adding those types of relationships can easily introduce identification problems.

RECOGNIZING IDENTIFICATION PROBLEMS Although identification issues underlie many estimation problems faced in SEM modeling, there are few certain indicators of the existence and source of identification problems. Many times the software programs will provide some form of solution even in the presence of identification issues. Therefore, researchers must consider a wide range of symptoms to assist in recognizing identification problems. It should be noted that occasionally error warnings or messages will suggest that a single parameter is not identified. Although the researcher can attempt to solve the problem by deleting the offending variable, many times this does not address an underlying cause and the problem persists. Regrettably, the SEM programs provide minimal diagnostic measures for identification problems. Thus, researchers must typically rely on other means of recognizing identification problems by the symptoms described in the following list:

- Very large standard errors for one or more coefficients.
- An inability of the program to invert the information matrix (no solution can be found).
- Wildly unreasonable or impossible estimates such as negative error variances, or very large parameter estimates, including standardized factor loadings and correlations among the constructs outside the range of +1.0 to −1.0.
- Models that result in differing parameter estimates based on the use of different starting values. In SEM programs, the researcher can specify an initial value for any estimated parameter as a starting point for the estimation process. Model estimates, however, should be comparable given any set of reasonable starting values. When questions about the identification of any single parameter occur, a second test can be performed. You first estimate a CFA model and obtain the parameter estimate. Next, fix the coefficient to its estimated value and rerun the model. If the overall fit of the model varies markedly, identification problems are indicated.

As you can see, identification problems can be manifested in SEM results in many different ways. The researcher should never rely only on the software to recognize identification problems, but must also diligently examine the results to ensure that no problems exist.

SOURCES AND REMEDIES OF IDENTIFICATION PROBLEMS Does the presence of identification problems mean your model is invalid? Although some models may need respecification, many times identification issues arise from common mistakes in specifying the model and the input data. In the discussion that follows, we will not only discuss the typical types of sources for identification problems, but also offer suggestions for dealing with the problems where possible. Some of the most common issues leading to problems with identification include the following.

Incorrect Indicator Specification. A researcher can easily make mistakes such as (1) not linking an item to any construct, (2) linking an indicator to two or more constructs, (3) selecting an indicator variable twice in the same model, or (4) not creating and linking an error term for each indicator. Although all of these mistakes seem obvious, specifying the measurement model is a process that requires complete accuracy. Programs such as LISREL, which may require command syntax, are perhaps the easiest for mistakes to occur in several areas. Something as simple as listing a variable twice in the SELECT command in LISREL or mistakenly dragging a variable from the AMOS variable list more than once creates an error that is hard to recognize at first glance. Even in programs such as AMOS, overlooking a loading between an indicator and construct or an error term is quite easy in a complicated model. We encourage the researcher to carefully examine the model specification if identification problems are indicated.

"Setting the Scale" of a Construct. A second common mistake that creates identification problems is not "setting the scale" of each construct. As discussed earlier, each construct must have one value specified (either a loading of an indicator or the construct variance). Failure to do this for any construct will create a problem and the model will not estimate. This type of problem occurs in initially specifying the model, but also in model respecification when indicators may be eliminated from the model. If an indicator with the fixed loading is deleted for some reason, then another indicator must be fixed. Even AMOS, does not automatically fix another parameter if that indicator is eliminated from the model at a later stage.

Too Few Degrees of Freedom. This problem is likely accompanied by a violation of the three-indicator rule. Small sample size (fewer than 200) increases the likelihood of problems in this situation. The simplest solution is to avoid this situation by including enough measures to avoid violating these rules. If this is not possible, the researcher can try to add some constraints that will free up degrees of freedom [39]. One possible solution is imposing **tau-equivalence** assumptions, which require all factor loadings on a particular factor to be equal. Tau-equivalence can be done for one or more factors. A second solution is to fix the error variances to a known or specified value. Third, the correlations between constructs can be fixed if some theoretical value can be assigned. The researcher should remember that each of these solutions, however, has implications for the construct validity of the constructs involved and should be undertaken with great care and an understanding of their impact on the constructs.

Identification problems must be solved before the results can be accepted. Although careful model specification using the guidelines discussed earlier can help avoid many of these problems, researchers must always be vigilant in scrutinizing the results to recognize identification problems wherever they occur.

Problems in Estimation

Even with no identification problems, SEM models may result in the estimation of parameters that are logically impossible. By this we mean that the estimated values are nonsensical. Rather than not provide results, most SEM programs will complete the estimation process in spite of these issues. It then becomes the responsibility of the researcher to identify the illogical results and correct the model to obtain acceptable results. We will discuss the two most common types of estimation problems as well as potential causes and solutions.

ILLOGICAL STANDARDIZED PARAMETERS   The most basic estimation problem with SEM results is when correlation estimates (i.e., standardized estimates) between constructs exceed |1.0| or even standardized path coefficients exceed |1.0|. These estimates are theoretically impossible and many times identification problems are the cause. But they also may occur from data issues (e.g., highly correlated indicators or violations of the underlying statistical assumptions) or even poorly specified constructs (e.g., extremely low reliability or other issues in construct validity). After examining the data, researchers should also examine each construct involved to remedy issues that relate to this situation.

HEYWOOD CASES   A SEM solution that produces an error variance estimate of less than zero (a negative error variance) is termed a **Heywood case.** Such a result is logically impossible because it implies a less than 0 percent error in an item, and by inference, it implies that more than 100 percent of the variance in an item or a construct is explained. Heywood cases are particularly problematic in CFA models with small samples or when the three-indicator rule is not followed [58]. Models with sample size greater than 300 that adhere to the three-indicator rule are unlikely to produce Heywood cases. Even when a Heywood case(s) is present, the SEM program may produce a solution, but it is an improper solution in which the model did not fully converge. This is usually accompanied by a warning or error message indicating that an error variance estimate is not identified and cautioning that the solution may not be reliable.

Several options are possible when Heywood cases arise. The first solution should be to ensure construct validity. This may involve the elimination of an offending item, but the researcher may be

---

RULES OF THUMB 12-4

Designing a Study to Provide Empirical Results

- The scale of a latent construct can be set by either:
  - Fixing one loading and setting its value to 1, or
  - Fixing the construct variance to 1
- Congeneric, reflective measurement models in which all constructs have at least three item indicators are statistically over-identified in models with two or more constructs
- The researcher should check for errors in the specification of the measurement model when identification problems are indicated
- Models with large samples (more than 300) that adhere to the three-indicator rule generally do not produce Heywood cases

---

limited if this creates a violation of the three-indicator rule. An alternative is to try and add more items if possible or assume tau-equivalence (all loadings in that construct are equal). Each of these is preferable to the "last resort" solution, which is to fix the offending estimate to a very small value, such as .005 [30]. Although this value may identify the parameter, it can lead to lower fit because the value is not likely to be the true sample value. It also means that the underlying cause cannot be remedied in the model specification, but must be addressed in an "ad hoc" fashion.

STAGE 4: ASSESSING MEASUREMENT MODEL VALIDITY

Once the measurement model is correctly specified, a SEM model is estimated to provide an empirical measure of the relationships among variables and constructs represented by the measurement theory. The results enable us to compare the theory against reality as represented by the sample data. In other words, we see how well the theory fits the data.

Assessing Fit

Fit was discussed in detail in Chapter 11. Recall that the sample data are represented by a covariance matrix of measured items, and the theory is represented by the proposed measurement model. Equations are implied by this model as discussed earlier in this chapter and in Chapter 11. These equations enable us to estimate reality by computing an estimated covariance matrix based on our theory. Fit compares the two covariance matrices.

The guidelines for good fit provided in Chapter 11 apply. Here the researcher attempts to examine all aspects of construct validity through various empirical measures. The result is that CFA enables us to test or confirm whether a theoretical measurement model is valid. It is quite different from EFA, which explores data to identify potential constructs. Many researchers conduct EFA on one or more separate samples before reaching the point of trying to confirm a model. EFA is the appropriate tool for identifying factors among multiple variables. As such, EFA results can be useful in developing theory that will lead to a proposed measurement model. It is here that CFA enters the picture. It can be used to confirm the measurement model developed using EFA.

Path Estimates

One of the most fundamental assessments of construct validity involves the measurement relationships between items and constructs (i.e., the *path estimates* linking constructs to indicator variables). When testing a measurement model, the researcher should expect to find relatively high loadings. After all, once CFA is used, a good conceptual understanding of the constructs and its items should exist. This knowledge, along with preliminary empirical results from exploratory studies, should provide these expectations.

SIZE OF PATH ESTIMATES AND STATISTICAL SIGNIFICANCE   Earlier, we provided rules of thumb suggesting that loadings should be at least .5 and ideally .7 or higher. Loadings of this size or larger confirm that the indicators are strongly related to their associated constructs and are one indication of construct validity. Note that these guidelines apply to the standardized loadings estimates, which remove effects due to the scale of the measures much like the differences between correlation and covariance. Thus, the researcher must be certain that they are included in the output. The default output often displays the unstandardized maximum likelihood estimates, which are more difficult to interpret with respect to these guidelines.

Researchers should also assess the statistical significance of each estimated (free) coefficient. Nonsignificant estimates suggest an item should be dropped. Conversely, a significant loading alone does not indicate an item is performing adequately. A loading can be significant at impressive levels of significance (i.e., $p < .01$) but still be considerably below $|.5|$. Low loadings suggest that a variable is a candidate for deletion from the model.

SEM models also typically display the **squared multiple correlations** for each measured variable. In a CFA model, this value represents the extent to which a measured variable's variance is explained by a latent factor. From a measurement perspective, it represents how well an item measures a construct. Squared multiple correlations are sometimes referred to as *item reliability, communality,* or *variance extracted* (more discussion in following section on construct validity). We do not provide specific rules for interpreting these values here because in a congeneric measurement model they are a function of the loading estimates. Recall that a congeneric model is one in which no measured variable loads on more than one construct. The rules provided for the factor loading estimates tend to produce the same diagnostics.

IDENTIFYING PROBLEMS   Loadings also should be examined for offending estimates as indications of overall problems. One often overlooked task is to make sure the loadings make sense. For instance, items with the same valence (e.g., positive or negative wording) should produce the same sign. If an attitude scale consists of responses to four items—good, likeable, unfavorable, bad—then two items should carry positive loadings and two should carry negative loadings (unless they have previously been recoded). If the signs of the loadings are not opposite, the researcher should not have confidence in the results.

As discussed earlier, standardized loadings above 1.0 or below −1.0 are out of the feasible range and an important indicator of a problem with the model. The reader can refer to the discussion of problems in parameter estimation to examine what this situation may mean for the model overall. It is important to point out that the problem may not reside solely in the variable with the out-of-range loading. So simply dropping this item may not provide the best solution. To summarize, the loading estimates can suggest either dropping an individual item or that some offending estimate indicates a larger overall problem.

Construct Validity

Recall that in Chapter 3 validity was defined as the extent to which research is accurate, and the discussion centered on validating summated scales. CFA eliminates the need to summate scales because the SEM programs compute latent construct scores for each respondent. This process allows relationships between constructs to be automatically corrected for the amount of error variance that exists in the construct measures.

One of the primary objectives of CFA/SEM is to assess the construct validity of a proposed measurement theory. **Construct validity** is the extent to which a set of measured items actually reflects the theoretical latent construct those items are designed to measure.

Thus, it deals with the accuracy of measurement. Evidence of construct validity provides confidence that item measures taken from a sample represent the actual true score that exists in the population.

Construct validity is made up of four components. These components were introduced in Chapter 3 along with summated scales. Here, we expand on those ideas and discuss them in terms more appropriate for CFA.

CONVERGENT VALIDITY   The items that are indicators of a specific construct should converge or share a high proportion of variance in common, known as **convergent validity.** Several ways are available to estimate the relative amount of convergent validity among item measures.

Factor Loadings.   The size of the factor loading is one important consideration. In the case of high convergent validity, high loadings on a factor would indicate that they converge on a common point, the latent construct. At a minimum, all factor loadings should be statistically significant [1]. Because a significant loading could still be fairly weak in strength, a good rule of thumb is that standardized loading estimates should be .5 or higher, and ideally .7 or higher. In most cases, researchers should interpret standardized parameter estimates because they are constrained to range between −1.0 and +1.0. Unstandardized loadings represent covariances and thus have no upper or lower bound. SEM programs provide standardized estimates, although it is usually an option that must be requested.

The rationale behind this rule can be understood in the context of an item's **communality** (see Chapter 3). The square of a standardized factor loading represents how much variation in an item is explained by the latent factor and is termed the **variance extracted** of the item. Thus, a loading of .71 squared equals .5. In short, the factor is explaining half the variation in the item with the other half being error variance. As loadings fall below .7, they can still be considered significant, but more of the variance in the measure is error variance than explained variance.

Average Variance Extracted.   With CFA, the **average variance extracted (AVE)** is calculated as the mean variance extracted for the items loading on a construct and is a summary indicator of convergence [33]. This value can be calculated using standardized loadings:

$$\text{AVE} = \frac{\sum\limits_{i=1}^{n} L_i^2}{n}$$

The L_i represents the standardized factor loading and i is the number of items. So for n items, AVE is computed as the total of all squared standardized factor loadings (squared multiple correlations) divided by the number of items.[1] In other words, it is the average squared completely standardized factor loading or average communality. Using this same logic, an AVE of .5 or higher is a good rule of thumb suggesting adequate convergence. An AVE of less than .5 indicates that, on average, more error remains in the items than variance explained by the latent factor structure imposed on the measure. An AVE measure should be computed for each latent construct in a measurement model. In Figure 12-1 an AVE estimate is needed for both the Supervisor Support and Work Environment constructs.

Reliability.   Reliability is also an indicator of convergent validity. Considerable debate centers around which of several alternative reliability estimates is best [6]. Coefficient alpha remains a commonly applied estimate although it may understate reliability. Different reliability coefficients do not produce dramatically different reliability estimates, but a slightly different **construct reliability (CR)** value is often used in conjunction with SEM models. It is computed

---

[1]SEM programs offer several different types of standardization. Where we use the term standardized, we refer to completely standardized estimates unless otherwise noted.

from the squared sum of factor loadings (L_i) for each construct and the sum of the error variance terms for a construct (e_i) as:

$$CR = \frac{\left(\sum_{i=1}^{n} L_i \right)^2}{\left(\sum_{i=1}^{n} L_i \right)^2 + \left(\sum_{i=1}^{n} e_i \right)}$$

The rule of thumb for either reliability estimate is that .7 or higher suggests good reliability. Reliability between .6 and .7 may be acceptable provided that other indicators of a model's construct validity are good. High construct reliability indicates that internal consistency exists, meaning that the measures all consistently represent the same latent construct.

DISCRIMINANT VALIDITY   **Discriminant validity** is the extent to which a construct is truly distinct from other constructs. Thus, high discriminant validity provides evidence that a construct is unique and captures some phenomena other measures do not. CFA provides two common ways of assessing discriminant validity.

First, the correlation between any two constructs can be specified (fixed) as equal to one. In essence, it is the same as specifying that the items making up two constructs could just as well make up only one construct. If the fit of the two-construct model is significantly different from that of the one-construct model, then discriminant validity is supported [1, 7]. In practice, however, this test does not provide strong evidence of discriminant validity, because high correlations, sometimes as high as .9, can still produce significant differences in fit between the two models.

Referring back to Figure 12-1, note that there are two constructs each with four indicator variables. Discriminant validity could be assessed by setting the value of the relationship between the two constructs to 1.0. Or the researcher could change that model to one in which all eight measured items are indicators of only one latent construct. In either case, the researcher could then test a model with this specification and compare its fit to the fit of the original two-construct model. If the model fits were significantly different, this would suggest that the eight items represent two separate constructs.

A more rigorous test is to compare the average variance-extracted values for any two constructs with the square of the correlation estimate between these two constructs [33]. The variance-extracted estimates should be greater than the squared correlation estimate. The logic here is based on the idea that a latent construct should explain more of the variance in its item measures that it shares with another construct. Passing this test provides good evidence of discriminant validity.

In addition to distinctiveness between constructs, discriminant validity also means that individual measured items should represent only one latent construct. The presence of cross-loadings indicates a discriminant validity problem. If high cross-loadings do indeed exist, and they are not represented by the measurement model, the CFA fit should not be good.

NOMOLOGICAL VALIDITY AND FACE VALIDITY   Constructs also should have face validity and nomological validity. The processes for testing these properties are the same whether using CFA or EFA, so the reader is referred to Chapter 3 for more detailed clarification. **Face validity** must be established *prior* to any theoretical testing when using CFA. Without an understanding of every item's content or meaning, it is impossible to express and correctly specify a measurement theory. Thus, in a very real way, face validity is the most important validity test. **Nomological validity** is then tested by examining whether the correlations among the constructs in a measurement theory make sense. The matrix of construct correlations can be useful in this assessment.

Researchers often test a measurement theory using constructs measured by multi-item scales developed in previous research. For instance, if HBAT wished to measure customer satisfaction with their services, it could do so by evaluating and selecting one of several customer satisfaction scales in the marketing literature. Handbooks exist in many social science disciplines that catalog multi-item scales [9, 68]. Similarly, if HBAT wanted to examine the relationship between cognitive dissonance and customer satisfaction, a previously applied cognitive dissonance scale could be used.

Any time previously used scales are in the same model, even if they have been applied successfully with adequate reliability and validity in other research, the researcher should pay careful attention that the item content of the scales does not overlap. In other words, when using borrowed scales, the researcher should still check for face validity. It is quite possible that when two borrowed scales are used together in a single measurement model, face validity issues become apparent that were not seen when the scales were used individually.

Model Diagnostics

CFA's ultimate goal is to obtain an answer as to whether a given measurement model is valid. But the process of testing using CFA provides additional diagnostic information that may suggest modifications for either addressing unresolved problems or improving the model's test of measurement theory.

Model respecification, for whatever reason, always impacts the underlying theory upon which the model was formulated. If the modifications are minor, then the theoretical integrity of a measurement model may not be severely damaged and the research can proceed using the prescribed model and data after making suggested changes. If the modifications are more than minor, then the researcher must be willing to modify the measurement theory, which will result in a new measurement model and potentially require a new data sample. Given the strong theoretical basis for CFA, the researcher should avoid making changes based solely on empirical criteria such as the diagnostics provided by CFA. Moreover, other concerns should be considered before making any change, including the theoretical integrity of the individual constructs and overall measurement model and the assumptions and guidelines that go along with good practice, much of which have already been discussed.

What diagnostic cues are provided when using CFA? They include fit indices such as those discussed and analyses of residuals as well as some specific diagnostic information provided in most CFA output. Many diagnostic cues are provided and we focus here on those that are both useful and easy to apply. Some areas that can be used to identify problems with measures are standardized residuals, modification indices, and specification search.

STANDARDIZED RESIDUALS   The standard output produced by most SEM programs includes residuals. **Residuals** refer to the individual differences between observed covariance terms and the fitted (estimated) covariance terms. The better the fit, the smaller are the residuals. Thus, a residual term is associated with every unique value in the observed covariance matrix. The **standardized residuals** are simply the raw residuals divided by the standard error of the residual. They are not dependent on the actual measurement scale range, which makes them useful in diagnosing problems with a measurement model.

Residuals can be either positive or negative, depending on whether the estimated covariance is under or over the corresponding observed covariance. Researchers can use these values to identify item pairs for which the specified measurement model does not accurately predict the observed covariance between those two items. Typically, standardized residuals less than |2.5| do not suggest a problem. Conversely, residuals greater than |4.0| raise a red flag and suggest a potentially unacceptable degree of error. It should be remembered that some large standardized residuals may occur just because of sampling error. The value of |4.0| just cited relates to a significance level of .001. Thus, we may accept one or two of these large residuals in many instances. What is of concern is a consistent pattern of large standardized residuals, associated either with a single variable and a number of other variables or residuals for several of the variables within a construct. Either occurrence suggests problems. The most likely, but not automatic, response is dropping one of the items associated with a residual greater than |4.0|. Standardized residuals between |2.5| and |4.0| deserve some attention, but may not suggest any changes to the model if no other problems are associated with those two items.

MODIFICATION INDICES   Typical SEM output also provides modification indices. A **modification index** is calculated for every possible relationship that is *not* estimated in a model. For example, in Figure 12-1 variable X_1 has a loading on the Supervisor Support construct, but not on the Work Environment construct. That is, the loading of X_1 on the Work Environment construct is fixed at zero. There would then be a modification index value for the possible loading of X_1 on the other construct. The modification index value would show how much the overall model χ^2 value would be reduced by also estimating a loading for X_1 to the Work Environment construct. Likewise, there would be modification indices calculated for the remainder of the items that loaded on Supervisor Support and not Work Environment, as well as vice versa (those items that loaded on Work Environment and not on Supervisor Support).

Modification indices of approximately 4.0 or greater suggest that the fit could be improved significantly by freeing the corresponding path to be estimated. But making model changes based solely on modification indices is not recommended. Doing so would be inconsistent with the theoretical basis of CFA and SEM in general. Modifications do provide important diagnostic information about the potential cross-loadings that could exist if specified. As such, they assist the researcher in assessing the extent of model misspecification without estimating a large number of new models. This is an important tool for identifying problematic indicator variables if they exhibit the potential for cross-loadings. Modification indices are estimated for all nonestimated parameters, so they are also generally provided for diagnosing error term correlations and also correlational relationships between constructs that may not be initially specified in the CFA model. Researchers should consult other residual diagnostics for a change suggested by a modification index and then take appropriate action, if justified by theory.

SPECIFICATION SEARCHES   A **specification search** is an empirical trial-and-error approach that uses model diagnostics to suggest changes in the model. In fact, when we make changes based on any diagnostic indicator, we are performing a type of specification search [67]. SEM programs such as AMOS and LISREL can perform specification searches automatically. These searches identify the set of "new" relationships that best improve the overall model fit. This process is based on freeing fixed (nonestimated) relationships with the largest modification index. Specification searches are fairly easy to implement.

Although it may be tempting to rely largely on specification searches as a way of finding a model with a good fit, this approach is not recommended [48, 50]. The biggest problem is its inconsistency with the intended purpose and use of procedures such as CFA. Namely, CFA tests theory and is generally applied in a confirmatory approach, not as an exploratory tool. Second, the results for one parameter depend on the results of estimating other parameters, which makes it difficult to be certain that the true problem with a model is isolated in the variables suggested by a modification index. Third, empirical research using simulated data has shown that mechanical specification searches are unreliable in identifying a true model and thus can provide misleading results. Therefore, CFA specification searches should involve identifying only a small number of major problems. A researcher in exploratory mode can use specification searches to identify a plausible measurement theory. But new construct structures suggested by specification searches must be confirmed using a new data set.

CAVEATS IN MODEL RESPECIFICATION   What types of modifications are more than minor? The answer to this question is not simple or clear-cut. If model diagnostics indicate the existence of some new factor not suggested by the original measurement theory, verifying such a change would require a new data set. When more than 20 percent of the measured variables are dropped or changed with respect to the factor they indicate, then a new data set should be used for further verification. In contrast, dropping one or two items from a large battery of items is less consequential and the confirmatory test may not be jeopardized.

Because CFA tests a measurement theory, changes to the model should be made only after careful consideration. The most common change would be the deletion of an item that does not perform well with respect to model integrity, model fit, or construct validity. At times, however, an item may be retained even if diagnostic information suggests that it is problematic. For instance, consider an item with high content validity (e.g., "I was very satisfied," in a satisfaction scale) within an overall CFA model with good overall fit and strong evidence for construct validity. Dropping it would not seem to accomplish much. It might buy a little fit at the expense of some conceptual consistency. In sum, a poorly performing item may be retained at times to satisfy statistical identification requirements, to meet the minimal number of items per factor, or based on face validity considerations. In the end, however, theory should always be prominently considered in making model modifications.

Summary Example

We will now illustrate not only how to assess the overall model fit of a CFA model, but also the use of several diagnostic measures. The measures will include standardized loadings, standardized residuals, and modification indices. Table 12-1 shows selected output from testing a CFA model that extends the model shown in Figure 12-4. Another construct labeled Job Satisfaction (JS) has been added to the model. The two constructs represent employee Empowerment and Job Satisfaction. The model fit as indicated by the CFI (.99) and the RMSEA (.04) appears good. The model χ^2 is significant, which is to be expected given the large sample size ($N = 800$), but the normed χ^2 is within suggested guidelines at 2.83.

We begin by looking at the standardized loadings. All of the loadings estimates are statistically significance, thus providing initial evidence of convergent validity. Given the sample size of 800, however, it is not unexpected that all of the loadings are statistically significant, and that should not be the only criteria used. Three of the estimates for Job Satisfaction fall below the .7 cutoff, although only one falls below the less conservative .5 cutoff (X_9). Thus, X_9 becomes a prime candidate for deletion. The loadings for X_5 and X_8 are lower than preferred, but unless some other

---

RULES OF THUMB 12-5

Assessing the Measurement Model Validity

- Loading estimates can be statistically significant but still be too low to qualify as a good item (standardized loadings below |.5|); in CFA, items with low loadings become candidates for deletion
- Completely standardized loadings above 1.0 or below −1.0 are out of the feasible range and can be an important indicator of some problem with the data
- Typically, standardized residuals less than |2.5| do not suggest a problem:
  - Standardized residuals greater than |4.0| suggest a potentially unacceptable degree of error that may call for the deletion of an offending item
  - Standardized residuals for any pair of items between |2.5| and |4.0| deserve some attention, but may not suggest any changes to the model if no other problems are associated with those two items
- The researcher should use the modification indices only as a guideline for model improvements of those relationships that can theoretically be justified
- Specification searches based on purely empirical grounds are discouraged because they are inconsistent with the theoretical basis of CFA and SEM
- CFA results suggesting more than minor modification should be reevaluated with a new data set (e.g., if more than 20% of the measured variables are deleted, then the modifications cannot be considered minor)

TABLE 12-1 Model Fit Measures, Loadings, Standardized Residuals, and Modification Indices in CFA

Overall Model Fit Measures

$\chi^2 = 68.0$ with 26 degrees of freedom ($p = .000013$)
CFI $= .99$
RMSEA $= .04$

> Fit Indices

Standardized Loadings (AMOS = Regression Weights)

	EMPOWERMENT	JOB SATISFACTION
X_1	0.78	—
X_2	0.89	—
X_3	0.83	—
X_4	0.87	—
X_5	—	0.58
X_6	—	0.71
X_7	—	0.69
X_8	—	0.52
X_9	—	0.46

> New Construct Job Satisfaction has five indicator variables.

Largest Negative Standardized Residuals

RESIDUAL	FOR X_3	AND X_1	−3.12
RESIDUAL	FOR X_4	AND X_2	−3.04
RESIDUAL	FOR X_6	AND X_4	−2.70
RESIDUAL	FOR X_9	AND X_5	−3.76

Largest Positive Standardized Residuals

RESIDUAL	FOR X_2	AND X_1	3.05
RESIDUAL	FOR X_4	AND X_3	3.90
RESIDUAL	FOR X_5	AND X_1	3.08
RESIDUAL	FOR X_5	AND X_2	2.72

Modification Indices for Cross-Loading Estimates

	EMPOWERMENT	JOB SATISFACTION
X_1	—	0.00
X_2	—	5.04
X_3	—	0.01
X_4	—	5.29
X_5	4.09	—
X_6	2.72	—
X_7	0.04	—
X_8	2.30	—
X_9	2.06	—

> Modification indices for each cross-loading not estimated above.

> Modification indices for each possible error term correlation.

Modification Indices for Error Term Estimates

	X_1	X_2	X_3	X_4	X_5	X_6
X_1	—					
X_2	9.30	—				
X_3	9.72	0.90	—			
X_4	0.01	9.26	15.17	—		
X_5	10.04	2.40	2.62	1.86	—	
X_6	0.28	0.00	1.40	2.73	0.86	—
X_7	2.04	0.09	0.17	0.28	0.16	0.07
X_8	0.00	0.84	3.82	0.06	6.62	0.26
X_9	0.78	0.08	2.14	0.00	14.15	4.98

evidence suggests they are problematic, they will likely be retained to support content validity. For all practical purposes, X_7's loading is adequate given it is only .01 below .70.

The next step is to calculate the construct reliabilities of both constructs. The reliability for Empowerment is .91 and the reliability of Job Satisfaction is .73. Both exceed the suggested threshold of .70. In terms of discriminant validity, we need to compare the AVEs for each construct with the square of the estimated correlation between these constructs. The AVEs are .71 and .36 for Empowerment and Job Satisfaction, respectively. Note that Job Satisfaction's AVE falls below the suggested level of .50, another indicator of perhaps improvement of the construct by eliminating an item. The correlation between constructs is .48 and its squared value is .23. Thus, discriminant validity of the two constructs is also supported because the AVE of both constructs is greater than the squared correlation.

In terms of other diagnostic measures, we next examine the standardized residuals. In the table all standardized residuals greater than |2.5| are shown. Two residuals approach but do not exceed 4.0. The largest, between X_3 and X_4 (3.90), suggests that the covariance estimate between these indicator variables could be more accurate. In this case, no change will be made based on the residual between X_3 and X_4 because the fit remains good despite the high residual. Deleting either variable would leave fewer than four indicator variables for this construct. Also, freeing the parameter representing the error covariance between these two would be inconsistent with the congeneric properties of the measurement model. Thus, it appears "we can live with" this somewhat high residual for now. The second highest residual is between X_5 and X_9 (−3.76). It provides further evidence (in addition to the low standardized loading) that X_9 may need to be dropped.

Finally, we examine the modification index associated with each of the loadings of the indicators. As we see, the values represent the cross-loadings of items if they were estimated. Here the information is consistent with that obtained from the residuals, leading to much the same conclusion. First, none of the values among loadings are high enough to indicate that a cross-loading is required. In looking at the modification indices for the error terms, we see that the value for the covariance between X_3 and X_4 error terms is 15.17. Although we do not recommend adding this relationship to the model, it does indicate a high degree of covariance between these two items that is not captured by the construct. But given the high loading estimates for each, no change is made. Looking further, X_9 has a high value of 14.15, just one more indication of a poorly performing item.

Given that there is a high standardized residual associated with X_9 (−3.76), there is a high modification index between X_9 and X_5 (14.15), and its loading is below .5, X_9 thus becomes a candidate for deletion. The final decision should be made based not only on model fit improvement, but the extent to which deleting X_9 would diminish the content validity of the construct.

CFA ILLUSTRATION

We now illustrate CFA using HBAT as an example. In this section we apply the first four stages of the six-stage process to a problem faced by management. We begin by briefly introducing the context for this new HBAT study.

HBAT employs thousands of workers in different operations around the world. Like many firms, one of its biggest management problems is attracting and keeping productive employees. The cost to replace and retrain employees is high. Yet the average new person hired works for HBAT less than 3 years. In most jobs, the first year is not productive, meaning the employee is not contributing as much as the costs associated with employing him/her. After the first year, most employees become productive. HBAT management would like to understand the factors that contribute to employee retention. A better understanding can be gained by learning how to measure the key constructs. Thus, HBAT is interested in developing and testing a measurement model made up of constructs that affect employees' attitudes and behaviors about remaining with HBAT.

Stage 1: Defining Individual Constructs

With the general research question defined, the researcher now selects the specific constructs that represent the theoretical framework to be tested and will be included in the analysis. The indicators used to operationalize the constructs may come from prior research or be developed specifically for this project.

HBAT initiated a research project to study the employee turnover problem. Preliminary research discovered that a large number of employees are exploring job options with the intention of leaving HBAT should an acceptable offer be obtained from another firm. To conduct the study, HBAT hired consultants with a working knowledge of the organizational behavior theory dealing with employee retention. Based on published literature and some preliminary interviews with employees, a study was designed focusing on five key constructs. The consulting team and HBAT management also agreed on construct definitions based on how they have been used in the past. The five constructs along with a working definition are as follows:

- *Job Satisfaction (JS).* Reactions resulting from an appraisal of one's job situation.
- *Organizational Commitment (OC).* The extent to which an employee identifies and feels part of HBAT.
- *Staying Intentions (SI).* The extent to which an employee intends to continue working for HBAT and is not participating in activities that make quitting more likely.
- *Environmental Perceptions (EP).* Beliefs an employee has about day-to-day, physical working conditions.
- *Attitudes Toward Coworkers (AC).* Attitudes an employee has toward the coworkers he/she interacts with on a regular basis.

The consultants proposed a set of multiple-item reflective scales to measure each construct. Face validity appears evident, and the conceptual definitions match well with the item wordings. Additionally, a pretest was performed in which three independent judges matched items with the construct names. No judge had difficulty matching items to constructs, providing further confidence the scales contain face validity. Having established face validity, HBAT proceeded to finalize the scales. Scale purification based on item-total correlations and EFA results (as in Chapter 3) from a pretest involving 100 HBAT employees resulted in the measures shown in Table 12-2. The job satisfaction scale contains multiple measures, each assessing the degree of satisfaction felt by respondents with a different type of scale. The complete questionnaire is available on the text's Web sites at www.pearsonglobaleditions.com/hair or www.mvstats.com.

Stage 2: Developing the Overall Measurement Model

With the constructs specified, the researcher next must specify the measurement model to be tested. In doing so, not only are relationships among constructs defined, but the nature of each construct (reflective versus formative) is also specified.

A visual diagram depicting the measurement model is shown in Figure 12-5. The model displays 21 measured indicator variables and five latent constructs. Without a reason to think the constructs are independent, all constructs are allowed to correlate with all other constructs. All measured items are allowed to load on only one construct each. Moreover, the error terms (not shown in the illustration) are not allowed to relate to any other measured variable, and the measurement model is congeneric. Four constructs are indicated by four measured items and one (JS) is indicated by five measured items. Every individual construct is identified. The overall model has more degrees of freedom than paths to be estimated. Therefore, in a manner consistent with the rule of thumb recommending a minimum of three indicators per construct but encouraging at least four, the order condition is satisfied. In other words, the model is overidentified. Given the number

TABLE 12-2   Observed Indicators Used in HBAT CFA of Employee Behavior

Item	Scale Type	Description	Construct
JS_1	0–10 Likert Disagree–Agree	All things considered, I feel very satisfied when I think about my job.	JS
OC_1	0–10 Likert Disagree–Agree	My work at HBAT gives me a sense of accomplishment.	OC
OC_2	0–10 Likert Disagree–Agree	I am willing to put in a great deal of effort beyond that normally expected to help HBAT be successful.	OC
EP_1	0–10 Likert Disagree–Agree	I am comfortable with my physical work environment at HBAT.	EP
OC_3	0–10 Likert Disagree–Agree	I have a sense of loyalty to HBAT.	OC
OC_4	0–10 Likert Disagree–Agree	I am proud to tell others that I work for HBAT.	OC
EP_2	0–10 Likert Disagree–Agree	The place I work in is designed to help me do my job better.	EP
EP_3	0–10 Likert Disagree–Agree	There are few obstacles to make me less productive in my workplace.	EP
AC_1	5-point Likert	How happy are you with the work of your coworkers? ____ Not happy ____ Somewhat happy ____ Happy ____ Very happy ____ Extremely happy	AC
EP_4	7-point Semantic Differential	What term best describes your work environment at HBAT? Too hectic _____ Very soothing	EP
JS_2	7-point Semantic Differential	When you think of your job, how satisfied do you feel? Not at all satisfied _____ Very much satisfied	JS
JS_3	7-point Semantic Differential	How satisfied are you with your current job at HBAT? Very unsatisfied _____ Very satisfied	JS
AC_2	7-point Semantic Differential	How do you feel about your coworkers? Very unfavorable _____ Very favorable	AC
SI_1	5-point Likert Disagree–Agree	I am not actively searching for another job. Strongly disagree _____ Strongly agree	SI
JS_4	5-point Likert	How satisfied are you with HBAT as an employer? ____ Not at all ____ Little ____ Average ____ A lot ____ Very much	JS
SI_2	5-point Likert Disagree–Agree	I seldom look at the job listings on monster.com. Strongly disagree _____ Strongly agree	SI
JS_5	Percent Satisfaction	Indicate your satisfaction with your current job at HBAT by placing a percentage in the blank, with 0% = Not satisfied at all, and 100% = Highly satisfied. _____	JS
AC_3	5-point Likert	How often do you do things with your coworkers on your days off? ____ Never ____ Rarely ____ Occasionally ____ Often ____ Very often	AC
SI_3	5-point Likert Disagree–Agree	I have no interest in searching for a job in the next year. Strongly disagree _____ Strongly agree	SI
AC_4	6-point Semantic Differential	Generally, how similar are your coworkers to you? Very different _____ Very similar	AC
SI_4	5-point Likert	How likely is it that you will be working at HBAT one year from today? ____ Very unlikely ____ Unlikely ____ Somewhat likely ____ Likely ____ Very likely	SI

of indicators and a sufficient sample size, no problems with the rank condition are expected either. Any such problems should emerge during the analysis.

In the proposed model, all of the measures are hypothesized as reflective. That is, the direction of causality is from the latent construct to the measured items. For instance, an employee's

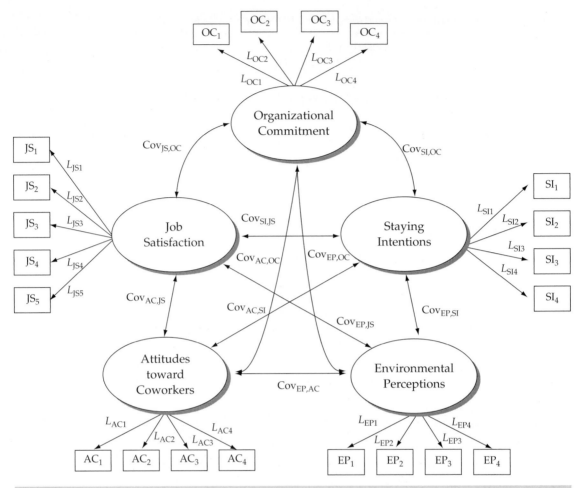

FIGURE 12-5   Measurement Theory Model (CFA) for HBAT Employees

Note: Measured variables are shown with a box with labels corresponding to those shown in the questionnaire. Latent constructs are shown with an oval. Each measured variable has an error term (*e*) associated with it, but they are not shown in the exhibit for simplicity. Two-headed connections indicate covariance between constructs (Cov). One-headed connectors indicate a causal path from a construct to an indicator (*L*).

desire to quit would tend to cause low scores on each of four indicators loading on the Staying Intentions (SI) construct. Each construct also has a series of indicators that share a similar conceptual basis, and empirically they would tend to move together. That is, we would expect that when one changes, systematic change will occur in the other.

Stage 3: Designing a Study to Produce Empirical Results

The next step requires that the study be designed and executed to collect data for testing the measurement model. The researcher must consider issues such as sample size and model specification, particularly in establishing the identification of the model.

HBAT next designed a study to test the measurement model. HBAT's interest was among its hourly employees, not its management team. Therefore, the HBAT personnel department supplied a random sample of 500 employees. The 500 represent employees from each of HBAT's divisions, including their operations in the United States, Europe, Asia, and Australia. Four hundred completed responses were obtained on the scale items described in Table 12-2.

If the model is overidentified, then based on pretests it is expected the communalities will exceed .5, and may exceed .6, and the sample size should be adequate. If the model contained some underidentified factors or if some communalities fall below .5, then a larger sample may be required. The sample size is also sufficient to enable maximum likelihood estimation. Several classification variables also were collected with the questionnaire. Employees were allowed to respond to the questionnaires while they were at work and return them anonymously. Initial screening showed no problems with missing data. Only two responses included any missing data. In one case, an out-of-range response was given, which is treated as a missing response. Using our rule of thumb from the previous chapter, the effective sample size using pairwise deletion (otherwise known as *all-available treatment*) is 399 because it is the minimum number of observations for any observed covariance.

SPECIFYING THE MODEL   Depending on the software you use, different approaches are required at this point. Two of the most popular software packages will be discussed, although many other software packages can be used to obtain identical results.

If you choose to use AMOS, then you begin by using the graphical interface to draw the model depicted in Figure 12-5. Once the model is drawn, you can drag the measured variables into the model and run the software. In contrast, if you choose to use LISREL, you can either use the drop-down menus to generate the syntax that matches the measurement model, or write the appropriate code into a syntax window, and/or generate a path diagram from which models can be estimated.

IDENTIFICATION   Once the measurement model is specified, the researcher is ready to estimate the model. The SEM software will provide a solution for the specified model if everything is properly specified. The default estimation procedure is maximum likelihood, which will be used in this case because preliminary analysis with the data leads HBAT to believe that the distributional properties of the data are acceptable for this approach. The researcher must now choose the remaining options that are needed to properly analyze the results. A more complete discussion of options is included on the text's Web sites at www.pearsonglobaleditions.com/hair or www.mvstats.com.

Table 12-3 shows an initial portion of an output from the CFA results for this model. It provides an easy way to quickly identify the parameters to be estimated and the degrees of freedom for the model. In this case, 52 parameters are to be estimated. Of the 52 free parameters, 16 are factor loadings, 15 represent factor variance and covariance terms, and 21 represent error variance terms. The total number of unique variance and covariance terms is:

$$(21 \times 22)/2 = 231$$

Because 231 is greater than 52, the model is identified with respect to the order condition. It includes more degrees of freedom than free parameters. No problems emerge with the rank condition for identification because we have at least four indicators for each construct. Furthermore, our sample size is sufficient, so we believe the model will converge and produce reliable results. It is an important way of checking the specification to avoid or spot potential identification problems.

Stage 4: Assessing Measurement Model Validity

We now examine the results of testing this measurement theory by comparing the theoretical measurement model against reality, as represented by this sample. Both the overall model fit and the criteria for construct validity must be examined. Therefore, we will review key fit statistics and the parameter estimates.

TABLE 12-3   Parameters to be Estimated in the HBAT CFA Model

Indicator Variable Loadings

	JS	OC	SI	EP	AC
JS_1	0	0	0	0	0
JS_2	1	0	0	0	0
JS_3	2	0	0	0	0
JS_4	3	0	0		0
JS_5	4	0	0	0	
OC_1	0	0	0	0	0
OC_2	0	5	0	0	0
OC_3	0	6	0	0	0
OC_4	0	7	0	0	0
SI_1	0	0	0	0	0
SI_2	0	0	8	0	0
SI_3	0	0	9	0	0
SI_4	0	0	10	0	0
EP_1	0	0	0	0	0
EP_2	0	0	0	11	0
EP_3	0	0	0	12	0
EP_4	0	0	0	13	0
AC_1	0	0	0	0	0
AC_2	0	0	0	0	14
AC_3	0	0	0	0	15
AC_4	0	0	0	0	16

> Note that one loading per construct is not estimated as it will be set to a value of 1.0 to "set the scale." Thus, 16 parameters are estimated for loadings.

Construct Variances and Covariances

	JS	OC	SI	EP	AC
JS	17				
OC	18	19			
SI	20	21	22		
EP	23	24	25	26	
AC	27	28	29	30	31

> Fifteen estimated parameters—one variance for each construct plus the 10 unique covariances among constructs

Error Terms for Indicators (one per indicator) = 21
Total number of estimated parameters: 16 + 15 + 21 = 52

OVERALL FIT   CFA output includes many fit indices. We did not present all possible fit indices. Rather, we will focus on the key GOF values using our rules of thumb to provide some assessment of fit. Each SEM program (AMOS, LISREL, EQS, etc.) includes a slightly different set, but they all contain the key values such as the χ^2 statistic, the CFI, and the RMSEA. They may appear in a different order or perhaps in a tabular format, but you can find enough information to evaluate your model's fit in any SEM program.

Table 12-4 includes selected fit statistics from the CFA output. The overall model χ^2 is 236.62 with 179 degrees of freedom. The p-value associated with this result is .0061. This p-value is significant using a type I error rate of .05. Thus, the χ^2 goodness of fit statistic does not indicate that the observed covariance matrix matches the estimated covariance matrix within sampling variance. However, given the problems associated with using this test alone, and the effective sample size of 399, we examine other fit statistics closely as well.

TABLE 12-4   The HBAT CFA Goodness-of-Fit Statistics

Chi-square (χ^2)
  Chi-square = 236.62 (p = 0.0061)
  Degrees of freedom = 179

Absolute Fit Measures
  Goodness-of-fit index (GFI) = 0.95
  Root mean square error of approximation (RMSEA) = 0.027
  90 percent confidence interval for RMSEA = (0.015; 0.036)
  Root mean square residual (RMR) = 0.086
  Standardized root mean residual (SRMR) = 0.035
  Normed chi-square = 1.32

Incremental Fit Indices
  Normed fit index (NFI) = 0.97
  Non-normed fit index (NNFI) = 0.99
  Comparative fit index (CFI) = 0.99
  Relative fit index (RFI) = 0.97

Parsimony Fit Indices
  Adjusted goodness-of-fit index (AGFI) = 0.93
  Parsimony normed fit index (PNFI) = 0.83

Next we look at several other fit indices. Our rule of thumb suggests that we rely on at least one absolute fit index and one incremental fit index, in addition to the χ^2 results. The value for RMSEA, an absolute fit index, is 0.027. This value appears quite low and is below the .08 guideline for a model with 21 measured variables and a sample size of 399. Using the 90% confidence interval for this RMSEA, we conclude the true value of RMSEA is between 0.015 and 0.036. Thus, even the upper bound of RMSEA is low in this case. The RMSEA therefore provides additional support for model fit. Next we see the standardized root mean square residual (SRMR) with a value of .035, below even the conservative cutoff value of .05. The third absolute fit statistic is the normed χ^2, which is 1.32. This measure is the chi-square value divided by the degrees of freedom (236.62/179 = 1.32). A number smaller than 2.0 is considered very good and between 2.0 and 5.0 is acceptable. Thus, the normed χ^2 suggests an acceptable fit for the CFA model.

Moving to the incremental fit indices, the CFI is the most widely used index. In our HBAT CFA model, CFI has a value of 0.99, which, like the RMSEA, exceeds the CFI guidelines of greater than .90 for a model of this complexity and sample size. The other incremental fit indices also exceed suggested cutoff values. Although this model is not compared to other models, the parsimony index of AGFI has a value (.93), which reflects good model fit.

The CFA results suggest the HBAT measurement model provides a reasonably good fit and thus it is suitable to proceed to further examination of the model results. Issues related to construct validity will be examined next and then the focus shifts to model diagnostics aimed at improving the specified model.

CONSTRUCT VALIDITY   To assess construct validity, we examine convergent, discriminant, and nomological validity. Face validity, as noted earlier, was established based on the content of the corresponding items.

Convergent Validity.   CFA provides a range of information used in evaluating convergent validity. Even though maximum likelihood factor loading estimates are not associated with a specified range of acceptable or unacceptable values, their magnitude, direction, and statistical significance should be evaluated.

We begin by examining the unstandardized factor loading estimates in Table 12-5. In LISREL these are termed lambda values, whereas in AMOS the estimates are referred to as regression weights and are shown under the "Estimates" portion of the output. Loading estimates that are statistically significant provide a useful start in assessing the convergent validity of the measurement model. The results confirm that all loadings in the HBAT model are highly significant as required for convergent validity.

Maximum likelihood estimates are the default option for most SEM programs, including AMOS and LISREL. Unstandardized loadings are provided but they offer little diagnostic information other than directionality and statistical significance. Instead, we examine standardized loadings because they are needed to calculate discriminant validity and reliability estimates. For construct validity, our guidelines are that individual standardized factor loadings (regression weights) should be at least .5, and preferably .7. Moreover, variance extracted measures should equal or exceed 50 percent, and .70 is considered the minimum threshold for construct reliability, except when conducting exploratory research.

Table 12-6 displays standardized loadings (standardized regression weights using AMOS terminology). When we refer to loading estimates, we refer to the standardized values unless otherwise noted. The lowest loading obtained is .58, linking organizational commitment (OC) to item OC1. Two other loading estimates fall just below the .7 standard. The AVE estimates and the construct reliabilities are shown at the bottom of Table 12-6. The AVE estimates range from 51.9 percent for JS to 68.1 percent for AC. All exceed the 50-percent rule of thumb. Construct reliabilities range from .83 for the OC construct to .89 for both SI and AC. Once again, these exceed .7, suggesting adequate reliability. These values were computed using the formulas shown earlier in the chapter when convergent validity was discussed. As of this date, SEM programs do not routinely provide these values.

TABLE 12-5   HBAT CFA Factor Loading Estimates and *t*-values

Indicator	Construct	Estimated Loading	Standard Error	*t*-value
JS_1	JS	1.00	_a	_a
JS_2	JS	1.03	0.08	13.65
JS_3	JS	0.90	0.07	12.49
JS_4	JS	0.91	0.07	12.93
JS_5	JS	1.14	0.09	13.38
OC_1	OC	1.00	_a	_a
OC_2	OC	1.31	0.11	12.17
OC_3	OC	0.78	0.08	10.30
OC_4	OC	1.17	0.10	11.94
SI_1	SI	1.00	_a	_a
SI_2	SI	1.07	0.07	16.01
SI_3	SI	1.06	0.07	16.01
SI_4	SI	1.17	0.06	19.18
EP_1	EP	1.00	_a	_a
EP_2	EP	1.03	0.07	14.31
EP_3	EP	0.80	0.06	13.68
EP_4	EP	0.90	0.06	14.48
AC_1	AC	1.00	_a	_a
AC_2	AC	1.24	0.06	18.36
AC_3	AC	1.04	0.06	18.82
AC_4	AC	1.15	0.06	18.23

[a]Not estimated when loading set to fixed value (i.e., 1.0).

TABLE 12-6   HBAT Standardized Factor Loadings, Average Variance Extracted, and Reliability Estimates

	JS	OC	SI	EP	AC
JS_1	0.74				
JS_2	0.75				
JS_3	0.68				
JS_4	0.70				
JS_5	0.73				
OC_1		0.58			
OC_2		0.88			
OC_3		0.66			
OC_4		0.84			
SI_1			0.81		
SI_2			0.86		
SI_3			0.74		
SI_4			0.85		
EP_1				0.70	
EP_2				0.81	
EP_3				0.77	
EP_4				0.82	
AC_1					0.82
AC_2					0.82
AC_3					0.84
AC_4					0.82
Average Variance Extracted	51.9%	56.3%	66.7%	60.3%	68.1%
Construct Reliability	0.84	0.83	0.89	0.86	0.89

Computed using the formula above as the average squared factor loading (squared multiple correlation).

Computed using the formula from above and the squared sum of the factor loadings.

Taken together, the evidence supports the convergent validity of the measurement model. Although three loading estimates are below .7, two of these are just below the .7 and the other does not appear to be significantly harming model fit or internal consistency. The average variance-extracted estimates all exceed .5 and the reliability estimates all exceed .7. In addition, the model fits relatively well. Therefore, all the items are retained at this point and adequate evidence of convergent validity is provided.

Discriminant Validity. We now turn to discriminant validity. First, we examine the interconstruct covariances. After standardization, the covariances are expressed as correlations. All SEM programs provide the construct correlations whenever standardized results are requested. Some (LISREL) will have a default text output that prints them as an actual correlation matrix. Others (i.e., AMOS) may simply list them in text output. The information is the same.

The conservative approach for establishing discriminant validity compares the AVE estimates for each factor with the squared interconstruct correlations associated with that factor. All AVE estimates from Table 12-6 are greater than the corresponding interconstruct squared correlation estimates in Table 12-7 (above the diagonal). Therefore, this test indicates that there are no problems with discriminant validity for the HBAT CFA model.

TABLE 12-7   HBAT Construct Correlation Matrix (Standardized)

	JS	OC	SI	EP	AC
JS	1.00	.04	.05	.06	.00
OC	0.21***	1.00	.30	.25	.09
SI	0.23***	0.55***	1.00	.31	.10
EP	0.24***	0.50***	0.56***	1.00	.06
AC	0.05	0.30***	0.31***	0.25***	1.00

Significance Level: * = .05, ** = .01, **** = .001

Note: Values below the diagonal are correlation estimates among constructs, diagonal elements are construct variances, and values above the diagonal are squared correlations.

The congeneric measurement model also supports discriminant validity because it does not contain any cross-loadings among either the measured variables or the error terms. This congeneric measurement model provides a good fit and shows little evidence of substantial cross-loadings. Taken together, these results support the discriminant validity of the HBAT measurement model.

Nomological Validity. Assessment of nomological validity is based on the approach outlined in Chapter 3 for EFA. The correlation matrix provides a useful start in this effort to the extent that the constructs are expected to relate to one another. Previous organizational behavior research suggests that more favorable evaluations of all constructs are generally expected to produce positive employee outcomes. For example, these constructs are expected to be positively related to whether an employee wishes to stay at HBAT. Moreover, satisfied employees are more likely to continue working for the same company. Most important, this relationship simply makes sense.

Correlations between the factor scores for each construct are shown in Table 12-7. The results support the prediction that these constructs are positively related to one another. Specifically, satisfaction, organizational commitment, environmental perceptions, and attitudes toward coworkers all have significant positive correlations with staying intentions. In fact, only one correlation is inconsistent with this prediction. The correlation estimate between AC and JS is positive, but not significant (p = 0.87). Because the other correlations are consistent, this one exception is not a major concern.

Nomological validity can also be supported by demonstrating that the constructs are related to other constructs not included in the model in a manner that supports the theoretical framework. Here the researcher must select additional constructs that depict key relationships in the theoretical framework being studied. In addition to the measured variables used as indicators for the constructs, several classification variables such as employee age, gender, and years of experience were also collected. Moreover, the performance of each employee was evaluated by management on a 5-point scale ranging from 1 = "Very Low Performance" to 5 = "Very High Performance." Management provided this information to the consultants who then entered it into the database.

These other measures are helpful in establishing nomological validity. Previous research suggests that job performance is determined by an employee's working conditions [3, 64]. The job performance–job satisfaction relationship is generally positive, but typically not a strong relationship. A positive organizational commitment–job performance relationship also is expected. In contrast, the relationship between job performance and staying is not as clear. Better-performing employees tend to have more job opportunities, which can cancel out the effects of "employees who perform better are more comfortable on the job." A positive environmental perceptions–job performance relationship is expected because one's working conditions directly contribute to how one performs a job. We also expect that experience will be associated with staying intentions. Thus, when intentions to stay are higher, an employee is more likely to remain with an organization. Age and staying intentions are not likely to be highly related. Employees approaching retirement are relatively older and could possibly report lower intentions to stay. This result would interfere with a positive age–staying intentions relationship that might otherwise exist.

Correlations between these three items and the factor scores for each measurement model construct are shown in Table 12-8. Correlations corresponding to the predictions made in the previous paragraph can be compared with the results. This comparison shows that the correlations are consistent with the theoretical expectations as described. Therefore, the analysis of the correlations among the measurement model constructs and the analysis of correlations between these constructs and other variables both support the nomological validity of the model.

MODIFYING THE MEASUREMENT MODEL In addition to evaluating goodness-of-fit statistics, the researcher must also check a number of model diagnostics. They may suggest some way to further improve the model or perhaps some specific problem area not revealed to this point. The following diagnostic measures from CFA should be checked: path estimates, standardized residuals, and modification indices.

Path Estimates. Evaluation of the loadings of each indicator on a construct provides the researcher with evidence of the indicators that may be candidates for elimination. Loadings below the suggested cutoff values should be evaluated for deletion, but the decision is not made based just on the loadings, but on the other diagnostic measures as well.

Results are positive to this point. Even with good fit statistics, however, HBAT should check the model diagnostics. The path estimates were examined earlier. One loading estimate—the .58 associated with OC_1—was noted because it fell below the ideal loading cutoff of .7. It did not appear to be causing problems, however, because the fit remained high. If other diagnostic information suggests a problem with this variable, action may be needed.

Standardized Residuals. The next diagnostic measures are the standardized residuals. The LISREL output shows the highest standardized residuals (e.g., greater than $|2.5|$), which prevents the researcher from having to search through all of the residuals. This can be a substantial task because a residual term is computed for every covariance and variance term in the observed covariance matrix.

The HBAT CFA model has 231 residuals (remember, this was the number of unique elements in the observed covariance matrix). We will not display them all here. In Table 12-9 we show all standardized residuals greater than $|2.5|$. No standardized residuals exceed $|4.0|$, the benchmark value that may indicate a problem with one of the measures. Those between $|2.5|$ and $|4.0|$ also may deserve attention if the other diagnostics indicate a problem. The largest residual is 3.80 for the covariance between SI_2 and SI_1. Both of these variables have a loading estimate greater than .8 on the SI construct. This residual may be explained by the content of the items. In this case, SI_2 and SI_1 may have slightly more in common with each other contentwise than they do with SI_3 and SI_4, the other two items representing SI.

The HBAT analyst decides not to take action in this case given the high reliability and high variance extracted for the construct. In addition, the model fit does not suggest a great need for improvement. Three of the highest negative residuals are associated with variable OC_1, which also

TABLE 12-8 **Correlations Between Constructs and Age, Experience, and Job Performance**

	JS	OC	SI	EP	AC
Job Performance (JP)	.15 (.003)	.27 (.000)	.10 (.041)	.29 (.000)	.06 (.216)
Age	.14 (.005)	.12 (.021)	.06 (.233)	−.01 (.861)	.15 (.003)
Experience (EXP)	.08 (.110)	.07 (.159)	.15 (.004)	.01 (.843)	.12 (.018)

*Note: p-*values shown in parentheses.

TABLE 12-9   Model Diagnostics for the HBAT CFA Model

Standardized Residuals (all residuals greater than |2.5|)

Negative Standardized Residuals

SI_3	and	OC_1	-2.68
SI_4	and	OC_1	-2.74
EP_3	and	OC_1	-2.59

Positive Standardized Residuals

SI_2	and	SI_1	3.80
SI_4	and	SI_3	3.07
EP_2	and	OC_3	2.98
EP_4	and	OC_3	2.88
EP_4	and	EP_3	3.28

Modification Indices for factor loadings

	JS	OC	SI	EP	AC
JS_1	—	0.19	1.44	2.71	0.69
JS_2	—	2.11	0.32	0.53	2.55
JS_3	—	0.00	0.29	0.16	0.00
JS_4	—	0.59	0.09	0.40	0.10
JS_5	—	3.20	2.59	1.38	4.96
OC_1	0.64	—	10.86	3.02	2.75
OC_2	0.07	—	10.84	0.51	7.14
OC_3	1.01	—	3.15	7.59	1.86
OC_4	0.00	—	0.07	0.02	1.02
SI_1	0.00	0.00	—	0.29	0.02
SI_2	1.89	0.08	—	1.66	0.59
SI_3	0.15	1.85	—	0.10	0.00
SI_4	2.78	0.55	—	2.46	0.37
EP_1	0.10	1.85	1.74	—	0.05
EP_2	0.11	3.48	0.78	—	0.53
EP_3	0.31	0.17	3.00	—	0.00
EP_4	0.17	0.17	0.11	—	0.85
AC_1	0.70	0.38	0.02	0.07	—
AC_2	0.43	2.45	0.84	0.22	—
AC_3	1.59	0.07	0.02	0.89	—
AC_4	1.29	3.70	0.89	3.01	—

is the variable with the lowest loading estimate (.58). Again, no action is taken at this point given the overall positive results. If a residual associated with OC_1 exceeded |4.0|, however, or if the model fit was marginal, OC_1 would be a prime candidate for being dropped from the model. In this case, the congeneric representation, which meets the standards of good measurement practice, appears to hold quite well.

Modification Indices. Modification indices (MI) are calculated for every fixed parameter (i.e., all of the possible parameters that were not estimated in the model). The two sets of MIs most useful in a CFA are for the factor loadings and the error terms between items. Note that there are generally not any MIs for the relationships between constructs because each construct has an estimated path to every other construct. As you would expect, a full listing of all modification indices is quite extensive and will not be provided here. Instead, we will identify the largest MI and also examine the MIs for the factor loadings.

First, the largest modification index is 14.44 for the covariance of the error terms of SI_1 and SI_2. (the full output can be found at the text's Web sites at www.pearsonglobaleditions.com/hair or www.mvstats.com). Although the modification indices for the error term correlations are somewhat useful in diagnosing problems with specific items, the researcher should avoid making model respecifications that involve correlated error terms.

The second type of MI that is quite useful in a CFA is for the factor loadings. As you can see in Table 12-9, each item has a modification index for all the constructs except the one it is hypothesized to relate to. This provides the researcher with an empirical estimate of how strongly each item is associated with other constructs (i.e., the potential for cross-loadings). As you can see, most of the values above 4.0 are associated with the items in the OC construct, which have fairly large values for the SI, EP, and AC constructs. OC_2 may be most problematic in that it has high values for both SI and AC, although OC_1 and OC_3 also have high values with at least one other construct. This may indicate some lack of unidimensionality for these two items, and this is reinforced by the fact that they have the two lowest standardized loadings across all items. But elimination of both items would violate the three-indicator rule, so they will be retained at this time.

A further specification search is not needed because the model has a solid theoretical foundation and the CFA is testing rather than developing a model. If the fit were poor, however, a specification search could take place as described earlier in the chapter. Such an effort would rely heavily on the combined diagnostics provided by the factor loading estimates, the standardized residuals, and the modification indices. At this point, HBAT can proceed with confidence that the questionnaire measures these key constructs well.

HBAT CFA Summary

Four SEM stages are complete. The CFA results generally support the measurement model. The χ^2 statistic is significant above the .01 level, which is not unusual given a total sample size of 400 (with an effective sample size of 399 using the all-available [PD] approach). Both the CFI and RMSEA appear quite good. Overall, the fit statistics suggest that the estimated model reproduces the sample covariance matrix reasonably well. Further, evidence of construct validity is present in terms of convergent, discriminant, and nomological validity. Thus, HBAT can be fairly confident at this point that the measures behave as they should in terms of the unidimensionality of the five measures and in the way the constructs relate to other measures. Remember, however, that even a good fit is no guarantee that some other combination of the 21 measured variables would not provide an equal or better fit. The fact that the results are conceptually consistent is of even greater importance than are fit results alone.

PART 2: WHAT IS A STRUCTURAL MODEL?

In the previous section, we learned that the goal of measurement theory is to produce ways of measuring concepts in a reliable and valid manner. Measurement theories are tested by how well the indicator variables of theoretical constructs relate to one another. The relationships between the indicators are captured in a covariance matrix. CFA tests a measurement theory by providing evidence on the validity of individual measures based on the model's overall fit and other evidence of construct validity. CFA alone is limited in its ability to examine the nature of relationships between constructs beyond simple correlations. A measurement theory then is often a means to an end of examining relationships between constructs, not an end in itself.

A **structural theory** is a conceptual representation of the **structural relationships** between constructs. It can be expressed in terms of a **structural model** that represents the theory with a set of structural equations and is usually depicted with a visual diagram. The structural relationship between any two constructs is represented empirically by the **structural parameter estimate,** also known as a **path estimate.** Structural models are referred to by several terms, including a theoretical model or, occasionally, a **causal model.** A causal model infers that the relationships meet the conditions necessary for causation. The conditions for causality were discussed in Chapter 11 and the researcher should be careful not to depict the model as having causal inferences unless all of the conditions are met.

A SIMPLE EXAMPLE OF A STRUCTURAL MODEL

The transition from a measurement model to a structural model is strictly the application of the structural theory in terms of relationships among constructs. Recall from Part 1 of this chapter that a measurement model typically represents all constructs with noncausal or correlational relationships among them. The structural model applies the structural theory by specifying which constructs are related to each other and the nature of each relationship. We will revisit our simple measurement model example from Part 1 to illustrate this point.

Figure 12-6 shows a two-construct structural model. The assumption now is that the first construct, *Supervisor Support,* is related to *Job Satisfaction* in a way that the relationship can be expressed as a regression coefficient. In the figure this relationship is shown as a structural relationship and labeled with a P = path estimate. In a causal theory, the model would imply that Supervisor Support causes or helps bring about Job Satisfaction.

The diagram in Figure 12-6 is similar to the CFA model. But when we move from measurement (CFA) to structural models (SEM), there are some changes in abbreviations, terminology, and notation. No changes are made to the left side of the diagram representing the Supervisor Support construct. But there are changes in other areas including the following:

- The relationship between the Supervisor Support and Job Satisfaction constructs in a CFA would be represented by a two-headed curved arrow (correlational relationship). This relationship changes to a dependence relationship and is now represented in Figure 12-6 by a single-headed arrow (P = path estimate). This arrow can be thought of as a relationship that is represented in a multiple regression model and estimated by a regression coefficient. This path shows the direction of the relationship in a structural model and represents the structural relationship that will be estimated. However, it would rarely have the same value using SEM because more information is used in deriving its value, including information that allows a correction for measurement error. Structural models in SEM differ from CFA models, therefore, because the emphasis moves from the relationship between latent constructs and measured indicator variables to the nature and magnitude of the relationships between constructs.
- The constructs are now identified differently. In CFA exogenous and endogenous constructs are not distinguished, but with a structural model we must distinguish between exogenous and endogenous constructs. The traditional independent variables are now labeled exogenous and

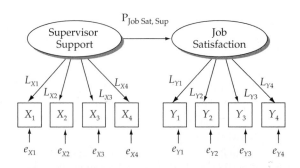

FIGURE 12-6 Visual Representation (Path Diagram) of a Simple Structural Theory

are still connected by correlations (double-headed curved arrows). Traditional dependent variables are now labeled endogenous. Theory is tested by examining the effect of exogenous constructs (predictors) on endogenous constructs (outcomes). Also, with most SEM models there is more than one endogenous construct so you have one endogenous construct predicting another. In such cases, one or more of the endogenous constructs operate as both predictor and outcome variables. This situation is not represented in Figure 12-6 but will be shown in later examples.

- The measured indicator variables (items) are no longer all represented by the letter X. Only the indicator variables for the exogenous construct are still represented by X. In contrast, the indicator variables for the endogenous construct are now represented by Y. This is a typical distinction in structural models and is consistent with the approach used in other multivariate procedures (X associated with predictors and Y associated with outcomes).
- The error variance terms now also have a notation that matches the exogenous–endogenous distinction. Error terms for all the variables are now labeled by the appropriate item (i.e., X variables or Y variables and the item number).
- The loading estimates are also changed to indicate exogenous or endogenous constructs. Variable loading estimates for exogenous constructs are represented by X and the item number, whereas variable loading estimates for endogenous constructs are represented by Y and the number.

Structural models differ from measurement models in that the emphasis moves from the relationship between latent constructs and measured variables to the nature and magnitude of the relationships between constructs. Measurement models are tested using CFA. The CFA model is then altered based on the nature of relationships among constructs. The result is a structural model specification that is used to test the hypothesized theoretical model.

With these theoretical distinctions between CFA and SEM represented in the path diagram, we now move on to estimate the structural model using SEM procedures. Note that in this type of situation the observed covariance model does not change between models. Differences in model fit are based solely on the different relationships represented in the structural model.

AN OVERVIEW OF THEORY TESTING WITH SEM

Given that the measurement model has already been examined and validated in a CFA analysis, the focus in a SEM analysis is testing structural relationships by examining two issues: (1) overall and relative model fit as a measure of acceptance of the proposed model and (2) structural parameter estimates, depicted with one-headed arrows on a path diagram.

The theoretical model shown in Figure 12-7 is evaluated based on how well it reproduces the observed covariance matrix and on the significance and direction of the hypothesized paths. Note that in this figure we are also identifying each of the hypothesized relationships in the path diagram (H_{number}). If the model shows good fit, and if the hypothesized paths are significant and in the direction hypothesized (H_1, H_2, and H_3 are positive and H_4 is negative), then the model is supported. But good fit does not mean that some alternative model might not fit better or be more accurate. Thus, further verification may be needed to ensure nomological validity, which is simply a theoretical plausibility test. In other words, the researcher checks to see if the relationships make sense. In short, the relationships must be consistent with the theory that suggests they should be positive or negative. If the relationships do not make sense, they should not be relied upon.

The estimation of the structural parameter estimates is through the same process used in CFA models. The primary distinction is that the structural model does not have all the constructs related to each other as is done in CFA. Thus, the structural model replaces the correlational relationships with dependence relationships. In doing so we introduce the concept of direct and indirect effects. The derivation of the path estimates and the identification of direct and indirect effects are described in the

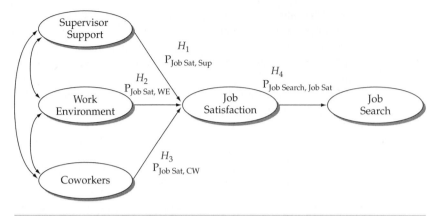

FIGURE 12-7 Expanded "Theoretical Model" of Job Search

Basic Stats appendix and materials available on the text's Web sites (www.pearsonglobaleditions.com/hair or www.mvstats.com). The reader may find this information helpful in understanding the full impact of any dependence relationship.

STAGES IN TESTING STRUCTURAL THEORY

Theory testing with SEM closely follows measurement theory testing using CFA. The process is similar conceptually in that a theory is proposed and then tested based on how well it fits the data. Now as we deal with the theoretical relationships between constructs, greater attention is focused on the different types of relationships that may exist.

One-Step Versus Two-Step Approaches

Even though SEM has the advantage of simultaneously estimating the measurement model and the structural model, our six-stage overall process is consistent with a **two-step SEM process** [1]. By two steps, we mean that in the first step we test the fit and construct validity of the proposed measurement model. Once a satisfactory measurement model is obtained, the second step is to test the structural theory. Thus, two key tests—one measurement and one structural—assess fit and validity. In Chapter 11, we referred to this concept as the two parts of SEM testing. Thus, the measurement model fit provides a basis for assessing the validity of the structural theory [2].

Some argue for the superiority of a one-step approach in which the overall fit of a model is tested without regard to separate measurement and structural models [34]. Yet a one-step model provides only one key test of fit and validity. It does not separate the measurement model assessment from the structural model assessment.

The authors recommend separate testing of the measurement model via a two-step approach as essential because valid structural theory tests cannot be conducted with bad measures. A valid measurement model is essential because with poor measures we would not know what the constructs truly mean. Therefore, if a measurement model cannot be validated, researchers should first refine their measures and collect new data. If the revised measurement model can be validated, then and only then do we advise proceeding with a test of the full structural model. A more detailed discussion of this issue is presented later in the chapter in the section on interpretational confounding.

The six SEM stages now continue. Stages 1–4 covered the CFA process from identifying model constructs to assessing the measurement model validity (see Part 1 of the chapter). If the measurement is deemed sufficiently valid, then the researcher can test a structural model composed of these measures, bringing us to stages 5 and 6 of the SEM process. Stage 5 involves specifying the structural model and stage 6 involves assessing its validity.

STAGE 5: SPECIFYING THE STRUCTURAL MODEL

We turn now to the task of specifying the structural model. This process involves determining the appropriate unit of analysis, representing the theory visually using a path diagram, clarifying which constructs are exogenous and endogenous, and several related issues such as sample size and identification.

Unit of Analysis

One issue not visible in a model is the **unit of analysis.** The researcher must ensure that the model's measures capture the appropriate unit of analysis. For instance, organizational researchers often face the choice of testing relationships representing individual perceptions versus the organization or business unit as a whole. Marketing researchers also study organizations, but they sometimes look at an exchange dyad (buyer and seller), retail store or advertisement as the unit of analysis. Individual perceptions represent each person's opinions or feelings. Organizational factors represent characteristics that describe an individual organization. A construct such as employee *esprit de corps* may well exist at both the individual and organizational levels. *Esprit de corps* can be thought of as how much enthusiasm an employee has for the work and the firm. In this way, one employee can be compared to another. But it also can be thought of as a characteristic of the firm overall. In this way, one firm can be compared to another and the sample size is now determined by how many firms are measured rather than by the number of individual respondents. The choice of unit of analysis determines how a scale is treated.

For example, a multiple-item scale could be used to assess the *esprit de corps* construct. If the desired unit of analysis is at the individual level and we want to understand relationships that exist among individuals, the research can proceed with individual responses. However, if the unit of analysis is the organization, or any other group, responses must be aggregated over all individuals responding for that particular group. Thus, organizational-level studies require considerably more data, because multiple responses must be aggregated into one group. Organizations also possess many non-latent characteristics such as size, location, and organizational structure.

Once the unit of analysis is decided and data are collected, the researcher must aggregate the data if group-level responses are used to set up the appropriate SEM. If the unit of analysis is the individual, the researcher can proceed as before. Sometimes, multiple units of analysis are included in the same model. For instance, organizational culture may cause an individual's job satisfaction. The term multilevel model refers to these analyses.

Model Specification Using a Path Diagram

We now consider how a theory is represented by visual diagrams. Paths indicate relationships. *Fixed parameters* are relationships that will not be estimated by the SEM routine; they are typically assumed to be set at zero and are not shown on a visual diagram. *Free parameters* are relationships that will be estimated; they are generally depicted by an arrow in a visual diagram.

Figure 12-7 included both fixed and free parameters. For example, no relationship is specified between Supervisor Support and Job Search. Therefore, no arrow is shown, and the theory assumes that this path is fixed at zero. However, there is a path between Job Satisfaction and Job Search that represents the relationship between these two constructs and for which a parameter will be estimated.

The parameters representing structural relationships between constructs are now our focus. These are in many ways the equivalent of regression coefficients and can be interpreted in a similar way. With SEM these parameters are divided into two types: (1) relationships between exogenous and endogenous constructs and (2) relationships between two endogenous constructs. Some software programs make a distinction between these two types (e.g., LISREL), whereas others (e.g., AMOS) consider all structural relationships similarly. For more detail on LISREL notation, refer to materials on the text's Web sites (www.pearsonglobaleditions.com/hair or www.mvstats.com).

STARTING WITH A MEASUREMENT MODEL   Once a theory is proposed, the first step is specifying the measurement theory and validating it with CFA. This allows for the full focus to be on establishing construct validity for all of the constructs. Only after that is the structural theory represented by specifying the set of relationships between constructs. Some relationships will be estimated, meaning that the theory states that two constructs are related to one another. In contrast, some relationships will be fixed, meaning that the theory states that the two constructs are not related to one another.

Figure 12-8 shows a CFA model that is converted into a subsequent structural model. The constructs are based on the previous employee Job Satisfaction example. Figure 12-8a shows a CFA that tests the measurement model. Each construct is indicated by four indicator items. Thus, four latent constructs are measured by 16 measured indicator variables (X_1–X_{16}). The error variance terms are not shown in the exhibit, but each of the 16 indicator items also has a corresponding error variance term that is estimated in the CFA model. Relationships between constructs are estimated by correlational relationships (Cov). In this case, there are six covariance/correlation terms between constructs.

TRANSFORMING TO A STRUCTURAL MODEL   The primary objective is to specify the structural model relationships as replacements for the correlational relationships found in the CFA model. This process, however, also involves a series of other changes, some just in terms of notation and others of more substantive issues (e.g., changing from exogenous to endogenous constructs). The following section describes the theoretical and notational changes involved in this process.

Theoretical Changes.   Specifying the structural model based on the measurement model necessitates the use of structural relationships (single-headed arrows for the hypothesized causal relationships)

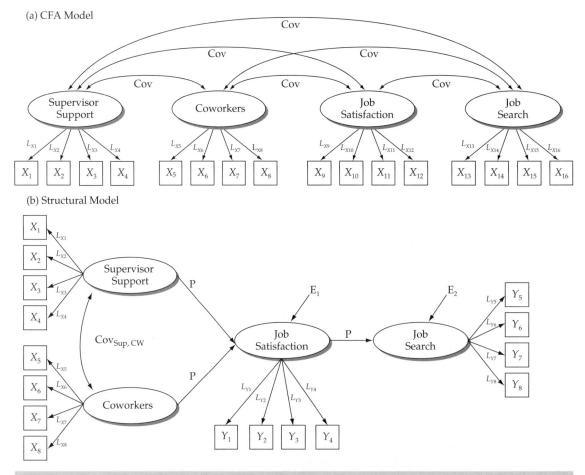

FIGURE 12-8   Changing a CFA Model to a Structural Model

Note: The error variance terms for the loadings are omitted from the diagram for simplicity. However, in SEM path diagrams, particularly those used in AMOS, each error term should be included as shown in Chapter 10.

in place of the correlational relationships among constructs used in CFA. The structural theory is specified by using free parameters (those to be estimated) and fixed parameters (those fixed at a particular value, usually zero) to represent hypothesized relationships. However, in specifying the structural relationships two other significant changes occur. First, a distinction must be made between endogenous and exogenous constructs. Recall that in the CFA model this distinction is not made. But now those constructs that act as outcomes (i.e., structural relationships predict them) must be specified as endogenous. Endogenous constructs are easily recognized in the path diagram because they have one or more arrows depicting structural relationships pointing toward them. A second change is that endogenous constructs are not fully explained and so each is associated with an error term (E). The user should be careful to insert an error term for endogenous constructs when using AMOS. LISREL includes the error term for endogenous constructs by default.

Note that although it may seem that changing a relationship from a correlational to a dependence relationship only involves changing a double-headed arrow to a single-headed one, the implications for the estimation of parameters in the model may be substantial. As described in the Basic Stats appendix, parameter estimates are calculated based on the direct and indirect effects in the model. Specifying a relationship as a dependence relationship specifies constraints on which effects are used in estimating the path estimate. Thus, specifying a relationship as causal has theoretical and practical implications and should only be done within an established theoretical model.

Notational Changes. The second type of change is one that is primarily notational to reflect either the change in the type of relationship (correlational to structural) or type of construct (exogenous versus endogenous). As we have discussed earlier, the underlying indicators do not change, but their notation may change. In a CFA model, all indicators used the designation of X. However, indicators of endogenous constructs are distinguished by Y. This impacts not only the item labeling but also the notation used for factor loadings and error terms. Remember, the underlying observed measures do not change, just their notation in the model.

We will now demonstrate the changes that occur in translating the CFA model described earlier in the chapter into a SEM model with hypothesized structural relationships. Let's assume the employee job satisfaction theory hypothesized that Supervisor Support and Coworkers are related to Job Satisfaction. This implies a single structural relationship with Job Satisfaction as a function of Supervisor Support and Coworkers. When Job Search is included in the theoretical model, it is viewed as an outcome of Job Satisfaction. Supervisor Support and Coworkers are not hypothesized as being directly related to Job Search.

Figure 12-8b corresponds to this structural theory. Several changes can be seen in transforming the CFA measurement model into the SEM structural model:

Theoretical Changes

1. Based on the proposed theory, there are two exogenous constructs and two endogenous constructs. Supervisor Support and Coworkers are exogenous based on our theory, and the model therefore has no arrows pointing at them. Job Satisfaction is a function of Supervisor Support and Coworkers and is therefore endogenous (arrows pointing at it). Job Search is a function of Job Satisfaction and therefore is also endogenous. Thus, the representation of Supervisor Support and Coworkers is not changed.
2. The hypothesized relationships between Supervisor Support and Job Satisfaction and Coworkers and Job Satisfaction, as well as the relationship between Job Satisfaction and Job Search, are all represented by a P (path coefficient).
3. No relationships are shown between Supervisor Support and Job Search or Coworkers and Job Search because they are fixed at zero. That is, the theory does not hypothesize a direct relationship between either Supervisor Support and Job Search or Coworkers and Job Search.
4. The hypothesized relationship between Supervisor Support and Coworkers remains a correlational relationship and is still a two-headed arrow and represented by Cov.

Notational Changes

1. The measured indicator variables for the exogenous constructs are still identified as X_1 to X_8. But the measured indicator variables for the endogenous constructs are now identified as Y_1 to Y_8.
2. The parameter coefficients representing the loading paths for exogenous constructs are still identified as L_{X1} to L_{X8}. In contrast, the parameter coefficients representing the loading paths for endogenous constructs are now identified as L_{Y1} to L_{Y8}.
3. Two new terms appear: E_1 and E_2. They represent the error variance of prediction for the two endogenous constructs. After these error variances are standardized, they can be thought of as the opposite of an R^2. That is, they are similar to the residual in regression analysis.

Degrees of Freedom.   Computation of the degrees of freedom in a structural model proceeds in the same fashion as the CFA model, except that the number of estimated parameters is generally smaller. First, because the number of indicators doesn't change, neither does the total number of degrees of freedom available. What changes in most situations is the number of structural relationships between constructs rather than the full set of correlational relationships among constructs in CFA.

In our example, 16 indicators result in a total of 136 unique values in the covariance matrix $[(N \times N + 1)/2 = (16 \times 17)/2 = 136]$. In Part 1 we learned that there were 38 estimated parameters for a value of 98 degrees of freedom in the CFA model.

For the structural model, we still have the same 136 unique values in the covariance matrix, but the degrees of freedom now differ in the relationships between constructs. We still have 32 estimated parameters for the indicators (a loading and error term for each of the 16 indicators). Now, however, instead of the six correlational relationships between the four constructs, we have one correlational relationship (Supervisor Support $\leftrightarrow$ Coworkers) and three structural relationships (Supervisor Support $\rightarrow$ Job Satisfaction; Coworkers $\rightarrow$ Job Satisfaction; Job Satisfaction $\rightarrow$ Job Search). This gives a total of 36 free parameters or 100 degrees of freedom $(136 - 36 = 100)$. The two additional degrees of freedom come from not specifying (i.e., constraining to 0) the direct relationships from Supervisor Support and Coworkers to Job Search. Rather, they both are directly related to Job Satisfaction, which in turn directly predicts Job Search.

The process of making these changes can be somewhat substantial in most software programs. In LISREL, additional matrices are introduced (e.g., gamma, beta, and zeta). With AMOS, the user needs to change the arrows using the graphical interface to show how the correlational CFA relationships become SEM dependence relationships and add error terms to each endogenous construct. Materials on the text's Web sites (www.pearsonglobaleditions.com/hair or www.mvstats.com) describe this process in more detail.

Recursive Versus Nonrecursive Models. One final distinction that must be made when specifying the structural model is if it is to be a recursive or nonrecursive model. A model is considered recursive if the paths between constructs all proceed only from the predictor (antecedent) construct to the dependent or outcome construct (consequences). In other words, a recursive model does not contain any constructs that are both determined by some antecedent and help determine that antecedent (i.e., no pair of constructs has arrows going both ways between them). This is the type of model we have used in examples in this and past chapters. Recursive SEM models will never have fewer degrees of freedom than a CFA model involving the same constructs and variables.

In contrast, a **nonrecursive model** contains feedback loops. A **feedback loop** exists when a construct is seen as both a predictor and an outcome of another single construct. The feedback loop can involve direct or even indirect relationships. In the indirect relationship the feedback occurs through a series of paths or even through correlated error terms.

Figure 12-9 shows a structural model that is nonrecursive. Notice that the construct Job Search is both determined by and determines Job Satisfaction. The parameters for the model include the path coefficients corresponding to both of these paths (P). If the model included a path from Job Search back to Coworkers, the model would also be nonrecursive. This is because Job Search would

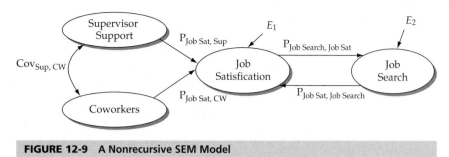

FIGURE 12-9   A Nonrecursive SEM Model

be determined indirectly by Coworkers through Job Satisfaction, and Coworkers would be directly determined by Job Search with the new path.

A theoretical interpretation of a nonrecursive relationship between two constructs is that one is both a cause and effect of the other. Although this situation is unlikely with cross-sectional data, it becomes more plausible with longitudinal data. It is difficult to produce a set of conditions that support a reciprocal relationship with cross-sectional data.

For instance, both intelligence and success in school can be thought of as latent constructs measured by multiple items. Does intelligence cause success in school or does success in school cause intelligence? Could it be that both are causal influences on each other? Longitudinal data may help sort out this issue because the time sequence of events can be taken into account.

Nonrecursive models many times have problems with statistical identification. By including additional constructs and/or measured variables, we can help ensure that the order condition is met. The rank condition for identification could remain problematic, however, because a unique estimate for a single parameter may no longer exist (see earlier discussion in this chapter). Therefore, we recommend avoiding nonrecursive models, particularly with cross-sectional data.

Designing the Study

Whenever SEM is used, sample size and identification are both important issues. Earlier we covered conditions for identification with an adequate sample size across various situations. If these conditions are satisfied for the CFA model, they are likely satisfied for the structural model too, especially for recursive structural models. A structural model is nested within a CFA model and is more parsimonious because it contains fewer estimated paths with the same number of constructs and items. Therefore, if the CFA model is identified, the structural model also should be identified, but only as long as the model is recursive, no interaction terms are included, the sample size is adequate, and a minimum of three measured items per construct is used. We now turn to some other issues that may occur in transitioning from a measurement model to a structural model.

SINGLE-ITEM MEASURES   Occasionally a structural model will involve a single-item measure. That is, structural relationships are hypothesized between a single variable and latent constructs. The problem is that a single-item measure's reliability and validity cannot be computed. The question then becomes, how can a single-item measure be represented within a CFA/SEM framework? How is it specified? Because its measurement characteristics are unknown, it requires the researcher's best judgment to fix the measurement parameter associated with the single item.

Many variables can be measured by single items of interest. Many times specific behaviors or outcomes (e.g., purchase/nonpurchase, amount spent, compensation levels, sales, etc.) have only a single measure. Moreover, specific characteristics may have quite specific measures that are widely accepted (e.g., household income). Finally, many times summated scale values are available and need to be included in the analysis. All of these variables are widely used in regression models, yet have only single measures. The question arises: How do we incorporate them into our SEM model?

The variable representing the single item would be incorporated into the observed covariance matrix just like any other item. The difference, however, is that it will be the only item associated with its construct. Thus, one measurement path links them together. As noted, the primary problem with single-item measures is that they are underidentified and their loading and error terms cannot be estimated. So the researcher must specify both of these values.

Factor loadings and error terms for single-item constructs should be set based on the best knowledge available. The construct can be thought of as the "true" value and the variable as the observed value. If the researcher feels there is very little measurement error (i.e., the "true" score is very close to the observed score), then a high loading value and corresponding low error term will be specified. For example, if the researcher felt that there was no error in the observed value, then the loading would be set to 1.0 and the error term to zero. We should note that this is the situation in multiple regression and other multivariate techniques where we assumed no measurement error. But the assumption of no error may not hold, so the researcher may need to specify some amount of error. In that case, the factor loading is then set (fixed) to the square root of the estimated reliability. The corresponding error term is set to 1, which is the reliability estimate.

As noted in the discussion of identification, single-item measures can create identification problems in SEM models; thus, we suggest their use be limited. Given the nature of SEM, latent constructs represented by multiple items are the preferred approach. However, in instances in which single-item measures are required, then the procedures discussed here are available. We should note that the "conservative" approach is to assume low or no measurement error, because this will result in little impact on the observed covariances.

MODELING THE CONSTRUCT LOADINGS WHEN TESTING STRUCTURAL THEORY   The CFA model in Figure 12-8a is modified to test the structural model shown in the bottom portion. The measurement portion of the structural model consists of the loading estimates for the measured items and the correlation estimates between exogenous constructs. The factor loading estimates from CFA can be treated several different ways in the structural model.

One argument suggests that, with the CFA model already estimated at this point, the factor loading estimates are known. Therefore, their values should be fixed and specified to the loading estimates obtained from the CFA model. In other words, they should no longer be free parameter estimates. Similarly, because the error variance terms are provided from the CFA, their values can also be fixed rather than estimated.

The rationale for fixing these values is that they are "known" and should not be subject to change because of relationships specified in the structural model. If they would change, this would be evidence of **interpretational confounding,** which means that the measurement estimates for one construct are being significantly affected by the pattern of relationships between constructs. In other words, the loadings for any given construct should not change noticeably just because a change is made to the structural model. An advantage to this approach is that the structural model is easier to estimate because so many more parameters have values that are fixed. A disadvantage is that the change in fit between the CFA and the structural model may be due to problems with the measures instead of with the structural theory.

Another approach is to use the CFA factor pattern and allow the coefficients for the loadings and the error variance terms to be estimated along with the structural model coefficients. It simplifies the transition from the CFA to the structural testing stage by eliminating the need to go through the process of fixing all the construct loading estimates and error variance terms to the CFA values. The process also can reveal any interpretational confounding by comparing the CFA loading estimates with those obtained from the structural model. If the standardized loading estimates vary substantially, then evidence of interpretational confounding exists. Small fluctuations are expected (.05 or less). As inconsistencies increase in size and number, however, the researcher should

---

RULES OF THUMB 12-6

Specifying the Structural Model

- CFA is limited in its ability to examine the nature of relationships between constructs beyond simple correlations
- A structural model should be tested after CFA has validated the measurement model
- The structural relationships between constructs can be created by:
  - Replacing the two-headed arrows from CFA with single-headed arrows representing a cause-and-effect type relationship
  - Removing the two-headed curved arrows connecting constructs that are not hypothesized to be directly related
- Recursive SEM models cannot be associated with fewer degrees of freedom than a CFA model involving the same constructs and variables
- Nonrecursive models involving cross-sectional data should be avoided in most instances:
  - It is difficult to produce a set of conditions that could support a test of a reciprocal relationship with cross-sectional data
  - Nonrecursive models yield more problems with statistical identification
- When a structural model is being specified, it should use the CFA factor pattern corresponding to the measurement theory and allow the coefficients for the loadings and the error variance terms to be estimated along with the structural model coefficients
- Measurement paths and error variance terms for constructs measured by only a single item (single measured variables or summated construct scores) should be based on the best knowledge available. When measurement error for a single item is modeled:
  - The loading estimate between the variable and the latent construct is set (fixed) to the square root of the best estimate of its reliability
  - The corresponding error term is set (fixed) to 1 minus the reliability estimate

---

examine the measures more closely. Another advantage of this approach is that the original CFA model fit becomes a convenient basis of comparison in assessing the fit for the structural model. This approach is used most often in practice, and it is the one recommended here.

STAGE 6: ASSESSING THE STRUCTURAL MODEL VALIDITY

The final stage of the decision process evaluates the validity of the structural model based on the a comparison of the structural model fit compared to the CFA model as well as an examination of model diagnostics. The comparison of structural model fit to the CFA model assesses the degree to which the structural model decreases model fit due to its specified relationships. Here the researcher also determines the degree to which each specified relationship is supported by the estimated model (i.e., the statistical significance of each hypothesized path). Finally, model diagnostics are used to determine if any model respecification is indicated.

Understanding Structural Model Fit from CFA Fit

This stage assesses the structural model's validity. The observed data are still represented by the observed sample covariance matrix, which will be compared to the estimated covariance matrix. In CFA, the estimated covariance matrix is computed based on the restrictions (pattern of free and fixed parameter estimates) corresponding to the measurement theory. As long as the structural theory is recursive, then it cannot include more relationships between constructs than can the CFA model from which it is developed. Therefore, a recursive structural model cannot have a lower χ^2 value than that obtained in CFA. The practical implication is that a structural model will not improve model fit when compared to a CFA model. Researchers are mistaken if they hope to "fix" a poorly performing CFA model in their structural model. If adequate fit was not found in the CFA model, the focus should be on improving that model before moving on to the structural model.

SATURATED THEORETICAL MODELS If the SEM model specifies the same number of structural relationships as are possible construct correlations in the CFA, the model is considered a **saturated structural model.** Saturated theoretical models are not generally interesting because they usually cannot reveal any more insight than the CFA model. The fit statistics for a saturated theoretical model should be the same as those obtained for the CFA model, which is a useful point. One way researchers can check to see whether the transition from a CFA model setup to a structural model setup is correct is to test a saturated structural model. If its fit does not equal the CFA model fit, a mistake has been made.

ASSESSING OVERALL STRUCTURAL MODEL FIT The structural model fit is assessed as was the CFA model fit. Therefore, good practice dictates that more than one fit statistic be used. Recall from Chapter 11 that, at a minimum, we recommended one absolute index, one incremental index, and the model χ^2 be used. Once again, no magic set of numbers suggests good fit in all situations. Even a CFI equal to 1.0 and an insignificant χ^2 may not have a great deal of practical meaning in a simple model. Therefore, only general guidelines are given for different situations. Those guidelines remain the same for evaluating the fit of a structural model.

COMPARING THE CFA FIT AND STRUCTURAL MODEL FIT The CFA fit provides a useful baseline to assess the structural or theoretical fit. A recursive structural model cannot fit any better (have a lower χ^2) than the overall CFA. Therefore, one can conclude that the structural theory lacks validity if the structural model fit is substantially worse than the CFA model fit [2]. A structural theory seeks to explain all of the relationships between constructs as simply as possible. The standard CFA model assumes a relationship exists between each pair of constructs. Only a saturated structural model would make this assumption. Thus, SEM models attempt to explain interconstruct relationships more simply and precisely than does CFA. When they fail to do so, the failure is reflected with relatively poor fit statistics. Conversely, a structural model demonstrating an insignificant $\Delta\chi^2$ value with its CFA model is strongly suggestive of adequate structural fit.

EXAMINING HYPOTHESIZED DEPENDENCE RELATIONSHIPS Recall that assessment of CFA model validity was based not only on fit, but on construct validity as well. Likewise, good fit alone is insufficient to support a proposed structural theory. The researcher also must examine the individual structural parameter estimates against the corresponding hypotheses. Theory validity increases to the extent that the parameter estimates are:

- *Statistically significant and in the predicted direction.* That is, they are greater than zero for a positive relationship and less than zero for a negative relationship.
- *Nontrivial.* This aspect can be checked using the standardized loading estimates. The guideline here is the same as in other multivariate techniques.

The researcher also can examine the variance-explained estimates for the endogenous constructs, which are essentially an analysis of the R^2. The same general guidelines apply for these values, as applied with multiple regression.

Also remember, particularly in structural models, that good fit does not guarantee that the SEM model is the single best representation of the data. Like CFA models, alternative models can often produce the same empirical results. The concept of equivalent models takes on much more meaning when examining a structural model in the sense that an equivalent model is an "alternative theory" that has identical fit to the model being tested. Once again, theory becomes essential in assessing the validity of a structural model.

Examine the Model Diagnostics

The same model diagnostics are provided for SEM as for the CFA model. For example, the pattern and size of standardized residuals can be used to identify problems in fit. We can assume the CFA model has sufficient validity if we have reached this stage, so the focus is on the diagnostic information about relationships between constructs. Particular attention is paid to path estimates, standardized residuals, and modification indices associated with the possible relationships between constructs in any of three possible forms (exogenous → endogenous constructs, endogenous → endogenous constructs, and error covariance among endogenous constructs). For instance, if a problem with model fit is due to a currently fixed relationship between an exogenous construct and an endogenous construct, it likely will be revealed through a standardized residual or a high modification index.

Consider the structural model in Figure 12-10. The model does not include a path linking Supervisor Support and Job Search. If the model were tested and a relationship between the two really exists, a high standardized residual or pattern of residuals would likely be found between indicator variables that make up these two constructs (X_1–X_4 and Y_5–Y_8 in this case). It would be telling us that the covariance between these sets of items has not been accurately reproduced by our initial theory. In addition, a high modification index might exist for the path that would be labeled $P_{\text{Job Search, Supervisor Support}}$ (the causal relationship from Supervisor Support to Job Search). The modification indices for paths

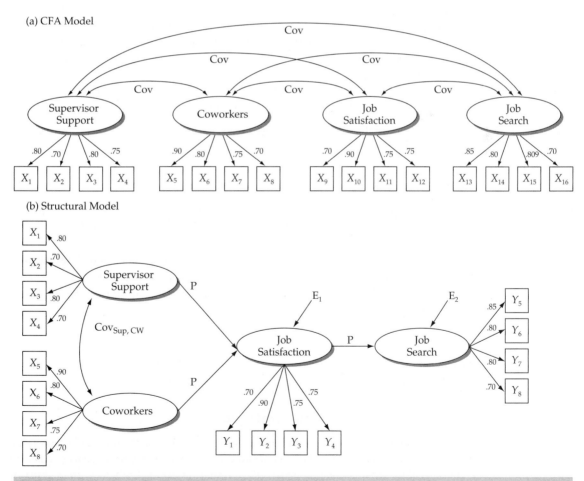

FIGURE 12-10   CFA Loading Estimates in a Structural Model

that are not estimated are shown in the standard SEM output. They can also be requested to be shown on a path diagram by using the appropriate drop-down menus. Generally speaking, model diagnostics are examined in the same manner as they are for CFA models.

Should a model be respecified based on this diagnostic information? It is fairly common practice to conduct **post hoc analyses** following the theory test. Post hoc analyses are after-the-fact tests of relationships for which no hypothesis was theorized. In other words, a path is tested where the original theory did not contain a path. Recall that SEM provides an excellent tool for *testing* theory. Therefore, any relationship revealed in a post hoc analysis provides only empirical evidence, not theoretical support. For this reason, those relationships identified post hoc should not be relied on in the same way as the original theoretical relationships. Only when all of the caveats of a model respecification strategy are noted should these changes be considered. Post hoc structural analyses are useful only in specifying potential model improvements that *must* make both theoretical sense and be cross-validated by testing the model with new data drawn from the same population. Thus, post hoc analyses are not useful in theory testing, and any such attempt should be discouraged.

SEM ILLUSTRATION

The CFA illustrations in Part 1 began by testing a measurement theory. The end result was validation of a set of construct indicators that enable HBAT to study relationships among five important constructs. HBAT would like to understand why some employees stay on the job longer than others. They know they can improve service quality and profitability when employees stay with the company longer. The six-stage SEM process begins with this goal in mind. For this illustration, we use the HBAT_SEM data set, available on the text's Web sites (www.pearsonglobaleditions.com/hair or www.mvstats.com).

The full measurement model was tested in Part 1 and was shown to have adequate fit and construct validity. Recall that the CFA fit statistics for this model were:

- χ^2 is 236.62 with 179 degrees of freedom (.05)
- CFI = .99
- RMSEA = 0.027

To refresh your memory, the five constructs are defined here:

- *Job Satisfaction (JS).* Reactions resulting from an appraisal of one's job situation.
- *Organizational Commitment (OC).* The extent to which an employee identifies and feels part of HBAT.
- *Staying Intentions (SI).* The extent to which an employee intends to continue working for HBAT and is not participating in activities that make quitting more likely.
- *Environmental Perceptions (EP).* Beliefs an employee has about day-to-day, physical working conditions.
- *Attitudes Toward Coworkers (AC).* Attitudes an employee has toward the coworkers he/she interacts with on a regular basis.

The analysis will be conducted at the individual level. HBAT is now ready to test the structural model using SEM.

Stage 5: Specifying the Structural Model

With the construct measures in place, researchers now must establish the structural relationships among the constructs and translate them into a form suitable for SEM analysis. The following sections detail the structural theory underlying the analysis and the path diagram used for estimation of the relationships.

DEFINING A STRUCTURAL THEORY The HBAT research team proposes a theory based on the organizational literature and the collective experience of key HBAT management employees. They agree that it is impossible to include all the constructs that might possibly relate to employee retention (staying intentions). It would be too costly and too demanding on the respondents based on the large number of survey items to be completed. Thus, the study is conducted with the five constructs listed previously.

The theory leads HBAT to expect that EP, AC, JS, and OC are all related to SI, but in different ways. For example, a high EP score means that employees believe their work environment is comfortable and allows them to freely conduct their work. This environment is likely to create high job satisfaction, which in turn will facilitate a link between EP and SI. Because it would require a fairly extensive presentation of key organizational concepts and findings, we will not develop the theory in detail here.

HBAT management wants to test the following hypotheses:

H_1: Environmental perceptions are positively related to job satisfaction.

H_2: Environmental perceptions are positively related to organizational commitment.

H_3: Attitudes toward coworkers are positively related to job satisfaction.

H_4: Attitudes toward coworkers are positively related to organizational commitment.

H_5: Job satisfaction is related positively to organizational commitment.

H_6: Job satisfaction is related positively to staying intentions.

H_7: Organizational commitment is related positively to staying intentions.

VISUAL DIAGRAM The theory can be expressed visually. Figure 12-11 shows the diagram corresponding to this theory. For simplicity, the measured indicator variables and their corresponding paths and errors have been left off of the diagram. If a graphical interface is used with a SEM program, then all measured variables and error variance terms would have to be shown on the path diagram.

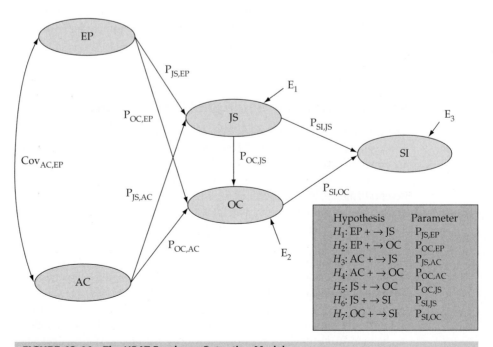

FIGURE 12-11 The HBAT Employee Retention Model

Exogenous Constructs. EP and AC are exogenous constructs in this model. They are considered to be determined by things outside of this model. In practical terms, this means that no hypothesis predicts either of these constructs. Like independent variables in regression, they are used only to predict other constructs.

The two exogenous constructs—EP and AC—are drawn at the far left. No single-headed arrows enter the exogenous constructs. A curved two-headed arrow is included to capture any covariance between these two constructs ($Cov_{AC,EP}$). Although there is no hypothesis between these two, there is no reason to suspect that they are independent constructs. So, if the measurement model estimates a path coefficient between constructs not involved in any hypothesis, then that parameter should also be estimated in the SEM model.

Endogenous Constructs. JS, OC, and SI are each endogenous constructs in this model. Each is determined by constructs included in the model, and so each is also seen as an outcome based on the hypotheses. Notice that both JS and OC are used as outcomes in some hypotheses and as predictors in others. This is perfectly acceptable in SEM, and a test for all hypotheses can be provided with one structural model test. This would not be possible with a single regression model because we would be limited to a single dependent variable.

The structural path model begins to develop from the exogenous constructs. A path should connect any two constructs linked theoretically by a hypothesis. Therefore, after drawing the three endogenous constructs (JS, OC, and SI), single-headed arrows are placed connecting the predictor (exogenous) constructs with their respective outcomes based on the hypotheses. The legend in the bottom right of Figure 12-11 lists each hypothesis and the path to which it belongs. Each single-headed arrow represents a direct path and is labeled with the appropriate parameter estimate. For example, H_2 hypothesizes a positive EP–OC relationship. A parameter estimate linking an exogenous construct to an endogenous construct is designated by the symbol P. The convention is that the subscript first lists the number (or abbreviation) of the construct to which the path points and then the subscript for the construct from which the path begins. So, H_1 is represented by $P_{JS,EP}$. Similarly then, H_7, linking SI with OC, is represented by $P_{SI,OC}$.

As discussed earlier, the CFA model must be transformed into a structural model for all of the software programs. Although the issues are specific to each program, the user must essentially redefine construct types (exogenous to endogenous), replace the correlational paths with the structural relationships, and change the notation associated with these changes. Materials on the text's Web sites (www.pearsonglobaleditions.com/hair or www.mvstats.com) describe this process in more detail.

Stage 6: Assessing the Structural Model Validity

The structural model shown in the path diagram in Figure 12-11 can now be estimated and assessed. To do so, the emphasis first will be on SEM model fit and then whether the structural relationships are consistent with theoretical expectations.

The information in Table 12-10 shows the overall fit statistics from testing the Employee Retention model. The χ^2 is 283.43 with 181 degrees of freedom ($p < .05$) and the normed chi-square is 1.57. The model CFI is .99, with a RMSEA of .036 and 90% confidence interval of .027 to .045. All of these measures are within a range that would be associated with good fit. These diagnostics suggest the model provides a good overall fit (see Chapter 11 for a review of fit guidelines). We also see that overall model fit changed very little from the CFA model (Table 12-10). The only substantive difference is a chi-square increase of 46.81 and a difference of two degrees of freedom. The standardized path coefficients are shown in Figure 12-12.

We next examine the path coefficients and loadings estimates to make sure they have not changed substantially from the CFA model (see Table 12-11). The loadings estimates are virtually unchanged from the CFA results. Only three estimated standardized loadings change and the

TABLE 12-10 Comparison of Goodness-of-Fit Measures Between HBAT Employee Retention and CFA Models

GOF Index	Employee Retention Model	CFA Model
Absolute Measures		
χ^2 (chi-square)	283.43	236.62
Degrees of freedom	181	179
Probability	0.00	0.00
GFI	.94	.95
RMSEA	.036	.027
Confidence interval of RMSEA	.027–.045	.015–.036
RMR	.110	.085
SRMR	.060	.035
Normed chi-square	1.57	1.32
Incremental Fit Measures		
NFI	.96	.97
NNFI	.98	.99
CFI	.99	.99
RFI	.96	.97
Parsimony Measures		
AGFI	.92	.93
PNFI	.83	.83

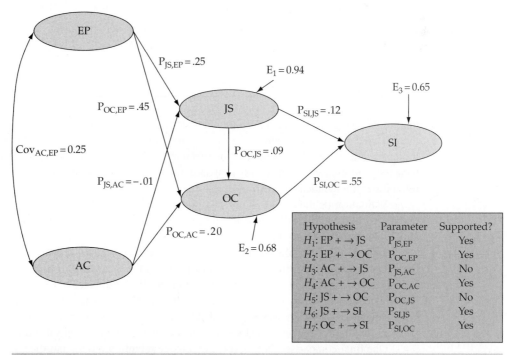

Hypothesis	Parameter	Supported?
H_1: EP + → JS	$P_{JS,EP}$	Yes
H_2: EP + → OC	$P_{OC,EP}$	Yes
H_3: AC + → JS	$P_{JS,AC}$	No
H_4: AC + → OC	$P_{OC,AC}$	Yes
H_5: JS + → OC	$P_{OC,JS}$	No
H_6: JS + → SI	$P_{SI,JS}$	Yes
H_7: OC + → SI	$P_{SI,OC}$	Yes

FIGURE 12-12 Standardized Path Estimates for the HBAT Structural Model

TABLE 12-11 Comparison of Standardized Factor Loadings and Construct Reliabilities for HBAT Employee Retention and CFA Models

INDICATOR	CONSTRUCT	EMPLOYEE RETENTION MODEL	CFA MODEL
		Standardized Factor Loading	
JS_1	JS	0.74	0.74
JS_2	JS	0.75	0.75
JS_3	JS	0.68	0.68
JS_4	JS	0.70	0.70
JS_5	JS	0.73	0.73
OC_1	OC	0.58	0.58
OC_2	OC	0.88	0.88
OC_3	OC	0.66	0.66
OC_4	OC	0.83	0.84
SI_1	SI	0.81	0.81
SI_2	SI	0.87	0.86
SI_3	SI	0.74	0.74
SI_4	SI	0.85	0.85
EP_1	EP	0.69	0.70
EP_2	EP	0.81	0.81
EP_3	EP	0.77	0.77
EP_4	EP	0.82	0.82
AC_1	AC	0.82	0.82
AC_2	AC	0.82	0.82
AC_3	AC	0.84	0.84
AC_4	AC	0.82	0.82
		Construct Reliabilities	
	JS	0.84	0.84
	OC	0.83	0.83
	SI	0.89	0.89
	EP	0.86	0.86
	AC	0.89	0.89

maximum change is .01. Thus, if parameter stability had not already been tested in the CFA stage, there is now evidence of stability among the measured indicator variables. In technical terms, this indicates that no problem is evident due to interpretational confounding and further supports the measurement model's validity. As we would expect with so little change in loadings, the construct reliabilities are identical as well.

Validation of the model is not complete without examining the individual parameter estimates. Are they statistically significant and meaningful? All of these answers must be addressed along with assessing model fit.

Table 12-12 shows the estimated unstandardized and standardized structural path estimates. All but two structural path estimates are significant and in the expected direction. The exceptions are the estimates between AC and JS and between JS and OC. Both estimates have significance below the critical t-value for a Type I error of .05. Therefore, although the estimate is in the hypothesized direction, it is not supported. Overall, however, given that five of seven estimates are consistent with the hypotheses, these results support the theoretical model with a caveat for the two paths that are not supported.

One final comparison between the employee retention model and the CFA model is in terms of the structural model estimates. Table 12-13 contains the standardized parameter estimates for all seven of the structural relationships as well as the correlational relationship among EP and AC. As

TABLE 12-12   Structural Parameter Estimates for HBAT Employee Retention Model

Structural Relationship	Unstandardized Parameter Estimate	Standard Error	t-value	Standardized Parameter Estimate
H_1: EP → JS	0.20	0.05	4.02	0.25
H_2: EP → OC	0.52	0.08	6.65	0.45
H_3: AC → JS	−0.01	0.05	−0.17	−0.01
H_4: AC → OC	0.26	0.07	3.76	0.20
H_5: JS → OC	0.13	0.08	1.60	0.09
H_6: JS → SI	0.09	0.04	2.38	0.12
H_7: OC → SI	0.27	0.03	8.26	0.55
EP correlated AC	0.37	0.09	4.19	0.25

noted earlier, five of these seven relationships were supported with significant path estimates. But what about the relationships not in the hypothesized model? Although we will examine model diagnostics in the next section, we can also compare the correlational relationships from the CFA model with the structural relationships. As we see in Table 12-13, the estimated parameters are quite comparable between the two models. But we also see that the two possible excluded structural relationships (EP → SI and AC → SI) correspond to significant relationships in the CFA model. This would suggest that model improvement might be possible with the addition of one or more of these relationships.

EXAMINING MODEL DIAGNOSTICS   As discussed earlier, several diagnostic measures are available for researchers to evaluate SEM models. They range from fit indices to standardized residuals and modification indices. Each of these will be examined in the following discussion to determine if model respecification should be considered.

The first comparison in fit statistics is the chi-square difference between the hypothesized model and the measurement model where we see a $\Delta\chi^2$ of 46.81 with two degrees of freedom ($p < .001$). The difference in degrees of freedom is two, which is due to the fact that all but two of the possible structural paths are estimated. Because the difference is highly significant, it suggests that fit may be improved by

TABLE 12-13   Comparison of Structural Relationships with CFA Correlational Relationships

HBAT Employee Retention Model		HBAT CFA Model	
Structural Relationship	Standardized Parameter Estimate	Comparable Correlational Relationship	Standardized Parameter Estimate
H_1: EP → JS	0.25	EP correlated JS	0.24
H_2: EP → OC	0.45	EP correlated OC	0.50
H_3: AC → JS	−0.01	AC correlated JS	0.05
H_4: AC → OC	0.20	AC correlated OC	0.30
H_5: JS → OC	0.09	JS correlated OC	0.21
H_6: JS → SI	0.12	JS correlated SI	0.23
H_7: OC → SI	0.55	OC correlated SI	0.55
EP correlated AC	0.25	EP correlated AC	0.25
Not estimated	—	EP correlated SI	0.56
Not estimated	—	AC correlated SI	0.31

estimating another structural path. The possibility of another meaningful structural path should be considered, particularly if other diagnostic information points specifically to a particular relationship. All of the other fit statistics are also supportive of the model and there were no substantive changes in the other fit indices between the CFA and structural model.

The next step is to examine the standardized residuals and modification indices for the structural model. As before, patterns of large standardized residuals and/or large modification indices indicate changes in the structural model that may lead to model improvement. Table 12-14 contains the standardized residuals greater than $|2.5|$. In looking for patterns of residuals for a variable or set of variables, one pattern is obvious: each item of the EP construct (EP_1 to EP_4) has significant standardized residual with at least three of the four items in the SI construct. This indicates that there may be a substantial relationship omitted between these two constructs. At the moment, there is no direct relationship between these two constructs, only indirect relationships ($EP \rightarrow JS \rightarrow SI$ and $EP \rightarrow JS \rightarrow OC \rightarrow SI$).

TABLE 12-14   Model Diagnostics for HBAT Employee Retention Model

Standardized Residuals (all residuals greater than $|2.5|$)

Largest Negative Standardized Residuals

SI_2	and	OC_1	−2.90
SI_3	and	OC_1	−2.88
SI_4	and	OC_1	−2.99
EP_3	and	OC_1	−2.90

Largest Positive Standardized Residuals

SI_2	and	SI_1	3.45
SI_4	and	SI_3	3.47
EP_1	and	SI_1	3.78
EP_1	and	SI_2	4.27
EP_1	and	SI_3	3.78
EP_1	and	SI_4	4.50
EP_2	and	OC_3	2.63
EP_2	and	SI_1	3.41
EP_2	and	SI_2	4.20
EP_2	and	SI_3	3.70
EP_2	and	SI_4	5.84
EP_3	and	SI_2	2.69
EP_3	and	SI_3	2.73
EP_3	and	SI_4	3.76
EP_4	and	SI_1	4.52
EP_4	and	SI_2	3.60
EP_4	and	SI_3	3.97
EP_4	and	SI_4	4.27
EP_4	and	EP_3	3.01
AC_3	and	SI_4	2.73
AC_4	and	SI_4	2.95

Modification Indices for Structural Relationships

Structural Relationship (not estimated)	Modification Index
$EP \rightarrow SI$	40.12
$AC \rightarrow SI$	8.98
$SI \rightarrow JS$	38.66
$SI \rightarrow OC$	45.12

Examination of the two modification indices for the direct paths of EP → SI and AC → SI shows that both have values over 4.0, although the EP → SI value is much higher (40.12 versus 8.98). This strongly supports the addition of the EP → SI relationship, if it can be supported theoretically. This also corresponds to the pattern of residuals described above between the indicators of these two constructs. It also casts doubt on the premise that JS mediates the relationship between EP and SI.

Note that modification indices can be estimated for the "second half" of the recursive relationships between SI → JS and SI → OC. As we can see, both indicate substantial improvement in model fit. But more important, because they have no theoretical basis for inclusion in the model, they highlight the potential dangers of making model respecifications based solely on improvement of model fit without regard to a theoretical basis.

The researcher must evaluate the information provided by the model fit measures and other diagnostics to determine (1) the level of theoretical support provided by the results and (2) any potential model respecifications that would provide improvement in the model while also having theoretical support.

MODEL RESPECIFICATION   Many times the diagnostic measures in SEM indicate model respecification should be considered. Any respecification must have strong theoretical as well as empirical support. Model respecification should not be the result of searching for relationships, but for improving model fit that is theoretically justified. Based on the residuals and modification indices from the initial SEM model, we examine a respecification of our HBAT example.

To further assess the SEM model, the HBAT research team conducts a post hoc analysis, adding a direct relationship between EP and SI. The SEM program is instructed to free this path, and the model is re-estimated. Figure 12-13 shows the model including a free path corresponding to EP → SI ($P_{SI,EP}$). Table 12-15 compares the GOF measures for the "original" and revised models. The resulting standardized parameter estimate for $P_{SI,EP}$ is 0.37 ($p < .001$). In addition, the overall fit reveals a χ^2 value of 242.2 with 180 degrees of freedom and a normed χ^2 value of 1.346. The CFI remains .99 and the RMSEA is .029, which is practically the same as the value for the CFA model. This is a better fit than the original structural model because the $\Delta\chi^2$ is 41.2 with one degree of freedom, which is significant ($p < .001$).

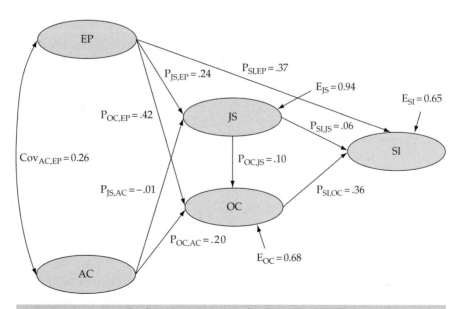

FIGURE 12-13   Standardized Path Estimates for the Revised HBAT Structural Model

TABLE 12-15 Comparison of Goodness-of-Fit Measures Between HBAT Employee Retention and Revised Employee Retention Models

GOF Index	Revised Employee Retention Model	Employee Retention Model
Absolute Measures		
χ^2 (chi-square)	242.23	283.43
Degrees of freedom	180	181
Probability	0.00	0.00
GFI	.95	.94
RMSEA	.029	.036
Confidence interval of RMSEA	.018–.038	.027–.045
RMR	.090	.110
SRMR	.040	.060
Normed chi-square	1.346	1.57
Incremental Fit Measures		
NFI	.97	.96
NNFI	.99	.98
CFI	.99	.99
RFI	.96	.96
Parsimony Measures		
AGFI	.93	.92
PNFI	.83	.83

Several of the path estimates from the original model have changed slightly as would be expected (see Table 12-16). Most notably, the JS–SI relationship ($P_{SI,JS} = .06$) is no longer significant and the SI–OC relationship ($P_{SI,OC} = .36$) remains significant but is substantially smaller than before.

The $\Delta\chi^2$ value between the revised SEM and CFA models is 5.61 with one degree of freedom, which is significant at a type I error rate of .05. The squared multiple correlation (i.e., R^2) for

TABLE 12-16 Comparison of Structural Relationships for the Original and Revised HBAT Employee Retention Models

HBAT Employee Retention Model		Revised HBAT Employee Retention Model	
Structural Relationship	**Standardized Parameter Estimate**	**Structural Relationship**	**Standardized Parameter Estimate**
H_1: EP → JS	0.25[*]	H_1: EP → JS	0.24[*]
H_2: EP → OC	0.45[*]	H_2: EP → OC	0.42[*]
H_3: AC → JS	−0.01	H_3: AC → JS	−0.01
H_4: AC → OC	0.20[*]	H_4: AC → OC	0.20[*]
H_5: JS → OC	0.09	H_5: JS → OC	0.10
H_6: JS → SI	0.12[*]	H_6: JS → SI	0.06
H_7: OC → SI	0.55[*]	H_7: OC → SI	0.36[*]
EP correlated AC	0.25[*]	EP correlated AC	0.26[*]
		EP → SI	0.37[*]

[*]Statistically significant at .05 level.

SI also improves from .35 to .45 with the addition of this relationship. These findings suggest that the structural model does a good, but not perfect, job in explaining the observed covariance matrix. Thus, we can proceed to interpret the precise nature of the relationships with a fair degree of confidence.

At this point, HBAT has tested its original structural model. The results showed reasonably good overall model fit and the hypothesized relationships were generally supported. However, the large difference in fit between the structural model and CFA model and several key diagnostics, including the standardized residuals, suggested one improvement to the model. This change improved the model fit. Now, HBAT must consider testing this model with new data to examine its generalizability.

PART 3: EXTENSIONS AND APPLICATIONS OF SEM

The widespread use of SEM models in almost every discipline has heightened interest in using SEM methods for more advanced issues. Whether it be higher-order factor models, testing mediation or moderation, or assessing construct invariance across groups, SEM researchers seek to utilize the flexibility of SEM models to address all of these questions and more. This final section of the chapter provides both the theoretical foundations for these approaches as well as the issues of estimation and interpretation that are involved. As SEM models become more accepted their use for these more specific research questions will become more widespread as well.

Among the topics covered in this part are the distinction between a formative and reflective measurement model; higher-order factor analysis; multigroup analysis, with a particular focus on measurement invariance testing; measurement bias; testing for mediation and moderation; and some discussion of PLS, an alternative approach to SEM modeling. It is difficult to address all of the issues concerning these advanced topics, but our goal is to provide some key points that will help in understanding the issues and modeling benefits from these extensions of the basic SEM modeling.

REFLECTIVE VERSUS FORMATIVE MEASURES

The issue of causality (i.e., correlational versus dependence relationships) has played a key role in our specification of the structural model. For example, relationships between constructs were assumed to be correlational in CFA. In contrast, with SEM the relationships between constructs were assumed to be dependent, except those between only exogenous constructs. Up to this point we have assumed a measurement theory wherein latent factors (constructs) are thought to cause measured variables. At times, however, the causality between constructs and indicator variables may be reversed. This contrasting direction of causality leads to a different measurement approach known as *formative measurement models*.

Reflective Versus Formative Measurement Theory

Until now, our discussion of constructs and the measurement model has assumed a reflective measurement theory. A *reflective measurement theory* is based on the idea that latent constructs cause the measured variables and that the error results in an inability of the construct to fully explain these measured variables. Thus, the direction of the arrows is from latent constructs to measured variables and error terms are associated with each measured variable. As such, reflective measures are consistent with classical test theory [60]. Construct validity of a reflective latent construct ensures that the "meaning" of the construct will remain consistent given the measures used and should not vary when associated with other constructs.

Because a reflective measure dictates that all indicator items are caused by the same latent construct, items within a construct should be highly correlated with each other. Individual items should be interchangeable and any single item can be left out without changing the construct as long as two conditions are satisfied: (1) the construct must have sufficient reliability and (2) at least three items must be specified to avoid identification problems [16]. Reflective indicators can be viewed as a sample of all the possible items available within the conceptual domain of the construct [26]. As a consequence, reflective indicators of a given construct are expected to move together, meaning that changes in one are associated with proportional changes in the other constructs.

Reflective indicator models are the predominant measurement theory used in the social sciences [14]. Typical social science constructs such as attitudes, personality, and behavioral intentions fit the reflective measurement model well [16]. Likewise, a study of medical symptoms typically would be reflective. For example, symptoms such as shortness of breath, tiring easily, wheezing, and reduced lung functioning would be considered indicators that would reflect the latent factor of emphysema. The symptoms *do not cause* the disease. Rather, the disease causes the symptoms.

In contrast, a *formative measurement theory* is based on the assumption that the measured variables cause the construct. A typical example would be a social class index [29]. Social class often is viewed as a composite of one's educational level, occupational prestige, and income (or sometimes wealth). Social class does not cause these indicators as in the reflective case. Rather, each formative indicator is considered a partial cause of the measure. In a business setting, investors often compute a bankruptcy index that indicates how close an individual or company is to financial bankruptcy. Key financial measures (e.g., total sales, assets, liabilities, expenses, retained earnings, and interest, among others) could be thought of as causing bankruptcy and thus they would be appropriate as formative indicators. Finally, using a health-related example, a formative emphysema factor might specify indicators such as cigarette consumption, exposure to toxins, chronic bronchitis, and others. These indicators would *form* rather than reflect the probability of an individual having emphysema. The fact that one smokes cigarettes has little connection to the other indicators.

Although a formative construct may seem quite simple—just reverse the arrows—it also reverses the way we think about constructs. A key assumption is that formative constructs are not considered latent and the indicators need not have a consistent inherent meaning [45]. Formative constructs are better viewed as indices where each indicator is a potential contributing cause. In a way, the loadings of individual items, which ultimately determine the meaning of the construct, are identified by other relationships in the model. This is because a formative measure requires at least two separate reflective items or other endogenous constructs to act as "outcome" measures to be identified and estimated (see later discussion) [40]. The result is that the construct could take on different "meanings" depending on the constructs used as outcomes. In some sense selecting the outcome measures is as important as the construct indicators themselves [28]. The problem is similar to the issue of interpretational confounding discussed earlier [19], but in this case it is inherent in all formative measures.

Operationalizing a Formative Construct

Figure 12-14 illustrates a formative indicator model. Each indicator (X) is an index item that causes the composite construct. A correlation is shown among the index items (X_1–X_3), and E is a parameter indicating the amount of error variance in the index. Notice that the error is now in the factor and not in the measured items. Similarly, because the causality is from the items to the factor (construct), and not the reverse, the factor does not explain the item intercorrelations. These differences lead to some changes in scale testing and usage as discussed in the next sections.

The implications of collinearity and dropping indicator items are different in reflective and formative models. Reflective items are presumed to each be representative of the same conceptual domain. Therefore, dropping reflective items does not change the latent construct's meaning. Items with low factor loadings can be dropped from reflective models without serious consequences as long as a construct retains a sufficient number of indicators. Moreover, collinearity is expected in reflective measures as an indication of convergent validity. But in formative models, the items define the construct, so dropping or adding an item can have profound changes in the meaning of the construct. Conceptually, a formative factor should be represented by the entire population of

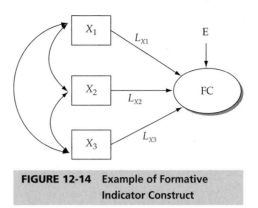

FIGURE 12-14   **Example of Formative Indicator Construct**

indicators that form it [45]. Also, because there is no "common cause" for the items in the construct, there is not any requirement for the items to be correlated, and they may even be completely independent. As a matter of fact, collinearity among formative indicators can present significant problems because the loadings of the formative indicators to the construct can become unreliable in estimation (similar to the impact of multicollinearity in multiple regression discussed in Chapter 4). If these parameters are unreliable, then it becomes impossible to validate the item. Thus, the researcher can face a dilemma, because dropping an item may make the index incomplete, but keeping it may make an estimate(s) unreliable. These issues that are associated with formative indicator models have yet to be resolved fully [28, 29].

Formative measurement models also require a different validation process. Because formative indicators do not have to be correlated, internal consistency (reliability) is not an appropriate validation criterion for formative indicators. Indeed, formative items may even be mutually exclusive [45]. Because the error is in the factor, the most important validation criteria relate to criterion or predictive validity. As noted earlier, the "validity" of a formative construct is contingent on the other constructs or variables it is related to in the model. Guidelines for validating formative factors are not as easily determined as with reflective models [29, 70].

Identification and estimation of a formative construct, as noted earlier, requires a relationship with two reflective measures or constructs that act as "outcomes." These outcomes become, in a simple sense, the dependent variables that the formative construct predicts. Figure 12-15 depicts three methods in which these outcome measures can be specified. In Figure 12-15a we see that two reflective indicators have been added to the formative construct similar to the MIMIC model [46]. In Figure 12-15b the formative construct is related to other multiple item reflective constructs. Finally, in Figure 12-15c a combination of one reflective construct and one reflective indicator provides identification for the formative construct. Each of these approaches has advantages and disadvantages that is the topic of continued discussion and research [27]. Because of these issues, formative indicator models present greater difficulties with statistical identification [49].

Distinguishing Reflective from Formative Constructs

Meaningful differences separate reflective and formative measurement models, but differentiating between the two is not always easy. Reflective models are generally easier to work with, have traditionally been more commonly employed in the social sciences, and are thought to best represent many individual difference characteristics and perceptual measures. Modeling a factor incorrectly, however, could cause misinterpretation and lead to questionable conclusions. The ultimate decision on the type of measurement model should be based on the true nature of the construct being studied [17, 45].

a. Inclusion of Two Reflective Indicators

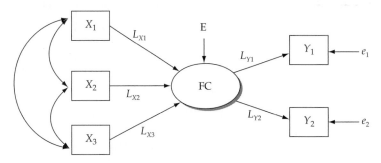

b. Inclusion of Two Reflective Constructs

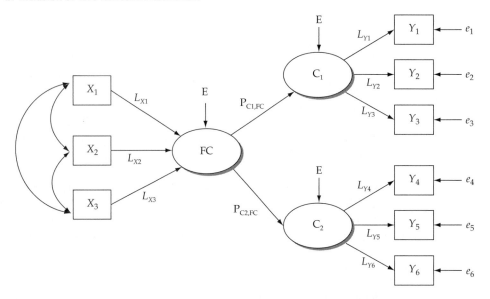

c. Inclusion of One Reflective Construct and One Reflective Indicator

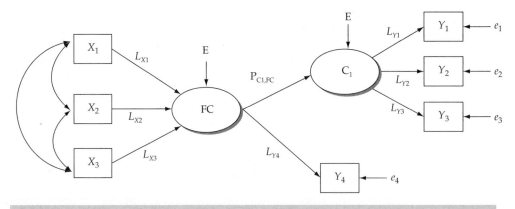

FIGURE 12-15 Three Approaches to Identification of Formative Constructs

Table 12-17 presents a series of characteristics that may assist in selecting the appropriate measurement model form. As discussed earlier, the reflective and formative measurement theories are directly opposite approaches. In the reflective model, the items are caused by the construct and should be "outcomes" of the latent construct. All of the items should have some conceptual linkage

TABLE 12-17   Distinguishing Between Reflective and Formative Constructs

Characteristic	*Indicative of:*	
	Reflective	Formative
Causality of construct	Items are caused by construct.	Construct is formed from items.
Conceptual relationship among items	All items are related conceptually because they have a common cause.	No requirement of conceptual linkage to other items.
Domain of items	Representative sample of potential items.	Exhaustive inventory of all possible items.
Covariance among items	Expected collinearity among items.	No expectation of collinearity. High collinearity among formative items can be problematic.
Internal consistency	Required.	Not required.
Forms of construct validity	Internal and external.	Only external.

and should covary together. The items in a reflective model need only to be a representative sample, and items can be added or dropped as long as the set of items provides coverage of the domain and the construct is identified. Reflective constructs are required to exhibit internal consistency as a requirement for validity, whereas validity must be established internally (convergent validity) as well as externally (discriminant and predictive/criterion validity).

Formative constructs are best characterized as indices rather than latent constructs because there is nothing "unobservable" when the items define the construct. Specification of the complete domain of the construct through an exhaustive set of possible items is required, thus raising the possibility of violating content validity if essential items are omitted or dropped. Because items are not required to be conceptually linked except in their relationship to other constructs there is also no requirement for collinearity among the items, and thus no level of internal consistency. In terms of construct validity, the lack of any internal validity measures requires that validity only be established through criterion or predictive validity and is contingent on the constructs used in the validation process.

Which to Use—Reflective or Formative?

Formative constructs have received considerable attention in recent years [27], particularly in light of the potential consequences of measurement model misspecification by incorrectly using reflective rather than formative measures. Research has estimated that a third or more of the constructs in the marketing, management, and strategic management journals have misspecified constructs as reflective when they are formative [45, 63]. In doing so, research indicates that the general effect is to increase the size of the structural parameters versus those found when formative constructs are used [52]. SEM procedures can appropriately be used to model formative constructs as long as the models are statistically identified.

The trend toward more widespread use of formative constructs has not been without concerns. Perhaps the most widespread concern has been the inherent lack of internal validity in formative constructs and their potential for interpretational confounding based on the constructs and/or approach selected for identification and estimation purposes (see [71] for a comprehensive review). Additional issues of both a conceptual and empirical nature have been raised to the extent that some researchers call into question any use of formative measures [17, 18, 61, 71].

At this point, there is no definitive answer to the question of which to use. Research is continuing to address the questions raised by both proponents and opponents. What can be said with certainty is that the content domain of the construct is most crucial, no matter which approach is used. At times, a construct can be represented by either approach with careful consideration of the

---

RULES OF THUMB 12-7

Formative Measurement Models

- Formative factors are not latent and are not validated, as are conventional reflective factors (internal consistency and reliability are not important)
- The variables that make up a formative factor should explain the largest portion of variation in the formative construct itself and should relate highly to other outcomes that are conceptually related (minimum correlation of .5)
- Formative factors present greater difficulties with statistical identification
- At least two additional reflective variables or constructs with reflective variables must be included, along with a formative construct, in order to achieve an overidentified model
- A formative factor should be represented by the entire population of items that form it; therefore, items should not be dropped because of a low loading
- SEM is appropriate for analyzing models with formative factors so long as statistical identification is possible.

---

items selected. Bollen and Ting [17] even suggest that the same set of indicators may be reflective in one formulation of a construct and formative in another. Thus, it is the researcher who controls the formulation of the construct and its implementation as either reflective or formative.

HIGHER-ORDER FACTOR ANALYSIS

The CFA model described in Part 1 is a **first-order factor model.** A first-order factor model means that the covariances between measured items are explained with a single latent factor layer. For now, think of a layer as one level of latent constructs.

Researchers increasingly are employing higher-order factor analyses although this aspect of measurement theory is not new. Higher-order CFAs most often test a **second-order factor** structure that contains two layers of latent constructs. They introduce a second-order latent factor(s) that causes multiple first-order latent factors, which in turn cause the measured variables (x). Theoretically this process can be extended to any number of multiple layers. Thus, the term *higher-order factor analysis.* Researchers seldom examine theories beyond a second-order model. Figure 12-16 contrasts path diagrams between a conventional first-order factor model with one layer in part (a) and a second-order factor model (b) with two layers in part (b).

Empirical Concerns

Both theoretical and empirical considerations are associated with higher-order CFA. All CFA models must account for the relationships among constructs. In a first-order CFA model, these covariance terms are typically free (estimated) unless the researcher has a strong theoretical reason to hypothesize independent dimensions. Therefore, all of the factors are interrelated but without a specific causal construct (i.e., thus the correlational relationships among the constructs). Higher-order factors can be thought of as explicitly representing the causal constructs that impact the first-order factors.

Another way to view a higher-order factor is that it accounts for covariance between constructs just as first-order factors account for covariation between observed variables [1]. In other words, the first-order factors now act as indicators of the second-order factor. All the considerations and rules of thumb (items per factor, identification, scale, etc.) apply to second-order factors just as they do to first-order factors. The difference is that the researcher must consider the first-order constructs as indicators of the second-order construct.

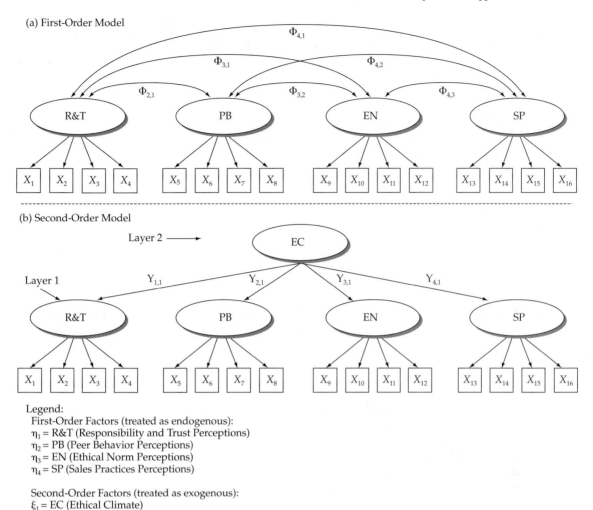

FIGURE 12-16   **Contrasting Path Diagrams for a First- and Second-Order Measurement Theory**

Figure 12-16a shows a conventional factor model with six unique covariances between four latent factors. Figure 12-16b depicts a CFA model where a second-order factor (Ethical Climate, EC) is introduced as the cause of the four first-order factors (R&T, PB, EN, and SP), each measured by four reflective items. It is important to note that the introduction of a second-order factor changes the designation of the constructs. First, the first-order factors from the CFA model (which were originally exogenous constructs) now become endogenous constructs (note the arrows point from the higher-order construct toward the first-order constructs). The second-order factor is now the specified cause of the four constructs versus using the correlational relationships among constructs to represent an unspecified common cause as was done in the CFA. Second, the higher-order construct is now the exogenous construct and it has no measured variables as indicators. Because it represents a relationship among constructs, the first-order factors act as its "indicators" through the structural model relationships. Finally, just as was required in specifying each first-order construct, the scale must be set for the second-order construct as well. The same two approaches are available that we discussed when using measured variables. First, one structural path from the second-order factor to a first-order factor can be fixed at 1.0 to set the scale. Alternatively, all four loading estimates can be estimated if the variance of the second-order factor is fixed at 1.0.

Theoretical Concerns

Theoretically, constructs sometimes can be operationalized at different levels of abstraction. Each layer in Figure 12-16b refers to a different level of abstraction. We will discuss two examples that illustrate the role of second-order factors.

Psychological constructs are often defined at different levels of abstraction. Personality can be represented by numerous related first-order factors. Each can be measured using dozens of multiple-item scales tapping a specific personality dimension. Psychological constructs representing first-order factors include scales for anxiety, pessimism, creativity, imaginativeness, and self-esteem, among many others. Alternatively, the first-order factors can be viewed as indicators of a smaller set of more abstract higher-order factors that reflect broader, more abstract personality orientations, such as extraversion, neuroticism, conscientiousness, agreeableness, and intellect [12, 66]. These more abstract personality constructs sometimes are referred to as the "Big 5" personality factors.

Similarly, one can imagine that many different factors might indicate how well one would do in graduate school. Multiple indicators from a standardized test could be used to represent verbal performance and quantitative performance, among other exam characteristics. Multiple items also could be used to assess how well a candidate performs in school, including GPAs in college, GPAs in high school, and perhaps several other grade-related scores. We also could use multiple-item scales to assess how motivated one is to succeed in graduate school. Once we have identified all the factors related to performance in graduate school, we may end up with a few dozen indicator variables for several factors such as reading comprehension, quantitative ability, problem solving, school performance, and desire. Each of these aspects is in itself a factor. However, they may all be driven by a higher-order factor that we could label "Likelihood of Success." It may be difficult to look at one's credentials and directly assess likelihood of success. However, it may be indicated quite well by more tangible factors such as problem-solving ability. In the end, key decisions may be made based on the more abstract success factor, and hopefully these decisions are better than relying on the individual more specific factors. Thus, the individual factors are first-order factors, and Likelihood of Success could be thought of as a second-order factor. This type of situation calls for the testing of a second-order CFA model.

It cannot be emphasized enough that the ultimate criterion in deciding to form a second-order measurement model is theory. Does it make theoretical sense? What logical reason leads us to expect layers of constructs? The increasing number of second-order factor models seen in the literature is partially the result of more researchers learning how to use SEM to represent and test a higher-order factor structure. The ability to conduct a second-order test does not justify doing so. The need for theory is particularly true when trying to decide between a first- and second-order factor configuration for a given measurement theory.

Using Second-Order Measurement Theories

The specification of a second-order CFA model is actually quite similar to a first-order model if we view the first-order constructs as indicators. Considering Figure 12-16a, the first-order model estimates a relationship (two-headed path in this case) for each potential covariance. The higher-order model in Figure 12-16b accounts for these six relationships with four factor loadings. Although the comparison between a first- and second-order measurement model is generally nested, the empirical comparison using a $\Delta\chi^2$ statistic is not as useful as it is when comparing competing measurement models of the same order [53]. The first-order model should fit better in absolute terms, because it uses more paths to capture the same amount of covariance.

In contrast, the higher-order model is more parsimonious (it consumes fewer degrees of freedom). Thus, it should perform better on indices that reflect parsimony (PNFI, RMSEA, etc.). Note, however, that even though a higher-order model is more parsimonious from the standpoint of degrees of freedom, it is not "simpler" because it involves multiple levels of abstraction. This complicates empirical comparisons and thus places more weight on theoretical and pragmatic concerns.

Higher-order measurement models are also still subject to construct validity standards. In particular, second-order factors should be rigorously examined for nomological validity because it is possible that various confounding explanations may exist for a higher-order factor. For example, if all item measures use the same type of rating scale, there could be a common methods factor influencing all first-order constructs. The second-order factor could be interpreted as common measurement bias in this case. If the second-order factor reacts to other theoretical constructs as expected, the chance of it being of this type is lower. More specifically, if the higher-order factor explains theoretically related outcomes such as organizational commitment and job satisfaction as well or better than does the combined set of first-order factors, then evidence in favor of the higher-order representation is provided [53]. Thus, a primary validation criterion becomes how well a higher-order factor explains theoretically related constructs. When comparing measurement models of different orders, a second-order model is supported to the extent that it shows greater nomological validity than a first-order model.

When to Use Higher-Order Factor Analysis

Although higher-order measurement models might seem to have many advantages, we must also consider the disadvantages. In general, they are conceptually more complicated. A construct can become so abstract that it is difficult to adequately describe its meaning. The added complexity also can diminish the diagnostic value of a construct as it becomes further removed from the tangible measured items. Higher-order CFA models also create more potential for unidentified or improper CFA solutions. For instance, researchers may have one or more higher-order factors with fewer than three indicators. Note that a minimum of three first-order (first-level) constructs is required in order to assess a single second-order construct.

With a reflective second-order or higher factor model, all first-order factors, which are now indicators of the second-order factor, are expected to move together (covary), just as with the measured items indicating first-order factors. When multiple first-order factors are used as indicators of a second-order factor, the researcher gives up the ability to test for relationships between these first-order factors and other key constructs. Thus, a drawback of the measurement model shown in Figure 12-16b is that we cannot investigate, for example, direct relationships between peer behavior (PB) and other key job outcomes such as turnover. Thus, the presumption is that all four first-order indicators would influence any other construct (e.g., turnover) the same way. If a conceptual case can be made that anyone of these first-order factors would affect another key construct differently, then perhaps a second-order measurement theory should not be used. This case is typified when one of a set of related first order constructs would be expected to affect some other construct positively, whereas other first-order constructs would affect it negatively.

Some questions that can help determine whether a higher-order measurement model is appropriate are listed here:

1. Is there a theoretical reason to expect that multiple conceptual layers of a construct exist?
2. Are all the first-order factors expected to influence other nomologically related constructs in the same way?
3. Are the higher-order factors going to be used to predict other constructs of the same general level of abstraction (i.e., global personality–global attitudes)?
4. Are the minimum conditions for identification and good measurement practice present in both the first-order and higher-order layers of the measurement theory?

> ### RULES OF THUMB 12-8
>
> #### Higher-Order Factor Models
>
> - Higher-order factors must have a theoretical justification and should be used only in relationships with other constructs of the same general level of abstraction
> - All of the first-order factors should be expected to influence other related constructs in the same way
> - At least three first-order constructs should be used to meet the minimum conditions for identification and good measurement practice

If the answer to each of these questions is yes, then a higher-order measurement model becomes applicable. After empirically testing higher-order models, the following questions should be addressed.

1. Does the higher-order factor model exhibit adequate fit?
2. Do the higher-order factors predict other conceptually related constructs adequately and as expected?
3. When comparing to a lower-order factor model, does the higher-order model exhibit equal or better predictive validity?

Once again, if the answer to these questions is yes, then a higher-order measurement theory would be supported.

MULTIPLE GROUPS ANALYSIS

Numerous SEM applications involve analyzing groups of respondents. Groups are sometimes formed from an overall sample by dividing it by meaningful characteristic such as a respondent's gender. For example, we may expect that men and women do not respond similarly across a wide range of social science issues. Alternatively, a large sample may be broken randomly into two subsamples so that a cross-validation can take place. But groups are not always separated after the fact. Many times different populations are sampled with the ultimate aim of testing for similarities and differences between those populations. For example, the populations may involve people from different cultures.

Multiple group analysis is a SEM framework for testing any number or type of differences between similar models estimated for different groups of respondents. The general objective is to see if there are differences between individual group models. This procedure is different from testing models with different specifications for the same sample of respondents. Here we are comparing the same model across different samples of respondents. Although very specific tests of differences can be performed for unique research questions, a general framework has emerged for comparing the measurement models and then structural models across groups. We will first discuss measurement model invariance because it is generally assumed to be a prerequisite for making comparisons at the structural model level.

Measurement Model Comparisons

A key benefit of achieving construct validity is that a construct will meet all of the requirements of reliability and validity not only in one situation, but hopefully across all of the potential situations in which it can be applied. Although rigorous testing in the development stage may support construct validity, researchers have long been aware of the need to reassess a construct, particularly when comparisons are made within the same study [31, 42, 43, 44, 56]. Increased applications of SEM to cross-cultural studies, longitudinal studies, assessment of differences based on personal differences (e.g., gender), and even context (e.g., type of workplace setting) have brought to our attention the need for a formalized process of making group comparisons [22, 23, 69, 70].

Today we see measurement model comparisons in two related areas. The first is broadly known as **measurement invariance** (or **measurement equivalence**). The primary objective is to ensure that measurement models conducted under different conditions yield equivalent representations of the same construct. As just discussed, the types of situations using this approach has become quite extensive. Yet even with the increased awareness and availability of SEM programs, in many areas of research that require measurement invariance (e.g., cross-cultural research) it is still infrequently used [73]. A more specific instance of measurement model comparisons is **cross-validation,** the attempt to reproduce the results found in one sample using data from a different sample. Generally, cross-validation uses two samples drawn from the same population. In other words, the sampling units in each group would have the same characteristics. Perhaps the most basic application is providing a second confirmation of a measurement theory that survived initial testing. One way of accomplishing this task is to split a large sample randomly into two groups so that each sample meets the minimum size requirements discussed earlier.

A SIX-STEP PROCESS OF GROUP COMPARISONS   What both of these areas have in common is their joint use of what has become known as **multisample confirmatory factors analysis (MCFA).** CFA, as discussed in Part 1, provides the basis for establishing construct validity through the measurement model. MCFA now extends CFA into a multigroup situation where separate samples are collected for each group and then comparisons made to determine their invariance (or equivalence). As might be expected, there are numerous aspects of a CFA model for comparison. Recent research has converged to identify a systematic framework for evaluating all of these aspects in a progressively more rigorous set of comparisons that addresses the most elemental aspects in the earlier stages [43, 55, 56].

The foundation of the process is a series of empirical comparisons of models with increasingly restrictive constraints. The fundamental measure of difference used is the chi-square difference ($\Delta\chi^2$). This measure allows for an overall comparison between two model specifications (e.g., one with and one without constraints). The basic logic is that if a set of constraints is applied and the model fit (as measured by chi-square) does not show a significant increase (meaning worse fit) from a less constrained model, then the constraints can be accepted [10, 37, 51].

General practice is to start with the most unconstrained model (i.e., a separate and unique CFA model is estimated for each group). Then **between-group constraints** are added to reflect specific measurement model comparisons. A between-group constraint estimates a single parameter for the relationship rather than estimating a unique parameter for each group. Thus, it represents the hypothesis that the relationship being tested is invariant (equal) across the groups. If imposing this constraint does not significantly increase model fit, then it can be accepted and the researcher can assume invariance for that relationship in the measurement model.

One feature of the MCFA approach is that while basic measures of fit are provided for each group model, all of the model fit measures are provided for the collective set of group models. So in simple terms the chi-squares for all the group models are added together and measures such as CFI, RMSEA, and others are calculated for the entire set. In this way comparisons can be made on model fit measures (e.g., $\Delta\chi^2$) across MCFA models with differing sets of constraints.

Although all model fit indices are available for the set of group models, the primary measure used for comparison is the **chi-square difference ($\Delta\chi^2$)** because it can be assessed with a statistical significance level. The degrees of freedom for any model comparison are the number of parameter estimates constraints that are added from one stage to the next. For example, assume that we have three groups. Before any comparison is made we would have separate models for each group, meaning three unique loading estimates each variable in each group. Now assume that the loading estimates are assumed to be equal. Instead of three estimates, we would have only one estimate that is the best estimate for the three groups combined. Now we would have two fewer loadings to estimate for each variable. So in general the number of degrees of freedom for the difference test is equal to the number of equality constraints multiplied by one less than the number of groups. This will determine the degrees of freedom used to test the significance of each $\Delta\chi^2$ test.

We will first describe the six stages in terms of both the measurement model issues they address as well as the nature of the constraints used. It is important to note that at each stage a new set of constraints is "added on" to those in the previous model. For example, the model at stage 3 will have all of the constraints imposed in stage 2 plus those added in stage 3. So the chi-square difference test can be made between models at each stage rather than the initial or baseline model. It should also be noted that there are various levels of "passing" the tests at each step, and these will be addressed in the following section.

Stage 1: Configural Invariance. The first stage confirms **configural invariance**—that the same basic factor structure exists in all of the groups. Researchers should confirm that each group CFA model has the same number of constructs and items associated with each construct. Moreover, it must be shown that each group model meets appropriate levels of model fit and construct validity. In measurement theory terms, we are now ensuring that the constructs are congeneric across groups. This model is sometimes referred to as the **totally free multiple group model (TF)** because all free parameters are estimated separately and therefore free to take on different values in each group. The TF model also becomes the baseline model for comparison.

Stage 2: Metric Invariance. The second stage provides the first empirical comparison between MCFA models groups and involves the equivalence of the factor loadings. **Metric invariance** establishes the equivalence of the basic "meaning" of the construct because the loadings denote the relationship between indicators and latent construct. This is a critical test of invariance and the degree to which this is met determines cross-group validity beyond the basic factor structure. Constraints are set so that the factor loadings are equal across groups (e.g., $L_{X1,\text{Group1}} = L_{X1,\text{Group2}}$, $L_{X2,\text{Group1}} = L_{X2,\text{Group2}}, \dots$). Note that although the loadings for X_1 are equal across groups, each measured variable has its own unique loading estimate. The $\Delta\chi^2$ is computed between this model and the TF model with the degrees of freedom equaling the number of constrained loading estimates across the groups.

Stage 3: Scalar Invariance. The third stage is **scalar invariance,** which tests for the equality of the measured variable intercepts (i.e., means) on the construct. Support for scalar invariance is required if any comparisons of level (e.g., mean scores) are made across groups. Scalar invariance allows the relative amounts of latent constructs to be compared between groups.

Stage 4: Factor Covariance Invariance. In the fourth stage the covariances between constructs are constrained. **Factor covariance invariance** tests if constructs are related to each other in a similar fashion across groups. Note that in this stage the degrees of freedom come from constraining the factor covariances rather than loadings for each measured variable.

Stage 5: Factor Variance Invariance. Now the test is for **factor variance invariance,** which assesses the equality of the variances of the constructs across the groups. If factor variances and covariances are equivalent across groups, then latent construct correlations are equal as well.

Stage 6: Error Variance Invariance. The final stage tests for the **error term invariance** for each measured variable across the groups. This test is for the amount of measurement error present in the indicators and the extent to which it is equivalent across models.

FULL VERSUS PARTIAL INVARIANCE   The chi-square difference test is a test for full invariance, meaning that constraining all the parameters relative to that type of invariance to be the same in each group does not significantly worsen fit. In the case of metric invariance, this would mean constraining each corresponding loading to be the same in each group. Yet full invariance becomes more difficult to achieve as models become complex and the tests progress to later stages [42]. **Partial invariance** is a less conservative standard involving at least multiple estimates per construct to be equivalent across groups [20]. A general consensus has developed that if two parameters per construct (e.g., loadings in metric invariance, intercepts in scalar invariance, or even error terms in error variance invariance) are found to be invariant, then partial invariance is found and the process can extend to the next stage.

If full invariance is not supported, the researcher can systematically "free" the constraints on each factor that have the greatest differences in the hope that the $\Delta\chi^2$ will become nonsignificant with at least two constraints per construct. One approach to identifying the constraints that should be eliminated first is to examine the modification indices for the fully constrained model. As you will remember, modification indices suggest the change in χ^2 associated with estimating a relationship. Thus, equality constraints with the largest modification indices should be freed first. An approach using the specification search feature of SEM programs to identify and free the most restrictive constraints has also been proposed [74]. No matter what approach is taken, the objective is to free as few constraints as possible to achieve invariance.

WHAT LEVEL OF INVARIANCE IS NEEDED?   With the six stages and the levels of invariance now defined, the final question is: What level of invariance is needed? Is it necessary to achieve at least partial invariance for all six stages? The answer to that question is dependent on the type of research question being addressed. But for most research questions involved with comparing constructs means across groups, achieving partial scalar invariance is sufficient. Metric invariance is relevant in establishing relationships among constructs, which is more likely needed done in testing structural relationships differences such as in tests of moderation. Moreover, it has been found that error variance invariance is rarely achieved and it is only necessary to demonstrate equal construct reliability across the groups.

Table 12-18 provides guidelines for the level of invariance needed for different types of research questions [23, 69, 70]. Focusing on the measurement model issues, we can see that the most common issue, equivalence of the basic structure of the construct, requires at configural invariance only. A tests of whether the constructs relate to indicators in the same way requires partial metric invariance (i.e., at least two loadings per construct must be invariant). Only when comparisons of the latent construct means are made is scalar invariance required. Note that a full invariance is always a stronger support of the types of constraints being tested, partial invariance at any of the stages except configural invariance is a practical standard that is acceptable, particularly as a model becomes more complex.

HBAT INVARIANCE ANALYSIS   During the course of interviews between HBAT management and the consultants, numerous issues arose suggesting a need to compare full-time versus part-time employees. The concern was that just taking an overall perspective on employees might overlook

TABLE 12-18 Suggested Minimum Levels of Invariance by Type of Research Question

Level of Invariance	*Measurement Model Comparisons*[a]		*Structural Model Comparisons*[a]	
	Basic Structure: Is the construct perceived and used in a similar manner?	**Mean Levels: Do the groups have equal amounts of latent constructs?**	**Theoretical Relationship Equivalence: Is the relationship between constructs the same across groups?**	**Theoretical Relationship Equivalence: Are correlations or standardized loadings the same across groups?**
Configural	Full	Full	Full	Full
Metric	Partial	Partial	Partial	Partial
Scalar		Partial		Partial
Factor covariance				Partial
Factor variance				
Error term				

[a]Minimum levels of invariance required.

noticeable differences between these two groups. It was felt that the first step should be to make sure that these two groups had common perceptions about the workplace attitudes that HBAT considered important in their employee retention efforts. This led to a call for an empirical comparison of the two groups on the five constructs in the Employee Retention model (see Parts 1 and 2 for more complete description of the constructs and model).

Groups were formed from the respondents to the HBAT employee survey based on work status—full-time or part-time. In this case the groups were of almost equal size (191 part-time employees, 209 full-time employees). With the groups and their responses defined, the invariance testing process can begin.

Six-Stage Invariance Testing Process. Group models were specified based on the six-stage process and then estimated. Table 12-19 contains the model fit statistics for each model and the chi-square differences test for each model comparison. In the first stage of configural invariance, the separate models for full-time and part-time employees both exhibit acceptable levels of model fit, as does the combined MCFA model ($\chi^2 = 438.1$, $df = 358$, $p = .002$, RMSEA = .024, and CFI = .98). Note that the MCFA χ^2 is equal to the sum of the two employee group models and the other fit

TABLE 12-19 Measurement Invariance Tests for Full-Time Versus Part-Time Employees

Model Tested	*Model Fit Measures*					*Model Differences*		
	χ^2	*df*	*p*	RMSEA	CFI	$\Delta\chi^2$	Δdf	*p*
Separate groups								
Part-time employees	259.8	179	.000	.049	.96			
Full-time employees	178.3	179	.500	.000	1.00			
Configural invariance	438.1	358	.002	.024	.98			
Metric invariance	450.8	374	.004	.023	.98	12.7	16	.69
Scalar invariance	521.8	395	.000	.028	.97	71.0	21	.00
Factor covariance invariance	535.6	405	.000	.028	.97	13.8	10	.19
Factor variance invariance	536.2	410	.000	.028	.97	.6	5	.98
Error variance invariance	587.7	431	.000	.030	.96	51.5	21	.00
Partial scalar invariance	457.4	384	.006	.022	.98	6.6	10	.77

measures signify acceptable fit across the two groups—indicating configural invariance. The next test is for metric invariance and involves constraining each matching loading, to be equal across the groups. We can see that the $\Delta\chi^2$ is only 12.7 with 16 degrees of freedom, which indicates a nonsignificant difference. The 16 degrees of freedom represent the 16 free factor loadings that were constrained to be equal to the other group (remember that one parameter was already constrained to 1.0 to set the scale on each construct, thus leaving 16 free parameters across the measured variables). Thus, the two models exhibit full metric invariance.

The next stage is to test for scalar invariance. Here the $\Delta\chi^2$ is 71.0 with 21 degrees of freedom. This difference is statistically significant, indicating that full scalar variance is not supported. For illustrative purposes we will continue to the next stage of the process and then return to assess if at least partial scalar invariance can be achieved. The next stage tests for factor covariance invariance, and the $\Delta\chi^2$ of 13.8 with 10 degrees of freedom is nonsignificant. From this result we can support invariance in the covariances among matching constructs. The next test for factor variance invariance shows very little difference for this constraint ($\Delta\chi^2 = .6$, $df = 5$), indicating that factor variances are almost identical between the two models. The final invariance test is for equivalence of the error terms of the indicators. As expected this test had a significant chi-square difference ($\Delta\chi^2 = 51.5$, $df = 21$). If this would have been supported, then equal reliabilities would have been found for constructs in each group.

We now return to see if we can achieve at least partial scalar invariance. Modification indices for the scalar invariance model were examined to identify the constraints of item intercepts that could be freed to most reduce the chi-square difference. As a result, 10 items (two per construct) were identified to retain their constraints in a test of partial scalar invariance. The items were JS_2, JS_4, OC_3, OC_4, SI_1, SI_3, AC_1, AC_4, EP_2, and EP_4. A model constraining each of these parameters to be equal to one another in each group produced a $\Delta\chi^2$ of only 6.6 ($df = 10$) from the metric invariance model, which was nonsignificant. Thus, partial scalar invariance can be supported as well. Thus, comparisons between construct means are possible.

Measurement Invariance Conclusions. The measurement invariance testing process demonstrated that the five constructs use in the Employee Retention model meet the criteria for configural invariance, full metric invariance and partial scalar invariance. As a result, most any form of group comparison can be made without concern that the differences are due to differing measurement properties between the two groups.

Structural Model Comparisons

The process of group comparisons for structural model parameters first builds upon the measurement model process and then performs similar types of comparisons to assess the differences in the structural model. Any type of structural model comparison first requires at least partial metric invariance of the measurement model to ensure that the constructs are comparable. If metric invariance is not achieved, then the researcher cannot be sure whether the differences seen in a structural model parameter are due to a group idiosyncracy or truly represent a differing structural relationship.

Structural model comparisons provide a specific test for addressing any number of research hypotheses, but the most common use is the test of moderation. Discussed in more detail in a following section, moderation assesses the differences in structural relationships between groups formed on a third variable. Group model comparisons can identify the extent of the differences either for an entire model or for a specific relationship. We will have a complete discussion of moderation and an example using the HBAT Employee Retention model in a later section.

The other research question concerning the structural model is to compare the means of latent constructs. In making these comparisons, SEM programs compare means only in a relative sense. In other words, they can tell you whether the mean is higher or lower relative to another group [57]. To do so requires that latent construct means are fixed to zero in one group (group 1). Values are then

---

RULES OF THUMB 12-9

Multiple-Group Models and Measurement Invariance Testing

- Multigroup models provide a comprehensive framework for comparing any model parameter between two or more samples of respondents
- Multisample confirmatory factor analysis (MCFA), a form of multigroup analysis, is the framework for assessing measurement invariance
- The chi-square difference test is the empirical means of assessing if a between-group constraint is statistically significant
- If a between-group constraint is nonsignificant, then the parameter being evaluated does not vary between groups
- Metric invariance, involving the equivalence of factor loadings, is needed for making model comparisons of relationships between constructs
- Scalar invariance, the equality of indicator intercepts, allows for comparing latent construct means across groups
- If full invariance cannot be achieved, partial invariance is acceptable if two indicators per construct are found invariant

---

estimated for the other group(s) and are interpreted as to how much higher or lower the latent construct means are in this group relative to group 1. Interested readers can find examples of comparison of latent construct means on the text's Web sites (www.pearsonglobaleditions.com/hair or www.mvstats.com).

MEASUREMENT BIAS

Researchers sometimes become concerned that survey responses are biased based on the way the questions are asked. For instance, it could be argued that the order in which questions are asked could be responsible for the covariance among items that are grouped close together. Similarly, researchers often are faced with resolving the question of constant methods bias. **Constant methods bias** would imply that the covariance among measured items is influenced by the fact that some or all of the responses are collected with the same type of scale. Thus, the covariance could be explained by the way respondents use a certain scale type, in addition to or instead of the content of the scale items.

Constant methods bias is an example of what is known as a nuisance factor. A **nuisance factor** is something that may affect the responses but is not of primary interest to the research question. For many effects of this type it is assumed that they are just represented in the error terms. But if the impact of a nuisance factor is substantial or systematic enough to impact the results, then it should be included in the model. We first discuss the model specification issues involved in assessing the impact of a nuisance factor and then examine model estimates to assess the extent of the effect.

Model Specification

The concept of a nuisance factor is widely used in experimental designs, where such factors in administration of the experiment (time of day, room conditions, or even administrator characteristics) may be thought to have some impact and thus need to be "controlled for" in the design. The most common approach is to introduce blocking factors to represent these factors when estimating effects. Then, these factors do not have to be considered in interpreting the results because they have already been accounted for. In SEM we follow a similar approach. To do so, we create a latent construct to represent the nuisance factor and then account for it in the model. In constant methods bias the focus is on effects related to the questionnaire design and administration, but the approach could be extended to any type of nuisance effect.

The rationale of a constant methods effect is that the use of a certain type of question impacts the results. In SEM terms, this means that an external effect (e.g., the scale type) impacts the measured variables. To represent this external effect we create an additional latent construct and relate it to the measured variables it impacts. Because the "cause" impacts the measured variables, the construct is reflective (i.e., arrows go from construct to measured variables) and operates as any other construct of this type. Note that a construct of this type violates the principles of good measurement theory because it creates cross-loadings for the measured variables involved and thus impacts the unidimensionality of the construct.

Let's use the HBAT employee questionnaire as an example. In gathering responses several different types of rating scales were used. Although it could be argued that respondents prefer a single format on any questionnaire, several advantages come with using a small number of different formats. One advantage is that the researcher can assess the extent to which any particular scale type may be biasing the results.

In this case, HBAT is concerned that the semantic differential items are causing measurement bias. The analyst suggests that respondents may have consistent patterns of responses to semantic differential scales no matter what the subject of the item is. Therefore, a semantic differential factor may help explain results. A CFA model can be used to test this proposition. To do so, an additional construct is created to represent the effect causing the semantic differential items. In this case, items EP_4, JS_2, JS_3, AC_2, and SI_4 are all measured with semantic differential scales. Thus, the model needs to estimate paths between this new construct and these measured items. Remember that factors of this type that are added will not have congeneric measurement properties.

We will modify the original HBAT CFA model (see Figure 12-5) by adding a sixth construct to represent the constant methods bias with dependence paths (loadings) to the five measured variables (EP_4, JS_2, JS_3, AC_2, and SI_4). These five variables will now have two factor loadings, one to their original latent construct (e.g., EP_4 loading on the EP construct) as well as a loading to the constant methods bias construct. The measurement model no longer exhibits a simple structure because each of the five measured variables is now determined both by its conceptual factor and by the new construct. We provide instructions on the text's Web sites (www.pearsonglobaleditions.com/hair or www.mvstats.com) on how to do this using either LISREL or AMOS. When all items are measured with the same scale, the equivalent addition would be a single factor causing all measured items.

Model Interpretation

Assessing the impact of the constant methods bias (or any other nuisance factor) is similar to any other set of competing models. First, overall model fits are compared to see if the additional factor has a significant impact. This is done through a chi-square difference ($\Delta\chi^2$) test and examination of the model fit indices. If the model fit significantly improves with the addition of the nuisance factor construct, then the researcher would know that the effect was substantial and should proceed to understanding the nature of the effect. Examination of the factor loadings provides a more precise estimate of the extent and magnitude of the effects on the individual items. Significant loadings indicate the presence of another cause for the measured variables, in addition to the original latent construct it represents.

Researchers should always be cautious in interpreting these effects because they are based on the assumption that the nuisance factor actually represents that effect (e.g., constant methods bias) and does not in actuality represent some other factor. Because the nuisance factor is many times not directly observed, the researcher must be careful to not allow spurious or other nonspecified effects to confound the nuisance factor.

To test for a constant methods effect due to the semantic differential scale, the CFA model is estimated with the addition of the sixth construct. The following fit statistics are obtained: χ^2 is 232.6 with 174 degrees of freedom and the RMSEA, PNFI, and CFI are .028, .80, and .99, respectively. The

$\Delta\chi^2$ of 4.0 (236.6 − 232.6) with five (179 − 174) degrees of freedom is insignificant. In addition, all of the estimates associated with the nuisance construct are nonsignificant. Also, the values for the original parameter estimates remain virtually unchanged as well.

Based on the model fit comparisons, the insignificant parameter estimates, and the parameter stability, no evidence supports the proposition that responses to semantic differential items are biasing results, and in this case the responses are not subject to measurement bias. Another factor could be added to act as a potential nuisance cause for the items representing another scale type, such as all Likert items. The test would proceed in much the same way.

A final note of caution is suggested regarding the concept of constant methods bias. In the design of a questionnaire researchers must be careful not to change the types of scales too frequently in an effort to eliminate this type of bias. Each time a new scale is used respondents must think about how to answer the new scale type. This effort requires additional learning, and therefore more in-depth respondent engagement in the process of answering the questions. Changing scale types too frequently in a questionnaire also produce problems. Thus, when designing questionnaires it is best to minimize changes in scale types unless there is an underlying logic to doing it. That is, if changing scale types is likely to enhance the quality of responses then do it. If not, use scale types that have demonstrated good results in previous research.

RELATIONSHIP TYPES: MEDIATION AND MODERATION

In Chapter 11, we introduced two basic types of relationships: correlational and dependence. As we have seen in the models using these relationship types, they are the "building blocks" of our structural models. We now discuss two variations of these basic relationships: mediation and moderation. For each new relationship type the discussion will first focus on the theoretical nature of the relationship and how it can be incorporated into our SEM models. Then we discuss the process whereby we incorporate these relationships into our existing models.

Mediation

In simple terms, a **mediating effect** is created when a third variable/construct intervenes between two other related constructs [8]. To understand how mediating effects are shown in our structural model, let us examine a model in terms of direct and indirect effects. **Direct effects** are those relationships that link two constructs with a single arrow. **Indirect effects** are those relationships that involve a sequence of relationships with at least one intervening construct involved. Thus, an indirect effect is a sequence of two or more direct effects (compound path) and is represented visually by multiple arrows. The following diagram shows both a direct effect (K → E) and an indirect effect of K on E in the form of a K → M → E sequence.

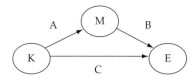

The indirect effect (K → M → E) represents the mediating effect of construct M on the relationship between K and E.

CONCEPTUAL BASIS FOR MEDIATION   From a theoretical perspective, the most common application of mediation is to "explain" why a relationship between two constructs exists. We may observe a relationship between constructs (e.g., K and E), but not know "why" it exists. We can then posit some explanation in terms of an intervening variable that operates to take the "inputs" from K and translate them into the "output" E. As such, the mediator (M) clarifies or explains the relationship between the two original constructs. A simple example illustrates these points.

The construct K could be a student's intelligence and E could be classroom performance. What is interesting is not just that the relationship exists between intelligence and classroom performances, but how it actually works. Can we explain how students translate their intelligence into performance? We may find that sometimes a student exhibits high intelligence, but does not always perform well. Moreover, some students with lower intelligence scores perform extremely well. So, is there some other process going on that translates student intelligence into actual classroom performance?

The intervening process is the mediating effect. In this case, we could propose a construct termed study effectiveness. Here we refer to such characteristics as the ability of students to focus their efforts on their class work, organize their class-related and other activities to provide them with sufficient time to complete their homework, and other "good" study habits. If a student is intelligent, this quality may encourage the student to study longer and better, which could result in higher classroom performance. In such a case, the significant correlation between K and E would be explained by the K → M → E sequence of relationships. We could then say that study effectiveness mediates the relationship between student's intelligence and classroom performance.

In broader terms the concept of mediation is common in structural models in general. If we view exogenous constructs as the "inputs" to our model explaining some final "outcome" represented by an endogenous construct, then any constructs going between these correspond to a theory involving at least some mediation. Hopefully these mediating relationships are successful in representing those mediating effects and we have good model fit.

TESTING FOR MEDIATION   Mediation requires significant correlations among all three constructs. Theoretically, a mediating construct facilitates the relationship between the other two constructs involved. If the mediating construct completely explains the relationship between the two original constructs (e.g., K and E), then we term this **complete mediation.** But if we find that there is still some of the relationship between K and E that is not explained away by the mediator, then we denote this as **partial mediation.**

A researcher can determine if mediation exists, and whether it is complete or partial, in several ways. First, if the path labeled C is expected to be zero due to mediation (representing complete mediation), a SEM model can represent mediation by including only the paths A and B in the model. It would not include a path directly from K to E. If the estimated model suggests the sequence K → M → E provides a good fit, it would support a mediating role for M. In addition, the fit of this model could be compared with the SEM results of a model including the K–E path (C). If the addition of path C improves fit significantly as indicated by the $\Delta\chi^2$, then complete mediation is not supported. If the two models produce similar χ^2 then mediation is supported.

Because relationships are not always clear, a series of steps can be followed to evaluate mediation. These steps apply whether using SEM or any other general linear model (GLM) approach, including multiple regression analysis. Using the previous mediation diagram, the steps are [25]:

1. Establish that the necessary individual relationships have statistically significant relationships:
   a. *K is related to E:* We establish that the direct relationship does exist.
   b. *K is related to M:* We establish that the mediator is related to the "input" construct.
   c. *M is related to E:* We establish that the mediator does have a relationship with the outcome construct.

2. Estimate an initial model with only the direct effect (C) between K and E. Then estimate a second model adding in the mediating variable (M) and the two additional path estimates (A and B). Then assess the extent of mediation as follows:

   a. If the relationship between K and E (C) remains *significant and unchanged* once M is included in the model as an additional predictor (K and M now predict E), then mediation is not supported.

   b. If C is *reduced but remains significant* when M is included as an additional predictor, then *partial mediation* is supported.

   c. If C is reduced to a point where it is *not statistically significantly* after M is included as a mediating construct, then **full mediation** is supported.

HBAT ILLUSTRATION OF MEDIATION   The HBAT model shown in Figure 12-17 hypothesizes several mediating effects. The relationships of both exogenous constructs (Environmental Perceptions [EP] and Attitudes Toward Coworkers [AC]) with Staying Intentions (SI) are hypothesized to be fully mediated by two endogenous constructs—Job Satisfaction (JS) and Organizational Commitment (OC). The model hypothesizes that the effects of EP and AC can be fully explained through these two mediating constructs. In Part 2 we examined model fit as an explanation of the overall set of relationships and confirmed that model. But we can also explicitly test for mediation.

Let us examine the possibility that the relationship between EP and SI is mediated by the two constructs of JS and OC. To do so, we will follow the two-step process described above:

Step 1: Establish Significant Relationships Between the Constructs.   In this situation, we can refer back to Table 12-7, which provided the correlations between constructs in the CFA model. From that analysis, we can first see that EP was significantly related SI (.56), ensuring that the direct, unmediated relationship was significant. We also find that EP was significantly related to both JS (.24) and OC (.50), establishing a relationship with both potential mediators. Finally, SI was significantly related to both JS (.23) and OC (.55), thus supporting relationships between the mediators and the outcome variable.

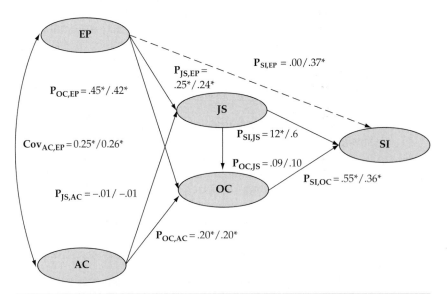

FIGURE 12-17   **Adding Direct Effects for Testing Mediation in the HBAT Employee Retention Model**

Values represent parameter estimates for initial/respecified (direct effect added) models.
*Statistically significant at .05 level.

Step 2: Estimate the Mediated Model and Assess the Level of Mediation. In this analysis, we will first use the original Employee Retention model (see Figure 12-17), which did not estimate the direct effect from EP to SI. To assess if adding the direct effect would substantially change the model fit, we can estimate a revised HBAT model (dotted path in Figure 12-17), which adds the direct path between EP and SI.

The revised model with the direct relationship had a significant decrease in chi-square ($\Delta \chi^2 =$ 41.2, $df = 1$, $p = .00$), a substantive improvement in model fit and a significant path estimate for the EP $\rightarrow$ SI relationship (see Table 12-20). These results suggest that there is not complete mediation. But is there partial mediation? To establish partial mediation, we will need to identify a significant indirect effect leading from EP to SI through the mediating variables. There are a number of potential compound paths, but only three reflect indirect causal mediated effects: (1) EP $\rightarrow$ OC $\rightarrow$ SI; (2) EP $\rightarrow$ JS $\rightarrow$ OC $\rightarrow$ SI; and (3) EP $\rightarrow$ JS $\rightarrow$ SI. If one or more of these indirect effects contains paths that are all significant, then the model supports partial mediation. By this we mean that there is a significant direct relationship between EP $\rightarrow$ SI, but that there is also a significant indirect effect through the mediator.

In the revised model the path estimates between EP and both mediators (JS and OC) are still significant (Table 12-20). But whereas one mediator still has a significant relationship with SI (OC $\rightarrow$ SI is significant), the other relationship from a mediator to SI (JS $\rightarrow$ SI) becomes nonsignificant. Does this mean that there is not partial mediation? No, it just means that although JS does not act as a mediator, OC can still have a mediating effect. Note that in the three indirect mediated effects, the nonsignificant relationship (JS $\rightarrow$ SI) is only a part of the third effect. The other two indirect mediating effects still have statistically significant individual paths. Thus, the model supports the finding that OC provides partial mediation of the relationship between EP and SI.

The magnitude of the mediating effects is demonstrated by breaking down the total effects into direct and indirect effects. Most software programs provide this decomposition of effects. But if not, the researcher can calculate the effects. The interested reader is referred to the Basic Stats appendix on the text's Web sites (www.pearsonglobaleditions.com/hair or www.mvstats.com) where this process is described in more detail. Table 12-21 provides a breakdown of the effects of EP $\rightarrow$ SI both in the original HBAT model (no direct effects from EP $\rightarrow$ SI) and the revised HBAT model (direct effect added for EP $\rightarrow$ SI). As we can see in the original model, substantial indirect

TABLE 12-20   Testing for Mediation in the HBAT Employee Retention Model

Model Element	HBAT Employee Retention Model	Revised Model with Direct Effect
Model fit		
χ^2 (chi-square)	283.43	242.23
Degrees of freedom	181	180
Probability	0.00	0.00
RMSEA	.036	.029
CFI	.99	.99
Standardized parameter estimates		
EP $\rightarrow$ JS	0.25*	0.24*
EP $\rightarrow$ OC	0.45*	0.42*
JS $\rightarrow$ SI	0.12*	0.06
JS $\rightarrow$ OC	0.09	0.10
OC $\rightarrow$ SI	0.55*	0.36*
EP $\rightarrow$ SI	Not estimated	0.26*

*Statistically significant at .05 level.

TABLE 12-21   Assessing Direct and Indirect Effects in a Mediated Model

Effects[a] of EP → SI	Original HBAT Model (Only Indirect Effects)	Revised HBAT Model (Indirect and Direct Effects)
Total effects	.29	.55
Direct effects	.00	.37
Indirect effects	.29	.18

[a]Values in the table represent standardized effects.

effects are present, thus supporting the presence of mediating effects of JS and OC. The remaining question is: Do those effects remain when the direct path is added in the revised model? As we can see, although the indirect effects do decrease, they are still significant and represent a substantial portion of the total effects. We should also note that the direct effect is significant, adds considerably to the total effects, and constitutes the majority of the total effects, making this a partial mediation situation.

The magnitude of any individual mediating effect can be calculated using the process for calculating indirect effects. If this is done, you will find that most of the mediating effect comes from the EP → OC → SI mediating relationship. Given this post hoc theoretical analysis, HBAT needs to cross-validate this result with new data before considering it reliable. However, managerial implications for each of the supported hypotheses can be developed based on the overall positive results.

Moderation

A **moderating effect** occurs when a third variable or construct changes the relationship between two related variables/constructs [8]. For example, we would say that a relationship is moderated by gender if we found that the relationship between two variables differed significantly between males and females. For example, the relationship between two variables may be negative for males and positive for females or significant in one group and not the other. In this type of situation, we would need to know whether the respondents were males or females before we could accurately estimate and interpret the relationship.

We have discussed moderators in the other multivariate techniques. In multiple regression, for example, there were interaction terms where the regression coefficient changed based on the value of a second variable. And in ANOVA/MANOVA, interaction effects were used to assess whether the differences between groups were constant across the values of another variable. In both of these examples, the interaction effects are in fact moderators.

It is important to note that moderating variables must be chosen with strong theoretical support. The assumption of causality by the moderator is one that cannot be tested directly and becomes potentially confounded as the moderator becomes correlated with either of the variables in the relationship. Therefore, analysis of moderators is easiest when the moderator has no significant linear relationship with either of the constructs [8, 25, 36]. The lack of a relationship between the moderator and the other constructs helps distinguish moderators from mediators (remember that the mediator must be related to both constructs in the relationship being mediated). As multicollinearity increases between the mediator and the other constructs, the causal inference is called into question, making a valid interpretation increasingly difficult.

NONMETRIC MODERATORS   A moderator variable can be metric or nonmetric. Nonmetric, categorical variables often are hypothesized as moderators. These moderators typically are classification variables of some type. One common type of moderator is respondent characteristics, such as gender, age, or other characteristics. Differing situations or contexts are another type of categorical moderator. A common example would be cross-cultural studies where country-of-origin becomes a moderating variable. Similarly, dividing respondents into current customers versus noncustomers would be using customer status as a moderating variable.

As has been noted, theory is important in evaluating a moderator because a researcher should find some reason to expect that the moderator changes a relationship. Researchers have any number of ways for dividing the sample into groups, but the section of a moderator should not be based on whether it demonstrates significant moderating effects, but instead on its theoretical foundation.

Once the moderating variable is selected, groups of respondents can be defined and multigroup analysis applied. The procedure is similar to that described earlier in the discussion of invariance testing with the primary difference being that in moderation the focus is typically on structural model estimates rather than the measurement model.

METRIC MODERATORS   A moderator can also be a continuous/metric variable and evaluated using SEM. If the continuous variable can be categorized in a way that makes sense (i.e., is based on theory or logic), then groups can be created and the same procedures used for nonmetric moderators can be applied. For instance, if the continuous variable shows bimodality (i.e., the frequency distribution shows two clear peaks rather than one), then logical groups could be created around each mode. As an example, job satisfaction could be measured metrically, but if there is a highly satisfied group versus a moderately satisfied group then the two groups defined by their feelings about satisfaction would be used as a moderating variable. Cluster analysis also might be useful to form groups. However, if the moderator variable displays a clear unimodal distribution (one peak), then grouping is not justified. It is possible that some fraction (i.e., one-third) of the observations around the median value could be deleted and the remaining observations (which are likely now bimodal) used to create groups. An obvious drawback to this approach is the increased cost, time, and effort associated with the need to gather a larger sample. The advantage is that multigroup analysis can be used providing an intuitive way of testing and demonstrating moderation.

Researchers also can model a metric moderator by creating interaction terms as when using a regression approach. Using regression terminology, the independent variable can be multiplied by the moderator to create an interaction term. However, taking this approach with multiple-item constructs is complicated by numerous factors. This topic is beyond the scope of this text. A brief introduction is provided on the text's Web sites (www.pearsonglobaleditions.com/hair or www.mvstats.com). We encourage all but the advanced users to apply the nonmetric multigroup approach unless it cannot be justified.

USING MULTIGROUP SEM TO TEST MODERATION   Multigroup SEM is used to test moderating effects when the moderating variable is either nonmetric or a metric moderator has been transformed into a nonmetric variable (see previous description). Moderation typically involves the testing of structural model estimates. Thus, the process becomes an extension of the multigroup analysis for testing measurement invariance. As an initial step, some form of metric invariance must be established before examining any differences in structural model estimates.

With measurement invariance established, the structural model estimate is then assessed for moderation by a comparison of group models, much like invariance testing. The first group model is estimated with path estimates calculated separately for each group. This is identical to the TF (totally free) model described earlier. A second group model is then estimated where the path estimate of interest is constrained to be equal between the groups. Comparison of the differences between models with a chi-square difference test ($\Delta\chi^2$) indicates if the model fit significantly decreased (i.e., an increase in chi-square) when the estimates were constrained to be equal. A statistically significant difference between models indicates that the path estimates were different (i.e., model fit was significantly better when separate path estimates were made) and that moderation does exist. If the models are not significantly different, then there is no support for moderation (because the path estimates were not different between groups). When testing for moderation the researcher is looking for significant differences in the two models to support the hypothesis of differences in the path estimates. The researcher should also examine the path estimates in question to assess if the differences in both group models are theoretically consistent.

HBAT ILLUSTRATION OF MODERATION   The HBAT management team suspected that men and women may not exhibit the same relationships in the role that the variable Attitudes Toward Coworkers may play in creating job satisfaction. Theory would suggest that there could be a gender difference in this relationship where the effect would be greater among women relative to men. This was reinforced when the Employee Retention model discussed earlier found a nonsignificant relationship between AC → JS. Questions arose that perhaps the "common" path estimate from combining both groups was not reflecting the actual differences between groups. As a result, the HBAT research team decided to conduct a multigroup analysis using the gender classification variable.

The multigroup CFA established metric invariance as described earlier in this chapter. This was sufficient to now test for moderation in the relationships between constructs, specifically the AC → JS relationship.

Following the same steps used to specify the two-group CFA model testing for differences based on gender, a two-group structural model was set up. The TF structural model estimates an identical structural model in both groups simultaneously. The model fit statistics and path estimates for the AC → JS relationship are shown in Table 12-22. Then a second group model is estimated, the only difference being that the AC → JS path estimates are constrained to be equal in both groups. These fit results and path estimates are also shown in Table 12-22.

Both models show acceptable fit indices (CFI and RMSEA), indicating their overall acceptability. The chi-square difference between models ($\Delta\chi^2$) is 11.1 with one degree of freedom. This is significant ($p < .001$), indicating that constraining the AC → JS path estimate to be equal between groups produces worse fit. Therefore, the unconstrained (TF) model in which the AC → JS relationship is freely estimated in both groups is supported. This result suggests that gender does moderate the relationship between AC and JS.

Looking at the standardized parameter estimates for the TF results, HBAT researchers find that the AC → JS relationship is significant in both groups. As predicted, the relationship is greater for women with a completely standardized estimate of 0.24, compared to a completely standardized estimate of −0.17 for men. Thus, it seems that attitudes toward coworkers is positively related to job satisfaction among women, but negatively related to job satisfaction among men. Moreover, the

TABLE 12-22   Testing for Gender as a Moderator in the HBAT Employee Retention Model

Model Characteristic	Unconstrained Group Model (TF for Each Group)	Constrained Group Model (AC → JS Equal Across Groups)	Model Differences
Model fit			
Chi-square	401.1	412.2	11.1
df	360	361	1
CFI	0.99	0.99	—
RMSEA	0.024	0.027	—
Path estimate ($P_{JS,AC}$)	.24 (female)*	−.01 (combined)	
	−.17 (male)*		

*Significant at .05 level.

RULES OF THUMB 12-10

Mediation and Moderation

- Mediation involves the comparison of a direct effect between two constructs while also including an indirect effect through a third construct
- Full mediation is found when the direct effect becomes nonsignificant in the presence of the indirect effect, whereas partial mediation occurs when the direct effect is reduced, but still significant
- Although indirect effects that are small (less than .08) are generally not of interest because they are likely trivial relative to direct effects, the combined total of all indirect effects may be substantial
- Moderation by a classification variable can be tested with multigroup SEM:
  - A multigroup SEM first allows all hypothesized parameters to be estimated freely
  - Then, a second model is estimated in which the relationships that are thought to be moderated are constrained to be equal in all groups
  - If the second model fits as well as the first, moderation is not supported
  - If its fit is significantly worse, then moderation is evident
- The multigroup model is convenient for testing moderation:
  - If a continuous moderating variable can be collapsed into groups in a way that makes sense, then groups can be created and the procedures described previously can be used to test for moderation
  - Cluster analysis may be used to identify groups for multigroup comparisons
  - Unimodal data should not be split into groups based on a simple median split
- When using a continuous variable moderator formed as a construct interaction:
  - The direct relationships between the predictor construct and the outcome construct and between the moderator construct and the outcome construct should be estimated only if the relationship between the construct interaction and the outcome is insignificant
  - Larger sample sizes are needed to accommodate continuous variable interactions (i.e., $N > 500$)

---

nonsignificant path estimate from the combined model (males and females together) may lead to the incorrect conclusion that AC is not related to JS. The moderated relationship "cancels" out the different effects of males and females when estimated together. The result is a clear case of moderation where the nature of a relationship (AC → JS in this case) changes based on a third variable (gender).

LONGITUDINAL DATA

SEM is increasingly applied to longitudinal data. Given the added insight from tracking changes in constructs and relationships over time, the increasing use of longitudinal data may be beneficial in many fields. Because many different types of longitudinal study designs lead to many different SEM applications, this section provides only a brief introduction to some of the key differences in dealing with longitudinal data. The interested reader is referred to other sources for a more detailed discussion and review [32].

Additional Covariance Sources: Timing

One of the key issues in modeling longitudinal data with SEM involves added sources of covariance associated with taking measures on the same units over time. For instance, consider a model hypothesizing that reading ability causes math ability. The theoretical rationale may lie in the fact that one needs to be able to read to adequately study mathematics [62]. Suppose that longitudinal data are available and that math and reading ability are each treated as latent constructs measured by multiple indicators. Math ability in any given time (t) could be modeled as a function of reading ability in time (t), reading ability in the previous time period ($t-1$), and math ability in the previous time period ($t-1$). A significant reading ability in time period ($t-1$) → math ability in time period (t) relationship would help provide evidence of causality in that it would be consistent with a causal time sequence and establish covariation. It is easy to see how the model could be extended to more time periods.

One key issue is whether it is reasonable to expect that the matching indicator measures in different time periods are unrelated. In other words, if reading speed is an indicator of the reading ability construct in each time period, should the correlation between the reading speed indicator test in time $t-1$ and the reading speed indicator test in time t be modeled? Generally speaking, the answer is yes.

Using Error Covariances to Represent Added Covariance

Including a measurement error covariance term or an additional construct that is seen as another cause of the matching indicators will represent the added covariance. Figure 12-18 shows a path diagram illustrating the use of error covariance parameters to capture the additional source of commonality. Each path that must be estimated is indicated with an arrow. Notice that two-headed

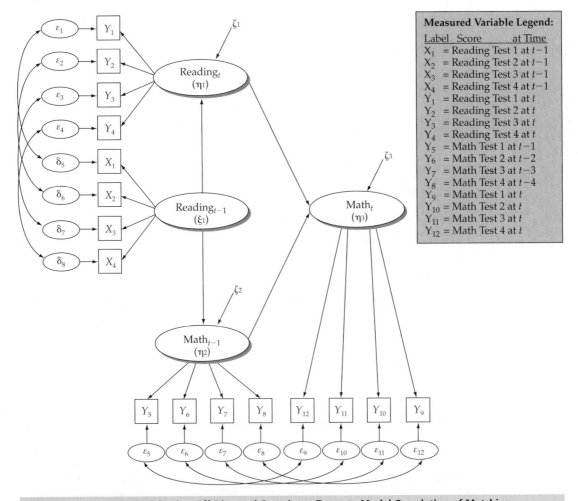

Measured Variable Legend:

Label	Score	at Time
X_1	= Reading Test 1 at $t-1$	
X_2	= Reading Test 2 at $t-1$	
X_3	= Reading Test 3 at $t-1$	
X_4	= Reading Test 4 at $t-1$	
Y_1	= Reading Test 1 at t	
Y_2	= Reading Test 2 at t	
Y_3	= Reading Test 3 at t	
Y_4	= Reading Test 4 at t	
Y_5	= Math Test 1 at $t-1$	
Y_6	= Math Test 2 at $t-2$	
Y_7	= Math Test 3 at $t-3$	
Y_8	= Math Test 4 at $t-4$	
Y_9	= Math Test 1 at t	
Y_{10}	= Math Test 2 at t	
Y_{11}	= Math Test 3 at t	
Y_{12}	= Math Test 4 at t	

FIGURE 12-18 SEM Model Using Off-Diagonal Covariance Terms to Model Correlation of Matching Measures Across Different Time Periods

curved arrows are now shown from each measured variable error term between matching tests. In other words, the first reading test (reading test 1) is the same test applied in both time periods t and $t - 1$, so a student's score on each should be correlated. The same can be said for each of the four separate math tests. In this way, an attempt can be made to control for the additional sources of covariation that emerge from the use of longitudinal data. These attempts at control become complicated quickly as the number of constructs and variables increases and across the different types of situations involving longitudinal data. But the models enable a closer examination of trend effects and can help establish the time sequence condition for causality. A final caution is that the increased number of parameters estimated may at times lead to problems with statistical identification.

Another way in which longitudinal analyses are conducted with SEM models is to track changes in correlations over different time periods. Assume that data for a SEM model has been collected for three different time periods. A multigroup analysis can be conducted over the three groups (time periods) to track potential changes in construct means or relationships. The process follows a sequence much like the multigroup process for moderation described earlier. For instance, a model setting a particular relationship to be equal across all three time periods can be tested against a model allowing the relationship to be estimated in different time periods. Then the fits for the two models can be compared to see whether the relationship is stable over time.

PARTIAL LEAST SQUARES

Partial least squares (PLS) has become increasingly popular as an alternative to SEM (e.g., LISREL, AMOS, or similar programs). Originally developed for econometrics [47, 72], it first gained widespread application in chemometric research and has more recently been adopted in research in business, education, and the social sciences. Although there are some differences between terminology in PLS and SEM programs, the basic specification of the structural model can be similar.

Although the structural models might look identical, there are substantive differences in terms of almost every aspect of developing, estimating, and interpreting a proposed model. Both PLS and SEM have advantages and disadvantages, which we will discuss. As you will see, there is no "clear-cut" winner in terms of estimating models with latent constructs. It is beyond the scope of this text to provide an exhaustive comparison between the methods. What we will provide is some perspective on the most suitable applications and the caveats found in each approach.

Characteristics of PLS

Let us begin with a brief description of PLS, focusing on some of its distinguishing characteristics. First, PLS specifies relationships in terms of measurement and structural models, which are termed *outer* and *inner models,* respectively. It can handle all types of data, from nonmetric to metric, with very minimal assumptions about the characteristics of the data. PLS handles both reflective and formative constructs and all **recursive models** are identified, even with single-item constructs. It differs, as implied in the name, in that PLS is estimated with regression-based methods rather than MLE. PLS focuses on explanation of variance (prediction of the constructs) rather than covariance (explanation of the relationships between items), and significance testing of parameter estimates is not possible without using bootstrapping methods.

PLS is now available in graphical-interface programs as well as modules in major software packages (SAS, SPSS) and even SEM programs such as LISREL. For example, a variation of PLS available within the SAS statistical software is easy to use. All the researcher needs to specify is the intended outcome variable, the set of measured variables that may predict it, and the number of factors that exist within that set of items. The researcher does not specify a factor pattern in this type of application.

At the core of the differences between PLS and SEM programs is its fundamental objective. PLS statistically produces parameter estimates that maximize explained variance much as was the case with OLS multiple regression. Therefore, the focus is much more on prediction. SEM, in contrast, tries to reproduce the observed covariation among measures, which makes it more oriented toward how well a given theory, as represented by a SEM model, explains these observations. Thus, SEM is more concerned with explanation [54] and is a more appropriate tool for theory testing. PLS does not provide a test of theoretical fit. This distinction is hard to conceptualize. We will therefore review some specific advantages and disadvantages of PLS as well as suggestions on its use.

Advantages and Disadvantages of PLS

PLS has several advantages as well as disadvantages relative to SEM. The advantages lie mainly in its robustness, meaning that it will provide a solution even when problems exist that may prevent a solution in SEM. First, poor measurement is one of the major obstacles to getting a SEM solution. For instance, when a researcher is attempting to test a structural model with single-item measures or a mix of several single- and two-item measures, PLS may be an option because of the identification problems that may occur with SEM. As has been noted, all recursive models are identified (do not exhibit statistical identification problems) in PLS, even with single-item measures. Thus, whereas validating one- and two-item measures in the context of a measurement theory has little meaning with SEM, PLS is uninhibited by such concerns.

PLS also readily handles both formative and reflective constructs. Many researchers have utilized PLS for just this reason, given the perceived difficulties in formative model specification in SEM. PLS also can be a useful way of quickly exploring a large number of variables to identify sets of variables (principal components) that can predict some outcome variable. PLS does not face the issues of model complexity that SEM does, and therefore it is able to handle large numbers of measured variables and/or constructs easily. Finally, PLS is insensitive to sample size considerations. Its estimation approach handles both very small and very large samples with more ease than does SEM. PLS is particularly useful in generating estimates even with very small samples (as low as 30 observations or less) where SEM programs are just not applicable.

But with these advantages come some disadvantages. The primary disadvantage of PLS is its focus on prediction rather than explanation. As noted earlier, this is a difficult concept to understand, but it is reflected in the fact that PLS will obtain solutions when SEM will fail. Some individuals might value a technique that always gives results. But the important consideration here is that we must understand what the results mean and how they are to be interpreted. We will use a simple example to highlight these differences.

Chin [24] provides a comparison of LISREL and PLS in an example with four measured variables and two constructs (see Figure 12-19). As he demonstrates, vastly different solutions are obtained using LISREL versus PLS methods. The LISREL solution produces a path estimate between the constructs of .83, which is much higher than any of the observed correlations, whereas the PLS path estimate of .22 is much closer to the observed correlations. Moreover, the factor loadings are much higher in the PLS model, whereas none of the loadings in the LISREL model meet acceptable levels. Thus, construct validity is questionable.

A case can be made that the PLS model is useful when measurement is poor. But we must make that statement considering the purpose of PLS, which is prediction. To understand the differences with LISREL and its emphasis on explanation, look at the measurement model results. The LISREL results indicate that both constructs do not meet the criteria for convergent validity discussed earlier. All of the LISREL loadings are below acceptable levels. If we calculated reliability and average variance extracted values they would be below acceptable standards as well.

So is this just a failure of LISREL in comparison to PLS? The answer comes from an examination of the measurement model. From a measurement theory perspective, we would conclude that the two constructs are unacceptable as latent constructs and would not proceed. Looking at the

Correlation Matrix	X_1	X_2	Y_1	Y_2
X_1	1.00			
X_2	.087	1.00		
Y_1	.140	.080	1.00	
Y_2	.152	.143	.272	1.00

Partial Least Squares Results   LISREL Results

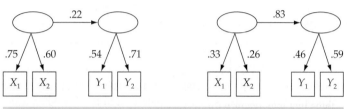

FIGURE 12-19 **Comparison of LISREL and PLS Estimates**

Source: Chin [15].

inter-item correlations, we see that they are .087 and .272, neither of which meet the suggested levels. This would lead us to conclude that the indicators do not support construct validity of the latent constructs and thus the model should not be estimated. The failure is not in the technique, but in the fact that the specified constructs do not meet the requirements of latent constructs as required by the SEM approach. As discussed earlier in Chapter 11 concerning the role of reliability in the SEM estimation process, it is assumed that the constructs exhibit adequate reliability and validity. This is an example of just such a situation, where a poor measurement model impacts the structural relationship.

But consider the following question: PLS provides loading estimates that are in the acceptable range, thus aren't the latent constructs valid? The difference arises from the fact that the loadings of the measured variables for the exogenous constructs in PLS are based on their prediction of the endogenous measured variables, not their shared variance among measured variables on the same construct. Thus, the loadings in the PLS solutions are in a way their contribution to the path estimates. This may seem similar to SEM models, but it is quite different in that the loadings are now dependent on which exogenous constructs they are related to. As such, they are much more susceptible to variation when moved from one model to another if the other constructs in the model change. This is in contrast to SEM approaches in which the emphasis is on construct validity irrespective of the other constructs in the model. PLS provides no omnibus test of the measurement model as is the case with CFA.

Choosing PLS Versus SEM

Conceptually and practically, PLS is similar to using multiple regression analysis to examine possible relationships with less emphasis on the measurement model as a separate and distinct submodel. PLS certainly can address a wider range of problems given its ability to work efficiently with a much wider range of sample sizes and model complexity and its less strict assumptions about the underlying data. Moreover, the lack of emphasis on the measurement properties of the constructs makes it more amenable to the use of constructs with fewer items (e.g., one or two) than are required for SEM. Thus, in certain cases, particularly when measures are problematic and/or the emphasis is more on exploration than confirmation, PLS may is an attractive alternative to SEM.

RULES OF THUMB 12-11

Longitudinal Data and PLS

- Longitudinal data is one situation in which correlated error terms are justified to accommodate covariance of the same indicator across time periods
- PLS is an alternative approach to SEM modeling and is well suited for:
  - Sample sizes either much smaller or larger than generally accommodated by SEM approaches
  - Situations where emphasis is more on prediction versus explanation
  - The measurement model properties do not permit the use of SEM
- PLS can provide results with samples too small to use in other approaches. Caution is needed in generalizing results from any small sample no matter the technique.

Although PLS can be applied to a wider range of situations, the researcher must always be aware of the differences in interpretation of the results, particularly as they relate to the measurement properties of the constructs. Is it appropriate to turn to PLS when a measurement theory fails to stand the scrutiny of a CFA and subsequent test of convergent validity? PLS still provides estimates of relationships between the model's constructs, but it is incumbent on the researcher in these situations to qualify the results based on the adequacy of the measures. As the concern for good measurement quality increases and as multi-item measures become available for latent constructs, PLS is not recommended as an alternative to SEM. Similarly, PLS can produce results even with a very small sample (even less than the number of variables). These may be useful for exploratory purposes but the generalizability of these results is limited by the small sample no matter what statistical approach is used.

In conclusion, the authors do not recommend one technique versus another in all situations. However, we strongly suggest that the researcher understand the differences for which each approach was developed and use them accordingly. Thankfully, it has been our experience that models with good measurement properties seems to reach very comparable results. Moreover, issues such as the appropriate use and interpretation of formative versus reflective measures are still found with each approach. It is in those instances in which the measurement properties are in doubt that the results may diverge and thus require a reasoned judgment by the researcher on which approach is most appropriate.

Summary

The widespread use of SEM techniques has greatly improved quantitative measurement in the social sciences. Researchers now have a tool that provides a strong test of one's theory from both the measurement and theoretical relationship perspectives. The key advantage is that the researcher can analytically test a conceptually grounded theory explaining how different measured items represent important psychological, sociological, or business measures. Then these constructs can be related in dependence relationships that provide strong support for the underlying theoretical models. Once the basic structural model has been examined, SEM provides a number of alternative analyses and specifications that provide further empirical support on such topics as higher-order models, measurement invariance, the use of formative measures, and tests for mediation and moderation.

In this chapter we have attempted to provide some insight into the broad range of topics covered under SEM modeling. Our hope is that you now have a greater understanding of the basic principles underlying SEM techniques, particularly those relating to the objectives of the chapter:

Distinguish between exploratory factor analysis and confirmatory factor analysis. CFA cannot be conducted appropriately unless the researcher can specify both the number of constructs that exist within the data to be analyzed and which specific measures should be assigned to each of these constructs. In contrast, EFA is conducted without knowledge of either of these things. EFA does not provide an assessment of fit. CFA provides an assessment of fit.

Assess the construct validity of a measurement model. Construct validity is essential in confirming a measurement model. Multiple components of construct validity include convergent validity, discriminant validity, face validity, and nomological validity. Construct reliabilities and variance-extracted estimates are useful in establishing convergent validity. Discriminant validity is supported when the average variance extracted (AVE) for a construct is greater than the shared variance between constructs. Face validity is established when the measured items are conceptually consistent with a construct definition. Nomological validity is supported to the extent that a construct relates to other constructs in a theoretically consistent way.

Know how to represent a measurement model using a path diagram.   Visual diagrams, or path diagrams, are useful tools in helping to translate a measurement theory into something that can be tested using standard CFA procedures. SEM programs make use of these path diagrams to show how constructs are related to measured variables. Good measurement practice suggests that a measurement model should be congeneric, meaning that each measured variable should load on only one construct. Unless some strong theoretical reason indicates doing otherwise, all constructs should be linked with a two-headed, curved arrow in the path diagram, indicating that the correlation between constructs will be estimated.

Understand the basic principles of statistical identification and know some of the primary causes of SEM identification problems.   Statistical identification is extremely important in obtaining useful CFA results. Underidentified models cannot produce reliable results. Overidentified models with an excess number of degrees of freedom are required for statistical identification. In addition, each estimated parameter should be statistically identified. Many problems associated with CFA and SEM in general, including identification and convergence problems, result from two sources: insufficient sample size and an insufficient number of indicator variables per construct. The researcher is strongly encouraged to provide for an adequate sample based on the model conditions and to plan on at least three or four measured items for every construct.

Understand the concept of fit as it applies to measurement models and be able to assess the fit of a confirmatory factor analysis model.   CFA is a multivariate tool that computes a predicted covariance matrix using the equations that represent the theory tested. The predicted covariance matrix is then compared to the actual covariance matrix computed from the raw data. Generally speaking, models fit well as these matrices become more similar. Multiple fit statistics should be reported to help understand how well a model truly fits.

They include the χ^2 goodness-of-fit statistic and degrees of freedom, one absolute fit index (such as the GFI or SRMR), and one incremental fit index (such as the TLI or CFI). One of these indices should also be a badness-of-fit indicator such as SRMR or RMSEA. No absolute value for the various fit indices suggests a good fit, only guidelines are available for this task. The values associated with acceptable models vary from situation to situation and depend considerably on the sample size, number of measured variables, and the communalities of the factors.

Distinguish a measurement model from a structural model.   The key difference between a measurement model and a structural model is the way the relationships between constructs are treated. In CFA, a measurement model is tested that usually assumes each construct is related to each other construct. No distinction is made between exogenous and endogenous constructs, and the relationships are represented as simple correlations with a two-headed curved arrow. In the structural model, endogenous constructs are distinguished from exogenous constructs. Exogenous constructs have no arrows entering them. Endogenous constructs are determined by other constructs in the model as indicated visually by the pattern of single-headed arrows that point to endogenous constructs.

Describe the similarities between SEM and other multivariate techniques.   Although CFA has much in common with EFA, the structural portion of SEM is similar to multiple regression. The key differences lie in the fact that the focus is generally on how constructs relate to one another instead of how variables relate to one another. Also, it is quite possible for one endogenous construct to be used as a predictor of another endogenous construct within the SEM model.

Depict a theoretical model with dependence relationships using a path diagram.   The chapter described procedures for converting a CFA path diagram into a structural path diagram. In a path diagram, the relationships between constructs are represented with one-headed arrows. Also, the common abbreviations change. Measured indicator items for endogenous constructs are generally referred to with a *y,* and the exogenous construct indicators are referred to with an *x.*

Test a structural model using SEM.   The CFA setup can be modified and the structural model tested using the same SEM program. Models are supported to a greater extent as the fit statistics suggest that the observed covariances are reproduced adequately by the model. The same guidelines that apply to CFA models apply to the structural model fit. Also, the closer the structural model fit is to the CFA model fit, then the more confidence the researcher can have in the model. Finally, the researcher also must examine the statistical significance and direction of the relationships. The model is supported to the extent that the parameter estimates are consistent with the hypotheses that represented them prior to testing.

Diagnose problems with the SEM results.   The same diagnostic information can be used for structural model fit as for CFA model fit. The statistical significance, or lack thereof, of key relationships, the standardized residuals, and modification indices all can be used to identify problems with a SEM model.

Understand the differences between reflective and formative constructs.   Up to this time, all of the measurement models were based on the reflective model—latent constructs cause the indicators, so the arrows point from construct to indicator. But an alternative perspective has emerged in recent years termed the formative model whereby the measurement model is reversed—the arrows point from indicator to construct. Numerous researchers have identified constructs that have supposedly been mistakenly estimated as reflective when they actually were formative, thus creating an increased awareness of the possible need to specify a construct as formative. But as with so many other issues in SEM, theory must guide the researcher in this decision because the ability to assess construct validity is quite limited for formative constructs.

Be able to specify formative constructs in SEM models. Formative constructs can be estimated in any of the SEM computer programs, although it generally takes more complicated model specification than with reflective constructs. A unique feature of the formative construct is that it is not identified until it is associated with at last two other reflective indicators or constructs. These reflective measures act as the "outcome" or dependent variable for the formative construct. At issue is the fact that the loadings of the formative construct are determined based on the "outcome" measures used. As such, loadings can vary markedly if different outcomes are used. This lack of internal consistency is a fundamental difference between reflective and formative constructs.

Identify the situation in which higher-order factor analysis models are appropriate.   The most common higher-order model is the second-order model. A second-order latent factor(s) causes multiple first-order latent factors, which in turn cause the measured variables (x). In a simple sense, the first-order latent constructs become the "indicators" of the second-order latent construct. As one would expect, the higher-order construct is more abstract because it has only latent constructs as its indicators. When theoretical support can be found for a higher-order model, then any SEM program can estimate the higher-order model.

Know how SEM can be used to compare results between groups.   Multigroup comparisons can be useful for examining both the measurement and structural models. They require that researchers test their hypotheses of group differences with between-group constraints representing the various degrees of measurement model invariance or equality of structural relationships between the groups. The $\Delta\chi^2$ is a primary statistic for testing invariance and for drawing conclusions about the differences between groups.

Perform an invariance measurement analysis using multigroup methods.   Invariance testing is performed through a six-stage process. At each stage, similar models are estimated for each group, and measures of model fit are calculated for the collective set of models. For each stage, a specific type of between-group constraint is specified, adding a new type of constraint to the previous model. So each model becomes more constrained. The between-group constraints test for the equivalence across groups of a specific element of the measurement model. For example, metric invariance examines the equivalence of the factor loadings across groups. If a group model does not differ significantly from the previous, less restrictive model, then we can say that the constraints were appropriate. So in the case of metric invariance, if the chi-square difference from the prior model was not statistically significant, then we would say we have full invariance and that the factor loadings were the same across the groups. This chi-square difference test is tested at each stage where a more restrictive model is specified. If the chi-square difference test is significant (meaning that full invariance is not supported), then the researcher can try and achieve partial invariance. In this case, a subset of the parameters (e.g., factor loadings) are constrained to be equivalent across groups. If this provides a nonsignificant difference, then partial invariance is achieved.

Understand the concepts of statistical mediation and moderation.   Several different types of relationships are discussed. In particular, the concepts of mediation and moderation are explained. Mediation involves a sequencing of relationships so that some construct intervenes in a sequence between two other constructs. Moderation involves changes in relationships based on the influence of some third variable or construct. Moderation was discussed in the context of multigroup SEM models and continuous variable interactions. Whenever possible, the multigroup approach is recommended.

Appreciate the differences between SEM and PLS approaches to structural model estimation. PLS is becoming a widely used approach to estimating path models and provides an alternative method to the SEM researcher. Partial least squares (PLS), as the name implies, is more of a "regression-based" approach that focuses on prediction rather than explanation. As such, when compared to true SEM approaches, it is more robust with fewer identification issues, works with both much smaller and much larger samples, and can incorporate formative as well as reflective constructs. But with these advantages come some disadvantages, primarily in the focus of PLS on the structural model relationships, many times to the detriment of the measurement model. When PLS will provide such estimates for measurement models that are lacking sufficient construct validity by SEM standards, the researcher must question what the "latent constructs" of a PLS model represent. Thus, the selection of PLS should be done with a slightly different perspective than other SEM methods.

Questions

1. How does CFA differ from EFA?
2. List and define the components of construct validity.
3. What are the steps in developing a new construct measure?
4. What are the properties of a congeneric measurement model? Why do they represent the properties of good measurement?
5. What is a Heywood case, and how is it treated using SEM?
6. Is it possible to establish precise cutoffs for CFA fit indices? Explain.
7. Find an article in a business journal that reports a CFA result. Does the model show evidence of adequate construct validity? Explain.
8. In what ways is a measurement theory different from a structural theory? What implications do these differences have for the way a SEM model is tested? How does the visual diagram for a measurement model differ from that of a SEM model?
9. How can a measured variable represented with a single item be incorporated into a SEM model?
10. What is the distinguishing characteristic of a nonrecursive SEM model?
11. How is the validity of a SEM model estimated?
12. Why is it important to examine the results of a measurement model before proceeding to test a structural model?
13. What conditions make a second-order factor model appropriate?
14. What conditions must be satisfied in order to draw valid conclusions about differences in relationships and differences in means between three different groups of respondents—one from Canada, one from Italy, and one from Japan? Explain.
15. An interviewer collects data on automobile satisfaction. Ten questions are collected via a personal interview. Then, the respondent responds to another 20 items by marking the items using a pencil. How can CFA be used to test whether the question format has biased the results?
16. What is meant by a formative construct? Can programs like LISREL, AMOS and Mplus accommodate formative constructs?
17. How do formative constructs differ from reflective constructs, both in specification/estimation and interpretation?
18. What conditions make a second-order factor model appropriate?
19. What conclusions can be drawn from measurement invariance testing?
20. What are the most common uses for multigroup testing?
21. Describe the process of multigroup testing in general terms.
22. What is a major concern when using SEM techniques with longitudinal data?
23. What is PLS, and how does it from SEM?
24. Draw a structural model hypothesizing that three exogenous constructs, X, Y, and Z, each affects a mediating construct, M, which in turns determines two other outcomes, P and R.
25. How can SEM test for a moderating effect?

Suggested Readings

A list of suggested readings illustrating issues and applications of multivariate techniques in general is available on the Web at www.pearsonglobaleditions.com/hair or www.mvstats.com.

References

1. Anderson, J. C., and D. W. Gerbing. 1988. Structural Equation Modeling in Practice: A Review and Recommended Two-Step Approach. *Psychological Bulletin* 103: 411–23.

2. Anderson, J. C., and D. W. Gerbing. 1992. Assumptions and Comparative Strengths of the Two-Step Approach. *Sociological Methods and Research* 20 (February): 321–33.

3. Babin, B. J., and J. B. Boles. 1998. Employee Behavior in a Service Environment: A Model and Test of Potential Differences Between Men and Women. *Journal of Marketing* 62 (April): 77–91.

4. Babin, B. J., J. B. Boles, and D. P. Robin. 2000. Representing the Perceived Ethical Work Climate Among Marketing Employees. *Journal of the Academy of Marketing Science* 28 (Summer): 345–59.

5. Babin, Barry J., and Mitch Griffin. 1998. The Nature of Satisfaction: An Updated Examination and Analysis. *Journal of Business Research* 41 (February): 127–36.

6. Bacon, D. R., P. L. Sauer, and M. Young. 1995. Composite Reliability in Structural Equations Modeling. *Educational and Psychological Measurement* 55 (June): 394–406.

7. Bagozzi, R. P., and L. W. Phillips. 1982. Representing and Testing Organizational Theories: A Holistic Construal. *Administrative Science Quarterly,* 27(3): 459–89.

8. Baron, R. M., and D. A. Kenny. 1986. The Moderator-Mediator Variable Distinction in Social Psychological Research: Conceptual, Strategic and Statistical Considerations. *Journal of Personality and Social Psychology* 51: 1173–82.

9. Bearden, W. O., R. G. Netemeyer, and M. Mobley. 1993. *Handbook of Marketing Scales: Multi-Item Measures for Marketing and Consumer Behavior.* Newbury Park, CA: Sage.

10. Bentler, P. M. 1980. Multivariate Analysis with Latent Variables: Causal Modeling. *Annual Review of Psychology* 31: 419–56.

11. Bergkvist, L., and J. R. Rossiter. 2007. The Predictive Validity of Multiple-Item Versus Single-Item Measures of the Same Constructs. *Journal of Marketing Research* 44: 175–84.

12. Blaha, John, S. P. Merydith, F. H. Wallbrown, and T. E. Dowd. 2001. Bringing Another Perspective to Bear on the Factor Structure of the Minnesota Multiphasic Personality Inventory-2. *Measurement & Evaluation in Counseling & Development* 33 (January): 234–43.

13. Blalock, H. M. 1964. *Causal Inferences in Nonexperimental Research.* Chapel Hill: University of North Carolina Press.

14. Bollen, K. A. 2002. Latent Variables in Psychology and the Social Sciences. *Annual Review of Psychology* 53: 605–34.

15. Bollen, K. A., and K. G. Jöreskog. 1985. Uniqueness Does Not Imply Identification. *Sociological Methods and Research* 14: 155–63.

16. Bollen, K., and R. Lennox. 1991. Conventional Wisdom on Measurement: A Structural Equation Perspective. *Psychological Bulletin* 110: 305–14.

17. Bollen, K. A., and K. Ting. 2000. A Tetra Test for Causal Indicators. *Psychological Methods* 5: 3–32.

18. Borsboom, D., G. J. Mellenbergh, J. van Heerden. 2003. The Theoretical Status of Latent Variables. *Psychological Review* 110: 203–19.

19. Burt, R. S. 1976. Interpretational Confounding of Unobserved Variables in Structural Equations Models. *Sociological Methods Research* 5: 3–52.

20. Byrne, B. M., R. J. Shavelson, and B. Muthén. Testing For The Equivalence of Factor Covariance and Mean Structures: The Issue of Partial Measurement Invariance. *Psychological Bulletin* 105: 456–66.

21. Carmines, E. G., and J. P. McIver. 1981. Analyzing Models with Unobserved Variables: Analysis of Covariance Structures. In G. W. Bohrnstedt and E. F. Borgotta (eds.), *Social Measurement: Current Issues.* Beverly Hills, CA: Sage, pp. 65–115.

22. Cattell, R. B. 1956. Validation and Intensification of the Sixteen Personality Factor Questionnaire. *Journal of Clinical Psychology* 12: 205–14.

23. Cheung, G. W., and R. B. Rensvold. 2002. Evaluating Goodness-of-Fit Indexes for Testing Measurement Invariance. *Structural Equation Modeling* 9(2): 233–55.

24. Chin, W. W. 2000. Partial Least Squares for Researchers: An Overview and Presentation of Recent Advances Using the PLS Approach. http://disc-nt.cba.uh.edu/chin/icis2000plstalk.pdf.

25. Cohen, J., and P. Cohen. 1983. *Applied Multiple Regression/Correlation Analysis for the Behavioral Sciences,* 2nd ed. Mahwah, NJ: Lawrence Erlbaum Associates.

26. DeVellis, R. F. 1991 *Scale Development: Theory and Applications.* Newbury Park, CA: Sage.

27. Diamantopoulos, A., P. Riefler, and K. P. Roth. 2008. Advancing Formative Measurement Models. *Journal of Business Research* 61(12): 1203–18.

28. Diamantopoulos, A., and J. Siguaw. 2006. Formative Versus Reflective Indicators in Organizational Measure Development: A Comparison and Empirical Illustration. *British Journal of Management* 17(4): 263–82.

29. Diamantopoulos, A., and H. M. Winklhofer. 2001. Index Construction with Formative Indicators: An Alternative to Scale Development. *Journal of Marketing Research* 38 (May): 269–77.

30. Dillon, W., A. Kumar, and N. Mulani. 1987. Offending Estimates in Covariance Structure Analysis—Comments on the Causes and Solutions to Heywood Cases. *Psychological Bulletin* 101: 126–35.

31. Drasgow, F. 1984. Scrutinizing Psychological Tests: Measurement Equivalence and Equivalent Relations with

External Variables Are the Central Issues. *Psychological Bulletin* 95: 134–35.

32. Ferrer, E., F. Hamagami, and J. J. McArdle. 2004. Modeling Latent Growth Curves with Incomplete Data Using Different Types of SEM and Multi-Level Software. *Structural Equation Modeling* 11(3): 452–83.

33. Fornell, C., and D. F. Larcker. 1981. Evaluating Structural Equations Models with Unobservable Variables and Measurement Error. *Journal of Marketing Research* 18 (February): 39–50.

34. Fornell, C., and Y. Yi. 1992. Assumptions of the Two-Step Approach to Latent Variable Modeling. *Sociological Methods and Research* 20 (February): 291–320.

35. Gerbing, D. W., and J. C. Anderson. 1984. On the Meaning of Within-Factor Correlated Measurement Errors. *Journal of Consumer Research* 11: 572–80.

36. Gogineni, A., R. Alsup, and D. F. Gillespie. 1995. Mediation and Moderation in Social Work Research. *Social Work Research* 19 (March): 57–63.

37. Griffin, M., B. J. Babin, and D. Modianos. 2000. Shopping Values of Russian Consumers: The Impact of Habituation in a Developing Economy. *Journal of Retailing* 76 (Spring): 33–52.

38. Hair, J. F., B. J. Babin, A. Money, and P. Samouel. 2003. *Essentials of Business Research.* Indianapolis, IN: Wiley.

39. Hayduk, L. A. 1987. *Structural Equation Modeling with LISREL.* Baltimore, MD: Johns Hopkins University Press.

40. Heise, D. R. 1972. Employing Nominal Variables, Induced Variables, and Block Variables in Path Analysis. *Sociological Research Methods* 1: 147–73.

41. Herting, J. R., and H. L. Costner. 1985. Respecification in Multiple Response Indicator Models. In *Causal Models in the Social Sciences,* 2nd ed. New York: Aldine, pp. 321–93.

42. Horn, J. L. 1991. Comments on "Issues in Factorial Invariance" in *Best Methods for the Analysis of Change.* Eds. L. M. Collins and J. L. Horn. Washington, D.C.: American Psychological Association, pp. 114–25.

43. Horn, J. L., and J. J. McArdle. 1992. A Practical and Theoretical Guide to Measurement Invariance in Aging Research. *Experimental Aging Research* 18 (Fall–Winter): 117–44.

44. Hui, C. H., and H. C. Triandis. 1985. Measurement in Cross-Cultural Psychology: A Review and Comparison of Strategies. *Journal of Cross-Cultural Psychology* 16 (June): 131–52.

45. Jarvis, C. B., S. B. Mackenzie, and P. M. Padsakoff. 2003. A Critical View of Construct Indicators and Measurement Model Misspecification in Marketing and Consumer Research. *Journal of Consumer Research* 30 (September): 199–218.

46. Landis, R. S., D. J. Beal, and P. E. Tesluk. 2000. A Comparison of Approaches to Forming Composite Measures in Structural Equation Models. *Organizational Research Methods* 3: 186–207.

47. Lohmoller, J. B. 1989. *Latent Variable Path Modeling with Partial Least Squares.* Heidelberg: Physica-Verlag.

48. MacCallum, R. C. 2003. Working with Imperfect Models. *Multivariate Behavioral Research* 38(1): 113–39.

49. MacCallum, R. C., and M. W. Browne. 1993. The Use of Causal Indicators in Covariance Structure Models: Some Practical Issues. *Psychological Bulletin* 114(3): 533–41.

50. MacCallum, R. C., M. Roznowski, and L. B. Necowitz. 1992. Model Modification in Covariance Structure Analysis: The Problem of Capitalization on Chance. *Psychological Bulletin* 111: 490–504.

51. MacCallum, R., M. Rosnowski, C. Mar, and J. Reith. 1994. Alternative Strategies for Cross-Validation of Covariance Structure Models. *Multivariate Behavioral Research* 29: 1–32.

52. MacKenzie, S. B., P. M. Podsakoff, and C. B. Jarvis. 2005. The Problem of Measurement Model Misspecification in Behavioral and Organizational Research and Some Recommended Solutions. *Journal of Applied Psychology* 90(July): 710–30.

53. Marsh, H. W., and S. Jackson. 1999. Flow Experience in Sport: Construct Validation of Multidimensional, Hierarchical State and Trait Responses. *Structural Equations Modeling* 6(4): 343–71.

54. McDonald, R. P. 1996. Path Analysis with Composite Variables. *Multivariate Behavioral Research* 31(2): 239–70.

55. Meade, A. W., and G. J. Lautenschlager. 2004. A Monte-Carlo Study of Confirmatory Factor Analytic Tests of Measurement Equivalence/Invariance. *Structural Equation Modeling* 11(1): 60–72.

56. Meredith, W. 1993. Measurement Invariance, Factor Analysis, and Factorial Invariance. *Psychometrika* 58 (December): 525–43.

57. Millsap, R. E., and H. Everson. 1991. Confirmatory Measurement Model Using Latent Means. *Multivariate Behavioral Research* 26: 479–97.

58. Nasser, F., and J. Wisenbaker. 2003. A Monte-Carlo Study Investigating the Impact of Item Parceling on Measures of Fit in Confirmatory Factor Analysis. *Educational and Psychological Measurement* 63 (October): 729–57.

59. Netemeyer, R. G., W. O. Bearden, and S. Sharma. 2003. *Scaling Procedures: Issues and Applications.* Thousand Oaks, CA: Sage.

60. Nunnally, J. C. 1978. *Psychometric Theory.* New York: McGraw-Hill.

61. Ping, R. A. 2004. On Assuring Valid Measurement for Theoretical Models Using Survey Data. *Journal of Business Research* 57: 125–41.

62. Plewis, I. 2001. Explanatory Models for Relating Growth Processes. *Multivariate Behavioral Research* 36(2): 207–25.

63. Podsakoff, N. P., W. Shen, and P. M. Podsakoff. 2006. The Role of Formative Measurement Models in Strategic Management Research: Review, Critique, and Implications for Future Research. *Research Methodology in Strategic Management* 3: 197–252.

64. Quinones, M. A., and K. J. Ford. 1995. The Relationship Between Work Experience and Job Performance: A Conceptual and Meta-Analytic Review. *Personnel Psychology* 48 (Winter): 887–910.

65. Rigdon, E. E. 1995. A Necessary and Sufficient Identification Rule for Structural Models Estimated in Practice. *Multivariate Behavior Research* 30(3): 359–83.

66. Shafer, A. B. 1999. Relation of the Big Five and Factor V Subcomponents to Social Intelligence. *European Journal of Personality* 13 (May–June): 225–40.

67. Silvia, E., M. Suyapa, and R. C. MacCallum. 1988. Some Factors Affecting the Success of Specification Searches in Covariance Structure Modeling. *Multivariate Behavioral Research* 23 (July): 297–326.

68. Smitherman, H. O., and S. L. Brodsky. 1983. *Handbook of Scales for Research in Crime and Delinquency.* New York: Plenum Press.

69. Steenkamp, J., and H. Baumgartner. 1998. Assessing Measurement Invariance in Cross-Cultural Research. *Journal of Consumer Research* 25 (June): 78–79.

70. Vandenberg, R. J., and C. E. Lance. 2000. A Review and Synthesis of the Measurement Invariance Literature: Suggestions, Practices, and Recommendations for Organizational Research. *Organizational Research Record* 3: 4–69.

71. Wilcox, J. B., R. D. Howell, and E. Breivik. 2008. Questions About Formative Measurement. *Journal of Business Research* 61(12): 1219–28.

72. Wold, H. 1966. Estimation of Principal Components and Related Models by Iterative Least Squares. In P. R. Krishnaiah, ed. *Multivariate Analysis.* New York: Academic Press, pp. 391–420.

73. Yi He, Y., M. D. Merz, and D. L. Alden. 2008. Diffusion of Measurement Invariance Assessment in Cross-National Empirical Marketing Research: Perspectives from the Literature and a Survey of Researchers. *Journal of International Marketing* 16(2): 64–83.

74. Yoon, M., and R. E. Millsap. 2007. Detecting Violations of Factorial Invariance Using Data-Based Specification Searches: A Monte Carlo Study. *Structural Equation Modeling* 14(3): 435–563.

INDEX

Page numbers followed by f indicate figures; those followed by t indicate tables.